Engineering Mechanics: STATICS and DYNAMICS

Engineering Mechanics: STATICS and DYNAMICS

R.C. Hibbeler

Department of Engineering Mechanics
Illinois Institute of Technology

Macmillan Publishing Co., Inc.
New York
Collier Macmillan Publishers
London

The two parts of this book are also published separately as *Engineering Mechanics: Statics* and *Engineering Mechanics: Dynamics,* copyright © 1974 by R. C. Hibbeler.

Macmillan Publishing Co., Inc.
866 Third Avenue, New York, New York 10022

Collier-Macmillan Canada, Ltd.

Library of Congress Cataloging in Publication Data

Hibbeler, R. C.
   Engineering mechanics: statics and dynamics.
   "The two parts of this book are also published separately as Engineering mechanics: statics, and Engineering mechanics: dynamics."
   1. Mechanics, Applied. I. Title.
TA350.H48   620.1   73-1048
ISBN 0-02-354050-8

Printing: 1 2 3 4 5 6 7 8   Year: 4 5 6 7 8 9 0

Preface

The purpose of this book is to provide the student with a clear and thorough presentation of the theory and application of engineering mechanics as it is currently required for the engineering curriculum. One of the major aims of the book is to help develop the student's ability to analyze problems—a most important skill for any engineer.

Although engineering mechanics is based upon relatively few principles, it is these principles which provide a necessary means for the solution of many problems relating to present-day engineering design and analysis. In this book, emphasis is given to both the understanding and the application of these principles, so that the student will have a firm basis for understanding how these principles are utilized both in more advanced courses of study and in engineering practice.

For teaching purposes, as each new principle is introduced, it is first applied to simple situations. Specifically, each principle is applied first to a particle, then to a rigid body subjected to a system of coplanar forces (or a body having planar motion), and finally to the most general case of three-dimensional force systems acting on a rigid body (or a body having spatial motion). This presentation allows the instructor some flexibility for teaching the material, since each of the three cases is elaborated upon in separate chapters.

The contents of each chapter are separated into well-defined units. Each unit contains the development and explanation of a specific topic, illustrative example problems, and a set of problems designed to test the student's ability to apply the theory. In any set, the problems are generally arranged in order of increasing difficulty, and the answers to all even-numbered problems are given in the back of the book. Many of the problems depict realistic situations encountered in engineering practice. It is hoped that this realism will stimulate student interest in engineering mechanics and furthermore will help the student to develop his skills in

reducing any such problem from its physical description to a model or symbolic representation whereby the principles of mechanics may be applied. All numerical problems are stated in terms of the currently used British system of units; however, if it is deemed necessary, some of the problems may be worked with metric units, using the table of conversion factors given in Appendix C.

Diagrams are used extensively throughout the book, since they provide an important means for obtaining a complete understanding of both the theory and its applications. By emphasizing the use of diagrams in the examples, the student is given a logical and orderly procedure for the solution of problems. Furthermore, when diagrams are included in the analysis, physical insight relating to the problem is maintained. For example, the free-body diagram is introduced early in the book, and it is used whenever a force analysis is required for a problem solution.

Mathematics provides a systematic means of applying the principles of mechanics. The student is expected to have prior knowledge of algebra, geometry, trigonometry, and some calculus. Vector analysis is introduced in this book at points where it is most applicable. Its use often provides a convenient means for presenting concise derivations of the theory, and it makes possible a simple and systematic solution of many complicated three-dimensional problems. Occasionally, the example problems are solved in several ways in an effort to compare the use of a vector analysis with other mathematical techniques based upon a conventional scalar approach. In this way, the student develops the ability to use mathematics as a tool whereby the solution of any problem may be carried out in the most direct and effective manner.

The contents of the book are divided into 22 chapters. The subject of statics is presented in the first eleven chapters.* In particular, the fundamental concepts of rigid-body mechanics are outlined in Chapter 1. The notion of a vector and the properties of a concurrent force system are introduced in Chapter 2, and this theory is then applied to the equilibrium of particles in Chapter 3. Chapter 4 contains a general discussion of both concentrated and distributed force systems and methods used to simplify them. The principles of rigid-body equilibrium are developed in Chapter 5. These principles are applied to specific problems involving the equilibrium of trusses, frames, and machines in Chapter 6 and to the analysis of internal forces in beams and cables in Chapter 7. Applications to problems involving frictional forces are discussed in Chapter 8. By introducing the methods for reducing a distributed load to a concentrated resultant force early in Chapter 4, this theory may be incorporated with problems involving equilibrium. At the discretion of the instructor, however, this material may be covered later in conjunction with topics relating to centroids and center of gravity, given in Chapter 9. This can be done

---

*This book is also available in two separate volumes, *Engineering Mechanics: Statics* and *Engineering Mechanics: Dynamics*.

by simply omitting the problems involving distributed force systems in the intervening chapters. On the other hand, since the theory in Chapter 9 does not depend upon the methods of equilibrium, this chapter may also be covered before rigid-body equilibrium (Chapter 5). If time permits, sections concerning more advanced topics, indicated by stars, may be studied. Some topics in Chapter 10 ("Moment of Inertia of an Area") and all of Chapter 11 ("Virtual Work") may be omitted without any loss of continuity from the basic course.

The concepts of dynamics are presented in Chapters 12 through 22. Due to the nature of this subject, this portion of the book provides the instructor with a greater versatility in presenting the material. Two suggested teaching sequences follow.

	Chapter
Sequence I	
Kinematics of Particles and Rigid Bodies	12, 13
Kinetics of Particles and Planar	
Kinetics of Rigid Bodies	
Equations of Motion	14, 15
Work and Energy	16, 17
Impulse and Momentum	18, 19
Sequence II	
Kinematics of Particles	12
Kinetics of Particles	
Equations of Motion	14
Work and Energy	16
Impulse and Momentum	18
Kinematics of Rigid Bodies	13
Planar Kinetics of Rigid Bodies	
Equations of Motion	15
Work and Energy	17
Impulse and Momentum	19

Time permitting, some of the material involving three-dimensional rigid-body motions may be included in the course. The kinematics of this motion are discussed in Sections 13-10 through 13-13. Detailed treatment of mass moment of inertia of a body is given in Chapter 20. (The mass moment of inertia concept is briefly introduced in Section 15-4 and forms a suitable basis for solving planar motion problems.) The kinetics of the spatial motions of a rigid body are given in Chapter 21. Chapter 22 ("Vibrations") may be included, provided the student has had the necessary mathematical background. Sections of the book which are beyond the basic mechanics course are indicated by a star and may be omitted. (Note that this more advanced material also provides a suitable reference for basic principles when it is covered in more advanced courses.)

The author has endeavored to write a textbook that will appeal to both

student and instructor alike. In doing so, it must be admitted that the development of this book was not accomplished singlehandedly. Through the years, many people have given both their support and their help. Although I cannot list them all, I wish to acknowledge the valuable suggestions and comments made by Professors G. Mavrigian, Youngstown State University; W. G. Plumtree, California State University at Los Angeles; M. E. Raville, Georgia Institute of Technology; A. Pytel, Pennsylvania State University; G. W. Washa, University of Wisconsin; R. Schaefer, University of Missouri at Rolla; and P. K. Mallick, Illinois Institute of Technology. Gratitude is extended to the engineering students at Youngstown State University and Illinois Institute of Technology for their assistance both in the preparation and classroom testing of the manuscript. I wish also to thank my secretaries, Bernadine, Joanne, Judy, and Leslie, who tirelessly typed and retyped the manuscript. Lastly, I would like to praise John Beck, J. Edward Neve, and the publishing staff for presenting the manuscript in such fine artistic form.

It is hoped that all instructors and students using this textbook will find it most suitable for the purpose for which it is intended. Since they can provide the most critical review of this work, the author welcomes and will gladly acknowledge any comments they make regarding its contents.

Russell C. Hibbeler

Contents

Engineering Mechanics: STATICS and DYNAMICS

1

General Principles

1-1. Definitions

Mechanics, in general, is that branch of the physical sciences concerned with the state of rest or motion of bodies which are subjected to the action of forces. The theory of mechanics forms the basis of most present-day engineering design and analysis. For example, a comprehensive study of the engineering sciences relating to structural members, machine elements, fluid flow, electrical instrumentation, and even molecular and atomic behavior requires a thorough knowledge of the principles of mechanics.

Besides being one of the most important physical sciences in engineering, mechanics is also considered one of the oldest physical sciences. Writings as early as those of Archimedes (287–212 B.C.) record how he developed the principle of buoyancy in hydrostatics, and explained the equilibrium of the lever. Simon Stevin (1548–1620) not only formulated the parallelogram law used to combine forces vectorally, but he is also recognized for the development of many important concepts in the subject of statics. Galileo Galilei (1564–1642) was one of the first major contributors to the field of dynamics. His work consisted of experiments made with pendulums and falling bodies. The most significant contributions to the field of mechanics, however, were made by Sir Isaac Newton (1642–1727). He is noted for his formulation of the laws of motion and the law of universal gravitational attraction. Some of the other notable pioneers in the science of mechanics who made important contributions were Bernoulli, Euler, Lagrange, and Hamilton.

Mechanics is generally subdivided into five branches: *rigid-body mechanics, deformable-body mechanics, fluid mechanics, relativistic mechanics,* and *quantum mechanics*. The subject of engineering mechanics as treated in this book is concerned with the laws of mechanics applicable to the

study of rigid-body behavior. The basic laws governing the behavior of rigid bodies are formulated from the three laws of motion first postulated by Sir Isaac Newton. For this reason, rigid-body mechanics is sometimes referred to as *Newtonian* or *classical mechanics*. A thorough study of rigid-body mechanics is a necessary prerequisite for the study of the mechanics of deformable bodies and the mechanics of fluids.

In 1905 Albert Einstein presented his general theory of relativity which placed limitations on classical mechanics. In this regard, the findings of relativistic mechanics differ significantly from classical mechanics when the speed of the body approaches the speed of light (186,000 mi/sec or 3.0×10^8 m/sec). Hence, the theory of relativity is most often applied to study the motions of celestial bodies and subatomic size particles, which move at fantastically high rates of speed. The theory of quantum mechanics was developed during 1925 and 1926 after it was realized that classical mechanics failed to predict the behavior of particles moving within atomic distances of one another. Because man, however, is macroscopic in size, we directly observe and obey the laws of classical mechanics. Therefore, this science forms a suitable basis for the design and analysis of the majority of engineering problems, since the speed of the body studied is generally much smaller than the speed of light, and the body is many times greater than atomic size.

The study of rigid-body mechanics is generally divided into two subject areas: statics and dynamics. *Statics* deals with the equilibrium of bodies under the action of forces, while *dynamics* is concerned with the accelerated motion of bodies. Dynamics is further subdivided into *kinematics*, which is the study of the geometry of motion without reference to the forces which cause the motion, and *kinetics*, which relates the imposed forces to the resulting motions. Basically, statics can be considered as a special case of dynamics in which the body is in equilibrium and therefore is either at rest or moves with constant velocity. Most structures are designed with the intention that they remain in equilibrium. Furthermore, even structural members having small accelerations, such as some aircraft and machine components, can be designed under pseudoequilibrium conditions, provided these accelerations are neglected. Clearly, therefore, many engineers need a thorough working knowledge of the principles of statics.

1-2. Fundamental Quantities, Units

Four fundamental quantities commonly used in mechanics are length, time, force, and mass. In general, each of these quantities is defined on the basis of an arbitrarily chosen *unit* or standard. By applying a simple experimental process to compare quantities of the same kind to the

standard unit, we can then form a basis for defining the standard quantities.

The concept of *length* is needed to locate the position of a point in space and thereby describe the size of a physical system. The unit of length measurement is the standard *meter* (m) which was originally defined by the distance between parallel lines marked on a platinum-iridium bar kept at standard pressure and temperature and located in the International Bureau of Weights and Measures at Sèvres, France. A more current definition of the meter,* which makes this unit readily accessible to scientists throughout the world, is taken to be 1,650,763.73 wave lengths in vacuum of the orange-red line of the spectrum of krypton-86. All other units of length are defined in terms of the standard meter. For example, 1 centimeter (cm) is equal to 0.01 m and 1 foot (ft) is equal to 0.3048m.

Time is conceived by a succession of events. Some convenient units for measurements of time are the second, minute, and hour. By international agreement, each of these measurements is a fraction of a standard unit of time measured by an atomic clock, having a time unit based upon the duration of 9,192,631,770 cycles of radiation associated with a specified transition of the isotope cesium-133.

Force is generally considered as a "push" or "pull" exerted by one body on another. A man pushing on a wall is exerting a force on the wall; the wall in turn is pushing back on the man with an equal, but opposite force. In the same manner, the man standing on a floor exerts a force on the floor, which is equal and opposite to the force of the floor on the man. Forces acting between objects always occur in equal and opposite pairs; no single force can act alone. This fact follows from Newton's third law of motion—for every action there is an equal and opposite reaction. Force interaction can occur when there is direct contact between bodies, such as a man pushing on a wall, or it can occur through a distance by which the bodies are physically separated. Examples of the latter type include gravitational, electrical, and magnetic forces. Engineers define the standard unit of force using either the *pound* (lb) or the kilogram-force (kg_f). Each of these two units is measured by the amount of pull exerted by the earth upon an object at a specified latitude and height above sea level. This force or gravitational attraction by the earth on a body is defined as the *weight* of the body. Comparison of weights can be made using a spring balance. Since the effect of gravity changes with respect to the distance from the center of the earth, it is important to make comparisons at a specified latitude and height above sea level.

Mass is regarded as a property which is used to determine the resistance of matter to a change in its motion. Unlike the concept of weight, *mass is constant regardless of position*. For this reason, comparison of masses

*11th International Conference on Weights and Measures, Paris, 1960.

is usually made by means of a lever-arm balance. The international standard of mass, as adopted by the 11th International Conference on Weights and Measures, is the *kilogram* (kg) defined by a bar of platinum alloy kept at the International Bureau of Weights and Measures. Occasionally grams (g) are also used to measure mass, 1 g = 0.001 kg.

The four fundamental quantities, length, time, force, and mass are all related using Newton's law of motion—force is proportional to the product of mass and acceleration, i.e., $F = kma$. For convenience, if we select the constant of proportionality $k = 1$, the *units* used to define force, mass, length, and time cannot *all* be selected arbitrarily. Instead, the equality $F = ma$ is maintained if three of the four units (called *primary units*) are arbitrarily defined and the fourth unit is derived from the equation.

Four important systems of units used to describe the four fundamental quantities are listed in Table 1-1. In particular, the MKS and CGS systems are both defined using length, time, and mass as primary units. These two systems are called *absolute,* since the defined measurements of the primary units can be made at *any location.* For example, the meter has the same (or absolute) length regardless of where the measurement is made. The unit of force in these systems is a *derived* unit. Hence, in the MKS system, a *newton* is the amount of force needed to give 1 kg mass an acceleration of 1 m/sec² ($F = ma$). Whereas in the CGS system, a *dyne* is defined as the amount of force which gives a mass of 1 g an acceleration of 1 cm/sec². These two absolute systems of units are used almost universally by scientists throughout the world.

Table 1–1   System of Units

Type of System	Name of System	Length	Mass	Force	Time
Absolute	Metric (MKS)	meter (m)	kilogram (kg)	newton* $\left(\dfrac{\text{kg-m}}{\text{sec}^2}\right)$	second (sec)
Absolute	(CGS)	centimeter (cm)	gram (g)	dyne* $\left(\dfrac{\text{g-cm}}{\text{sec}^2}\right)$	second (sec)
Gravitational	Metric Gravitational (MKGFS)	meter (m)	metric slug* $\left(\dfrac{\text{kg}_f\text{-sec}^2}{\text{m}}\right)$	kilogram force (kg_f)	second (sec)
Gravitational	British Gravitational (FPS)	foot (ft)	slug* $\left(\dfrac{\text{lb-sec}^2}{\text{ft}}\right)$	pound (lb)	second (sec)

*Derived unit

Since most experiments involve a direct measurement of force, engineers prefer to use a *gravitational system* of units in preference to an absolute system. In a gravitational system, the three primary units are force, length, and time. Mass, therefore, becomes a derived unit, Table 1-1. In formulating the MKGFS system, the units of length (m) and time (sec) are the same as in the MKS system. The MKGFS system, however, is often confused with the MKS system since the unit of mass in the MKS system and the unit of force in the MKGFS system have the same name—the kilogram. Due to frequent misunderstanding in solving problems, it should therefore be kept firmly in mind that the kilogram (kg) is a unit of mass in the MKS system and the kilogram-force (kg_f) is a unit of force in the MKGFS system. Specifically, the unit of mass in the MKGFS system is the *metric slug* defined as the amount of mass to which a force of 1 kg_f imparts an acceleration of 1 m/sec². On the other hand, in the FPS system of units, the unit of mass is called a *slug*, which is equal to the amount of mass which is accelerated at 1 ft/sec² when acted upon by a force of 1 lb. By comparison, 1 lb = 0.4536 kg_f.

Currently the FPS system of units is predominantly used by American engineers; however, by act of Congress, the MKGFS units of measurement are intended, in time, to replace the FPS system. By adopting the MKGFS units, American engineers will be in a more suitable position to communicate in a quantitative sense with engineers located in other nations of the world. Although this change will come in due time, in this book we will use the FPS system of units as a means for applying the principles of mechanics to solving problems. If deemed necessary, the problems for solution may be worked in metric units using the appropriate conversion factors, 1 ft = 0.3048 m, and 1 lb = 0.4536 kg_f.

1-3. Fundamental Idealizations

If a physical body is represented in its natural state, application of the principles of mechanics becomes exceedingly complex. In the physical world all matter is composed of atoms. Thus, when a force is applied to a solid body, a series of interatomic forces is established among all the atoms of the solid to prevent the body from collapsing. In an effort to find the actual amount of force resisted by the body, at some point inside the body, it would be necessary to analyze each of the atomic interactions to arrive at an answer—a nearly impossible task. To simplify this problem, we may view the body as having a macroscopic form and consider each element of the body to be made up of a continuous "spread" or *continuum* of matter, rather than a finite number of discrete atomic particles. The behavior of this assumed continuum can then be studied experimentally in order to find relations between deformations of the solid

and the forces applied to it. This deformable-continuum model of a physical body is an important idealization used throughout the study of deformable-body mechanics and fluid mechanics.

We may further simplify the model representing the physical body by assuming the body to be entirely rigid. For a *rigid body* each small portion or particle of the body remains at a constant distance away from every other particle in the body both before and after applying a load to the body. This simplification eliminates the need for experimental testing, since the body cannot deform. The concept of the rigid-body assumption is an important idealization in the study of engineering mechanics. In solving problems, however, we should be clearly aware of its limitations, for, in reality, all bodies deform. Since many of the principles of engineering mechanics are related to the shape of the body, it is important to recognize problems where applied loadings might cause large deformations of the body. If the loads severely deform a body such that the orientation of the loading *after* it is applied is *not known,* the rigid-body model must be replaced by the deformable-continuum model. However, if the deformed shape of the body is *known,* so that the final position of the applied loads can be determined with sufficient accuracy, then the rigid-body assumption may be justified. In many cases the deformations occurring in engineering structures, machines, mechanisms, etc. are relatively small, and the rigid-body assumption is suitable for analysis. Thus, with the exception of springs, the principles of mechanics as discussed in this book will be based on the assumption that all materials are rigid.

If the size or shape of a body has no consequence in the problem solution, we may use a particle to represent a model of the body. A *particle* essentially has no size but has a mass. This idealization is useful in statics for studying the equilibrium of a body subjected to a concurrent force system. In dynamics its use is often justified in studying the motion of bodies having trajectories larger than the dimensions of the body, such as the motion of a planetary body about the sun. Since geometry of the body is not involved, the principles of mechanics reduce to a particularly simplified form when applied to a particle. For this reason, throughout this book specific principles of mechanics are first applied to a particle before they are applied to a rigid body.

Another important fundamental idealization used in mechanics is the concept of a concentrated force. When any external load is applied to the surface of a body, it is actually distributed over a finite surface area. The intensity of these *distributed forces* is measured in terms of pressure or force per unit area. In some cases, however, the distributed load can be simplified to a *concentrated force.* This can be done provided the area over which the load is applied is small compared to the finite dimensions of the body. When a distributed load can be represented as a concentrated force, computations regarding the effect of the load on the body are greatly simplified.

1–4. Fundamental Principles

The entire structure of rigid-body mechanics is formulated on the basis of relatively few fundamental principles. These principles or laws were conceived on the basis of *experimental evidence*. We will briefly state these principles here and discuss them in greater detail at points in the book where they are needed.

Newton's Three Laws of Motion. In the late seventeenth century, Sir Isaac Newton formulated his three fundamental laws governing the motion of a particle. These laws are as follows:

First Law. A particle originally at rest, or moving at a constant velocity, will continue to remain at rest or move at constant velocity in a straight line, provided there is no unbalanced force acting on the particle.

Second Law. A particle acted upon by an unbalanced force **F** receives an acceleration **a** in the direction of the force. The acceleration is directly proportional to the force and inversely proportional to the particle mass m. This law is commonly expressed in mathematical terms as

$$\mathbf{F} = m\mathbf{a}$$

Third Law. For every force acting on a particle, the particle exerts an equal, opposite, and collinear reactive force.

Newton's Law of Gravitational Attraction. Shortly after formulating his three laws of motion for a particle, Newton postulated a law governing the mutual attraction between any two particles. In mathematical form this law can be expressed by the equation

$$F = G\frac{m_1 m_2}{r^2} \tag{1-1}$$

where F = the force of attraction between the two particles
  G = the universal constant of gravitation; according to experimental evidence $G = 3.44 \times 10^{-8}$ lb–ft^2/slug^2 $(6.67 \times 10^{-11}$ m^3/(kg–sec^2))
  m_1, m_2 = the masses of each of the two particles
  r = the distance between the centers of both particles

Any two particles or bodies have a mutual attractive (gravitational) force acting between them. In the case of a particle located at or near the surface of the earth, however, the only attractive force having any sizable magnitude is that of the earth's gravitation. Thus, for our purpose, this will be the only gravitational force considered.

The Parallelogram Law for Finding the Force Resultant. Originally developed by Simon Stevin back in the late sixteenth century, the paral-

lelogram law is used to find a single resultant force equivalent to two forces acting at a point. This is done by constructing a parallelogram whose sides represent the given forces. The diagonal of the parallelogram then represents the magnitude of the resultant force.

The Principle of Transmissibility. This principle states that a force, applied to a rigid body, may act at any point along its line of action without altering the conditions of equilibrium or the conditions of motion of the body.

1–5. Method of Problem Solution

One of the most effective ways of studying engineering mechanics is to work problems. Merely studying concepts or general principles about some theory will be of little use if this knowledge is not applied in a meaningful way. In working problems it is very important to present the work in a *logical* and *orderly* manner. It is strongly advised to follow these steps:

1. Read the problem carefully. List the data given and the results required.
2. Draw and label any necessary diagrams needed for the solution. In solving equilibrium problems, for example, this step will include drawing the necessary free-body diagrams.
3. List all the relevant principles, generally in mathematical form.
4. Think about the problem in terms of the actual *physical situation*. Try to correlate this knowledge with each mathematical expression that is written out.
5. Solve the necessary equations algebraically as far as is practical, then complete the solution numerically. For engineering work, a slide rule in most cases provides the desired degree of accuracy for solution.
6. Check through the problem making sure that the equations used are dimensionally homogeneous and that the units of the numerical data used in the solution are consistent.*
7. Study the answer with technical judgment and common sense to determine whether or not the answer seems reasonable.
8. Once the solution has been completed, review the problem. Try to think of other ways of obtaining the same solution.

---

*The terms of any equation used to describe a physical process must be dimensionally homogeneous, that is, each term must be expressed in the same units. Provided this is the case, all the terms of the equation can then be combined if numerical data is substituted for the variables. For later applications, the example on page 11 illustrates the method for converting from one set of units to another.

In applying the above procedure, do the work as neatly as possible the first time. Being neat generally stimulates clear and orderly thinking and vice versa.

Example

Given a velocity $v = 10$ ft/sec, convert the units to mi/hr.

Solution

Since 5,280 ft = 1 mi, 60 sec = 1 min, and 60 min = 1 hr, then

$$v = 10 \text{ ft/sec} = \frac{10 \text{ ft}}{1 \text{ sec}} \frac{60 \text{ sec}}{1 \text{ min}} \frac{60 \text{ min}}{1 \text{ hr}} \frac{1 \text{ mi}}{5,280 \text{ ft}}$$

$$= \frac{36,000}{5,280} \text{ mi/hr} = 6.82 \text{ mi/hr} \qquad\qquad Ans.$$

2
Force and Position Vectors

2-1. Scalars and Vectors

Many of the physical quantities in mechanics can be expressed mathematically by means of scalars and vectors. A quantity possessing only a magnitude is called a *scalar*. Some scalars are always positive—for example, mass, energy, or volume. Others, classified as algebraic scalars, such as temperature, may be either positive or negative. In this book scalars will be indicated by letters of ordinary type. The mathematical operations involving scalars follow the same rules as those of elementary algebra.

A *vector* is a quantity having a magnitude, direction, and sense. It is necessary that two vectors be combined using the parallelogram law. This law utilizes a particular form of construction which accounts for the combined magnitude, direction, and sense of the vector. Vector quantities which are encountered in the study of statics are the position, force, and moment vectors.

For analytical work, a vector is often represented by a letter with an arrow written over it, such as $\vec{A}$, and its magnitude is designated by $|\vec{A}|$ or simply A. In this book vectors will be designated in boldface type; for example **A** is used to designate the vector "A," while its magnitude will be indicated by A.

Graphically, a vector is represented by an arrow, used to define its magnitude and direction. The *magnitude* of the vector is indicated by the *length* of the arrow. The vector **A** shown in Fig. 2–1 has a magnitude of four units. The line of action has a *direction* of 20° above the horizontal with a *sense* indicated by the arrowhead. The point O is commonly

Fig. 2–1

13

A = B

Fig. 2–2

Fig. 2–3

referred to as the *initial point, origin,* or *tail* of the vector, and the point P is referred to as the *terminal point, terminus,* or *tip*.

The two vectors **A** and **B** shown in Fig. 2–2 are considered *equal* provided they have the same magnitude, direction, and sense. Although it is not necessary that they originate from the same point, it is required that a pair of equal vectors be of the same *type;* that is, the magnitude of the vectors must have the same units. For example, the magnitudes of two force vectors may both be expressed in units of pounds. Borrowing the equal sign from scalar algebra, we may write the equality of vectors **A** and **B** as

$$\mathbf{A = B}$$

A *negative vector* has a direction opposite to its positive counterpart but has the same magnitude, as shown in Fig. 2–3.

There are three *kinds* of vectors which are important to us in our study of statics. They are the bounded or fixed vector, the sliding vector, and the free vector. A *fixed vector* is one that acts at a particular point in space. For example, a force vector **F** acting on the particle P, shown in Fig. 2–4a, must be a fixed vector, since a particle itself is a well-defined point of application. A *sliding vector* is a vector which may be applied anywhere along its line of action. A force applied to a rigid body may be thought of as a sliding vector when the effect of the force on the entire body is being considered (refer to the principle of transmissibility in Sec. 1–4). Provided this is the case, the force vector **F** shown in Fig. 2–4b may be applied at any point $P(P')$ lying along its line of action. *Free vectors* can act at any point in space; it is only necessary that they preserve their same magnitude, direction, and sense. In some instances the moment vector of a couple is treated as a free vector. This concept will be discussed later; however, if this is the case, the couple vector **M** acting on the rigid body shown in Fig. 2–4c may then be placed at *any point P* on the body.

(a)

(b)

(c)

Fig. 2–4

2–2. Addition and Subtraction of Vectors

The law for addition of two vectors **A** and **B**, Fig. 2–5a, is defined using either the "triangle" or "parallelogram" rules. To form the sum $\mathbf{R =}$ $\mathbf{A + B}$ by means of the *triangle rule,* vector **B** is added to vector **A** by placing the initial point of **B** at the terminal point of **A** and then joining

the initial point of **A** to the terminal point of **B** (see Fig. 2–5b). The vectors **A** and **B** are referred to as *component vectors,* and their sum yields the *resultant vector* **R.** In a similar manner, vector **R** can also be obtained by adding **A** to **B,** Fig. 2–5c. Both ways of addition form the basis for the *parallelogram law* of vector addition. According to this method, vectors **A** and **B** each form an adjacent side of a parallelogram. The resultant vector **R** is then determined by constructing an arrow from the initial point of **A** and **B** along the diagonal of the parallelogram to the opposite corner. The construction is shown in Fig. 2–5d. From this figure, or Figs. 2–5b and 2–5c, it can be seen that vector addition is commutative; i.e., **A** + **B** = **B** + **A.** It is important to note that when adding vectors, the addition must be carried out using either the "triangle" or "parallelogram" laws, as shown here. Furthermore, the vectors added must have magnitudes measured with a consistent set of units.

When adding two vectors *graphically,* the vector arrows must be drawn to scale to make the operation meaningful. The resultant vector's magnitude and its direction can then be measured directly from the graph. See Example 2–1.

In many cases the *resultant of two vectors* can be determined conveniently through analytical methods. Since the triangle rule forms the basis for vector addition, we can use trigonometry to find the resultant's magnitude and direction. In some problems the resultant vector may already be known, in which case the problem unknowns involve the magnitudes and/or the directions of the component vectors. (See Examples 2–3 and 2–4.) In any case, the law of sines and the law of cosines may be helpful in solving for the unknowns. These formulas are given in Fig. 2–6 for the triangle *ABC.*

When adding *three* or *more vectors* which are *coplanar,* i.e., lying in the same plane, the polygon rule, which is a simple extension of the triangle rule, may be used. This rule is illustrated in Fig. 2–7 for the addition of vectors **A, B,** and **C.** In forming the sum **R** = **A** + **B** + **C,** we arrange the vectors in a "tip-to-táil" fashion. The resultant **R** is thus

Fig. 2–5

Sine law:
$$\frac{A}{\text{Sin } a} = \frac{B}{\text{Sin } b} = \frac{C}{\text{Sin } c}$$

Cosine law:
$$C = \sqrt{A^2 + B^2 - 2AB \text{ Cos } c}$$

Fig. 2–6

Fig. 2–7

15

formed by connecting the "tail" or origin of the first vector **A** with the "tip" of the last one **C.** This procedure is equivalent to applying the triangle rule twice; that is, adding **B** to **A,** we form the resultant (**A** + **B**), shown as a dashed arrow in Fig. 2–7. **C** is then added to this vector, which yields the resultant **R.**

Noting the dashed arrows in Fig. 2–7, we see that the addition of three vectors is associative; i.e.,

$$R = (A + B) + C = A + (B + C)$$

The associative rule can easily be extended to include more than three vectors. Since vector addition is commutative, the order of addition of the vectors according to the polygon rule is immaterial; that is, the vectors may be added together in any order.

In defining the (resultant) *difference* of two vectors **A** and **B,** one may write

$$R = A - B = A + (-B)$$

and then form the vector sum. It is seen that subtraction is defined as a special case of addition, and the rules that apply for vector addition will also apply for vector subtraction. A graphic example of vector subtraction is given in Fig. 2–8.

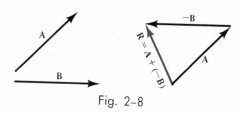

Fig. 2–8

If **A** = **B** then the difference **R** is defined as a zero or *null vector,* written as

$$R = A - B = 0$$

Null vectors have zero magnitude and arbitrary direction.

2–3. Multiplication of a Vector by a Scalar

The product of a vector **A** by a scalar m, yielding m**A,** is defined as a vector having a magnitude $|m|\ A$. The sense of direction is the same as **A** provided m is positive, or opposite to **A** if m is negative. (If $m = 0$, then m**A** is a null vector.) The negative of a vector can thus be defined on the basis of multiplication of the vector by the scalar -1. By definition,

the vector *magnitude* is always a *positive quantity. A minus sign in front of a vector simply means that its directional sense is reversed* (see Fig. 2–3).

Multiplication of a scalar by a vector can be shown to be associative; i.e.,

$$m(n\mathbf{A}) = (mn)\mathbf{A}$$

commutative; i.e.,

$$m\mathbf{A} = \mathbf{A}m$$

and distributive; i.e.,

$$(m + n)\mathbf{A} = m\mathbf{A} + n\mathbf{A}$$

and

$$m(\mathbf{A} + \mathbf{B}) = m\mathbf{A} + m\mathbf{B}$$

Note that these laws apply only when scalars are multiplied by vectors. The product of two vectors shall be defined in Chapter 4.

Division of a vector by a scalar can be defined by means of the laws of multiplication, since

$$\frac{\mathbf{A}}{m} = \frac{1}{m}\mathbf{A} \qquad m \neq 0$$

Graphic examples of multiplication and division of vectors by scalars are shown in Fig. 2–9.

Fig. 2–9

2–4.   Force as a Vector Quantity

As mentioned in Chapter 1, *force* is considered as a push or pull that one body exerts on another. In other words, force might be regarded as a directed action which can be classified as a vector quantity. The *force magnitude* is commonly measured in terms of pounds (lb) or kilogram-force (kg$_f$). Large forces may be represented in units of kilopounds (kips) or tons (a *kip* and a *ton* are equivalent to 1,000 lb and 2,000 lb, respectively).

To represent a force as a vector graphically, one would arbitrarily choose a scale of measuring units to represent the units of force. A line segment having a scaled length equivalent to the force *magnitude* would then be constructed in a *direction* acting along the line of action of the force. The *sense* of the force would be indicated by an arrowhead.

Since forces are vector quantities, they not only have a magnitude, direction, and sense, but they also must add according to the triangle or parallelogram laws of vector addition. The fact that forces do indeed obey these laws has been proven by *experimental observation.* If forces F_1 and F_2 act at the same point P, their effect on P is the same as if these forces are replaced by their vector resultant $F_R = F_1 + F_2$, Fig. 2–10.

When a system of two or more forces acts through the *same point,* the vectors are known as a system of *concurrent* forces. Furthermore, if all these forces act in the *same plane,* such as F_1 and F_2 in Fig. 2–10, the force system is both concurrent and *coplanar.* From this and the previous discussion of vectors, we will now give some examples involving concurrent–coplanar force systems.

Fig. 2–10

Example 2-1

The "eyebolt" in Fig. 2–11a is subjected to two forces F_1 and F_2. Determine the magnitude and direction of the resultant force acting on the "eyebolt."

Solution

Graphical Procedure. Choosing an arbitrary scale of 50 lb $= \frac{1}{2}$ in., the force parallelogram is constructed as shown in Fig. 2–11b. Using a scale

(a)

(b)

(c)

Fig. 2–11

and protractor, we can measure the force resultant $\mathbf{F}_R$ directly from this drawing. It is found that

$$F_R = 213 \text{ lb} \qquad\qquad \textit{Ans.}$$

$$\alpha = 55° \qquad\qquad \textit{Ans.}$$

Analytical Procedure. The two component force vectors may be added together in a "tip-to-tail" fashion according to the triangle law of vector addition, as shown in Fig. 2–11c. The magnitude of the resultant may be determined from the triangle by means of the law of cosines:

$$F_R = \sqrt{(100)^2 + (150)^2 - 2(100)\,150\,\cos 115°}$$
$$= \sqrt{10{,}000 + 22{,}500 - 30{,}000\,(-0.423)}$$
$$= 212.6 \text{ lb} \qquad\qquad \textit{Ans.}$$

Applying the law of sines, using the computed value of F_R, we find the angle θ:

$$\frac{150}{\sin\theta} = \frac{212.6}{\sin 115°}$$

$$\sin\theta = \frac{150}{212.6}\,(0.906) = 0.639$$

$$\theta = 39.8$$

Thus,

$$\alpha = 39.8° + 15.0° = 54.8° \qquad\qquad \textit{Ans.}$$

Example 2–2

The gusset plate in Fig. 2–12a is subjected to four forces which are concurrent at point O. Determine the resultant of these forces using the polygon method of vector addition.

Solution

Choosing an arbitrary scale of 50 lb $= \frac{1}{4}$ in. and starting with $\mathbf{F}_1$, we add all the vectors together in a "tip-to-tail" fashion, as shown in Fig. 2–12b. The resultant force $\mathbf{F}_R$ is then drawn from the initial point of $\mathbf{F}_1$ to the head of $\mathbf{F}_4$. Direct measurements from the figure give

$$F_R = 408 \text{ lb} \qquad\qquad \textit{Ans.}$$

$$\theta = 26° \qquad\qquad \textit{Ans.}$$

(a)

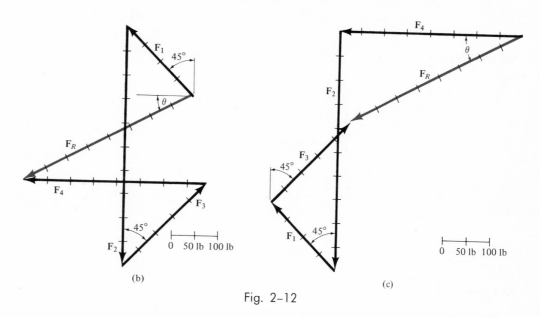

(b)

(c)

Fig. 2–12

Since vector addition is commutative, the same results are obtained from the construction shown in Fig. 2–12c. In this case $\mathbf{F}_R = \mathbf{F}_4 + \mathbf{F}_2 + \mathbf{F}_1 + \mathbf{F}_3$.

Example 2–3

Resolve the 200-lb force resultant shown acting on the pin, Fig. 2–13a, into components in (a) the x and y directions, (b) the x' and y' directions, and (c) the x' and y directions.

Fig. 2–13

Solution

Part (a). Using the parallelogram law, we resolve **F** into x and y components, Fig. 2–13b. From this figure,

$$F_x = 200 \cos 40° = 153.2 \text{ lb} \qquad \textit{Ans.}$$
$$F_y = 200 \sin 40° = 128.6 \text{ lb} \qquad \textit{Ans.}$$

Part (b). The x' and y' components of **F** are shown in Fig. 2–13c. The magnitudes of these components are

$$F_{x'} = 200 \cos 70° = 68.4 \text{ lb} \qquad \textit{Ans.}$$
$$F_{y'} = 200 \sin 70° = 187.9 \text{ lb} \qquad \textit{Ans.}$$

Part (c). The two components $\mathbf{F}_{x'}$ and $\mathbf{F}_y$ are shown in Fig. 2–13d. Note how the construction is based on the parallelogram law. Using the law of sines, to obtain the magnitude of each component,

$$\frac{F_y}{\sin 70°} = \frac{200 \text{ lb}}{\sin 60°}$$
$$F_y = 217.0 \text{ lb} \qquad \textit{Ans.}$$
$$\frac{F_{x'}}{\sin 50°} = \frac{200 \text{ lb}}{\sin 60°}$$
$$F_{x'} = 177.0 \text{ lb} \qquad \textit{Ans.}$$

Example 2–4

The ring shown in Fig. 2–14a is subjected to two forces $\mathbf{F}_1$ and $\mathbf{F}_2$. If it is required that the resultant of these two applied forces have a magni-

(a)   (b)   (c)   (d)

Fig. 2–14

tude of 1,000 lb and be directed vertically downward, determine (a) the magnitude of F_1 and F_2 provided $\theta = 30°$, and (b) the magnitude of F_1 and F_2 when F_2 is a minimum.

Solution

Part (a). A sketch of the vector addition, using the parallelogram law, is given in Fig. 2–14b. The unknown magnitudes F_1 and F_2 can be found using the law of sines; namely,

$$\frac{F_1}{\sin 30°} = \frac{1,000 \text{ lb}}{\sin 130°}$$

$$F_1 = 652.7 \text{ lb} \qquad\qquad Ans.$$

$$\frac{F_2}{\sin 20°} = \frac{1,000 \text{ lb}}{\sin 130°}$$

$$F_2 = 446.5 \text{ lb} \qquad\qquad Ans.$$

Part (b). As shown in Fig. 2–14c, using the triangle law, vector F_2 may be added to F_1 in various ways to yield the net resultant F_R. In particular, the *minimum* length or magnitude of F_2 will occur when the line of action of F_2 is perpendicular to F_1. Any other direction, such as OA or OB, yields a larger value for the magnitude of F_2. Hence, when $\theta = 90° - 20° = 70°$, the value of F_2 is minimum. From the triangle shown in Fig. 2–14d, it is seen that

$$F_1 = 1,000 \sin 70° = 939.7 \text{ lb} \qquad\qquad Ans.$$

$$F_2 = 1,000 \sin 20° = 342.0 \text{ lb} \qquad\qquad Ans.$$

Problems

2-1. Find the resultant force acting on the pin.

Prob. 2-1

2-2. Using the polygon law for vector addition, determine graphically the resultant vector of the four coplanar forces.

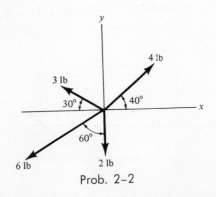

Prob. 2-2

2-3. The riveted bracket supports two forces. Determine the angle θ so that the line of action of the resultant force is horizontal. What is the magnitude of this resultant force?

Prob. 2-3

2-4. Two forces are applied at the end of a "screw eye" in order to remove the post. Determine the angle θ so that the resultant force acting on the post is directed vertically upward and has a magnitude of 100 lb.

Prob. 2-4

23

2-5. Using a graphical procedure, determine the magnitude of the resultant force and its direction measured from the horizontal.

Prob. 2-5

2-6. Find the x and y components of the 500-lb force acting on the bracket.

Prob. 2-6

2-7. Resolve the force **F** into components acting parallel and perpendicular to the *aa* axis of the boom. **F** is coplanar with *aa* and the vertical.

Prob. 2-7

2-8. Using the polygon law for vector addition, determine the resultant of the three forces acting at the end of the boom.

Prob. 2-8

2-9. Using trigonometry, determine (a) $\mathbf{F}_1 + \mathbf{F}_2$, and (b) $\mathbf{F}_1 - \mathbf{F}_2$.

Prob. 2-9

2-10. The force **F** has a magnitude of 500 lb and is to be resolved into two components acting along the axes of booms *AB* and *AC*. Determine the angle θ so that the component of **F** directed along *AC* acts from *A* to *C* and has a magnitude of 200 lb.

Prob. 2-10

(c) $2\mathbf{v}_1 - 2\mathbf{v}_2 - \mathbf{v}_3$, where $v_1 = 0$, $v_2 = 1$ ft/sec, $v_3 = 5$ ft/sec.

Prob. 2-12

2-13. Two forces act at the end of the boom. Determine the magnitude of force $\mathbf{F}$ and its direction θ such that (a) the resultant of the two forces, $\mathbf{F}_R$, has a magnitude of 20 lb and is directed along the negative x axis, (b) the resultant $\mathbf{F}_R$ is a null vector.

2-11. The boom collar A resists the pull of the 1,000-lb applied loading. Resolve this force into components acting along the x and y axes.

Prob. 2-11

Prob. 2-13

2-12. Given the three velocity vectors $\mathbf{v}_1$, $\mathbf{v}_2$, and $\mathbf{v}_3$ having the directions shown, use the polygon rule to find the following sums:
(a) $\mathbf{v}_1 + \mathbf{v}_2 + \mathbf{v}_3$, where $v_1 = 7$ ft/sec, $v_2 = 8$ ft/sec, $v_3 = 12$ ft/sec.
(b) $\mathbf{v}_1 - \frac{1}{2}(\mathbf{v}_2 + \mathbf{v}_3)$, where $v_1 = 4$ ft/sec, $v_2 = 200$ ft/ min, $v_3 = 24$ in./sec.

2-14. The jet aircraft is subjected to the towing forces developed in the cables AB and AC. If it is required that the resultant of these two forces have a magnitude of 500 lb and be directed along the axis of the fuselage of the aircraft, determine (a) the magnitude of the tow-

25

ing forces, provided $\theta = 15°$, and (b) the value of θ such that the force in cable *AB* is a minimum.

Prob. 2–14

2–15. Resolve the 175-lb force acting on the support into components acting in (a) the x' and y' directions, and (b) the x' and x directions.

Prob. 2–16

Prob. 2–15

2–16. Using trigonometry, determine the magnitude and direction θ of force **F**, which passes through the center of the 100-lb sphere so that the resultant of the two forces acting on the sphere has a magnitude of 210 lb and is directed along the positive y axis.

2–17. The 500-lb force acting on the frame is to be resolved into two components acting along the axes of the struts *AB* and *AC*. If the component of force along member *AC* is required to be 300 lb, directed from *A* to *C*, determine the magnitude of force acting along *AB* and the angle θ of the 500-lb force.

Prob. 2–17

26

2-18. A vertical force of 200 lb is applied to the end of the horizontal beam *AB*. If it is known that the component of this force acting along *CA* is 500 lb, directed from *C* to *A*, find the component of force acting along *AB*, and the angle θ.

Prob. 2–18

2-5. Unit Vectors

A *unit vector* is a vector having a magnitude of unity. If **A** is a vector having a magnitude $A \neq 0$, then a unit vector having the *same direction* as **A** is represented by

$$\mathbf{u}_A = \frac{\mathbf{A}}{A} \tag{2-1}$$

Rewriting this equation,

$$\mathbf{A} = A\mathbf{u}_A \tag{2-2}$$

Since vector **A** is of a certain kind, for example, a force vector, it is customary to assign to it the proper set of units used to describe it. The magnitude *A* also has this same set of units; hence, from Eq. 2–1, the *unit vector will be dimensionless.* Equation 2–2 indicates that vector **A** may therefore be expressed in terms of both its magnitude and direction *separately;* that is, *A* (a scalar) expresses the *magnitude* of **A**, and $\mathbf{u}_A$ (a dimensionless vector) expresses the *directional sense* of **A**.

2–6. Components of a Vector and Cartesian Unit Vectors

Any vector $\mathbf{A}$ may be represented by the sum of three component vectors acting along the x, y, and z axes of a rectangular coordinate system. The proof of this statement can be established with reference to Fig. 2–15.

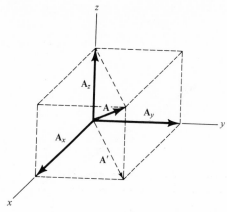

Fig. 2–15

Using the parallelogram law of vector addition, we have

$$\mathbf{A}' = \mathbf{A}_x + \mathbf{A}_y$$

and

$$\mathbf{A} = \mathbf{A}' + \mathbf{A}_z$$

Combining these equations,

$$\mathbf{A} = \mathbf{A}_x + \mathbf{A}_y + \mathbf{A}_z \qquad (2\text{–}3)$$

In this book only right-hand Cartesian coordinate systems will be used. A coordinate system is said to be *right-handed* provided the thumb of the right hand points in the direction of the positive z axis, while the "curl" of the right-hand fingers is drawn from the positive x to the positive y axis, as shown in Fig. 2–16.

A set of unit vectors $\mathbf{i}$, $\mathbf{j}$, and $\mathbf{k}$ will be used to define the *positive direction* of the x, y, and z axes, respectively, as shown in Fig. 2–17. Using this set of unit vectors and our definition of a unit vector as expressed by Eq. 2–2, we may recast the vector components of Eq. 2–3 into a more convenient form:

$$\mathbf{A} = A_x\mathbf{i} + A_y\mathbf{j} + A_z\mathbf{k}$$

There is a distinct advantage to writing vectors in this manner, that is, in terms of their $\mathbf{i}$, $\mathbf{j}$, and $\mathbf{k}$ (or x, y, and z) components. In doing so,

Fig. 2-16

Fig. 2-17

the magnitude and direction of each component vector are *separated,* and as a result, this simplifies the operations of vector algebra. For example, to form the vector sum $\mathbf{R} = \mathbf{A} + \mathbf{B}$, we must add the corresponding vectors $\mathbf{A}$ and $\mathbf{B}$ according to the parallelogram law in the manner shown shaded in Fig. 2-18*a*. However, an easier method of vector addition consists of forming the resultant by adding together the magnitudes of each of the corresponding $\mathbf{i}$, $\mathbf{j}$, and $\mathbf{k}$ components of vectors $\mathbf{A}$ and $\mathbf{B}$. From Fig. 2-18*a*,

$$\mathbf{A} = A_x\mathbf{i} + A_y\mathbf{j} + A_z\mathbf{k}$$

and

$$\mathbf{B} = B_x\mathbf{i} + B_y\mathbf{j} + B_z\mathbf{k}$$

Thus,

$$\mathbf{R} = \mathbf{A} + \mathbf{B} = A_x\mathbf{i} + A_y\mathbf{j} + A_z\mathbf{k} + B_x\mathbf{i} + B_y\mathbf{j} + B_z\mathbf{k}$$

Since $A_x\mathbf{i}$ and $B_x\mathbf{i}$ are collinear vectors, their resultant is determined by the scalar addition of the components. Adding the $\mathbf{j}$ and $\mathbf{k}$ components in a similar manner, we obtain the necessary resultant

$$\mathbf{R} = (A_x + B_x)\mathbf{i} + (A_y + B_y)\mathbf{j} + (A_z + B_z)\mathbf{k}$$

This vector sum is illustrated graphically in Fig. 2-18*b*.

2-7. Magnitude and Direction of Cartesian Vectors

It is always possible to obtain the magnitude and direction of a vector when the vector is expressed in terms of its Cartesian components. If

(a)

(b)

Fig. 2-18

29

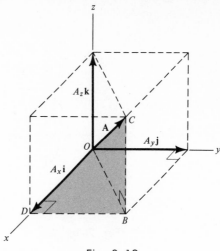

Fig. 2–19

$\mathbf{A} = A_x\mathbf{i} + A_y\mathbf{j} + A_z\mathbf{k}$, the *magnitude* of $\mathbf{A}$ can be found by applying the Pythagorean theorem. Consider the construction shown in Fig. 2–19. From the right triangle ODB, the magnitude of line segment OB may be represented as

$$OB = \sqrt{(OD)^2 + (DB)^2}$$

Similarly, from the right triangle OBC,

$$OC = \sqrt{(OB)^2 + (BC)^2}$$

Combining both equations,

$$OC = \sqrt{(OD)^2 + (DB)^2 + (BC)^2}$$

Recognizing that each of these line segments represents the magnitude of vector $\mathbf{A}$ and its components, we may express the magnitude of $\mathbf{A}$ as

$$A = \sqrt{(A_x)^2 + (A_y)^2 + (A_z)^2} \qquad (2\text{--}4)$$

Hence, *the magnitude of a vector is equal to the positive square root of the sum of the squares of the vector's Cartesian components*. The positive square root is chosen since the magnitude represents the measured *length* of the vector.

The *direction* of vector $\mathbf{A}$ is represented by *coordinate direction angles* α, β, and γ which the tail of vector $\mathbf{A}$ makes with the positive x, y, and z axes, as shown in Fig. 2–20. Note that α, β, or γ can never be greater than 180°. Referring to Fig. 2–20a, it is necessary that ODB be a right triangle, having a right angle at point D. Thus, provided the magnitude of $\mathbf{A}$ is calculated (see Eq. 2–4) the angle α can be determined from

$\cos \alpha = OD/OB = A_x/A$. In a similar manner, using right triangles OCB and OBE, shown in Figs. 2–20b and 2–20c, we obtain $\cos \beta = A_y/A$ and $\cos \gamma = A_z/A$. The numbers $\cos \alpha$, $\cos \beta$, and $\cos \gamma$ are known as the *direction cosines* of vector **A**. The angles α, β, and γ are therefore defined from the relations:

$$\alpha = \cos^{-1} \frac{A_x}{A}$$

$$\beta = \cos^{-1} \frac{A_y}{A} \qquad (2\text{--}5)$$

$$\gamma = \cos^{-1} \frac{A_z}{A}$$

Knowing the magnitude of **A**, it is possible to form the unit vector $\mathbf{u}_A$ which, by definition, has a magnitude of one (dimensionless unit) and acts in the direction of **A**. Using Eq. 2–1, this unit vector is

$$\mathbf{u}_A = \frac{\mathbf{A}}{A} = \frac{A_x}{A}\mathbf{i} + \frac{A_y}{A}\mathbf{j} + \frac{A_z}{A}\mathbf{k} \qquad (2\text{--}6)$$

From the definition of the direction cosines given above, it is seen that the **i**, **j**, *and* **k** *components, namely A_x/A, A_y/A and A_z/A, of the unit vector $\mathbf{u}_A$ represent the direction cosines of* **A**. Hence, we may also write,

$$\mathbf{u}_A = \cos \alpha \mathbf{i} + \cos \beta \mathbf{j} + \cos \gamma \mathbf{k} \qquad (2\text{--}7)$$

(a)

(b)

(c)

Fig. 2–20

Since the magnitude of a vector is equal to the square root of the sum of the squares of its Cartesian components, and $\mathbf{u}_A$ has a magnitude of unity, using Eq. 2–7, we obtain an important relation between the direction cosines:

$$\cos^2 \alpha + \cos^2 \beta + \cos^2 \gamma = 1 \qquad (2\text{–}8)$$

Hence, when two coordinate direction angles of a vector are known, the third direction angle is specified by using Eq. 2–8.

2–8. Position Vectors and Unit Vector Operations

In general, the coordinate axes are used as a space frame. Hence, the units along the three axes are measured in terms of length. By definition, a *position vector* is used to locate one point in space relative to another point. The position vector $\mathbf{r}_1$, Fig. 2–21, extending *from the origin to an arbitrary point* $P_1(x_1, y_1, z_1)$ in space is therefore

$$\mathbf{r}_1 = x_1\mathbf{i} + y_1\mathbf{j} + z_1\mathbf{k}$$

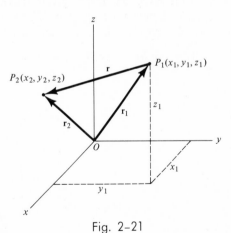

Fig. 2–21

A position vector may also be constructed between two arbitrary points in space. For example, consider finding the position vector $\mathbf{r}$ having an initial point $P_1(x_1, y_1, z_1)$ and terminal point $P_2(x_2, y_2, z_2)$, Fig. 2–21. Forming the position vectors $\mathbf{r}_1$ and $\mathbf{r}_2$ yields

$$\mathbf{r}_1 = x_1\mathbf{i} + y_1\mathbf{j} + z_1\mathbf{k}$$
$$\mathbf{r}_2 = x_2\mathbf{i} + y_2\mathbf{j} + z_2\mathbf{k}$$

Using the triangle rule for vector addition, Fig. 2–21, we require

$$\mathbf{r}_1 + \mathbf{r} = \mathbf{r}_2$$

Hence,

$$\mathbf{r} = \mathbf{r}_2 - \mathbf{r}_1 = (x_2\mathbf{i} + y_2\mathbf{j} + z_2\mathbf{k}) - (x_1\mathbf{i} + y_1\mathbf{j} + z_1\mathbf{k})$$

or

$$\mathbf{r} = (x_2 - x_1)\mathbf{i} + (y_2 - y_1)\mathbf{j} + (z_2 - z_1)\mathbf{k} \qquad (2\text{–}9)$$

Thus, *the components of a position vector acting between any two points in space may be formed by taking the coordinates of the terminus or head of the vector $P_2(x_2, y_2, z_2)$ and subtracting from them the corresponding coordinates $P_1(x_1, y_1, z_1)$ of its initial point or tail.*

Following the method used in Sec. 2–7, the *magnitude* of $\mathbf{r}$ is

$$r = \sqrt{(x_2 - x_1)^2 + (y_2 - y_1)^2 + (z_2 - z_1)^2} \qquad (2\text{–}10)$$

The *direction* of $\mathbf{r}$ can be determined by *first* constructing a unit vector $\mathbf{u}_r$ which is directed from point P_1 to point P_2. This vector is determined by using Eqs. 2–9 and 2–10, i.e.,

$$\mathbf{u}_r = \frac{\mathbf{r}}{r} = \frac{(x_2 - x_1)}{r}\mathbf{i} + \frac{(y_2 - y_1)}{r}\mathbf{j} + \frac{(z_2 - z_1)}{r}\mathbf{k}$$

See Fig. 2–22. On the basis of Eq. 2–7, the angles α, β, and γ which $\mathbf{r}$ makes with respect to the x, y, and z axes are found by using the *components* of this unit vector:

$$\alpha = \cos^{-1}\frac{(x_2 - x_1)}{r}$$

$$\beta = \cos^{-1}\frac{(y_2 - y_1)}{r}$$

$$\gamma = \cos^{-1}\frac{(z_2 - z_1)}{r}$$

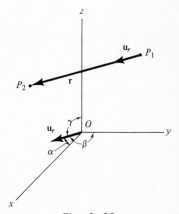

Fig. 2–22

The unit vector $\mathbf{u}_r$ can also be used to construct a vector of a *different kind* but having the same direction as $\mathbf{r}$. As an example, suppose that points P_1 and P_2 act as fixed end supports for a cord which has a tension of T lb, as shown in Fig. 2–23a. When the cord is removed, it exerts a force of $\mathbf{T}_{P_1 P_2}$, which acts on point P_1 toward point P_2; and, similarly, an equal and opposite force $\mathbf{T}_{P_2 P_1}$ acts on point P_2 toward point P_1, Fig. 2–23b. Both forces have a magnitude of T lb. The *direction* of force $\mathbf{T}_{P_1 P_2}$ may be specified by using the unit vector $\mathbf{u}_r$. Since $\mathbf{u}_r$ has a

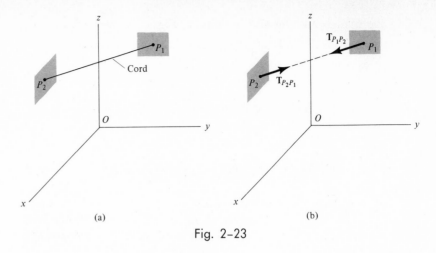

Fig. 2-23

magnitude of one (dimensionless) unit, we may form the force vector $\mathbf{T}_{P_1 P_2}$ by applying Eq. 2-2. Thus,

$$\mathbf{T}_{P_1 P_2} = (T \text{ lb})\mathbf{u}_r$$

$$= \left\{ T\frac{(x_2 - x_1)}{r}\mathbf{i} + T\frac{(y_2 - y_1)}{r}\mathbf{j} + T\frac{(z_2 - z_1)}{r}\mathbf{k} \right\} \text{lb}$$

Also,

$$\mathbf{T}_{P_2 P_1} = -\mathbf{T}_{P_1 P_2} = T \text{ lb} (-\mathbf{u}_r) = \{-T\mathbf{u}_r\} \text{ lb}$$

$$= \left\{ -T\frac{(x_2 - x_1)}{r}\mathbf{i} - T\frac{(y_2 - y_1)}{r}\mathbf{j} - T\frac{(z_2 - z_1)}{r}\mathbf{k} \right\} \text{lb}$$

The following examples further illustrate the methods used to apply the above theory to the solution of problems.

Example 2-5

The rope in Fig. 2-24a is subjected to a tension force of 60 lb when used to pull the crate up the 20° incline. Express this force as a Cartesian vector having components parallel and perpendicular to the incline.

Solution I

As shown in Fig. 2-24b, the 60-lb force is resolved into its Cartesian components using the parallelogram law. From this figure,

$$F_x = 60 \cos 30° = 52.0 \text{ lb}$$
$$F_y = 60 \sin 30° = 30.0 \text{ lb}$$

(a) (b) (c)

Fig. 2–24

Since the magnitude of the **i** and **j** components of **F** are now known, the force may be expressed as

$$\mathbf{F} = \{52.0\mathbf{i} + 30.0\mathbf{j}\} \text{ lb} \qquad \textit{Ans.}$$

Solution II

The result may be obtained by using a unit vector approach. As shown in Fig. 2–24c, the unit vector $\mathbf{u}_F$ acting in the direction of **F** has direction cosines along the x and y axes of $\cos \alpha = \cos 30°$ and $\cos \beta = \cos 60°$, respectively. Thus, from Eq. 2–6, or simply from the components,

$$\mathbf{u}_F = \cos \alpha \mathbf{i} + \cos \beta \mathbf{j} = \cos \alpha \mathbf{i} + \sin \alpha \mathbf{j} = \sin \beta \mathbf{i} + \cos \beta \mathbf{j}$$
$$= 0.866\mathbf{i} + 0.500\mathbf{j}$$

Since the force has a magnitude of 60 lb, from Eq. 2–2,

$$\mathbf{F} = (60 \text{ lb})\mathbf{u}_F = \{52.0\mathbf{i} + 30.0\mathbf{j}\} \text{ lb} \qquad \textit{Ans.}$$

Example 2-6

The gusset plate shown in Fig. 2–25a is subjected to four forces which are concurrent at point O. Using Cartesian vectors, determine the magnitude and direction of the resultant of these forces.

Solution

The resultant force is the vector sum of all the forces. Hence, it is first necessary to express each force in Cartesian vector form. This can be accomplished by first establishing a unit vector in the direction of each force, and then multiplying the components of the unit vector by the magnitude of the force in accordance with Eq. 2–2. Thus,

$$\mathbf{F}_1 = (400 \text{ lb})(-\mathbf{i}) = \{-400\mathbf{i}\} \text{ lb}$$
$$\mathbf{F}_2 = (250 \text{ lb})(\sin 45°\mathbf{i} + \cos 45°\mathbf{j}) = \{176.8\mathbf{i} + 176.8\mathbf{j}\} \text{ lb}$$

35

(a)

(b)

Fig. 2-25

$$\mathbf{F}_3 = (500\ \text{lb})(-\mathbf{j}) = \{-500\mathbf{j}\}\ \text{lb}$$
$$\mathbf{F}_4 = (200\ \text{lb})(-\sin 45°\mathbf{i} + \cos 45°\mathbf{j}) = \{-141.4\mathbf{i} + 141.4\mathbf{j}\}\ \text{lb}$$

The resultant force is therefore

$$\mathbf{F}_R = \mathbf{F}_1 + \mathbf{F}_2 + \mathbf{F}_3 + \mathbf{F}_4$$
$$= -400\mathbf{i} + (176.8\mathbf{i} + 176.8\mathbf{j}) + (-500\mathbf{j}) + (-141.4\mathbf{i} + 141.4\mathbf{j})$$
$$= \{-364.6\mathbf{i} - 181.8\mathbf{j}\}\ \text{lb}$$

The magnitude of this vector is

$$F_R = \sqrt{(-364.6)^2 + (-181.8)^2} = 407.4\ \text{lb} \qquad\qquad \textit{Ans.}$$

A unit vector acting in the direction of $\mathbf{F}_R$ is

$$\mathbf{u}_R = \frac{\mathbf{F}_R}{F_R} = -\frac{364.6}{407.4}\mathbf{i} - \frac{181.8}{407.4}\mathbf{j} = -0.895\mathbf{i} - 0.446\mathbf{j}$$

The direction cosines of $\mathbf{u}_R$ are therefore

$$\cos\alpha = -0.895, \qquad \text{so that } \alpha = 153.5° \qquad\qquad \textit{Ans.}$$
$$\cos\beta = -0.446, \qquad \text{so that } \beta = 116.5° \qquad\qquad \textit{Ans.}$$

The resultant $\mathbf{F}_R$ is illustrated graphically in Fig. 2-25b. Note, by definition, the direction angles α and β must be measured from the *positive* x and y axes to the tail of the force.

Example 2-7

Using the hoisting rig shown in Fig. 2-26a, a man lifts a 100-lb crate by pulling on the cable ABC. The pulley located at B is free to rotate,

(a)

(b)   (c)

Fig. 2–26

and therefore the entire cable *ABC* is subjected to a tension of 100 lb. Determine the magnitude and direction of the resultant force which this cable exerts on the hoisting rig at point *B*.

Solution

As shown in Fig. 2–26*b*, there are two 100-lb forces acting on the hoisting rig which represent the effect of the cable after the cable is *removed*. It is the resultant of these two forces which is to be found. One

of these forces is caused by the weight of the crate. Since this force acts in the $-\mathbf{k}$ direction and has a magnitude of 100 lb,

$$\mathbf{F}_{BC} = \{-100\mathbf{k}\} \text{ lb}$$

The second force $\mathbf{F}_{BA}$ is directed from point B (0 ft, 0 ft, 16 ft) to point A (6 ft, -3 ft, 4 ft). The direction of $\mathbf{F}_{BA}$ is computed using this fact. A position vector *from point B to point A* is calculated by Eq. 2-9.

$$\mathbf{r}_{BA} = (6-0)\mathbf{i} + (-3-0)\mathbf{j} + (4-16)\mathbf{k} = \{6\mathbf{i} - 3\mathbf{j} - 12\mathbf{k}\} \text{ ft}$$

The magnitude of this vector (which represents the distance from point B to point A) is

$$r_{BA} = \sqrt{(6)^2 + (-3)^2 + (-12)^2} = 13.75 \text{ ft}$$

Forming the unit vector $\mathbf{u}_{BA}$, which then specifies the *direction* of both $\mathbf{r}_{BA}$ and $\mathbf{F}_{BA}$,

$$\mathbf{u}_{BA} = \frac{\mathbf{r}_{BA}}{r_{BA}} = \frac{6}{13.75}\mathbf{i} - \frac{3}{13.75}\mathbf{j} - \frac{12}{13.75}\mathbf{k}$$

or

$$\mathbf{u}_{BA} = 0.436\mathbf{i} - 0.218\mathbf{j} - 0.873\mathbf{k}$$

Knowing the magnitude of $\mathbf{F}_{BA}$ is 100 lb, $\mathbf{F}_{BA}$ may therefore be expressed in Cartesian vector form as

$$\mathbf{F}_{BA} = (100 \text{ lb}) \, \mathbf{u}_{BA} = \{43.6\mathbf{i} - 21.8\mathbf{j} - 87.3\mathbf{k}\} \text{ lb}$$

The resultant force $\mathbf{F}_R$ acting at point B is then the vector sum:

$$\mathbf{F}_R = \mathbf{F}_{BA} + \mathbf{F}_{BC} = \{43.6\mathbf{i} - 21.8\mathbf{j} - 187.3\mathbf{k}\} \text{ lb}$$

The magnitude of this resultant is

$$F_R = \sqrt{(43.6)^2 + (-21.8)^2 + (-187.3)^2} = 193.5 \text{ lb} \qquad \textit{Ans.}$$

A unit vector acting in the direction of $\mathbf{F}_R$ is

$$\mathbf{u}_R = \frac{\mathbf{F}_R}{F_R} = \frac{43.6}{193.5}\mathbf{i} - \frac{21.8}{193.5}\mathbf{j} - \frac{187.3}{193.5}\mathbf{k}$$

or

$$\mathbf{u}_R = 0.225\mathbf{i} - 0.113\mathbf{j} - 0.968\mathbf{k}$$

The $\mathbf{i}$, $\mathbf{j}$, and $\mathbf{k}$ components of $\mathbf{u}_R$ represent the direction cosines of $\mathbf{F}_R$. Therefore,

$$\alpha = \cos^{-1}(0.225) = 77.0° \qquad \textit{Ans.}$$
$$\beta = \cos^{-1}(-0.113) = 96.5° \qquad \textit{Ans.}$$
$$\gamma = \cos^{-1}(-0.968) = 165.4° \qquad \textit{Ans.}$$

The resultant vector $\mathbf{F}_R$ is shown in Fig. 2–26c. Note that each of the direction angles is measured from the *positive* coordinate axis to the tail of the force.

Example 2–8

Three guy wires are attached to the top of a mast as shown in Fig. 2–27a. The tension in each of these wires is as follows: $T_{AB} = 200$ lb, $T_{AC} = 300$ lb, and $T_{AD} = 500$ lb. Determine the (x, y) coordinate position of point C so that the resultant of these three forces is directed downward along the axis of the mast, acting from point A to point O. What is the magnitude of the resultant force?

(a)   (b)

Fig. 2–27

Solution

When each cable is removed, the tension force in each cable is directed away from point A, as shown in Fig. 2–27b. Each of these three forces may be expressed in Cartesian vector form. This is done by *first* forming unit vectors along the cables. With reference to Fig. 2–27a,

$$\mathbf{r}_{AB} = \{-20\mathbf{j} - 100\mathbf{k}\}\ \text{ft};\ r_{AB} = \sqrt{(-20)^2 + (-100)^2} = 102.0\ \text{ft}$$

so that,

$$\mathbf{u}_{AB} = \frac{\mathbf{r}_{AB}}{r_{AB}} = -0.196\mathbf{j} - 0.980\mathbf{k}$$

$$\mathbf{r}_{AC} = \{x\mathbf{i} + y\mathbf{j} - 100\mathbf{k}\}\ \text{ft};\ r_{AC} = \sqrt{(x)^2 + (y)^2 + (-100)^2}$$

so that,

$$\mathbf{u}_{AC} = \frac{\mathbf{r}_{AC}}{r_{AC}} = \frac{x\mathbf{i} + y\mathbf{j} - 100\mathbf{k}}{\sqrt{x^2 + y^2 + (-100)^2}}$$

$$\mathbf{r}_{AD} = \{-25\mathbf{i} + 15\mathbf{j} - 100\mathbf{k}\} \text{ ft};$$
$$r_{AD} = \sqrt{(-25)^2 + (15)^2 + (-100)^2} = 104.2 \text{ ft}$$

so that,

$$\mathbf{u}_{AD} = \frac{\mathbf{r}_{AD}}{r_{AD}} = -0.240\mathbf{i} + 0.144\mathbf{j} - 0.960\mathbf{k}$$

In accordance with Eqs. 2–1 and 2–2, the force vectors acting along the cables are therefore

$$\mathbf{T}_{AB} = (200 \text{ lb})\mathbf{u}_{AB} = \{-39.2\mathbf{j} - 196.0\mathbf{k}\} \text{ lb}$$
$$\mathbf{T}_{AC} = (300 \text{ lb})\mathbf{u}_{AC} = \left\{ \frac{300x\mathbf{i} + 300y\mathbf{j} - 30{,}000\mathbf{k}}{\sqrt{x^2 + y^2 + (-100)^2}} \right\} \text{ lb}$$
$$\mathbf{T}_{AD} = (500 \text{ lb})\mathbf{u}_{AD} = \{-120.0\mathbf{i} + 72.0\mathbf{j} - 480.0\mathbf{k}\} \text{ lb}$$

Since the *resultant force* $\mathbf{F}_R$ is to be directed downward along the mast axis, $\mathbf{F}_R$ acts in the $-\mathbf{k}$ direction (Fig. 2–27b). If the unknown magnitude of this force is represented as F_R, then

$$\mathbf{F}_R = -F_R\mathbf{k}$$

It is required that

$$\mathbf{F}_R = \mathbf{T}_{AB} + \mathbf{T}_{AC} + \mathbf{T}_{AD}$$

or

$$-F_R\mathbf{k} = (-39.2\mathbf{j} - 196.0\mathbf{k}) + \left(\frac{300x\mathbf{i} + 300y\mathbf{j} - 30{,}000\mathbf{k}}{\sqrt{x^2 + y^2 + (-100)^2}} \right)$$
$$+ (-120.0\mathbf{i} + 72.0\mathbf{j} - 480.0\mathbf{k})$$

To satisfy this vector equation, the corresponding magnitudes of each of the **i, j,** and **k** components of the vector on the left-hand side of the equation must be *equal* to the corresponding magnitudes of the **i, j,** and **k** components of the vectors on the right. Hence,

$$0 = \frac{300x}{\sqrt{x^2 + y^2 + (-100)^2}} - 120.0 \qquad (1)$$

$$0 = -39.2 + \frac{300y}{\sqrt{x^2 + y^2 + (-100)^2}} + 72.0 \qquad (2)$$

$$-F_R = -196.0 - \frac{30{,}000}{\sqrt{x^2 + y^2 + (-100)^2}} - 480.0 \qquad (3)$$

After placing the constants in Eqs. (1) and (2) on the left-hand side of the equal sign, squaring and rearranging terms, we have

$$y^2 = 5.25x^2 - 10,000$$

and

$$x^2 = 82.7y^2 - 10,000$$

Substituting one equation into the other and solving yields $x = \pm 44.0$ ft and $y = \pm 12.0$ ft. The squaring of Eqs. (1) and (2) has introduced extraneous roots in the solution. By substitution, however, the only two of the four roots which satisfy Eqs. (1) and (2) are

$$x = 44.0 \text{ ft} \qquad \qquad Ans.$$
$$y = -12.0 \text{ ft} \qquad \qquad Ans.$$

Substituting these values into Eq. (3) yields

$$F_R = 949.0 \text{ lb} \qquad \qquad Ans.$$

Problems

2-19. For the bracket connection, express the forces **T** and **F** in Cartesian vector notation with respect to the x and y axes. Determine the magnitude of the resultant force and its direction with respect to the x axis.

2-21. Express each of the three forces acting on the bracket in Cartesian vector notation, and compute the magnitude and direction of the resultant force.

Prob. 2-19

Prob. 2-21

2-20. Determine the magnitude and direction of (a) the force vector $\mathbf{F} = \{10\mathbf{i} + 25\mathbf{j} - 30\mathbf{k}\}$ lb, and (b) the position vector $\mathbf{r} = (-10 \text{ in.})\mathbf{i} + (2 \text{ ft})\mathbf{j}$.

2-22. Represent the position vector acting from points (3 ft, 5 ft, 6 ft) to (5 ft, -2 ft, 1 ft) in Cartesian vector notation. Give its direction cosines, and find the distance between these two points.

2-23. Find the magnitude and the direction of **h**, where **h** is the resultant of the Cartesian unit vectors; i.e.,

$$\mathbf{h} = \mathbf{i} + \mathbf{j} + \mathbf{k}$$

2-24. Given the three position vectors

$$\mathbf{r}_1 = \{4\mathbf{i} + 2\mathbf{j} + 5\mathbf{k}\} \text{ ft}$$
$$\mathbf{r}_2 = \{2\mathbf{i} + 1\mathbf{k}\} \text{ ft}$$
$$\mathbf{r}_3 = \{-24\mathbf{i} + 12\mathbf{j}\} \text{ in.}$$

find the magnitude and direction of (a) $\mathbf{r}_3$, and (b) $\mathbf{r}_1 - \mathbf{r}_2 + \tfrac{1}{2}\mathbf{r}_3$.

2-25. Determine the length of the sides AB, BC, and AC of the triangular plate.

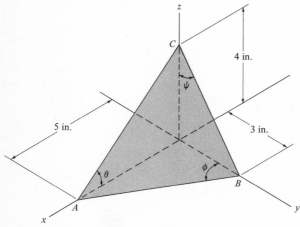

Prob. 2-25

2-26. Determine the angles θ, ϕ, and ψ which the edges of the triangular plate ABC, shown in Prob. 2–25, make with the x, y, and z axes.

2-27. If $\mathbf{r}_1 = \{3\mathbf{i} - 2\mathbf{j} + 3\mathbf{k}\}$ ft, $\mathbf{r}_2 = \{4\mathbf{i} - 3\mathbf{k}\}$ ft, $\mathbf{r}_3 = \{\mathbf{i} - 2\mathbf{j} + 10\mathbf{k}\}$ ft, determine the magnitude and direction of (a) $\mathbf{r}_1 - \mathbf{r}_2 + \mathbf{r}_3$, and (b) $-\mathbf{r}_1 + \mathbf{r}_2 - \mathbf{r}_3$.

2-28. Determine the magnitude and direction of the resultant of the three forces in Prob. 2–8 using Cartesian vector notation.

2-29. The cable OA exerts a force on point O of $\mathbf{F} = \{4\mathbf{i} + 3\mathbf{j} + 10\mathbf{k}\}$ lb, directed from O to A. If the length of the cable is 5 ft, what are the coordinates (x, y, z) of point A?

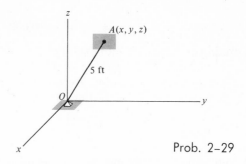

Prob. 2-29

2-30. A position vector is drawn from the origin to the point (4 ft, 3 ft, 10 ft). Determine the angles α, β, and γ which this vector makes with the x, y, and z axis, respectively.

2-31. The antenna cables CA and CB exert forces of 50 lb and 63 lb, respectively on point C. If these forces act away from point C, express the resultant of these two forces, acting on point C, as a vector. What is the magnitude and direction of the resultant?

Prob. 2-31

42

2–32. The boom OA and the two guy wires support a 100-lb lamp. If the cables AB and AC exert forces of 50 and 70 lb, respectively on point A, determine the magnitude and direction of the resultant force of the weight of the lamp and the two cables acting on the boom at point A. The cable forces are directed from A to B and A to C.

Prob. 2–32

2–33. The magnitude of the y and z components of the force F are known to be 30 and 10 lb, respectively, as shown. The angle α which F makes with the x axis is 45°. Determine the magnitude and direction cosines of F.

Prob. 2–33

2–34. A position vector r has a magnitude of 10 ft and a direction defined by the angles $\alpha = 30°$, $\gamma = 75°$. If it is known that the y component of r is negative, determine the angle β and the components of r.

2–35. The guy line OA used to hold the umbrella tent up is 6 ft long and exerts a supporting force of $F = \{4i - 3j - 12k\}$ lb acting in a direction from point A to O on the tent. Determine the height h of point A relative to the ground.

Prob. 2–35

2–36. A wire is bent into the shape shown. If it is known that a position vector acting from point A to point B is represented as $r_{AB} = \{i + 3j - 5k\}$ in., determine the length OA and its direction with respect to the x, y, and z axes.

Prob. 2–36

43

2-37. The shaft *S* exerts three force components on the die *D*. Find the magnitude and direction of the resultant force.

Prob. 2–37

2-38. The door is held open by means of two chains *AB* and *CD*. If the tension in *AB* and *CD* is 15 and 20 lb, respectively, express each of these forces in Cartesian vector notation. Consider the forces as acting on the door and directed from point *D* to *C* and point *B* to *A*.

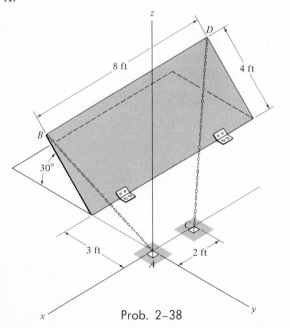

Prob. 2–38

2-39. The ring is subjected to the force of three springs, each spring having a tension of 50 lb. Express these forces as Cartesian vectors directed outward from point *O*, and find the resultant force acting on point *O*.

Prob. 2–39

2-40. Because of the action of the 200-lb weight, the boom *AB* exerts a 250-lb force on the pin at *A* directed from *B* to *A*. The cable *AC* exerts a force of 175 lb on *A*, acting from *A* to *C*. Determine the magnitude of force in the horizontal cord *AD* and the height *h* so that the resultant of the four forces is zero.

Prob. 2–40

Equilibrium of a Particle

3–1. Conditions for the Equilibrium of a Particle

In this section we will develop both the necessary and sufficient conditions required for the equilibrium of a particle. A *particle* may be defined as being a relatively small portion of matter such that its dimensions or size will be of no consequence in the analysis of the physical problem. (The concept of a particle used in mechanics is therefore analogous to the point used in geometry.) Thus, objects considered as particles can only be subjected to a system of *concurrent forces*. Since a body can be thought of as a combination of many particles, the study of the equilibrium of particles is a necessary prerequisite to the study of body equilibrium.

The conditions for *particle equilibrium* are based on a balance of forces. In particular, if a concurrent force system acts on a particle such that the resultant of the force system is zero, the particle will either remain motionless or continue to move in a straight line at constant speed, depending upon the initial condition of motion. In a physical sense, equilibrium means that there is as much force acting on the particle in one direction as in the opposite direction. This concept, which is both necessary and sufficient for equilibrium, is formally stated as Newton's first law of motion: If the *resultant force* acting on a particle is *zero,* the particle is in *equilibrium*.

As a consequence of equilibrium, motion of the particle is static, i.e., unchanging. Hence, as mentioned, equilibrium requires that the particle either be at rest, if originally at rest, or move in a straight line with constant speed, if originally in motion. Most often, however, the term "equilibrium" or more specifically, "static equilibrium," is used to describe an object at rest.

From a graphical point of view, equilibrium requires that the force

polygon, constructed for the system of forces acting on the particle, form a *closed* polygon. Why? Thus, if the particle P, shown in Fig. 3–1a, is in equilibrium when subjected to the applied (coplanar) force system the forces, when added vectorially in a "head-to-tail" fashion, form a closed polygon. For example, starting with force $\mathbf{F}_1$ and adding the other forces, the head of vector $\mathbf{F}_4$ coincides with the starting point, O, and therefore the resultant of the applied force system is zero, Fig. 3–1b.

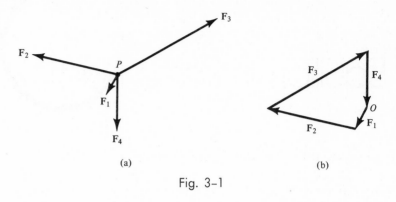

(a)   (b)

Fig. 3–1

The condition for particle equilibrium may be formulated mathematically as

$$\Sigma \mathbf{F} = 0 \tag{3–1}$$

where $\Sigma \mathbf{F}$ represents the *vector sum* of *all forces* acting on the particle. If each of the concurrent forces is resolved into its respective $\mathbf{i}, \mathbf{j}$, and $\mathbf{k}$ components, Eq. 3–1 may then be written in the form

$$\Sigma F_x \mathbf{i} + \Sigma F_y \mathbf{j} + \Sigma F_z \mathbf{k} = 0 \tag{3–2}$$

where ΣF_x, ΣF_y, and ΣF_z represent the *algebraic sum* of the vector components acting along the x, y, and z axis, respectively. Since the $\mathbf{i}, \mathbf{j}$, and $\mathbf{k}$ directions are mutually *orthogonal* or perpendicular to each other, vectors acting along any one of these directions cannot have components acting in any of the other two directions. Therefore, to satisfy Eq. 3–2, the force summations along each of the unit vector directions must be equal to zero, i.e.,

$$\Sigma F_x = 0$$
$$\Sigma F_y = 0$$
$$\Sigma F_z = 0 \tag{3–3}$$

Vector Eq. 3–1, or the scalar Eqs. 3–3, provide both necessary and sufficient conditions for the equilibrium of a particle.

3–2. Free-Body Diagrams

When solving equilibrium problems, it is strongly advised to draw a *free-body diagram* as the *first step* of the problem solution. Provided the free-body diagram is correctly drawn, the effect of all the forces acting on the particle is then taken into account when the equations of equilibrium are written. To construct a free-body diagram for a particle, the particle is isolated from its surroundings, an outlined shape of the particle is sketched, and *all* the forces acting *on the particle* are labeled on this sketch. All forces are either *active forces,* which tend to set the particle in motion, for example, weight, magnetic, and electrostatic interaction; or *reactive forces,* such as those caused by the surroundings, which tend to prevent motion. The forces on the free-body diagram which are known should be labeled with their proper magnitudes and directions. Letters are used to represent the magnitudes and directions of forces which are unknown.

Cables are often encountered in equilibrium problems. Except for Sec. 7–4, throughout this book all cables will be assumed to have negligible weight and to be inextensible, that is, they cannot be stretched. Such cables carry only tension forces, and these forces are always directed along the axes of the cables. In Chapter 6, it will be shown that the tension developed in a continuous cable, which passes over a frictionless pulley, must be *constant* to keep the cable in equilibrium.* For example, for *any angle θ,* a 5-lb force is needed to lift the 5-lb block shown in Fig. 3–2.

$\mathbf{T} = 5$ lb   5 lb

Fig. 3–2

If an *elastic spring* is attached to a cable, the length of the cable and spring will change in proportion to the tensile force developed in the cable. A characteristic which defines the elasticity of a spring is the

---

*If the pulley has an angular acceleration, this statement is no longer true. This case is discussed in *Engineering Mechanics: Dynamics.*

stiffness k. The magnitude of force, developed by an elastic spring which has a stiffness k and is deformed (compressed or elongated) a distance x, measured from its unloaded position, is $F = kx$. The spring shown in Fig. 3–3 has a stiffness of $k = 5$ lb/in. A 10-lb force is therefore needed to hold the spring at a distance of $x = \pm 2$ in. from its unstretched position. [$F = kx = (5$ lb/in.)$(2$ in.$) = 10$ lb]

Fig. 3–3

A special case of equilibrium occurs when a *rigid body is subjected to a system of concurrent forces.* As an example, consider the 10-lb sphere shown in Fig. 3–4a, which is supported by the rope AB and the smooth plane. A free-body diagram of the sphere is shown in Fig. 3–4b. There

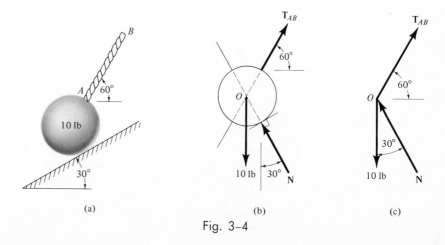

(a)   (b)   (c)

Fig. 3–4

are three forces acting *on* the sphere. The tensile force $\mathbf{T}_{AB}$ has an un-known magnitude of T_{AB} lb and acts in the direction of the rope. The reactive force of the plane has an unknown magnitude of N lb. Since the inclined surface is *smooth*, this force will act *normal to the tangent of the surface at the point of contact*.† Finally, there is the 10-lb weight of the sphere which represents the effect of gravity, and therefore acts downward toward the center of the earth. As you will recall from your study of physics, the weights for each particle of the sphere contribute to the total weight of the sphere. Provided the material of the sphere is homogeneous, the total weight acts through the geometric center O. (See Chapter 5 for further discussion.) As shown in Fig. 3–4*b*, the *lines of action* of these three forces intersect at the common point O. In Sec. 1–4, it was pointed out that a force acting on a rigid body may be considered as a sliding vector when analyzing the conditions of equilibrium for the rigid body (principle of transmissibility). Hence, for equilibrium of the sphere, the force system acting on the sphere may be considered *concurrent* at point O, as shown in Fig. 3–4*c*, and must therefore satisfy the same equilibrium condition as for a particle, i.e., Eq. 3–1.

Two other examples of free-body diagrams for a particle are given in Figs. 3–5 and 3–6. Besides its weight, the 5-lb ball in Fig. 3–5*a* is acted upon by the unknown force in the spring, $\mathbf{F}_s$, and the 6-lb vertical force. The free-body diagram, showing all these forces, is given in Fig. 3–5*b*. Point A located on the top of the mast in Fig. 3–6*a* is acted upon by four forces, namely, the tension forces $\mathbf{F}_{AB}$ and $\mathbf{F}_{AC}$ due to the cables AB and AC, the force of the banner, $\mathbf{F}_B$, and the reaction $\mathbf{F}_P$, which the pole AF exerts on A. The free-body diagram is shown in Fig. 3–6*b*. The force $\mathbf{F}_P$ acts at an angle with the vertical since the pole restrains *all* spatial movements of point A. Note that the force which the cable DE exerts on the pole is *not* shown on the free-body diagram of point A. This force acts on the pole at D, not at A.

(a)   (b)

Fig. 3–5

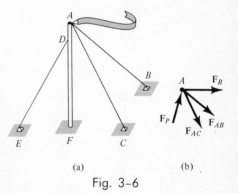

(a)   (b)

Fig. 3–6

3–3. Two-Dimensional Force Systems Acting on a Particle

In some instances, the forces acting on a particle may all lie in the *same plane*, that is, the force system is *coplanar*. If this is the case, only *two* scalar equations must be satisfied for equilibrium. For an (x, y) Cartesian coordinate system,

$$\Sigma F_x = 0$$
$$\Sigma F_y = 0 \tag{3-4}$$

†Rough surfaces of contact contribute frictional forces which act tangent to the surface at points of contact. This type of force is discussed in Chapter 8.

These two equations can be solved for, at most, two unknowns, generally represented as angles and/or magnitudes of forces shown on the free-body diagram of the particle.

When constructing a free-body diagram, the sense of direction of a force having an unknown magnitude can be assumed. The correctness of the assumed sense of direction will become apparent after solving the equilibrium equations for the unknown quantity. By definition, the *magnitude* of a force is *always positive,* hence, if the solution yields a positive magnitude, the sense of direction of the unknown force was assumed correctly; whereas, if the solution yields a negative magnitude, the minus sign indicates that the sense of direction is opposite to that which was assumed. As a simple example to illustrate this point again consider the free-body diagram of the 5-lb ball shown in Fig. 3–5b. The force $\mathbf{F}_s$ on this diagram represents the effect of the spring *on the ball.* The sense of direction of this force has been assumed downward as shown. Applying the equilibrium equation in the vertical direction to the force system, with the positive direction *assumed upward,**

$$+\uparrow\Sigma F_y = 0; \qquad\qquad -F_s - 6\,\text{lb} - 5\,\text{lb} = 0$$

Hence, $$F_s = -11\,\text{lb}$$

The negative sign indicates that the sense of direction of $\mathbf{F}_s$ is actually *upward* and not downward as shown on the free-body diagram in Fig. 3–5b.

Coplanar equilibrium problems may also be solved by using a graphical procedure. To satisfy vector Eq. 3–1, it is required that all forces acting on the particle form a closed polygon. Under the action of only three forces, the force polygon closes to form a triangle. When constructed, the unknowns may be found using either the *law of sines* and/or the *law of cosines* (see Example 3–2). In either case, a vector analysis (using Eq. 3–1), a scalar analysis (using Eqs. 3–4), or the force-polygon method leads to the same results.

Example 3–1

A 10-lb block is held in equilibrium by two cords of negligible weight, as shown in Fig. 3–7a. The cord *ABC* is continuous and passes over a frictionless pulley at *B.* Determine the tension in cord *AD* and the weight *W* needed to hold the 10-lb block in the equilibrium position.

Solution I   (Vector Analysis)
The first step in the solution is to draw a free-body diagram of point

*The downward direction can also be assumed positive when applying $\Sigma F_y = 0$. Check and see that you obtain the same result.

Fig. 3–7

A. Why? There are three concurrent forces acting *on* this point. Cable *AE* exerts a downward pull on *A* having a magnitude of 10 lb. The force in cable *AD* has an unknown magnitude T_{AD} and is directed from *A* to *D*. Since cable *ABC* is continuous, an unknown tension force having a magnitude of *W* lb acts on *A* and is directed from *A* to *B*. The free-body diagram for point *A* is shown in Fig. 3–7*b*.

Expressing each of the three forces in Cartesian vector form, we have

$$\mathbf{T}_{AD} = -T_{AD}\mathbf{i}$$
$$\mathbf{W} = W(\cos 30°\mathbf{i} + \sin 30°\mathbf{j}) = 0.866\,W\mathbf{i} + 0.500\,W\mathbf{j}$$
$$\mathbf{F} = \{-10\mathbf{j}\}\text{ lb}$$

Applying Eq. 3–1, in reference to the free-body diagram,

$$\Sigma\mathbf{F} = 0; \qquad\qquad \mathbf{T}_{AD} + \mathbf{W} + \mathbf{F} = 0$$
$$-T_{AD}\mathbf{i} + 0.866\,W\mathbf{i} + 0.500\,W\mathbf{j} - 10\mathbf{j} = 0$$

Equating the **i** and **j** components to zero, we have

$$\xrightarrow{+}\Sigma F_x = 0; \qquad\qquad -T_{AD} + 0.866\,W = 0$$
$$+\uparrow\Sigma F_y = 0; \qquad\qquad 0.500\,W - 10 = 0$$

Solving these two equations yields

$$W = 20.0\text{ lb} \qquad\qquad\qquad Ans.$$
$$T_{AD} = 17.3\text{ lb} \qquad\qquad\qquad Ans.$$

Solution II   (Scalar Analysis)

After the free-body diagram for point *A* has been established, Eqs. 3–4 may be applied *directly* to the forces shown on this diagram, since vector components are easily obtained in two dimensions, Fig. 3–7*b*. Hence,

$$\xrightarrow{+}\Sigma F_x = 0; \qquad\qquad -T_{AD} + W\cos 30° = 0$$
$$+\uparrow\Sigma F_y = 0; \qquad\qquad W\sin 30° - 10 = 0$$

These equations are the same as those previously obtained.

If we compare the two solutions, it is clearly seen that expressing each force as a Cartesian vector (Solution I) prior to applying the equations of equilibrium has no added advantage when solving problems in two dimensions. Because of the extra labor involved, it is suggested that problems involving coplanar force systems be solved using a scalar analysis (Solution II). The real advantage of vector analysis will be realized when problems in three dimensions are considered.

Example 3–2

Two cables are used to support the 50-lb crate shown in Fig. 3–8a. Determine the tensile force developed in the cables.

Solution

The free-body diagram of point A is shown in Fig. 3–8b. There are

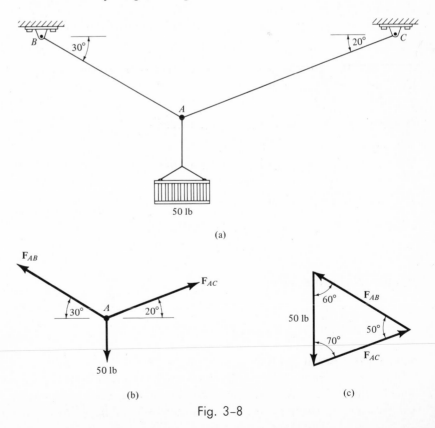

50 lb

(a)

F_{AB}

F_{AC}

30°   20°

A

50 lb

(b)

50 lb

60° F_{AB}

50°

70°

F_{AC}

(c)

Fig. 3–8

three forces which act *on* this point—the cable forces $\mathbf{F}_{AB}$ and $\mathbf{F}_{AC}$, and the 50-lb weight. The two unknowns F_{AB} and F_{AC} can be obtained in the usual manner of applying the two equations of equilibrium. (See Example 3–1.) Here, however, we will illustrate the use of the force-polygon method for the solution.

Drawing the force polygon, the vectors on the free-body diagram are added in a "head-to-tail" fashion forming the *closed* triangle shown in Fig. 3–8c. Applying the law of sines to this triangle, we have

$$\frac{F_{AC}}{\sin 60^\circ} = \frac{50}{\sin 50^\circ}$$

$$F_{AC} = 50\left(\frac{0.866}{0.766}\right) = 56.5 \text{ lb} \qquad\qquad Ans.$$

$$\frac{F_{AB}}{\sin 70^\circ} = \frac{50}{\sin 50^\circ}$$

$$F_{AB} = 50\left(\frac{0.940}{0.766}\right) = 61.3 \text{ lb} \qquad\qquad Ans.$$

Example 3–3

The spring-loaded supporting rods AB and AC in Fig. 3–9a are used as shock absorbers for transporting fragile loads. Knowing that the elastic spring constants for the rods AB and AC are $k_{AB} = 30$ lb/in. and $k_{AC} = 50$ lb/in., respectively, determine how far point A moves vertically when the 1,000-lb load is removed. Neglect the weight of the rods.

Solution

If we know the forces $\mathbf{F}_{AB}$ and $\mathbf{F}_{AC}$ developed in rods AB and AC, we can then determine the *extension* of each rod when the rods support the 1,000-lb load. Using geometry, we may then arrive at the undeflected position of point A when the load is removed.

The free-body diagram of point A is shown in Fig. 3–9b. Applying the scalar equations of equilibrium, using the geometry to determine the sines and cosines of the angles, we have

$$\xrightarrow{+}\Sigma F_x = 0; \qquad \frac{4}{\sqrt{(4)^2 + (5)^2}}F_{AB} - \frac{6}{\sqrt{(6)^2 + (5)^2}}F_{AC} = 0$$

or
$$0.625 F_{AB} - 0.768 F_{AC} = 0 \qquad\qquad (1)$$

$$+\uparrow\Sigma F_y = 0;$$

$$\frac{5}{\sqrt{(4)^2 + (5)^2}}F_{AB} + \frac{5}{\sqrt{(6)^2 + (5)^2}}F_{AC} - 1,000 = 0$$

or
$$0.781 F_{AB} + 0.640 F_{AC} - 1,000 = 0 \qquad\qquad (2)$$

The tension forces in the rods can be determined by solving Eqs. (1) and (2) simultaneously. In particular, solving for F_{AB} in Eq. (1),

$$F_{AB} = 1.229\, F_{AC} \qquad (3)$$

Substituting this into Eq. (2),

$$0.781(1.229\, F_{AC}) + 0.640\, F_{AC} - 1,000 = 0$$

or

$$1.600\, F_{AC} = 1,000$$

Thus,

$$F_{AC} = 625 \text{ lb} \qquad \textit{Ans.}$$

Using Eq. (3),

$$F_{AB} = 1.229(625 \text{ lb})$$
$$F_{AB} = 768 \text{ lb} \qquad \textit{Ans.}$$

The forces shown on the free-body diagram of point A represent the effect of the rods and the weight *on* point A. According to Newton's third law of motion, point A must exert *equal, but opposite forces* on the rod and weight, as shown in Fig. 3–9c. (Note that the weight and rods are *not* represented as complete free-body diagrams in this figure.) The *extension* of each rod created by forces $\mathbf{F}_{AB}$ and $\mathbf{F}_{AC}$ is therefore

$$F_{AB} = k_{AB}\,\Delta x_{AB}; \quad \Delta x_{AB} = \frac{768 \text{ lb}}{30 \text{ lb/in.}} = 25.6 \text{ in.} = 2.13 \text{ ft}$$

$$F_{AC} = k_{AC}\,\Delta x_{AC}; \quad \Delta x_{AC} = \frac{625 \text{ lb}}{50 \text{ lb/in.}} = 12.5 \text{ in.} = 1.04 \text{ ft}$$

Using the Pythagorean theorem, we may determine the *extended length* of each rod when subjected to the 1,000-lb load. Referring to Fig. 3–9a, we have

$$l_{AB} = \sqrt{(4)^2 + (5)^2} = 6.40 \text{ ft}$$
$$l_{AC} = \sqrt{(6)^2 + (5)^2} = 7.81 \text{ ft}$$

Thus, the *unloaded length* of the rods becomes

$$l'_{AB} = l_{AB} - \Delta x_{AB} = 6.40 - 2.13 = 4.27 \text{ ft}$$
$$l'_{AC} = l_{AC} - \Delta x_{AC} = 7.81 - 1.04 = 6.77 \text{ ft}$$

With reference to Fig. 3–9d, when the weight is *removed,* point A moves to location (x, y). The distance y may be related to the distance x by the equations

$$y^2 + x^2 = (6.77)^2 \qquad (4)$$
$$y^2 + (10 - x)^2 = (4.27)^2 \qquad (5)$$

three forces which act *on* this point—the cable forces $\mathbf{F}_{AB}$ and $\mathbf{F}_{AC}$, and the 50-lb weight. The two unknowns F_{AB} and F_{AC} can be obtained in the usual manner of applying the two equations of equilibrium. (See Example 3–1.) Here, however, we will illustrate the use of the force-polygon method for the solution.

Drawing the force polygon, the vectors on the free-body diagram are added in a "head-to-tail" fashion forming the *closed* triangle shown in Fig. 3–8c. Applying the law of sines to this triangle, we have

$$\frac{F_{AC}}{\sin 60°} = \frac{50}{\sin 50°}$$

$$F_{AC} = 50\left(\frac{0.866}{0.766}\right) = 56.5 \text{ lb} \qquad\qquad \textit{Ans.}$$

$$\frac{F_{AB}}{\sin 70°} = \frac{50}{\sin 50°}$$

$$F_{AB} = 50\left(\frac{0.940}{0.766}\right) = 61.3 \text{ lb} \qquad\qquad \textit{Ans.}$$

Example 3–3

The spring-loaded supporting rods AB and AC in Fig. 3–9a are used as shock absorbers for transporting fragile loads. Knowing that the elastic spring constants for the rods AB and AC are $k_{AB} = 30$ lb/in. and $k_{AC} = 50$ lb/in., respectively, determine how far point A moves vertically when the 1,000-lb load is removed. Neglect the weight of the rods.

Solution

If we know the forces $\mathbf{F}_{AB}$ and $\mathbf{F}_{AC}$ developed in rods AB and AC, we can then determine the *extension* of each rod when the rods support the 1,000-lb load. Using geometry, we may then arrive at the undeflected position of point A when the load is removed.

The free-body diagram of point A is shown in Fig. 3–9b. Applying the scalar equations of equilibrium, using the geometry to determine the sines and cosines of the angles, we have

$$\overset{+}{\rightarrow}\Sigma F_x = 0; \quad \frac{}{\sqrt{(4)^2}} F_{AB} - \frac{6}{\sqrt{(6)^2 + (5)^2}} F_{AC} = 0$$

or

$$0.768 F_{AC} = 0 \qquad\qquad (1)$$

$$+\uparrow\Sigma F_y = 0; \quad \frac{5}{\sqrt{(4)^2 + (5)^2}} F_{A}$$

$$0.781 F_{AB} \quad F_{AC} - 1{,}000 = 0$$

or

$$00 = 0 \qquad\qquad (2)$$

The tension forces in the rods can be determined by solving Eqs. (1) and (2) simultaneously. In particular, solving for F_{AB} in Eq. (1),

$$F_{AB} = 1.229\, F_{AC} \tag{3}$$

Substituting this into Eq. (2),

$$0.781(1.229\, F_{AC}) + 0.640\, F_{AC} - 1,000 = 0$$

or

$$1.600\, F_{AC} = 1,000$$

Thus,

$$F_{AC} = 625 \text{ lb} \hspace{3cm} \textit{Ans.}$$

Using Eq. (3),

$$F_{AB} = 1.229(625 \text{ lb})$$
$$F_{AB} = 768 \text{ lb} \hspace{3cm} \textit{Ans.}$$

The forces shown on the free-body diagram of point A represent the effect of the rods and the weight *on* point A. According to Newton's third law of motion, point A must exert *equal, but opposite forces* on the rod and weight, as shown in Fig. 3–9c. (Note that the weight and rods are *not* represented as complete free-body diagrams in this figure.) The *extension* of each rod created by forces $\mathbf{F}_{AB}$ and $\mathbf{F}_{AC}$ is therefore

$$F_{AB} = k_{AB}\,\Delta x_{AB}; \quad \Delta x_{AB} = \frac{768 \text{ lb}}{30 \text{ lb/in.}} = 25.6 \text{ in.} = 2.13 \text{ ft}$$

$$F_{AC} = k_{AC}\,\Delta x_{AC}; \quad \Delta x_{AC} = \frac{625 \text{ lb}}{50 \text{ lb/in.}} = 12.5 \text{ in.} = 1.04 \text{ ft}$$

Using the Pythagorean theorem, we may determine th*xtended length* of each rod when subjected to the 1,000-lb load. Ref ^{4g} to Fig. 3–9a, we have

$$l_{AB} = \sqrt{(4)^2 + (5)^2} = 6.4$$
$$l_{AC} = \sqrt{(6)^2 + (5)^2} =$$

Thus, the *unloaded length* of the rods $= 4.27 \text{ ft}$

$$l'_{AB} = l_{AB} - \Delta x_{AB} = \qquad = 6.77 \text{ ft}$$
$$l'_{AC} = l_{AC} - \Delta x_A \text{elated to the distance } x \text{ by the}$$

With reference to Fig. 3–9d $(6.77)^2$
to location (x, y). The dis $)^2 = (4.27)^2$
equations

$$\tag{4}$$
$$\tag{5}$$

(a)

(b)

(c)

(d)

Fig. 3–9

Solving for y^2 in Eq. (4), and substituting into Eq. (5) yields

$$[(6.77)^2 - x^2] + (10 - x)^2 = (4.27)^2$$

or

$$20x - 127.6 = 0$$

Hence,

$$x = 6.38 \text{ ft}$$

From Eq. (4),

$$y^2 = (6.77)^2 - (6.38)^2 = 5.13$$

Choosing the positive square root for the solution of y, since it represents the physical situation,

$$y = 2.26 \text{ ft}$$

From Fig. 3–9d, point A therefore moves vertically a distance of

$$\Delta y = 5 \text{ ft} - y = 5 \text{ ft} - 2.26 \text{ ft}$$
$$\Delta y = 2.74 \text{ ft}$$ *Ans.*

when the 1,000 lb load is removed.

Problems

3-1. Particle P is subjected to the concurrent force system shown. Determine the magnitude of **F** and its direction θ which is required to keep the particle in equilibrium.

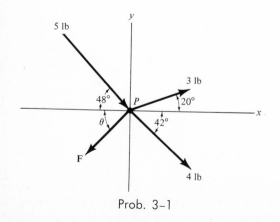

Prob. 3–1

3-2. The joint of a light metal truss is formed by riveting four angles to the gusset plate G. Knowing the force in members A and C, determine the compressive forces F_B and F_D acting on members B and D which are required for equilibrium. The force system is concurrent at point O.

Prob. 3–2

3-3. The particle P is subjected to the action of four coplanar forces. Determine the magnitude of $\mathbf{F}$ and the angle θ for equilibrium.

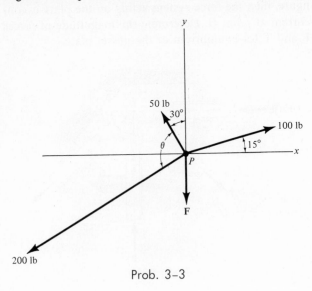

Prob. 3-3

3-4. Provided $\theta = 30°$, determine the forces $\mathbf{F}_A$ and $\mathbf{F}_B$ required to hold the 5-lb sphere in equilibrium. At what angle θ will the force $\mathbf{F}_B$ be a minimum? Assume that the line of action of each force is directed toward the center point O of the sphere.

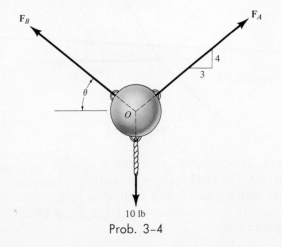

Prob. 3-4

3-5. Determine the length of string l so that $\theta = 20°$ and the spring AB remains horizontal. The unstretched length of AB is 1 ft.

Prob. 3-5

3-6. The links AB, BC, and CD support the loading shown. Determine the tension in AB and the angle θ for equilibrium. The force in each link acts along the axis of the link.

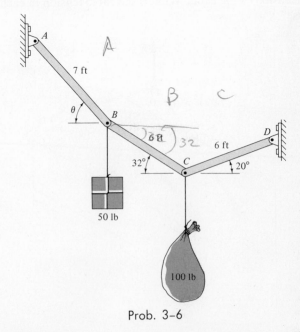

Prob. 3-6

57

3-7. The 25-lb pipe rests between the two smooth inclined planes. If plane A exerts a reaction of 8 lb on the pipe, determine the magnitude of the applied force P and the reaction that plane B exerts on the pipe.

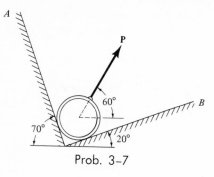

Prob. 3-7

3-8. Determine the magnitude of the maximum force P required to hold the 25-lb pipe of Prob. 3-7 in equilibrium. Assume that P acts at $60°$, as shown.

3-9. The 200-lb crate is suspended by means of a 6-ft long cable attached to the sides of the crate and which passes over the small pulley located at O. If the cable can be attached at either points A and B, or C and D, determine which attachment produces the least amount of tension in the cable. What is the cable tension in each case?

Prob. 3-9

3-10. The members of a truss are connected to the gusset plate. If the force acting in each member is directed along the axis of the member, as shown in the figure, then the force system acting on the plate is concurrent at point O. Determine the magnitude of forces F and T for equilibrium of the gusset plate.

Prob. 3-10

3-11. Determine the magnitude of R and the angle θ required for equilibrium of particle P.

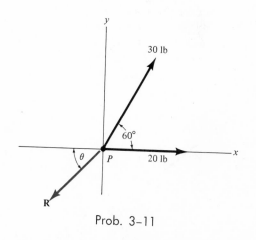

Prob. 3-11

3-12. Two small blocks A and B rest on the smooth inclines. The blocks are connected by a light inextensible cord which passes over a frictionless pulley. If block A weighs 15 lb, determine the weight W_B of block B for equilibrium.

Prob. 3-12

3-13. The 10-lb roller mechanism is supported by two springs. If the springs are unstretched when the roller is at the equilibrium position A, determine the stiffness k so that the roller will be in equilibrium when the height $h = 1$ ft. Assume that the surfaces of contact are frictionless. Neglect the size of the roller.

Side View

Top View

Prob. 3-13

3-14. Determine the distance s for the location of the horizontal strut AB so that a tension of 140 lb is developed in each of the links OA and OB. The force in each link acts along the axis of the link.

Prob. 3-14

3-15. The two spheres A and B have an equal weight and are electrostatically charged, such that the repulsive force acting between them has a magnitude of 0.6 lb and is directed along the dashed line. Determine the angle θ, the tension in the cables, and the weight W of each sphere.

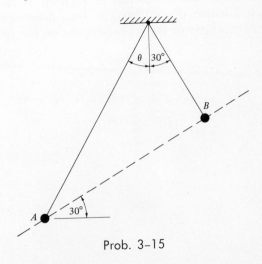

Prob. 3-15

59

3–16. The roller mechanism AB is at rest between the smooth inclines of the 30° wedge. If each of the rollers weighs 5 lb, and the undeformed length of each spring is 4 in., determine the distance h for equilibrium of the mechanism. Assume that both rollers remain in the same horizontal plane aa. The spring constant for each spring is $k = 1$ lb/in. Neglect the size of the roller.

Side View   Top View

Prob. 3–16

3–17. The 50-lb sphere rests between two smooth inclined planes and is attached to a spring. If the unstretched length of the spring is 4 in., determine the normal reactions of the inclines acting on the sphere.

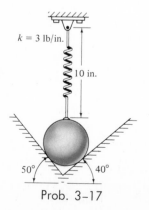

Prob. 3–17

3–18. Three weights are connected by the light inextensible cords shown. If the cords pass over small frictionless pulleys, determine the distance s for equilibrium.

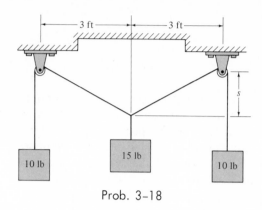

Prob. 3–18

3–19. A continuous cable of total length 40 in. is wrapped around the *small* frictionless pulleys at A, B, C, and D. If the spring constants are both $k = 3$ lb/in. and the springs are each stretched 6 in., determine the weight W of each block. Neglect the weight of the pulleys and cords. The springs are unstretched when $d = 20$ in.

Prob. 3–19

3-20. A small 5-lb sphere is resting on the smooth parabolic surface. Determine the force it exerts on the surface and the weight W needed to hold it in the equilibrium position shown.

Prob. 3-20

3-4. Three-Dimensional Force Systems Acting on a Particle

The general force system acting on the particle is, of course, three-dimensional. *After* the free-body diagram for the particle is established, all the forces acting on the particle should be expressed in Cartesian vector form. Equation 3-1 may then be applied to solve for the unknown quantities. This vector equation actually represents the *three* scalar Eqs. 3-3. Thus, for a complete solution, there can be at most three unknown force magnitudes and/or directions on the free-body diagram. (In problems where two direction angles for a force are known, the third angle is specified by using Eq. 2-8, that is, $\cos^2 \alpha + \cos^2 \beta + \cos^2 \gamma = 1$.) The resulting equilibrium equations generally must be solved simultaneously to obtain a complete solution. If these equations form a set of linear algebraic equations, which is most often the case, it may be expedient to solve them by means of the computer program given in Appendix *A*.

Example 3-4

The small 50-lb block shown in Fig. 3-10*a* rests on a smooth inclined plane and is prevented from slipping down the plane by the two guy

61

Fig. 3–10

wires AC and AB. Determine the tension developed in these wires and the reactive force at the contacting surface.

Solution

A free-body diagram of the block is shown in Fig. 3–10b. Why do we have to construct this diagram first? There are four forces acting *on* the block: the tensile forces $\mathbf{F}_{AC}$ and $\mathbf{F}_{AB}$ developed by the wires, the weight $\mathbf{W}$ of the block acting vertically downward, and the normal force $\mathbf{N}$ which the smooth plane exerts on the block. Having accounted for all the forces that act on the block, we may express each force as a vector.

$$\mathbf{F}_{AC} = F_{AC}\frac{\mathbf{r}_{AC}}{r_{AC}} = F_{AC}\left(\frac{\mathbf{i} - 2\mathbf{j} + 8\mathbf{k}}{\sqrt{69}}\right) = \frac{F_{AC}}{\sqrt{69}}\mathbf{i} - \frac{2F_{AC}}{\sqrt{69}}\mathbf{j} + \frac{8F_{AC}}{\sqrt{69}}\mathbf{k}$$

$$\mathbf{F}_{AB} = F_{AB}\frac{\mathbf{r}_{AB}}{r_{AB}} = F_{AB}\left(\frac{-5\mathbf{i} + 6\mathbf{k}}{\sqrt{61}}\right) = -\frac{5F_{AB}}{\sqrt{61}}\mathbf{i} + \frac{6F_{AB}}{\sqrt{61}}\mathbf{k}$$

$$\mathbf{N} = N(\sin 20°\mathbf{j} + \cos 20°\mathbf{k}) = (N\sin 20°)\mathbf{j} + (N\cos 20°)\mathbf{k}$$

$$\mathbf{W} = 50 \text{ lb}(-\mathbf{k}) = \{-50\mathbf{k}\} \text{ lb}$$

Applying Eq. 3–1 gives

$$\Sigma\mathbf{F} = 0; \qquad\qquad \mathbf{F}_{AC} + \mathbf{F}_{AB} + \mathbf{N} + \mathbf{W} = 0$$

or

$$\frac{F_{AC}}{\sqrt{69}}\mathbf{i} - \frac{2F_{AC}}{\sqrt{69}}\mathbf{j} + \frac{8F_{AC}}{\sqrt{69}}\mathbf{k} - \frac{5F_{AB}}{\sqrt{61}}\mathbf{i} + \frac{6F_{AB}}{\sqrt{61}}\mathbf{k}$$

$$+ (N\sin 20°)\mathbf{j} + (N\cos 20°)\mathbf{k} - 50\mathbf{k} = 0$$

Equating the components in the $\mathbf{i}$, $\mathbf{j}$, and $\mathbf{k}$ directions equal to zero, we obtain

$$\Sigma F_x = 0; \qquad \frac{1}{\sqrt{69}} F_{AC} - \frac{5}{\sqrt{61}} F_{AB} = 0 \qquad\qquad (1)$$

$$\Sigma F_y = 0; \qquad -\frac{2}{\sqrt{69}} F_{AC} + N \sin 20° = 0 \qquad\qquad (2)$$

$$\Sigma F_z = 0; \qquad \frac{8}{\sqrt{69}} F_{AC} + \frac{6}{\sqrt{61}} F_{AB} + N \cos 20° - 50 = 0 \qquad\qquad (3)$$

The three unknowns in the above equations may be obtained by simultaneous solution of the equations. For example, solve for F_{AB} in Eq. (1) and for N in Eq. (2)—both in terms of F_{AC}. Substitute these results in Eq. (3), thereby obtaining an equation in terms of the single unknown F_{AC}. After solving for F_{AC}, it is possible to obtain the values of F_{AB} and N by resubstituting into Eqs. (1) and (2) the value of F_{AC}. (Recognizing Eqs. (1–3) to be a set of three linear algebraic equations, the computer program in Appendix A may also be used for the solution.) The results are

$$F_{AC} = 28.3 \text{ lb} \qquad\qquad Ans.$$
$$F_{AB} = 5.31 \text{ lb} \qquad\qquad Ans.$$
$$N = 19.9 \text{ lb} \qquad\qquad Ans.$$

Example 3–5

A 1,000-lb crate is supported by three struts CF, CD, and CE and the cable arrangement shown in Fig. 3–11a. Determine the forces developed in the struts and the tension in the cables. Assume that the force developed in each strut passes along the axis of the strut. These forces may either be tensile or compressive.

Solution

It is first necessary to draw a free-body diagram of point C in order to expose the unknown forces in the struts and the cable BC. As shown in Fig. 3–11b, the four forces acting *on* this point are: the cable tension F_{CB}, and the forces in the three struts F_{CD}, F_{CE}, and F_{CF}. Due to geometry, the line of action (or direction) of each of the four forces is *known*. However, the sense of direction of the forces acting in the struts is not known, and it has been assumed that F_{CD} and F_{CE} are tensile forces; that is, struts CD and CE *pull* on point C. Similarly, the force F_{CF} is assumed to be compressive so that strut CF *pushes* on point C. Since only *three* scalar equilibrium equations can be applied to the force system acting on the free-body diagram at point C, it is necessary to solve for one of the four unknown force magnitudes with the aid of *another* free-body diagram. A free-body diagram of point B is shown in Fig. 3–11c. (Why is this a free-body diagram?) Since the force system at B is *coplanar*, the

forces in the cables $\mathbf{F}_{BA}$ and $\mathbf{F}_{BC}$ may be determined using the *two* scalar equations of equilibrium, namely,

$$\xrightarrow{+}\Sigma F_y = 0; \qquad\qquad F_{BC}\cos 30° - F_{BA} = 0$$
$$+\uparrow\Sigma F_z = 0; \qquad\qquad F_{BC}\sin 30° - 1{,}000 = 0$$

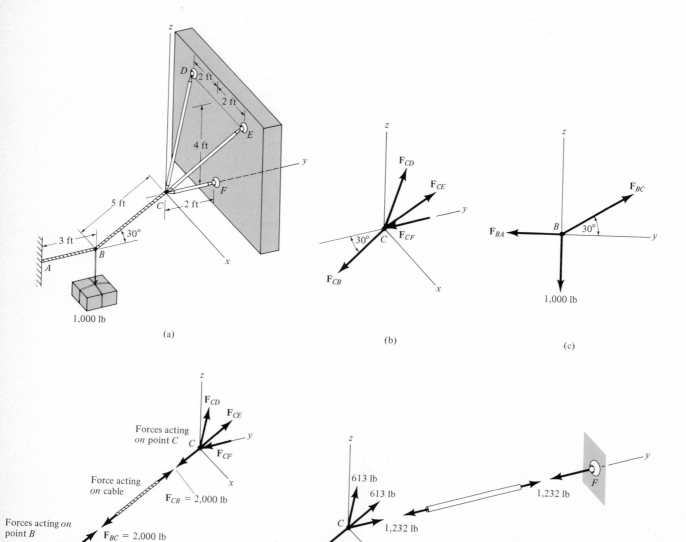

Fig. 3–11

Solving, we have

$$F_{BA} = 1,732 \text{ lb} \qquad \text{Ans.}$$
$$F_{BC} = 2,000 \text{ lb} \qquad \text{Ans.}$$

For every action there is an equal and opposite reaction. Therefore, force F_{BC} has a directional sense, as shown on each of the three free-body diagrams in Fig. 3–11d. (Notice how a *pull* on point B creates a *tension* in the rope.)

Since the magnitude of $\mathbf{F}_{BC}$ is known, the magnitudes of the other three forces acting at point C may now be computed. Expressing each of the forces acting on the free-body diagram of point C (Fig. 3–11b) as a vector, we have

$$\mathbf{F}_{CB} = (2,000 \text{ lb})\mathbf{u}_{CB} = 2,000 \text{ lb}(-\cos 30°\mathbf{j} - \sin 30°\mathbf{k})$$
$$= \{-1,732\mathbf{j} - 1,000\mathbf{k}\} \text{ lb}$$

$$\mathbf{F}_{CD} = F_{CD}\frac{\mathbf{r}_{CD}}{r_{CD}} = -\frac{2F_{CD}}{\sqrt{24}}\mathbf{i} + \frac{2F_{CD}}{\sqrt{24}}\mathbf{j} + \frac{4F_{CD}}{\sqrt{24}}\mathbf{k}$$

$$\mathbf{F}_{CE} = F_{CE}\frac{\mathbf{r}_{CE}}{r_{CE}} = \frac{2F_{CE}}{\sqrt{24}}\mathbf{i} + \frac{2F_{CE}}{\sqrt{24}}\mathbf{j} + \frac{4F_{CE}}{\sqrt{24}}\mathbf{k}$$

$$\mathbf{F}_{CF} = F_{CF}(-\mathbf{j}) = -F_{CF}\mathbf{j} \quad \text{(By inspection, Fig. 3–11b, note that } \mathbf{F}_{CF} \text{ acts}$$
$$\text{in the } -\mathbf{j} \text{ direction.)}$$

Applying Eq. 3–1 yields

$$\Sigma\mathbf{F} = 0; \qquad \mathbf{F}_{CB} + \mathbf{F}_{CD} + \mathbf{F}_{CE} + \mathbf{F}_{CF} = 0$$

or

$$-1,732\mathbf{j} - 1,000\mathbf{k} - \frac{2F_{CD}}{\sqrt{24}}\mathbf{i} + \frac{2F_{CD}}{\sqrt{24}}\mathbf{j} + \frac{4F_{CD}}{\sqrt{24}}\mathbf{k} + \frac{2F_{CE}}{\sqrt{24}}\mathbf{i} + \frac{2F_{CE}}{\sqrt{24}}\mathbf{j}$$
$$+ \frac{4F_{CE}}{\sqrt{24}}\mathbf{k} - F_{CF}\mathbf{j} = 0$$

Thus, equating the respective $\mathbf{i}$, $\mathbf{j}$, and $\mathbf{k}$ components equal to zero,

$$\Sigma F_x = 0; \qquad -\frac{2}{\sqrt{24}}F_{CD} + \frac{2}{\sqrt{24}}F_{CE} = 0$$

$$\Sigma F_y = 0; \qquad -1,732 + \frac{2}{\sqrt{24}}F_{CD} + \frac{2}{\sqrt{24}}F_{CE} - F_{CF} = 0$$

$$\Sigma F_z = 0; \qquad -1,000 + \frac{4}{\sqrt{24}}F_{CD} + \frac{4}{\sqrt{24}}F_{CE} = 0$$

The first and third of these equations may be solved simultaneously for F_{CD} and F_{CE}. Substituting these values into the second equation gives F_{CF}. The results are

$$F_{CD} = 612.4 \text{ lb} \qquad \qquad \textit{Ans.}$$
$$F_{CE} = 612.4 \text{ lb} \qquad \qquad \textit{Ans.}$$
$$F_{CF} = -1{,}232 \text{ lb} \qquad \qquad \textit{Ans.}$$

Since $\mathbf{F}_{CF}$ is calculated as a *negative* quantity, this force, as drawn on the free-body diagrams in Figs. 3–11*b* and 3–11*d*, acts in the opposite sense of direction. Strut *CF* must therefore exert a force of 1,232 lb on point *C*, acting as shown in Fig. 3–11*e*. With reference to this figure, given the equal but opposite reactions, strut *CF* is subjected to a tensile (elongation) force of 1,232 lb because of the weight of the crate. (Can you explain why the supporting bracket at *F* in Fig. 3–11*e* does not represent a complete free-body diagram?)

Problems

3–21. The small sphere having a weight of 10 lb is forced up against the corner of three smooth mutually perpendicular planes *A*, *B*, and *C*. The applied force **F** has a line of action which passes from point *P* (2 ft, 2 ft, 3 ft) through the center of the sphere. Determine the normal reactions of the three planes acting on the sphere.

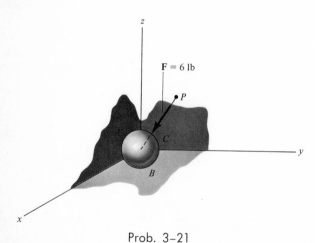

Prob. 3–21

3–22. A 1,500-lb homogeneous crate is supported in the horizontal plane by means of four cables. Determine the magnitude of force developed in each of these cables if all the cables have an equal length.

Prob. 3–22

3-23. The joint of a space frame is subjected to three forces. Strut OA lies in the xy plane and strut OB lies in the yz plane. Determine the forces acting in each of the struts required for equilibrium of the joint.

Prob. 3–23

3-24. Determine the magnitude and direction of the force **P** required to keep the concurrent force system in equilibrium.

Prob. 3–24

3-25. Determine the tension developed in each of the three cables and the strut OC used to support the 500-lb crate. The spring cable OA has an unstretched length of 2 ft and a stiffness $k_{OA} = 10$ lb/in. The force in the strut acts along the axis of the strut.

Prob. 3–25

3-26. Three electrostatically charged balls, each having a weight of 4.4×10^{-4} lb, are suspended from a common point. If the electrostatic repulsive force acting between any two balls is 1.0×10^{-3} lb and is directed along a line of action joining these two balls, determine the tension developed in each cord for equilibrium.

Prob. 3–26

3–27. Three 10-lb blocks and one 15-lb block are suspended from the pulley and cable system. If the pulleys are frictionless and the weights of the cables are negligible, determine the sag *s* for equilibrium of the system.

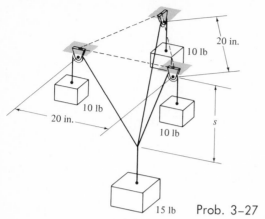

15 lb   Prob. 3–27

3–28. Determine the tensile forces acting along the axis of struts *DE* and *DF* required to support the 500-lb load. The force acts along the axis of each strut.

Prob. 3–28

3–29. A 5-lb sphere rests between the 45° grooves of a 30° incline and a vertical wall. If all the surfaces of contact are smooth, determine the reactions of the surfaces on the sphere.

Prob. 3–29

3–30. A small peg *P* rests on a spring which is contained inside the smooth cylindrical pipe. When the spring is compressed so that $s = 12$ in, the spring exerts an upward force of 60 lb on the peg. Determine the point of attachment $(x, y, 0)$ of cable *PA* so that the tensions in cables *PB* and *PC* equal 20 and 30 lb, respectively.

Prob. 3–30

3–31. Determine the force acting along the axis of struts *OA* and *OB* and the tension developed in cables *OC*, *OD*, and *DE*.

68

3-33. The boom OA is used to support a 1,500-lb crate. If the force of the boom acting on point A is directed from O to A, determine this force and the tension forces $\mathbf{F}_{AC}$ and $\mathbf{F}_{AB}$ acting in the cables.

Prob. 3-33

2,500 lb

Prob. 3-31

3-32. If the tension acting in cable AD is 73 lb, determine the tension in BD and CD and the weight W of the bucket.

3-34. Determine the tension in cables AB and CB and the compression in strut OB for equilibrium of the automobile engine. Assume that the force in the strut acts along the axis of the strut.

250 lb

Prob. 3-34

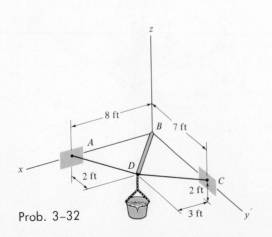

Prob. 3-32

69

3-35. The triangular frame ABC can be adjusted vertically between the three cords. Determine the distance s so that the tension in each of the slanted cords equals 40 lb. The frame remains in the horizontal plane.

Prob. 3-35

3-36. Determine the magnitude of forces **P**, **R**, and **F** for equilibrium of the concurrent force system. The force **F** is located in the positive x, y, z octant.

Prob. 3-36

3-37. For the tripod structure shown, determine the angle θ such that the tensile force developed in the legs OB and OC are equal. Neglect the weight of the tripod and assume that the force acting in each leg is directed along the axis of the leg. What is the force developed in each leg? Leg OA lies in the zy plane and force **F** lies in the xy plane.

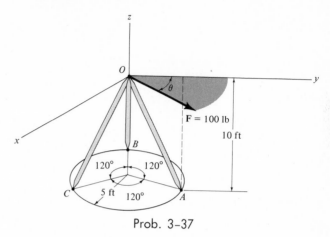

Prob. 3-37

3-38. Knowing that cord OB has been stretched 2 in., and fixed in place as shown, determine the tension developed in each of the other three cables in order to hold the 225-lb weight in equilibrium. Cable OD lies in the xy plane.

Prob. 3-38

70

3-39. Determine the tension developed in the three cables required to support the 60-lb traffic light.

3-41. A force of 100 lb holds the 400-lb crate in equilibrium. Determine the coordinates (O, y, z) of point A. The tension in cords AC and AB is 700 lb each.

Prob. 3-41

Prob. 3-39

3-40. Determine the magnitude of **P** and the direction of the 200-lb force required to keep the concurrent force system in equilibrium. Assume that the 200-lb force acts within the octant shown.

Prob. 3-40

71

4

Equivalent Force Systems

4-1. Force Systems Acting on a Rigid Body

In the last two chapters we have studied only the properties of a concurrent force system. This system of forces acts on a particle since geometrically a particle is represented as a point. As stated in Chapter 1,
a *rigid body* can be considered as a combination of a large number of
particles in which all the particles remain at a fixed distance from one
another, both before and after applying a load. The size and shape of
a rigid body often plays an important part in the force analysis since
the applied force system will not necessarily be concurrent. Indeed, each
force may act at *different points* on the rigid body. Clearly the two forces
F_1 and F_2 acting on the block in Fig. 4–1 are not a system of concurrent
forces; rather, the forces are noncollinear yet parallel to each other.

Vectors F_1 and F_2, shown in Fig. 4–1, represent *concentrated* forces
applied at some point and along a unique line of action. This, of course,
is an idealization. When any external loading is applied to the surface
of a body it is actually *distributed* over a finite area. This distribution
of force can be represented as a concentrated force only when the area
over which the load is applied is rather small compared to the total surface
area of the body. There are some cases, however, where this simplification
cannot be made, and hence, a variation of the force distribution over
the *surface area* must be taken into account. Such loadings may be caused
by wind, hydrostatic pressure, or simply the weight of material supported
by the body. Forces may also be distributed throughout the *volume* of
the body. Weight and magnetic forces act in this manner.

Fig. 4–1

In the last chapter it is shown that the condition for the equilibrium of a particle or a concurrent force system, simply requires that the resultant of the force system be equal to zero. In Chapter 5, it is shown that such a restriction is necessary but not sufficient for the equilibrium of rigid bodies. A further restriction must be made with regard to the nonconcurrency of the applied force system, giving rise to the concept of *moment*. In this chapter, we will develop a formal definition of a moment and discuss ways of finding the moments caused by concentrated forces and distributed loadings about points and axes. We will also develop methods of reducing nonconcurrent concentrated force systems and distributed loadings to equivalent, yet simpler, systems. Before doing any of this, however, we must first extend our knowledge of vector algebra.

4–2. Vector Cross Product

As shown in Sec. 2–3, there is only one way to multiply a vector by a scalar. In vector algebra, however, it is possible to *define* several ways of multiplying two vectors together. In this section we will discuss the *cross-product* method of vector multiplication. Since this operation yields a vector, it is often referred to as the *vector product*.

The vector cross product of two vectors **A** and **B** which yield the resultant vector **C** is written as

$$C = A \times B$$

Fig. 4–2

and read as "**C** equals **A** cross **B**." The *magnitude* of vector **C** is defined as the product of the magnitudes of the two vectors **A** and **B** and the sine of the angle θ made between them ($0° \leq \theta \leq 180°$). Thus, $C = AB \sin \theta$. Vector **C** has a *direction* which is perpendicular to the plane containing the two vectors **A** and **B** such that **A, B,** and **C** form a *right-handed system;* that is, the *sense* of the vector **C** is found by using the right-hand rule, curling the fingers of the right hand from vector **A** (cross) to vector **B**. The thumb then points in the direction of **C**, as shown in Fig. 4–2.

Knowing both the magnitude and direction of **C** it is possible to write this vector in vector notation as

$$C = A \times B = (AB \sin \theta)u_C \qquad (4\text{–}1)$$

Fig. 4–3

Here $AB \sin \theta$ defines the *magnitude* of **C** and the unit vector u_C defines the *directional sense* of **C**, which is determined by the right-hand rule. The terms of Eq. 4–1 are illustrated graphically in Fig. 4–3.

If the resultant of the vector product **A × B** is zero (provided vectors **A** and **B** are not null vectors), then **A** must be *parallel* to **B.** This follows

from the definition of the cross product given by Eq. 4–1, i.e., $AB \sin \theta$ $\mathbf{u}_C = 0$. Thus, $\sin \theta = 0$, or $\theta = 0°, 180°$, so that $\mathbf{A}$ is parallel to $\mathbf{B}$. In a similar manner

$$\mathbf{A} \times \mathbf{A} = 0 \qquad (4\text{–}2)$$

Three important relations valid for the cross product are as follows:

1. The cross product is *not* commutative, i.e.,

$$\mathbf{A} \times \mathbf{B} \neq \mathbf{B} \times \mathbf{A}$$

Rather,

$$\mathbf{A} \times \mathbf{B} = -\mathbf{B} \times \mathbf{A} \qquad (4\text{–}3)$$

This can be shown from Fig. 4–2 by using the right-hand rule. The cross product $\mathbf{B} \times \mathbf{A}$ yields a vector which acts in the opposite direction to $\mathbf{C}$, i.e., $\mathbf{B} \times \mathbf{A} = -\mathbf{C}$.

2. The distributive law:

$$\mathbf{A} \times (\mathbf{B} + \mathbf{D}) = (\mathbf{A} \times \mathbf{B}) + (\mathbf{A} \times \mathbf{D}) \qquad (4\text{–}4)$$

Given vectors $\mathbf{A}$, $\mathbf{B}$, and $\mathbf{D}$, the terms on the left and right-hand side of this equation are constructed as shown in Fig. 4–4. The proof is left as an exercise for the reader (see Prob. 4–4). It is important to note that proper order of the cross products in Eq. 4–4 must be maintained, since the cross product is not commutative. In accordance with Eq. 4–3, reversing the cross-product operation in any term requires a change in the sign of that term.

1. Add $\mathbf{B} + \mathbf{D}$

2. Form $\mathbf{A} \times (\mathbf{B} + \mathbf{D})$

1. Form $\mathbf{A} \times \mathbf{B}$

2. Form $\mathbf{A} \times \mathbf{D}$

3. Add $(\mathbf{A} \times \mathbf{B}) + (\mathbf{A} \times \mathbf{D})$

$$\mathbf{A} \times (\mathbf{B} + \mathbf{D}) = (\mathbf{A} \times \mathbf{B}) + (\mathbf{A} \times \mathbf{D})$$

(a)

(b)

Fig. 4–4

3. Multiplication by a scalar:

$$m(\mathbf{A} \times \mathbf{B}) = (m\mathbf{A}) \times \mathbf{B} = \mathbf{A} \times (m\mathbf{B}) = (\mathbf{A} \times \mathbf{B})m \qquad (4\text{–}5)$$

This property is easily shown, since the magnitude of the resultant vector ($|m|AB \sin \theta$) and its direction are the same in each case.

In order to conveniently use the vector cross product, this operation will be formulated by using Cartesian component notation. To do this, we must first determine the cross products of each of the unit vectors **i**, **j**, and **k**. This can readily be done by using Eq. 4–1. For example, to find the cross product **i** × **j**, the magnitude of the resultant vector is $(i)(j)(\sin 90°) = (1)(1)(1) = 1$, and its direction is determined using the right-hand rule. As shown in Fig. 4–5, the resultant vector points in the +**k** direction. Thus,

$$\mathbf{i} \times \mathbf{j} = (1)\mathbf{k}$$

In a similar manner,

$$\mathbf{i} \times \mathbf{j} = \mathbf{k} \qquad \mathbf{i} \times \mathbf{k} = -\mathbf{j} \qquad \mathbf{i} \times \mathbf{i} = 0$$
$$\mathbf{j} \times \mathbf{k} = \mathbf{i} \qquad \mathbf{j} \times \mathbf{i} = -\mathbf{k} \qquad \mathbf{j} \times \mathbf{j} = 0$$
$$\mathbf{k} \times \mathbf{i} = \mathbf{j} \qquad \mathbf{k} \times \mathbf{j} = -\mathbf{i} \qquad \mathbf{k} \times \mathbf{k} = 0$$

Fig. 4–5

These results should not be memorized; instead, you should comprehend how each "cross" product was derived, using Eq. 4–1.

In the general case $\mathbf{A} = A_x\mathbf{i} + A_y\mathbf{j} + A_z\mathbf{k}$ and $\mathbf{B} = B_x\mathbf{i} + B_y\mathbf{j} + B_z\mathbf{k}$, represent any two arbitrary vectors. We can form the cross product

$$\mathbf{A} \times \mathbf{B} = (A_x\mathbf{i} + A_y\mathbf{j} + A_z\mathbf{k}) \times (B_x\mathbf{i} + B_y\mathbf{j} + B_z\mathbf{k})$$

using the distributive law, Eq. 4–4, and the property of scalar multiplication, Eq. 4–5. Hence,

$$\mathbf{A} \times \mathbf{B} = A_xB_x(\mathbf{i} \times \mathbf{i}) + A_xB_y(\mathbf{i} \times \mathbf{j}) + A_xB_z(\mathbf{i} \times \mathbf{k})$$
$$+ A_yB_x(\mathbf{j} \times \mathbf{i}) + A_yB_y(\mathbf{j} \times \mathbf{j}) + A_yB_z(\mathbf{j} \times \mathbf{k})$$
$$+ A_zB_x(\mathbf{k} \times \mathbf{i}) + A_zB_y(\mathbf{k} \times \mathbf{j}) + A_zB_z(\mathbf{k} \times \mathbf{k})$$

Carrying out the cross-product operations and combining terms yields

$$\mathbf{A} \times \mathbf{B} = (A_yB_z - A_zB_y)\mathbf{i} - (A_xB_z - A_zB_x)\mathbf{j} + (A_xB_y - A_yB_x)\mathbf{k}$$

This equation may also be written in determinant form as

$$\mathbf{A} \times \mathbf{B} = \begin{vmatrix} \mathbf{i} & \mathbf{j} & \mathbf{k} \\ A_x & A_y & A_z \\ B_x & B_y & B_z \end{vmatrix} \qquad (4\text{–}6)$$

Thus, to find the cross product of any two Cartesian vectors **A** and **B,** we expand a determinant whose first row of elements are the unit vectors

i, j, and **k** and whose second and third rows represent the x, y, and z components of the two vectors **A** and **B**, respectively.* The order of arrangement is important because the commutative law for the cross product does not hold.

4–3. Principle of Transmissibility

The *principle of transmissibility* is an important concept often used in mechanics for studying the action of forces on rigid bodies. This principle states that the conditions of equilibrium or the motion of a rigid body remain *unchanged* when a force **F**, acting at a given point on the rigid body, is replaced to another point *lying on the line of action of the force.* As a simple example of this, consider the beam shown in Fig. 4-6. The force **F** acting at point P_1 represents the effect of some loading applied to the top of the beam. The beam is supported by a roller at A and a pin at B. Using the methods developed in the next chapter, we can calculate the forces at these supports. Clearly these reactions represent *external forces* acting on the beam. Rather than applying the force **F** at

Fig. 4–6

---

*A determinant having three rows and three columns can be expanded using three minors, each of which is multiplied by one of the three terms in the first row. There are four elements in each minor, for example,

$$\begin{vmatrix} A_{11} & A_{12} \\ A_{21} & A_{22} \end{vmatrix}$$

By *definition*, this notation represents the terms $(A_{11}A_{22}\text{-}A_{12}A_{21})$ which is simply the product of the two elements of the arrow slanting downwards to the right $(A_{11}A_{22})$ minus the product of the two elements intersected by the arrow slanting downwards to the left $(A_{12}A_{21})$. For a three-by-three determinant, such as Eq. 4-6, the three minors can be generated in accordance with the following scheme:

For element **i**: $\begin{vmatrix} \mathbf{i} & \mathbf{j} & \mathbf{k} \\ A_x & A_y & A_z \\ B_x & B_y & B_z \end{vmatrix} = \mathbf{i}(A_yB_z - A_zB_y)$

For element **j**: $\begin{vmatrix} \mathbf{i} & \mathbf{j} & \mathbf{k} \\ A_x & A_y & A_z \\ B_x & B_y & B_z \end{vmatrix} = -\mathbf{j}(A_xB_z - A_zB_x)$

For element **k**: $\begin{vmatrix} \mathbf{i} & \mathbf{j} & \mathbf{k} \\ A_x & A_y & A_z \\ B_x & B_y & B_z \end{vmatrix} = \mathbf{k}(A_xB_y - A_yB_x)$

Adding the results, and noting that the **j** element must include the minus sign, yields the expanded form of **A** × **B** given above Eq. 4-6.

the beam's top surface, when computing the external forces we can hang the load from the bottom of the beam at point P_3, or drill a hole in the beam to point P_2 and apply the load at this point. In any case, since points P_1, P_2, and P_3 lie on the line of action of $\mathbf{F}$, the principle of transmissibility states that $\mathbf{F}$ may be applied at *any* of these three points and the reactive forces developed at the supports A and B will remain the same. For our purposes, the principle of transmissibility is used for computing the *external reactions;* however, the *internal forces* developed in the beam will depend upon where $\mathbf{F}$ acts. Provided $\mathbf{F}$ acts at P_1, the internal forces in the beam have a high intensity around point P_1. When $\mathbf{F}$ acts at P_3, the effects of $\mathbf{F}$ will influence particles surrounding P_1 to a lesser degree than those particles surrounding point P_3.

When computing the *external effects* of forces acting on a rigid body, it follows from the principle of transmissibility that any force acting on the rigid body may be considered as a *sliding vector;* that is, *the force may act at any point along its line of action.* This is contrary to the notion of a force being a fixed vector, which applies to any force acting on a particle.

4–4. Moment of a Force About a Point

We can illustrate the concepts dealing with the moment of a force by using a simple example. Consider force $\mathbf{F}$ applied at the end of a pipe wrench, Fig. 4–7. The effect of this force tends to cause the pipe to twist about the *aa* axis. Experience teaches that to produce the greatest turning effect, the force should be applied at right angles to the handle, and the distance d should be made as large as possible. This twisting or rotational effect of a force about the *aa* axis is defined as the *moment of a force* or simply the *moment*. In all cases, the tendency for rotation will occur about an axis *perpendicular* to a plane containing both the force $\mathbf{F}$ and the distance line d, as shown in Fig. 4–7.

Consider now a force $\mathbf{F}$ and point O in space. The line of action of $\mathbf{F}$ and the point are contained within the shaded plane shown in Fig. 4–8a. We will define the moment of force $\mathbf{F}$ about point O as a *vector quantity.* This vector has a *magnitude* which is the product of the magnitude of the force vector $\mathbf{F}$ and the perpendicular distance d from the point to the line of action of the force, i.e., $M_O = Fd$. The units of moment magnitude are thus force times distance, e.g., lb-ft or kg_f-m. The *direction* of the moment vector will be specified using the right-hand rule. The fingers are curled in the sense of rotation as caused by the force acting about the point; the *thumb* then *points* in the *directive sense* of the moment vector, which is *perpendicular* to the shaded plane containing $\mathbf{F}$ and d.

In three dimensions, the moment vector is illustrated by a regular vector with a curl on it to *distinguish* it from a force vector. See Figs. 4–7 and

Fig. 4–7

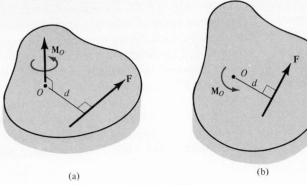

(a)

(b)

Fig. 4-8

4-8a. Many problems in mechanics involve coplanar force systems. In such cases, these problems may be conveniently viewed in two dimensions. For example, a projected view of Fig. 4–8*a* is given in Fig. 4–8*b*. The moment $\mathbf{M}_O$ is simply represented here by the counterclockwise curl, which indicates the action of $\mathbf{F}$. This curl is used to show the sense of rotation caused by the moment $\mathbf{M}_O$. Using the right-hand rule, realize that the *actual direction* of the moment vector in Fig. 4–8*b* points *out* of the page, since the fingers follow the curl.

The definition of moment of a force about point *O* may be expressed in terms of the vector cross product. With reference to Fig. 4–9, this expression is

$$\mathbf{M}_O = \mathbf{r} \times \mathbf{F} \qquad\qquad (4\text{-}7)$$

As shown, $\mathbf{r}$ represents the vector drawn from point *O* to *any point A* lying on the line of action of the force $\mathbf{F}$. We will now show that the moment vector determined by Eq. 4–7 is in accordance with the definition of a moment. From the definition of the vector cross product, Eq. 4–1, the *magnitude* of $\mathbf{M}_O$ is

$$M_O = rF \sin \theta = F(r \sin \theta) = Fd$$

Here θ defines the angle made between the *tails* of the two vectors $\mathbf{F}$ and $\mathbf{r}$ (see Fig. 4–3). This angle, shown in Fig. 4–9, is constructed by sliding vector $\mathbf{r}$ to point *A*. The *direction* of $\mathbf{M}_O$ is determined by means of the right-hand rule, based upon the definition of the cross product as given by Eq. 4–7. After extending $\mathbf{r}$ to the dotted position, we form the cross product $\mathbf{r} \times \mathbf{F}$ by curling the right-hand fingers from the direction of $\mathbf{r}$ toward the direction of $\mathbf{F}$ (vector $\mathbf{r}$ cross vector $\mathbf{F}$). The thumb is then directed upward or perpendicular to the plane in the direction of $\mathbf{M}_O$, Fig. 4–9. Since the cross product is not commutative, it is important that the *proper order* of vector multiplication be maintained in Eq. 4–7.

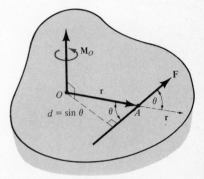

Fig. 4-9

Fig. 4–10

If we reverse the order so that $\mathbf{M}_O = \mathbf{F} \times \mathbf{r}$, the sense of direction of $\mathbf{M}_O$ would be *opposite* to the *actual* sense of direction shown in Fig. 4–8a.

As stated before, $\mathbf{r}$ in Eq. 4–7 can be drawn from point O to *any point* lying on the line of action of $\mathbf{F}$. This definition for $\mathbf{r}$ is due to the transmissibility of $\mathbf{F}$, which was discussed qualitatively in Sec. 4–3. We will now show mathematically that $\mathbf{F}$ is transmissible; that is, $\mathbf{F}$ is a *sliding vector*. In Fig. 4–10, $\mathbf{F}$ is applied at two points, P_1 and P_2, lying along its line of action. These points lie at a distance $\mathbf{r}_1$ and $\mathbf{r}_2$, respectively, from the origin of coordinates O. For transmissibility, it is necessary that the moment of $\mathbf{F}$ about O be the *same* when $\mathbf{F}$ acts at either point P_1 or P_2, i.e.,

$$\mathbf{r}_1 \times \mathbf{F} = \mathbf{r}_2 \times \mathbf{F}$$

From the figure,

$$\mathbf{r}_2 = \mathbf{r}_1 + \mathbf{r}_{P_1 P_2}$$

Substituting gives

$$\mathbf{r}_1 \times \mathbf{F} = (\mathbf{r}_1 + \mathbf{r}_{P_1 P_2}) \times \mathbf{F}$$

Using the distributive law, Eq. 4–4, we obtain

$$\mathbf{r}_1 \times \mathbf{F} = \mathbf{r}_1 \times \mathbf{F} + \mathbf{r}_{P_1 P_2} \times \mathbf{F}$$

or

$$(\mathbf{r}_1 - \mathbf{r}_1) \times \mathbf{F} = \mathbf{r}_{P_1 P_2} \times \mathbf{F}$$
$$0 \equiv 0$$

The left-hand side of this equation is zero since $(\mathbf{r}_1 - \mathbf{r}_1)$ forms a null vector. The right-hand side is zero since $\mathbf{r}_{P_1 P_2}$ and $\mathbf{F}$ are parallel to one another; refer to Eq. 4–2. Thus, $\mathbf{F}$ is a sliding vector, and by definition $\mathbf{r}$ in Eq. 4–7 may be directed to *any point* lying on the line of action of $\mathbf{F}$.

To summarize the concepts covered in this section, consider the rigid body shown in Fig. 4–11.* Force $\mathbf{F}$ is applied at point $P_2(x_2, y_2, z_2)$. The moment of this force about point $P_1(x_1, y_1, z_1)$ may be computed on the basis of Eq. 4–7. The position vector $\mathbf{r}$ extends from point P_1 *to* point P_2 (or any other point lying on the line of action of the force). Hence,

$$\mathbf{r} = (x_2 - x_1)\mathbf{i} + (y_2 - y_1)\mathbf{j} + (z_2 - z_1)\mathbf{k}$$

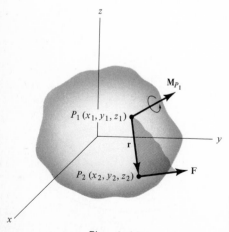

Fig. 4–11

*The shaded object shown in Fig. 4–11 represents a rigid body having an *arbitrary shape*. Similarly, the object shown in Fig. 4–17 represents the plane view of a rigid body. Such objects are simply *graphical representations* used to explain the theory. However, it should be kept in mind that the arbitrary shape of these objects may in reality represent an airplane, beam, bridge, stone, etc.

If **F** has Cartesian components having magnitudes F_x, F_y, and F_z, we may express **F** in vector notation as

$$\mathbf{F} = F_x\mathbf{i} + F_y\mathbf{j} + F_z\mathbf{k}$$

Thus,

$$\mathbf{M}_{P_1} = \mathbf{r} \times \mathbf{F} = \begin{vmatrix} \mathbf{i} & \mathbf{j} & \mathbf{k} \\ (x_2 - x_1) & (y_2 - y_1) & (z_2 - z_1) \\ F_x & F_y & F_z \end{vmatrix} \qquad (4\text{-}8)$$

As shown in Fig. 4–11, $\mathbf{M}_{P_1}$ is *perpendicular* to the shaded plane containing vectors **r** and **F**. Hence, the force **F** tends to rotate the body about an axis defined by the line of action of $\mathbf{M}_{P_1}$.

4–5. Varignon's Theorem

A theorem often used in mechanics is *Varignon's theorem,* which states that the sum of the moments of all the forces of a concurrent force system about a given point is equal to the same moment created by the resultant force of the concurrent force system about the point.

The proof of this theorem follows directly from the distributive property of the vector cross product, Eq. 4–4. For example, consider the system of three concurrent forces shown in Fig. 4–12. The resultant force is determined by the vector sum $\mathbf{F}_R = \mathbf{F}_1 + \mathbf{F}_2 + \mathbf{F}_3$. Hence, the total moment of the three forces about point O is

$$\mathbf{M}_R = \mathbf{r} \times \mathbf{F}_1 + \mathbf{r} \times \mathbf{F}_2 + \mathbf{r} \times \mathbf{F}_3 = \mathbf{r} \times (\mathbf{F}_1 + \mathbf{F}_2 + \mathbf{F}_3)$$

or

$$\mathbf{M}_R = \mathbf{r} \times \mathbf{F}_R$$

Fig. 4–12

where **r** is the position vector drawn from O to P.

This property was originally developed by the French mathematician Varignon (1654–1722). It has important application to the solution of problems and proofs of theorems which follow, since it allows us to consider the moment of a force's two components rather than the force itself. Used in this way, Varignon's theorem states that the moment of a force about a point is equivalent to the moment of the force's two components about the point.

Example 4–1

A force **F** having a magnitude of 5 lb acts at the end of the block shown in Fig. 4–13a. Determine the x, y, and z components of the moment at point O created by this force.

(a)

(b)

Fig. 4-13

(c)

Solution I

We can compute the moment at point O using Cartesian unit vectors in accordance with Eq. 4-7, i.e., $\mathbf{M}_O = \mathbf{r} \times \mathbf{F}$. From Fig. 4-13a, $\mathbf{F} = \{-5\mathbf{i}\}$ lb. Either vector $\mathbf{r}_{OA} = \{5\mathbf{i} + 4\mathbf{j} - 3\mathbf{k}\}$ ft or $\mathbf{r}_{OB} = \{4\mathbf{j} - 3\mathbf{k}\}$ ft may be used in the cross product. Why? From Eq. 4-7,

$$\mathbf{M}_O = \mathbf{r}_{OA} \times \mathbf{F} = \begin{vmatrix} \mathbf{i} & \mathbf{j} & \mathbf{k} \\ 5 & 4 & -3 \\ -5 & 0 & 0 \end{vmatrix} = \mathbf{r}_{OB} \times \mathbf{F} = \begin{vmatrix} \mathbf{i} & \mathbf{j} & \mathbf{k} \\ 0 & 4 & -3 \\ -5 & 0 & 0 \end{vmatrix}$$

Expanding the determinants, yields the same result, namely,

$$\mathbf{M}_O = \{15\mathbf{j} + 20\mathbf{k}\} \text{ lb-ft} \qquad \textit{Ans.}$$

Hence, the moment has only two components $\mathbf{M}_y$ and $\mathbf{M}_z$, as shown in Fig. 4-13b. Each component represents the effect of $\mathbf{F}$ in tending to rotate the block about the y and z axis.

Solution II

According to the principle of transmissibility, the force $\mathbf{F}$ may be applied to the block at point B, as shown in Fig. 4-13c, without altering the moment of the force produced about point O. When applied at point B, the *perpendicular distance d* from the line of action of the force to point O is 5 ft. Hence, the magnitude of the moment $\mathbf{M}_O$ about point O is $M_O = F d = (5 \text{ lb})(5 \text{ ft}) = 25$ lb-ft. By the right-hand rule this vector

acts *perpendicular* to the shaded diagonal plane containing d and **F**, Fig. 4–13c. The magnitude of the two components of $\mathbf{M}_O$ are therefore

$$M_y = \tfrac{3}{5}(25) = 15 \text{ lb-ft}$$

and

$$M_z = \tfrac{4}{5}(25) = 20 \text{ lb-ft}$$

Thus, the moment at O can be expressed in Cartesian component form as

$$\mathbf{M}_O = \{15\,\mathbf{j} + 20\,\mathbf{k}\}\text{lb-ft} \qquad\qquad Ans.$$

Note that even though **F** acts at a perpendicular distance of 5 ft from the x axis (see Fig. 4–13c) this force does not create a tendency for rotation about the x axis since the force is *parallel* to the x axis.

Example 4–2

A 50-lb force acts on the handle of a lever, as shown in Fig. 4–14a. Determine the moment created by this force about point A.

Solution I

In Fig. 4–14b, the *perpendicular distance* from the line of action of the force to the moment point is computed using trigonometry. Since $\angle BCD$ is 120° then $\angle DBC$ is 15° so that

$$\frac{BC}{\sin 45°} = \frac{2 \text{ in.}}{\sin 15°}$$

$$BC = 2\left(\frac{0.707}{0.259}\right) = 5.46 \text{ in.}$$

$$AB = 10.00 - 5.46 = 4.54 \text{ in.}$$

Hence, from $\triangle ABE$,

$$d = 4.54 \sin 15° = 1.174 \text{ in.}$$

Therefore,

$$M_A = Fd = (50 \text{ lb})(1.174 \text{ in.}) = 58.7 \text{ lb-in.}$$

According to the right-hand rule, for the coordinate system shown counterclockwise rotation in the plane of the figure will be considered positive ($+\mathbf{k}$ direction). Since the 50-lb force tends to rotate the lever *clockwise* about point A, then

$$\mathbf{M}_A = \{-58.7\mathbf{k}\}\text{lb-in.} \qquad\qquad Ans.$$

Solution II

Following the principle of transmissibility, force $\mathbf{F}$ may be applied at point B. The force may then be resolved into components acting parallel and perpendicular to the lever, as shown in Fig. 4–14c. Only the component acting perpendicular to the lever produces a moment about point A. Why? Using the length of AB computed in Solution I, we have

$$M_A = F_\perp d_{AB} = (50 \sin 15°)4.54 = 58.7 \text{ lb-in.}$$

Since the moment is clockwise about point A,

$$\mathbf{M}_A = \{-58.7\mathbf{k}\} \text{ lb-in.} \qquad Ans.$$

The line of action of $\mathbf{F}_\parallel$ passes through point A and hence creates zero moment about this point.

Solution III

The 50-lb force is resolved into its x and y components, as shown in Fig. 4–14d. Applying Varignon's theorem, the total moment computed

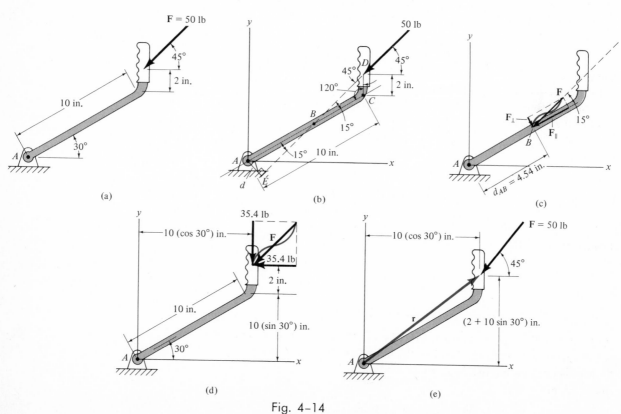

(a)   (b)   (c)

(d)   (e)

Fig. 4–14

about point A is equivalent to the sum of the moments produced by the two components. Assuming counterclockwise rotation as positive, that is, $+\mathbf{k}$ direction,

$$\mathbf{M}_A = -35.4(10 \cos 30°)\mathbf{k} + 35.4(2 + 10 \sin 30°)\mathbf{k}$$

or

$$\mathbf{M}_A = \{-58.7\mathbf{k}\} \text{ lb-in.} \qquad \textit{Ans.}$$

Solution IV

Using a vector approach, we can represent the required force and position vectors as shown in Fig. 4–14e. Expressing each of these vectors in Cartesian vector notation yields

$$\mathbf{r} = 10 \cos 30°\mathbf{i} + (2 + 10 \sin 30°)\mathbf{j} = \{8.66\mathbf{i} + 7.00\mathbf{j}\} \text{ in.}$$
$$\mathbf{F} = -50 \cos 45°\mathbf{i} - 50 \sin 45°\mathbf{j} = \{-35.4\mathbf{i} - 35.4\mathbf{j}\} \text{ lb}$$

Hence, the required moment becomes

$$\mathbf{M}_A = \mathbf{r} \times \mathbf{F} = \begin{vmatrix} \mathbf{i} & \mathbf{j} & \mathbf{k} \\ 8.66 & 7.00 & 0 \\ -35.4 & -35.4 & 0 \end{vmatrix} = \{-58.7\mathbf{k}\} \text{ lb-in.} \quad \textit{Ans.}$$

From the analysis, it is seen that Solution III yields the most direct solution to this problem. This method, which involves dividing forces into their rectangular components *before* computing the moment, is generally recommended for solving problems involving coplanar force systems. Note, in particular, that the vector approach, Solution IV, becomes tedious when working problems represented in two dimensions. Its use is generally recommended for solving problems in three dimensions.

Example 4–3

Determine the point of application P and the direction for a 20-lb force, which lies in the plane of the square plate shown in Fig. 4–15a, so that this force creates the greatest counterclockwise moment about point O. What is this moment?

Solution

Since the maximum moment created by the force is required, the force must act on the plate at a distance *farthest* from point O. As shown in Fig. 4–15b, the point of application of $\mathbf{F}$ must therefore be at the diagonal corner. In order to produce counterclockwise rotation of the plate, the force $\mathbf{F}$ must act at an angle $45° < \theta < 225°$. The greatest moment is produced when the line of action of $\mathbf{F}$ is *perpendicular* to $\mathbf{r}_{OP}$, i.e., $\theta = 135°$, Fig. 4–15c. The maximum moment is therefore

(a)   (b)   (c)

Fig. 4-15

$$M_O = Fr_{OP} = (20\text{lb})(2\sqrt{2}\text{ in.}) = 56.6 \text{ lb-in.} \mathord{\downharpoonright}$$ *Ans.*

By the right-hand rule, $\mathbf{M}_O$ has a sense of direction which is pointing out of the page.

Example 4-4

The cord BC shown in Fig. 4-16a exerts a force of 60 lb on pole AB. Determine the magnitude and the direction of the moment created by this force about point A.

Solution

Since this problem is expressed in three dimensions, it is recommended that a vector approach be used for the solution. For convenience, the

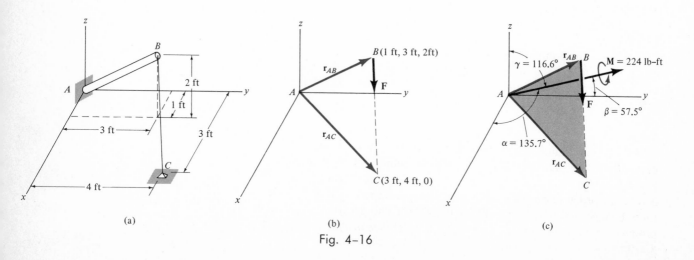

(a)   (b)   (c)

Fig. 4-16

origin of coordinates is established at point A, and the coordinates of points B and C are determined as shown in Fig. 4-16b.

The moment of force $\mathbf{F}$ about point A is $\mathbf{M}_A = \mathbf{r} \times \mathbf{F}$. $\mathbf{r}$ represents a position vector drawn from point A to *any point* on the line of action of the force. Two convenient position vectors which may be used are $\mathbf{r}_{AB}$ and $\mathbf{r}_{AC}$, as indicated in Fig. 4-16b. These vectors are represented as

$$\mathbf{r}_{AB} = \{\mathbf{i} + 3\mathbf{j} + 2\mathbf{k}\} \text{ ft} \qquad \text{and} \qquad \mathbf{r}_{AC} = \{3\mathbf{i} + 4\mathbf{j}\} \text{ ft}$$

The force $\mathbf{F}$, having a magnitude of 60 lb, may be expressed in Cartesian vector form as

$$\mathbf{F} = (60 \text{ lb}) \frac{\mathbf{r}_{BC}}{r_{BC}} = (60)\left(\frac{2\mathbf{i} + \mathbf{j} - 2\mathbf{k}}{\sqrt{(2)^2 + (1)^2 + (-2)^2}}\right) = \{40\mathbf{i} + 20\mathbf{j} - 40\mathbf{k}\} \text{ lb}$$

Therefore,

$$\mathbf{M}_A = \mathbf{r}_{AB} \times \mathbf{F} = \begin{vmatrix} \mathbf{i} & \mathbf{j} & \mathbf{k} \\ 1 & 3 & 2 \\ 40 & 20 & -40 \end{vmatrix} = \mathbf{r}_{AC} \times \mathbf{F} = \begin{vmatrix} \mathbf{i} & \mathbf{j} & \mathbf{k} \\ 3 & 4 & 0 \\ 40 & 20 & -40 \end{vmatrix}$$

or

$$\mathbf{M}_A = \{-160\mathbf{i} + 120\mathbf{j} - 100\mathbf{k}\} \text{ lb-ft}$$

The *magnitude* of $\mathbf{M}_A$ is therefore

$$M_A = \sqrt{(-160)^2 + (120)^2 + (-100)^2} = 224 \text{ lb-ft} \qquad Ans.$$

A unit vector acting in the direction of $\mathbf{M}_A$ is

$$\mathbf{u}_{M_A} = \frac{\mathbf{M}_A}{M_A} = -0.716\mathbf{i} + 0.537\mathbf{j} - 0.447\mathbf{k}$$

The components of $\mathbf{u}_{M_A}$ are the *direction cosines* of $\mathbf{M}_A$. Hence,

$$\alpha = \cos^{-1}(-0.716) = 135.7° \qquad Ans.$$
$$\beta = \cos^{-1}(0.537) = 57.5° \qquad Ans.$$
$$\gamma = \cos^{-1}(-0.444) = 116.6° \qquad Ans.$$

As shown in Fig. 4-16c, $\mathbf{M}_A$ acts perpendicular to the shaded plane containing vectors $\mathbf{F}$, $\mathbf{r}_{AB}$, and $\mathbf{r}_{AC}$. Why?

Problems

4-1. Find the resulting vector for each of the following:
(a) $(2\mathbf{i} - 2\mathbf{k}) \times 3\mathbf{j}$
(b) $5\mathbf{j} \times (3\mathbf{j} + 4\mathbf{k})$
(c) $(-10\mathbf{i} - 2\mathbf{k}) \times (3\mathbf{j} - \mathbf{k})$.

4-2. Given $\mathbf{A} = 3\mathbf{i} + 4\mathbf{j} + 2\mathbf{k}$, $\mathbf{B} = 3\mathbf{i} + 2\mathbf{k}$, and $\mathbf{C} = 2\mathbf{i} - 4\mathbf{j} + 3\mathbf{k}$, find (a) $(\mathbf{A} \times \mathbf{B}) \times \mathbf{C}$, and (b) $\mathbf{A} \times (\mathbf{B} \times \mathbf{C})$.

4-3. Determine a unit vector perpendicular to the plane containing the two vectors $\mathbf{A} = 3\mathbf{i} - 2\mathbf{j}$ and $\mathbf{B} = 2\mathbf{k}$.

4-4. If $\mathbf{A}$, $\mathbf{B}$, and $\mathbf{C}$ are given vectors, prove the distributive law for the vector cross product, i.e., $\mathbf{A} \times (\mathbf{B} + \mathbf{C}) = (\mathbf{A} \times \mathbf{B}) + (\mathbf{A} \times \mathbf{C})$. *Suggestion:* Use Cartesian vectors.

4-5. Determine the cross product $\mathbf{r}_1 \times \mathbf{r}_2$ for the vectors shown in the figure.

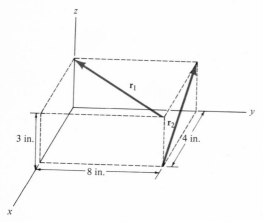

Prob. 4–5

4-6. A force of 30 lb is applied to the handle of the wrench. Determine the moment of this force about point O. Use both a scalar and vector approach to solve the problem.

Prob. 4–6

4-7. Resolve the 100-lb force into its x and y components, and determine the moment of each of these components about point O. Solve the problem using both a scalar and a vector approach.

Prob. 4–7

4-8. If the tension in the cable BC is 30 lb, determine the moment created by this force acting on the flag pole at the base A.

Prob. 4–8

4-9. Determine the angle α at which the 50-lb force must act so that the moment of this force about point B is equal to zero.

Sept 24 = hour exam   probs 10-16-18

Prob. 4-9

4-10. A 20-lb force is applied 90° to the handle of the socket wrench. Determine the magnitude and direction of the moment created by this force about point O.

Prob. 4-10

4-11. To prevent the bell crank from rotating about the pin connection at O, the total moment created by forces T and R about this point must be zero. If $R = 20$ lb, determine the magnitude of T to prevent rotation.

4-12. Determine the magnitude of the force F which is applied at the end of the shift lever such that this force creates a clockwise moment of 20 lb-ft about point O when $\alpha = 30°$.

Prob. 4-12

4-13. If the maximum force F which can be developed at the end of the shift lever in Prob. 4-12 is 50 lb, determine the angle α so that the clockwise moment at O created by this force is 20 lb-ft.

4-14. A force F having a magnitude of 90 lb acts at a point located along the diagonal of the parallelepiped. Using vector notation, show that the moment created by F about point A is the same, regardless of whether F is applied at point B or point C.

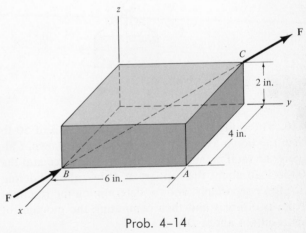

Prob. 4-11

Prob. 4-14

89

4-15. The moment created about the winch at O must be zero or else it will rotate. To prevent rotation due to the 100-lb load W, a powl at A exerts a horizontal force $\mathbf{T}$ on the ratchet of the winch. Determine the magnitude of this force which is required to produce zero moment at O.

Prob. 4-16

Prob. 4-15

4-16. The cable AB is subjected to a tension of 50 lb when the boom assembly is in the position shown. Calculate the total moment produced by this force and the 100-lb load about the support at D by (a) considering each force separately, and (b) determining the resultant of the two forces and then computing the moment of this resultant about D.

4-17. Find the moment created by the 200-lb force acting at the end of the pipe assembly about each of the joints at A, B, and C.

Prob. 4-17

90

4-18. The three struts AB, AC, and AD hold the 10-lb lantern in equilibrium. Determine the force in each of these struts and show that the total moment of these three forces about point P is zero. Assume that the force in each strut acts along the axis of the strut. Is the total moment of these three forces also zero about point P'? Why or why not?

4-19. The moment created at point P by the applied force $\mathbf{F}$ acting on the block is $\mathbf{M} = \{-\mathbf{i} - 3\mathbf{j} - 9\mathbf{k}\}$ lb-in. Determine the dimensions b and c of the block if $a = 3$ in.

Prob. 4–19

Prob. 4–18

4-20. Determine the direction of the force $\mathbf{F}$ applied at the end of the pipe section such that the moment created by $\mathbf{F}$ about point A is zero.

Prob. 4–20

4-6. Moment of a Couple

A *couple* is defined as two parallel forces having the same magnitude, opposite direction, and acting at some perpendicular distance d apart. Since the resultant force of the two forces composing the couple is zero, the entire effect of a couple is to produce a pure *moment,* or tendency of rotation in a specified direction.

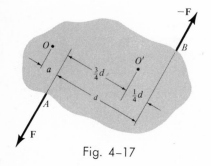

Fig. 4-17

As shown in Fig. 4-17, the net effect of the two forces acting on the plate lying in the xy plane will produce a rotation of the plate about an axis *perpendicular* to the plate. By definition, these two forces constitute a couple. The moment produced by this couple, when computed about any point lying in the plane of the plate, can be calculated by using the definition of the moment of a force and the principle of transmissibility of a force. Hence, the moment about point O, Fig. 4-17, is equal to the moment of the force at A about point O plus the moment of the force at B about point O. Assuming positive moment counterclockwise, that is, directed out of the page,

$$M_O = -(a)F + (a + d)F = Fd \ \text{↺}$$

In a similar manner, the moment of the couple computed about the point O' is

$$M_{O'} = (\tfrac{3}{4}d)F + (\tfrac{1}{4}d)F = Fd \ \text{↺}$$

Since points O and O' were arbitrarily selected in the xy plane, these two calculations illustrate an important property regarding the moment of a couple: The magnitude of the moment and the direction of its rotation are the *same* when the couple moment is computed about *any point* lying in the plane of the couple. Hence, it may be concluded that the moment of a couple is a *free vector,* preserving its magnitude and direction but having no specific point of application. This free vector is obviously a moment vector. The *magnitude* of the couple moment is $M_C = Fd$, where F is the magnitude of one of the forces and d is the *perpendicular distance* between the two forces. The *direction* (and sense) of the couple moment is determined by the right-hand rule and acts perpendicular to the plane containing the two forces.

Since it was shown in Sec. 4-4 that a moment is represented in vector notation by the vector cross product, the moment of a couple can be defined by

$$\mathbf{M}_C = \mathbf{r} \times \mathbf{F} \tag{4-9}$$

where $\mathbf{r}$ represents the position vector, drawn from *any point* on the line

of action of one force to *any point* on the line of action of the other force. As shown in Figs. 4–18a and 4–18b, the force **F** in Eq. 4–9 represents that force to which the vector **r** is *directed*. The *magnitude* of the couple moment is

$$M_C = rF \sin \theta = F(r \sin \theta) = Fd$$

In accordance with the cross product, note how the angle θ is defined in Fig. 4–18. The *direction* of **M**$_C$ is defined by the right-hand rule. In both cases, the directional sense is upward, as shown in the figure. Recall that **M**$_C$ is a free vector, and can therefore act through *any point* in the plane, Fig. 4–18.

Two couples are said to be *equal* if they produce a moment vector having the same magnitude and direction. Since the moment produced by a couple is always perpendicular to the plane containing the couple forces, it is therefore necessary that the forces of equal couples lie either in the same plane or in corresponding planes that are *parallel* to one another. In this way, the line of action of each couple moment will be the same, that is, perpendicular to the parallel planes. For example, the two couples shown in Fig. 4–19a are equivalent. One couple is produced by forces of 10 lb acting at a distance of $d = 5$ ft, and the other is produced by forces of magnitude of 20 lb at a distance of 2.5 ft. Since the

(a)

(b)

Fig. 4–18

(a)

(b)

Fig. 4–19

planes in which the forces act are parallel, the moment produced by each of these couples may be expressed in vector form as

$$\mathbf{M}_C = \{50\mathbf{k}\}\text{lb-ft}$$

Since couple moments are free vectors, they may be applied at some *arbitrary point P,* and added vectorially. As shown in Fig. 4–19*b,* the resultant of both couple moments is therefore,

$$\mathbf{M}_R = 50\mathbf{k} + 50\mathbf{k} = \{100\mathbf{k}\}\text{lb-ft}$$

The moments of two couples acting on nonparallel planes are also added using the rules of vector addition. For example, the two couples acting on different planes of the rigid body in Fig. 4–20*a* may be replaced by their corresponding moment vectors $\mathbf{M}_{C_1}$ and $\mathbf{M}_{C_2}$, shown in Fig.

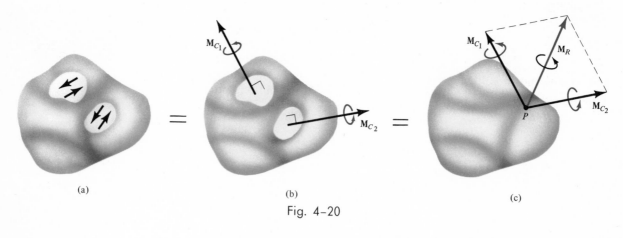

(a)

(b)

(c)

Fig. 4–20

4–20*b.* These free vectors may be moved to an *arbitrary point P* and added vectorially to obtain their resultant vector sum $\mathbf{M}_R = \mathbf{M}_{C_1} + \mathbf{M}_{C_2}$, shown in Fig. 4–20*c.* The effect of the two couples in Fig. 4–20*a* imparts a total twist or couple moment $\mathbf{M}_R$ to the body acting in the direction of $\mathbf{M}_R$.

4–7. Resolution of a Force into a Force and a Couple

Before the methods of rigid-body equilibrium can be developed, we must first study what effects are caused by moving a force from one point to another on a rigid body. For example, consider moving the force $\mathbf{F}$ acting at point *A* on the constrained rigid body in Fig. 4–21*a* to the arbitrary point *O* which does *not* lie along the line of action of $\mathbf{F}$. Using the principle of transmissibility, $\mathbf{F}$ may first be applied at point *A'*, which lies at a perpendicular distance *d* from point *O* to the line of action of the force,

(a)   (b)   (c)   (d)

Fig. 4–21

as shown in Fig. 4–21*b*. Applying equal and opposite forces **F** *at point O*, as shown in Fig. 4–21*c*, in no way alters the equilibrium state of the body. Two of the three forces (indicated by a slash across them) in this figure form a couple. The moment of this couple has a magnitude $M_C = Fd$ and tends to rotate the body in a counterclockwise direction. Since the couple moment is a free vector, it may be applied at *any point P* on the rigid body, as shown in Fig. 4–21*d*, and still produce the *same* external effects on the rigid body. In addition to this couple, force **F** now acts at point *O*. Thus, when **F** acts at *A*, Fig. 4–21*a*, it will produce *the same reactions* at the three roller supports as when **F** is applied at point *O* and a couple moment $\mathbf{M}_C$ is applied to the body, Fig. 4–21*d*. *The magnitude and direction of the couple moment is determined by taking moments of* **F** *about point O when the force is located at its original point A (or point A′). The line of action of* $\mathbf{M}_C$ *is thus perpendicular to the plane containing* **F** *and d.*

For the general case where the rigid body is viewed in three dimensions, we can formulate the above procedure using vector notation. With refer-

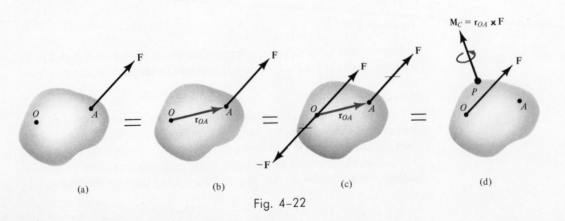

(a)   (b)   (c)   (d)

Fig. 4–22

ence to Fig. 4–22, force vector $\mathbf{F}$ is replaced from point A to point O, and a couple moment vector $\mathbf{M}_C$ is *added*, where

$$\mathbf{M}_C = \mathbf{r}_{OA} \times \mathbf{F}$$

The position vector $\mathbf{r}_{OA}$ *is drawn from point* O *to any point lying along the line of action of the force. According to the right-hand rule, the moment vector is perpendicular to the plane made by force* $\mathbf{F}$ *and position vector* $\mathbf{r}_{OA}$. (The constructions in Fig. 4–22 are equivalent to those used for the two-dimensional case shown in Fig. 4–21.) As before, since $\mathbf{M}_C$ is a free vector, it can act at *any point P* on the body.

Whenever one has a resulting force and a resulting couple moment which are *mutually perpendicular*, it is always possible to move the force from its original line of action to a unique parallel line of action, thereby eliminating the moment. To do this, simply reverse the procedure as just discussed. For example, the force $\mathbf{F}$ and moment $\mathbf{M}_C$, shown in Fig. 4–21d, are mutually perpendicular. (The direction of $\mathbf{M}_C$ is perpendicular to the page in accordance with the right-hand rule.) If $\mathbf{F}$ is moved so that its line of action passes through points AA', Fig. 4–21b, the moment $\mathbf{M}_C$ will be eliminated. In three dimensions, vectors should be used for this operation, Fig. 4–22. A further discussion of this is given in Sec. 4–10. See Example 4–6.

The concepts presented in this section regarding the movement of a force may be summarized by the following two statements:

1. If the force is to be moved to a *point located on its line of action,* by the principle of transmissibility, simply move the force to the point.
2. If the force is to be moved to a *point which is not located on its line of action,* move the force to the point and *add* a couple moment. The magnitude and direction of the couple moment are determined by the cross product of a *position vector,* directed from the point to any point lying on the line of action of the force, and the *force vector.*

Example 4–5

Replace the three couples acting on the block shown in Fig. 4–23a by a single resultant couple moment. Determine its magnitude and direction.

Solution

Each pair of equal and opposite forces may be replaced by its equivalent couple-moment vector, as shown in Fig. 4–23b. The sense of direction of these vectors is determined by using the right-hand rule. In particular, the inclined plane has a slope of $\frac{4}{3}$; hence, the unit vector $\mathbf{u}_n$ acting *perpendicular* to this plane, Fig. 4–23b, may be represented by

$$\mathbf{u}_n = \tfrac{4}{5}\mathbf{j} + \tfrac{3}{5}\mathbf{k}$$

Fig. 4–23

The (10 lb)(6 in.) = 60 lb-in. couple moment may therefore be written in Cartesian component notation as

$$\mathbf{M}_C = (60 \text{ lb-in.})\mathbf{u}_n = \{48\mathbf{j} + 36\mathbf{k}\} \text{ lb-in.}$$

Since couple moments are free vectors, they may be moved to some arbitrary point P on the block and added vectorially, Fig. 4–23c. The resultant moment becomes

$$\mathbf{M}_{C_R} = 120\mathbf{i} - 60\mathbf{j} + 48\mathbf{j} + 36\mathbf{k}$$
$$= \{120\mathbf{i} - 12\mathbf{j} + 36\mathbf{k}\} \text{ lb-in.} \qquad \textit{Ans.}$$

The magnitude of this resultant is

$$M_{C_R} = \sqrt{(120)^2 + (-12)^2 + (36)^2} = 125.9 \text{ lb-in.} \qquad \textit{Ans.}$$

The direction of $\mathbf{M}_{C_R}$ is determined from the unit vector

$$\mathbf{u}_C = \frac{\mathbf{M}_{C_R}}{M_{C_R}} = \frac{120}{125.9}\mathbf{i} - \frac{12}{125.9}\mathbf{j} + \frac{36}{125.9}\mathbf{k}$$

$$= 0.954\mathbf{i} - 0.095\mathbf{j} + 0.286\mathbf{k}$$

Thus,

$$\alpha = \cos^{-1}(0.954) = 17.4° \qquad \textit{Ans.}$$
$$\beta = \cos^{-1}(-0.095) = 95.5° \qquad \textit{Ans.}$$
$$\gamma = \cos^{-1}(0.286) = 73.4° \qquad \textit{Ans.}$$

Example 4-6

Replace the force and couple acting on the beam shown in Fig. 4–24a by an equivalent force and couple-moment system acting at point A. Can this system be further reduced to a single force? If so, specify the location of the force.

Solution

Since this problem is represented in two dimensions, the solution is most easily obtained using a scalar approach. The couple moment formed by the two 10-lb forces acting at B and C has a magnitude of $M_1 = (3\text{ ft})(10\text{ lb}) = 30$ lb-ft and acts in the $+\mathbf{k}$ direction (counterclockwise). Because $\mathbf{M}_1$ is a *free vector*, it may be moved directly to point A, as shown in Fig. 4–24b. Since point A does *not* lie on the line of action of the 20-lb force, when this force is moved to point A a couple moment of magnitude $M_2 = (8\text{ ft})(20\text{ lb}) = 160$ lb-ft *must be added* to the beam, Fig. 4–24b. By the right-hand rule, this moment acts *clockwise*, i.e. in the $-\mathbf{k}$ direction. (The force, when acting at point D, tends to rotate the beam about point A clockwise.) Adding the two moments at point A, the total moment acting at A is thus

(a)   (b)   (c)   (d)

Fig. 4–24

$$M = M_1 + M_2 = 30\,k - 160\,k = \{-130\,k\}\text{ lb-ft}$$

The result is shown in Fig. 4–24c.

Since the resultant force and moment acting at A are mutually *perpendicular* to one another, this system may indeed be further reduced to a single resultant force. To do this the 20-lb force must be moved to the *right* a distance d to point A such that the couple moment created by the 20-lb force, when the force is moved, has a magnitude of 130 lb-ft. Hence,

$$130\text{ lb-ft} = d(20\text{ lb})$$

or

$$d = 6.5\text{ ft} \qquad\qquad Ans.$$

The result is illustrated in Fig. 4–24d, where it is shown that when the 20-lb force acts to the *right* of point A, the force imparts the *required* 130 lb-ft *clockwise* rotation to the beam about point A.

Example 4–7

Replace the couple moment M and forces F_1 and F_2 acting on the rigid body shown in Fig. 4–25a by an equivalent force and couple-moment system acting at the origin of coordinates.

Solution

Expressing the forces and couple moment in Cartesian vector notation, we have

$$F_1 = (50\text{ lb})\frac{r_{AB}}{r_{AB}} = 50\left(\frac{2i + j + 4k}{\sqrt{21}}\right) = \{21.8i + 10.9j + 43.6k\}\text{ lb}$$

$$F_2 = (75\text{ lb})\frac{r_{CO}}{r_{CO}} = 75\left(\frac{-3i - 9k}{\sqrt{90}}\right) = \{-23.7i - 71.2k\}\text{ lb}$$

$$M = (200\text{ lb-ft})\frac{r_{BC}}{r_{BC}} = 200\left(\frac{i - 3j + 5k}{\sqrt{35}}\right)$$
$$= \{33.8i - 101.4j + 169.0k\}\text{ lb-ft}$$

Since M is a *free vector,* it may be moved directly to the origin O, as shown in Fig. 4–25b. Because force is a *sliding vector,* F_2 may be placed at the origin, since the origin is a point lying on the line of action of F_2. Force F_1 must be replaced at the origin using the method discussed in Sec. 4–7. This is accomplished by moving F_1 directly to the origin and *adding* a couple moment M_1, as shown in Fig. 4–25b. The moment is perpendicular to the shaded plane and defined by the cross product:

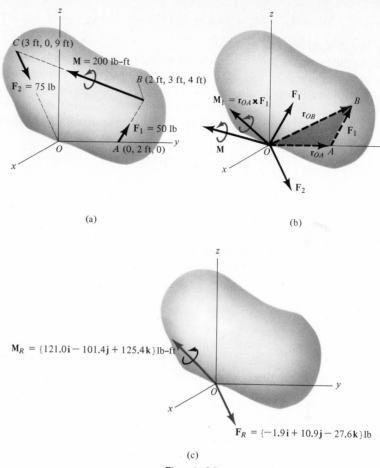

(a)

(b)

(c)

Fig. 4–25

$$\mathbf{M}_1 = \mathbf{r}_{OA} \times \mathbf{F}_1 = 2\mathbf{j} \times (21.8\mathbf{i} + 10.9\mathbf{j} + 43.6\mathbf{k})$$

or

$$\mathbf{M}_1 = \{87.2\mathbf{i} - 43.6\mathbf{k}\} \text{ lb-ft}$$

(From Fig. 4–25a, note also that $\mathbf{M}_1 = \mathbf{r}_{OB} \times \mathbf{F}_1$ since point B is on the line of action of $\mathbf{F}_1$.) The resultant force and moment system acting at the origin is thus

$$\mathbf{F}_R = \mathbf{F}_1 + \mathbf{F}_2 = (21.8 - 23.7)\mathbf{i} + (10.9)\mathbf{j} + (43.6 - 71.2)\mathbf{k}$$
$$= \{-1.9\mathbf{i} + 10.9\mathbf{j} - 27.6\mathbf{k}\} \text{ lb} \qquad \qquad Ans.$$

$$\mathbf{M}_R = \mathbf{M}_1 + \mathbf{M} = (87.2 + 33.8)\mathbf{i} + (-101.4)\mathbf{j} + (-43.6 + 169.0)\mathbf{k}$$
$$= \{121.0\mathbf{i} - 101.4\mathbf{j} + 125.4\mathbf{k}\} \text{ lb-ft} \qquad \qquad Ans.$$

These results are shown in Fig. 4–25c.

Problems

4-21. Determine the resultant couple moment at points A and B caused by the two couples acting on the block assembly.

Prob. 4-21

4-22. The resultant moment created by the two couples acting on the disk is 10 kip-in. acting counterclockwise. Determine the magnitude of force **T**.

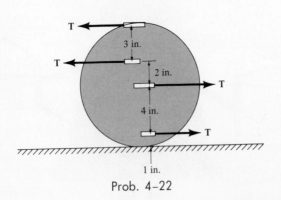

Prob. 4-22

4-23. When the engine of the plane is running, the vertical reaction which the ground exerts at A is measured as 1,100 lb. On the other hand, when the engine is turned off, the vertical reactions at A and B are 900 lb each. The difference in readings at A is caused by the couple acting on the propeller when the engine is running. This couple tends to overturn the plane in the sense opposite to the counterclockwise rotation of the propeller. Determine the magnitude of the couple moment and the magnitude of the vertical force exerted at B when the engine is running.

Prob. 4-23

4-24. Determine the magnitude and direction of the resultant couple of the three couples acting on the disk and bar assembly. At what point does this resultant couple act when you consider the effects of external forces acting on the assembly?

Prob. 4-24

101

4-25. Determine the magnitude and direction of the resultant couple acting on the pipe assembly. The 60-lb forces lie in the horizontal plane.

Prob. 4-25

4-26. If the resultant couple moment of the three couples acting on the triangular block is zero, determine the magnitude of forces **F** and **P**.

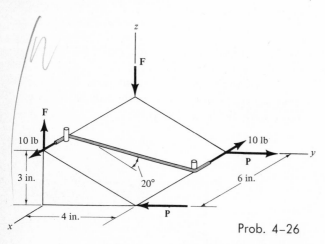

Prob. 4-26

4-27. The cord passing over the two small pegs *A* and *B* of the board is subjected to a tension of 10 lb. Determine the *minimum* tension **P** and the direction α of the cord passing over the pegs *C* and *D*, so that the resultant couple moment produced by the two cords is 20 lb-in., acting clockwise.

Prob. 4-27

4-28. If the resultant of the two couples acting on the fire hydrant is $\mathbf{M}_R = \{-32\mathbf{i} + 20\mathbf{j}\}$ lb-ft, determine the magnitude of the applied force **P**.

Prob. 4-28

4-29. A 40 lb-ft couple moment is developed by the electric motor. Determine the reactive forces caused by this couple at the supports *A* and *B*. Neglect the weight of the motor, and assume that the supports at *A* and *B* transmit only vertical forces.

M = 40 lb-ft

15 in.

A

60°

B

R

R

Prob. 4-29

4-31. The wheel is subjected to a force of 10 lb and a couple moment of 6 lb-in. Determine an equivalent system of a force and a couple moment acting at point A; repeat for point B.

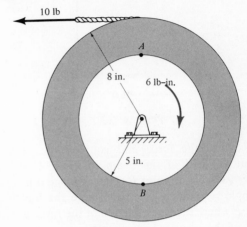

10 lb

A

8 in.

6 lb–in.

5 in.

B

Prob. 4-31

4-30. Replace the 100-lb vertical load acting on the bracket by a force and a couple moment at point P; then distribute the moment into a couple having a pair of horizontal forces acting along the axis of the rivets at A and B.

4-32. Replace the force and couple-moment system acting on the rigid frame by an equivalent force and couple-moment system acting at point O.

6 in.

A

3 in.

100 lb

P

3 in.

B

Prob. 4-30

10 lb–in.

35°

6 in.

10 lb–in.

5 in.

10 lb

6 in.

O

Prob. 4-32

103

4-33. Replace the force and couple-moment system acting at the end of the beam by an equivalent system acting at point A.

Prob. 4–33

4-34. Replace the force and couple-moment system acting at the end of the beam in Prob. 4–33 by an equivalent system at point B. Then replace this loading by a single resultant force. Determine the location, x, of point P where the line of action of the force intersects the axis of the beam.

4-35. Replace the couple moment and force acting on the welded pipe section by an equivalent force and couple-moment system acting at point O.

Prob. 4–36

4-37. Replace the 200-lb force acting at the end of the beam by an equivalent force and couple-moment system at point O.

Prob. 4–35

4-36. A force of 100 lb acts at the end of the beam. Replace this force by an equivalent force and couple-moment system at (a) point A; (b) at point O. Determine the magnitude and direction of the resultant couple vector in each case.

Prob. 4–37

4-38. Replace the two forces and couple moment by an equivalent force and couple-moment system acting at point O.

Prob. 4-38

4-39. Replace the loading acting on the cantilever shaft by an equivalent force and couple-moment system acting at point O.

Prob. 4-39

4-40. Replace the system of three forces shown acting on the beam cross section by an equivalent force and couple-moment system acting at point A.

Prob. 4-40

4-8. Dot or Scalar Product

The *dot product* of two vectors **A** and **B**, written **A · B** (read **A** dot **B**), is defined as the product of the magnitude of vectors **A** and **B** and the cosine of the angle between them. Expressed in equation form,

$$\mathbf{A} \cdot \mathbf{B} = AB \cos \theta \tag{4-10}$$

where $0° \leq \theta \leq 180°$. The dot product is often referred to as the *scalar product* of vectors since, unlike the cross product, the result is a *scalar* and not a vector. If the resultant of the dot product **A · B** is *zero* (provided vectors **A** and **B** are not null vectors), then **A** must be *perpendicular* to **B**. This follows from the definition of the dot product given by Eq. 4–10, i.e., $AB \cos \theta = 0$. Thus $\cos \theta = 0$, or $\theta = 90°$, so that **A** is perpendicular to **B**.

Three important laws of operation which are valid for the dot product are:

1. Commutative law:

$$\mathbf{A} \cdot \mathbf{B} = \mathbf{B} \cdot \mathbf{A} \tag{4-11}$$

2. Multiplication by a scalar:

$$m(\mathbf{A} \cdot \mathbf{B}) = (m\mathbf{A}) \cdot \mathbf{B} = \mathbf{A} \cdot (m\mathbf{B}) = (\mathbf{A} \cdot \mathbf{B})m \tag{4-12}$$

3. Distributive law:

$$\mathbf{A} \cdot (\mathbf{B} + \mathbf{D}) = (\mathbf{A} \cdot \mathbf{B}) + (\mathbf{A} \cdot \mathbf{D}) \tag{4-13}$$

Equations 4–11 and 4–12 can be easily proven by using the definition of the dot product, Eq. 4–10. The proof of Eq. 4–13 is left as an exercise for the reader (see Prob. 4–42).

Equation 4–10 may be used to find the dot product of each of the Cartesian component unit vectors. For example,

$$\mathbf{i} \cdot \mathbf{i} = (1)(1) \cos 0° = 1$$

In a similar manner,

$$\begin{array}{lll} \mathbf{i} \cdot \mathbf{i} = 1 & \mathbf{i} \cdot \mathbf{j} = 0 & \mathbf{i} \cdot \mathbf{k} = 0 \\ \mathbf{j} \cdot \mathbf{j} = 1 & \mathbf{j} \cdot \mathbf{k} = 0 & \mathbf{j} \cdot \mathbf{i} = 0 \\ \mathbf{k} \cdot \mathbf{k} = 1 & \mathbf{k} \cdot \mathbf{i} = 0 & \mathbf{k} \cdot \mathbf{j} = 0 \end{array} \tag{4-14}$$

You should not memorize these results; rather, you should understand clearly how each is derived using Eq. 4–10.

The above results can be used to determine the dot product of two vectors **A** and **B** when these vectors are written in Cartesian component form. If

$$\mathbf{A} = A_x\mathbf{i} + A_y\mathbf{j} + A_z\mathbf{k}$$

and

$$\mathbf{B} = B_x\mathbf{i} + B_y\mathbf{j} + B_z\mathbf{k}$$

then

$$\mathbf{A} \cdot \mathbf{B} = (A_x\mathbf{i} + A_y\mathbf{j} + A_z\mathbf{k}) \cdot (B_x\mathbf{i} + B_y\mathbf{j} + B_z\mathbf{k})$$

Using the scalar and distributive laws of operation, Eqs. 4–12 and 4–13, we have

$$
\begin{aligned}
\mathbf{A} \cdot \mathbf{B} = {} & A_xB_x(\mathbf{i} \cdot \mathbf{i}) + A_xB_y(\mathbf{i} \cdot \mathbf{j}) + A_xB_z(\mathbf{i} \cdot \mathbf{k}) \\
& + A_yB_x(\mathbf{j} \cdot \mathbf{i}) + A_yB_y(\mathbf{j} \cdot \mathbf{j}) + A_yB_z(\mathbf{j} \cdot \mathbf{k}) \\
& + A_zB_x(\mathbf{k} \cdot \mathbf{i}) + A_zB_y(\mathbf{k} \cdot \mathbf{j}) + A_zB_z(\mathbf{k} \cdot \mathbf{k})
\end{aligned}
$$

In accordance with Eqs. 4–14, the final result becomes

$$\mathbf{A} \cdot \mathbf{B} = A_xB_x + A_yB_y + A_zB_z \qquad (4\text{–}15)$$

Thus, to determine the dot product of two vectors, multiply the corresponding x, y, and z components together and sum their products.

The dot product may be applied to determine the *angle formed between two vectors*. This angle may be determined by solving for θ in Eq. 4–10,

$$\theta = \cos^{-1}\frac{\mathbf{A} \cdot \mathbf{B}}{AB}, \quad 0 \le \theta \le \pi$$

The dot product may also be used to find the *projected magnitude of a vector in a given direction*. For example, the magnitude of the projection of vector $\mathbf{A}$ in the direction of $\mathbf{B}$, as shown in Fig. 4–26 is defined by $A\cos\theta$, where θ is the angle formed between vectors $\mathbf{A}$ and $\mathbf{B}$. Using Eq. 4–10, this magnitude may be expressed as

$$A\cos\theta = \frac{1}{B}(\mathbf{A} \cdot \mathbf{B})$$

or

$$A\cos\theta = \mathbf{A} \cdot \frac{\mathbf{B}}{B} = \mathbf{A} \cdot \mathbf{u}_B$$

Here $\mathbf{u}_B$ is a unit vector acting in the direction of $\mathbf{B}$, Fig. 4–26.

Fig. 4–26

4–9. Moment of a Force About a Specified Axis

A useful application of the dot product, in statics, consists of finding the projection of a moment vector along a given axis. As an example, let us consider what effect the force $\mathbf{F}$, acting at the end of the socket

(a)

(b)

Fig. 4-27

wrench shown in Fig. 4–27a, has on turning the bolt *B*. The effective action of **F** in this regard depends upon the *moment* which **F** creates about the *z* axis (since a rotation about this axis will cause the bolt to turn). Separating **F** into its rectangular components F_x, F_y, and F_z, as shown in Fig. 4–27a, it is seen that *only* component F_x causes a rotation of the wrench about the *z* axis. (Since the line of action of F_y passes through a point lying on the *z* axis and the line of action of F_z is parallel to the *z* axis both of these forces create zero moment about the *z* axis.) The moment created by the force component F_x has a magnitude $M_z = F_x l$ and a direction defined by the right-hand rule, that is, in the $-\mathbf{k}$ direction. Thus, $\mathbf{M}_z = -F_x l \mathbf{k}$.

(a)   (b)

Fig. 4–28

The problem of determining the moment component of a force along a given axis may be greatly simplified by using vector analysis. Consider the rigid body, shown in Fig. 4–28a, which is subjected to a force $\mathbf{F}$. This force imparts a tendency for rotation of the body about the fixed aa axis (provided, of course, the line of action of $\mathbf{F}$ does not intersect aa). If $\mathbf{r}$ represents a vector which is directed from the arbitrary point A *lying on the axis* to *any point B*, lying on the line of action of force $\mathbf{F}$, then the moment produced by $\mathbf{F}$ about point A is

$$\mathbf{M}_A = \mathbf{r} \times \mathbf{F}$$

In accordance with the definition of the cross product, the direction of $\mathbf{M}_A$ is perpendicular to the shaded plane containing vectors $\mathbf{r}$ and $\mathbf{F}$, as shown in Fig. 4–28a.

The vector $\mathbf{M}_A$ can be separated into two rectangular *components*, $\mathbf{M}_a$ and $\mathbf{M}_\perp$, acting parallel and perpendicular to the aa axis, as shown in Fig. 4–28a. In particular, the magnitude of $\mathbf{M}_a$ represents the *projection* of $\mathbf{M}_A$ onto the aa axis and can be found using the dot product, as discussed in Sec. 4–8. Hence,

$$M_a = \mathbf{u}_a \cdot \mathbf{M}_A$$

where $\mathbf{u}_a$ is a unit vector defining the *direction* of the *aa* axis. Combining the last two equations yields

$$M_a = \mathbf{u}_a \cdot (\mathbf{r} \times \mathbf{F}) \qquad (4\text{-}16)$$

In vector algebra, this combination of dot and cross products, yielding the scalar M_a, is termed the *mixed triple product*. Knowing the Cartesian components of each of the vectors, the mixed triple product may be written in determinant form as

$$M_a = (u_{a_x}\mathbf{i} + u_{a_y}\mathbf{j} + u_{a_z}\mathbf{k}) \cdot \begin{vmatrix} \mathbf{i} & \mathbf{j} & \mathbf{k} \\ r_x & r_y & r_z \\ F_x & F_y & F_z \end{vmatrix}$$

or simply

$$M_a = \mathbf{u}_a \cdot (\mathbf{r} \times \mathbf{F}) = \begin{vmatrix} u_{a_x} & u_{a_y} & u_{a_z} \\ r_x & r_y & r_z \\ F_x & F_y & F_z \end{vmatrix} \qquad (4\text{-}17)$$

Here u_{a_x}, u_{a_y}, and u_{a_z} represent the Cartesian components of the unit vector defining the direction of the *aa* axis; r_x, r_y, and r_z represent the Cartesian components of the position vector drawn from *any point* on the *aa* axis to *any point* on the line of action of the force; and F_x, F_y, and F_z represent the Cartesian components of the force vector.

The fact that $\mathbf{M}_a$ represents the tendency created by force $\mathbf{F}$ for rotating the body about the *aa* axis can be shown with reference to Fig. 4–28*b*. The shaded plane is perpendicular to the *aa* axis and intersects this axis at point O. Both the position vector $\mathbf{r}$ and the force $\mathbf{F}$ are resolved into two components which lie in and perpendicular to this plane. The perpendicular component of force, $\mathbf{F}_2$, does not contribute a moment about the *aa* axis, since this force is parallel to the axis. $\mathbf{M}_a$ results only from the action of $\mathbf{F}_1$, which lies at a perpendicular distance d from the axis. Thus, the magnitude of $\mathbf{M}_a$ is $M_a = F_1 d = \mathbf{r}_{OB} \times \mathbf{F}_1 = \mathbf{u}_a \cdot (\mathbf{r} \times \mathbf{F})$.

The socket-wrench problem discussed earlier may be solved using Eq. 4–17. As shown in Fig. 4–27*b*, the axis of rotation is defined by the unit vector $\mathbf{u}_a = \mathbf{k}$. Both vectors $\mathbf{r}_1 = l\mathbf{j} + a\mathbf{k}$ and $\mathbf{r}_2 = l\mathbf{j}$ satisfy the requirements for $\mathbf{r}$, defined in Eq. 4–17. Why? Hence,

$$M_z = \mathbf{u}_a \cdot (\mathbf{r}_1 \times \mathbf{F}) = \begin{vmatrix} 0 & 0 & 1 \\ 0 & l & a \\ F_x & F_y & F_z \end{vmatrix} = -lF_x$$

or

$$M_z = \mathbf{u}_a \cdot (\mathbf{r}_2 \times \mathbf{F}) = \begin{vmatrix} 0 & 0 & 1 \\ 0 & l & 0 \\ F_x & F_y & F_z \end{vmatrix} = -lF_x$$

Since both the magnitude and direction of $\mathbf{M}_z$ are known, $\mathbf{M}_z$ may be written in vector form as

$$\mathbf{M}_z = M_z(-\mathbf{k}) = -F_x l \mathbf{k}$$

4-10. Reduction of a Force and Couple System

For purposes of simplification, it is possible to reduce a system of forces and couple-moment vectors acting on a rigid body to a *single* resultant force acting at an arbitrary point and a *single* resultant moment. Consider, for example, the rigid body subjected to the system of force and couple moment vectors shown in Fig. 4–29a. To simplify this loading system, each vector is transferred to some *arbitrary point O* on the body. The couple moments $\mathbf{M}$ and $\mathbf{M}'$ are free vectors; hence they may be applied

Fig. 4–29

directly at point O, as shown in Fig. 4–29b. The forces $\mathbf{F}_1$, $\mathbf{F}_2$, and $\mathbf{F}_3$ are sliding vectors, and since point O does *not* lie along the line of action of any of these forces, each force must be replaced at point O in accordance with the procedure outlined in Sec. 4–7. For example, when vector $\mathbf{F}_1$ is applied to point O, a corresponding couple-moment vector $\mathbf{M}_1 = \mathbf{r}_1 \times \mathbf{F}_1$ must be *applied* to the body. The *equivalent* system of force and moment vectors applied at point O is shown in Fig. 4–29b. Using the laws of vector addition, we may sum this concurrent system of vectors to a single resultant force and resultant moment.

$$\mathbf{F}_R = \Sigma\mathbf{F} = \mathbf{F}_1 + \mathbf{F}_2 + \mathbf{F}_3$$
$$\mathbf{M}_R = \Sigma\mathbf{M}_o = \mathbf{M} + \mathbf{M}' + (\mathbf{r}_1 \times \mathbf{F}_1) + (\mathbf{r}_2 \times \mathbf{F}_2) + (\mathbf{r}_3 \times \mathbf{F}_3)$$

The results are shown in Fig. 4–29c. Note that if the force and couple-moment system in Fig. 4–29a was replaced at some other reference point, other than O, one would obtain the *same* force resultant $\mathbf{F}_R$ but a *different* resultant moment $\mathbf{M}_R$. The force resultant is simply the sum of all the force vectors of the system, whereas the moment resultant depends upon the location of the reference point.

It was discussed in Sec. 4–7 that if $\mathbf{F}_R$ is *perpendicular* to $\mathbf{M}_R$, the system of a force and a moment may be reduced to a *single force* acting through some unique point. Provided $\mathbf{F}_R$ and $\mathbf{M}_R$ are not null vectors, perpendicularity is guaranteed if

$$\mathbf{F}_R \cdot \mathbf{M}_R = 0 \tag{4–18}$$

A procedure for eliminating the moment vector in this case follows the steps outlined in Fig. 4–21, going from Fig. 4–21d to Fig. 4–21a; or in Fig. 4–22, going from Fig. 4–22d to Fig. 4–22a. In particular, if a force system is *coplanar* or all the forces act *parallel* to one another, the system of forces can always be resolved into a single resultant force. In *both* these cases, if the force systems are reduced to a single resultant force and moment acting at some *arbitrary point* on the body, the moment vector will *always* be perpendicular to the force and, therefore, one can always locate the force at some unique point in order to eliminate the moment vector. These two specific cases involving coplanar and parallel force systems are discussed further in Examples 4–10 and 4–11.

For the general case, the resultant force $\mathbf{F}_R$ will not be perpendicular to the moment $\mathbf{M}_R$, Fig. 4–29c. As shown in Fig. 4–29d, however, $\mathbf{M}_R$ may be resolved into two components; one perpendicular, $\mathbf{M}_\perp$, and the other parallel, $\mathbf{M}_\parallel$, to the line of action of the force resultant $\mathbf{F}_R$. The magnitude of component $\mathbf{M}_\parallel$ may be found using the dot product:

$$M_\parallel = M_R \cos\theta = \frac{\mathbf{F}_R}{F_R} \cdot \mathbf{M}_R = \mathbf{u}_R \cdot \mathbf{M}_R$$

Here $\mathbf{u}_R$ denotes a unit vector in the direction of $\mathbf{F}_R$. In vector notation,

$$\mathbf{M}_{\parallel} = M_{\parallel}\mathbf{u}_R$$

Since the component $\mathbf{M}_{\perp}$ is *perpendicular* to $\mathbf{F}_R$, $\mathbf{M}_{\perp}$ may be *eliminated* by *moving* $\mathbf{F}_R$ to point O', as shown in Fig. 4–29e. Why? The point O' lies along the line segment bb at a distance d from point O. Because bb is perpendicular to both $\mathbf{M}_{\perp}$ and $\mathbf{F}_R$, then

$$M_{\perp} = F_R d$$

or

$$d = \frac{M_{\perp}}{F_R}$$

It should be clearly understood that $\mathbf{F}_R$ must be located at point O' and *not* O'', as shown in Fig. 4–29e. If $\mathbf{F}_R$ is applied at point O', the moment of $\mathbf{F}_R$ tending to cause rotation of the body *about* point O is in the *same* direction as $\mathbf{M}_{\perp}$ (shown dashed in the figure).

The resulting force $\mathbf{F}_R$ and *collinear* moment $\mathbf{M}_{\parallel}$, Fig. 4–29f, is called a *wrench*. The *axis of the wrench* has the same line of action as the force. The wrench tends to cause both a translation along and a rotation about this axis. Comparing Fig. 4–29a with Fig. 4–29f, it is therefore possible to simplify a general system of force and couple-moment vectors, acting on a rigid body, to a wrench. The axis of the wrench and a point through which this axis passes is unique and can always be determined.

Example 4–8

Given the two position vectors

$$\mathbf{r}_1 = \{3\mathbf{i} + 2\mathbf{j} - 4\mathbf{k}\} \text{ ft and } \mathbf{r}_2 = \{-1\mathbf{i} + 3\mathbf{j} + 5\mathbf{k}\} \text{ ft.}$$

Determine (a) the angle θ formed between the two vectors $\mathbf{r}_1$ and $\mathbf{r} = \mathbf{r}_2 - \mathbf{r}_1$, and (b) the angle β which $\mathbf{r}_2$ makes with the y axis.

Solution

Part (a). Vectors $\mathbf{r}_1$ and $\mathbf{r}_2$ are represented graphically in Fig. 4–30a. (Note that $\mathbf{r}_1 + \mathbf{r} = \mathbf{r}_2$, so that $\mathbf{r} = \mathbf{r}_2 - \mathbf{r}_1$.) θ is defined as the angle formed between the *tails* of vectors $\mathbf{r}_1$ and $\mathbf{r}$ ($0° < \theta < 180°$). This angle can be determined using the dot product of vectors $\mathbf{r}_1$ and $\mathbf{r}$. Computing the required quantities,

$$\mathbf{r} = \mathbf{r}_2 - \mathbf{r}_1 = \{-4\mathbf{i} + 1\mathbf{j} + 9\mathbf{k}\} \text{ ft}$$
$$r = \sqrt{(-4)^2 + (1)^2 + (9)^2} = 9.90 \text{ ft}$$
$$r_1 = \sqrt{(3)^2 + (2)^2 + (-4)^2} = 5.39 \text{ ft}$$

Using Eq. 4–15, we write

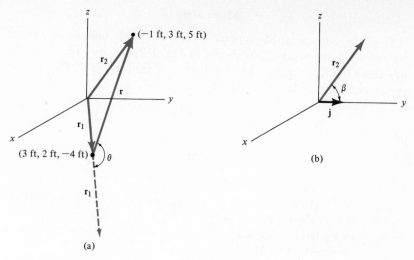

Fig. 4–30

$$\mathbf{r} \cdot \mathbf{r}_1 = (-4\mathbf{i} + 1\mathbf{j} + 9\mathbf{k}) \cdot (3\mathbf{i} + 2\mathbf{j} - 4\mathbf{k})$$
$$= -4(3) + 1(2) + 9(-4) = -46 \text{ ft}^2$$

Thus, by definition of the dot product

$$\mathbf{r} \cdot \mathbf{r}_2 = rr_1 \cos \theta$$

or

$$\theta = \cos^{-1}\frac{\mathbf{r} \cdot \mathbf{r}_1}{rr_1} = \cos^{-1}\frac{-46}{9.90(5.39)}$$
$$= \cos^{-1}(-0.862) = 149.6° \qquad \textit{Ans.}$$

Part (b). The required angle β is shown in Fig. 4–30*b*. The direction of the y axis is represented by the $\mathbf{j}$ unit vector. Since,

$$\mathbf{r}_2 = \{-1\mathbf{i} + 3\mathbf{j} + 5\mathbf{k}\} \text{ ft}, \; r_2 = \sqrt{(-1)^2 + (3)^2 + (5)^2} = 5.92 \text{ ft}$$
$$\mathbf{r}_2 \cdot \mathbf{j} = (-1\mathbf{i} + 3\mathbf{j} + 5\mathbf{k}) \cdot (\mathbf{j}) = -1(0) + 3(1) + 5(0) = 3 \text{ ft}$$

Then,

$$\mathbf{r}_2 \cdot \mathbf{j} = r_2 j \cos \beta$$

or

$$\beta = \cos^{-1}\frac{\mathbf{r}_2 \cdot \mathbf{j}}{(r_2)(j)} = \cos^{-1}\frac{3 \text{ ft}}{(5.92 \text{ ft})(1)}$$
$$= \cos^{-1}(0.507) = 59.5° \qquad \textit{Ans.}$$

We can obtain this same result by determining the *direction cosine* of r_2 using the unit vector $\mathbf{u}_2$ acting in the direction of r_2.

$$\mathbf{u}_2 = \frac{\mathbf{r}_2}{r_2} = -\frac{1}{\sqrt{35}}\mathbf{i} + \frac{3}{\sqrt{35}}\mathbf{j} + \frac{5}{\sqrt{35}}\mathbf{k} = -0.169\mathbf{i} + 0.507\mathbf{j} + 0.845\mathbf{k}$$

Hence, from the $\mathbf{j}$ component of this vector,

$$\cos \beta = 0.507$$
$$\beta = \cos^{-1}(0.507) = 59.5° \qquad \textit{Ans.}$$

Example 4–9

Find the projections of the force vector $\mathbf{F} = \{40\mathbf{i} + 30\mathbf{j}\}$ lb in the direction of the position vectors $\mathbf{r}_1 = \{2\mathbf{j} + 4\mathbf{k}\}$ ft and $\mathbf{r}_2 = \{1\mathbf{i} + 5\mathbf{j} + 2\mathbf{k}\}$ ft, shown in Fig. 4–31a.

Solution

The projection of a vector in a given direction can be determined using the dot product. The unit vectors in Fig. 4–31a, which define the directions of $\mathbf{r}_1$ and $\mathbf{r}_2$, are

$$\mathbf{u}_1 = \frac{\mathbf{r}_1}{r_1} = \frac{2\mathbf{j} + 4\mathbf{k}}{\sqrt{20}} = 0.447\mathbf{j} + 0.894\mathbf{k}$$

$$\mathbf{u}_2 = \frac{\mathbf{r}_2}{r_2} = \frac{1\mathbf{i} + 5\mathbf{j} + 2\mathbf{k}}{\sqrt{30}} = 0.183\mathbf{i} + 0.913\mathbf{j} + 0.365\mathbf{k}$$

These unit vectors, along with the force vector $\mathbf{F}$, are shown in Figs. 4–31b and 4–31c. It is assumed that $\mathbf{F}$ makes an angle of ψ with $\mathbf{u}_1$ and ϕ with $\mathbf{u}_2$. If we consider the x', y', and z' axes to be measured in units of pounds, the *magnitude* of each projected vector becomes

Fig. 4–31

$$F \cos \psi = \mathbf{F} \cdot \mathbf{u}_1 = (40\mathbf{i} + 30\mathbf{j}) \cdot (0.447\mathbf{j} + 0.894\mathbf{k})$$
$$= 40(0) + 30(.447) + 0(.894) = 13.41 \text{ lb}$$
$$F \cos \phi = \mathbf{F} \cdot \mathbf{u}_2 = (40\mathbf{i} + 30\mathbf{j}) \cdot (0.183\mathbf{i} + 0.913\mathbf{j} + 0.365\mathbf{k})$$
$$= 40(0.183) + 30(0.913) + 0(0.365) = 34.71 \text{ lb}$$

Hence, the projected *vectors* are

$$\mathbf{F} \cos \psi = (F \cos \psi)\mathbf{u}_1 = (13.41 \text{ lb})(0.447\mathbf{j} + 0.894\mathbf{k})$$
$$= \{6.00\mathbf{j} + 12.00\mathbf{k}\} \text{ lb} \qquad\qquad\qquad \textit{Ans.}$$
$$\mathbf{F} \cos \phi = (F \cos \phi)\mathbf{u}_2 = (34.7 \text{ lb})(0.183\mathbf{i} + 0.913\mathbf{j} + 0.365\mathbf{k})$$
$$= \{6.35\mathbf{i} + 31.70\mathbf{j} + 12.67\mathbf{k}\} \text{ lb} \qquad\qquad \textit{Ans.}$$

The results are shown in Figs. 4–31*b* and 4–31*c*.

Example 4–10

The beam *AE* in Fig. 4–32*a* is subjected to a series of concentrated loads acting along its length. Determine the magnitude, the direction, and the location of a single resultant force which is equivalent to the given system of forces.

Solution I

The force system can be reduced to a single resultant force and couple moment acting at point *E* (or any other arbitrary point along the axis of the beam). Position vectors, shown in Fig. 4–32*b*, are drawn *from* point *E* to *any point* lying on the line of action of each of the forces. Each force **F** is then moved to point *E*, and a moment $\mathbf{M}_E = \Sigma \mathbf{r} \times \mathbf{F}$ must be *added* to the beam. Why? The resultant force and resultant moment acting at point *E* is therefore

$$\mathbf{F}_R = \mathbf{F}_1 + \mathbf{F}_2 + \mathbf{F}_3$$
$$= (50 \cos 60°\mathbf{i} - 50 \sin 60°\mathbf{j}) + 20\mathbf{j} + 10\mathbf{i}$$

or

$$\mathbf{F}_R = \{35.0\mathbf{i} - 23.3\mathbf{j}\} \text{ lb}$$

and

$$\mathbf{M}_E = (\mathbf{r}_{EB} \times \mathbf{F}_1) + (\mathbf{r}_{EC} \times \mathbf{F}_2) + (\mathbf{r}_{ED} \times \mathbf{F}_3)$$
$$= -8\mathbf{i} \times (50 \cos 60°\mathbf{i} - 50 \sin 60°\mathbf{j}) + (-5\mathbf{i}) \times 20\mathbf{j}$$
$$+ (-3\mathbf{i} + \mathbf{j}) \times 10\mathbf{i}$$

or

$$\mathbf{M}_E = \{236.4\mathbf{k}\} \text{ lb-ft}$$

These results are shown in Fig. 4–32*c*.

Since the force system, Fig. 4–32*a*, is *coplanar*, $\mathbf{F}_R$ is *perpendicular to* $\mathbf{M}_E$, i.e., $\mathbf{M}_E \cdot \mathbf{F}_R = 0$. Thus, the resultant force $\mathbf{F}_R$ can be located at a

Fig. 4-32

unique point P on the axis of the beam and thereby eliminate the moment. If $\mathbf{r}_{EP} = x\mathbf{i}$ denotes a position vector drawn *from* point E *to* point P, it is necessary that

$$\mathbf{M}_E = \mathbf{r}_{EP} \times \mathbf{F}_R$$
$$236.4\mathbf{k} = x\mathbf{i} \times (35.0\mathbf{i} - 23.3\mathbf{j}) \qquad (1)$$

or

$$236.4 = -23.3x$$

Hence,

$$x = -\frac{236.4}{23.3} = -10.1 \text{ ft} \qquad \textit{Ans.}$$

The negative sign indicates that $\mathbf{F}_R$ must be placed 10.1 ft to the *left* of

point E, as shown in Fig. 4–32d. The required (counterclockwise) moment M_E will then be produced when $\mathbf{F}_R$ is moved from point P to point E.*

The magnitude of $\mathbf{F}_R$ is

$$F_R = \sqrt{(35.0)^2 + (-23.3)^2} = 42.0 \text{ lb} \qquad Ans.$$

Measuring θ from *below* the x axis, we have

$$\theta = \tan^{-1} \frac{23.3}{35.0} = 33.6° \qquad \theta\, \mathbf{F}_R \qquad Ans.$$

Solution II

Since the problem involves a coplanar force system, the solution is facilitated using a semi-scalar approach. For convenience, the 50-lb force acting at B is separated into its two rectangular components as shown in Fig. 4–32e. The resultant force, acting on the beam, is the vector sum of all the forces acting on the beam. Thus,

$$\mathbf{F}_R = 50 \cos 60°\mathbf{i} - 50 \sin 60°\mathbf{j} + 20\mathbf{j} - 10\mathbf{i}$$

or

$$\mathbf{F}_R = \{35.0\mathbf{i} - 23.3\mathbf{j}\} \text{ lb}$$

So that

$$F_R = \sqrt{(35.0)^2 + (-23.3)^2} = 42.0 \text{ lb}$$

and

$$\theta = \tan^{-1} \frac{23.3}{35.0} = 33.6° \qquad \theta\, \mathbf{F}_R$$

The resultant force $\mathbf{F}_R$ is shown acting on the beam in Fig. 4–32f.

The location of the force along the axis of the beam, P, can be determined by realizing that the moment of the *force resultant* $\mathbf{F}_R$ about point E, $(\mathbf{M}_R)_E$, must be equivalent to the moment of *all the forces* shown in Fig. 4–32e about point $E(\Sigma\mathbf{M}_E)$. (The 35-lb component of $\mathbf{F}_R$ does not create a moment about point E. Why?) Hence, assuming moments are positive counterclockwise, i.e., $+\mathbf{k}$ direction, we have

$$\zeta+ \qquad\qquad (\mathbf{M}_R)_E = \Sigma\mathbf{M}_E$$

$$d(23.3 \text{ lb}) + 0(35.0 \text{ lb}) = (8 \text{ ft})(50 \sin 60° \text{ lb}) - (5 \text{ ft})(20 \text{ lb}) - (1 \text{ ft})(10 \text{ lb})$$

Thus,

$$d = 10.1 \text{ ft} \qquad Ans.$$

---

*Except for purposes of explanation, as stated here, it is not strictly necessary to replace all the forces at point E. Essentially what is required is to realize that the *moment of all the forces* in Fig. 4–32b must be *equivalent* to the *moment produced by the force resultant* $\mathbf{F}_R$ acting at P, Fig. 4–32d. In this way Eq. (1) can be written *directly* without the intermediate step which involves Fig. 4–32c. See Solution II.

Since d is a positive quantity, $\mathbf{F}_R$ acts 10.1 ft to the *left* of E as shown in Fig. 4–32*f*.

Example 4–11

The slab in Fig. 4–33*a* is subjected to a series of five parallel forces. Determine the magnitude and direction of a single resultant force equivalent to the given force system, and locate its point of application on the slab.

Solution

Since the force system is represented in three dimensions, a vector approach will be used for the solution. The resultant of the force system is the sum of all the forces, that is,

$$\mathbf{F}_R = \Sigma \mathbf{F} = -600\mathbf{k} + 100\mathbf{k} - 300\mathbf{k} - 400\mathbf{k} - 500\mathbf{k}$$

or

$$\mathbf{F}_R = \{-1,700\mathbf{k}\} \text{ lb}$$

The moment $\mathbf{M}_O$ created by the force system about the origin O can be obtained by means of the position vectors shown in Fig. 4–33*b*. Each position vector is drawn from O to a point on the line of action of each force. Thus,

$$\mathbf{M}_O = \Sigma \mathbf{r} \times \mathbf{F} = [\mathbf{r}_{OA} \times (-600\mathbf{k})] + [\mathbf{r}_{OB} \times 100\mathbf{k}] + [\mathbf{r}_{OC} \times (-300\mathbf{k})]$$
$$+ [\mathbf{r}_{OD} \times (-400\mathbf{k})]$$

or

$$\mathbf{M}_O = [8\mathbf{i} \times (-600\mathbf{k})] + [(6\mathbf{i} + 5\mathbf{j}) \times 100\mathbf{k}] + [(2\mathbf{i} + 5\mathbf{j}) \times (-300\mathbf{k})]$$
$$+ [10\mathbf{j} \times (-400\mathbf{k})]$$

$$= \{-5,000\mathbf{i} + 4,800\mathbf{j}\} \text{ lb-ft} \tag{1}$$

Fig. 4–33

Since all the force vectors are *parallel* to one another, $\mathbf{F}_R$ is perpendicular to $\mathbf{M}_O$, i.e., $\mathbf{F}_R \cdot \mathbf{M}_O = 0$. Therefore, the resultant force $\mathbf{F}_R$ can be placed at a unique point $P(x, y, 0)$ on the slab and produce the equivalent moment $\mathbf{M}_O$. If $\mathbf{r}_{OP} = x\mathbf{i} + y\mathbf{j}$ denotes a position vector acting from the origin to point P on the surface of the slab, Fig. 4–33c, then

$$\mathbf{M}_O = \mathbf{r}_{OP} \times \mathbf{F}_R = \begin{vmatrix} \mathbf{i} & \mathbf{j} & \mathbf{k} \\ x & y & 0 \\ 0 & 0 & -1{,}700 \end{vmatrix} = \{-1{,}700y\mathbf{i} + 1{,}700x\mathbf{j}\} \text{ lb-ft} \quad (2)$$

Hence, equating Eqs. (1) and (2)*

$$-5{,}000\mathbf{i} + 4{,}800\mathbf{j} = -1{,}700y\mathbf{i} + 1{,}700x\mathbf{j}$$

The $\mathbf{i}$ and $\mathbf{j}$ components on the left and right-hand sides of the equation must be equal. Thus,

$$-5{,}000 = -1{,}700y; \qquad y = 2.94 \text{ ft} \qquad \qquad Ans.$$
$$4{,}800 = 1{,}700x; \qquad x = 2.82 \text{ ft} \qquad \qquad Ans.$$

The single resultant force of 1,700 lb acting on the slab at point P (2.82 ft, 2.94 ft, 0) is therefore equivalent to the system of parallel forces in Fig. 4–33a.

Example 4–12

The rectangular block, shown in Fig. 4–34a, is subjected to a system of force and couple-moment vectors. Determine (a) an equivalent force-moment system acting at point A; and (b) the projection of moment of this system acting along the diagonal AB.

Solution

Part (a). By the principle of transmissibility, the force $\mathbf{F}_1$ may be moved directly to point A, since A lies on the line of action of $\mathbf{F}_1$. The couple-moment vectors may be moved directly to point A because they are free vectors. Moving the force $\mathbf{F}_2$ to point A requires adding an additional moment $\mathbf{M}_{AB} = \mathbf{r}_{AB} \times \mathbf{F}_2$. Hence, the force and moment resultants are

$$\mathbf{F}_R = \Sigma\mathbf{F} = \mathbf{F}_1 + \mathbf{F}_2 = (-10\mathbf{k}) + (-3\mathbf{i} + 2\mathbf{j} + 5\mathbf{k})$$
$$= \{-3\mathbf{i} + 2\mathbf{j} - 5\mathbf{k}\} \text{ lb} \qquad \qquad Ans.$$

---

*This can be done directly. In words, the moment due to the *force system* about point O, Fig. 4–33b, is *equivalent* to the moment due to the *resultant force* about point O, Fig. 4–33c. (This principle applies, as well, to any arbitrary point lying in the plane of the slab.)

Fig. 4-34

$$\mathbf{M}_A = \mathbf{M}_1 + \mathbf{M}_2 + (\mathbf{r}_{AB} \times \mathbf{F}_2)$$
$$= [-20\mathbf{i}] + [50\mathbf{i}] + [(-2\mathbf{i} + 6\mathbf{j} + 2\mathbf{k}) \times (-3\mathbf{i} + 2\mathbf{j} + 5\mathbf{k})]$$
$$= \{56\mathbf{i} + 4\mathbf{j} + 14\mathbf{k}\} \text{ lb-ft} \qquad\qquad Ans.$$

The results are shown in Fig. 4-34b.

Part (b). As shown in Fig. 4-34b, a unit vector which defines an axis directed from A to B is

$$\mathbf{u}_{AB} = \frac{\mathbf{r}_{AB}}{r_{AB}} = \frac{-2\mathbf{i} + 6\mathbf{j} + 2\mathbf{k}}{\sqrt{(-2)^2 + (6)^2 + (2)^2}} = -0.302\mathbf{i} + 0.905\mathbf{j} + 0.302\mathbf{k}$$

The projection of $\mathbf{M}_A$ acting along the diagonal is thus

$$M_{AB} = M_A \cos \theta = \mathbf{M}_A \cdot \mathbf{u}_{AB}$$
$$= (56\mathbf{i} + 4\mathbf{j} + 14\mathbf{k}) \cdot (-0.302\mathbf{i} + 0.905\mathbf{j} + 0.302\mathbf{k})$$
$$= (56)(-0.302) + (4)(0.905) + (14)(0.302)$$
$$= -9.06 \text{ lb-ft}$$

The negative sign indicates that the moment projection acts in the *opposite direction* to $\mathbf{u}_{AB}$. Expressed as a vector, this projection is

$$\mathbf{M}_{AB} = M_{AB}\mathbf{u}_{AB} = (-9.06 \text{ lb-ft})(-0.302\mathbf{i} + 0.905\mathbf{j} + 0.302\mathbf{k})$$

or

$$\mathbf{M}_{AB} = \{2.74\mathbf{i} - 8.20\mathbf{j} - 2.74\mathbf{k}\} \text{ lb-ft} \qquad\qquad Ans.$$

4-41. If $\mathbf{A} = (3 \text{ ft})\mathbf{i} + (24 \text{ in.})\mathbf{j}$ and $\mathbf{B} = (1 \text{ ft})\mathbf{i} + (2 \text{ ft})\mathbf{j} + (30 \text{ in.})\mathbf{k}$, determine $\mathbf{A} \cdot \mathbf{B}$.

4-42. Given the three vectors **A**, **B**, and **C**, show that $\mathbf{A} \cdot (\mathbf{B} + \mathbf{C}) = \mathbf{A} \cdot \mathbf{B} + \mathbf{A} \cdot \mathbf{C}$. *Suggestion:* Use Cartesian vectors.

4-43. Given the three nonzero vectors **A**, **B**, and **C**, show that if $\mathbf{A} \cdot (\mathbf{B} \times \mathbf{C}) = 0$, the three vectors *must* then lie in the same plane.

4-44. Find the equation of a plane which contains the three points $(2, 4, -1)$, $(3, 1, 1)$, and $(2, -2, 4)$. (*Hint:* Construct direction vectors to each of these points and some arbitrary point (x, y, z) lying in the plane, and use the result of Prob. 4-43.)

4-45. Determine the angle made between the two forces $\mathbf{F}_1$ and $\mathbf{F}_2$ shown acting on the plate. What is the moment of these two forces about the z axis?

Prob. 4-46

4-47. Determine the angle which the force **F** makes with the diagonal AB of the crate. The force **F** lies in the plane $DBEC$ and makes an angle of $10°$ with the extended line DB as shown.

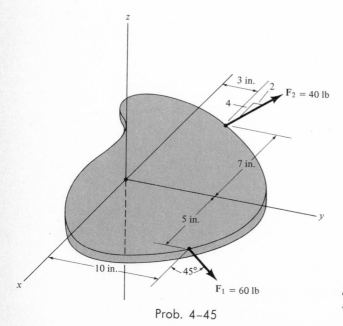

Prob. 4-45

4-46. Determine the projection of the moment created by the 20-lb force which acts along the aa axis of the socket wrench. The force lies in the horizontal plane.

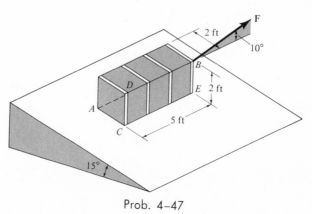

Prob. 4-47

4-48. For the crate in Prob. 4-47, determine the angle which **F** makes with the vertical.

4-49. The ropes AB and CD exert a tension force of 20 lb each on the plane. Determine the resultant force and moment created by these forces at the midsection point O of the plane wing.

Prob. 4-49

4-51. A 70-lb force acts vertically on the "Z" bracket. Determine the moment of this force about the bolt axis (z axis).

Prob. 4-51

4-50. The door is held open by two chains AB and AC. These chains exert a force of 20 and 40 lb, respectively, on the door. Determine the moment created by these two forces about point O and the direction of the axis about which this moment acts. What is the projection of this moment about axis OD?

4-52. The chain AB exerts a force of 20 lb on the door at B. Determine the projection of the moment of this force acting along the hinged axis x of the door.

Prob. 4-50

Prob. 4-52

4-53. Determine the total moment created by forces **P** and **W** about the *AB* axis of the hoist.

Prob. 4–53

4-54. Force **F,** having a magnitude of 50 lb and a direction specified by **u** = 0.200**i** + 0.141**j** + 0.200**k,** acts at point *G* of the block. (a) Determine the *x*, *y*, and *z* components of **F** when this force is moved to point *B*. (b) Using the scalar product, show that the projections of **F** along the *x*, *y*, and *z* axes equal the respective *x*, *y*, and *z* components of **F**. Explain why this is so.

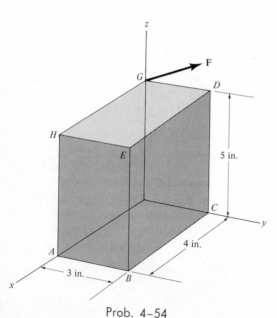

Prob. 4–54

4-55. Replace the given loading system acting on the beam by an equivalent force and couple moment acting at *O*.

Prob. 4–55

4-56. Replace the force system acting on the plate by an equivalent single resultant force. Specify the location of this force.

Prob. 4–56

4-57. Three forces act at the end of the cantilever beam. Determine the moment created by each of these forces about the x axis. Solve the problem using both a vector and a scalar approach.

Prob. 4-57

4-59. Two forces F_1 and F_2 are applied at the end of the pipe assembly. If F_1 acts in the direction shown and has a magnitude of 90 lb, and F_2 has a magnitude of 50 lb, determine the appropriate direction angles β and γ for F_2 so that the resultant moment of both forces about point O is $M_R = \{-400i + 240k\}$ lb-in. Require $\alpha = 90°$ for F_2.

Prob. 4-59

4-58. Replace the two wrenches and the 100-lb force acting on the pipe assembly by an equivalent force and couple-moment system acting at point O.

Prob. 4-58

4-60. Replace the two forces shown acting on the rectangular block by an equivalent force and couple-moment system acting at point A.

Prob. 4-60

125

4-61. Determine the moment of force **F** about the centerline of the shaft. In what direction does the moment tend to twist the shaft? The centerline of the shaft lies in the *zy* plane.

F = {4i + 3j + 3k} lb

2 in.

10 in.

20°

Prob. 4-61

4-62. Specify the location $(x, y, 0)$ for the resultant of the three parallel column forces acting on the circular slab.

700 lb

900 lb

500 lb

10°

20° 7 ft

Prob. 4-62

4-11. Reduction of a Distributed Loading

In the previous sections of this chapter, we have considered ways of simplifying a system of *concentrated forces* which act on a body. In many practical situations, however, the surface of a body may be subjected to *distributed loadings*. Such loadings are caused by wind, hydrostatic pressure, or simply the weight of material supported by the body. The distribution of such surface loadings is described by a *loading function*. In general, at any point on the surface of the body, this loading function defines the intensity of load measured as a *force per unit area* or *pressure* acting on the surface. The direction of the intensity of distributed loading is indicated by arrows acting on the load intensity diagram. For example, the loading distribution or pressure acting on the plate shown in Fig. 4-35 is described by the loading function $p = p(x,y)$ lb/ft². Knowing this function, it is possible to determine the magnitude of *force d***F** acting on an elemental surface area dA ft² of the plate located at the arbitrary point (x, y). This force magnitude is $dF = (p(x, y)$ lb/ft²$)(dA$ ft²$) = dV$ lb, where dV represents the small *differential volume element* shown shaded in Fig. 4-35. We can obtain the magnitude of the *total resultant force* by *integrating* the loading function over the entire surface area A of the plate, i.e.,

p

dF

$p = p(x, y)$ lb/ft²

dA ft²

Fig. 4-35

$$F_R = \int_A p(x,y)\, dA = \int_V dV = V$$

126

Hence, the *magnitude of the resultant force is equal to the volume under the distributed loading diagram.* Using the methods of Chapter 9, it will be shown that *the line of action of the force resultant,* $\mathbf{F}_R$, *passes through the geometric-center or centroid of this volume.*

A simple example which illustrates these principles is shown in Fig. 4–36a. The plate is subjected to a uniform load of 10 lb/ft². Hence, the *resultant force* of this distributed loading is equal to the *volume* of the rectangular block,

$$F_R = V = (10 \text{ lb/ft}^2)(2 \text{ ft})(4 \text{ ft}) = 80 \text{ lb}$$

The line of action of this force acts through the centroid or geometric center of the block and therefore is located on the plate at point

$$x = 1 \text{ ft, and } y = 2 \text{ ft}$$

If the loading function is *symmetric* about any axis, it can be *simplified geometrically* and thereby represented as an *area* rather than a volume. The loading intensity is then measured as a *force per unit length,* rather than a force per unit area. Consider again the uniformly loaded plate, Fig. 4–36a. The loading function has two axes of symmetry. If the distributed loading is multiplied by the width of the plate, the loading function is rectangular and represents the distribution of load *along the length* of the plate. Hence,

$$w_l = (10 \text{ lb/ft}^2)(2 \text{ ft}) = 20 \text{ lb/ft}$$

This distributed loading is uniform and is shown in Fig. 4–36b. In a

(a)

(b)

(c)

Fig. 4–36

Fig. 4–37

similar manner, the loading *along the width* of the plate, Fig. 4–36*c*, is

$$w_w = (10 \text{ lb/ft}^2)(4 \text{ ft}) = 40 \text{ lb/ft}$$

As a second example, consider the beam shown in Fig. 4–37*a* which is subjected to the distributed loading $p = 8x$ lb/ft². In particular, where $x = 9$ ft, the pressure intensity is $p = 8(9) = 72$ lb/ft², as shown. Note that the beam has a constant width of $\frac{1}{2}$ ft and the load is *symmetric* with respect to the *width*. The loading diagram can therefore be viewed in two dimensions along the *length* of the beam, provided

$$w = (8x \text{ lb/ft}^2)(\tfrac{1}{2} \text{ ft}) = 4x \text{ lb/ft}$$

This loading is shown in Fig. 4–37*b* where at $x = 9$ ft, $w = 4(9) = 36$ lb/ft. Throughout the rest of this section, we will consider only loading functions which are represented as an area, i.e., $w = w(x)$ lb/ft. Methods for simplifying the more general loading function, $p = p(x,y)$ lb/ft², are discussed further in Sec. 9–5.

The magnitude and location of the resultant force of the distributed loading $w = w(x)$ lb/ft can be determined using the methods of Sec. 4–10. Specifically, the method used is the *same* as that used for finding the magnitude and location of the resultant force of a system of coplanar parallel forces. For example, consider the beam shown in Fig. 4–38*a* which is subjected to the arbitrary distributed loading $w = w(x)$ lb/ft. An element of the beam, dx, located at a distance x from the left support, is subjected to a force having a magnitude of $dF = (w(x) \text{ lb/ft})(dx \text{ ft}) = dA$ lb. This force represents the shaded differential area segment dA under the loading curve. By integration, the magnitude of the total resultant force $\mathbf{F}_R$, supported by the beam, is determined by integration, i.e.,

$$F_R = \int_0^L w(x) \, dx = \int_A dA = A \qquad (4\text{–}19)$$

Hence, it is seen that *the magnitude of the total resultant force is equal to the area under the loading diagram.*

In order to replace the distributed loading system $w(x)$ by $\mathbf{F}_R$, it is necessary to determine *where* on the beam $\mathbf{F}_R$ must act. To obtain this location, first move the incremental force $d\mathbf{F}$, shown in Fig. 4–38a, to the origin O. When this is done, a clockwise couple moment having a magnitude of $dM_O = x\,dF = x(w(x)\,dx)$ must be added to the beam. Provided *all* the differential forces dF acting from $x = 0$ to $x = L$ are replaced at point O, the resultant moment acting at the origin will be

$$M_O = \int_0^L x\,w(x)\,dx$$

(a)

From this derivation, the resultant force $\mathbf{F}_R$ and moment $\mathbf{M}_O$ shown in Fig. 4–38b are therefore *equivalent* to the distributed loading $w(x)$ shown in Fig. 4–38a. Since $\mathbf{M}_O$ is *perpendicular* to $\mathbf{F}_R$, the force-moment system may be reduced to the single resultant force $\mathbf{F}_R$ acting at a distance $\bar{x}$ from point O, where

$$M_O = \bar{x}F_R$$

or

$$\bar{x} = \frac{M_O}{F_R} = \frac{\displaystyle\int_0^L x\,w(x)\,dx}{\displaystyle\int_0^L w(x)\,dx} = \frac{\displaystyle\int_A x\,dA}{\displaystyle\int_A dA} \qquad (4\text{–}20)$$

(b)

(c)

Fig. 4–38

See Fig. 4–38c. The distance $\bar{x}$, computed from Eq. 4–20, may be interpreted as the coordinate distance from the origin O to the geometric center or centroid of the area under the loading curve.

A detailed treatment for locating the centroid of an area is given in Chapter 9. Essentially the centroid is located using integration formulas similar to Eq. 4–20, provided the boundaries of the area can be described by mathematical functions. The centroids of common area shapes such as triangles, trapezoids, etc. have been computed using the techniques of Sec. 9–2 and are tabulated in Appendix B. If the area shape of the loading diagram is contained in Appendix B, the centroidal distance can be obtained *directly* from this table without the need for integration.

To illustrate the techniques used to reduce a distributed area loading to single resulting force, consider again the uniform loading shown in Fig. 4–36b. The loading diagram is represented geometrically as a rectangle. The *area of this rectangle* is equal to the *magnitude of the resultant force*, i.e.,

$$F_R = (20 \text{ lb/ft})(4 \text{ ft}) = 80 \text{ lb}$$

The *line of action* of this force acts through the *centroid* of the rectangle, i.e., $\bar{y} = 2$ ft. In a similar manner, F_R in Fig. 4–36c is

$$F_R = (40 \text{ lb/ft})(2 \text{ ft}) = 80 \text{ lb}$$

and $\bar{x} = 1$ ft as shown. Notice that the above results are equivalent to those obtained previously for F_R in Fig. 4–36a.

As a second example to illustrate these techniques, consider the triangular load distribution shown in Fig. 4–37b. The *area of the triangle* represents the *magnitude of the resultant force*, i.e.,

$$F_R = \tfrac{1}{2}(9 \text{ ft})(36 \text{ lb/ft}) = 162 \text{ lb}$$

The line of action of this force passes through the centroid of the triangle. Thus, using Appendix B, the centroid is $\tfrac{1}{3}(9 \text{ ft}) = 3$ ft from the fixed support (see Fig. 4–37b).

Besides being distributed over a surface area, forces can also be distributed throughout the *volume* of a body. These forces are termed *body forces*. Their loading intensity is given in units of *force per unit volume*. The most common body force encountered in statics is that due to the earth's gravitational field. In the case of a *particle*, the magnitude of this force of attraction is termed the *weight* of the particle. The weight can be found using Newton's universal law of gravitation, Eq. 1–1. This force acts along a line directed from the particle to the center of the earth. Thus, *the resultant weight of a body is the resultant of the weights of all the particles that compose it.* Provided we deal with bodies that are small compared to the size of the earth, it is reasonable to assume that all the particles of the body contribute weight forces which are parallel to one another. It is shown in Chapter 9 that the resultant weight of a body can be represented as a *concentrated force* which acts through a unique point in the body, commonly referred to as the *center of gravity*. Provided the body is composed of *homogeneous matter,* the center of gravity *coincides* with the geometric center or centroid of the volume of the body. The techniques for integration used for finding the center of gravity and centroids of volume shapes are given in Sec. 9–2. The volume centroids for geometrically simple shapes, however, are *tabulated* in Appendix B. These data provide a useful means for locating the line of action of the resultant weight of homogeneous bodies encountered in some of the problems given in the following chapters.

Example 4-13

Determine the magnitude and location of the resultant of the distributed load acting on the beam shown in Fig. 4–39a.

Solution

As shown in Fig. 4–39b, the loading may be divided into three seg-

ments: the parabolic loading $(0 \leq x < 4$ ft), the triangular loading $(4$ ft $< x < 7$ ft), and the trapezoidal loading $(7$ ft $< x \leq 13$ ft). The *area* under each of the three distributed loading curves equals the *resultant force* for that loading curve.

Parabolic loading. Using Eq. 4–19 in conjunction with the differential element shown in Fig. 4–39a, the area is found by integration.

$$F_P = \int_0^4 w(x)\,dx = \int_0^4 12x^2\,dx = \frac{12}{3}x^3\Big]_0^4 = 4(4^4 - 0^4) = 256 \text{ lb}$$

Triangular loading. The area of the triangle is

$$F_T = \tfrac{1}{2}(3 \text{ ft})(200 \text{ lb/ft}) = 300 \text{ lb}$$

(a)

(b)

(c)

Fig. 4–39

Trapezoidal loading. The area of the trapezoid can be found by using the formula in Appendix B:

$$F_{Tr} = \tfrac{1}{2}h(a + b) = \tfrac{1}{2}(6 \text{ ft})(50 \text{ lb/ft} + 200 \text{ lb/ft}) = 750 \text{ lb}$$

Each of these resultant forces *acts* through the *centroid* of its respective loading curve area.

Parabolic loading. Referring to the differential element shown in Fig. 4–39a, and using Eq. 4–20, we have

$$\bar{x}_P = \frac{\displaystyle\int_0^L x\, w(x)\, dx}{\displaystyle\int_0^L w(x)\, dx} = \frac{\displaystyle\int_0^4 x(12x^2)\, dx}{\displaystyle\int_0^4 12x^2\, dx} = \frac{\left.\dfrac{12}{4}x^4\right|_0^4}{\left.\dfrac{12}{3}x^3\right|_0^4} = \frac{3(4^4 - 0^4)}{4(4^3 - 0^3)} = 3 \text{ ft}$$

measured from the left support.

Triangular loading. Using Appendix B, we have

$$\bar{x}_T = \frac{h}{3} = \frac{3}{3} = 1 \text{ ft}$$

measured from the right side of the triangle.

Trapezoidal loading. Using Appendix B, we have

$$\bar{x}_{Tr} = \frac{1}{3}\left(\frac{2a + b}{a + b}\right)h = \frac{1}{3}\left(\frac{2(50 \text{ lb/ft}) + 200 \text{ lb/ft}}{50 \text{ lb/ft} + 200 \text{ lb/ft}}\right)(6 \text{ ft}) = 2.40 \text{ ft}$$

measured from the left side of the trapezoid.

These results are all shown in Fig. 4–39b. By means of the method discussed in Sec. 4–10, this system of three *coplanar parallel forces* may be replaced by a *single concentrated force* acting at a unique point on the beam. Why? As shown in Fig. 4–39c, the resultant concentrated load for these three forces is

$$F_R = \Sigma F = 256 + 300 + 750 = 1{,}306 \text{ lb} \qquad \textit{Ans.}$$

Assuming that this force acts at a distance x_R from point O, we require

$$F_R\, x_R = \Sigma M_O$$
$$1{,}306\, x_R = 256(3) + 300(6) + 750(9.40)$$

Thus,

$$x_R = 7.36 \text{ ft} \qquad \textit{Ans.}$$

Problems

4-63. Determine the magnitude and location of the resultant force acting on the beam.

Prob. 4-63

4-64. Determine the equivalent force and couple-moment system acting at point O on the beam.

Prob. 4-64

4-65. The beam supports forty sandbags, each weighing 90 lb. Approximate the magnitude and the location of the resultant force due to the loading.

Prob. 4-65

4-66. Replace the loading system shown acting on the beam by a single resultant force. Where is this force located on the beam?

Prob. 4-66

4-67. Determine the resultant force acting on the beam.

Prob. 4-67

4-68. Determine the magnitude and location of the resultant force of the distributed loading shown acting on the beam.

Prob. 4-68

133

4-69. Determine the magnitude and location of the resultant force acting on the beam.

Prob. 4-69

4-70. The lifting force acting across the span of a wing on an airplane has a parabolic distribution (symmetrical about the w axis). The wing weighs 15,654 lb with a center of gravity at point P_1. The engine weighs 13,000 lb with a center of gravity at point P_2. Replace the force system by an equivalent force and couple-moment system at point A.

Prob. 4-70

4-71. The beam is subjected to an upward triangular load and a downward uniform load as shown in the figure. Determine the length b of uniform load and its position a on the beam such that the total resultant force and couple-moment acting on the beam is zero.

Prob. 4-71

4-72. The curved beam is subjected to a uniform load of 2 kip/ft. Determine an equivalent force and couple-moment system acting at point A.

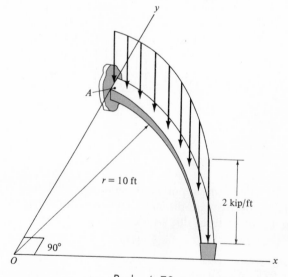

Prob. 4-72

5
Equilibrium of a Rigid Body

In this chapter we will discuss the fundamental concepts which underlie the equilibrium of a rigid body. In particular, it was pointed out in Chapter 3 that the methods of particle equilibrium pertain as well to the equilibrium of a rigid body subjected to a system of *concurrent forces*. (Recall that according to the principle of transmissibility, a system of concurrent forces may be represented as acting on a specific particle of the body.) In the first part of this chapter, we will consider the equilibrium of a body subjected to a *coplanar force system*. By this we assume that the body and the forces which are applied to it can be projected onto a single plane. Consequently, the reactions needed to maintain equilibrium of the body may also be represented in this same plane. The more general discussion of a *three-dimensional force system* acting on a rigid body will be given in the second part of this chapter.

Two-Dimensional Force Systems Acting on a Rigid Body

5-1. Free-Body Diagrams and Support Reactions

The free-body diagram concept, which was first introduced in Chapter 3 in connection with problems involving the equilibrium of a particle, plays an important role in the solution of equilibrium problems for a rigid body. Provided the free-body diagram is correctly drawn, the effects of all the applied forces and couples acting on the rigid body can be accounted for when the equations of equilibrium are written. Any force or couple

moment omitted from the diagram will obviously lead to errors in the solution. *For this reason, a comprehensive understanding in drawing correct free-body diagrams is of primary importance for solving problems in mechanics.* It is the preliminary step before the equations of equilibrium can be applied.

To construct a free-body diagram for a rigid body, the body (or group of bodies, considered as a single system) is *isolated* from all its supports and adjacent contacting bodies. An outlined shape of the body is drawn, and all the forces and couple moments acting *on the body* are labeled on this sketch. Included also are dimensions of the body necessary for computing the moments of forces. All *known* quantities should be drawn and labeled with their proper magnitude and direction. Letters are used to represent the magnitudes and directions of forces and couple moments which are *unknown*. The vector sense (represented by the arrowhead) of a force or couple moment having an unknown magnitude but known direction can be assumed. The correctness of the assumed sense will become apparent after solving the equilibrium equations for the unknown quantity. A positive sign indicates that the assumed sense was correct; a minus sign indicates that the sense is opposite to that which was assumed.

Forces and couple moments generally encountered are those due to (1) applied external loadings, (2) reactions occurring at the supports or at points of contact between bodies, and (3) the weight of the isolated body. (Forces which are *internal* to the body, such as the forces occuring between any two particles of the body, are *not shown* on the free-body diagram of the body since these forces always occur in equal and opposite collinear pairs. The net effect of these forces on the body will then be zero.)

Before we are able to draw a free-body diagram for a rigid body, it is necessary to first consider the various types of *reactions* that occur at supports and points of contact between rigid bodies. For two-dimensional problems, that is, bodies subjected to coplanar force systems, the types of support reactions commonly encountered are given in Table 5–1. You should *carefully study* the symbols used to represent these reactions, and you should clearly *understand* how the forces and/or couple moments are developed at all the supports. In particular, a *force* is developed on a connecting member by the support if the support *prevents translation* of the connecting member, and a couple moment is developed by the support if it *prevents rotation* of the connecting member. For example, a body in contact with the smooth surface (6) is prevented from translating *only* in a direction (downward) *normal* to the surface. Hence the surface exerts only a *normal force* on the body at the point of contact. The magnitude of this force represents *one unknown*. Since the body is free to rotate on the surface, no couple moment is developed by the surface on the body at its point of contact. The pin or hinge support (8) prevents translation of the connecting member at its point of connection. Unlike

Table 5-1 Support Reactions for Rigid Bodies Subjected to Two-Dimensional Force Systems

Types of connection	Reaction	Number of unknowns
(1) light cable		One unknown: magnitude of force **F**; this force acts in the direction of the cable so that the cable remains in tension
(2) weightless link	or	One unknown: magnitude of force **F**; this force acts in the direction of the link
(3) roller		One unknown: magnitude of force **F**; this force acts normal to the supporting surface
(4) roller or pin in confined smooth slot	or	One unknown: magnitude of force **F**; this force acts in either direction normal to the confined slot
(5) rocker		One unknown: magnitude of force **F**; this force acts normal to the inclined surface
(6) smooth contacting surface		One unknown: magnitude of force **F**; this force acts perpendicular to the tangent at the point of contact
(7) collar on smooth rod	or	One unknown: magnitude of force **F**; this force acts normal to the axis of the supporting rod

137

Table 5-1 (Contd.)

Types of connection	Reaction	Number of unknowns
(8) smooth pin or hinge		Two unknowns: the magnitudes of the two force components F_x and F_y, or the magnitude of the force F and its direction θ; note that β and θ are not necessarily equal [usually not, unless the rod shown is a link as in (2)]
(9) fixed support		Three unknowns: the magnitudes of the couple M and two force components F_x and F_y, or the magnitudes of the couple M and force F and the direction of the force, θ

the smooth surface, the translation is prevented in all directions. Hence, an unknown force F must be developed at the support. This force has *two unknowns,* namely, its magnitude F and direction θ, or the magnitudes of its two components F_x and F_y. Since the connecting member is allowed to rotate freely in the plane about the support at its point of contact, a pin connection does not resist a couple moment acting perpendicular to this plane. On the other hand, the fixed support (9) prevents *both* planar translation and rotation of the connecting member at the point of connection. Therefore, this type of support exerts both a force and a couple moment on the member. Note that the couple moment acts *perpendicular* to the plane of the page since rotation is prevented in the plane. Hence, there are *three unknowns* at a fixed support.*

In reality, all supports actually exert only *distributed surface loads* on their contacting members. The concentrated force and/or couple-moment vectors shown in Table 5-1 represent the *resultants* of this load distribution. This representation is, of course, an idealization. However, it can be used here since the surface area over which the distributed load acts on the support is considerably *smaller* than the *total* surface area of the connecting member.

Before proceeding further, carefully review this section. Then attempt to draw the free-body diagrams for the following example problems before "looking" at the solution. Further practice in drawing free-body diagrams may be gained by solving the problems given at the end of this section.

*A free body diagram for a body having a fixed support is shown in Fig. 5-1b.

(a)

(b)

Fig. 5-1

Example 5-1

Draw the free-body diagram for the uniform beam shown in Fig. 5-1a. The beam weighs 150 lb.

Solution

The free-body diagram for the beam is shown in Fig. 5-1b. Since the support is a fixed wall, there are three reactions acting *on the beam* at A, denoted as A_x, A_y, and M_A. The magnitudes of these vectors are *unknown* and their sense of direction have all been *assumed*. (How does one realize the *correct sense* of direction for these unknown magnitudes?)

For convenience in computing the reactions at A, the distributed loading is reduced to a concentrated force acting on the beam. Using the methods of Sec. 4–11, the resultant of the distributed loading is equal to the area under the triangular loading diagram, i.e., $\frac{1}{2}(6 \text{ ft})(100 \text{ lb/ft}) = 300 \text{ lb}$. This force acts through the centroid of the triangle, i.e., 2 ft from A. The 150-lb weight of the beam acts 3 ft from A since the beam has a uniform weight.

(a)

(b)

Fig. 5-2

Example 5–2

Draw the free-body diagram for the bell crank ABC shown in Fig. 5–2a, and determine the number of unknown force components.

Solution

The free-body diagram is shown in Fig. 5–2b. The pin support at B exerts forces $\mathbf{B}_x$ and $\mathbf{B}_y$ on the crank, each component having a known line of direction, but unknown magnitude. The link at C exerts an unknown magnitude of force $\mathbf{F}$ acting in the direction of the link. The dimensions of the crank are also labeled on the free-body diagram since this information will be useful in computing the moments of the forces. As usual, the sense of direction of the three unknown forces have been assumed. The correct sense of direction will become apparent after solving the equilibrium equations.

(a)

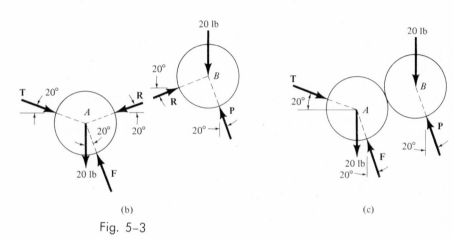

(b)   (c)

Fig. 5–3

Example 5–3

Two smooth balls A and B, each weighing 20 lb, rest between the smooth inclined planes shown in Fig. 5–3a. Draw the free-body diagrams for (a) ball A and ball B individually, and (b) balls A and B combined.

Solution

Part (a). The free-body diagrams for balls A and B individually are shown in Fig. 5–3b. Since all the contacting surfaces are *smooth*, the reactive forces $\mathbf{P}$, $\mathbf{T}$, $\mathbf{R}$, and $\mathbf{F}$ act in a direction *normal* to the tangent at their surface of contact. In particular, the reactive force $\mathbf{R}$ acting *on ball A* represents the effect which ball B exerts on ball A. Using Newton's third law, the free-body diagram of ball B must include an *equal but opposite force* $\mathbf{R}$ which ball A *exerts on ball B*.

Part (b). The free-body diagram for both balls A and B combined is shown in Fig. 5-3c. The reactions at the surfaces of contact with the inclined planes must be the *same* as those listed on the separate free-body diagrams given in Part (a). These unknowns represent the magnitudes of the forces which the planes exert *on* the balls. Since the contact force **R**, which acts between balls A and B, occurs as an equal and opposite collinear pair of forces, it is considered here as an *internal force* and hence is not shown on the free-body diagram in Fig. 5-3c.

Problems

Draw the free-body diagram in each of the following problems and determine the total number of unknown force and couple magnitudes and/or directions. Neglect the weight of the members unless otherwise stated.

5-1. The 10-lb sphere resting between the smooth inclined planes.

S ET

Prob. 5-1

5-2. The uniform 8-ft long beam pinned at A and supported by a cord at B; the total weight of the beam is 40 lb.

Prob. 5-2

5-3. The "spanner wrench" subjected to the 20-lb force; the support at A is pinned and the surface of contact at B is smooth.

Prob. 5-3

5-4. The "C" bracket supported at A, B, and C by rollers.

Prob. 5-4

5-5. The beam supported at A by a smooth roller and at B by a fixed foundation.

Prob. 5-5

5-6. The homogeneous triangular plate supported by cables at A and B and having a weight of 60 lb.

Prob. 5-6

5-7. The uniform beam supported at A by a pin, at B by a spring, and at C by a cord; the beam weighs 150 lb.

Prob. 5-7

5-8. The bell-crank supported at A by a pin and subjected to cable forces at points B and C.

Prob. 5-8

5-9. The bent beam supported at A by a fixed support and at B by a roller, and subjected to the weight of the 100-lb load.

Prob. 5-9

5-10. The beam supported at A by a smooth roller and at B by a pin.

Prob. 5-10

142

5-11. The homogeneous concrete slab weighing 3,000 lb and supported on its platform by the cables at A and B.

5-12. The uniform rod supported by a pin at A and a short link BC.

Prob. 5-11

Prob. 5-12

5-2. Equations of Coplanar Equilibrium

In Sec. 4-10 it was shown that a system of *coplanar forces* acting on a rigid body may be simplified to a single resultant force $\mathbf{F}_R = \Sigma\mathbf{F}$, acting through some arbitrary point O on the body, and a single resultant couple moment $\mathbf{M}_R = \Sigma\mathbf{M}_O$. The resultant force is equal to the vector sum of all the forces of the system and will therefore have the *same* magnitude and direction regardless of the location of point O. The magnitude and sense of direction of the couple moment, however, *depends* upon the location of point O. For a coplanar force system, $\mathbf{M}_R$ will always be directed *perpendicular* to the plane containing the system of forces.

In particular, if the *resultant force and moment* of the external forces acting on a body is *zero*, at any arbitrary point O, the body is said to be in *static equilibrium*. Physically, this means that equilibrium maintains a balance of force and moment. Equations which express these two conditions of equilibrium are

$$\Sigma\mathbf{F} = 0$$
$$\Sigma\mathbf{M}_O = 0 \tag{5-1}$$

$\Sigma\mathbf{F}$ represents the vector sum of all the forces acting on the body. If the free-body diagram is drawn in the xy plane, the applied system of forces may be expressed in terms of their respective $\mathbf{i}$ and $\mathbf{j}$ force *components*, i.e., $\Sigma\mathbf{F} = \Sigma F_x\mathbf{i} + \Sigma F_y\mathbf{j}$. The resultant moment vector acts in either the plus or minus $\mathbf{k}$ direction and has a magnitude ΣM_O. Thus, $\Sigma\mathbf{M}_O = \Sigma M_O\mathbf{k}$. Since the $\mathbf{i}, \mathbf{j},$ and $\mathbf{k}$ directions are mutually perpendicular, Eqs. 5-1 may be expressed in the form of *three independent scalar equations:*

143

$$\Sigma F_x = 0$$
$$\Sigma F_y = 0 \qquad\qquad (5\text{-}2)$$
$$\Sigma M_O = 0$$

In these equations, ΣF_x and ΣF_y represent the summation of the *components* of all the forces shown on the free-body diagram in the x and y directions, respectively. ΣM_O represents the sum of the applied couple moments acting on the body and the moments of all the forces about the arbitrary point O.

Although Eqs. 5-2 are *most often* used for solving statics problems involving coplanar force systems, certain alternate sets of three equilibrium equations may be used instead. One such set is

$$\Sigma F_y = 0$$
$$\Sigma M_A = 0 \qquad\qquad (5\text{-}3)$$
$$\Sigma M_B = 0$$

In using these equations, it is required that the moment points A and B do *not* lie on a line *perpendicular* to the y axis. To show that Eqs. 5-3 form a necessary set of three equilibrium conditions, consider the body shown in Fig. 5-4. $\mathbf{F}_R$ represents the *resultant* of *all* the external forces acting on the body.* For $\mathbf{F}_R$ to satisfy the equation $\Sigma F_y = 0$, $\mathbf{F}_R$ must be perpendicular to the y axis, as shown. Also, the equation $\Sigma M_A = 0$ is satisfied provided point A is chosen so that it lies along the line of action of $\mathbf{F}_R$. If point B is chosen such that line AB does not coincide with the line of action of $\mathbf{F}_R$ (which is perpendicular to the y axis), then $\Sigma M_B = 0$ is satisfied only if $\mathbf{F}_R$ is equal to zero.

In a similar manner, it can be shown that

$$\Sigma F_x = 0$$
$$\Sigma M_A = 0 \qquad\qquad (5\text{-}4)$$
$$\Sigma M_B = 0$$

form a complete set of three necessary equilibrium equations provided A and B do *not* lie on a line *perpendicular* to the x axis.

A third alternate set of equilibrium equations is

$$\Sigma M_A = 0$$
$$\Sigma M_B = 0 \qquad\qquad (5\text{-}5)$$
$$\Sigma M_C = 0$$

Here it is required that points A, B, and C do not lie along the *same*

$\mathbf{F}_R$ satisfies $\begin{cases} \Sigma F_y = 0 \\ \Sigma M_A = 0 \end{cases}$

Require $\mathbf{F}_R = 0$ to satisfy $\Sigma M_B = 0$

Fig. 5-4

---

*It was shown in Sec. 4–10 that a system of forces can always be replaced by a resultant force $\mathbf{F}_R$ and resultant moment $\mathbf{M}_R$. For a *coplanar force system,* however, $\mathbf{F}_R$ will *always* be perpendicular to $\mathbf{M}_R$, hence, the force-moment system may be further reduced to the *single resultant force* $\mathbf{F}_R$ acting through some unique point.

line. If both $\Sigma M_A = 0$ and $\Sigma M_B = 0$, then the resultant force $\mathbf{F}_R$, if it exists, must pass through the line connecting points A and B as shown in Fig. 5-5. Provided point C is not collinear with A and B, $\mathbf{F}_R$ must equal zero in order to satisfy $\Sigma M_C = 0$. Thus, Eqs. 5-5 form a necessary set of complete equilibrium equations.

If one *violates* the stated restriction for using the sets of equilibrium Eqs. 5-3, 5-4, or 5-5, one obtains equations which *are not linearly independent,* that is, one equation will differ from another by a constant multiple. It will then be necessary to choose other points or axes about which to sum moments or forces, until three independent equations are finally written.

An essential point to keep in mind is that one can write only *three independent scalar equations for the equilibrium of a rigid body subjected to a coplanar force system.* In solving equilibrium problems in two dimensions, it is best to *orient* the x and y axes in a direction which will facilitate *reduction* of all the forces into their x and y components. Similarly, moments should be summed about points which lie along the line of action of as *many unknown forces* as possible. Doing this reduces the number of unknowns in the moment equations and will generally eliminate the need for solving a set of simultaneous equations.

It is *not* recommended that equilibrium problems involving coplanar force systems be solved by direct application of vector Eqs. 5-1 (see Example 5-4). It is true that a vector approach to the solution is mathematically more sophisticated than a scalar approach; however, a vector solution really becomes advantageous when the geometry of the problem becomes complicated, as in the case of three dimensions. A scalar solution is more direct and lends greater insight into the physical effects caused by each of the applied forces and couple moments.

The three independent scalar equations of equilibrium provide the *necessary conditions* of equilibrium for a coplanar force system acting on a rigid body. If the body is properly supported so that all the forces acting on the body are determined solely from the equilibrium equations, then these equations are also *sufficient conditions* for equilibrium. Those types of problems for which the equilibrium equations do not provide sufficient conditions for equilibrium will be discussed in Sec. 5-5.

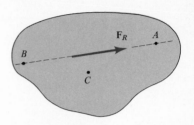

$\mathbf{F}_R$ satisfies $\begin{cases} \Sigma M_A = 0 \\ \Sigma M_B = 0 \end{cases}$

Require $\mathbf{F}_R = 0$ to satisfy $\Sigma M_C = 0$

Fig. 5-5

Example 5-4

Determine the support reactions for the loaded beam shown in Fig. 5-6a. Neglect the weight of the beam in the calculations.

Solution I

The *first step* is to draw a free-body diagram of the beam. Why? This diagram, shown in Fig. 5-6b, accounts for all the forces that act *on the beam.* The reaction at A is "vertical," since there is a smooth *roller* at

(a)

(b)

Fig. 5–6

this support. The *pin support* at *B* exerts an unknown force, represented by the two components $\mathbf{B}_x$ and $\mathbf{B}_y$. The directional sense of the three vectors $\mathbf{A}_y$, $\mathbf{B}_x$, and $\mathbf{B}_y$ have been assumed. The correct sense of these vectors will become apparent after solving the equations of equilibrium. For simplicity in applying the equilibrium equations, the 60-lb force is represented by its two orthogonal components as shown.

Applying the equilibrium equation $\Sigma F_x = 0$ to the force system acting on the free-body diagram in Fig. 5–6b yields

$$\xrightarrow{+}\Sigma F_x = 0;$$

$$60 \cos 45° \text{ lb} + B_x = 0, \qquad B_x = -42.4 \text{ lb}$$

Since the result is a negative quantity, $\mathbf{B}_x$ has a sense of direction *opposite* to that shown (assumed) in Fig. 5–6b.

Equating to zero the sum of moments about point *B* of *all* the forces shown on the free-body diagram we can directly calculate the unknown $\mathbf{A}_y$. Why? As stated in Sec. 4–4, the moment of a force is determined by multiplying the magnitude of the force by its *perpendicular distance* from the line of action of the force to point *B*. Considering counterclockwise rotation about point *B* to be positive, ($+\mathbf{k}$ direction) we have

$$\zeta+\Sigma M_B = 0;$$

$$10 \text{ lb } (3 \text{ ft}) + (60 \sin 45° \text{ lb})(9 \text{ ft}) - A_y(13 \text{ ft}) = 0, \quad A_y = 31.7 \text{ lb} \quad \textit{Ans.}$$

(Show that this same result would be obtained if moments had been assumed positive clockwise.) It should be apparent that the moments of forces 60 cos 45° lb, $\mathbf{B}_x$ and $\mathbf{B}_y$ do not appear in the moment equation, since the line of action of these force vectors passes through point B.

Summing forces in the y direction, using the result $A_y = 31.7$ lb obtained above, yields

$+\uparrow\Sigma F_y = 0;$

$$31.7 \text{ lb} - 60 \sin 45° \text{ lb} - 10 \text{ lb} + B_y = 0, \qquad B_y = 20.7 \text{ lb}$$

The magnitude of force $\mathbf{B}_y$ may also be obtained by summing moments about point A, i.e.,

$\zeta + \Sigma M_A = 0;$

$$-(60 \sin 45° \text{ lb})(4 \text{ ft}) - 10 \text{ lb} (10 \text{ ft}) + B_y(13 \text{ ft}) = 0, \qquad B_y = 20.7 \text{ lb}$$

The first solution for B_y is the least complicated; however, the second solution is desirable, since, for accuracy, it does not depend upon prior knowledge of the magnitude of $\mathbf{A}_y$.

The resultant force at B may be computed by adding vectorially the components $\mathbf{B}_x$ and $\mathbf{B}_y$, i.e.,

$$F_B = \sqrt{(B_x)^2 + (B_y)^2} = \sqrt{(42.4)^2 + (20.7)^2} = 47.2 \text{ lb} \qquad Ans.$$

Realizing that $\mathbf{B}_x$ acts in the opposite sense to that shown in Fig. 5–6b,

$$\theta = \tan^{-1}\frac{B_y}{B_x} = \tan^{-1}\frac{20.7}{42.4} = 26.0° \qquad {}^\theta \!\searrow^{F_B} \qquad Ans.$$

Solution II

After the free-body diagram for the beam is drawn, using the methods of vector algebra, we may apply Eqs. 5–1 directly to obtain the solution. With reference to the x and y axes shown in Fig. 5–6b, each of the forces acting on the free-body diagram must be expressed in Cartesian vector form.

$$\mathbf{F}_A = A_y\mathbf{j}$$
$$\mathbf{F}_1 = \{60 \cos 45° \,\mathbf{i} - 60 \sin 45° \,\mathbf{j}\} \text{ lb}$$
$$\mathbf{F}_2 = \{-10\mathbf{j}\} \text{ lb}$$
$$\mathbf{F}_B = B_x\mathbf{i} + B_y\mathbf{j}$$

Thus,

$\Sigma\mathbf{F} = 0;$

$$A_y\mathbf{j} + 60 \cos 45° \,\mathbf{i} - 60 \sin 45° \,\mathbf{j} - 10\mathbf{j} + B_x\mathbf{i} + B_y\mathbf{j} = 0$$

To satisfy this vector equation, the scalar components in the $\mathbf{i}$ and $\mathbf{j}$ directions must be equated to zero. Hence,

$$\xrightarrow{+}\Sigma F_x = 0; \qquad\qquad 60\cos 45° + B_x = 0 \tag{1}$$
$$+\uparrow\Sigma F_y = 0; \qquad A_y - 60\sin 45° - 10 + B_y = 0 \tag{2}$$

Summing moments about point B yields

$$\Sigma M_B = 0; \qquad (\mathbf{r}_{BD} \times \mathbf{F}_2) + (\mathbf{r}_{BC} \times \mathbf{F}_1) + (\mathbf{r}_{BA} \times \mathbf{F}_A) = 0$$
$$(-3\mathbf{i}) \times (-10\mathbf{j}) + (-9\mathbf{i}) \times (60\cos 45°\,\mathbf{i} - 60\sin 45°\,\mathbf{j})$$
$$+ (-13\mathbf{i}) \times (A_y\mathbf{j}) = 0$$

In this equation, each vector cross product yields a moment vector acting in the positive or negative $\mathbf{k}$ direction. After computing the cross products, the above equation reduces to

$$\zeta+\Sigma M_B = 0; \qquad 10(3) + 60\sin 45°\,(9) - A_y(13) = 0 \tag{3}$$

By comparison, note that Eqs. (1) through (3) are identical to those obtained in Solution I.

Although the vector approach is mathematically more elegant than the scalar approach, the extra labor needed to obtain the above results generally does not justify the use of Cartesian vectors for the solution of equilibrium problems involving *coplanar force systems*. For this reason, the remaining example problems will be solved using a scalar analysis.

Example 5–5

Determine the reactions at the fixed support A for the loaded frame shown in Fig. 5–7a.

(a)

(b)

Fig. 5–7

Solution

The free-body diagram for the frame is shown in Fig. 5–7b. There are three unknowns at the fixed support, represented by the magnitudes of A_x, A_y, and M_A. The distributed load is simplified using the methods of Sec. 4–11. Specifically, the resultant force of this loading is equal to the area under the loading curve, that is (6 ft)(5 lb/ft) = 30 lb. This force acts through the centroid, or geometric center of the loading curve, 3 ft from B or C.

Applying the equilibrium equations to the system of forces yields,

$\xrightarrow{+} \Sigma F_x = 0;$ $\qquad\qquad A_x = 0$ $\qquad\qquad$ *Ans.*

$+\uparrow\Sigma F_y = 0;$ $\quad A_y - 30$ lb $- 20$ lb $= 0,$ $\quad A_y = 50$ lb $\qquad$ *Ans.*

$(+\Sigma M_A = 0;$ $\quad M_A - 30$ lb (3 ft) $- 20$ lb (6 ft) $= 0,$

$\qquad\qquad\qquad M_A = 210$ lb-ft $\qquad\qquad$ *Ans.*

Rather than summing moments about point A, any other point (*on or off* the free-body diagram) might have been used. However, point A was chosen since the lines of action of forces $\mathbf{A}_x$ and $\mathbf{A}_y$ pass through point A, and therefore these forces were not included in the moment summation about this point. Note in particular that the couple-moment $\mathbf{M}_A$ must be *included* in the moment summation. (Actually $\mathbf{M}_A$ is a free vector and can therefore be placed at any point on the free-body diagram.)

Although only *three* independent equilibrium equations can be written for a rigid body, it is a good practice to check all calculations using a fourth equilibrium equation. This later equation may be obtained from one of the other sets of equilibrium Eqs. 5–3, 5–4, or 5–5. For example, the above computations may be verified by summing moments about point C:

$(+\Sigma M_C = 0;$ $\quad 30$ lb (3 ft) $- 50$ lb (6 ft) $+ 210$ lb-ft $\equiv 0$

$\qquad\qquad 90$ lb-ft $- 300$ lb-ft $+ 210$ lb-ft $\equiv 0$

Example 5–6

The uniform beam shown in Fig. 5–8a weighs 50 lb and is subjected to a concentrated force and couple moment. If the beam is supported at A by a smooth wall and at B and C by rollers, determine the reactions at the supports.

Solution

The free-body diagram for the beam is shown in Fig. 5–8b. All the support reactions act normal to the surface of contact since the contacting surfaces are smooth. The reactions at B and C are shown acting in the positive y' direction. This assumes that only the rollers located on the bottom of the beam are used for support. Since the beam is uniform, its 50-lb weight acts at the midpoint. The 800-lb-ft couple moment is a

(a)

(b)

Fig. 5–8

free vector, and it can therefore be shown acting at *any point* on the free-body diagram without affecting the computed results.

Using the xy coordinate system in Fig. 5–8b and applying the equations of equilibrium, we have

$$\xrightarrow{+}\Sigma F_x = 0; \qquad C_{y'} \sin 30° + B_{y'} \sin 30° - A_x = 0 \qquad (1)$$

$$+\uparrow\Sigma F_y = 0; \quad -30 \text{ lb} + C_{y'} \cos 30° - 50 \text{ lb} + B_{y'} \cos 30° = 0 \qquad (2)$$

$$\zeta +\Sigma M_A = 0; -B_{y'}(2 \text{ ft}) + (50 \cos 30° \text{ lb})(4 \text{ ft}) + 800 \text{ lb-ft}$$
$$- C_{y'}(6 \text{ ft}) - (30 \cos 30° \text{ lb})(8 \text{ ft}) = 0 \qquad (3)$$

In writing the moment equation, it should be noticed that the lines of action of the force components $30 \sin 30°$ lb and $50 \sin 30°$ lb pass through point A, and therefore these forces are not included in the moment equation.

If we solve Eqs. (2) and (3) simultaneously,

$$B_{y'} = -52.8 \text{ lb} \qquad\qquad\qquad Ans.$$
$$C_{y'} = 145.2 \text{ lb} \qquad\qquad\qquad Ans.$$

Since $B_{y'}$ is a negative quantity, its sense of direction is opposite to that shown on the free-body diagram in Fig. 5–8b. Therefore, the top roller at B serves as the support rather than the bottom roller.

When the values of $B_{y'}$ and $C_{y'}$ are substituted into Eq. (1) we obtain

$$145.2 \sin 30° \text{ lb} - 52.8 \sin 30° \text{ lb} - A_x = 0, \qquad A_x = 46.2 \text{ lb} \quad Ans.$$

Note that the negative sign of $B_{y'}$ must be *retained* when substituting into this equation, since the equations of equilibrium were *originally* written with the assumption that the sense of direction of all the forces acting on the free-body diagram was correct.

(a)

(b)

(c)

Fig. 5–9

Example 5–7

The uniform beam AB shown in Fig. 5–9a weighs 100 lb and is supported at A by a pin and at B and C by a continuous cable which wraps

around a frictionless pulley located at D. If a maximum tension force of 80 lb can be developed in the cable before it breaks, determine the greatest length b of a uniform 20-lb/ft distributed load which can be placed on the beam. The uniform load begins at the left-hand support. What is the reaction at A just before the cable breaks?

Solution

A free-body diagram of the beam is shown in Fig. 5–9b. Two unknown force components act at A since the beam is supported there by a pin. As stated in the problem, the maximum tensile force which can be developed in the cable is 80 lb. Since the cable is continuous and passes over a frictionless pulley, the entire cable is subjected to this force. Hence, the cable exerts an 80-lb force at points C and B *on the beam* in the direction of the cable.

The fact that the cable maintains a *constant tensile force* of 80 lb throughout its length when passing over the pulley at D can be easily proved by considering rotational equilibrium of the pulley about the axis of the pin. Assuming that the radius of the pulley is r and using the free-body diagram of the pulley shown in Fig. 5–9c, we see that

$$\zeta + \Sigma M_D = 0; \qquad r(80\text{ lb}) - r(80\text{ lb}) \equiv 0$$

Notice that this result is *independent* of the *direction* of the ends of the cable. It indicates that the cable must always sustain the same force of 80 lb at its end points. (If needed, the magnitude of the force components $\mathbf{D}_x$ and $\mathbf{D}_y$ can be found using the force equilibrium equations.)

The uniformly distributed load is reduced to a concentrated force in accordance with the methods of Sec. 4–11. The magnitude of this force is equivalent to the area under the diagram, that is $20b$ lb. The force acts through the centroid of the area, a distance of $b/2$ from point A. Summing moments of the force system about point A (Fig. 5–9b) we can obtain a direct solution for the dimension b. Why?

$$\zeta + M_A = 0; \quad -(20b\text{ lb})\left(\frac{b}{2}\text{ ft}\right) - 100\text{ lb (5 ft)} + (80\sin 60°\text{ lb})(8\text{ ft}) +$$

$$80\text{ lb (10 ft)} = 0, \qquad b = 9.24\text{ ft} \quad Ans.$$

Using this result and summing forces in the x and y direction, we have

$$\xrightarrow{+}\Sigma F_x = 0; \qquad -A_x + 80\cos 60°\text{ lb} = 0, \qquad A_x = 40\text{ lb}$$

$$+\uparrow\Sigma F_y = 0; \quad A_y - 20\text{ lb/ft (9.24 ft)} - 100\text{ lb} + 80\sin 60°\text{ lb}$$

$$+\ 80\text{ lb} = 0, \qquad A_y = 135.5\text{ lb}$$

By vector addition of the two force components at A, the total reaction at A becomes

$$F_A = \sqrt{(A_x)^2 + (A_y)^2} = \sqrt{(40)^2 + (135.5)^2} = 141.4\text{ lb} \qquad Ans.$$

$$\theta = \tan^{-1}\frac{A_y}{A_x} = \tan^{-1}\frac{135.5}{40} = 73.6° \qquad \theta\searrow^{F_A} \qquad Ans.$$

Problems

Neglect the weight of the members in the following problems unless specified.

5-13. Determine the reactive force components **T**, **R**, and **F** acting on ball *A* shown in Fig. 5-3*b*, p. 140.

5-14. Determine the tensile force acting in the vertical cables which supports the 60-lb triangular plate in Prob. 5-6.

5-15. Determine the horizontal and vertical components of force developed by the pin at *A*, and the force in link *BC*, of the rod in Prob. 5-12.

5-16. Three smooth rollers, each having a weight of 50 lb, rest on the 20° inclined plane. Determine the reaction of this plane on each of the rollers.

Prob. 5-16

5-17. Determine the horizontal and vertical components of reaction acting at the pin *A* and the reaction at roller *B* for the loaded beam.

Prob. 5-17

5-18. The bent beam is supported by a pin at *A* and by a smooth surface at *B*. Determine the horizontal and vertical components of reaction at *A* and the reaction at *B*. The spring is stretched three inches.

152

Prob. 5-18

5-19. Determine the reactions at the wall, *A*, for the beam loaded as shown.

Prob. 5-19

5-20. A 150-lb man stands uniformly on a smooth ladder rung at *A*. Assuming the man's leg bone is only subjected to a compressive force F_B, determine this force and the muscle force **T**. Discuss the validity of the solution. The man stands with both feet on the ladder.

Prob. 5-20

5-21. Determine the bone force F_B and muscle force **T** necessary to hold the 10-lb ball in equilibrium.

Prob. 5-21

5-22. Determine the tension in the cable and the horizontal and vertical components of reaction of the pin A. The pulley at D is frictionless.

Prob. 5-22

5-23. Determine the components of reaction at pin A and roller B for the beam loaded as shown.

Prob. 5-23

5-24. Determine the reactions at A and B for the beam loaded as shown.

Prob. 5-24

5-25. Determine the horizontal and vertical components of reaction at the pin support A and the reaction at the roller B for the loaded frame.

Prob. 5-25

5-26. Determine the horizontal and vertical components of reaction at the pin support A and the reaction at the smooth wall at B for the beam loaded as shown.

153

Prob. 5-26

Prob. 5-28

5-27. Determine the horizontal and vertical components of reaction at the pin A and the tension in the cable BC for the boom. Two of these booms are used to support the 1,000-lb lifeboat. *Hint:* the pulley at D exerts a total downward reaction of 500 lb on the boom.)

5-29. Calculate the components of reaction at the fixed wall A for the loaded beam. The cable is fixed at B and passes over a smooth peg at C.

Prob. 5-27

Prob. 5-29

5-28. Determine the horizontal and vertical components of reaction at the pin and the reaction at the roller support for the loaded semicircular arch.

5-30. Two spheres, each weighing 20 lb, are suspended from the same point by equal length cords. Determine

154

the tension in each cord and the reactive force acting between the spheres.

Prob. 5-30

5-31. The forces acting on the jet aircraft, which is flying level at constant velocity, must be in equilibrium. If the aircraft weighs 18,000 lb, and the jet provides a thrust **T** of 1,200 lb, determine the lift provided by both the wings, L_w, and tail L_t, and the drag **D**. The center of gravity of the aircraft is at B as shown.

Prob. 5-31

5-32. Determine the reactions at A and B for the loaded truss.

Prob. 5-32

5-33. Determine the horizontal and vertical components of reaction at the pin support A and the force in the link CE for the loaded frame.

Prob. 5-33

5-34. A uniform rod AB having a weight of 70 lb supports a 50-lb block. Determine the angles α and β of the supporting cables, and the necessary weight W

155

which is required to hold the rod in the horizontal position. The pulleys are frictionless.

75 lb

50 lb

Prob. 5-34

around a fixed drum at *B* and passes over a frictionless pulley located at *C*.

8 ft

13 ft

5 ft

C

B 5 ft

20°

A

4 ft

100 lb

250 lb

Prob. 5-36

5-35. Determine the horizontal and vertical components of reaction at the pin *A* and the tension in the cable *BCD* required to support the 250-lb beam. The cord is attached to a fixed drum at *B* and passes over a smooth peg at *C*. The center of gravity for the beam and drum is located at *G*.

5-37. Two smooth tubes *A* and *B*, each having the same weight *W*, are suspended from a common point by means of equal length cords. A third tube, *C*, is placed between *A* and *B*. Determine the greatest weight of *C* without upsetting equilibrium.

D

45°

B

r = 1.5 ft

C

Smooth peg

2 ft

1 ft

G

1.5 ft

8 ft

30° 10 ft

A

Prob. 5-35

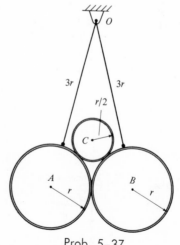

O

3*r*   3*r*

r/2

C

A   *B*

r   *r*

Prob. 5-37

5-36. Calculate the tension in the cable at *B* required to hold the boom in the position shown. The cable wraps

5-38. A uniform beam having a weight *W* and length *l* rests horizontally on the surface of two smooth inclined planes. If *α* is known, show that *θ* = *α* is required for equilibrium. Also, compute the reactions at the ends of the beam.

Prob. 5-38

5-39. Determine the tension in the supporting cables *BC* and *DEF*, and the reaction at the smooth incline *A* of the loaded beam.

Prob. 5-39

5-40. Determine the least force **F** and the angle θ required to move the 250-lb roller over the 3-in. block.

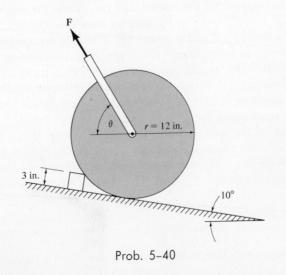

Prob. 5-40

5-41. The bracket is used to support a smooth cylindrical tube having a weight of 40 lb. If the bracket is pinned at *D*, determine the angle θ such that the normal reaction at *A* is 10 lb. What are the normal reactions at points *B* and *C*?

Prob. 5-41

5-42. Determine the maximum load *W* which can be supported by the crane. For the calculation, assume that the cable *BC* is attached at both ends. The tractor portion weighs 20,000 lb and has a center of gravity at *G*. Neglect the weight of the boom *AB*.

Prob. 5-42

Three-Dimensional Force Systems Acting on a Rigid Body

5–3. Free-Body Diagrams and Support Reactions

The *first step* in solving equilibrium problems involving three-dimensional force systems, as in the case of two dimensions, is to draw a free-body diagram of the body (or group of bodies considered as a system). Why would you want to do this? The procedure for constructing the free-body diagram is identical with that given for two dimensional problems. In drawing a free-body diagram, it is important to include *all* forces and couple moments which act *on* the body or group of bodies in question. These interactions are commonly caused by the external applied loadings, contact forces exerted by adjacent bodies on the body, support reactions, and the weight of the body. Once the free-body diagram is drawn, the equations of equilibrium can then readily be applied.

The reactive forces and moments acting at various types of supports and connections, when the members are viewed in three dimensions, are listed in Table 5–2. It is important both to recognize the symbols used to represent each of the supports and to clearly understand how the forces and/or couple moments are developed by each support. In particular, a *force* is developed by a support which restricts the *translation* of the connecting member in a given direction. A *couple moment* is developed when *rotation* of the connecting member is prevented. For example, in Table 5–2, the ball-and-socket joint (4) prevents any translation of the connecting member; therefore, a force must act on the member at the point of connection. This force has three force components acting in orthogonal directions and having unknown magnitudes, F_x, F_y, F_z. Provided these three components are known, one can obtain the magnitude of the force, $F = \sqrt{F_x^2 + F_y^2 + F_z^2}$, and its direction, defined by $\alpha = \cos^{-1} F_x/F$, $\beta = \cos^{-1} F_y/F$, and $\gamma = \cos^{-1} F_z/F$.* Since the connecting member is allowed to rotate freely about *any* axis, no moment is resisted at this support.† The smooth bearing (5) allows both translation and rotation of its connecting shaft only along the *y* axis. However, motion is restricted in the *xz* plane. As a result, force and couple-moment components act on the shaft along the *x* and *z* axes.

In some cases, not all the reactive forces existing at a support can be developed because of other physical constraints acting on the rigid body. For example, the single hinge (7) may develop a total of five unknowns (three force components and two moment components) when it acts as

---

*The three unknowns may also be represented as an unknown force magnitude F, and two unknown direction cosines for the force. The third direction cosine is obtained using the identity $\cos^2 \alpha + \cos^2 \beta + \cos^2 \gamma = 1$ (Eq. 2–8).

†A free-body diagram for a body having a ball-and-socket joint is shown in Fig. 5–15*b*.

Table 5-2

Types of connection	Reaction	Number of unknowns
(1) light cable		One unknown: magnitude of **F**; this force acts in the direction of the cable
(2) smooth surface support		One unknown: magnitude of **F**; this force acts normal to the surface at the point of contact
(3) roller on a smooth surface		One unknown: magnitude of **F**; this force acts normal to the surface at the point of contact
(4) ball and socket		Three unknowns: magnitudes of three force components F_x, F_y, and F_z
(5) smooth bearing		Four unknowns: magnitudes of two force components F_x, F_z and the two couples M_x, M_z
(6) smooth pin		Five unknowns: magnitudes of three force components F_x, F_y, F_z and two couples M_y, M_z

159

Table 5–2 (Contd.)

Types of connection	Reaction	Number of unknowns
(7) single hinge		Five unknowns: magnitudes of three force components F_x, F_y, F_z and two couples M_x, M_z
(8) fixed support		Six unknowns: magnitudes of three force components F_x, F_y, F_z and three couples M_x, M_y, M_z

a *primary single means of support*. However, when two hinges are present and aligned properly along the *same axis,* only three unknown components of force may be developed. Moment reactions cannot occur in this case, since rotation about the axis of one hinge is prevented by constraining forces developed at the other hinge. (See Examples 5–8 and 5–11.)

All reactions which *can exist* at a support must be included on the free-body diagram. Those unknowns which are zero will automatically be accounted for when the equations of equilibrium are applied. As usual, when an unknown is represented as the magnitude of a force or couple moment, the sense of direction for the vector may be *assumed*. If the solution yields a negative answer for the unknown, the vector sense of direction is then opposite to the assumed direction.

5–4. Equations of Equilibrium

A necessary condition for the equilibrium of a rigid body subjected to a three-dimensional force system requires that the total *resultant* force and *resultant* couple moment acting on the body be equal to *zero*. By reducing the system of forces acting on a rigid body to a single resultant force $\Sigma \mathbf{F}$ acting through a specified point O and a resultant moment $\Sigma \mathbf{M}_O$, these necessary conditions for equilibrium may be postulated in vector component form as

$$\Sigma \mathbf{F} = \Sigma F_x \mathbf{i} + \Sigma F_y \mathbf{j} + \Sigma F_z \mathbf{k} = 0 \tag{5-6a}$$

$$\Sigma \mathbf{M}_O = \Sigma M_x \mathbf{i} + \Sigma M_y \mathbf{j} + \Sigma M_z \mathbf{k} = 0 \tag{5-6b}$$

Since the components in the **i, j,** and **k** directions are independent of one another, the necessary conditions for the equilibrium of a rigid body may be expressed in the form of *six independent scalar equations:*

$$\Sigma F_x = 0$$
$$\Sigma F_y = 0 \qquad\qquad (5\text{--}7a)$$
$$\Sigma F_z = 0$$

and

$$\Sigma M_x = 0$$
$$\Sigma M_y = 0 \qquad\qquad (5\text{--}7b)$$
$$\Sigma M_z = 0$$

These six equations may be used to solve for at most six unknowns shown on the free-body diagram.

Equations 5–7a express the fact that the sum of the force components acting in the x, y, and z directions, respectively, must be zero. The coordinate axes x, y, and z should be oriented in a direction which will enable one to easily calculate the force components to be used in Eqs. 5–7a. Equations 5–7b require the sum of the moment components directed along the x, y, and z *axes,* respectively, to be zero. It is *not necessary* that the set of axes chosen for moment summation *coincide* with the set of axes chosen for force summation. It is recommended that one *choose the direction of one of the moment axes such that it intersects the line of action of as many unknown forces as possible.* The moment of forces passing through points on this axis or parallel to the axis will then be zero.

It will be stated without proof that *any three nonorthogonal axes* may be chosen for either force or moment summations. These axes must, however, *not be parallel* to one another or else linearly dependent equations may result. By the proper choice of axes, it may be possible to solve directly for an unknown quantity, or at least reduce the need for solving a large number of simultaneous equations for the unknowns. In some cases, however, after applying the equations of equilibrium, one may unavoidably obtain a complicated system of linear algebraic equations which must then be solved simultaneously, e.g., six equations with six unknowns. In such cases the equations may be solved using the computer program listed in Appendix A.

Often, the three-dimensional geometry of a problem may be complicated. In such circumstances, it is recommended that the (two) vector equations of equilibrium (Eqs. 5–6) *first* be used to obtain the necessary (six) scalar equations. A vector approach to the solution is very methodical and eliminates the need for having to visualize the direction of moments created by the force components.

5–5. Sufficient Conditions for Equilibrium

Equations 5–2 and 5–7 provide the necessary conditions for equilibrium of a rigid body subjected to two- and three-dimensional force systems, respectively. In all problems presented thus far, these equations are also sufficient conditions for equilibrium. Such problems are termed *statically determinate*. There are, however, certain types of problems in which the equilibrium equations are not sufficient for determining all the forces acting on a rigid body. These types of problems are called *statically indeterminate*.

(a)

(b)

Fig. 5–10

Statical indeterminacy can arise if a body has more supports than are necessary (redundant) to maintain the equilibrium configuration. For example, the two-dimensional problem, Fig. 5–10a, and the three-dimensional problem, Fig. 5–10b, shown together with their free-body diagrams, are both statically indeterminate because of redundancy in the number of support reactions. In the two-dimensional case, there are five unknowns, that is, A_x, A_y, B_y, C_y, and M_A, for which only three equilibrium equations can be written (Eqs. 5–2). Two more equations are needed for a complete solution to the problem. This problem is termed indeterminate to the second degree (five unknowns minus three equations). The three-dimensional problem has nine unknowns, for which only six equilibrium

equations can be written (Eqs. 5–7). It is indeterminate to the third degree.
The additional equations needed to solve indeterminate problems of the type shown in Fig. 5–10 are obtained from the conditions of deformation which occur between the loads and the internal movements within the body. These relations involve the physical properties of the body which are studied on subjects relating to mechanics of deformation, such as strength of materials.

In some cases, there may be an equal number of unknown forces as there are equations of equilibrium; however, *instability* can develop because of *improper constraining* action by the supports. This may occur in two ways. In the case of three-dimensional problems, the body is improperly constrained if the support reactions all intersect a common axis. In two-dimensional problems, the axis is *perpendicular* to the figure, and appears as a point. When the reactive forces are concurrent at this point, the body is improperly constrained. Examples of both cases are given in Fig. 5–11. From the free-body diagram of each body it is seen that the summation of moments about axis AB, Fig. 5–11a, or point O, Fig. 5–11b, will *not* be equal to zero; thus rotation about axis AB or point O will take place.* It becomes impossible to solve completely for all the unknowns, since one can write a moment equation which does not involve any of the unknown support reactions. (This limits by one the number of available equilibrium equations.)

(a)

(b)

Fig. 5–11

*For the three dimensional problem $\Sigma M_{AB} = (100 \text{ lb})(6 \text{ ft}) \neq 0$, and for the two dimensional problem $\Sigma M_O = (16 \text{ lb})(2 \text{ ft}) \neq 0$.

(a)

(b)

Fig. 5-12

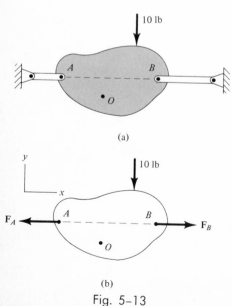

(a)

(b)

Fig. 5-13

Another way in which partial constraining leads to instability occurs when the *reactive forces* are all parallel. Three- and two-dimensional examples of this are shown in Figs. 5-12a and b. In both cases the summation of forces along the horizontal *aa* axis will *not* be equal to zero. The problem is unstable since a force equilibrium equation in this direction *will not* involve any unknown *reactive* forces.

In some cases, a body may have *fewer* reactive forces than equations of equilibrium which must be satisfied. Under these conditions, the body is only *partially constrained*. Consider the body shown in Fig. 5-13a with its corresponding free-body diagram in Fig. 5-13b. Although the equations $\Sigma F_x = 0$ and $\Sigma M_O = 0$ (O lies at some point not on the line of action AB) will be satisfied by proper choice of the reactions $\mathbf{F}_A$ and $\mathbf{F}_B$, the equation $\Sigma F_y = 0$ will not be satisfied under the loading conditions and therefore equilibrium will not be maintained.

To ensure equilibrium of a rigid body, it is therefore necessary to properly constrain the body. This can be accomplished provided: (1) there are at least as many reactive forces as equations of equilibrium, (2) the lines of action of the reactive forces do not intersect points on a common axis, and (3) the reactive forces are not parallel to one another. When the number of reactive forces needed to properly constrain the body in question

(a)

(b)

Fig. 5-14

is a minimum, the equations of equilibrium are *both* necessary and sufficient conditions for determining all the forces acting on the body.

Example 5-8

The platform shown in Fig. 5-14a has negligible weight and is subjected
to two concentrated forces. It is supported at A and B by hinges and at C
by a vertical cable. Draw a free-body diagram of the platform.

Solution

The free-body diagram of the platform is shown in Fig. 5-14b. This
diagram includes the dimensions of the body which will be necessary for
computing the moments of the forces. Provided the hinges are *properly
aligned,* three components of force act at each hinge. Because of the
constraints caused by *both* hinges, rotation about the x or z axis of a
single hinge is prevented; hence, unknown moment components of the
hinges do not act on the body and are therefore excluded from the
free-body diagram (see (7) in Table 5-2). For example, any rotation of
the hinge at A about the z axis is prevented by the moment of force B_x
developed at B, that is, $(B_x)(3 \text{ ft})$, *not* by a moment component $(M_A)_z$. On
the free-body diagram there is a total of *seven* unknown magnitudes of
force: $A_x, A_y, A_z, B_x, B_y, B_z,$ and C_z. Only *six* scalar equilibrium equations
can be written to relate these unknowns and therefore only a partial solution to the problem can be obtained (see Prob. 5-43). Since there is one
more unknown than equations of equilibrium, this problem is statically
indeterminate to the first degree.

165

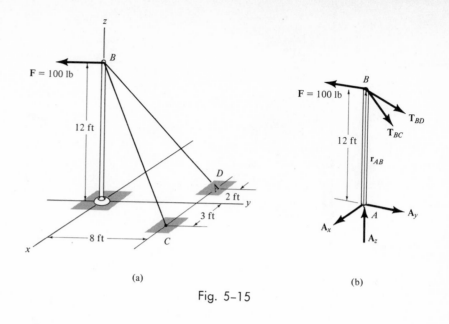

(a)

(b)

Fig. 5-15

Example 5-9

Determine the tensions in the cables BC and BD, and the reaction at the support A for the vertical mast shown in Fig. 5-15a. Neglect the weight of the mast.

Solution

The first step is to draw the free-body diagram of the mast, Fig. 5-15b. Why? Since the support at A is a ball-and-socket joint, three unknown force components act at this point. The cable tension forces $\mathbf{T}_{BC}$ and $\mathbf{T}_{BD}$ have a known direction and unknown magnitude. There are, therefore, five unknown force magnitudes shown on the free-body diagram.

Expressing each force acting on the free-body diagram in Cartesian vector notation,

$$\mathbf{F} = \{-100\mathbf{j}\} \text{ lb}$$

$$\mathbf{F}_A = A_x\mathbf{i} + A_y\mathbf{j} + A_z\mathbf{k}$$

$$\mathbf{T}_{BC} = T_{BC}\left(\frac{\mathbf{r}_{BC}}{r_{BC}}\right) = \frac{3T_{BC}}{\sqrt{217}}\mathbf{i} + \frac{8T_{BC}}{\sqrt{217}}\mathbf{j} - \frac{12T_{BC}}{\sqrt{217}}\mathbf{k}$$

$$= 0.204T_{BC}\mathbf{i} + 0.543T_{BC}\mathbf{j} - 0.815T_{BC}\mathbf{k}$$

$$\mathbf{T}_{BD} = T_{BD}\left(\frac{\mathbf{r}_{BD}}{r_{BD}}\right) = \frac{-2T_{BD}}{\sqrt{212}}\mathbf{i} + \frac{8T_{BD}}{\sqrt{212}}\mathbf{j} - \frac{12T_{BD}}{\sqrt{212}}\mathbf{k}$$

$$= -0.137T_{BD}\mathbf{i} + 0.549T_{BD}\mathbf{j} - 0.824T_{BD}\mathbf{k}$$

Applying the two vector equations of equilibrium gives

$$\Sigma F = 0; \qquad\qquad F + F_A + T_{BC} + T_{BD} = 0$$

$$-100j + A_x i + A_y j + A_z k + 0.204 T_{BC} i + 0.543 T_{BC} j - 0.815 T_{BC} k$$
$$-0.137 T_{BD} i + 0.549 T_{BD} j - 0.824 T_{BD} k = 0$$

or

$$(A_x + 0.204 T_{BC} - 0.137 T_{BD})i + (-100 + A_y + 0.543 T_{BC}$$
$$+ 0.549 T_{BD})j + (A_z - 0.815 T_{BC} - 0.824 T_{BD})k = 0$$

Equating to zero the **i**, **j**, and **k** components we obtain the three scalar equilibrium equations:

$$\Sigma F_x = 0; \qquad\qquad A_x + 0.204 T_{BC} - 0.137 T_{BD} = 0 \qquad\qquad (1)$$

$$\Sigma F_y = 0; \qquad\qquad A_y + 0.543 T_{BC} + 0.549 T_{BD} - 100 = 0 \qquad (2)$$

$$\Sigma F_z = 0; \qquad\qquad A_z - 0.815 T_{BC} - 0.824 T_{BD} = 0 \qquad\qquad (3)$$

Summing moments about axes passing through point A to eliminate the three unknown force components passing through this point,

$$\Sigma M_A = 0; \qquad (r_{AB} \times F) + (r_{AB} \times T_{BC}) + (r_{AB} \times T_{BD}) = 0$$

The position vector r_{AB} in this equation extends from point A to point B. Since this vector is a common term in the cross product, by the distributive law, it may be factored out, yielding,

$$r_{AB} \times (F + T_{BC} + T_{BD}) = 0$$

Substituting, we have

$$12k \times (-100j + 0.204 T_{BC} i + 0.543 T_{BC} j - 0.815 T_{BC} k$$
$$-0.137 T_{BD} i + 0.549 T_{BD} j - 0.824 T_{BD} k) = 0$$

or

$$-12(0.543 T_{BC} + 0.549 T_{BD} - 100)i + 12(0.204 T_{BC} - 0.137 T_{BD})j = 0$$

Setting the **i** and **j** components equal to zero yields

$$\Sigma M_x = 0; \qquad\qquad 0.543 T_{BC} + 0.549 T_{BD} - 100 = 0 \qquad\qquad (4)$$

$$\Sigma M_y = 0; \qquad\qquad 0.204 T_{BC} - 0.137 T_{BD} = 0 \qquad\qquad\qquad (5)$$

(The equilibrium equation about the z axis, $\Sigma M_z = 0$, is automatically satisfied for the given loading system since the lines of action of all the forces acting on the body pass through the z axis.)

Solving Eqs. (4) and (5) for T_{BC} and T_{BD} and substituting these values into Eqs. (1) through (3) yield

$$A_x = 0.0 \text{ lb} \qquad\qquad\qquad Ans.$$
$$A_y = 0.0 \text{ lb} \qquad\qquad\qquad Ans.$$
$$A_z = 150.0 \text{ lb} \qquad\qquad\quad Ans.$$
$$T_{BC} = 73.5 \text{ lb} \qquad\qquad\quad Ans.$$
$$T_{BD} = 109.4 \text{ lb} \qquad\qquad Ans.$$

Fig. 5-16

Example 5-10

The homogeneous plate shown in Fig. 5-16a weighs 100 lb/ft². The plate is suspended in the horizontal plane by means of three parallel cords. Determine the tension in each of these cords.

Solution

A free-body diagram of the plate is shown in Fig. 5-16b. The weight acts at the center of gravity of the plate (which coincides with the centroid). The magnitude of this force is the product of 100 lb/ft² times the area of the plate. Hence $W = (100 \text{ lb/ft}^2)(4 \text{ ft})(6 \text{ ft}) = 2{,}400$ lb. Therefore $\mathbf{W} = \{-2{,}400\mathbf{k}\}$ lb.

Orientating the x, y, and z axes as shown in the figure, we can write the vector force equilibrium equation as

$$\Sigma \mathbf{F} = 0; \qquad T_A\mathbf{k} + T_B\mathbf{k} + T_C\mathbf{k} - 2{,}400\mathbf{k} = 0$$

or

$$\Sigma F_z = 0; \qquad T_A + T_B + T_C - 2{,}400 = 0 \tag{1}$$

Summing moments of all the forces about the origin yields

$$\Sigma \mathbf{M}_A = 0; \quad (\mathbf{r}_{AB} \times T_B\mathbf{k}) + (\mathbf{r}_{AC} \times T_C\mathbf{k}) + (\mathbf{r}_{AD} \times \mathbf{W}) = 0$$

The position vectors can be determined from the geometry of the plate. Hence,

$$[6\mathbf{j} \times T_B\mathbf{k}] + [(4\mathbf{i} + 6\mathbf{j}) \times T_C\mathbf{k}] + [(2\mathbf{i} + 3\mathbf{j}) \times (-2{,}400\mathbf{k})] = 0$$

or

$$(6T_B + 6T_C - 7{,}200)\mathbf{i} + (-4T_C + 4{,}800)\mathbf{j} = 0$$

Equating to zero the corresponding $\mathbf{i}$ and $\mathbf{j}$ components gives

$$\Sigma M_x = 0; \qquad 6T_B + 6T_C - 7{,}200 = 0 \tag{2}$$

$$\Sigma M_y = 0; \qquad -4T_C + 4{,}800 = 0 \tag{3}$$

These two scalar equations represent the moment summation of forces about the x and y axes, respectively. The moment of a force about an axis is equal to the product of the force magnitude and the perpendicular distance from the line of action of the force to the axis. Hence, for example, taking moments of the forces $\mathbf{T}_C$ and $\mathbf{W}$ about the y axis shown in Fig. 5–16b yields the scalar Eq. (3).

Solving Eqs. (1) through (3), we obtain

$$T_A = 1{,}200 \text{ lb} \qquad\qquad Ans.$$
$$T_B = 0 \qquad\qquad Ans.$$
$$T_C = 1{,}200 \text{ lb} \qquad\qquad Ans.$$

Note that these reactions provide the necessary *vertical support* for the loading. If a force acting in the xy plane is applied to the plate, other supports would be required to maintain equilibrium. For this reason, the plate is only partially constrained.

Example 5–11

Determine the magnitude of the vertical force **P**, which must be applied to the handle of the windlass shown in Fig. 5–17a, to maintain equilibrium of the 150-lb weight. Also calculate the reactions at the smooth bearings A and B.

Solution

The free-body diagram of the windlass is shown in Fig. 5–17b. Provided the bearings at A and B are aligned correctly, *only* force reactions occur at these supports. (Moment reactions are not developed at these supports since rotation about the axis of one bearing is prevented by a *reactive force* occurring at the other bearing. For example, there is no moment

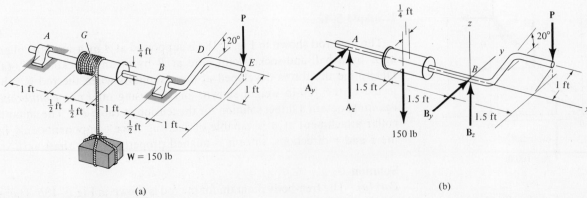

(a)

(b)

Fig. 5–17

component $(\mathbf{M}_A)_z$ developed at A since rotation about A is prevented by the reactive force $\mathbf{B}_y$ acting at B.)

For this particular problem, each of the five unknown force magnitudes shown on the free-body diagram can easily be obtained by direct application of the scalar equations of equilibrium. Summing moments about the x axis, we can obtain a direct solution for P. Why? For the scalar moment summation, we must compute the moment of each force as the product of the *perpendicular distance* from the x axis to the line of action of the force. Assuming positive moments act in the $+\mathbf{i}$ direction,

$$\Sigma(M_B)_x = 0; \quad 150 \text{ lb } (\tfrac{1}{4} \text{ ft}) - P(1 \cos 20° \text{ ft}) = 0, \qquad P = 39.9 \text{ lb} \quad Ans.$$

Using this result for P and summing moments in a similar manner about the y and z axes passing through point B we have,

$$\Sigma(M_B)_y = 0;$$

$$-150 \text{ lb } (1.5 \text{ ft}) + A_z(3 \text{ ft}) + (39.9 \text{ lb})(1.5 \text{ ft}) = 0, \qquad A_z = 55.0 \text{ lb} \quad Ans.$$

and

$$\Sigma(M_B)_z = 0; \qquad -A_y(3 \text{ ft}) = 0, \qquad A_y = 0 \qquad Ans.$$

(Note that summation of moments was taken about axes which pass through the bearing support at B in order to eliminate the two unknown force components $\mathbf{B}_y$ and $\mathbf{B}_z$ from the moment equation.)

The reactions at B are obtained by a force summation, using the results computed above,

$$\Sigma F_y = 0; \qquad 0 + B_y = 0, \qquad B_y = 0 \qquad Ans.$$

$$\Sigma F_z = 0; \quad 55.0 - 150 + B_z - 39.9 = 0, \qquad B_z = 134.9 \text{ lb} \qquad Ans.$$

As shown on the free-body diagram, the supports do not provide resistance against movement of the windlass in the x direction. Hence, the windlass is only partially constrained.

Example 5-12

The bent rod shown in Fig. 5–18a is supported at A by a smooth collar, at E by a ball-and-socket joint, and at B by means of cable BC. (a) Determine the forces developed at these supports when the rod is subjected to the 100-lb weight shown; and (b) using only *one equilibrium equation,* obtain a direct solution for the tension in cable BC. The smooth collar attachment at A is capable of exerting force components only in the z and y directions, since it is aligned properly on the shaft axis.

Solution
Part (a). The free-body diagram for the rod is shown in Fig. 5–18b. There are six unknowns in this problem. These include the three unknown force

(a)

(b)

Fig. 5–18

components at the ball-and-socket joint, two at the smooth collar bearing, and an unknown tension force in the cable. Expressing each force on the free-body diagram in vector component notation, we have

$$\mathbf{W} = \{-100\mathbf{k}\}\ \text{lb}$$
$$\mathbf{F}_A = A_y\mathbf{j} + A_z\mathbf{k}$$
$$\mathbf{F}_E = E_x\mathbf{i} + E_y\mathbf{j} + E_z\mathbf{k}$$
$$\mathbf{T}_{BC} = T_{BC}\left(\frac{\mathbf{r}_{BC}}{r_{BC}}\right) = \frac{3T_{BC}}{\sqrt{98}}\mathbf{i} - \frac{5T_{BC}}{\sqrt{98}}\mathbf{j} + \frac{8T_{BC}}{\sqrt{98}}\mathbf{k}$$
$$= 0.303T_{BC}\mathbf{i} - 0.505T_{BC}\mathbf{j} + 0.808T_{BC}\mathbf{k}$$

Force equilibrium requires

$$\Sigma\mathbf{F} = 0; \qquad\qquad \mathbf{W} + \mathbf{F}_A + \mathbf{F}_E + \mathbf{T}_{BC} = 0$$
$$-100\mathbf{k} + (A_y\mathbf{j} + A_z\mathbf{k}) + (E_x\mathbf{i} + E_y\mathbf{j} + E_z\mathbf{k}) + (0.303T_{BC}\mathbf{i}$$
$$-0.505T_{BC}\mathbf{j} + 0.808T_{BC}\mathbf{k}) = 0$$

or

$$(E_x + 0.303T_{BC})\mathbf{i} + (A_y + E_y - 0.505T_{BC})\mathbf{j}$$
$$+ (A_z + E_z + 0.808T_{BC} - 100)\mathbf{k} = 0$$

Thus, equating the **i** and **j** components equal to zero,

$$\Sigma F_x = 0; \qquad\qquad E_x + 0.303T_{BC} = 0 \qquad\qquad\qquad (1)$$
$$\Sigma F_y = 0; \qquad\qquad A_y + E_y - 0.505T_{BC} = 0 \qquad\qquad (2)$$
$$\Sigma F_z = 0; \qquad\qquad A_z + E_z + 0.808T_{BC} - 100 = 0 \qquad (3)$$

Summing moments about point E, to eliminate the three components of force acting there,

$$\Sigma\mathbf{M}_E = 0; \qquad (\mathbf{r}_{EB} \times \mathbf{T}_{BC}) + (\mathbf{r}_{EG} \times \mathbf{W}) + (\mathbf{r}_{EA} \times \mathbf{F}_A) = 0$$

Expressing each of the cross products in determinant form,

$$\begin{vmatrix} \mathbf{i} & \mathbf{j} & \mathbf{k} \\ 0 & -12 & -14 \\ 0.303T_{BC} & -0.505T_{BC} & 0.808T_{BC} \end{vmatrix} + \begin{vmatrix} \mathbf{i} & \mathbf{j} & \mathbf{k} \\ -4 & -12 & -14 \\ 0 & 0 & -100 \end{vmatrix}$$
$$+ \begin{vmatrix} \mathbf{i} & \mathbf{j} & \mathbf{k} \\ -8 & -12 & -14 \\ 0 & A_y & A_z \end{vmatrix} = 0$$

Expanding and rearranging terms,

$$(-16.77T_{BC} - 12A_z + 14A_y + 1{,}200)\mathbf{i} + (-4.24T_{BC} + 8A_z - 400)\mathbf{j}$$
$$+ (3.64T_{BC} - 8A_y)\mathbf{k} = 0$$

Equating the **i**, **j**, and **k** components equal to zero yields

$$\Sigma M_x = 0; \qquad -16.77 T_{BC} - 12 A_z + 14 A_y + 1,200 = 0 \qquad (4)$$

$$\Sigma M_y = 0; \qquad -4.24 T_{BC} + 8 A_z - 400 = 0 \qquad (5)$$

$$\Sigma M_z = 0; \qquad 3.64 T_{BC} - 8 A_y = 0 \qquad (6)$$

The simultaneous solution of Eqs. (1) through (6) may be rather tedious. Consequently, the computer program listed in Appendix A may be used to solve these equations. The final solution is

$$A_y = 16.3 \text{ lb} \qquad \qquad Ans.$$
$$A_z = 69.0 \text{ lb} \qquad \qquad Ans.$$
$$E_x = -10.84 \text{ lb} \qquad \qquad Ans.$$
$$E_y = 1.79 \text{ lb} \qquad \qquad Ans.$$
$$E_z = 2.09 \text{ lb} \qquad \qquad Ans.$$
$$T_{BC} = 35.8 \text{ lb} \qquad \qquad Ans.$$

Part (b).   Without knowing the other reactions, the cable tension $\mathbf{T}_{BC}$ may be obtained *directly* by summing moments about an axis passing through points A and E, Fig. 5–18b. (Note that the line of action of the other five unknown force components passes through this axis and hence, these forces are eliminated from the moment summation.) The unit vector $\mathbf{u}_{AE}$ shown in Fig. 5–18a which defines the position of this axis is

$$\mathbf{u}_{AE} = \frac{\mathbf{r}_{AE}}{r_{AE}} = \frac{8}{\sqrt{404}}\mathbf{i} + \frac{12}{\sqrt{404}}\mathbf{j} + \frac{14}{\sqrt{404}}\mathbf{k}$$
$$= 0.398\mathbf{i} + 0.597\mathbf{j} + 0.696\mathbf{k}$$

Hence, the sum of moments about this axis is zero provided

$$\Sigma M_{AE} = \mathbf{u}_{AE} \cdot \Sigma(\mathbf{r} \times \mathbf{F}) = 0$$

In this equation **r** represents a position vector drawn from *any point* on the axis to *any point* on the line of action of the force, **F.** (See Eq. 4–17, p. 110.) Thus, with reference to Fig. 5–18b, we have

$$\mathbf{u}_{AE} \cdot [(\mathbf{r}_{AB} \times \mathbf{T}_{BC}) + (\mathbf{r}_{AG} \times \mathbf{W})] = 0$$

$$(0.398\mathbf{i} + 0.597\mathbf{j} + 0.696\mathbf{k}) \cdot [(8\mathbf{i}) \times (0.303 T_{BC}\mathbf{i}$$
$$-0.505 T_{BC}\mathbf{j} + 0.808 T_{BC}\mathbf{k}) + (4\mathbf{i}) \times (-100\mathbf{k})] = 0$$

or

$$-6.67 T_{BC} + 238.8 = 0, \qquad T_{BC} = 35.8 \text{ lb} \qquad Ans.$$

Since the cable tension is known, Eqs. (1) through (6) may be readily solved for the remaining unknown reactions.

Problems

Neglect the weight of the members in the following problems except when specified.

5-43. Using a single equilibrium equation, solve for the tension in the cable located at C, for the platform in Example 5-8.

5-44. Calculate the magnitude and direction of the resultant force and moment acting at the fixed wall of the cantilevered beam.

Prob. 5-44

5-45. Determine the tension developed in the three supporting cables. The plate is held in the horizontal xy plane by the three cables.

Prob. 5-45

5-46. The windlass is subjected to a load of 150 lb. Determine the horizontal force **P** needed to hold the handle in the position shown, and the components of reaction at the ball-and-socket joint A and the smooth bearing B. The bearing at B is in proper alignment and exerts only a force reaction on the windlass.

Prob. 5-46

5-47. The homogeneous block weighs 1,000 lb and is supported at A by a ball-and-socket and at B by a smooth roller. Determine the tensions in the cables CD and EF and the reactions at A and B. The cable CD lies in a plane which is parallel to the yz plane.

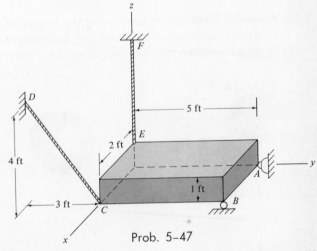

Prob. 5-47

173

5-48. The wing of the jet aircraft is subjected to the following force system: the weight of the 1,500-lb engine having a center of gravity point G_C, the weight of the wing and internal gas supply of 2,500 lb having a center of gravity G_W, and the vertical upward reaction of 3,000 lb exerted by the ground on *each* tire, T. For this loading system, calculate the magnitude and direction of the resultant force and moment exerted by the fuselage on the wing at the fixed point of attachment, A.

Prob. 5-48

5-49. A smooth 20-lb sphere rests along the surface of the arc support. Determine the force per unit length along the ring of contact which is exerted on the sphere.

Side View

Top View

Prob. 5-49

5-50. The composite rods are supported at points A and B by a smooth surface and at E by a ball-and-socket joint. Determine the reaction components if each rod weighs 2 lb/ft.

Prob. 5-50

5-51. Determine the reaction components at the two ball supports A and B and the ball-and-socket joint at C for the triangular plate.

Prob. 5-51

174

5-52. Determine the components of reaction at the smooth bearing supports A and B for the shaft. Both pulleys are firmly fixed to the shaft and lie in planes parallel to the xz plane. Is the shaft in complete equilibrium under the given loading system? The bearing supports are in proper alignment and exert only force reactions on the shaft.

Prob. 5-52

5-53. The uniform lid on a chest is propped open by means of the light rod CD. If the hinge at A prevents sliding of the lid along the axis AB while the hinge at B does not offer resistance in this direction, calculate the compressive force in the rod CD and the components of reaction at the hinges A and B. The lid weighs 25 lb. The hinges are in proper alignment and exert only force reactions on the lid.

Prob. 5-53

5-54. Determine the compressive force in the strut AB and the reactive components at the hinge C of the 3-ft-diameter hatch door. The door weighs 60 lb and has a center of gravity at its midpoint.

Prob. 5-54

5-55. The boom AC is subjected to a load of 400 lb and is supported at A by a ball-and-socket joint and by two cables BDC and CE. The cable BDC is continuous and passes around a frictionless pulley at D. Calculate the tension in the cables and the components of reaction at A.

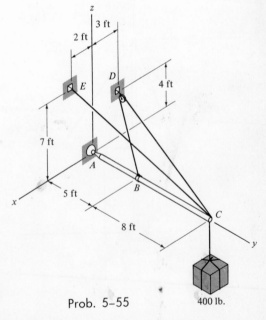

Prob. 5-55

175

5-56. The triangular plate is supported by a single hinge at *A* and a vertical cable *BC*. If the cable can support a maximum tension of 300 lb, determine the maximum force **F** which may be applied to the plate. Compute the reactive components at *A* for this loading.

Prob. 5-56

5-57. Determine the forces acting in the cables *AB* and *AC* and the components of reaction at the ball-and-socket joint *D* for the boom.

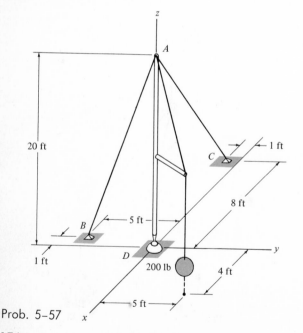

Prob. 5-57

5-58. The semicircular metal plate supports a 150-lb load. Determine the tensions in cables *BD* and *CD* and the components of reaction at the ball-and-socket joint at *A*.

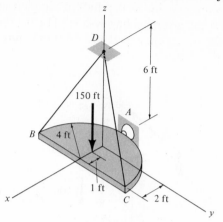

Prob. 5-58

5-59. The bent rod is supported at *A*, *B*, and *C* by smooth bearings. Compute the reaction components at the bearings if the rod is subjected to a 200-lb vertical force and a 30-lb-ft couple as shown. The bearings are in proper alignment and exert only force reactions on the rod.

Prob. 5-59

176

5-60. Determine the components of reaction at the ball supports B and C and the ball-and-socket A for the uniformly loaded plate.

Prob. 5-60

5-61. The uniform rod AB weighs 30 lb and is supported at A by a ball-and-socket joint and leans against the smooth vertical wall at B. A cable is attached at B to help support the rod. Determine the unknown force components acting on the rod.

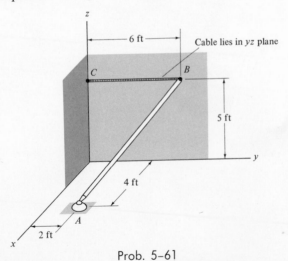

Prob. 5-61

5-62. The uniform door weighs 40 lb and is supported at A and B by hinges. The hinge at A supports only forces in the horizontal plane, while B supports both horizontal and vertical components of force. If the spring CE is fixed at the top of the door as shown in the figure, determine the magnitude of horizontal force **P** required to keep the door open at the position shown. What are the components of reaction at hinges A and B? The spring is unstretched when the door is closed.

Prob. 5-62

177

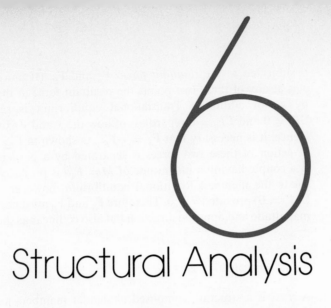

6

Structural Analysis

6-1. Two-Force Member

When forces are applied at only two points on a member, the member is commonly referred to as a *two-force member*. For *equilibrium,* the force acting at one point on the member must be of *equal magnitude, opposite direction, and collinear* with the force acting at the other point. To prove this statement, consider the two-force member shown in Fig. 6–1a which

(a)

(b)
Translational Equilibrium

(c)
Translational and
Rotational Equilibrium

Fig. 6–1

is subjected to the *resultant forces* $\mathbf{F}_A$ and $\mathbf{F}_B$. (If more than one force acts at any of these two points the resultant force at that point is found by vector addition.) Translational equilibrium is satisfied provided $\Sigma F_x = 0$ and $\Sigma F_y = 0$ regardless of how the x and y axes are orientated. Hence, it is necessary that $\mathbf{F}_A = -\mathbf{F}_B$. As shown in Fig. 6–1*b*, if the line of action of these two forces is separated by a perpendicular distance d, a couple having a magnitude of $M = F_A d$ is produced which tends to rotate the member. Rotational equilibrium, however, is satisfied (e.g., $\Sigma M_A = 0$) provided $d = 0$. Therefore $\mathbf{F}_A$ and $\mathbf{F}_B$ must not only be of equal magnitude and opposite direction but also collinear as shown in Fig. 6–1*c*.

6–2. Simple Trusses

A *truss* is a structure composed of slender members joined together at their end points. The members commonly used in construction consist of wood struts, metal bars, angles, or channels. The joint connections are usually formed by bolting or welding the ends of the members to a common plate (called a *gusset plate*), as shown in Fig. 6–2, or by simply

Fig. 6–2

passing large bolts or pins through each of the members. Most *planar* trusses are used for roof and bridge construction. The *Pratt truss, ABCDE,* shown in Fig. 6–3*a*, is an example of a typical roof-supporting truss. In this figure, the roof load is transmitted to the truss *at the joints* by means of a series of "purlins" such as beam *DD'*. Since the imposed loading acts in the same plane as the truss, Fig. 6–3*b*, the analysis of the forces developed in the truss members may be considered two dimensional.

To design both the members and the connections of a truss, it is *first necessary* to determine the *forces* developed in each of the members when the truss is subjected to a given loading. In this regard two important assumptions will be made in determining the forces acting in the members of a truss:

(a)

(b)

Fig. 6-3

1. The members are joined together by smooth pins. In cases where bolted or welded joint connections are used, this assumption is satisfactory provided the center lines of the joining members are concurrent at a point, as in the case of point A in Fig. 6-2.
2. All the loadings are applied at the joints. In most situations, such as for bridge and roof trusses, this assumption is true. The deck or floor of a bridge is usually supported by crossbeams which are connected to the side trusses at the joints. The loads acting on a roof are transmitted to the joints of the roof-supporting trusses by a series of purlins. Frequently in the force analysis the weight of the members is neglected, since the force supported by the members is large in comparison with their weight. If the weight of each of the members is to be included in the analysis, it is satisfactory to apply half the weight of each member as a vertical force acting at the end of each member.

In accordance with these assumptions *each truss member acts as a two-force member,* and therefore, the forces acting on a member must act axially along the member. If this force tends to *elongate* the member, it is a *tensile* force. When the force tends to *shorten* the member, the force

is *compressive*. In the actual design of a truss it is important to state whether the nature of the force is tensile or compressive. Most often, the compression members must be made thicker than the tension members, because of the buckling or column effect which occurs in compression members.

In this book we will be concerned only with the force analysis of simple trusses. To form a rigid stable unit lying in a single plane and consisting of pin-connected members, we must arrange the members as a series of connecting *triangular* elements. If we form polygons having more than three sides, the frame will collapse. For example, consider the four-sided frame shown in Fig. 6–4. When the load **P** is applied to this frame it will collapse unless a diagonal member *AB* is added to make the frame stable. When a truss is constructed entirely of triangular elements, it is known as a *simple truss*.

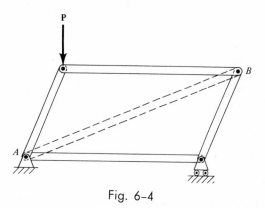

Fig. 6–4

The forces developed in simple truss members can be determined by means of the method of joints or the method of sections. Once the force analysis of the truss has been completed, the size of both the members and connections can then be determined using the theory of strength of materials.

6–3. Method of Joints

If a truss is in equilibrium, each of its joints must be in equilibrium. The *method of joints* consists of satisfying equilibrium for all the forces exerted *on the pin* at each joint. Since truss members are all straight two-force members, the force developed in each member is *directed along the axis of the member*. Hence, for planar trusses *the force system acting on the pin at each joint will be both coplanar and concurrent*. Equilibrium conditions for each joint therefore require $\Sigma F_x = 0$ and $\Sigma F_y = 0$.* Because

---

*Rotational or moment equilibrium is automatically satisfied since the force system is concurrent.

there are only two equations of equilibrium for each joint, the force analysis should generally begin at joints having at most two unknown forces. In this way one obtains two equations which have a unique solution. After having determined the forces at some joints, the forces at the other joints may be analyzed in turn. The following two examples illustrate the procedure.

Example 6-1

A plane truss is acted upon by the force system shown in Fig. 6–5a. Determine the force in each member of the truss. Indicate whether the members are in tension or compression. The truss is supported at A by means of rollers and at B by means of a pin.

Solution

If initially a free-body diagram of each joint were drawn, it would be found that no joint is subjected to *only* two unknown forces; therefore, it is first necessary to determine the external reactions at joints A and B. A free-body diagram of the truss is given in Fig. 6–5b. Applying the equations of equilibrium, we have

$$\xrightarrow{+} \Sigma F_x = 0; \qquad 500 \text{ lb} - B_x = 0, \qquad B_x = 500 \text{ lb}$$

$$\left(+\Sigma M_B = 0; \quad -A_y(20 \text{ ft}) + 500 \text{ lb} (10 \text{ ft}) = 0, \qquad A_y = 250 \text{ lb}\right.$$

$$+\uparrow \Sigma F_y = 0; \quad 250 \text{ lb} - 500 \text{ lb} - 1{,}000 \text{ lb} + B_y = 0, \qquad B_y = 1{,}250 \text{ lb}$$

Knowing these reaction forces, we can now begin the analysis by starting either at joints A or B, since there are two unknown member forces acting at each of these joints.

Joint A, Fig. 6–5c. As shown on the free-body diagram, there are three forces which act at joint A. In particular, force $\mathbf{F}_{AD}$ has been assumed to be directed *away* from the joint. Because for every action there is an equal but opposite reaction, this *"pulling on the pin"* causes a *tension* in member AD*. If, in the calculations, the numerical value of F_{AD} is positive, the assumed direction is correct; however, if it is negative, the force direction is reversed and the member is in compression.

In some cases it is possible to tell *by inspection* whether or not the sense of direction of an unknown member force is tensile or compressive. For example, the force in member AE must be compressive, Fig. 6–5c, since the vertical component of this force must act downward to balance the upward 250-lb force ($\Sigma F_y = 0$). The horizontal component of force $\mathbf{F}_{AE}$ then acts to the left, which requires member AD to be in tension ($\Sigma F_x = 0$). *In all cases,* the correct sense of an unknown force should be verified by drawing the free-body diagrams of each joint and then applying and solving the equations of equilibrium.

---

*To see this clearly, refer to Fig. 6–5h. Here it is likewise seen that a force which is *pushing on a pin* causes a corresponding *pushing* on the member and hence a *compression* in the member.

Fig. 6-5

The equations of equilibrium when applied to joint A yield

$\xrightarrow{+} \Sigma F_x = 0;$ $\qquad\qquad F_{AD} - F_{AE} \cos 45° = 0$

$+\uparrow \Sigma F_y = 0;$ $\qquad\qquad 250 \text{ lb} - F_{AE} \sin 45° = 0$

Solving, we have

$$F_{AE} = 354 \text{ lb} \quad \text{(compression)} \qquad\qquad \textit{Ans.}$$

$$F_{AD} = 250 \text{ lb} \quad \text{(tension)} \qquad\qquad \textit{Ans.}$$

Joint D, Fig. 6-5d. Having found the force in member AD, we may next analyze joint D, since there are only two unknown forces acting at this joint. The force $F_{AD} = 250 \text{ lb}$ is a *tensile* force and therefore acts *away* from joint D. This force is transferred from joint A to joint D via member AD, as shown by means of equal and opposite reactions in Fig. 6-5h. Applying the equations of equilibrium yields

$\xrightarrow{+} \Sigma F_x = 0;$ $\ F_{DC} - 250 \text{ lb} = 0,$ $\quad F_{DC} = 250 \text{ lb} \quad \text{(tension)} \qquad \textit{Ans.}$

$+\uparrow \Sigma F_y = 0;$ $F_{DE} - 500 \text{ lb} = 0,$ $\quad F_{DE} = 500 \text{ lb} \quad \text{(tension)} \qquad \textit{Ans.}$

Continuing in the same manner, we may analyze joints C and E next.

Joint C, Fig. 6-5e.

$\xrightarrow{+} \Sigma F_x = 0;$ $\qquad\qquad -250 \text{ lb} + F_{CE} \cos 45° = 0$

$+\uparrow \Sigma F_y = 0;$ $\qquad F_{CB} - F_{CE} \sin 45° - 1,000 \text{ lb} = 0$

Hence,

$$F_{CE} = 354 \text{ lb} \quad \text{(compression)} \qquad\qquad \textit{Ans.}$$

$$F_{CB} = 1,250 \text{ lb} \quad \text{(tension)} \qquad\qquad \textit{Ans.}$$

Joint E, Fig. 6-5f.

$\xrightarrow{+} \Sigma F_x = 0;$ $\ -F_{EB} - 354 \cos 45° \text{ lb} + 354 \cos 45° \text{ lb} + 500 \text{ lb} = 0$

$$F_{EB} = 500 \text{ lb} \quad \text{(compression)} \qquad\qquad \textit{Ans.}$$

$+\uparrow \Sigma F_y = 0;$ $\quad 2[354 \sin 45° \text{ lb}] - 500 \text{ lb} \equiv 0 \quad \text{(check)}$

All the forces in the members have been found; however, it is always prudent to analyze the last joint (joint B) in order to check the accuracy of the solution.

Joint B, Fig. 6-5g.

$\xrightarrow{+} \Sigma F_x = 0;$ $\qquad 500 \text{ lb} - 500 \text{ lb} \equiv 0 \quad \text{(check)}$

$+\uparrow \Sigma F_y = 0;$ $\qquad 1,250 \text{ lb} - 1,250 \text{ lb} \equiv 0 \quad \text{(check)}$

The results of this force analysis are summarized in Fig. 6–5h, which shows the complete free-body diagrams for each joint. Applying the principle of action and reaction, the forces acting in the members are equal and opposite to those acting at the joints, as shown in the figure. The truss members are labeled as being in either tension or compression.

Example 6–2

Using the method of joints, show that members QP, PG, HJ, JG, and GK of the *Fink truss,* shown in Fig. 6–6a, are subjected to zero force when the truss is loaded as shown.

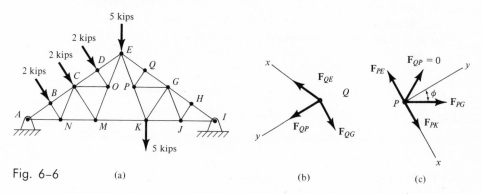

Fig. 6–6   (a)   (b)   (c)

Solution

At each joint the equations $\Sigma F_x = 0$ and $\Sigma F_y = 0$ must be satisfied. The orientation of the x and y axes is arbitrary for the application of these equations. Hence, if we consider a free-body diagram of joint Q so that the y axis is directed along member QP, as shown in Fig. 6–6b, the line of action of forces $\mathbf{F}_{QE}$ and $\mathbf{F}_{QG}$ are both along the x axis. Therefore the equation $\Sigma F_y = 0$ is satisfied provided $\mathbf{F}_{PQ} = 0$, i.e., QP is a zero-force member.

A free-body diagram of joint P is shown in Fig. 6–6c. By appropriate orientation of the x axis, equilibrium in the y direction is satisfied provided $\Sigma F_y = F_{PG} \cos \phi = 0$. Hence $F_{PG} = 0$. In a similar manner, by considering, in turn, free-body diagrams of joints H, J, and G, complete this example and show that the force in the respective members HJ, JG, and GK is zero.

Although the zero-force members are not necessary for supporting the loading shown in Fig. 6–6a, their use is important in the construction of the truss. Understand that zero-force members are actually determined on the basis of assuming the truss to be *rigid*. In reality, however, the truss may deform when it is loaded thereby transmitting, slightly, the applied loads to the zero-force members. Furthermore, these members add to the stability of the truss during construction and provide support if the applied loading is changed.

Problems

6–1. Determine the force in each member of the truss. Indicate whether the members are in tension or compression.

Prob. 6–1

6–2. Determine the force in each member of the truss. Indicate whether the members are in tension or compression.

Prob. 6–2

6–3. Compute the force in each member of the truss. Indicate whether the members are in tension or compression.

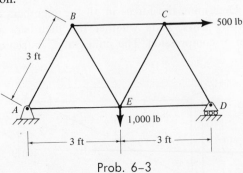

Prob. 6–3

6–4. Determine the force in each member of the truss. Indicate whether the members are in tension or compression.

6–4. Determine the force in each member of the truss. Indicate whether the members are in tension or compression.

Prob. 6–4

6–5. Determine the force in each member of the truss. Indicate whether the members are in tension or compression.

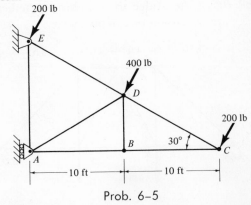

Prob. 6–5

6–6. The *Howe roof truss* supports the vertical loading shown. Determine the force in members *CL*, *MC*, and *CD*. Indicate whether these forces are tensile or compressive.

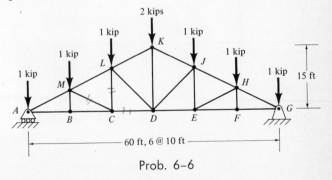

Prob. 6–6

187

6-7. The *Pratt bridge truss* is subjected to the loading shown. Determine the force in members *MD* and *BC*. Indicate whether the members are in tension or compression.

5 kips   5 kips   5 kips

Prob. 6-7

6-8. Determine the forces developed in members *AG*, *GF*, and *BC* of the bridge truss. Indicate whether the members are in tension or compression.

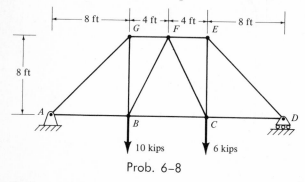

Prob. 6-8

6-9. Determine the force in members *AJ*, *DI*, and *CJ*, and indicate all the zero-force members of the *Howe bridge truss*. State whether members *AJ*, *DI*, and *CJ* are in tension or compression.

Prob. 6-9

6-10. Determine the forces in members *CD*, *AG*, and *BF* of the truss shown. Indicate whether the members are in tension or compression.

Prob. 6-10

6-11. Determine the forces in all the members of the truss. Indicate whether the members are in tension or compression.

Prob. 6-11

6-12. Determine the forces in all the members of the supporting truss. State whether the members are in tension or compression. The pulley at *C* is frictionless.

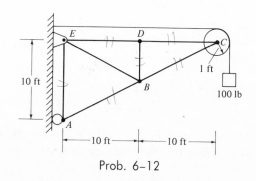

Prob. 6-12

6-13. Determine the force in members *CF, BH*, and *AH* of the *Warren roof truss*. Indicate whether the members are in tension or compression.

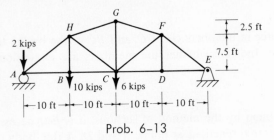

Prob. 6-13

6-14. Determine the force in each member of the truss in terms of the applied force **P.** Indicate whether the members are in tension or compression.

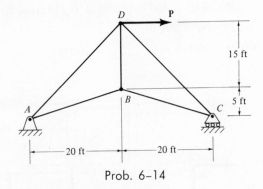

Prob. 6-14

6-15. Determine the forces in members *DC* and *HK* of the crane truss. Indicate whether the members are in tension or compression.

Prob. 6-15

6-4. Method of Sections

A second method for finding the forces in the members of a truss, the *method of sections,* consists of passing a section through the truss, thus cutting it into two parts. If the entire truss is in equilibrium, then each of the two parts must also be in equilibrium. Hence, after the truss has been sectioned, either part may be used to form a free-body diagram. Since only three equilibrium equations are available, the forces in all the members at the cut section may be determined provided the cutting section passes through not more than three members in which the forces are unknown. By choosing the proper points about which to sum moments, or directions for summation of forces, each of the three unknown forces may generally be solved for from a *single equation* rather than from a series of simultaneous equations.

When the forces in only a *few members* of a truss are to be found,

189

the method of sections combined with the method of joints generally provides the most direct means of obtaining these forces.

Example 6–3

Determine the force in members GE, GC, and BC of the bridge truss shown in Fig. 6–7a. Indicate whether the members are in tension or compression.

Solution

To solve this problem by the method of sections, it is *first* necessary to determine the external reactions at points A and D. A free-body diagram of the entire truss is shown in Fig. 6–7b. Applying the equations of equilibrium, we have

$$\xrightarrow{+}\Sigma F_x = 0; \qquad 400 \text{ lb} - A_x = 0, \qquad A_x = 400 \text{ lb}$$

$$(+\Sigma M_A = 0; \qquad -1{,}000 \text{ lb } (20 \text{ ft}) - 400 \text{ lb } (10 \text{ ft}) + D_y(30 \text{ ft}) = 0,$$

$$D_y = 800 \text{ lb}$$

$$+\uparrow\Sigma F_y = 0; \quad A_y - 1{,}000 \text{ lb} + 800 \text{ lb} = 0, \qquad A_y = 200 \text{ lb}$$

(a)

(b)

(c)

(d)

Fig. 6–7

Knowing these reactions, the truss can now be sectioned along the dotted line *aa*, cutting the three members in Fig. 6–7a. Either the right- or the left-hand side may be used to draw a free-body diagram of the cut section as shown in Figs. 6–7c and 6–7d. In constructing the two diagrams, note that the three unknown member forces acting on one side of the truss must be equal but opposite to those acting on the other side (Newton's third law of motion). Also, the line of action of each member force is specified from the *geometry of the truss* since the force in a member passes along the axis of the member.

In many cases the sense of direction of an unknown force may be determined by inspection *before* the equilibrium equations are applied. For example, the force in member *GE*, that is, $\mathbf{F}_{GE}$, must be compressive, as shown in Fig. 6–7c.* Summing moments about point *C*, this force will then provide a necessary counterclockwise moment which will balance the clockwise moment created by the 200-lb force—note that the line of action of $\mathbf{F}_{GC}$, $\mathbf{F}_{BC}$ and 400 lb passes through point *C*. Similarly, when moments are summed about point *G*, one comes to the conclusion that force $\mathbf{F}_{BC}$ must be tensile. To satisfy $\Sigma F_y = 0$, the force $\mathbf{F}_{GC}$ must be tensile so that its vertical component acts downward to balance the 200-lb reactive force acting upward at *A*. *In all cases,* the correct sense of an unknown member force will be verified after solving the equations of equilibrium. If the sense of direction of an unknown force was chosen *correctly,* the answer will be *positive* when the equations are solved for the force magnitude. A negative answer indicates that the force acts in the opposite direction to that shown on the free-body diagram.

By judicious application of the equations of equilibrium, we can solve directly for each of the three unknown forces at a cut section, rather than solving three equations simultaneously. Using the free-body diagram in Fig. 6–7c and summing moments about point *G*, as just discussed, eliminates the two unknown forces $\mathbf{F}_{GE}$ and $\mathbf{F}_{GC}$. Hence,

$$\zeta + \Sigma M_G = 0; \qquad -400 \text{ lb (10 ft)} - 200 \text{ lb (10 ft)} + F_{BC}(10 \text{ ft}) = 0,$$
$$F_{BC} = 600 \text{ lb} \quad \text{(tension)} \qquad\qquad Ans.$$

In the same manner, by summing moments about point *C* we obtain a direct solution for the force in *GE*.

$$\zeta + \Sigma M_C = 0; \qquad -200 \text{ lb (20 ft)} + F_{GE}(10 \text{ ft}) = 0,$$
$$F_{GE} = 400 \text{ lb} \quad \text{(compression)} \qquad\qquad Ans.$$

Since $\mathbf{F}_{BC}$ and $\mathbf{F}_{GE}$ have no vertical components of force, summing forces in the *y* direction yields F_{GC}, i.e.,

*By definition, a free-body diagram of the cut section of a truss must show *all* the forces acting *directly on the members located at the cut section.* Hence, the force acting on member *GE* in Figs. 6–7c or 6–7d is *compressive* since it tends to *push* on the member, likewise, the forces in members *GC* and *BC* are *tensile* since they tend to *pull* on the member.

$$+\uparrow \Sigma F_y = 0; \qquad 200\ \text{lb} - F_{GC}\sin 45° = 0,$$

$$F_{GC} = 282.9\ \text{lb} \quad \text{(tension)} \qquad \qquad Ans.$$

The same results can also be obtained by applying the equations of equilibrium to the free-body diagram shown in Fig. 6–7d. With this diagram how would you obtain each unknown force using a *single equation* of equilibrium for each calculation?

Example 6-4

Determine the force in member *GC* of the truss shown in Fig. 6–8a. Indicate whether the member is in tension or compression.

Solution

A free-body diagram of the entire truss is shown in Fig. 6–8b. We must first solve for the external reactions. Why?

(a)

(b)

(c)

(d)

Fig. 6-8

$\zeta + \Sigma M_A = 0;$

$-3{,}000 \text{ lb } (10 \text{ ft}) - 1{,}000 \text{ lb } (20 \text{ ft}) - 1{,}000 \text{ lb } (30 \text{ ft}) + E_y(40 \text{ ft}) = 0,$

$$E_y = 2{,}000 \text{ lb}$$

$\xrightarrow{+} \Sigma F_x = 0;$ $\qquad\qquad E_x = 0$

Using the computed value of E_y,

$+\uparrow \Sigma F_y = 0;$

$A_y - 1{,}000 \text{ lb} - 3{,}000 \text{ lb} - 1{,}000 \text{ lb} - 1{,}000 \text{ lb} + 2{,}000 \text{ lb} = 0,$

$$A_y = 4{,}000 \text{ lb}$$

According to the method of sections, any imaginary section which cuts through member GC will also have to cut through three other members whose forces are unknown. For example, the section bb shown in Fig. 6–8a cuts through members GJ, GC, HC, and BC. If we consider a free-body diagram of the left-hand portion of this section, (Fig. 6–8d) it is possible to determine the force in member GJ ($\mathbf{F}_{GJ}$) by summing moments about point C to eliminate the other three unknowns; however, the force in GC cannot be determined from the remaining two equilibrium equations. It is therefore necessary to determine the force in either HC or BC *before* using section bb. We may, for example, determine the force in member BC by considering the adjacent section aa, Fig. 6–8a. A free-body diagram of the truss on the left-hand side of section aa is given in Fig. 6–8c. Summing moments about point H, we may obtain the force in BC directly from one equation:

$\zeta + \Sigma M_H = 0; -4{,}000 \text{ lb } (10 \text{ ft}) + 1{,}000 \text{ lb } (10 \text{ ft}) + F_{BC}(10 \tan 30° \text{ ft}) = 0,$

$$F_{BC} = 5{,}200 \text{ lb} \quad \text{(tension)}$$

Knowing the magnitude and direction of this force we can now use a free-body diagram of the truss to the left of section bb, Fig. 6–8d. Applying the three equations of equilibrium, we have

$\zeta + \Sigma M_C = 0; \quad -4{,}000 \text{ lb } (20 \text{ ft}) + 1{,}000 \text{ lb } (20 \text{ ft}) + 3{,}000 \text{ lb } (10 \text{ ft})$
$$+ (F_{GJ} \cos 30°)(20 \tan 30° \text{ ft}) = 0,$$

$$F_{GJ} = 3{,}000 \text{ lb} \quad \text{(compression)}$$

(Note that the component ($F_{GJ} \sin 30°$) is not included in the moment equation since the line of action of this force at G passes through point C.)

$\zeta + \Sigma M_G = 0;$

$-4{,}000 \text{ lb } (20 \text{ ft}) + 1{,}000 \text{ lb } (20 \text{ ft}) + 3{,}000 \text{ lb } (10 \text{ ft}) + F_{BC}(20 \tan 30° \text{ ft})$
$$- (F_{HC} \cos 30°)(20 \tan 30° \text{ ft}) = 0$$

Substituting for F_{BC} and solving,

$$F_{HC} = 3{,}000 \text{ lb} \quad \text{(compression)}$$

(Note that F_{HC} might also be obtained by taking moments about point A, Fig. 6–8c.)

$$+\uparrow\Sigma F_y = 0; \qquad 4{,}000\text{ lb} - 1{,}000\text{ lb} - 3{,}000\text{ lb} - 1{,}000\text{ lb} - F_{GC}$$
$$+ (F_{GJ}\sin 30°) + (F_{HC}\sin 30°) = 0$$

Substituting for F_{GJ} and F_{HC} and solving,

$$F_{GC} = 2{,}000\text{ lb}\quad\text{(tension)}\qquad\qquad Ans.$$

Problems

6–16. Determine the forces in members EB and ED of the truss. Indicate whether the members are in tension or compression.

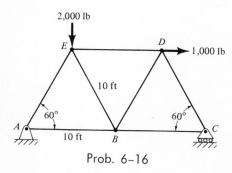

Prob. 6–16

6–17. Determine the forces in members EH and IH of the airplane supporting frame. Indicate whether the members are in tension or compression.

Prob. 6–17

6–18. Determine the forces in the members GF and BC of the truss. Indicate whether the members are in tension or compression.

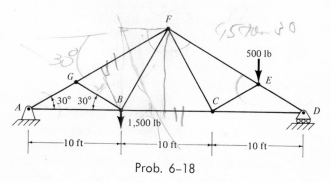

Prob. 6–18

6–19. Determine the forces in members BG and BF of the truss. Indicate whether the members are in tension or compression.

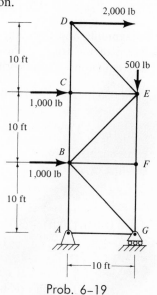

Prob. 6–19

6-20. Determine the force in members *GF*, *FC*, and *BC* for the truss in Prob. 6–10. Solve for each force directly using a single equilibrium equation. Indicate whether the members are in tension or compression.

6-21. Compute the force in members *EF* and *CB* for the truss in Prob. 6–8.

6-22. Determine the force in members *FE*, *FB*, and *BC* of the truss. Solve for each force directly using a single equilibrium equation. Indicate whether the members are in tension or compression.

Prob. 6–22

6-23. Determine the force in members *GF* and *CF* of the truss. Indicate whether the members are in tension or compression.

Prob. 6–23

6-24. Determine the force in members *FC*, *FB*, and *GA* of the crane truss. The pulley at *D* is frictionless. Indicate whether the members are in tension or compression.

Prob. 6–24

6-25. The crane truss supports a load of 1,000 lb. A 2,500-lb counterweight, having a center of gravity at *G*, is used to stabilize the truss from tipping. Determine the force in member *CE* for this loading. Indicate whether the members are in tension or compression.

Prob. 6–25

6-26. Compute the force in members *BG* and *HG* of the truss. Indicate whether the members are in tension or compression.

195

15 ft

10 ft

600 lb

800 lb

700 lb

|← 10 ft →|← 10 ft →|← 10 ft →|← 10 ft →|

Prob. 6-26

6-27. Determine the force in members *CD* and *GC* of the truss. Indicate whether the members are in tension or compression.

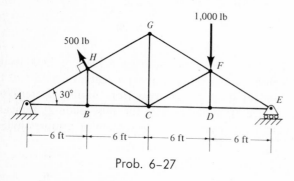

Prob. 6-27

6-28. Determine the force in members *DE* and *LE* of the *Baltimore truss*. Indicate whether the members are in tension or compression.

Prob. 6-28

6-29. Determine the forces in members *LM* and *MK* of the *K truss*. Indicate whether the members are in tension or compression.

Prob. 6-29

6-30. Determine the forces in members *JI* and *DM* of the truss. Indicate whether the members are in tension or compression.

Prob. 6-30

6-31. Determine the forces in members *SG*, *EF*, *ME*, and *SF* of the truss. Indicate whether the members are in tension or compression.

Prob. 6-31

196

*6-5. Space Trusses

A *space truss* consists of members joined together at their ends to form a stable three-dimensional structure. In Sec. 6–2 it was shown that the simplest form of a stable two-dimensional truss consists of the members arranged in the form of a triangle. We then built up the simple plane truss from this basic triangular element by adding two members at a time to form further triangular elements. In a similar manner, the simplest element of a stable space truss is a *tetrahedron,* formed by connecting six members together, as shown in Fig. 6–9. Any additional members added to this basic element would be redundant in supporting the force **P.** A simple space truss can be built from this basic tetrahedron element by adding three additional members, forming multiconnected tetrahedrons.

Fig. 6–9

Because truss members are slender, the external loading is assumed to act only at the joints. Assuming that each joint consists of a ball-and-socket connection, each member of a space truss may be treated as a two-force member. This assumption is justified provided the center lines of the joined members intersect at a common point and the weight of the member can be neglected. If the weight of each member is to be included in the analysis, it is generally satisfactory to apply half the weight of each member as a vertical force acting at the end of each member.

We may use either the method of joints or the method of sections to determine the forces developed in the members of a space truss. In using the method of joints, the three scalar equilibrium equations $\Sigma F_x = 0, \Sigma F_y = 0$, and $\Sigma F_z = 0$ must be satisfied at each joint. Rotational equilibrium is satisfied since the force system is concurrent. The force analysis must start at a joint where not more than three unknown forces exist. In using the method of sections, we can have no more than *six* unknown forces, shown on the free-body diagram of the isolated section, if all the forces at the cut sections are to be determined. If a section of the truss is isolated, the force system must satisfy the six scalar equilibrium equations $\Sigma F_x = 0, \Sigma F_y = 0, \Sigma F_z = 0, \Sigma M_x = 0, \Sigma M_y = 0$, and $\Sigma M_z = 0$. Because of the difficulty often encountered with three-dimensional geometry, it is recommended that you express the equilibrium equations in *vector form* when applying the method of joints or the method of sections in the force analysis of space trusses.

Since it is relatively easy to draw the free-body diagrams and apply the equations of equilibrium at each joint, the method of joints is very consistent in its application. However, as is often the case, one is left with a set of simultaneous equations to solve for the unknown forces.* When the method of sections is applied, however, many of the unknown forces may be computed *directly,* using a single equilibrium equation. The following example illustrates the use of both methods of solution.

*The computer program listed in Appendix A may be helpful for this type of analysis.

Example 6-5

Determine the forces acting in each of the members of the space frame shown in Fig. 6-10a. Indicate whether the members are in tension or compression.

Solution I (Method of Joints)

Since there is one known force and three unknown forces acting at joint A, we may begin the force analysis of this truss by applying the method of joints to joint A.

Joint A, Fig. 6-10b. Expressing each of the forces that acts on the free-body diagram of joint A in vector notation yields

$$\mathbf{P} = \{-4\mathbf{j}\} \text{ kips}$$
$$\mathbf{F}_{AB} = F_{AB}\mathbf{j}$$
$$\mathbf{F}_{AC} = -F_{AC}\mathbf{k}$$
$$\mathbf{F}_{AE} = F_{AE}\left(\frac{\mathbf{r}_{AE}}{r_{AE}}\right) = F_{AE}\frac{(\mathbf{i} + \mathbf{j} - \mathbf{k})}{\sqrt{3}}$$

For equilibrium,

$$\Sigma \mathbf{F} = 0; \qquad \mathbf{P} + \mathbf{F}_{AB} + \mathbf{F}_{AC} + \mathbf{F}_{AE} = 0$$

$$-4\mathbf{j} + F_{AB}\mathbf{j} - F_{AC}\mathbf{k} + \frac{1}{\sqrt{3}}F_{AE}\mathbf{i} + \frac{1}{\sqrt{3}}F_{AE}\mathbf{j} - \frac{1}{\sqrt{3}}F_{AE}\mathbf{k} = 0$$

$$\frac{1}{\sqrt{3}}F_{AE}\mathbf{i} + (-4 + F_{AB} + \frac{1}{\sqrt{3}}F_{AE})\mathbf{j} + (-F_{AC} - \frac{1}{\sqrt{3}}F_{AE})\mathbf{k} = 0$$

Equating the $\mathbf{i}$, $\mathbf{j}$, and $\mathbf{k}$ components to zero gives

$$\Sigma F_x = 0; \qquad \frac{1}{\sqrt{3}}F_{AE} = 0$$

$$\Sigma F_y = 0; \qquad -4 + F_{AB} + \frac{1}{\sqrt{3}}F_{AE} = 0$$

$$\Sigma F_z = 0; \qquad -F_{AC} - \frac{1}{\sqrt{3}}F_{AE} = 0$$

Solving these equations, we have

$$F_{AC} = F_{AE} = 0 \qquad\qquad\qquad\qquad Ans.$$
$$F_{AB} = 4 \text{ kips} \quad \text{(tension)} \qquad\qquad Ans.$$

Knowing the force $\mathbf{F}_{AB}$, we may proceed to joint B.

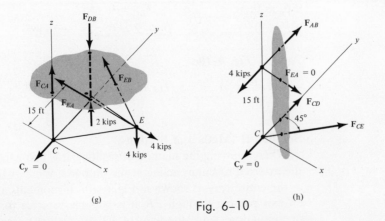

Fig. 6-10

Joint B, Fig. 6-10c.

$$\Sigma F = 0; \qquad 2k + F_{AB} + R_B + F_{BD} + F_{BE} = 0$$

$$2k - 4j - \frac{1}{\sqrt{2}} R_B i + \frac{1}{\sqrt{2}} R_B j + F_{BD} k + \frac{1}{\sqrt{2}} F_{BE} i - \frac{1}{\sqrt{2}} F_{BE} k = 0$$

$$\left(-\frac{1}{\sqrt{2}} R_B + \frac{1}{\sqrt{2}} F_{BE}\right) i + \left(-4 + \frac{1}{\sqrt{2}} R_B\right) j +$$

$$\left(2 + F_{BD} - \frac{1}{\sqrt{2}} F_{BE}\right) k = 0$$

Thus,

$$\Sigma F_x = 0; \qquad -\frac{1}{\sqrt{2}} R_B + \frac{1}{\sqrt{2}} F_{BE} = 0$$

$$\Sigma F_y = 0; \qquad -4 + \frac{1}{\sqrt{2}} R_B = 0$$

$$\Sigma F_z = 0; \qquad 2 + F_{BD} - \frac{1}{\sqrt{2}} F_{BE} = 0$$

Solving these equations yields

$$R_B = F_{BE} = 5.66 \text{ kips} \quad \text{(tension)} \qquad \textit{Ans.}$$
$$F_{BD} = 2 \text{ kips} \quad \text{(compression)} \qquad \textit{Ans.}$$

In a similar manner, applying the equations of equilibrium to the force system acting on the free-body diagrams of joints D and C, respectively, yields the following results.

Joint D, Fig. 6-10d.

$$\Sigma F_x = 0; \qquad F_{DE} = 0 \qquad\qquad\qquad \textit{Ans.}$$
$$\Sigma F_y = 0; \qquad F_{DC} = 0 \qquad\qquad\qquad \textit{Ans.}$$
$$\Sigma F_z = 0; \qquad D_z - 2 = 0; \quad D_z = 2 \text{ kips}$$

Joint C, Fig. 6-10e.

$$\Sigma F_x = 0; \qquad F_{CE} \sin 45° = 0; \quad F_{CE} = 0 \qquad \textit{Ans.}$$
$$\Sigma F_y = 0; \qquad\qquad C_y = 0$$

Solution II (Method of Sections)

When applying the method of sections, it is first necessary to calculate the external reactions acting at the supports. Why? A free-body diagram of the entire truss is shown in Fig. 6-10*f*. If moments are summed about an axis passing through EB, it is possible to solve for C_y directly from one equation. Since the lines of action of all the forces except C_y pass

through points lying on this axis, it is therefore necessary that $C_y = 0$ to insure $\Sigma M_{EB} = 0$. Summing moments about axis DE, we may obtain $\mathbf{R}_B$. Using the theory in Sec. 4-9, this requires

$$\Sigma M_{DE} = 0; \qquad \mathbf{i} \cdot [(15\mathbf{k}) \times (-4\mathbf{j}) + (15\mathbf{k}) \times R_B(-0.707\mathbf{i} + 0.707\mathbf{j})]$$

$$\mathbf{i} \cdot [60\mathbf{i} - 10.61\,R_B\mathbf{j} - 10.61\,R_B\mathbf{i}]$$

Thus

$$R_B = 5.66 \text{ kips}$$

so that

$$\mathbf{R}_B = \{-4\mathbf{i} + 4\mathbf{j}\} \text{ kips}$$

In a similar manner, the force $\mathbf{D}_z$ can be found using $\mathbf{R}_B$ and summing moments about an axis passing through CE. The result is

$$\mathbf{D}_z = \{2\mathbf{k}\} \text{ kips}$$

Finally, using the computed results, the force at E, $\mathbf{F}_E$, can be found by summing all the forces on the free-body diagram.

$$\Sigma \mathbf{F} = 0; \qquad -4\mathbf{j} + 2\mathbf{k} + (-4\mathbf{i} + 4\mathbf{j}) + (2\mathbf{k}) + \mathbf{F}_E = 0$$

$$\mathbf{F}_E = \{4\mathbf{i} - 4\mathbf{k}\} \text{ kips}$$

so that

$$E_x = 4 \text{ kips}, E_y = 0, E_z = -4 \text{ kips}$$

Having obtained these reactions, consider now the isolated section of the truss shown in Fig. 6-10g. Summing moments about axis CE eliminates the three unknown forces $\mathbf{F}_{EA}$, $\mathbf{F}_{EB}$ and $\mathbf{F}_{CA}$ and allows us to solve directly for $\mathbf{F}_{DB}$:

$$\Sigma M_{CE} = 0; \qquad \mathbf{u}_{CE} \cdot [15\mathbf{j} \times (2 - F_{DB})\mathbf{k}] = 0$$

$$\mathbf{u}_{CE} \cdot [15(2 - F_{DB})\mathbf{i}] = 0$$

$$15(2 - F_{DB})(\mathbf{u}_{CE} \cdot \mathbf{i}) = 0$$

Since

$$\mathbf{u}_{CE} \cdot \mathbf{i} \neq 0$$

then

$$15(2 - F_{DB}) = 0$$

Hence,

$$F_{DB} = 2 \text{ kips} \quad \text{(compression)} \qquad\qquad Ans.$$

or

$$\mathbf{F}_{DB} = \{-2\mathbf{k}\} \text{ kips}$$

Summing forces on the free-body diagram, we have

$$\Sigma\mathbf{F} = 0; \qquad 2\mathbf{k} + 4\mathbf{i} - 4\mathbf{k} + \mathbf{F}_{DB} + \mathbf{F}_{EB} + \mathbf{F}_{EA} + \mathbf{F}_{CA} = 0$$

Using the result for $\mathbf{F}_{DB}$,

$$2\mathbf{k} + 4\mathbf{i} - 4\mathbf{k} - 2\mathbf{k} - \frac{1}{\sqrt{2}} F_{EB}\mathbf{i} + \frac{1}{\sqrt{2}} F_{EB}\mathbf{k} - \frac{1}{\sqrt{3}} F_{EA}\mathbf{i} - \frac{1}{\sqrt{3}} F_{EA}\mathbf{j}$$
$$+ \frac{1}{\sqrt{3}} F_{EA}\mathbf{k} + F_{CA}\mathbf{k} = 0$$

or

$$\left(4 - \frac{1}{\sqrt{2}} F_{EB} - \frac{1}{\sqrt{3}} F_{EA}\right)\mathbf{i} + \left(-\frac{1}{\sqrt{3}} F_{EA}\right)\mathbf{j}$$
$$+ \left(2 - 4 - 2 + \frac{1}{\sqrt{2}} F_{EB} + \frac{1}{\sqrt{3}} F_{EA} + F_{CA}\right)\mathbf{k} = 0$$

Hence,

$$\Sigma F_x = 0; \qquad 4 - \frac{1}{\sqrt{2}} F_{EB} - \frac{1}{\sqrt{3}} F_{EA} = 0$$

$$\Sigma F_y = 0; \qquad -\frac{1}{\sqrt{3}} F_{EA} = 0$$

$$\Sigma F_z = 0; \qquad 2 - 4 - 2 + \frac{1}{\sqrt{2}} F_{EB} + \frac{1}{\sqrt{3}} F_{EA} + F_{CA} = 0$$

Solving these equations,

$$F_{EB} = 5.66 \text{ kips} \quad \text{(tension)} \qquad\qquad Ans.$$
$$F_{EA} = F_{CA} = 0 \qquad\qquad Ans.$$

To obtain the forces in the remaining members, consider the isolated section shown in Fig. 6–10h. Using the result that $F_{EA} = 0$ and summing moments about the x axis yields

$$\Sigma M_x = 0; \qquad \mathbf{i} \cdot [(15\mathbf{k}) \times (F_{AB} - 4)\mathbf{j}]$$
$$-15(F_{AB} - 4) = 0$$

Thus,

$$F_{AB} = 4 \text{ kips} \quad \text{(tension)} \qquad\qquad Ans.$$

Forces $\mathbf{F}_{CE}$ and $\mathbf{F}_{CD}$ may be obtained by the force equilibrium equations.

$$\Sigma F_x = 0; \qquad F_{CE} \sin 45° = 0; \quad F_{CE} = 0 \qquad Ans.$$
$$\Sigma F_y = 0; \qquad F_{CD} - 4 + 4 = 0; \quad F_{CD} = 0 \qquad Ans.$$

Problems

6-32. Determine the forces in members *AB*, *CB*, and *DB* of the space truss. Members *BD* and *BC* have the same length. Indicate whether the members are in tension or compression.

Prob. 6–32

6-33. A space truss in the form of a cube is subjected to two loads. Determine the forces acting in members *IC* and *EC*. Indicate whether the members are in tension or compression. Is this space truss stable enough to resist any type of loading?

Prob. 6–33

6-34. Determine the forces developed in all the members of the space truss. Indicate whether the members are in tension or compression.

Prob. 6–34

6-35. Determine the forces developed in all the members of the space truss. Indicate whether the members are in tension or compression.

Prob. 6–35

6-36. The triangular space frame is subjected to both a horizontal and vertical load at its apex *G*. Determine

the force in member *AD*. Indicate whether the member
is in tension or compression.

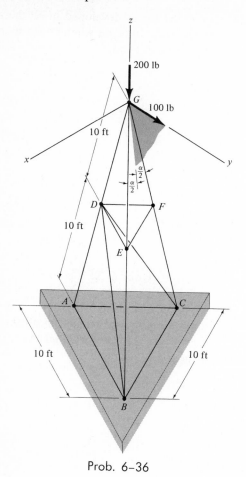

200 lb

100 lb

10 ft

$\frac{\alpha}{2}$

$\frac{\alpha}{2}$

10 ft

10 ft

10 ft

Prob. 6–36

6–6. Multiforce Members, Frames, and Machines

Since two-force members are subjected to loads only at the supporting
joints, the two forces have equal magnitudes, are collinear, and opposite
in direction. However, if a member is subjected to a loading *between* its
supporting joints, the forces acting *at* the joints will, in general, *not* be
directed along a line connecting the joints. These types of members are
often called *multiforce members*. Consider for example the beam shown
in Fig. 6–11*a* with its corresponding free-body diagram shown in Fig.
6–11*b*. The end reactions at *A* and *B* can be found by applying the three
equations of equilibrium, i.e.,

204

$$\zeta + \Sigma M_A = 0; \qquad -10 \sin 60° \text{ lb } (2 \text{ ft}) + 5 \text{ lb } (7 \text{ ft}) - B_y(10 \text{ ft}) = 0,$$
$$B_y = 1.77 \text{ lb}$$
$$\xrightarrow{+} \Sigma F_x = 0; \qquad A_x - 10 \cos 60° \text{ lb} = 0, \qquad A_x = 5.00 \text{ lb}$$
$$+\uparrow \Sigma F_y = 0; \qquad A_y - 10 \sin 60° \text{ lb} + 5 \text{ lb} - 1.77 \text{ lb} = 0,$$
$$A_y = 5.43 \text{ lb}$$

Hence,

$$F_A = \sqrt{(A_x)^2 + (A_y)^2} = \sqrt{(5.00 \text{ lb})^2 + (5.43 \text{ lb})^2} = 7.38 \text{ lb}$$

$$\theta = \tan^{-1} \frac{A_y}{A_x} = \tan^{-1} \frac{5.43 \text{ lb}}{5.00 \text{ lb}} = 47.5° \qquad \angle^{\theta} {}^{F_A}$$

(a)

(b)

(c)

Fig. 6–11

Obviously, the forces $\mathbf{F}_A$ and $\mathbf{B}_y$ acting at the two joints are *not* directed along the axis of the beam. As a consequence, if the beam has elastic properties, the loading will actually *bend* the beam into a shape similar to that shown in Fig. 6–11c.*

Frames and machines are two common types of structures which are often composed of pin-connected, multiforce members. *Frames* are generally stationary and are used to support loads; whereas, *machines* contain moving parts and are designed to transmit and alter the effect of forces. Both types of structures cannot be analyzed using the method of joints or the method of sections since the applied loading generally tends to bend the members as stated above. A partial analysis of the structure may be made, however, by removing the pin connections at the joints, and considering each member as a *separate* rigid body. The forces acting at the joints may then be determined by applying the equations of equilibrium to the free-body diagram of each member. Provided the structure is properly constrained and contains no more supports or members than are necessary to prevent collapse, the force analysis will be statically determinate. Hence, the equations of equilibrium for each member will provide both necessary and sufficient conditions for equilibrium. When the structure is subjected to a coplanar system of forces, *it is possible to write three independent equilibrium equations for the forces shown on the free-body diagram of each member.* By proper application of the equilibrium equations some of the unknowns may be solved for directly from a single equation rather than from a series of simultaneous equations.

A step-by-step procedure which may be useful in analyzing the *joint reactions* for frames and machines composed of multiforce members is as follows:

---

*Using the theory developed in Chapter 7, it will be shown that in order to hold any section of the beam in equilibrium it is necessary that both an internal force and bending moment be developed at the cut section.

1. Draw a free-body diagram of the entire structure, apply the three equations of equilibrium, and solve for as many of the unknown *external reactions* as possible. (It may not be possible to solve for all the external reactions on the free-body diagram since the whole structure is generally composed of many rigid bodies, consisting of struts, beams, pulleys, etc. See Examples 6–6 and 6–11.)

2. Separate the members of the structure and draw a free-body diagram for each member. Forces common to two members act with equal magnitude but opposite direction on the respective free-body diagrams of the members. Recall that all *two-force members* have only one unknown, that is, the magnitude of force which acts along a line joining the end points of the member. The unknown forces acting at the joints of multiforce members should be represented by their rectangular components. In many cases it is possible to tell "by inspection" or experience in which direction these unknown forces act. If this is impossible, a direction is assumed.

3. Apply the three equations of equilibrium to the force system shown on each of the free-body diagrams and solve for the unknown forces. If an unknown force magnitude is found to be a positive quantity, its assumed direction, as shown on the free-body diagram, is correct; if it is negative, the direction of the force is reversed. In particular, if a force is "negative", the negative sign must be *retained* with the force magnitude if the force magnitude is used in the solution of any other equations of equilibrium. The retention of the negative sign is necessary since the equilibrium equations were all written so as to account for the direction of the forces *originally* shown on the free-body diagrams.

The examples which follow should be *thoroughly understood* before proceeding to solve the problems. Specifically, each example illustrates the step-by-step procedure just discussed for obtaining the force which acts at the joints and connections of frames and machines. When these forces are obtained, it is possible to properly design the connections and supports using the theory of strength of materials. In particular, the design of a member follows from a force analysis of the member. This force analysis will be discussed in Chapter 7. It should be noted, however, that a force analysis of the member *first* requires a force analysis of the joints as discussed in this section.

Example 6–6

Determine the tension in the cables and also the force **P** required to support the 600-lb block, utilizing the frictionless pulley system shown in (a) Fig. 6–12*a*, and (b) Fig. 6–13*a*.

Fig. 6–12

(a)   (b)   (c)

Solution

Part (a). *Step 1:* A free-body diagram of the entire pulley system is shown in Fig. 6–12*b*. Applying the translational equilibrium equation in the vertical direction gives

$$+\uparrow\Sigma F_y = 0;\quad R - P - 600\text{ lb} = 0,\qquad R = 600\text{ lb} + P \tag{1}$$

Step 2: A free-body diagram of each pulley is given in Fig. 6–12*c*. Since the cable is *continuous* and the pulleys are frictionless, the cable has a *constant tension* of P lb acting throughout its length. (See Example 5–7.) The link connection between pulleys B and C is a two-force member and it therefore has an unknown tension of T lb acting in it. Notice that the *principle of action and reaction* must be carefully observed for forces **P** and **T** when drawing the *separate* free-body diagrams, as shown in Fig. 6–12*c*.

Step 3: If we apply the equilibrium equation $\Sigma F_y = 0$ to each pulley, we have for pulley A,

$$+\uparrow\Sigma F_y = 0;\qquad 3P - 600\text{ lb} = 0,\qquad P = 200\text{ lb} \qquad \textit{Ans.}$$

for pulley B,

$$+\uparrow\Sigma F_y = 0;\quad T - 2P = T - 400\text{ lb} = 0,\qquad T = 400\text{ lb}$$

and for pulley C,

$$+\uparrow\Sigma F_y = 0;\quad R - 2P - T = R - 2(200\text{ lb}) - 400\text{ lb} = 0,\quad R = 800\text{ lb}$$

This same result for R may also be obtained by using Eq. (1).

Fig. 6–13

(a)

(b)

(c)

Part (b). *Step 1:* A free-body diagram for the entire pulley system is shown in Fig. 6–13*b*. Since the cable passing over pulleys *B* and *C* is continuous, the tension in this cable is *P* lb. The other cables have tensions T_1 lb and T_2 lb acting in them. Applying the equation of equilibrium yields

$$+\uparrow\Sigma F_y = 0; \quad T_1 - 600\ \text{lb} + P + T_2 - P = 0, \quad T_1 + T_2 = 600\ \text{lb} \quad (2)$$

Step 2: A free-body diagram of each pulley is shown in Fig. 6–13*c*. Observe how the principle of action and reaction is applied to forces T_1 and **P** on the separate free-body diagrams.

Step 3: If we apply the equilibrium equation $\Sigma F_y = 0$ to each pulley, we have for pulley *A*,

$$+\uparrow\Sigma F_y = 0; \qquad 2T_1 - 600\ \text{lb} = 0, \qquad T_1 = 300\ \text{lb} \qquad \qquad Ans.$$

for pulley *B*,

$$+\uparrow\Sigma F_y = 0; \quad 2P - T_1 = 2P - 300\ \text{lb} = 0, \qquad P = 150\ \text{lb} \qquad Ans.$$

and for pulley *C*,

$$+\uparrow\Sigma F_y = 0; \quad T_2 - 2P = T_2 - 300\ \text{lb} = 0, \qquad T_2 = 300\ \text{lb} \qquad Ans.$$

The solution can be easily checked by using Eq. (2).

Example 6-7

Determine the forces which the pins at points *B* and *C* exert on member *ABCD* of the loaded frame shown in Fig. 6-14*a*.

208

Fig. 6–14

Solution

Step 1: A free-body diagram of the entire structure is shown in Fig. 6–14b. Applying the three equations of equilibrium yields

$$\zeta + \Sigma M_A = 0; \qquad -500 \text{ lb } (10 \text{ ft}) + D_x(14 \text{ ft}) = 0, \qquad D_x = 357 \text{ lb}$$
$$\overset{+}{\rightarrow}\Sigma F_x = 0; \qquad A_x - 357 \text{ lb} = 0, \qquad A_x = 357 \text{ lb}$$
$$+\uparrow\Sigma F_y = 0; \qquad A_y - 500 \text{ lb} = 0, \qquad A_y = 500 \text{ lb}$$

Since all the unknown forces are numerically positive, their assumed directions, indicated on the free-body diagram in Fig. 6–14b, are correct.

Step 2: A free-body diagram of each member of the frame is shown in Fig. 6–14c. Notice that member *BE* is a two-force member, and hence

209

only the magnitude of force acting at its two end points is *unknown* (see Sec. 6–1). The force acting at pin C has two unknowns represented by the two rectangular components $\mathbf{C}_x$ and $\mathbf{C}_y$.* As shown, the principle of action and reaction of forces must be applied at joints B, C, and E when the separate free-body diagrams are drawn.

Step 3: If we apply the three equations of equilibrium to member CEF, we have

$$\zeta + \Sigma M_C = 0; \quad -500 \text{ lb } (10 \text{ ft}) - (T_B \sin 45°) 8 \text{ ft} = 0,$$

$$T_B = -884 \text{ lb} \quad ^{45°}\!\!\nearrow T_B \quad Ans.$$

$$\xrightarrow{+} \Sigma F_x = 0; \quad -C_x - (-884 \cos 45° \text{ lb}) = 0, \quad C_x = 625 \text{ lb}$$

$$+\!\uparrow\! \Sigma F_y = 0; \quad C_y - (-884 \sin 45° \text{ lb}) - 500 \text{ lb} = 0, \quad C_y = -125 \text{ lb}$$

Since the magnitudes of forces $\mathbf{T}_B$ and $\mathbf{C}_y$ were calculated as negative quantities, they were assumed to be acting with the wrong sense of direction on the free-body diagrams shown in Fig. 6–14c. The correct directional sense of these forces might have been determined "by inspection" *before* applying and solving the equations of equilibrium for member CEF. As shown in Fig. 6–14c, moment equilibrium about point E on member CEF indicates that $\mathbf{C}_y$ must actually act *downward* to counteract the moment created by the 500-lb force about point E. Similarly, summing moments about point C, we can see that the vertical component of force $\mathbf{T}_B$ must actually act *upward*.

A free-body diagram of member $ABCD$, showing the magnitude and *correct sense of direction* of all the computed forces acting on it, is given in Fig. 6–14d. To check the previous calculations we may apply the three equations of equilibrium to this member:

$$\xrightarrow{+} \Sigma F_x = 0; \quad -357 \text{ lb} + 625 \text{ lb} - 884 \cos 45° \text{ lb} + 357 \text{ lb} \equiv 0$$

$$+\!\uparrow\! \Sigma F_y = 0; \quad 125 \text{ lb} - 884 \sin 45° \text{ lb} + 500 \text{ lb} \equiv 0$$

$$\zeta + \Sigma M_B = 0; \quad 357 \text{ lb } (10 \text{ ft}) - 625 \text{ lb } (8 \text{ ft}) + 357 \text{ lb } (4 \text{ ft}) \equiv 0$$

The total magnitude of force which the pin C exerts on member $ABCD$ is therefore

$$F_C = \sqrt{(C_x)^2 + (C_y)^2} = \sqrt{(625 \text{ lb})^2 + (125 \text{ lb})^2} = 637 \text{ lb} \quad Ans.$$

The direction of this force is

$$\theta = \tan^{-1} \frac{C_y}{C_x} = \tan^{-1} \frac{125 \text{ lb}}{625 \text{ lb}} = 11.3° \quad \measuredangle_\theta \mathbf{F}_C \quad Ans.$$

---

*The two rectangular components of an unknown force are considered here instead of the magnitude and direction of the force. It is generally easier to determine these two components using the equations of equilibrium.

Fig. 6–15

Example 6-8

The frame shown in Fig. 6–15a supports a load of 100 lb. If joints A, B, C, and D are all pin-connected, determine the horizontal and vertical components of force acting at each of these joints.

Solution

Step 1: A free-body diagram of the entire frame is shown in Fig. 6–15b. Applying the three equations of equilibrium gives

$$\zeta + \Sigma M_A = 0; \quad -100 \text{ lb } (5 \text{ ft}) - 60 \text{ lb-ft} + F_E(7 \text{ ft}) = 0, \quad F_E = 80 \text{ lb}$$

$$\xrightarrow{+} \Sigma F_x = 0; \quad A_x = 0 \qquad\qquad\qquad\qquad\qquad\qquad\qquad\qquad Ans.$$

$$+\uparrow \Sigma F_y = 0; \quad A_y - 100 \text{ lb} + 80 \text{ lb} = 0, \qquad A_y = 20 \text{ lb} \qquad Ans.$$

Step 2: Using these values for the reactions at A and E a free-body diagram of each member of the frame is shown in Fig. 6–15c. Since all the members are multiforce members, there are two unknown components of force acting at each of the joints B, C, and D. In accordance with the principle of action and reaction, these force components act with equal magnitude and opposite direction on adjacent free-body diagrams, as shown. (Each of the free-body diagrams should be carefully studied.)

The 60-lb-ft couple, being a "free vector," can act at any point on the free-body diagram of member *ABC*. This couple cannot, however, act on the adjacent members *CDE* or *BD*, since these members have been *separated* from member *ABC*, that is, *ABC* is a *separate* rigid body from the other members.

211

Step 3: Applying the three equations of equilibrium, we have for member *ABC*,

$$(\Sigma M_B = 0; \quad -20 \text{ lb (1 ft)} - 60 \text{ lb-ft} + C_x(8 \text{ ft}) + C_y(6 \text{ ft}) = 0 \quad (1)$$
$$\xrightarrow{+}\Sigma F_x = 0; \qquad\qquad\qquad -B_x - C_x = 0 \quad (2)$$
$$+\uparrow\Sigma F_y = 0; \qquad\qquad 20 \text{ lb} + B_y + C_y = 0 \quad (3)$$

and for member *BD*,

$$(+\Sigma M_B = 0; \qquad -100 \text{ lb (4 ft)} + D_y(6 \text{ ft}) = 0 \quad (4)$$
$$\xrightarrow{+}\Sigma F_x = 0; \qquad\qquad B_x - D_x = 0 \quad (5)$$
$$+\uparrow\Sigma F_y = 0; \qquad\qquad D_y - 100 \text{ lb} - B_y = 0 \quad (6)$$

Six *independent equations* have been written which contain six unknowns: B_x, B_y, C_x, C_y, D_x, and D_y. Due to the nature of these equations the solution is rather direct. For example, solve for D_y in Eq. (4) then proceed to obtain the remaining unknowns using Eqs. (6), (3), (1), (2), and (5), in that order. The final results are

$$B_x = 0.0 \qquad\qquad\qquad\qquad\qquad \textit{Ans.}$$
$$B_y = -33.3 \text{ lb} \qquad\qquad\qquad \textit{Ans.}$$
$$C_x = 0.0 \qquad\qquad\qquad\qquad\qquad \textit{Ans.}$$
$$C_y = 13.3 \text{ lb} \qquad\qquad\qquad\quad \textit{Ans.}$$
$$D_x = 0.0 \qquad\qquad\qquad\qquad\qquad \textit{Ans.}$$
$$D_y = 66.7 \text{ lb} \qquad\qquad\qquad\quad \textit{Ans.}$$

Since the magnitude of force $\mathbf{B}_y$ has been computed as a negative quantity, it was assumed acting in the wrong direction on the free-body diagrams in Fig. 6–15c. Hence, its direction is opposite to that shown on these diagrams. The above calculations may be checked in part by applying the equations of equilibrium to member *EDC*.

Example 6–9

Determine the horizontal and vertical force components acting at the pin connections *B* and *C* of the loaded frame shown in Fig. 6–16a.

Solution

Step 1: The free-body diagram of the entire frame is shown in Fig. 6–16b. Applying the three equations of equilibrium yields

$$(+\Sigma M_A = 0; \quad M_A - 100 \text{ lb (4 ft)} = 0, \qquad M_A = 400 \text{ lb-ft}$$
$$\xrightarrow{+}\Sigma F_x = 0; \quad A_x = 0$$
$$+\uparrow\Sigma F_y = 0; \qquad A_y - 100 \text{ lb} = 0, \qquad A_y = 100 \text{ lb}$$

Fig. 6-16

Step 2: Using these values for the reactions, the free-body diagram for each of the members is shown in Fig. 6-16c. Notice that member *CD* is a two-force member. Member *BED* and the pulley are considered to be one system that is, acting together as a single body. The cable, being continuous, is subjected to a tensile force of 100 lb throughout its entire length. Hence, when the cable is *removed*, a force of 100 lb pulls horizontally on member *ABC*, and both horizontally and vertically on the pulley.

Step 3: Applying the three equations of equilibrium to member *BED* yields

$$\zeta + \Sigma M_B = 0; \quad -100 \text{ lb (4 ft)} + 100 \text{ lb (1 ft)} + T_C \sin 45°\text{ (5 ft)} = 0,$$
$$T_C = 84.9 \text{ lb} \qquad\qquad Ans.$$

Using this result for T_C,

$\xrightarrow{+} \Sigma F_x = 0;$ $B_x - 100 \text{ lb} - 84.9 \cos 45° \text{ lb} = 0,$ $B_x = 160 \text{ lb}$ *Ans.*

$+\uparrow \Sigma F_y = 0;$ $84.9 \sin 45° \text{ lb} - 100 \text{ lb} + B_y = 0,$ $B_y = 40 \text{ lb}$ *Ans.*

It is suggested that the above calculations be checked by applying the three equations of equilibrium to member ABC.

An alternative approach may be used to solve for the forces acting on member BED. If the pulley is *removed,* it must be held in equilibrium by 100-lb force components exerted *on* it by the pin at point E, Fig. 6–16*d*. By inspection, both translational and rotational equilibrium of the pulley are satisfied. According to the principle of action and reaction, the force components exerted by the pulley on member BED are equal but opposite to those exerted by the member on the pulley. Hence, the free-body diagram of member BED shown in Fig. 6–16*d* may be used to obtain the reactions $\mathbf{B}_x$, $\mathbf{B}_y$, and $\mathbf{T}_C$. The results will be identical to those obtained previously. Can you show this?

Example 6–10

The frame shown in Fig. 6–17*a* is supported at A by a pin and at E by a smooth inclined plane. Determine the horizontal and vertical components of force that the pin connections exert on the frame. The pulley at C is frictionless.

Solution

Step 1: A free-body diagram of the entire frame is shown in Fig. 6–17*b*. Writing the three equations of equilibrium, we have

$\zeta + \Sigma M_A = 0;$ $-200 \text{ lb} (13 \text{ ft}) + F_E \cos 45°(12 \tan 30° \text{ ft}) = 0,$

$$F_E = 531 \text{ lb}$$

(Note that the component $F_E \sin 45°$ is not included in the moment equation since the line of action of this force passes through point A.) Using the computed value of F_E,

$\xrightarrow{+} \Sigma F_x = 0;$ $531 \cos 45° \text{ lb} - A_x = 0,$ $A_x = 375 \text{ lb}$ *Ans.*

$+\uparrow \Sigma F_y = 0;$ $-A_y + 531 \sin 45° \text{ lb} - 200 \text{ lb} = 0,$ $A_y = 175 \text{ lb}$ *Ans.*

Step 2: A free-body diagram of each member, including the pins at D and C, and the pulley is shown in Fig. 6–17*c*. The force $\mathbf{F}_E$ is shown on the free-body diagram of member EDC as two components having magnitudes of $(531 \text{ lb})(0.707) = 375 \text{ lb}$. In all cases, equal but opposite reactions are shown between two connected members, when the members are separated. You are strongly urged to study each of the free-body diagrams carefully. In particular, note that *three force interactions* occur

Fig. 6-17

on the pin located at C, namely, the force components $\mathbf{C}_x$ and $\mathbf{C}_y$ representing the force exerted by member ABC, the force components $\mathbf{C}'_x$ and $\mathbf{C}'_y$ representing the force exerted by member EDC, and finally the two 200-lb force components exerted by the pulley. Writing the equilibrium equations $\Sigma F_x = 0$ and $\Sigma F_y = 0$ for the pin at C will enable us to obtain two equations which relate the unknown forces magnitudes C_x, C'_x, and C_y, C'_y (Equations (4) and (5) in Step 3). When only *two force interactions* occur at a pin connection (such as pin D in Fig. 6-17c), to satisfy $\Sigma F_x = 0$ and $\Sigma F_y = 0$ the force components acting must be equal in magnitude but opposite in direction. For this reason the free-body diagrams for *pins* which connect *only two members* together were *not considered* in the force analysis of the previous examples. In all cases, the length of the pin is

considered *short* so that the moment created by the force interactions on the pin can be neglected.

Step 3: Applying the three equations of equilibrium, we have for member ABC,

$$\xrightarrow{+} \Sigma F_x = 0; \qquad -375 \text{ lb} + B_x + C_x = 0 \qquad (1)$$

$$+\uparrow \Sigma F_y = 0; \qquad -175 \text{ lb} + B_y - C_y = 0 \qquad (2)$$

$$\zeta + \Sigma M_B = 0; \qquad 175 \text{ lb} (6 \text{ ft}) - C_y(6 \text{ ft}) = 0 \qquad (3)$$

for pin C,

$$\xrightarrow{+} \Sigma F_x = 0; \qquad -C_x + C_x' - 200 \text{ lb} = 0 \qquad (4)$$

$$+\uparrow \Sigma F_y = 0; \qquad C_y + C_y' - 200 \text{ lb} = 0 \qquad (5)$$

and for member EDC,

$$\xrightarrow{+} \Sigma F_x = 0; \qquad 375 \text{ lb} - D_x - C_x' = 0 \qquad (6)$$

$$+\uparrow \Sigma F_y = 0; \qquad 375 \text{ lb} - D_y - C_y' = 0 \qquad (7)$$

$$\zeta + \Sigma M_D = 0; \qquad -375 \text{ lb} (6 \text{ ft}) + 375 \text{ lb} (6 \tan 30° \text{ ft}) - C_y'(6 \text{ ft})$$
$$+ C_x'(6 \tan 30° \text{ ft}) = 0 \quad (8)$$

Eight independent linear equations have been written which may be solved for the eight unknowns: B_x, B_y, C_x, C_y, C_x', C_y', D_x, and D_y.

Due to the nature of these equations, however, each equation may be solved directly for each unknown. This is done by first solving Eq. (3) for C_y then proceeding to obtain the remaining unknowns using Eqs. (2), (5), (8), (4), (6), (7), and (1), in that order. The final results are

$$B_x = 257 \text{ lb} \qquad \qquad \qquad Ans.$$
$$B_y = 350 \text{ lb} \qquad \qquad \qquad Ans.$$
$$C_x = 118 \text{ lb} \qquad \qquad \qquad Ans.$$
$$C_y = 175 \text{ lb} \qquad \qquad \qquad Ans.$$
$$C_x' = 318 \text{ lb} \qquad \qquad \qquad Ans.$$
$$C_y' = 25 \text{ lb} \qquad \qquad \qquad Ans.$$
$$D_x = 57 \text{ lb} \qquad \qquad \qquad Ans.$$
$$D_y = 350 \text{ lb} \qquad \qquad \qquad Ans.$$

It is suggested that the above calculations be checked by applying the equations of equilibrium to member BD.

Example 6–11

The piston and link mechanism shown in Fig. 6–18a is used for purposes of compacting material contained within the cylinder C. If a force

(a)

(b)

(c)

Fig. 6-18

of 80 lb is applied at a right angle to each handle of the mechanism, determine the compaction force exerted by the piston and the horizontal and vertical components of force at pins A and B. Assume that the surface of contact between the piston and the cylinder wall is smooth.

Solution

Step 1: A free-body diagram of the piston and link mechanism is shown in Fig. 6–18*b*. In particular, **N** represents the *resultant force* exerted by the walls of the cylinder on the piston and *d* is the horizontal distance at which this force acts from the pin at *A*. Applying the three equations of equilibrium yields

$$\xrightarrow{+}\Sigma F_x = 0; \qquad A_x - P - 2(80 \sin 15° \text{ lb}) = 0$$

$$P = A_x - 41 \text{ lb} \qquad\qquad (1)$$

$$(+\Sigma M_A = 0; \; -80 \text{ lb } (16 \text{ in.} - 4 \cos 30° \text{ in.}) +$$

$$80 \text{ lb } (16 \text{ in.} - 4 \cos 30° \text{ in.}) - Nd = 0$$

$$N = 0 \quad (\text{since } d \neq 0)$$

$$+\uparrow\Sigma F_y = 0; \quad -80 \cos 15° \text{ lb} + 80 \cos 15° \text{ lb} + A_y + 0 = 0,$$

$$A_y = 0$$

Note that we *cannot* obtain a direct solution for the compaction force **P** using the single free-body diagram, Fig. 6–18*b*. It is therefore necessary to separate the mechanism into its component parts and write equations of equilibrium for each rigid body.

Step 2: A free-body diagram of each member *including* pins *A* and *B* of the mechanism is shown in Fig. 6–18*c*. The pins at *A* and *B* are shown since *three connections* are made at each of these pins. Since both the loading and mechanism are symmetrical, forces having equal magnitudes occur in the corresponding (two-force) members *AE* and *AD*, and in members *GEB* and *HDB*. You should study each of the separate free-body diagrams carefully and note that the principle of action and reaction of forces must be observed when drawing the separate free-body diagrams.

Step 3: Applying the equations of equilibrium to member *GEB* and pin *A*; for member *GEB*,

$$(+\Sigma M_B = 0; \quad -S \sin 30°(4 \text{ in.}) + 80 \text{ lb } (20 \text{ in.}) = 0, \qquad S = 800 \text{ lb}$$

Using the computed value of *S*,

$$\xrightarrow{+}\Sigma F_x = 0; \qquad -80 \sin 15° \text{ lb} + 800 \cos 15° \text{ lb} - B_x = 0,$$

$$B_x = 752 \text{ lb} \qquad\qquad Ans.$$

$$+\uparrow\Sigma F_y = 0; \qquad -80 \cos 15° \text{ lb} + 800 \sin 15° \text{ lb} - B_y = 0,$$

$$B_y = 130 \text{ lb} \qquad\qquad Ans.$$

and for pin *A*, using the computed value of *S*,

$$\xrightarrow{+}\Sigma F_x = 0; \; A_x - 2(800 \cos 15° \text{ lb}) = 0, \qquad A_x = 1,545 \text{ lb} \qquad Ans.$$

$$+\uparrow\Sigma F_y = 0; \; A_y - 800 \sin 15° \text{ lb} + 800 \sin 15° \text{ lb} = 0, \quad A_y = 0 \quad Ans.$$

The magnitude of the compaction force **P** can be obtained from Eq. (1), using the value of A_x.

$$P = 1,545 \text{ lb} - 41 \text{ lb} \qquad P = 1,504 \text{ lb} \qquad \textit{Ans.}$$

This value can be checked by considering the equilibrium of the pin at B, namely,

$$\xrightarrow{+} \Sigma F_x = 0; \qquad\qquad 2B_x - P = 0,$$
$$2(752 \text{ lb}) - 1,504 \text{ lb} \equiv 0$$

Problems

Neglect the weight of the members unless stated otherwise.

6-37. Determine the horizontal and vertical components of force acting at A, B, and C of the frame.

Prob. 6-37

6-38. Determine the reactions at pins A and C for the two-member frame.

Prob. 6-38

6-39. Determine the horizontal and vertical components of the forces developed at pin connections A, B, and C for the two-member frame.

Prob. 6-39

6-40. Determine the horizontal and vertical components of force at pins A and C of the mechanism.

Prob. 6-40

6-41. Determine the angle α which the pulley-supporting link makes with the vertical. Neglect the weight of the pulleys.

Prob. 6-41

6-42. Determine the tension in member BC of the frame and the horizontal and vertical components of force acting on the pin at D.

Prob. 6-42

6-43. If a force of 40 lb is applied to the handles of the pliers, determine the force applied to the bolt B.

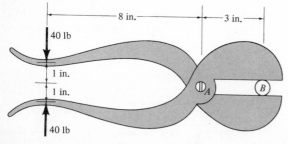

Prob. 6-43

6-44. Determine the horizontal and vertical components of force at pins A and C of the mechanism. The pulley at D is frictionless. The mechanism is pin connected at B.

Prob. 6-44

6-45. The jack shown supports a 800-lb automobile engine. Determine the compression in the hydraulic cylinder C and the reaction at the pin B.

Prob. 6-45

6-46. The 5,600-lb car, having a center of gravity at G, is supported by a hydraulic jack AD. Determine the compression in the cylinder C when the car is in the position shown.

Prob. 6-46

6-47. Determine the force **P** required to maintain equilibrium of the pulley mechanisms.

(a)   (b)   (c)

Prob. 6-47

6-48. Compute the reactions at supports *A*, *B*, *C*, and *D* of the compound beam.

Prob. 6-48

6-49. Determine the clamping force exerted on the pipe at *B* if a force of 50 lb is applied to each of the handles of the wrench.

Prob. 6-49

10.22

6-50. Determine the compressive force developed in member *BC* and the horizontal and vertical components of the force which the pin at *A* exerts on member *AB* of the frame.

Prob. 6-50

6-51. Determine the horizontal and vertical components of force acting at *A*, *B*, and *C* in the two-member arch.

Prob. 6-51

221

6-52. The tongs are used to lift the 100-lb crate. Determine the force developed at pins B and C. Due to friction, the shoes at E and F prevent the crate from slipping.

Prob. 6-52

6-53. The tractor shovel shown carries a 1,000-lb load of gravel, having a center of gravity at G. Compute the forces developed in the hydraulic cylinders BC and IJ due to this loading.

Prob. 6-53

6-54. Two blocks A and B are suspended from a ceiling by means of two cables. If block A weighs 50 lb and block B weighs 30 lb, determine the tension in the cables and the reactive force which block B exerts on block A.

Prob. 6-54

6-55. Determine the reactions at pins A and B of the frame. The large pulley is pinned at C and has a radius of $\frac{1}{2}$ ft. The small pulley at D has a radius of $\frac{1}{4}$ ft.

Prob. 6-55

6-56. The semicircular arch AB weighs 20 lb/ft, and the smooth cylinder D weighs 50 lb. Neglecting the weight of the brace BCD, determine the horizontal and vertical reactions at the pin B and the reaction of the roller C.

222

Prob. 6-56

6-57. The extension ladder rests against a smooth wall at *A* and is held from slipping at *B* by means of a friction shoe. If a 150-lb man climbs up the ladder to the position shown, determine the reactions at the (pin) connection *C* and the smooth contact *D*. Each section of the ladder weighs 25 lb and has a center of gravity at its midlength. Neglect the distance "*s*" in the calculations and assume that the only force which the man exerts on the ladder is his total weight, acting as a vertical reaction at *E*.

Prob. 6-57

6-58. The mechanism is subjected to a horizontal force of 80 lb. Determine the magnitude of the vertical force **P** which must be applied to the smooth block at *B*, required to maintain equilibrium.

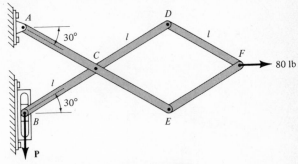

Prob. 6-58

6-59. If a vertical force of *P* = 60 lb is applied to the handle of the punch, determine the force exerted on the plate at *A*.

Prob. 6-59

6-60. Determine the horizontal and vertical components of the forces acting at pins *A*, *B*, and *C* of the frame.

Prob. 6-60

223

6-61. The drafting table is supported at B by a pin and a spring at A. If the table weighs 20 lb and has a center of gravity at G, determine the stiffness k of the spring so that the table remains in equilibrium for $\theta = 30°$. The spring is unstretched when $\theta = 90°$.

Prob. 6-61

6-62. Determine the horizontal and vertical components of force developed at pins A, B, and D of the "A" frame.

Prob. 6-62

6-63. The hoisting frame is supported at A by a pin and at C by a short link. The cable is attached to a drum at B and passes over frictionless pulleys at D and E. Determine the magnitude of force which the pins exert on members CD and AD and the force $\mathbf{P}$ needed for equilibrium.

Prob. 6-63

6-64. Determine the horizontal and vertical components of reaction at the pin supports A and D of the compound beam.

Prob. 6-64

224

6-65. Determine the tension in chain CD and the force which the pin at B exerts on member CB of the frame.

6-67. Determine the horizontal force **P** which must be applied to the lever of the mechanism in order to hold the 250-lb gate open. There is a smooth roller at F.

Prob. 6-65

Prob. 6-67

6-66. Determine the horizontal and vertical components of the forces acting on the pins A and C of the frame. The pulley at D is frictionless. Neglect the weight of the motor.

6-68. The mechanism for a tractor is shown in the figure. Determine the force created in the hydraulic cylinders AD and EF in order to hold the shovel in equilibrium. The load weighs 1,500 lb and has its center of gravity at G. All joints are pin connected.

Prob. 6-66

Prob. 6-68

225

Internal Forces

*7–1. Internal Forces Developed in Structural Members

In Secs. 6–3 and 6–6, we discussed methods for computing the equilibrium forces acting at supports and joint connections of rigid structural members. Given these forces, we may use the theory developed in strength of materials to design all the supports and connections required to hold the structure together. In a similar manner, before we can properly design a structural *member,* it is first necessary to know the *internal forces* acting within the member, when the member is subjected to a given system of loads.

The internal forces acting at any cross section along a member may be determined by using the *method of sections.** This method is generally applied after all the external forces acting on the member are computed. For example, consider the simply-supported beam loaded as shown in Fig. 7–1a. We can solve for the support reactions $\mathbf{A}_x$, $\mathbf{A}_y$, and $\mathbf{B}_y$ by applying the three equations of equilibrium to the force system on the free-body diagram shown in Fig. 7–1b. To determine the internal forces acting at location C along the axis of the beam (or any other location), the beam is sectioned perpendicular to its axis into two segments at that location, and the free-body diagrams shown in Fig. 7–1c are obtained. Since the two segments were prevented from both translating and rotating with respect to one another *before* the beam was sectioned, equilibrium is maintained when a *resultant* internal force, having rectangular components $\mathbf{A}_C$ and $\mathbf{V}_C$ and a *resultant* internal bending moment $\mathbf{M}_C$, are developed at the cut section. In structural mechanics, the force components $\mathbf{A}_C$ (acting normal to the section) and $\mathbf{V}_C$ (acting tangent to the section) are termed the *axial force* and *shear force,* respectively. In accor-

---

*We discussed the method of sections briefly in Sec. 6–4, where it was used in connection with obtaining the internal force in truss (two-force) members.

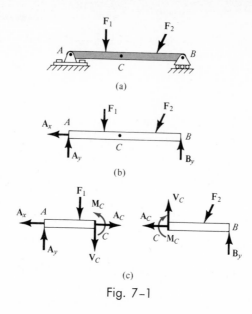

(a)

(b)

(c)

Fig. 7-1

dance with Newton's third law of motion, these internal forces and bending moment acting on an adjacent segment must be equal in magnitude and opposite in direction, as shown in Fig. 7-1c. The magnitudes of the internal forces and bending moment may be obtained by applying the three equations of equilibrium to *either* segment *AC* or *CB*.

The following examples further illustrate the application of the method of sections for obtaining the axial force, shear force, and bending moment at arbitrary points lying along the axis of a structural member.

Fig. 7-2

Example 7-1

Determine the axial force, shear force, and bending moment acting at points B and C of the cantilevered beam loaded as shown in Fig. 7-2a. The 50-lb axial load is applied along the central axis of the beam.

Solution

A free-body diagram of the entire beam is shown in Fig. 7-2b. Since the wall prevents both rotation and translation of the beam, an unknown moment and two unknown force components must act at this support. The distributed load has been replaced by its resultant force in accordance with the principles outlined in Sec. 4-11. Applying the three equations of equilibrium yields

$$\xrightarrow{+}\Sigma F_x = 0; \qquad -A_x + 50\text{ lb} = 0, \qquad A_x = 50\text{ lb}$$

$$+\uparrow\Sigma F_y = 0; \quad A_y - 300\text{ lb} - 100\text{ lb} = 0, \qquad A_y = 400\text{ lb}$$

$$\zeta+\Sigma M_A = 0; \quad -300\text{ lb (5 ft)} - 100\text{ lb-ft} - 100\text{ lb (20 ft)} + M_A = 0,$$

$$M_A = 3,600\text{ lb-ft}$$

These forces and end moment keep the *entire beam* in equilibrium.

Passing an imaginary cutting section, perpendicular to the axis of the beam, through point C, we obtain the free-body diagrams of the beam segments AC and CD shown in Fig. 7-2c. The magnitudes of the internal forces A_C and V_C and the internal moment M_C are obtained by applying the equations of equilibrium to either segment AC or CD. For segment CD we have

$$\xrightarrow{+}\Sigma F_x = 0; \qquad 50\text{ lb} - A_C = 0, \qquad A_C = 50\text{ lb} \qquad \textit{Ans.}$$

$$+\uparrow\Sigma F_y = 0; \qquad V_C - 100\text{ lb} = 0, \qquad V_C = 100\text{ lb} \qquad \textit{Ans.}$$

$$\zeta+\Sigma M_C = 0; \quad -M_C - 100\text{ lb (5 ft)} = 0, \qquad M_C = -500\text{ lb-ft} \qquad \textit{Ans.}$$

or for segment AC we have

$$\xrightarrow{+}\Sigma F_x = 0; \qquad A_C - 50\text{ lb} = 0, \qquad A_C = 50\text{ lb} \qquad \textit{Ans.}$$

$$+\uparrow\Sigma F_y = 0; \quad 400\text{ lb} - 300\text{ lb} - V_C = 0, \qquad V_C = 100\text{ lb} \qquad \textit{Ans.}$$

$$\zeta+\Sigma M_C = 0;$$

$$M_C - 100\text{ lb-ft} + 300\text{ lb (10 ft)} - 400\text{ lb (15 ft)} + 3,600\text{ lb-ft} = 0,$$

$$M_C = -500\text{ lb-ft} \qquad \textit{Ans.}$$

As shown, segment CD provides the *simplest* solution for the internal forces since there are *fewer loads* on the free-body diagram of this segment. Also, the results were obtained without having to first obtain the external reactions at A. Because the magnitude of $\mathbf{M}_C$ has been computed as a *negative* quantity, the sense of direction of $\mathbf{M}_C$ is *opposite* to that assumed on the free-body diagrams in Fig. 7-1c. Therefore, the correct direction of $\mathbf{M}_C$ is clockwise on segment AC and counterclockwise on segment CD.

Before the beam is sectioned, it is important that each load act at its exact point of application on the beam; otherwise erroneous free-body diagrams will result. For example, using the principle of transmissibility discussed in Sec. 4–3, the 50-lb horizontal load at D may be moved to point A *before* calculating the external reactions. However, if this is done *prior* to sectioning the beam at C, segment CD will *not* be subjected to the 50-lb load as it should be.

Passing a section through point B, we obtain the free-body diagrams of the two beam segments AB and BD shown in Fig. 7–2d. *After sectioning the beam,* the portion of the distributed load acting on each segment is replaced by its resultant force. Do you see why we must section the beam *first* before simplifying the distributed load? Applying the three equations of equilibrium to segment AB yields

$$\overset{+}{\rightarrow}\Sigma F_x = 0; \qquad A_B - 50 \text{ lb} = 0, \qquad A_B = 50 \text{ lb} \qquad\qquad Ans.$$

$$+\uparrow\Sigma F_y = 0; \quad 400 \text{ lb} - 150 \text{ lb} - V_B = 0, \qquad V_B = 250 \text{ lb} \qquad Ans.$$

$$\zeta+\Sigma M_B = 0; \quad M_B + 150 \text{ lb} (2.5 \text{ ft}) - 400 \text{ lb} (5 \text{ ft}) + 3{,}600 \text{ lb-ft} = 0,$$

$$M_B = -1{,}975 \text{ lb-ft} \qquad\qquad Ans.$$

Since the magnitude of $\mathbf{M}_C$ is a negative quantity, the sense of direction of $\mathbf{M}_C$ shown on the free-body diagrams in Fig. 7–2d must be reversed.

Example 7–2

Determine the axial force, shear force, and bending moment acting at points E, H, and D of the frame shown in Fig. 7–3a. Neglect the weights of the members.

Solution

The free-body diagram of the entire frame is shown in Fig. 7–3b. Applying the three equations of equilibrium yields

$$+\uparrow\Sigma F_y = 0; \qquad A_y - 200 \text{ lb} = 0, \qquad A_y = 200 \text{ lb}$$

$$\zeta+\Sigma M_A = 0; \qquad -200 \text{ lb} (4 \text{ ft}) + B_x(8 \text{ ft}) = 0, \qquad B_x = 100 \text{ lb}$$

$$\overset{+}{\rightarrow}\Sigma F_x = 0; \qquad 100 \text{ lb} - A_x = 0, \qquad A_x = 100 \text{ lb}$$

A free-body diagram of each member of the frame, including the pins at A and C, is shown in Fig. 7–3c. Since the weights of the members are neglected, GC and CA are both two-force members, and thus the forces acting at their respective endpoints must be equal, opposite, and collinear. (Are you able to understand why these end forces must be equal, opposite, and collinear? Refer to Sec. 6–1.) The 200- and 100-lb forces shown on the free-body diagram of the pin at A represent the effect of the *pin support* at A acting *on* the pin. (These are the same forces shown on the free-body diagram in Fig. 7–3b. There they represent the effect of the pin support on the entire frame.)

(a)

(b)

(c)

(d)

(e)

Fig. 7-3

(f)

Applying the equations of equilibrium to the pin at C gives

$$+\uparrow\Sigma F_y = 0; \quad R \sin 45° - 200 \text{ lb} = 0, \quad R = 283 \text{ lb}$$

$$\stackrel{+}{\rightarrow}\Sigma F_x = 0; \quad P - 283 \cos 45° \text{ lb} = 0, \quad P = 200 \text{ lb}$$

Knowing the magnitude of **P**, we may calculate A'_x and A'_y by applying the equations of equilibrium to member AGB:

$$+\uparrow\Sigma F_y = 0; \qquad\qquad A'_y = 0$$

$$\stackrel{+}{\rightarrow}\Sigma F_x = 0; \quad 100 \text{ lb} - 200 \text{ lb} + A'_x = 0, \quad A'_x = 100 \text{ lb}$$

$$\stackrel{\curvearrowleft}{+}\Sigma M_G = 0; \quad 100 \text{ lb} (4 \text{ ft}) - 100 \text{ lb} (4 \text{ ft}) \equiv 0 \quad \text{(check)}$$

231

The calculations may be *checked* by applying the magnitudes of the computed forces on the pin at A and showing that these forces satisfy the equations of equilibrium for the pin. (Note that a separate free-body diagram of the pins has been included here since there are *three* connections at each pin. For example, at pin A the forces are caused by members AC, AGB and the support.)

Member GC is sectioned at E and the free-body diagrams of the two segments are shown in Fig. 7–3d. Applying the three equations of equilibrium to either segment reveals that

$$\xrightarrow{+}\Sigma F_x = 0; \qquad\qquad A_E = 200 \text{ lb} \qquad\qquad Ans.$$
$$+\uparrow\Sigma F_y = 0; \qquad\qquad V_E = 0 \qquad\qquad Ans.$$
$$\zeta+\Sigma M_E = 0; \qquad\qquad M_E = 0 \qquad\qquad Ans.$$

The free-body diagrams shown in Fig. 7–3e are obtained when member AC is sectioned at H. Applying the three equations of equilibrium to section CH yields

$$\xrightarrow{+}\Sigma F_x = 0; \quad 283\cos 45°\text{lb} - V_H = 0, \quad V_H = 200 \text{ lb} \qquad Ans.$$
$$+\uparrow\Sigma F_y = 0; \quad -283\sin 45°\text{lb} + A_H = 0, \quad A_H = 200 \text{ lb} \qquad Ans.$$
$$\zeta+\Sigma M_H = 0; \quad 283\cos 45°\text{lb (2 ft)} - M_H, \quad M_H = 400 \text{ lb-ft} \qquad Ans.$$

The free-body diagrams of sections AD and DGB are shown in Fig. 7–3f. Applying the equations of equilibrium to segment AD gives

$$+\uparrow\Sigma F_y = 0; \qquad\qquad A_D = 0 \qquad\qquad Ans.$$
$$\xrightarrow{+}\Sigma F_x = 0; \quad 100\text{ lb} - V_D = 0, \quad V_D = 100 \text{ lb} \qquad Ans.$$
$$\zeta+\Sigma M_D = 0; \quad -100\text{ lb (2 ft)} + M_D = 0, \quad M_D = 200 \text{ lb-ft} \qquad Ans.$$

Fig. 7–4   (a)   (b)

Example 7-3

Determine the components of axial force, shear force, and bending moment at section H of the bent pipe shown in Fig. 7-4a.

Solution

The external reactions at A, B, and E have been solved for in Example 5-12. If the pipe is cut at section H and the segment HDE is isolated, the resulting free-body diagram is shown in Fig. 7-4b. Since the loading is three-dimensional, there are three unknown force components, and three unknown moment components acting at H. Using Cartesian vector notation, the vectors on this free-body diagram may be represented as

$$\mathbf{F}_H = F_{Hx}\mathbf{i} - F_{Hy}\mathbf{j} + F_{Hz}\mathbf{k}$$
$$\mathbf{M}_H = M_{Hx}\mathbf{i} - M_{Hy}\mathbf{j} + M_{Hz}\mathbf{k}$$
$$\mathbf{F}_E = \{-10.84\mathbf{i} + 1.79\mathbf{j} + 2.09\mathbf{k}\} \text{ lb}$$

Applying the two vector equations of equilibrium yields

$$\Sigma\mathbf{F} = 0; \quad \mathbf{F}_H + \mathbf{F}_E = 0$$
$$F_{Hx}\mathbf{i} - F_{Hy}\mathbf{j} + F_{Hz}\mathbf{k} - 10.84\mathbf{i} + 1.79\mathbf{j} + 2.09\mathbf{k} = 0 \qquad (1)$$

$$\Sigma\mathbf{M}_H = 0; \qquad \mathbf{M}_H + (\mathbf{r}_{HE} \times \mathbf{F}_E) = 0$$
$$M_{Hx}\mathbf{i} - M_{Hy}\mathbf{j} + M_{Hz}\mathbf{k} + (4\mathbf{j} + 14\mathbf{k}) \times (-10.84\mathbf{i} + 1.79\mathbf{j} + 2.09\mathbf{k}) = 0$$

or

$$M_{Hx}\mathbf{i} - M_{Hy}\mathbf{j} + M_{Hz}\mathbf{k} - 16.7\mathbf{i} - 151.8\mathbf{j} + 43.4\mathbf{k} = 0 \qquad (2)$$

Thus, equating the $\mathbf{i}$, $\mathbf{j}$, and $\mathbf{k}$ components of Eqs. (1) and (2) equal to zero, gives,

$\Sigma F_x = 0;$	$F_{Hx} - 10.84 = 0,$	$F_{Hx} = 10.84 \text{ lb}$	*Ans.*
$\Sigma F_y = 0;$	$-F_{Hy} + 1.79 = 0,$	$F_{Hy} = 1.79 \text{ lb}$	*Ans.*
$\Sigma F_z = 0;$	$F_{Hz} + 2.09 = 0,$	$F_{Hz} = -2.09 \text{ lb}$	*Ans.*
$\Sigma M_x = 0;$	$M_{Hx} - 16.7 = 0,$	$M_{Hx} = 16.7 \text{ lb-in.}$	*Ans.*
$\Sigma M_y = 0;$	$M_{Hy} - 151.8 = 0,$	$M_{Hy} = 151.8 \text{ lb-in.}$	*Ans.*
$\Sigma M_z = 0;$	$M_{Hz} + 43.4 = 0,$	$M_{Hz} = -43.4 \text{ lb-in.}$	*Ans.*

Are you able to recognize the significance of the negative signs of F_{Hz} and M_{Hz}?

Problems

7-1. Determine the internal axial force, shear force, and moment at point D for the frame in Prob. 6–37.

7-2. Determine the internal axial force, shear force, and moment developed at point D of the two-member frame in Prob. 6–39.

7-3. Determine the internal axial force, shear force, and moment developed at point E of the frame in Prob. 6–44.

7-4. Determine the internal axial force, shear force, and moment at points F and E of the compound beam.

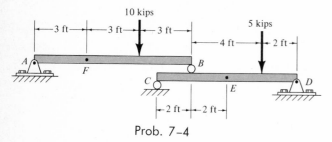

Prob. 7–4

7-5. Determine the internal axial force, shear force, and moment at points C, D, and E of the frame. Specifically, points D and E are located just to the left and just to the right of the concentrated force of 10 kips.

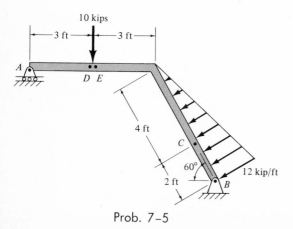

Prob. 7–5

7-6. Determine the internal axial force, shear force, and moment at points D and E of the beam.

234

Prob. 7–6

7-7. Determine the internal axial force, shear force, and moment at points D and E of the frame.

Prob. 7–7

7-8. Determine the internal axial force, shear force, and moment of points D and E of the frame.

Prob. 7–8

7-9. Determine the internal axial force, shear force, and moment at points C and D of the frame.

Prob. 7-9

7-10. Determine the components of internal axial force, shear force, and moment at point B of the cantilever rod.

Prob. 7-10

7-11. Determine the internal force and moment developed at point D in the frame.

Prob. 7-11

7-12. Determine the internal axial force, shear force, and moment developed at point E of the compound beam in Prob. 6-64.

7-13. Determine the internal axial force, shear force, and moment at points D and E of the frame.

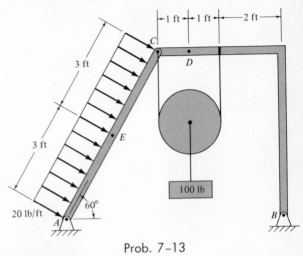

Prob. 7-13

7-14. The bent rod is subjected to two concentrated forces acting at its end. Determine the components of internal axial force, shear force, and moment acting at point B.

Prob. 7-14

235

7-15. Determine the internal axial force, shear force, and moment at the points E and F of the frame.

7-16. Determine the components of internal axial force, shear force, and moment at point A of the pipe. The pipe weighs 8 lb/ft.

Prob. 7-15

Prob. 7-16

*7-2. Shear and Bending-Moment Diagrams for a Beam

Beams are structural members which are designed to support transverse loadings, that is, loadings applied perpendicular to the axis of the member. In general, beams are long and straight bars, having a constant cross section. The design of such members requires a detailed knowledge of the internal shear forces and internal bending moment acting at *all points* along the longitudinal axis of the beam. When this force and bending moment analysis is complete, one can then use the theory of strength of materials to design the proper cross-sectional area of the beam.

In Example 7-1 we discussed a method for finding the internal axial force A, shear force V, and bending moment M at *specific points* along a beam axis. In that example it was shown that these internal forces and moment *vary* from one point to another, which is generally the case. In this section, we will discuss a general method for obtaining a variation of the internal shear force and bending moment in a beam as a function at any point, x along the beam axis. Graphs which show a variation of V and M as a function of x are termed the *shear diagram* and *bending-moment diagram*, respectively. We will not consider the internal axial force in the discussion for two reasons. In most cases, the loads applied to a beam act perpendicular to the axis of the beam and hence produce only internal shear and bending moment. For design purposes, the beam's resistance to shear, and particularly to bending, is more crucial than its ability to resist axial force.

Before presenting a method for determining the shear and bending moment as functions of x and later plotting these functions (shear and bending-moment diagrams), we must first establish a *sign convention* so as to define "positive" and "negative" shear force and bending moment acting in the beam. [This is analogous to assigning coordinate directions x positive to the right and y positive upward, for plotting a function $y = f(x)$.] The sign convention to be adopted here is illustrated in Fig. 7–5a. On the *right-hand face* (R.H.F.) of a beam segment, the internal shear force V acts upward and the internal moment M acts clockwise. In accordance with Newton's third law, an equal and opposite shear force and bending moment must act on the *left-hand face* (L.H.F.) of a segment. An easy way to remember this sign convention is to isolate a small beam segment and note that *positive shear tends to rotate the segment clockwise* (Fig. 7–5b) and a *positive bending moment tends to bend the segment, if it were elastic, concave upward* (Fig. 7–5c).

In determining the shear force and bending-moment functions, it is *first* necessary to compute all the external forces acting on the beam. The beam is then sectioned perpendicular to the axis of the beam at an arbitrary distance x measured from the *left end*. The unknown internal shear force and bending moment are indicated on the free-body diagram of the beam segment at the cut section. These internal loads are *always assumed to act in the positive direction* in accordance with the sign convention given in Fig. 7–5. Applying the two equations of equilibrium $\Sigma F_y = 0$ and $\Sigma M = 0$ to either of the two segmented portions of the beam, one obtains V and M, respectively, as functions of x. In general, the internal shear and bending moment functions will either be discontinuous or their slope will be discontinuous at points where a distributed load changes, and at points where concentrated loads or couples are applied. Because of this, the shear and bending-moment functions must be determined for *each segment* of the beam located between any two discontinuities of loading. As noted earlier, a graphical representation of the shear function (V versus x) is called the *shear diagram*. The *moment diagram* is a plot of M versus x. In plotting these functions, if computed values are positive, the magnitudes are plotted above the x axis, whereas negative values are plotted below the axis.

(a)

(b)

(c)

Fig. 7–5

Example 7–4

Draw the shear and bending-moment diagrams for the beam shown in Fig. 7–6a.

Solution

Using the equations of equilibrium, we can easily show that the support reactions are $A_y = C_y = 25$ kips, acting upward, and $A_x = 0$, Fig. 7–6d. If the beam is sectioned an arbitrary distance x from point A, extending within the region AB, the free-body diagram of the left-hand segment

237

Fig. 7–6

is shown in Fig. 7–6*b*. The unknowns **V** and **M** are indicated acting in the *positive direction* on the right-hand face of the element according to the established sign convention. Why? Summing forces in the vertical direction, we have

$$+\uparrow \Sigma F_y = 0; \qquad\qquad 25 - V = 0$$

or

$$V = 25 \text{ kips}, \qquad 0 \le x < 10 \text{ ft} \qquad (1)$$

Summing moments about point S yields

$$\zeta + \Sigma M_S = 0; \qquad\qquad M - 25(x) = 0$$

or

$$M = 25x \text{ kip-ft}, \qquad 0 \le x \le 10 \text{ ft} \qquad (2)$$

A free-body diagram for a left-hand segment of the beam extended to a distance x within the region BC is shown in Fig. 7–6*c*. As always, **V** and **M** are shown acting in the positive sense. Summing forces, we have

$$+\uparrow \Sigma F_y = 0; \qquad\qquad 25 - 50 - V = 0$$

or

$$V = -25 \text{ kips}, \qquad 10 \text{ ft} < x \le 20 \text{ ft} \qquad (3)$$

Summing moments about point S, we obtain

$$\zeta + \Sigma M_S = 0; \qquad M + 50(x - 10) - 25(x) = 0$$

or

$$M = (500 - 25x) \text{ kip-ft}, \qquad 10 \text{ ft} \le x \le 20 \text{ ft} \qquad (4)$$

Comparing the results, do you see why it was necessary to separate the beam into the two regions AB and BC?

When the above shear-force and bending-moment Eqs. (1) through (4) are plotted within the regions in which they are valid, we obtain the shear-force and bending-moment diagrams shown in Fig. 7–6*d*. The shear-force diagram indicates that the internal shear force is always 25 kips (positive) along the beam segment AB. Just to the right of point B, the shear force changes sign and remains at a constant value of -25 kips for segment BC. The moment diagram starts at zero, increases linearly to point C ($x = 10$ ft), where $M_{max} = 25$ kips (10 ft) $= 250$ kip-ft, and thereafter decreases back to zero.

It is seen in Fig. 7–6*d* that the graph of the shear and moment diagrams are discontinuous at points of concentrated force, that is, points A, B, and C. For this reason, as stated earlier, it is necessary to express the shear and bending-moment functions separately for regions between concentrated loads. It should be realized, however, that all loading discontinuities

are mathematical, arising from the *idealization* of a *concentrated force* and *couple*. Physically, all loads are applied over a finite area, and if this load variation could be accounted for, all shear and bending-moment diagrams would actually be continuous over the entire length of the beam.

Example 7–5

Draw the shear and bending-moment diagrams for the beam shown in Fig. 7–7a.

Solution

A free-body diagram of the entire beam is shown in Fig. 7–7b. To compute the external reactions, the distributed load has been temporarily reduced to two concentrated forces. Applying the three equations of equilibrium, we have

$$\xrightarrow{+}\Sigma F_x = 0; \qquad\qquad A_x = 0$$

$$\zeta + \Sigma M_A = 0; \quad -6 \text{ kips (4 ft)} - 16 \text{ kips (10 ft)} + C_y(14 \text{ ft}) = 0,$$
$$C_y = 13.14 \text{ kips}$$

$$+\uparrow\Sigma F_y = 0; \quad A_y - 6 \text{ kips} - 16 \text{ kips} + 13.14 \text{ kips} = 0, \quad A_y = 8.86 \text{ kips}$$

A free-body diagram for a segment of the beam extending within region *AB* is shown in Fig. 7–7c. The distributed loading acting on this segment has been replaced by a concentrated force *after* the segment is isolated as a free-body diagram. Since the distributed load now has a length x, the *magnitude* of the *concentrated force* is equal to the *area* under the distributed loading, i.e., $\frac{1}{2}(x)(\frac{1}{3}x) = x^2/6$. This force *acts* through the *centroid* of the distributed loading area, a distance $x/3$ from the right end (Appendix B). Applying the two equations of equilibrium yields

$$+\uparrow\Sigma F_y = 0; \qquad\qquad 8.86 - \frac{x^2}{6} - V = 0$$

$$V = \left(8.86 - \frac{x^2}{6}\right) \text{kips}, \qquad 0 \le x \le 6 \text{ ft} \qquad\qquad (1)$$

$$\zeta + \Sigma M_S = 0; \qquad M + \frac{x^2}{6}\left(\frac{x}{3}\right) - 8.86x = 0$$

$$M = \left(8.86x - \frac{x^3}{18}\right) \text{kip-ft}, \qquad 0 \le x \le 6 \text{ ft} \qquad\qquad (2)$$

Since the distributed loading diagram *changes slope* at point *B*, another section of the beam must be chosen within region *BC*. The free-body diagram for the *left-hand segment* which extends within this region is shown in Fig. 7–7d. Notice how the distributed loading has been simplified on this diagram. Applying the equilibrium equations gives

(a)

(b)

$\frac{x^2}{6}$ kips $\frac{x}{3}$ ft

$(\frac{x}{3})$ kip/ft

S $\downarrow$ $\mathbf{V}$ $)\mathbf{M}$

x

8.86 kips $0 \leqslant x \leqslant 6$ ft

(c)

6 kips $2(x-6)$ kips $(\frac{x-6}{2})$ ft

4 ft

2 kip/ft

A

6 ft S $\downarrow$ $\mathbf{V}$ $)\mathbf{M}$

x

8.86 kips 6 ft $\leqslant x \leqslant 14$ ft

(d)

$\frac{(14-x)}{2}$ ft $2(14-x)$ kips

2 kip/ft

$\mathbf{M}($ $\uparrow$

S C

$\mathbf{V}$

$(14-x)$ ft   13.14 kips

(e)

6 ft

2 kip/ft

A B C

8.86 kips   13.14 kips

V

8.86 kips

$V = (8.86 - \frac{x^2}{6})$ kips

(+)

$V = 0$

x

7.43 ft

(−)

13.14 kips

$(V = -2x + 14.86)$ kips

M $M = (8.86 x - \frac{x^3}{18})$ kip-ft

$M_{max} = 43.2$ kip-ft

$M = (-x^2 + 14.86x - 12)$ kip-ft

(+)

x

(f)

Fig. 7–7 6 ft $\leqslant x \leqslant 14$ ft

240

$$+\uparrow\Sigma F_y = 0; \qquad 8.86 - 6 - 2(x-6) - V = 0$$
$$V = (-2x + 14.86)\text{ kips}, \qquad 6\text{ ft} \le x \le 14\text{ ft} \qquad (3)$$

$$\zeta + \Sigma M_S = 0; \quad M + 2(x-6)\left(\frac{x-6}{2}\right) + 6(x-4) - 8.86x = 0$$

$$M = (-x^2 + 14.86x - 12)\text{ kip-ft}, \qquad 6\text{ ft} \le x \le 14\text{ ft} \qquad (4)$$

A second method for obtaining these *same* shear and bending-moment equations for segment *BC* consists of isolating the *right-hand segment* of the beam. Since the origin of the *x* axis has been established at point *A*, the length of this segment is $(14 - x)$ ft. The free-body diagram is shown in Fig. 7-7e. According to the established sign convention, note that the internal shear and bending moment acting at the cut section are indicated in the positive direction. For equilibrium, we again obtain

$$+\uparrow\Sigma F_y = 0; \qquad V - 2(14-x) + 13.14 = 0$$
$$V = (-2x + 14.86)\text{ kips}, \qquad 6\text{ ft} \le x \le 14\text{ ft} \qquad (3)$$

$$\zeta + \Sigma M_S = 0; \quad M - 2(14-x)\left(\frac{14-x}{2}\right) + 13.14(14-x) = 0$$

$$M = (-x^2 + 14.86x - 12)\text{ kip-ft} \qquad 6\text{ ft} \le x \le 14\text{ ft} \qquad (4)$$

The shear and bending-moment diagrams shown in Fig. 7-7f are obtained by plotting Eqs. (1) through (4) within the regions in which they are valid.

The point of *zero shear* can be found using Eq. (3):

$$V = -2x + 14.86 = 0$$
$$x = 7.43\text{ ft}$$

This value of *x* represents the point on the beam where the *maximum moment* occurs (see Sec. 7-3). Using Eq. (4), we have

$$M_{max} = (-(7.43)^2 + 14.86(7.43) - 12)\text{ kip-ft}$$
$$= 43.2\text{ kip-ft}$$

Problems

7-17. Draw the shear and bending-moment diagrams for the beam.

Prob. 7-17

7-18. Draw the shear and bending-moment diagrams for the beam.

Prob. 7-18

7-19. Draw the shear and bending-moment diagrams for the cantilever beam.

Prob. 7-19

7-20. Draw the shear and bending-moment diagrams for the beam.

Prob. 7-20

7-21. Draw the shear and bending-moment diagrams for the beam.

Prob. 7-21

7-22. Draw the shear and bending-moment diagrams for the cantilever beam.

Prob. 7-22

7-23. Draw the shear and bending-moment diagrams for the beam.

Prob. 7-23

7-24. Draw the shear and bending-moment diagrams for each of the two segments of the compound beam.

Prob. 7-24

7-25. Draw the shear and bending-moment diagrams for the beam.

Prob. 7-25

7-26. Draw the shear and bending-moment diagrams for the axle AB of the train wheels. The load transferred to the axle is 10 kips, distributed as shown.

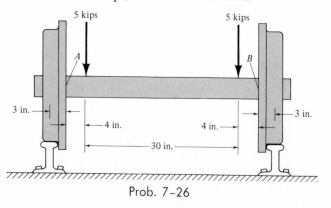

Prob. 7-26

7-27. Draw the shear and bending-moment diagrams for the beam.

Prob. 7–27

7-28. Draw the shear and bending-moment diagrams for the beam.

Prob. 7–28

7-29. Draw the shear and bending-moment diagrams for the beam.

Prob. 7–29

7-30. Draw the shear and bending-moment diagrams for the beam.

Prob. 7–30

7-31. Draw the shear and bending-moment diagrams for the beam *ABCD*. Neglect the weight of the beam and pulley arrangement.

Prob. 7–31

7-32. Draw the shear and bending-moment diagrams for the beam *ABCDE*. All pulleys have a radius of 1 ft. Neglect the weight of the beam and pulley arrangement.

Prob. 7–32

7-33. Determine the internal shear, axial load, and bending moment as a function of θ for the cantilevered semicircular arch. The total weight of the arch is 500 lb.

Prob. 7–33

243

*7-3. Relations Between Distributed Load, Shear, and Bending Moment

In cases where a beam is subjected to a series of several concentrated forces, couples, and distributed loads, the method of constructing the shear and bending-moment diagrams discussed in Sec. 7-2 may become quite tedious. In this section, we will discuss a simpler method for constructing these diagrams—a method based upon differential relations which exist between the load, shear, and bending moment.

Consider the beam AD shown in Fig. 7-8a, which is subjected to an arbitrary distributed loading $w = w(x)$ and a series of concentrated forces and moments. In the following discussion, *the distributed load will be considered positive when the loading acts upward,* as shown. The free-body diagram for a small segment of the beam having a length Δx is shown in Fig. 7-8b. This segment has been chosen at a point, x along the beam, which is *not* subjected to a concentrated force or couple. For this reason, any results obtained will not apply at points of concentrated loading. The internal shear force and bending moment, shown on the free-body diagram, are assumed to act in the *positive direction* according to the established sign convention. The shear force and moment acting on the right-hand face must be increased by a small finite amount in order to keep the segment in equilibrium. The distributed loading has been replaced by a force $w(x) \Delta x$, which acts at a fractional distance $\epsilon(\Delta x)$ from the right end, where $0 < \epsilon < 1$. Applying the equations of equilibrium,

$$+\uparrow \Sigma F_y = 0; \qquad V + w(x)\,\Delta x - (V + \Delta V) = 0$$
$$\Delta V = w(x)\,\Delta x$$

and

$$\zeta + \Sigma M_O = 0; \qquad -V\,\Delta x - M - w(x)\,\Delta x\,\epsilon(\Delta x) + (M + \Delta M) = 0$$

$$\Delta M = V\,\Delta x + w(x)\epsilon(\Delta x)^2$$

Dividing by Δx and taking the limit as $\Delta x \to 0$ these two equations become

$$\frac{dV}{dx} = w(x) \qquad (7\text{-}1)$$

and

$$\frac{dM}{dx} = V \qquad (7\text{-}2)$$

(a)

(b)

Fig. 7-8

The above equations provide a very useful means for plotting the shear-force and bending-moment diagrams of a loaded beam. For a

specific point on the beam, Eq. 7–1 states that the *slope* of the *shear diagram* is equal to the intensity of *loading* at the point, while Eq. 7–2 states that the *slope* of the *moment diagram* is equal to the corresponding intensity of *shear force* at the point. In particular, if the shear is equal to zero, $dM/dx = 0$, and therefore, points of zero shear correspond to points of maximum (or possibly minimum) moment.

Equations 7–1 and 7–2 may be integrated between points B and C of the beam, in which case

$$\Delta V_{BC} = \int_{x_B}^{x_C} w(x) \, dx \qquad (7\text{–}3)$$

and

$$\Delta M_{BC} = \int_{x_B}^{x_C} V \, dx \qquad (7\text{–}4)$$

Equation 7–3 states that the *change in shear* between points B and C is equal to the *area* under the *loading curve* between these points. Similarly, from Eq. 7–4, the *change in moment* between points B and C is equal to the *area* under the *shear diagram* from B to C.

From the given derivation and the discussion in Sec. 7–2, it should be noted that Eqs. 7–1 and 7–3 are invalid at points where a concentrated force acts, since these equations do not account for the sudden change in shear at these points. Likewise, because of a discontinuity of moment, Eqs. 7–2 and 7–4 are invalid at points where a couple is applied.

The following two examples illustrate the method used to apply these equations for the construction of the shear and bending-moment diagrams. After working through these examples, it is recommended that you solve Examples 7–4 and 7–5 using this method.

Example 7–6

Sketch the shear and bending-moment diagrams for the cantilever beam shown in Fig. 7–9a.

Solution

The reactions at A are first calculated by using the equations of equilibrium and are indicated on the free-body diagram shown in Fig. 7–9b. It is convenient to plot both the shear and bending-moment diagrams directly below the free-body diagram of the beam, as shown in the figure.

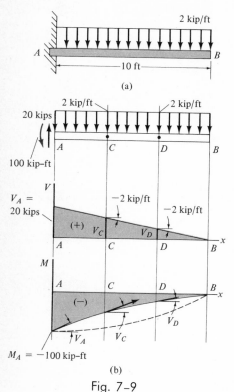

2 kip/ft

A   B

|——————10 ft——————|

(a)

2 kip/ft   2 kip/ft

20 kips

A   C   D   B

100 kip-ft

V

$V_A =$
20 kips

−2 kip/ft

(+) V_C −2 kip/ft
            V_D

A   C   D   B   → x

M

A   C   D   B   → x

(−)
                    V_D

V_A V_C

$M_A = -100$ kip-ft

(b)

Fig. 7–9

Shear Diagram. When constructing the shear diagram, the shear force at points A and B are plotted first, since their value can be taken directly from the free-body diagram of the beam. The shear force at A is $V_A = +20$ kips and at B, $V_B = 0$ (refer to Fig. 7–5). At an intermediate point C, the intensity of load acting on the beam is -2 kips/ft (a negative value since the load acts downward). In accordance with Eq. 7–1, the slope of the shear diagram at this point is $\dfrac{dV}{dx}\Big|_C = -2$ kips/ft. In a similar manner, the slope at point D is also $\dfrac{dV}{dx}\Big|_D = -2$ kips/ft. Since the distributed load is uniform, for other points along the beam we also obtain this same slope. Thus, the shear diagram has a *constant* negative slope of -2 kips/ft which extends from end A to end B.

Moment Diagram. From the free-body diagram, the moment diagram must start at a value of $M_A = -100$ kip-ft and terminate at a value of $M_B = 0$. Do you see why M_A is negative? Refer to Fig. 7–5. Using Eq. 7–2, we find that the slope of the moment diagram at point A, as shown in Fig. 7–9b, is $\dfrac{dM}{dx}\Big|_A = V_A = +20$ kip-ft/ft. The *slope* of the moment diagram will always be positive along the entire axis of the beam since the *shear* is always positive; however, the *slope* will *decrease* as the distance x increases because the *shear decreases*. The resulting moment function is thus a parabola, as shown in Fig. 7–9b. You may ask why this curve is not concave upward as shown by the dashed lines. This cannot be the case since the slope of the moment diagram at A is required to be greater than the slope at C; i.e., $V_A > V_C$ as shown in the shear diagram. Similarly, $V_C > V_D$ and more specifically, the slope at B must be zero.

From these diagrams, it is seen that when the distributed loading is constant, the shear diagram will vary linearly with x and the moment diagram will be proportional to x^2. The actual equations of these curves can be obtained using the methods discussed in Sec. 7–2. In a similar manner, if the distributed loading is triangular, that is, varies linearly with x, the shear diagram will be proportional to x^2 and the bending-moment diagram will be a function of x^3. This increase in the exponential power of x in going from the loading to the shear diagram to the bending-moment diagram is due to the integration expressed by Eqs. 7–3 and 7–4.

Example 7–7

Sketch the shear and bending-moment diagrams for the beam shown in Fig. 7–10a. Give values at the points where the slopes of the shear and bending-moment diagrams change.

(a)

(c)

(b)

(d)

(e)

Fig. 7-10

Solution

The free-body diagram of the beam is shown in Fig. 7–10*b*. For equilibrium, the vertical reactions at *B* and *E* are 3 kips and 1 kip, respectively. Why is it necessary to compute these external reactions?

Shear Diagram. Since the shear force at point *A* is zero, the shear diagram starts at zero. The triangular *load* acts *downward* and *increases* in magnitude from points *A* to *B*; hence, from Eq. 7–1, the *slope* of the *shear diagram* is *negatively increasing* from points *A* to *B*. For example, the slope at points *F* and *H* are $\left.\dfrac{dV}{dx}\right|_F = -w_F$ and $\left.\dfrac{dV}{dx}\right|_H = -w_H$, respectively, where $|w_H| > |w_F|$. Since the distributed loading within region *AB* is triangular (varies with *x*), the shear diagram within this region is parabolic (varies with x^2). The total resultant force of the distributed loading is $\frac{1}{2}$ (3 ft)(2 kips/ft) = 3 kips; hence, the internal shear force just to the *left* of point *B* is −3 kips. This follows from Eq. 7–3; that is, the change in shear ($V_B - 0$) is equal to the area under the loading diagram between points *A* and *B*. A free-body diagram of segment *AB* of the beam extending just to the *right* of point *B*, Fig. 7–10*c*, reveals that the shear force just to the *right* of support *B* must be zero for equilibrium. Since no load acts between points *B* and *C*, $dV/dx = 0$, and therefore, the shear force remains constant within this region, Fig. 7–10*b*. A free-body diagram for a segment of the beam extending just to the right of point *C*, Fig. 7–10*d*, indicates a jump in the shear load to − 1 kip. (Why is it negative?) Within region *CDE* of the beam the shear force again remains constant, since no concentrated force or distributed loads act within this region, Fig. 7–10*b*.

Moment Diagram. The moment acting at point *A* on the free-body diagram of the beam is zero, and therefore the moment diagram begins at zero. From Eq. 7–2, the *slope* of the moment diagram between points *A* and *B* becomes increasingly *negative*, since the *shear* within region *AB* is always *negative*. The curve within this region varies with x^3. Why? As shown on the free-body diagram in Fig. 7–10*c*, the required internal moment acting at point *B* is $M_B = -3$ kip-ft, for equilibrium. Since the shear jumps from a value of −3 kips to zero at point *B*, the slope of the moment diagram makes a corresponding change. The slope within region *BC* remains at a constant value of $\left.\dfrac{dM}{dx}\right|_{BC} = 0$. Why? We can verify the moment at point *C* by isolating segment *ABC* of the beam and determining the internal moment at point *C*, using the equations of equilibrium (see Fig. 7–10*d*). However, another method for determining this internal moment consists of using Eq. 7–4. The change in moment between points *B* and *C* is $\Delta M_{BC} = (4 \text{ ft})(0 \text{ kips}) = 0$ (the area under the

shear diagrams within region *BC*). The moment acting at point *C* is therefore $M_C = M_B + \Delta M_{BC} = -3$ kip-ft $+ 0 = -3$ kip-ft. The jump made in the shear diagram at point *C* causes the moment diagram to change slope from 0 to -1.0 (kip-ft)/ft. This slope like the shear remains constant to point *D*, as shown on the moment diagram. The internal moment acting at the left of point *D* is $M_D = M_C + \Delta M_{CD} = -3$ kip-ft $+ (-1$ kip$)(2$ ft$) = -5$ kip-ft. Using this value, a free-body diagram for a segment of the beam about point *D*, Fig. 7–10*e*, indicates that the internal moment acting to the right of the concentrated 7-kip-ft couple must be 2 kip-ft for equilibrium. Therefore, there is a jump in the moment diagram at this point, Fig. 7–10*b*. The moment diagram has a constant slope of -1 kip-ft/ft within region *DE* and closes to zero at point *E*. The moment must be zero at *E*, since the pin cannot support any moment.

Problems

7–34. Sketch the shear and bending-moment diagrams for Prob. 7–18.

7–35. Sketch the shear and bending-moment diagrams for Prob. 7–20.

7–36. Sketch the shear and bending-moment diagrams for Prob. 7–21.

7–37. Sketch the shear and bending-moment diagrams for Prob. 7–27.

7–38. Sketch the shear and bending-moment diagrams for Prob. 7–31.

7–39. Sketch the shear and bending-moment diagrams for Prob. 7–32.

7–40. Sketch the shear and bending-moment diagrams for the beam.

Prob. 7–40

7–41. Sketch the shear and bending-moment diagrams for the beam.

Prob. 7–41

7–42. Sketch the shear and bending-moment diagrams for the beam.

Prob. 7–42

7–43. Sketch the shear and bending-moment diagrams for the beam.

Prob. 7-43

7-44. Sketch the shear and bending-moment diagrams for the beam.

Prob. 7-44

7-45. Sketch the shear and bending-moment diagrams for the beam.

Prob. 7-45

7-46. Sketch the shear and bending-moment diagrams for the beam.

Prob. 7-46

*7-4. Cables

Flexible cables and chains are often used in engineering structures for support and to transport loads from one member to another. When used for supporting suspension bridges and trolley wheels, cables form the main load-carrying element of the structure. In the force analysis of such systems, the weight of the cable itself may frequently be neglected. However, when cables are used for transmission lines and guys for radio antennas and derricks, the cable weight may become important and must be included in the structural analysis. Regardless of which loading conditions are present, provided this loading is coplanar with the cable, the requirements for equilibrium are formulated in an identical manner. We will consider two cases in the analysis which follows: (1) a cable subjected to an external distributed load, and (2) a cable subjected to its own weight.

In deriving the necessary relations between the tensile force, sag, length, and span of a cable, we will make the assumption that the cable is *perfectly flexible* and *inextensible*. Because of the flexibility assumption, the cable offers no resistance to bending, and therefore, the tensile force acting in the cable is always tangent to the cable at points along the cable. Being inextensible, the cable has a constant length both before and after the load is applied. The derivations which follow are based on knowing the *final configuration* of the cable; that is, the geometry of the cable *after* the loading is applied. With the geometry known, a cable segment is then treated the same as a rigid body.

Case I: Cable Subjected to an External Distributed Load.   Consider the weightless cable shown in Fig. 7–11a, which is subjected to a loading function $w(x)$ pounds per unit length, *as measured in the x direction.* The free-body diagram of a small segment of the cable having a length Δs is shown in Fig. 7–11b. The tensile force in the cable changes continuously in both magnitude and direction along the length of the cable. Consequently, on the free-body diagram of the cable segment this change in cable tension is denoted by ΔT. The resultant of the distributed load is simply $w(x)(\Delta x)$. This force acts at a fractional distance $\epsilon \Delta x$ from point O, where $0 < \epsilon < 1$. Applying the equations of equilibrium to the element yields

$$\xrightarrow{+} \Sigma F_x = 0; \qquad -T\cos\theta + (T + \Delta T)\cos(\theta + \Delta\theta) = 0$$

$$+\uparrow\Sigma F_y = 0; \qquad -T\sin\theta - w(x)(\Delta x) + (T + \Delta T)\sin(\theta + \Delta\theta) = 0$$

$$\zeta+\Sigma M_O = 0; \qquad w(x)(\Delta x)\,\epsilon\,(\Delta x) - T\cos\theta\,\Delta y + T\sin\theta\,\Delta x = 0$$

If we divide each of the above equations by Δx and take the limit as $\Delta x \rightarrow 0$, and hence $\Delta y \rightarrow 0$ and $\Delta T \rightarrow 0$, we obtain

$$\frac{d(T\cos\theta)}{dx} = 0 \qquad\qquad (7\text{--}5)$$

$$\frac{d(T\sin\theta)}{dx} - w(x) = 0 \qquad\qquad (7\text{--}6)$$

and

$$\frac{dy}{dx} = \tan\theta \qquad\qquad (7\text{--}7)$$

Integrating Eq. 7–5, we have

$$T\cos\theta = \text{const} = F_H \qquad\qquad (7\text{--}8)$$

where F_H represents a horizontal component of tensile force at any point along the cable.

Integrating Eq. 7–6 gives

$$T\sin\theta = \int w(x)\,dx \qquad\qquad (7\text{--}9)$$

Dividing Eq. 7–9 by Eq. 7–8 eliminates T. Then, using Eq. 7–7 we obtain

$$\tan\theta = \frac{dy}{dx} = \frac{1}{F_H}\int w(x)\,dx$$

Performing a second integration gives

$$y = \frac{1}{F_H}\int\left(\int w(x)\,dx\right)dx \qquad\qquad (7\text{--}10)$$

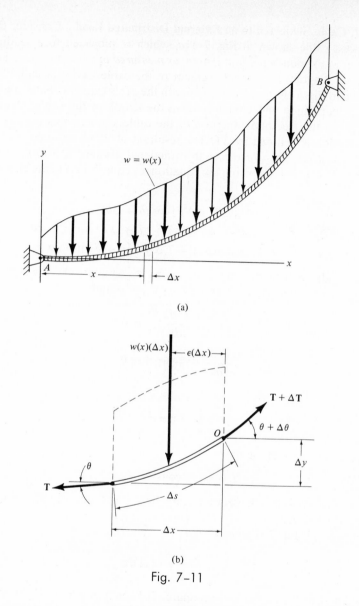

(a)

(b)

Fig. 7–11

This equation is used to determine the deflection curve for the cable. The two constants, say C_1 and C_2, resulting from the integration of Eq. 7–10 are determined by applying the boundary conditions for the cable. See Example 7–8.

Case II: Cable Subjected to its Own Weight. When the weight of the cable becomes important in the force analysis, for example, transmission lines, the loading function along the cable becomes a function of the arc

(a)

(b)

Fig. 7–12

length s, rather than the projected length x. A generalized loading function w = w(s) acting on the cable is shown in Fig. 7–12a. The free-body diagram for a segment of the cable is shown in Fig. 7–12b. Applying the equilibrium equations to the force system on this diagram, we obtain identical relationships as those given by Eqs. 7–5, 7–6, and 7–7, except with ds replacing dx. Therefore, it may be shown that

$$T \cos \theta = F_H$$

$$T \sin \theta = \int w(s) \, ds$$

$$\frac{dy}{dx} = \frac{1}{F_H} \left(\int w(s) \, ds \right) \tag{7-11}$$

To perform a direct integration of Eq. 7–11, we must first substitute dy/dx with ds/dx. Since

$$\frac{dy}{dx} = \sqrt{\left(\frac{ds}{dx} \right)^2 - 1}$$

We get

$$\frac{ds}{dx} = \left\{ 1 + \frac{1}{F_H{}^2} \left[\int w(s) \, ds \right]^2 \right\}^{1/2}$$

Separating the variables and integrating, we obtain

$$x = \int \frac{ds}{\left\{ 1 + \dfrac{1}{F_H{}^2} \left[\int w(s) \, ds \right]^2 \right\}^{1/2}} \tag{7-12}$$

The two constants of integration, say C_1 and C_2, are found using the boundary conditions for the cable.

The following examples illustrate the method for solving cable problems using these derived equations.

Example 7–8

The cable of a suspension bridge is used to support the uniform road surface between the two columns at A and B, as shown in Fig. 7–13a. If the road surface weighs w_o lb/ft, determine the maximum force developed in the cable and the length of the complete cable. The span length L and sag h are known.

Solution

For reasons of symmetry, the origin of coordinates has been placed at the center of the cable. The equation for the deflection curve of the cable can be found by using Eq. 7–10. Noting that $w(x) = w_o$,

$$y = \frac{1}{F_H} \int \left(\int w_o \, dx \right) dx$$

Integrating this equation twice,

$$y = \frac{1}{F_H} \left(\frac{w_o x^2}{2} + C_1 x + C_2 \right) \tag{1}$$

(a)

$$\sqrt{4F_H{}^2 + w_o{}^2L^2}$$

w_oL

θ_{max}

$2F_H$

(b)

Fig. 7–13

The constants of integration may be determined by using the boundary conditions $y = 0$ at $x = 0$, and $dy/dx = 0$ at $x = 0$. Substituting these conditions into Eq. (1) yields $C_1 = C_2 = 0$. The deflection curve then becomes

$$y = \frac{w_o}{2F_H} x^2 \qquad (2)$$

This is the equation of a *parabola*. The constant F_H may be obtained by using the boundary condition $y = h$ at $x = L/2$. Thus,

$$F_H = \frac{w_o L^2}{8h} \qquad (3)$$

Therefore, the deflection curve, Eq. (2), becomes

$$y = 4\frac{hx^2}{L^2} \qquad (4)$$

The maximum tension in the cable may be determined using Eq. 7–8. Since

$$T = \frac{F_H}{\cos \theta}, \quad \text{for} \quad 0 \le \theta \le \pi$$

T_{max} will occur when θ is maximum. From Eq. (2) this occurs when

$$\left. \frac{dy}{dx} \right|_{max} = \tan \theta_{max} = \frac{w_o}{F_H} x \bigg|_{x = L/2}$$

or

$$\theta_{max} = \tan^{-1} \frac{w_o L}{2 F_H}$$

Therefore,

$$T_{max} = \frac{F_H}{\cos \theta_{max}} \qquad (5)$$

Using the triangular relationship shown in Fig. 7–13b, we may write Eq. (5) as

$$T_{max} = \frac{\sqrt{4 F_H^2 + w_o^2 L^2}}{2}$$

Substituting Eq. (3) into the above equation yields

$$T_{max} = \frac{w_o L}{2} \sqrt{1 + \left(\frac{L}{4h} \right)^2} \qquad \textit{Ans.}$$

For a differential segment of cable length,

$$ds = \sqrt{(dx)^2 + (dy)^2} = \sqrt{1 + \left(\frac{dy}{dx} \right)^2}\, dx$$

Since $dy/dx = w_o x/F_H$, the total length of the cable $\mathcal{L}$ can be determined from

$$\mathcal{L} = 2 \int_0^{L/2} \sqrt{1 + \left(\frac{w_o x}{F_H} \right)^2}\, dx \qquad (6)$$

Integrating and substituting the limits yields

$$\mathcal{L} = \frac{L}{2} \sqrt{1 + \left(\frac{w_o L}{2 F_H} \right)^2} + \frac{F_H}{w_o} \sinh^{-1} \left(\frac{w_o L}{2 F_H} \right)$$

Using Eq. (3) and rearranging terms, we have

$$\mathcal{L} = \frac{L}{2} \left[\sqrt{1 + \left(\frac{4h}{L} \right)^2} + \frac{L}{4h} \sinh^{-1} \left(\frac{4h}{L} \right) \right] \qquad \textit{Ans.}$$

Example 7-9

Determine the deflection curve, the length of cable, and the maximum tension in the uniform cable shown in Fig. 7-14. The cable weighs $w_o = 5$ lb/ft.

Fig. 7-14

Solution

For reasons of symmetry, the origin of coordinates is located at the center of the cable. Applying Eq. 7-12 where $w(s) = w_o$ yields

$$x = \int \frac{ds}{\left[1 + \frac{1}{F_H{}^2}\left(\int w_o\, ds\right)^2\right]^{1/2}}$$

Integrating the term under the integral sign in the denominator we have

$$x = \int \frac{ds}{\left[1 + \frac{1}{F_H{}^2}\left(w_o s + C_1\right)^2\right]^{1/2}}$$

Substituting $u = 1/F_H\,(w_o s + C_1)$ so that $du = (w_o/F_H)\, ds$, a second integration yields

$$x = \frac{F_H}{w_o}\left[\sinh^{-1} u + C_2\right]$$

or

$$x = \frac{F_H}{w_o}\left\{\sinh^{-1}\left[\frac{1}{F_H}(w_o s + C_1)\right] + C_2\right\} \qquad (1)$$

From Eq. 7-11,

$$\frac{dy}{dx} = \frac{1}{F_H}\left(\int w_o\, ds\right)$$

or

$$\frac{dy}{dx} = \frac{1}{F_H}(w_o s + C_1)$$

Since $dy/dx = 0$ at $s = 0$, $C_1 = 0$ and

$$\frac{dy}{dx} = \tan\theta = \frac{w_o s}{F_H} \tag{2}$$

where θ is the angle of the cable. The constant C_2 may be evaluated by using the condition $s = 0$ at $x = 0$ in Eq. (1), in which case $C_2 = 0$. Solving for s in Eq. (1) yields

$$s = \frac{F_H}{w_o}\sinh\left(\frac{w_o}{F_H}x\right) \tag{3}$$

Substituting Eq. (3) into Eq. (2), we have

$$\frac{dy}{dx} = \sinh\left(\frac{w_o}{F_H}x\right)$$

Thus,

$$y = \frac{F_H}{w_o}\cosh\left(\frac{w_o}{F_H}x\right) + C_4 \tag{4}$$

If we use the boundary condition $y = 0$ at $x = 0$, the constant $C_4 = -F_H/w_o$, and therefore the deflection curve becomes

$$y = \frac{F_H}{w_o}\left[\cosh\left(\frac{w_o}{F_H}x\right) - 1\right] \qquad Ans.$$

This equation is often referred to as a *catenary curve*. The constant F_H is obtained by using the boundary condition that $y = h$ at $x = L/2$, in which case

$$h = \frac{F_H}{w_o}\left[\cosh\left(\frac{w_o L}{2F_H}\right) - 1\right] \tag{5}$$

Since $w_o = 5$ lb/ft, $h = 6$ ft, and $L = 20$ ft, Eqs. (4) and (5) become

$$y = \frac{F_H}{5\ \text{lb/ft}}\left[\cosh\left(\frac{5\ \text{lb/ft}}{F_H}x\right) - 1\right] \tag{6}$$

$$6\ \text{ft} = \frac{F_H}{5\ \text{lb/ft}}\left[\cosh\left(\frac{50\ \text{lb}}{F_H}\right) - 1\right] \tag{7}$$

Equation (7) can be solved for F_H by using a trial and error procedure. The result is

$$F_H = 45.8\ \text{lb}$$

and therefore Eq. (6) becomes

$$y = 9.16[\cosh(0.109x) - 1]\ \text{ft} \qquad Ans.$$

Using Eq. (3), we see that the half-length of the cable is ($x = 10$ ft)

$$\frac{\mathcal{L}}{2} = \frac{45.8\ \text{lb}}{5\ \text{lb/ft}}\sinh\left[\frac{5\ \text{lb/ft}}{45.8\ \text{lb}}(10\ \text{ft})\right] = 12.1\ \text{ft}$$

Hence,

$$\mathcal{L} = 24.2\ \text{ft} \qquad Ans.$$

Since $T = F_H/\cos\theta$, the maximum tension occurs when θ is maximum, i.e., at $s = \mathcal{L}/2 = 12.1$ ft. Using Eq. (2) yields

$$\frac{dy}{dx}\bigg|_{max} = \tan\theta_{max} = \frac{5 \text{ lb/ft}(12.1 \text{ ft})}{45.8 \text{ lb}} = 1.32$$

$$\theta_{max} = 52.8°$$

Thus,

$$T_{max} = \frac{45.8 \text{ lb}}{\cos 52.8°} = \frac{45.8 \text{ lb}}{0.604} = 75.8 \text{ lb} \qquad Ans.$$

Problems

7-47. The cable will break when the maximum tension reaches 2,000 lb. Determine the minimum length of cable and the sag h if the cable carries a uniform distributed load of 30 lb/ft of horizontal projection.

Prob. 7–49

7-50. The weightless cable is subjected to the triangular loading. If the slope of the cable at point A is zero, determine the deflection curve of the cable and the maximum tension developed in the cable. L, h and w_o are known.

Prob. 7–47

7-48. The platform has a total weight of 5,000 lb and is supported between walls at points A and B by means of two cables. Determine the maximum force developed in the cables.

Prob. 7–48

7-49. The cord is capable of sustaining a maximum tension force of 800 lb before it will snap. Determine the maximum loading w_o lb/ft which the cable can support.

Prob. 7–50

7-51. The cable is subjected to a uniform load of 60 lb/ft. Determine the maximum tension developed in the cable.

Prob. 7–51

7-52. The cable weighs 6 lb/ft and is 150 ft in length. Determine the sag h so that the cable spans 100 ft. Compute the minimum tension in the cable.

Prob. 7–52

7-53. The telephone cable weighs 0.1 lb per foot of length. At a temperature of 100° F, the cable sags 180 in. Determine the maximum tension in the cable and the length of the cable at this temperature.

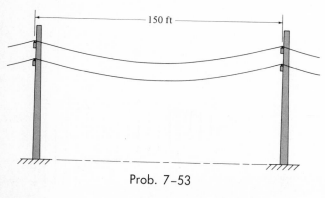

Prob. 7–53

7-54. The cable weighs 5 lb/ft. Determine the maximum and minimum tension in the cable and the length of the cable.

Prob. 7–54

7-55. A 100-lb wire is attached between two points at a distance of 50 ft apart, having equal elevations. If the maximum tension developed in the cable is 75 lb, determine the length of wire and the sag.

7-56. A 100-ft chain has a total weight of 150 lb and is suspended between two points 25 ft apart, at the same elevation. Determine the maximum tension and the sag in the chain.

7-57. A 50-ft cable is suspended between two points a distance 15 ft apart and at the same elevation. If the minimum tension in the cable is 200 lb, determine the total weight of the cable and the maximum tension developed in the cable.

7-58. The yacht is anchored with a 400-ft chain which weighs 8 lb/ft and makes a 60° angle with the horizontal. The tension in the chain at point A is 1,500 lb. Determine the length of the chain, l_d, which is lying at the bottom of the sea, and the depth d. Assume that the buoyancy effects of the water on the chain are negligible. (*Hint:* Establish the coordinate system at B as shown.)

Prob. 7–58

7-59. Show that the deflection curve of a cable discussed in Example 8–9 reduces to Eq. (4) in Example 8–8 when the *hyperbolic cosine function* is expanded in terms of a series, and only the first two terms are retained. (The answer indicates that the *catenary* may be replaced by a *parabola* in the analysis of problems in which the cable tension is very taut, that is, a small sag. In this case the cable weight is assumed uniformly distributed along the horizontal.

7-60. A uniform cord is suspended between two points having the same elevation. Determine the sag-to-span ratio so that the maximum tension in the cable equals the total weight of the cable.

7-61. The buoyant (or vertical) force acting on the weather balloon is 80 lb. If cable OB is 200 ft in length

and the cable weighs 0.25 lb/ft, determine the height h of the balloon.

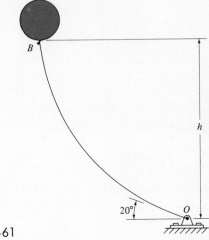

Prob. 7-61

261

Friction

8–1. Characteristics of Dry Friction

In the previous chapters we have considered the surface of contact between one body and another to be perfectly *smooth*. Because of this, the force of interaction between the bodies always acts *normal* to the surface at points of contact. In reality, however, all surfaces are *rough,* and depending upon the nature of the problem, the ability of a body to support a *tangential* as well as a *normal* force at its contacting surface must be taken into consideration. The tangential force developed at a surface of contact is caused by friction. Specifically, *friction* may be defined as a force of resistance acting between two contacting bodies which prevents or inhibits any possible slipping motion between the two bodies. This force always acts *tangent* to the surface at points of contact and is directed so as to oppose the possible or existing motion of the body at that point.

The effects of friction are both advantageous and detrimental. Without friction we would be unable to walk, or make use of any sort of transportation vehicle such as an automobile or boat. Furthermore, many engineering devices, such as belt drives and brakes, require frictional forces in order to function. The adverse effects of friction caused by the rubbing and wearing away of surfaces are reduced when bearings, rollers, and lubricants are used in machines.

In general, there are two types of friction which can occur between surfaces. *Fluid friction* exists when the contacting surfaces are separated by a thin film of fluid such as oil. When relative motion between the bodies occurs, frictional forces are developed between layers of fluid elements. This force depends upon the relative velocity of the fluid layers and viscosity or shearing resistance of the fluid. For this reason problems involving fluid friction are generally studied in fluid mechanics. In this

book we will study only the effects of *dry friction*. This type of friction is often called *Coulomb friction,* since its characteristics were extensively studied by C. A. Coulomb in 1781. Dry friction occurs between the contacting surfaces of rigid bodies in the absence of a lubricating fluid.

The theory of dry friction can best be explained by considering what effect a horizontal force **P** has on moving a block of weight **W** which is resting on a horizontal floor. The free-body diagram of the block is shown in Fig. 8–1*a*. Regardless of the magnitude of the applied force **P**, *vertical equilibrium* is always maintained provided the *resultant* normal force **N** is equal and opposite to **W**.* Consider now the variation of the applied force **P** with the *resultant* frictional force **F** developed at the floor. This variation is shown in Fig. 8–1*b*. Provided the magnitude of **P** is small enough, $\mathbf{F} = -\mathbf{P}$, and the block remains in *equilibrium*. Under these circumstances, the force **F** is called the *static-friction force,* and its value is obtained directly from the *equation of equilibrium* $\Sigma F_x = 0$, Fig. 8–2*a*.

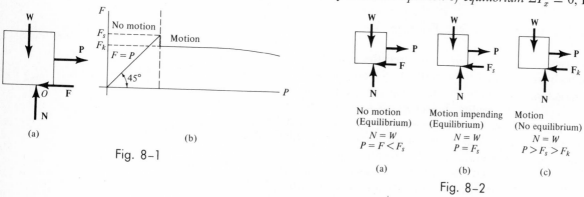

Fig. 8–1

Fig. 8–2

As the magnitude of force **P** is slowly increased, the magnitude of the friction force **F** attains a certain *maximum value* F_s, called the *limiting static-friction force* for which *motion impends,* Fig. 8–1*b*. This point causes a condition of *unstable equilibrium;* that is, a slight increase in *P* will cause sliding of the block, Fig. 8–2*b*. If *P* is increased slightly, *F* drops to a smaller value of F_k, Fig. 8–1*b*. F_k is now a function of velocity and is called the *kinetic-friction force*. As shown on the graph it remains essentially constant for a time, and then as *P* increases, due to aerodynamic effects, F_k decreases somewhat. This situation of loading on the block is shown in Fig. 8–2*c*. In summary then, the frictional force **F** is categorized in three different ways, namely, **F** is a static-frictional force if equilibrium is maintained, **F** is a limiting static-frictional force $\mathbf{F}_s$ when its magnitude reaches its maximum value needed to maintain equilibrium, and finally, **F** is termed a kinetic-frictional force $\mathbf{F}_k$ when sliding at the contacting surface occurs.

*Physically, note that the line of action of **N** must be to the *right* of the line of action of **W** so that moments about point *O* are balanced.

A close examination of the contacting surfaces between the floor and block, Fig. 8–3, reveals the nature of how frictional forces are developed. It can be seen that many microscopic irregularities exist between the two surfaces (those that are not theoretically smooth), and as a result, reactive forces $\Delta \mathbf{R}_n$ are developed at the protuberances. These forces act at points of contact, and must be directed both upward and in a direction opposed to the motion or tendency of motion of the block. In this way, each reactive force contributes both a frictional component of force $\Delta \mathbf{F}_n$ which tends to prevent or retard motion of the block, and a normal component $\Delta \mathbf{N}_n$. The drop made in the frictional force from F_s (static) to F_k (kinetic), Fig. 8–1b, occurs because the applied load $\mathbf{P}$ shears off some of the humps and the block "rides" on top of the humps. The contact forces $\Delta \mathbf{R}_n$ are then aligned slightly more in the vertical direction than before and hence contribute a smaller frictional component $\Delta \mathbf{F}_n$ as when the irregularities are meshed. For very large magnitudes of $\mathbf{P}$, Fig. 8–1b, the magnitude

Fig. 8–3

of **F** drops off ($F < F_k$). Such effects cause an even greater vertical alignment of $\Delta \mathbf{R}_n$ and consequently a smaller magnitude of $\Delta \mathbf{F}_n$.

As a result of *experiments* that pertain to this discussion, we may state the following rules which apply to bodies subjected to dry friction:

1. The frictional force acts *tangent* to the contacting surfaces in a direction *opposed* to the *relative motion*, (or tendency for motion) of one surface against another.

2. The limiting static-frictional force F_s which can be developed is independent of the area of contact, provided the normal pressure is not great enough to alter the contacting surface of the bodies.

3. The limiting static-frictional force F_s is generally greater than the kinetic frictional force F_k for any two surfaces of contact. At *extremely low* relative velocities, however, F_k becomes large, and hence, approximately equal to the limiting static-frictional force.

4. The limiting frictional force F_s is (approximately) proportional to the normal force N at the point of contact. This may be expressed in mathematical form as

$$\mu_s = \frac{F_s}{N} \tag{8-1}$$

where μ_s is called the *coefficient of static friction*.

5. The kinetic-frictional force F_k is (approximately) proportional to the normal force N at the point of contact. This may be expressed in mathematical form as

$$\mu_k = \frac{F_k}{N} \tag{8-2}$$

where μ_k is called the *coefficient of kinetic friction*.

Table 8-1 lists some typical values of the coefficients of static friction found in most engineering handbooks and journals. It should be noted that the last value gives a coefficient of static friction greater than 1. For surfaces consisting of soft metals, this is often the case. Corresponding

Table 8-1

Contact Materials	Coefficient of Static Friction (μ_s)
metal on ice	0.03–0.05
wood on wood	0.30–0.70
leather on wood	0.20–0.50
leather on metal	0.30–0.60
aluminum on aluminum	1.10–1.70

values of the coefficient of kinetic friction would be approximately 25 per cent smaller than those listed in the table. In general, it is seen that a wide range of values is given for each of the coefficients of friction. This is due to the variable conditions of normal pressure, roughness, and cleanliness of the contacting surfaces with which the investigators worked. It is therefore important that both caution and judgment be exercised when selecting a coefficient of friction for a given set of conditions. When a more exact determination of the frictional forces acting between two surfaces is required, the coefficient of friction should be determined directly by an experiment which involves the two materials to be used.

It should be observed that Eqs. 8–1 and 8–2 have a specific, yet limited, use in the solution of friction problems. In particular, the frictional force acting at a contacting surface is determined from Eq. 8–2 *only* if relative *motion* is occurring between the two surfaces. Furthermore, if two bodies are *stationary*, the magnitude of frictional force, *F*, acting at the contacting surfaces, is *not necessarily* equal to $\mu_s N$ (Eq. 8–1). *F* must, however, satisfy the inequality $F \le \mu_s N$. Only when impending motion occurs does *F* reach its upper limit, $F = F_s = \mu_s N$. To better understand the inequality, however, consider the block shown in Fig. 8–4a. In this case, the applied load has a magnitude of $P_s = F_s$. By equilibrium, this force and the weight of the block create a resultant reactive force $\mathbf{R}_s$ at the contacting surface having components $\mathbf{F}_s$ and $\mathbf{N}$. The block is thus *on the verge of sliding* (impending motion). The angle ϕ_s, which $\mathbf{R}_s$ makes with $\mathbf{N}$, is called the angle of static friction or the *angle of repose*. From the figure

$$\phi_s = \tan^{-1} \frac{F_s}{N} = \tan^{-1} \frac{\mu_s N}{N} = \tan^{-1} \mu_s$$

If the limiting force $\mathbf{F}_s$ is applied from *any direction* along the sides of the block, ϕ_s may be used to describe the locus of possible directions for the reaction $\mathbf{R}_s$. This locus will be a right circular cone, known as the *cone of static friction*, having a vertex angle $2\phi_s$, Fig. 8–4b. Provided the block is not in motion, any horizontal force $P < P_s = F_s$ causes a reactive force $\mathbf{R}$, having a line of action which lies *within* the cone of friction. Thus, it is necessary that the frictional force $F \le F_s$ if the block is not in motion.

(a)

(b)

Fig. 8–4

8–2. Problems Involving Dry Friction

If a rigid body is in equilibrium under a system of forces which includes the effect of friction, the force system must not only satisfy the equations of equilibrium but *also* the laws which govern the frictional forces. In general, there are three types of mechanics problems involving dry

friction. In the *first type,* it is known that either the *motion* of the body is *impending* along all its contacting surfaces, in which case $F = \mu_s N$, or the body is *sliding* with some velocity, so that $F = \mu_k N$. With the aid of the frictional equations, the equilibrium equations are applied in the usual manner to solve for any unknowns shown on the free-body diagram of the body (see Examples 8–1, 8–2, 8–5, and 8–6). In the *second type* of problem, the applied loads acting on the rigid body are *known,* and one has to determine whether or not the frictional forces developed at surfaces of contact are large enough to *hold the body in equilibrium.* At each rough point of contact there is a normal force **N** and associated frictional force **F.** One method for solving this type of problem is to use the equations of equilibrium and solve for both forces F and N, which are required to *hold the body in equilibrium.* Knowing the equilibrium normal force N at each contacting surface, the corresponding *maximum friction force* can be easily calculated using Eq. 8–1 ($F_s = \mu_s N$). By comparing this value of F_s with the friction force F needed for equilibrium it is possible to determine whether or not the body remains in equilibrium. Hence, if $F \leq F_s$, the body is in equilibrium; however, if $F > F_s$, slipping occurs at the contact point. (See Example 8–3.) The *third type* of problem is similar to the second type, that is, for a given loading one has to determine if the body or system of bodies remains in equilibrium. In this case, however, there may be several different ways in which the body or system of bodies may move out of equilibrium. For example, a block resting on a rough horizontal surface may either remain stationary, slide, or tip if it is subjected to a horizontal force. The solution follows the procedure of the second type of problem, except one must perform an analysis for each of the possible cases which may cause motion. (See Example 8–4.)

In Chapters 3 and 5, it was shown that one can *assume the sense of direction* of a force having an *unknown magnitude.* The correct sense of direction of the force is made known after the equations of equilibrium are applied and the magnitude of the force is determined. If the magnitude is positive, the sense of direction of the force was assumed correctly. However, if the magnitude is negative, the sense of direction of the force is the reverse of that which was assumed. This convenience of *assuming* the sense of direction prior to applying the equations of equilibrium is possible, since the equilibrium equations are *vector* equations. That is, the scalar equilibrium equations equate to zero the *components of vectors* acting in the *same direction.* For example, $\Sigma F_x = 0$ equates to zero the components of all force vectors acting in the x direction. The frictional equation $F_s = \mu_s N$, (or $F_k = \mu_k N$) however, relates the magnitudes of vectors acting in *perpendicular directions.* (**F**$_s$ is always perpendicular to **N.**) For this reason, when the frictional equation is applied and used *in conjunction with* the equilibrium equations to solve for any unknowns, it is necessary that the frictional forces acting on the free-body diagram

be shown acting in their *correct sense of direction*. The sense of direction of frictional forces acting on a body will always be such as to either *oppose the motion or impend the motion of the body*.

In most problems, the force system acting on the body is coplanar, in which case the scalar equations of equilibrium provide the most direct solution to the problem. When three-dimensional force systems are encountered, a vector approach is generally most convenient.

Example 8-1

The block shown in Fig. 8–5*a* weighs 200 lb and is subjected to a 100-lb force. Determine the angle θ such that the block is on the verge of moving (a) up the plane, and (b) down the plane. The coefficient of static friction is $\mu_s = 0.3$.

(a)

Solution

Part (a). The free-body diagram for the block is shown in Fig. 8–5*b*. Since motion is *impending* up the plane, the frictional force *must* act in the opposite direction to oppose the tendency for motion. The applied force system must satisfy *both* the equilibrium equations and the law of dry friction. *Impending motion* requires that $F = \mu_s N = 0.3N$. Why? Applying two of the equations of equilibrium, we write

$$+\nearrow\Sigma F_x = 0; \qquad 100\text{ lb} - 0.3N - 200\sin\theta\text{ lb} = 0 \qquad (1)$$

$$+\nwarrow\Sigma F_y = 0; \qquad N - 200\cos\theta\text{ lb} = 0 \qquad (2)$$

Solving for N in terms of θ in Eq. (2), substituting into Eq. (1), and simplifying yields,

$$100\text{ lb} - 60\cos\theta\text{ lb} - 200\sin\theta\text{ lb} = 0$$

The angle θ can be obtained by trial and error. The magnitude of the normal force **N** is then readily obtained from Eq. (2). The final results are

$$N = 195.7\text{ lb}$$

$$\theta = 12.0° \qquad\qquad Ans.$$

(b)

Knowing the height of the block, the distance d in Fig. 8–5*b* may be obtained by summing moments about point O, i.e.,

$$\zeta+\Sigma M_O = 0; \qquad N(d) - 0.3N(3\text{ in.}) = 0$$

N cancels from the equation so that $d = 0.9$ in. Note in particular, that the line of action of N lies to the *right* of point O because of the nature of the moment balance.

Part (b). The free-body diagram for the block having *impending motion*

(c)

Fig. 8–5

down the plane is shown in Fig. 8–5c. Again, $F = 0.3N$. The equations of equilibrium become

$$+\nearrow\Sigma F_x = 0; \qquad 100 \text{ lb} + 0.3N - 200 \sin\theta \text{ lb} = 0$$
$$+\nwarrow\Sigma F_y = 0; \qquad N - 200 \cos\theta \text{ lb} = 0$$

in which case

$$N = 140.6 \text{ lb}, \qquad \theta = 45.3° \qquad\qquad Ans.$$

Example 8–2

The uniform 60-lb plank shown in Fig. 8–6a rests against a floor and wall for which the coefficients of static friction are $(\mu_s)_f = 0.30$ and $(\mu_s)_w = 0.20$, respectively. Determine the distance s to which a 150-lb man can climb without causing the plank to slip.

Solution

The free-body diagram for the plank is shown in Fig. 8–6b. The frictional forces $\mathbf{F}_A$ and $\mathbf{F}_B$ oppose the motion of the plank. Since motion is impending at both surfaces,

$$F_A = (\mu_s)_f N_A = 0.30 N_A$$
$$F_B = (\mu_s)_w N_B = 0.20 N_B$$

Using these relations and applying the three equations of equilibrium, we obtain

$$\xrightarrow{+}\Sigma F_x = 0; \qquad\qquad 0.30 N_A - N_B = 0$$
$$+\uparrow\Sigma F_y = 0; \qquad N_A - 150 \text{ lb} - 60 \text{ lb} + 0.20 N_B = 0$$
$$\zeta+\Sigma M_A = 0; \quad -150 \text{ lb}(s\cos 40°) - 60 \text{ lb}(15\cos 40° \text{ ft})$$
$$+ 0.2 N_B(30\cos 40° \text{ ft}) + N_B(30\sin 40° \text{ ft}) = 0$$

Solving for the three unknowns yields

$$N_A = 198.1 \text{ lb}$$
$$N_B = 59.4 \text{ lb}$$
$$s = 6.34 \text{ ft} \qquad\qquad Ans.$$

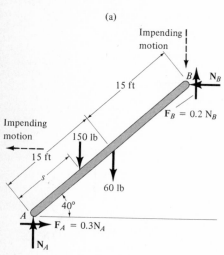

(a)

(b)

Fig. 8–6

Example 8–3

The brake shown in Fig. 8–7a bears against the outer surface of a 30-lb drum for which $\mu_s = 0.45$ and $\mu_k = 0.40$. Determine the magnitude of the force acting at pin O.

Solution

It is not known initially if the 20-lb force, applied to the brake, is

(a)

(b)

(c)

Fig. 8–7

adequate to prevent the wheel from rotating. This is important since the actual force developed at point O will depend upon whether or not the wheel is at rest or is rotating. In each of these two cases the frictional forces developed at A will be different.

Case I: Wheel Is Assumed at Rest. The free-body diagrams for both

the wheel and brake are shown in Fig. 8–7b. Due to the brake, the frictional force $\mathbf{F}_A$ tends to *prevent motion of the wheel* and so acts *downward on the wheel*. This force is assumed as an *equilibrium force* and is *not necessarily equal* to the *maximum frictional force* developed by the brake. Summing moments about point B on the wheel yields a direct solution for the magnitude of $\mathbf{F}_A$.

$$\zeta + \Sigma M_B = 0; \qquad 50\text{ lb }(6\text{ in.}) - F_A(12\text{ in.}) = 0; \qquad F_A = 25\text{ lb}$$

Summing moments about point O on the brake to obtain N_A gives

$$\zeta + \Sigma M_C = 0; \qquad 20\text{ lb }(20\text{ in.}) - 25\text{ lb }(3\text{ in.}) - N_A(8\text{ in.}) = 0,$$
$$N_A = 40.6\text{ lb}$$

The forces $\mathbf{F}_A$ and $\mathbf{N}_A$ as calculated here must be developed at point A to *hold the wheel in equilibrium*. It is *also* necessary that $\mathbf{F}_A$ satisfy the law of static friction at point A. Since $\mu_s = 0.45$, the *maximum* value of frictional force which can be developed at point A, where $N_A = 40.6$ lb, is

$$(F_s)_A = \mu_s N_A = 0.45(40.6\text{ lb}) = 18.3\text{ lb}$$

By comparison, it is seen that the force required for equilibrium, F_A, is greater than the maximum frictional force $(F_s)_A$ developed at the brake. Hence, the wheel is *rotating* so that our initial assumption was wrong.

Case II: Wheel Is Rotating. Since *motion* is occurring, the frictional force developed by the brake is related to the normal force by Eq. 8–2, i.e.,

$$(F_k)_A = \mu_k N_A' = 0.40 N_A'$$

The free-body diagrams of the wheel and brake are shown in Fig. 8–7c. Applying the three equations of equilibrium to the brake, since the brake does not move,

$$\xrightarrow{+} \Sigma F_x = 0; \qquad -O_x + N_A' - 20\text{ lb} = 0$$
$$+\uparrow \Sigma F_y = 0; \qquad -O_y + 0.40 N_A' = 0$$
$$\zeta + \Sigma M_C = 0; \qquad -N_A'(8\text{ in.}) - 0.40 N_A'(3\text{ in.}) + 20\text{ lb }(20\text{ in.}) = 0$$

Hence,

$$N_A' = 43.5\text{ lb}$$
$$O_x = 23.5\text{ lb}$$
$$O_y = 17.4\text{ lb}$$

The total force at pin O is therefore

$$F_O = \sqrt{O_x{}^2 + O_y{}^2} = \sqrt{(23.5\text{ lb})^2 + (17.4\text{ lb})^2} = 29.2\text{ lb} \qquad Ans.$$
$$\theta = \tan^{-1}\frac{O_y}{O_x} = \tan^{-1}\frac{17.4\text{ lb}}{23.5\text{ lb}} = 36.6° \quad \theta \diagup \mathbf{F}_C \qquad Ans.$$

Example 8-4

The homogeneous blocks A and B shown in Fig. 8-8a weigh 100 and 60 lb, respectively. Determine the greatest magnitude for the applied force **P** without causing the blocks to move. The static coefficients of friction between the contacting surfaces are given in the figure.

Fig. 8-8

Solution

There are three possible ways in which the blocks may begin to move: (1) impending motion of B on A with no motion of A; (2) tipping of B on A with no motion of A; and (3) no impending motion of B while motion of A is impending on the horizontal plane. (Since block A is relatively flat and long, tipping of this block is of no consequence in the analysis of the problem.) The magnitude of **P** needed to cause each of these motions will be computed. The least value of P is the required answer. Why?

Case I: Impending Motion of B; A Remains Stationary. The free-body

diagram of block B is shown in Fig. 8–8b. Impending motion of B implies that $F_{AB} = 0.3N_{AB}$. There are two unknown force magnitudes on the free-body diagram, N_{AB} and P.

Applying the equations of equilibrium* gives

$$\xrightarrow{+} \Sigma F_x = 0; \qquad\qquad 0.3N_{AB} - P = 0$$

$$+\uparrow \Sigma F_y = 0; \qquad\qquad N_{AB} - 60\text{ lb} = 0$$

Solving these equations, we obtain

$$N_{AB} = 60\text{ lb}$$

$$P = 18\text{ lb}$$

Case II: B tips; A Remains Stationary.   When block B is on the verge of tipping, the normal reaction $\mathbf{N}_{AB}$ acts at point O. This situation is shown in Fig. 8–8c. Applying the equations of equilibrium, we have

$$\xrightarrow{+} \Sigma F_x = 0; \qquad\qquad F_{AB} - P = 0$$

$$+\uparrow \Sigma F_y = 0; \qquad\qquad N_{AB} - 60\text{ lb} = 0$$

$$\zeta + \Sigma M_O = 0; \qquad P(6\text{ in.}) - 60\text{ lb }(3\text{ in.}) = 0$$

Hence,

$$N_{AB} = 60\text{ lb}$$

$$F_{AB} = 30\text{ lb}$$

$$P = 30\text{ lb}$$

Case III: A Has Impending Motion on the Horizontal Plane; B Remains Stationary on A.   The free-body diagram of both blocks as one system is shown in Fig. 8–8d. Since block A is on the verge of moving, $F = 0.25\ N$. Applying the equations of equilibrium yields

$$\xrightarrow{+} \Sigma F_x = 0; \qquad\qquad -25\text{ lb} + 0.25N - P = 0$$

$$+\uparrow \Sigma F_y = 0; \qquad\qquad N - 100\text{ lb} - 60\text{ lb} = 0$$

Hence,

$$N = 160\text{ lb}$$

$$P = 15\text{ lb} \qquad\qquad\qquad Ans.$$

The smallest magnitude of force $\mathbf{P}$ is developed under the conditions

---

*A third unknown d specifies the line of action of $\mathbf{N}_{AB}$. Since tipping is assumed not to occur, $0 < d \leq 3$ in. The magnitude of d may be computed by summing moments about point O, which yields

$$\zeta + \Sigma M_O = 0; \quad 60\text{ lb }(3\text{ in.}) - P(6\text{ in.}) - N_{AB}(d) = 0$$

Using the values of N_{AB} and P calculated above we obtain a value of $d = 1.20$ in.

of Case III. Therefore, motion will first occur when $P > 15$ lb, and as a result both blocks will initially slide together.

Example 8-5

The 30-lb block shown in Fig. 8–9a rests on an inclined plane for which $\mu_s = 0.35$. Determine the magnitude of minimum force **P** required to move the block and the direction of impending motion. The force **P** acts along the surface of the inclined plane.

(a)

Solution

A free-body diagram of the block is shown in Fig. 8–9b. Since the problem is three-dimensional, a vector approach to the solution will be considered. For simplicity, the x, y, and z axes are shown such that the x and y axes are located on the surface of the plane. Expressing each force in vector notation, we have

$$\mathbf{N} = N\mathbf{k}$$
$$\mathbf{P} = -P \sin 45°\mathbf{i} + P \cos 45°\mathbf{j} = -0.707P\mathbf{i} + 0.707P\mathbf{j}$$
$$\mathbf{W} = 30 \sin 15° \, \mathbf{j} - 30 \cos 15° \, \mathbf{k} = \{7.76\mathbf{j} - 29.0\mathbf{k}\} \text{ lb}$$
$$\mathbf{F} = F \sin \theta\mathbf{i} - F \cos \theta\mathbf{j}$$

(b)

Fig. 8-9

Using the vector force equation for equilibrium, we obtain

$$\Sigma \mathbf{F} = 0; \qquad \mathbf{N} + \mathbf{P} + \mathbf{W} + \mathbf{F} = 0$$

or

$$N\mathbf{k} - 0.707P\mathbf{i} + 0.707P\mathbf{j} + 7.76\mathbf{j} - 29.0\mathbf{k} + F \sin \theta\mathbf{i} - F \cos \theta\mathbf{j} = 0$$

Equating to zero each of the respective **i**, **j**, and **k** components gives

$$\Sigma F_x = 0; \qquad -0.707P + F \sin \theta = 0 \qquad (1)$$
$$\Sigma F_y = 0; \qquad 0.707P + 7.76 \text{ lb} - F \cos \theta = 0 \qquad (2)$$
$$\Sigma F_z = 0; \qquad N - 29.0 \text{ lb} = 0 \qquad (3)$$

Since the block is subjected to impending motion,

$$F = 0.35N \qquad (4)$$

Equation (3) may be solved for N and the value substituted into Eq. (4) to obtain F. When Eqs. (1) and (2) are added, P is eliminated and a solution for θ is obtained by trial and error. The value of P is then obtained using the computed value of θ in either Eq. (1) or (2). The final results are

$$N = 29.0 \text{ lb}$$
$$F = 10.14 \text{ lb}$$
$$P = 3.03 \text{ lb} \qquad \qquad Ans.$$
$$\theta = 12.2° \qquad \qquad Ans.$$

Since the frictional force acts in the opposite direction of impending motion, the block will move down the plane at an angle $\theta = 12.2°$ with the y axis when $P > 3.03$ lb.

Example 8-6

The 150-lb block A is used to lift the 200-lb block B by applying a horizontal force **P**, as shown in Fig. 8–10a. If the static coefficient of friction for all contacting surfaces is $\mu_s = 0.25$, determine the greatest magnitude of force **P** needed to cause impending motion of the blocks.

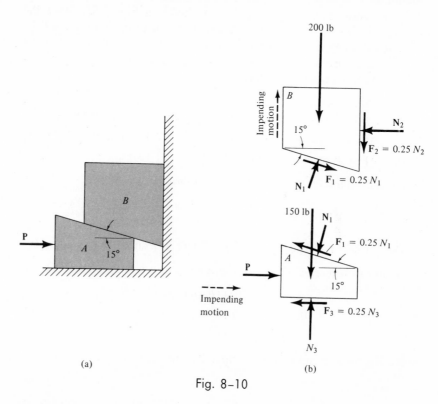

(a)

(b)

Fig. 8–10

Solution

The free-body diagram for each block is shown in Fig. 8–10b. Since motion is impending, Eq. 8–1 is used to relate the normal and frictional force at each contacting surface. As shown on block A, the friction forces $\mathbf{F}_1$ and $\mathbf{F}_3$ at each contacting surface act in the opposite direction to impending motion at their contacting surface. Equal but opposite reactions of $\mathbf{N}_1$ and $\mathbf{F}_1$ must act on block B as shown. Since this block is forced upward, $\mathbf{F}_2$ acts downward. Applying the two force equations of equilibrium to block B, we have

$$\xrightarrow{+}\Sigma F_x = 0; \qquad N_1 \sin 15° + 0.25N_1(\cos 15°) - N_2 = 0$$
$$+\uparrow\Sigma F_y = 0; \qquad N_1 \cos 15° - 0.25 N_1(\sin 15°) - 0.25 N_2 - 200 \text{ lb} = 0$$

Solving these equations simultaneously yields

$$N_1 = 258 \text{ lb}$$
$$N_2 = 129 \text{ lb}$$

Using the value of $N_1 = 258$ lb and applying the two force equations of equilibrium to block A we can write

$$\xrightarrow{+}\Sigma F_x = 0; \qquad P - 0.25 N_3 - 0.25(258 \cos 15° \text{ lb}) - 258 \sin 15° \text{ lb} = 0$$
$$+\uparrow\Sigma F_y = 0; \qquad N_3 - 258 \cos 15° \text{ lb} + 0.25(258 \sin 15° \text{ lb}) - 150 \text{ lb} = 0$$

Solving these equations simultaneously, we obtain

$$N_3 = 383 \text{ lb}$$
$$P = 225 \text{ lb} \qquad\qquad\qquad Ans.$$

Problems

Assume that $\mu = \mu_s = \mu_k$ in the following problems where specified.

8-1. The footing of a ladder rests on a concrete surface for which $\mu = 0.8$ for the two materials. If the magnitude of a force, directed along the axis of the ladder, is 600 lb, determine the friction force acting at the bottom of the ladder. What is the minimum coefficient of friction which will prevent the ladder from slipping?

8-2. An ax is driven into the block of wood. If the coefficient of friction between the ax and the wood is $\mu = 0.2$, determine the smallest angle α of the blade which will cause the ax to be "self-locking." Neglect the weight of the ax.

Prob. 8-1

Prob. 8-2

8-3. Determine the force **P** which must be applied to the wedge to lift the 1,000-lb block. The coefficient of friction for all contacting surfaces is $\mu = 0.3$. Neglect the weight of the wedge.

Prob. 8-3

8-4. The block weighs 60 lb and is subjected to a force **P** acting at an angle $\alpha = 20°$ with the horizontal. If $\mu_s = 0.5$, determine the magnitude of force **P** to just start the block moving down the plane.

Prob. 8-4

8-5. If the coefficient of kinetic friction between the inclined surface and the block in Prob. 8-4 is $\mu_k = 0.3$, determine the angle α of the force $P = 20$ lb which will allow the block to move down the plane with constant velocity.

8-6. The 100-lb block rests on the horizontal plane for which $\mu_s = 0.5$. If a force having a magnitude of 20 lb is applied to the block, determine the friction force developed between the block and the plane. Does the block slip?

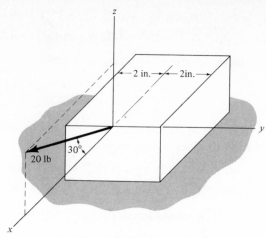

Prob. 8-6

8-7. The mine car has a uniform weight of 6,000 lb. If the coefficient of friction between the wheels and the tracks is $\mu_s = 0.4$ when the wheels are locked, find the normal force acting on each of the two wheels at points A and B when (a) the brakes at A are locked, and (b) the brakes at both A and B are locked. Does the car move when only the brakes at A are locked?

Prob. 8-7

8-8. The rectangular tank is filled with water having a density of 62.4 lb/ft³. If the tank weighs 20 lb when empty, determine the maximum height of water h which may be placed in the tank so that the tank will not tip when subjected to the horizontal force $P = 200$ lb.

278

3 ft

Top View

10 ft

h

3 ft

P

4 ft

Side View

Prob. 8-8

8-9. If block *B* is placed upon block *A,* determine the maximum angle α so that block *A* does not slip. Each block weighs 1,000 lb, and all surfaces of contact have a coefficient of friction $\mu = 0.2$.

B

α

A

Prob. 8-9

8-10. The 50-lb block is subjected to a horizontal force of 20 lb. Determine the smallest width *w* so that the block does not slide or tip.

w

20 lb

10 in.

$\mu_s = 0.8$

Prob. 8-10

8-11. Two blocks *A* and *B* each weighing 50 lb are joined by a light 3-ft rod having a pin at each end. Determine the maximum height *h* for which block *B* may be placed up the inclined plane without causing the assembly to slip down the plane. Neglect the size of the blocks in the calculations. $\mu = 0.15$ at the contacting surfaces.

B

h

A

30°

Prob. 8-11

8-12. Two blocks *A* and *B,* each weighing 50 lb, are connected by the linkage shown. If the coefficient of friction at the contacting surfaces is $\mu_s = 0.5$, determine the maximum vertical force **P** which may be applied to the linkage without causing the blocks to move.

B

30°

P

A

30°

Prob. 8-12

$\tan \phi = U_{ST}$

8–13. The hub of the 150-lb wheel bears against the vertical wall. If the coefficient of friction between all contacting surfaces is $\mu = 0.6$, determine the horizontal force **P** needed to rotate the wheel.

3 ft

P

5 ft

Prob. 8–13

P

3 in.

$\frac{1}{2}$ in.

$\frac{1}{2}$ in.

A

15°

F

5 in.

Prob. 8–15

Dec 1

19

8–14. Determine the minimum applied force **P** required to ~~hold wedge B~~ in place. The spring is compressed a distance of 6 in. Neglect the weight of the wedges and block *C*. The coefficient of friction for all contacting surfaces is $\mu = 0.35$. Neglect friction at the rollers.

wedge moves inward?

k = 100 lb/in.

C

P

B

10°

A

Prob. 8–14

8–15. If the vertical load **P** acting on the bracket is 500 lb, determine the horizontal force **F** required to move the wedge to the left. Neglect the weight of the wedge and the bracket. The bracket is pinned at *A*. The coefficient of friction between all the contacting surfaces of the wedge is $\mu = 0.2$.

8–16. A 100-lb weight rests on a 15° inclined plane. Determine the minimum force *P* which will cause impending motion of the weight up the plane; $\mu_s = 0.3$.

P

45°

15°

Prob. 8–16

8–17. A 100-lb disk rests on an inclined surface for which $\mu_s = 0.2$. Determine the maximum vertical load **P** which may be applied to the link *AB* without causing the disk to slip.

P

$\frac{1}{2}$ ft

2 ft

1 ft

3 ft

A

2 ft

30°

B

Prob. 8–17

8-18. Blocks A and B weigh 50 and 100 lb, respectively. If the coefficient of friction for all surfaces of contact is $\mu_s = 0.3$, calculate the magnitude of force P needed to cause impending motion of B to the right. What is the tension in the cord?

Prob. 8–18

8-19. Solve Prob. 8–18 assuming that the cord slopes upward to the left at an angle of 20°.

8-20. A 50-lb beam rests on two surfaces at points A and B. Determine the maximum distance x to which a girl can walk up the beam before the beam begins to slip. Assume that the girl weighs 120 lb and moves up the beam with a constant velocity.

Prob. 8–20

8-21. A 150-lb man squats 4 ft from the top of the ladder. If the ladder weighs 50 lb, with center of gravity at point G, and the coefficients of friction at points A and B are $(\mu_s)_A = 0.3$ and $(\mu_s)_B = 0.2$, respectively, determine the range of horizontal force which the man may exert on the ladder at point C without causing the ladder to move. Assume that the man squatting on the ladder exerts only a vertical force on the ladder at point D.

Prob. 8–21

8-22. Solve Prob. 8–21 given that the wall surface at point B is smooth.

8-23. Determine the minimum force F needed to push the two 150-lb cylinders up the plane. The coefficient of friction between all surfaces of contact is $\mu = 0.3$.

Prob. 8–23

8-24. A single force **P** is applied to the handle of the drawer. If friction is neglected at the bottom of the drawer, and the coefficient of friction along the sides is μ_s, determine the spacing s between the handles so that the drawer does not bind at the corner when the force **P** is applied.

Prob. 8-24

8-25. If the coefficient of friction for the drum and brake mechanism is $\mu_s = 0.6$, determine the reaction at the pin O. Does the 60-lb vertical force prevent the drum from rotating? Neglect the weight of the brake.

Prob. 8-25

8-26. The compound beam is adjusted into the horizontal position by means of a wedge located at its right support. If the coefficient of friction between the wedge and the surfaces of contact is $\mu = 0.25$, determine the horizontal force **P** required to remove the wedge.

Prob. 8-26

8-27. The 100-lb sphere rests between the two supports at A and B. Determine the minimum value of P and the coefficient of friction μ_s needed to roll the sphere off the edge B without causing the sphere to slip.

Prob. 8-27

8-28. If $\mu_s = 0.15$ for the contacting surfaces at A and B in Prob. 8-27, determine the magnitude of force **P** for impending motion of the sphere. What are the normal reactions at points A and B?

8-29. Two blocks A and B rest on the inclined planes. If the weight of block A is 60 lb, determine the weight of block B so that (a) block B is on the verge of sliding up the plane, and (b) block B moves at constant velocity down the plane. $\mu_s = 0.5$ and $\mu_k = 0.25$ for all surfaces.

Prob. 8–29

Prob. 8–31

8–30. A 50-lb block is attached to a light rod AB which pivots at pin A. If the coefficient of friction between the plane and the block is $\mu_s = 0.4$, determine the minimum angle θ at which the block may be placed without slipping on the plane.

8–32. The 100-lb crate C rests on a 10-lb dolly D. If the front casters of the dolly at A are locked to prevent them from rolling, determine the maximum force P which may be applied without causing motion of the crate. The coefficient of friction between the casters and the floor is $\mu_s = 0.4$, and between the dolly and the crate, $\mu_s = 0.3$.

Prob. 8–30

8–31. Two blocks are connected by a uniform 2-lb link making an angle of $\theta = 30°$ with the horizontal. If the coefficient of friction is $\mu_s = 0.35$ for the inclined plane and $\mu_s = 0.2$ for the vertical wall, determine the maximum weight of block B without upsetting the equilibrium position for the two blocks. Block A weighs 50 lb.

Prob. 8–32

283

8–33. The 10-ft-long slender beam weighs 100 lb and is connected at A by means of a ball-and-socket joint. If the beam rests against a vertical wall at B, determine the maximum angle θ_{max} at which the beam may be placed so that it does not rotate about the z axis. The coefficient of friction at the wall is $\mu_s = 0.3$. Is rotation of the beam also prevented about the x and y axes when $\theta = \theta_{max}$? Why is this so?

Prob. 8–33

8–34. If the coefficient of friction for all surfaces is $\mu_s = 0.5$, determine the maximum angle θ for which the uniform 100-lb beam AB may be placed against the wall and still not slip.

Prob. 8–34

8–35. The wheel C is subjected to the braking action caused by both wheel B and the brake arm A. If the coefficient of friction at all points of contact is $\mu_s = 0.3$, determine the least amount of force $\mathbf{P}$ which must be applied to the brake handle to prevent wheel C from slipping. Neglect friction at pins O and D. Wheel B weighs 100 lb.

Prob. 8–35

8–36. The 600-lb force applied to the wedge creates a compressive force at A. If friction at B is neglected, determine the compression at A if the coefficient of friction for the wedge is (a) $\mu = 0.3$, and (b) $\mu = 0$.

Prob. 8–36

8–37. A 100-lb cylinder rests against a wall and on the surface of a 20° wedge. Assuming that the cylinder, wedge, wall, and horizontal plane A all have the same surface characteristics, determine the coefficient of friction μ which will make the wedge "self-locking;" i.e., $\mathbf{P} = 0$.

Prob. 8–37

19 Dec. 1

8–39. A 50-lb wedge is placed in the grooved slot of an inclined plane. Determine the maximum angle θ for the inclined plane without causing the wedge to slip. The coefficient of friction for the wedge and surface is $\mu = 0.2$.

Prob. 8–39

19 Dec 1

8–38. The friction grip is used to lift steel plates. This is done by inserting a plate p between the loose fitting cylinder C and applying the lifting force **P** to the grip. If the coefficient of friction between the grip and the plate is μ, determine the angle α which will allow the plate to be lifted.

8–40. The two rods each have a weight W and are pinned together at point A. If the coefficient of friction at C is $\mu_s = 0.5$, determine the maximum angle θ for equilibrium.

Prob. 8–38

Prob. 8–40

8-41. The 100-lb cylindrical boiler rests between the two inclined planes. When $P = 15$ lb, the boiler is on the verge of impending motion. Determine the coefficient of friction μ between the surfaces of contact and the boiler.

8-42. The wedge used in the cotter joint has a coefficient of friction $\mu = 0.3$ with all contacting surfaces. Determine the angle α so that the wedge is self-locking, that is, that the wedge does not slip for any magnitude of **P**.

Prob. 8–41

Prob. 8–42

*8–3. Frictional Forces in Screws

A *screw* may be thought of simply as an inclined plane wrapped around a cylinder. A nut initially at position A on the screw shown in Fig. 8–11 will move to B when rotated 360° around the screw. This rotation is equivalent to translating the nut up an inclined plane of height p and length $l = 2\pi r$, where r is the mean radius of the thread. The rise p is often referred to as the *pitch* of the screw. The *pitch angle* is given by $\theta_p = \tan^{-1}(p/2\pi r)$.

In most cases screws are used as fasteners. In many types of machines screws are incorporated to transmit power or motion from one part of the machine to another. A *square-threaded screw* is most commonly used for this latter purpose, especially when large forces are applied along the

Fig. 8–11

axis of the screw. When subjected to large axial loads, the frictional forces developed at the thread of the screw become important in the analysis to determine the force needed to turn the screw. Consider, for example, the square-threaded jack screw shown in Fig. 8–12 which is subjected to an axial load **W** and a horizontal force **P**, acting at a right angle on the end of the lever of length L. Summing moments about the axis of the screw indicates that a horizontal force of magnitude $S = PL/r$, acting at the mean radius r of the thread, creates the same moment about the axis of the screw as the force **P**.

Fig. 8–12

Fig. 8–13

The reactive forces of the jack to loads S and W are distributed over the circumference of the screw thread in contact with the screw hole in the jack, that is, within region h shown in Fig. 8–12. For simplicity, we will consider this portion of thread as a simple block resting on an inclined plane having a pitch angle of θ_p, as shown in Fig. 8–13a. The inclined plane represents the inside *supporting thread* of the jack base. The block is subjected to the total axial load **W** acting on the jack, the horizontal force **S** (which is related to the magnitude of force **P**), and the *resultant force* **R**, which the plane exerts on the block. Although not shown, force **R** has components acting normal, **N**, and tangent, **F**, to the contacting surface. Provided force **P** is great enough, the thread (and hence the block) is either on the verge of impending motion, or motion is occurring. Under these conditions, **R** acts at an angle $(\phi + \theta_p)$ from

the vertical as shown in the figure, where $\phi = \tan^{-1} F/N = \tan^{-1} \mu N/N = \tan^{-1} \mu$. ($\mu_s$ is used here if impending motion occurs; otherwise μ_k is used in the case of motion.) Applying the two force equations of equilibrium to the block, we obtain

$$\xrightarrow{+} \Sigma F_x = 0; \qquad S - R\sin(\phi + \theta_p) = 0$$
$$+\uparrow \Sigma F_y = 0; \qquad R\cos(\phi + \theta_p) - W = 0$$

Eliminating R and solving for S, then substituting this value into the equation $P = Sr/L$, yields the force needed to *raise* the block.

$$P = \frac{Wr}{L}\tan(\phi + \theta_p) \qquad (8\text{-}3)$$

This equation gives the required value P necessary to cause upward impending motion of the screw when $\phi = \phi_s = \tan^{-1} \mu_s$ (the angle of static friction). If ϕ is replaced by $\phi_k = \tan^{-1} \mu_k$ (the angle of kinetic friction), Eq. 8-3 would give a smaller value P necessary to maintain uniform upward motion of the screw.

When the load W is to be *lowered*, the direction of P is reversed, in which case the angle ϕ (ϕ_s or ϕ_k) lies on the opposite side of the normal n to the plane supporting the block. This case is shown in Fig. 8-13b for $\phi < \theta_p$. Thus, Eq. 8-3 becomes

$$P = \frac{Wr}{L}\tan(\phi - \theta_p) \qquad (8\text{-}4)$$

If the load P is *removed*, the screw will remain self-locking; that is, it will support the weight W by friction forces alone, provided $\phi > \theta_p$. However, if $\phi = \phi_s = \theta_p$, Fig. 8-13c, the screw will be on the verge of rotating downward. When $\theta_p > \phi_s$, a restraining load is needed to prevent the screw from moving downward.

*8–4. Frictional Forces in Bearings and Disks

Rotational shafts or axles are generally supported by means of *bearings*. This type of support provides resistance to axial and lateral loads acting on the shaft without restricting the rotation of the shaft about its longitudinal axis. Well-lubricated bearings are subjected to the laws of fluid mechanics, in which the viscosity of the lubricant, the speed of shaft rotation, and the amount of clearance between the shaft and bearing are needed to determine the frictional resistance of the bearing. When the bearing is not lubricated, or only partially lubricated, a reasonable analysis of the frictional resistance may be based on the laws of dry friction.

Pivot or *collar* bearings are commonly used to support the *axial or normal loads* of a rotating shaft. These two types of support are shown in Fig. 8-14. If it is assumed that the normal pressure acting at the bearing

Pivot bearing

(a)

Collar bearing

(b)

Fig. 8–14

surface is *uniformly distributed* over the contact area, an analysis of the frictional forces developed by these supports can easily be performed. Consider, for example, the end of a collar-bearing shaft shown in Fig. 8–15, which is subjected to an axial thrust **P**. The total bearing or contact area is $\pi(R_2^2 - R_1^2)$. Since $\Sigma F_z = 0$, the uniform normal *pressure* or force per unit area acting at the bearing surface is

$$p = \frac{P}{\pi(R_2^2 - R_1^2)}$$

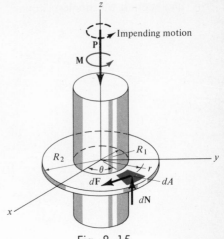

Fig. 8–15

An applied moment **M**, needed to cause impending rotation of the shaft, can be determined by considering the frictional forces developed at the bearing surface. A small area element $dA = (r\,d\theta)(dr)$, shown in Fig. 8–15, is subjected to a normal force

$$dN = p\,dA = \frac{P}{\pi(R_2^2 - R_1^2)}\,dA$$

and a frictional force

$$dF = \mu_s\,dN = \frac{\mu_s P}{\pi(R_2^2 - R_1^2)}\,dA$$

The normal force dN does not create a moment about the z axis of the shaft; however, the frictional force does, namely, $dM = r\,dF$. Integration is needed to compute the total moment created by all the frictional forces acting on differential areas dA. Therefore, for impending rotational motion,

$$\Sigma M_z = 0; \qquad M - \int_A r\,dF = 0$$

Integrating over the entire area $dA = (r\,d\theta)(dr)$, we have

$$M = \int_{R_1}^{R_2} \int_0^{2\pi} r\left(\frac{\mu_s P}{\pi(R_2^2 - R_1^2)}\right)(r\,d\theta\,dr) = \frac{\mu_s P}{\pi(R_2^2 - R_1^2)} \int_{R_1}^{R_2} r^2 dr \int_0^{2\pi} d\theta$$

or

$$M = \frac{2}{3}\mu_s P\left(\frac{R_2^3 - R_1^3}{R_2^2 - R_1^2}\right) \qquad (8\text{–}5)$$

Equation 8–5 gives the magnitude of moment **M** required for the impending motion of a shaft supported by a collar bearing. The frictional moment developed at the end of the shaft when the shaft is *rotating* at constant speed can be found by substituting μ_k for μ_s in Eq. 8–5. This equation may also be used for determining the moment transmitted between the surfaces of *disk-type brakes* or *clutches,* having a contact area described by an inner radius R_1 and an outer radius R_2. When $R_2 = R_0$

and $R_1 = 0$, as in the case of the pivot bearing, shown in Fig. 8-14a, Eq. 8-5 reduces to

$$M = \frac{2}{3}\mu_s PR_0$$

When a shaft or axle is subjected to *lateral loads,* a *journal bearing* is commonly used for support. A typical journal-bearing support is shown in Fig. 8-16. As the shaft rotates in the direction shown in the figure, it rolls up against the wall of the bearing to some point A where slippage occurs. If the lateral load acting at the end of the shaft is **W**, it is necessary that the bearing reactive force **R** acting at A be equal and opposite to **W**. The moment **M** needed to maintain the rotation of the shaft can be found by summing moments about the z axis of the shaft; that is,

$$\Sigma M_z = 0; \qquad\qquad -M + R(r\sin\phi_k) = 0$$

or

$$M = Rr\sin\phi_k$$

where ϕ_k is the angle of kinetic friction defined by $\tan\phi_k = \dfrac{F}{N} = \dfrac{\mu_k N}{N} = \mu_k$. If the bearing is partially lubricated, μ_k is small, and therefore

Fig. 8-16

$\mu_k = \tan \phi_k \approx \sin \phi_k \approx \phi_k$. Under these conditions, a reasonable approximation to the moment needed to overcome the frictional resistance becomes

$$M \approx Rr\mu_k$$

Example 8–9 illustrates a common application of this equation.

Example 8–7

The turnbuckle shown in Fig. 8–17 has a square thread with a mean radius of 0.3 in. and a pitch of 0.12 in. If the coefficient of friction between the screw and the turnbuckle is $\mu = 0.25$, determine the moment **M** which must be applied to the turnbuckle to draw the end screws closer together. Is the turnbuckle "self-locking"?

Solution

The moment M applied to the turnbuckle may be obtained by using Eq. 8–3. Since friction at two screws must be overcome, this requires

$$M = 2PL = 2Wr\tan(\phi + \theta_p) \qquad (1)$$

In this equation, the load $W = 200$ lb, $r = 0.3$ in., $\phi = \tan^{-1}\mu = \tan^{-1}(0.25) = 14.04°$, and $\theta_p = \tan^{-1}(p/2\pi r) = \tan^{-1}\{0.12 \text{ in.}/[2\pi(0.3 \text{ in.})]\} = 3.64°$. Substituting these values into Eq. (1) and solving gives

$$M = 2(200 \text{ lb})(0.3 \text{ in.})[\tan(14.04° + 3.64°)]$$

$$M = 38.3 \text{ lb-in.} \qquad \textit{Ans.}$$

When the moment M is *removed*, the turnbuckle will be self-locking since $\phi > \theta_p$.

Example 8–8

The uniform bar shown in Fig. 8–18a has a total weight of W lb. If it is assumed that the normal pressure acting at the contacting surface is uniformly distributed, determine the couple moment **M** required to rotate the bar. Assume that the width s is negligible in comparison to the length l. The coefficient of friction is equal to μ.

Solution

A free-body diagram of the bar is shown in Fig. 8–18b. Since the bar has a total weight of W lb, the weight per unit length along the x axis of the bar is $w = W/l$. The magnitude of force $d\mathbf{N}$ acting on a segment of area having a length dx is

$$dN = w\, dx = \frac{W}{l}\, dx$$

200 lb

M

200 lb

Fig. 8–17

(a)

(b)

Fig. 8-18

and therefore the frictional force acting on this same element of area is

$$dF = \mu \, dN = \frac{\mu W}{l} dx$$

The moment created by this frictional force about the z axis becomes

$$dM = x \, dF = \frac{\mu W}{l} x \, dx$$

Summing moments about the z axis of the bar gives

$$M = \int_{-l/2}^{l/2} \frac{\mu W}{l} x \, dx$$

or

$$M = \frac{\mu W l}{4} \qquad\qquad Ans.$$

Example 8-9

The 5-in.-diameter pulley shown in Fig. 8-19a fits loosely on a 1-in.-diameter shaft for which the coefficients of static and kinetic friction are $\mu = 0.26$. Determine the minimum tension $\mathbf{T}$ in the belt to (a) raise the 100-lb weight at constant velocity, and (b) lower the weight at constant velocity. Assume no slipping between the belt and pulley and neglect the weight of the pulley.

Solution
Part (a). A free-body diagram of the pulley is shown in Fig. 8-19b. When the pulley is subjected to cable tensions of 100 lb each, the pulley makes contact with the shaft at point P_1. As the tension $\mathbf{T}$ is increased, the pulley will roll around the shaft to point P_2 before motion impends. From the

(a)

(b)

(c)

Fig. 8-19

figure, $d = r \sin \phi$. As stated in the text, $\sin \phi \approx \mu$. Hence, $d \approx r\mu =$ ($\frac{1}{2}$ in.)0.26 = 0.13 in., so that summing moments about point P_2 gives

$$\zeta + \Sigma M_{P_2} = 0;$$
$$100 \text{ lb } (2.63 \text{ in.}) - T(2.37 \text{ in.}) = 0, \qquad T = 111.0 \text{ lb} \qquad Ans.$$

Part (b). When the weight is lowered, the resultant force **R** acting on the shaft passes through point P_3, as shown in Fig. 8–19*c*. Summing moments about this point yields

$$\zeta + \Sigma M_{P_3} = 0;$$
$$100 \text{ lb } (2.37 \text{ in.}) - T(2.63 \text{ in.}) = 0, \qquad T = 90.1 \text{ lb} \qquad Ans.$$

Problems

8-43. The threads of the jack screw shown in Fig. 8–12 have a mean radius of 1 in. and a pitch of $\frac{1}{4}$ in. If the static coefficient of friction is $\mu_s = 0.2$ and the lever arm has a length of 2 ft, determine the force **P** which must be initially applied to lift a weight $W = 5,000$ lb. What initial force **P** is required to lower the weight?

8-44. Prove that the pitch p must be less than $2\pi r\mu$ for the jack screw, shown in Fig. 8–12, to be "self-locking."

8-45. The square-threaded bolt is used to join two plates together. If the bolt has a mean diameter of $\frac{3}{4}$ in. and a pitch of $\frac{3}{16}$ in., determine the torque required to loosen the bolt if the tension in the bolt is 35,000 lb. The coefficient of friction between the threads and the bolt is $\mu_s = 0.15$.

8-46. The square-threaded screw of the vise clamps has a mean diameter of $\frac{1}{2}$ in. and a pitch of $\frac{1}{7}$ in. If $\mu_s = 0.25$ for the threads, and the load **P**, applied normal to the handle, is 50 lb, determine the compressive force in the block.

P = 50 lb

3 in.

6 in.

8 in.

Prob. 8–46

Prob. 8–45

$\frac{3}{4}''$

$\frac{1}{2}''$

8–47. The *universal testing machine* is used to measure the strength of a specimen S when the specimen is subjected to large tensile or compressive loads. These forces are applied by raising or lowering the movable head H of the machine. The head weighs 500 lb, and rests on two square threaded screws, each with a mean diameter of 3 in. and a pitch of $\frac{3}{8}$ in. The coefficient of static friction for the threads is $\mu_s = 0.2$. If the specimen is compressed in the machine such that the dial reads 3,000 lb, determine the torque T (lb-ft) which must be applied to each of the screws, by the motors of the machine, to increase the load. Neglect friction acting on the four guide posts G.

Prob. 8–49

8–50. The *double-collar bearing* is subjected to an axial load $P = 1,000$ lb. Assuming that collar A supports $\frac{3}{4}P$ and collar B supports $\frac{1}{4}P$, both with a uniform distribution of pressure, determine the maximum frictional moment $\mathbf{M}$ which may be resisted by the bearing. $\mu_s = 0.2$ for both collars.

Prob. 8–47

8–48. If a tensile specimen is placed in the testing machine of Prob. 8–47, determine the torque which must be applied to each of the screws by the machine to create a tensile force in the specimen of 6,000 lb.

8–49. The collar bearing supports an axial load of 1,500 lb. If a moment of 160 lb-ft is required to overcome the friction developed by the bearing, determine the coefficient of friction μ acting at the bearing surface.

Prob. 8–50

8-51. Determine the moment **M** which must be applied to the collar bearing in Prob. 8–50 to cause impending motion of the collar. Assume that the load $P = 1,000$ lb is distributed equally to each collar. The coefficient of friction for collar A is $\mu_s = 0.2$, and for collar B, $\mu_s = 0.3$.

8-52. The automobile jack is subjected to a vertical load of 6,000 lb. If the square-threaded screw has a pitch of $\frac{1}{4}$ in. and a mean diameter of $\frac{1}{2}$ in., determine the force which must be applied to the handle to (a) raise the load, and (b) lower the load; $\mu_s = 0.25$. The supporting plate exerts only vertical forces at A and B, and each cross-link has a length of 4 in.

Prob. 8–52

8-53. The pivot bearing has been bored out at its end. If the coefficient of friction at the bearing surface is $\mu_s = 0.35$, and a moment $M = 10$ lb-ft is required to rotate the shaft, determine the applied axial load P.

Prob. 8–53

8-54. The *disk clutch* is used in standard transmissions of automobiles. If four springs are used to force the two plates A and B together, determine the force in each spring required to transmit a moment of $M = 600$ lb-ft across the plates. The coefficient of friction is $\mu_s = 0.3$.

Prob. 8–54

8-55. The plate A is engaged to plate B using three springs. Each spring exerts a normal force of 500 lb on plate B. If $\mu_s = 0.35$ for the contacting surfaces, determine the maximum moment M which may be transmitted across the plates.

Prob. 8-55

8-56. Because of wearing at the edges, the pivot bearing is subjected to a conical-pressure distribution at its surface of contact. Determine the moment M required to turn the shaft which supports an axial load P. The coefficient of friction is μ.

Prob. 8-56

8-57. Determine the moment M required to rotate the "conical" pivot bearing when the bearing is subjected to an axial load P. The coefficient of friction for the contacting surfaces is μ. Assume that the pressure acting between the conical surface of contact is uniform.

Prob. 8-57

8-58. Determine the normal and frictional components of force developed on the collar bushing. The coefficient of friction is $\mu = 0.21$. Assume that the belt moves in the direction of force T.

Prob. 8-58

8-59. If a tension force $T = 215$ lb is required to pull the 200-lb force around the collar bushing in Prob. 8-58, determine the coefficient of friction at the contacting surface.

8-60. A 4-in.-diameter post is driven 12 ft into mud for which $\mu = 0.15$. If the normal pressure acting around the post varies linearly with depth as shown, determine the frictional moment **M** which must be overcome to rotate the post.

Prob. 8-60

8-61. Because of drying, the coefficient of friction of the mud acting on the post in Prob. 8-58 varies linearly from $\mu = 0.3$ at the top of the post to $\mu = 0.1$ at its base. Determine the frictional moment **M** needed to rotate the post. Use the data in Prob. 8-60.

8-62. A 12-in.-diameter disk fits loosely over a 3-in.-diameter shaft in which the coefficient of friction is $\mu = 0.15$. If the disk weighs 100 lb, determine the vertical tangential force which must be applied to the disk to cause it to slip over the shaft.

8-63. The 4-in.-diameter pulley fits loosely on a $\frac{1}{2}$-in.-diameter axle. If $\mu = 0.30$, determine the minimum tension **T** required to raise the 250-lb load. Neglect the weight of the pulley.

Prob. 8-63

8-64. The collar fits loosely around a 2-in.-diameter shaft. If the coefficient of friction between the shaft and the collar is $\mu = 0.29$, determine the maximum belt tension **T** needed to lower the 150-lb load.

Prob. 8-64

297

8-5. Frictional Forces in Flat Belts

In the design of belt drives or band brakes it is necessary to determine the frictional forces developed between a belt and its contacting surface. Consider the flat belt acting over a fixed drum of radius r, as shown in Fig. 8–20a. The *total angle of belt contact* is β, and the coefficient of friction between the two surfaces is μ. If it is known that the tension acting in the belt on the right of the drum is T_1, let it be required to find the tension T_2 needed to pull the belt counterclockwise over the surface of the drum. *Obviously, T_2 must be greater than T_1, since the belt must overcome the resistance of friction at the surface of contact.* A free-body diagram of the *entire belt* is shown in Fig. 8–20b. Because of the frictional effects acting over the curved surface, it is seen that the normal force N and the frictional force F, acting at various points along the contacting surface of the belt, will vary both in magnitude and direction. Due to this unknown force distribution, the analysis of the problem will proceed on the basis of initially studying the forces acting on a differential element of the belt.

A free-body diagram of an element of the belt having a length $r\, d\theta$ is shown in Fig. 8–20c. Assuming either impending motion, or motion of the belt, the magnitude of frictional force $dF = \mu\, dN$. This force opposes the sliding motion of the belt and thereby increases the magnitude of tensile force acting in the belt. Applying the two force equations of equilibrium, we have

$$\xrightarrow{+}\Sigma F_x = 0; \qquad T\cos\frac{d\theta}{2} + \mu\, dN - (T + dT)\cos\frac{d\theta}{2} = 0$$

$$+\uparrow\Sigma F_y = 0; \qquad dN - (T + dT)\sin\frac{d\theta}{2} - T\sin\frac{d\theta}{2} = 0$$

For an angle of infinitesimal size, $\sin d\theta/2$ and $\cos d\theta/2$ may be replaced by $d\theta/2$ and 1, respectively. Also, the *product* of two infinitesimals dT and $d\theta/2$ may be neglected when compared to infinitesimals of the first order. The above two equations therefore reduce to

$$\mu\, dN = dT$$

and

$$dN = T\, d\theta$$

Eliminating dN between these two relations yields

$$\frac{dT}{T} = \mu\, d\theta$$

We may integrate this equation between all the points of contact which the belt makes with the drum. Since $T = T_1$ at $\theta = 0°$, and $T = T_2$ at $\theta = \beta$, then

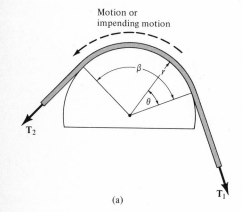

Motion or impending motion

T_2

β

θ

T_1

(a)

$N = N(\theta)$

$F = F(\theta)$

T_2

θ

T_1

(b)

y

$\frac{d\theta}{2}$ $r\,d\theta$ $d\mathbf{F} = \mu\, dN$

$\mathbf{T} + d\mathbf{T}$ dN $\mathbf{T}$

$\frac{d\theta}{2}$

$\frac{d\theta}{2}$ $\frac{d\theta}{2}$

$\frac{d\theta}{2}$

x

(c)

Fig. 8–20

$$\int_{T_1}^{T_2} \frac{dT}{T} = \mu \int_0^\beta d\theta$$

or

$$\ln \frac{T_2}{T_1} = \mu\beta$$

Solving for T_2, we obtain

$$T_2 = T_1 e^{\mu\beta} \tag{8-6}$$

This equation is valid for flat belts placed on *any shape* of contacting surface. In summary, the terms in this equation are defined as follows:

μ = coefficient of static or kinetic friction between the belt and the surface of contact

β = angle of belt and surface contact, measured in radians

T_2, T_1 = belt tensions; $\mathbf{T}_1$ opposes the direction of motion (or impending motion) of the belt, while $\mathbf{T}_2$ acts in the direction of belt motion (or impending motion); because of friction, $T_2 > T_1$.

e = 2.718, base of natural logarithm.

An application of Eq. 8-6 is given in Example 8-10.

*8-6. Rolling Resistance

If a *rigid* cylinder of weight W rolls at constant velocity along a *rigid* surface, the normal force exerted by the surface on the cylinder acts at the tangent point of contact, as shown in Fig. 8-21a. Under these condi-

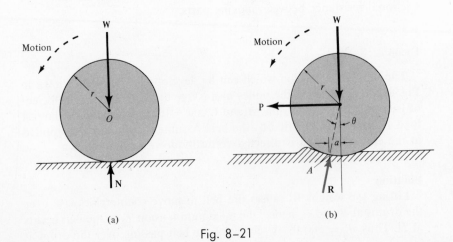

(a)

(b)

Fig. 8-21

tions, provided the cylinder does not encounter frictional resistance from the air, motion of the cylinder would continue indefinitely. Actually, however, no two materials are perfectly rigid. Hence, a normal reaction of the surface on the cylinder consists of a *distribution* of normal forces. Assuming that the cylinder is much harder than the surface, a small ridge is formed in front of the rolling cylinder caused by the indentation of the cylinder into the surface, Fig. 8–21b. Since the cylinder can never really "climb" over this "hill," the total reactive force **R** at the surface acts at some angle θ from the vertical. To keep the cylinder in equilibrium, that is, moving at constant velocity, it is necessary that **R** be *concurrent* with the driving force **P** and the weight **W.** Taking moments about point A gives $Wa = P(r\cos\theta)$. Since the deformations of the cylinder are generally small in relation to the radius, $\cos\theta \approx 1$; hence,

$$Wa \approx Pr$$

or

$$P \approx \frac{Wa}{r}$$

The distance "a" is termed the *coefficient of rolling resistance,* which has the dimension of length. Experimentally, this factor is difficult to measure, since it depends upon such parameters as the rate of rotation of the cylinder, the elastic properties of the contacting surfaces, and the frictional effects of the surfaces. For this reason, little reliance is placed on the data for determining a. The analysis presented here does, however, indicate why a heavy load offers greater resistance to motion than a light load under the same conditions.

The driving force **P** needed to roll a cylinder over a surface, however, is much less than that needed to slide the cylinder across the surface. For this reason roller or ball bearings are often used to minimize the frictional resistance between moving parts.

Example 8–10

The maximum tension which can be developed in the belt, shown in Fig. 8–22a, is 500 lb. If the pulley at A is free to rotate and the coefficient of friction at the fixed drums B and C is $\mu = 0.25$, determine the largest load W which can be lifted by the belt. Assume that the force **T** applied at the end of the belt is directed vertically downward, as shown.

Solution

Lifting the weight W causes the belt to move counterclockwise over the drums at B and C; hence, the maximum tension T_2 in the belt occurs at D. Thus, $T_2 = 500$ lb. A section of the belt passing over the drum at

(a)

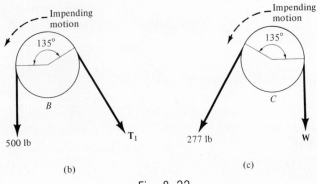

(b) (c)

Fig. 8–22

B is shown in Fig. 8–22b. The angle of contact between the drum and the belt is $\beta = (135°/180°)\pi = 3\pi/4$ rad. Using Eq. 8–6, we have

$$T_2 = T_1 e^{\mu\beta}$$
$$500 \text{ lb} = T_1 e^{0.25[(3/4)\pi]}$$

Hence,

$$T_1 = \frac{500 \text{ lb}}{e^{0.25[(3/4)\pi]}} = 277 \text{ lb}$$

Since the pulley at A is free to rotate, equilibrium requires that the tension in the belt remains the *same* on both sides of the pulley. Can you show this?

The section of the belt passing over the drum at C is shown in Fig. 8–22c. The load $W < 277$ lb. Why? Applying Eq. 8–6, we obtain

$$277 \text{ lb} = W e^{0.25[(3/4)\pi]}$$
$$W = 100 \text{ lb} \qquad\qquad Ans.$$

Problems

8-65. A "hawser" is wrapped around a fixed "capstan" to secure a ship for docking. If the tension in the rope, caused by the ship, is 1,500 lb, determine the least number of complete turns the rope must be wrapped around the capstan in order to prevent slipping of the rope. The greatest horizontal force that a longshoreman can exert on the rope is 50 lb. The coefficient of friction is $\mu_s = 1/\pi$.

1,500 lb

50 lb

Prob. 8-65

8-66. A flat belt is wrapped over the elliptical drum. If a weight of 10 lb is suspended from one end of the belt, determine the tension T in the belt needed to (a) raise the weight, and (b) lower the weight, at constant velocity. The coefficient of friction is $\mu = 1/\pi$.

3 in.

12 in.

10 lb

Prob. 8-66

T

8-67. Determine the maximum and minimum values of weight W which may be applied without causing the 50-lb block to slip. The coefficient of friction between the block and the plane is $\mu_s = 0.2$, and between the rope and the drum $D, \mu_s = 0.3$.

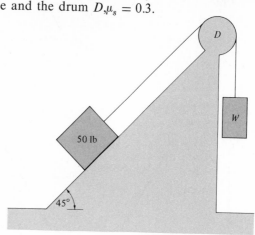

D

W

50 lb

45°

Prob. 8-67

8-68. If the applied moment M acting on the wheel is 200 lb-ft, determine the vertical force P which must be applied to the band brake to stop the wheel. The coefficient of friction for the brake is $\mu_k = 0.4$.

8 in.

M

4 in.

O

P

4 in.   12 in.

Prob. 8-68

20 12-8 8-2

8-69. The band brake is used to stop the wheel. If slipping of the belt occurs when a vertical load of 50 lb is applied to the lever, determine the reaction at the pin O. The coefficient of friction for the brake is $\mu_k = 0.25$.

Prob. 8–69

8-70. The blocks A and B weigh 100 and 50 lb respectively. The coefficient of friction at each contacting surface is indicated in the figure. Determine the maximum weight W which may be applied without causing motion of block B.

Prob. 8–70

8-71. The 14-ft uniform beam AB weighs 20 lb, and is supported by a rope which passes over a pulley at C and a fixed wheel at D. If the coefficient of friction at D is $\mu_s = 0.3$, determine the distance x from the end of the beam at which a 5-lb weight may be placed and not cause the beam to tip.

Prob. 8–71

8-72. Two 50-lb blocks are attached to a cable which passes over two drums. If $\mu = 0.3$ for the drums, determine the height h the blocks raise if a vertical load $P = 200$ lb is applied to the cable as shown. When $h = 0$ the tension in the cable is zero.

Prob. 8–72

303

8-73. A small motor is attached to the floor of the platform. If the total weight of the platform and its contents is 500 lb, determine the torque T which must be supplied by the motor to (a) raise the platform at constant velocity, and (b) lower the platform at constant velocity. In both cases, assume the pulley at A is suddenly fixed from rotating, in which case $\mu_k = 1/\pi$.

Prob. 8-73

8-74. Show that the frictional relation between the belt tensions, the coefficient of friction μ, and the angular contacts α and β for the V-belt is $T_2 = T_1 e^{[\mu\beta/\sin(\alpha/2)]}$.

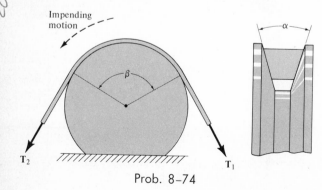

Prob. 8-74

8-75. A 60° V-fan belt of the automobile engine passes around the hub H of a generator G and over the housing F to a fan. If the generator locks, causing the hub to become fixed, determine the moment resisted by the generator as the belt slips over the hub. Assume that slipping of the belt only occurs at F and that the coefficient of friction for the hub is $\mu_s = 0.30$. The belt can sustain a maximum tension of 50 lb. (See Prob. 8-74.)

Prob. 8-75

8-76. Experimentally it is found that a 3-in.-diameter cylinder rolls at constant velocity down an inclined plane having a slope of $\frac{1}{16}$ in./ft. Determine the coefficient of rolling resistance for the cylinder.

8-77. The eight wheels on a tank car are each 30 in. in diameter. If the coefficient of rolling resistance is 0.002 in. with the tracks and the car weighs 30 tons, determine the amount of horizontal force **P** required to overcome the rolling resistance of the wheels.

8-78. The lawn roller weighs 300 lb. If the rod BA is held at an angle of 30° from the horizontal and the coefficient of rolling resistance for the roller is 2 in., determine the horizontal component of force, $\mathbf{F}_x$, needed to push the roller at constant speed. Neglect the friction developed at the axle and assume that the resultant force acting on the handle is applied along BA.

Prob. 8–78

8-79. The roller is subjected to the load of W lb. If the coefficient of rolling resistance for the top surface of the roller is a_1 and at the bottom surface is a_2, show that a force of $P = [W(a_1 + a_2)]/2r$ is required to move the load and the roller forward.

Prob. 8–79

8-80. The 3,000-lb steel beam is moved over a level surface by using a series of 3-in.-diameter rollers for which the coefficient of rolling resistance is $\frac{1}{8}$ in. at the ground and $\frac{1}{16}$ in. at the surface of the beam. Determine the horizontal force **P** needed to move the beam at constant speed. (*Hint:* See Prob. 8–79.)

Prob. 8–80

9

Center of Gravity and Centroids

9–1. Introduction

In Sec. 4–11, it was pointed out that distributed loads can act either on the surface of a body or throughout its volume. Since these types of loads often vary in some complex fashion, it becomes necessary to develop a general method for reducing distributed loads to a single resultant force having a specified location. Occasionally this general method is also used for finding the *centroid* or geometric center of a body. For example, it was stated in Sec. 4–11 that the center of gravity of a *homogeneous body* coincides with the volume centroid of the body. Recall that the weight of the body can then be replaced by a single resultant force acting through this point. In a similar manner, the resultant force of a distributed surface loading acts through the centroid of the distributed loading diagram. In the theory of strength of materials, the centroid of the cross-sectional *area* of a beam or column must be located in order to properly design these members.

In this chapter we will discuss a general method for determining the center of gravity of a body and the centroid of various geometric shapes. Before discussing this method as applied to continuous systems, however, it is instructive to first apply this same method to finding the center of gravity of a system of discrete particles.

9–2. Center of Gravity of a System of Particles

Consider the system of three particles shown in Fig. 9–1a. Each particle has a different weight, and the particles are distributed across the top

surface of an imaginary weightless beam in a manner shown in the figure. Using the methods developed in Sec. 4–10, this parallel system of forces may be reduced to an equivalent force-moment system acting at an arbitrary point O on the beam, as shown in Fig. 9–1b. From this figure,

$$W = W_1 + W_2 + W_3$$

and

$$\zeta + M_O = \tilde{x}_1 W_1 + \tilde{x}_2 W_2 + \tilde{x}_3 W_3$$

$\tilde{x}_1$, $\tilde{x}_2$, and $\tilde{x}_3$ represent the *algebraic distances* from the particles to the origin O. Since $\mathbf{M}_O$ is *perpendicular* to $\mathbf{W}$, the force-moment system may be further resolved to a single force resultant $\mathbf{W}$. To do this, $\mathbf{W}$ must be moved a distance $\bar{x}$ to the right of point O in order to create an equivalent moment $\mathbf{M}_O$ at point O, i.e., $M_O = \bar{x}W$. The distance $\bar{x}$ becomes

$$\bar{x} = \frac{M_O}{W} = \frac{\tilde{x}_1 W + \tilde{x}_2 W_2 + \tilde{x}_3 W_3}{W_1 + W_2 + W_3}$$

We may conclude from this result that the effect caused by a single particle having a weight W (equal to the total weight of all the particles) and placed at point G located a distance $\bar{x}$ to the right of point O, as shown in Fig. 9–1c, is *equivalent* to the effect caused by the weights of all three particles shown in Fig. 9–1a. The point G is termed the *center of gravity* of the system of particles. (For this special case G may be considered as a "balance point" for the unsupported weightless beam; that is, if a fulcrum were placed under the beam at point G as shown in Figs. 9–1a, 9–1b, or 9–1c, the beam would remain balanced in the horizontal position.)

The equation for $\bar{x}$ may be generalized to include n particles acting on the beam. For this case,

$$\bar{x} = \frac{\displaystyle\sum_{i=1}^{n} \tilde{x}_i W_i}{\displaystyle\sum_{i=1}^{n} W_i} \tag{9–1}$$

where W_i is the weight of the ith particle and $\tilde{x}_i$ is the *algebraic distance* from the origin to the ith particle.

If n particles, having weights $W_1, W_2, \ldots, W_n$, are all contained in the xy plane, the location of the ith particle may be specified using coordinates $(\tilde{x}_i, \tilde{y}_i)$. Summing moments of the weights of these particles about the x and y axes, respectively, in the same manner as above, one obtains the relations

$$\bar{x} = \frac{\displaystyle\sum_{i=1}^{n} \tilde{x}_i W_i}{\displaystyle\sum_{i=1}^{n} W_i}, \qquad \bar{y} = \frac{\displaystyle\sum_{i=1}^{n} \tilde{y}_i W_i}{\displaystyle\sum_{i=1}^{n} W_i} \tag{9–2}$$

(a)

(b)

(c)

Fig. 9–1

These equations locate the position $(\bar{x}, \bar{y})$ of the center of gravity for the system of particles.

In the most general case, the ith particle is located at a point $(\tilde{x}_i, \tilde{y}_i, \tilde{z}_i)$ in space. If the system of n particles is rotated with respect to any of the three axes, the weights of the various particles will not remain in the same relative positions; therefore, the resultant weight will be in a different position relative to the n particles. The line of action of the resultant weight, however, does act through the center of gravity $(\bar{x}, \bar{y}, \bar{z})$ regardless of how the system of particles is rotated. The location of this point can be determined by using the moment principle just described. The results are

$$\bar{x} = \frac{\sum\limits_{i=1}^{n} \tilde{x}_i W_i}{\sum\limits_{i=1}^{n} W_i}, \qquad \bar{y} = \frac{\sum\limits_{i=1}^{n} \tilde{y}_i W_i}{\sum\limits_{i=1}^{n} W_i}, \qquad \bar{z} = \frac{\sum\limits_{i=1}^{n} \tilde{z}_i W_i}{\sum\limits_{i=1}^{n} W_i} \qquad (9\text{--}3)$$

In order to study problems concerning the motion of *matter* under the influence of forces, that is, dynamics, it is necessary to locate the *center of mass*. In finding the location of this point for a system of particles one must consider the weighted average of the *mass* of each particle. In this way, equations analogous to Eq. 9–3 are used to locate the center of mass, only with m_i substituted for W_i in these expressions. Since particles have weight only when under the influence of a gravitational attraction, the center of gravity depends upon the existence of a gravitational field for its definition. The center of mass is independent of gravitational forces. For example, it would be meaningless to define the center of gravity of particles representing the planets of our solar system, whereas the center of mass of this system is important.

If all the particles are approximately the *same distance* from the center of the earth, the acceleration of gravity g for the particles may be considered a *constant*. Newton's second law of motion (see Chapter 1) then becomes $W_i = m_i g$ for all i. Substituting this expression into Eqs. 9–3, the constant g will cancel from both numerator and denominator; hence, for this case the points defining the center of mass and center of gravity for the system of particles will coincide.

(a)

(b)

Fig. 9–2

Example 9–1

Determine the location of the center of gravity G for the system of four particles shown in Fig. 9–2a with respect to the particle at B.

Solution

Locating the origin of coordinates at (the arbitrary) point B, as shown in Fig. 9–2a, the distance $\bar{x}$ from B to the center of gravity can be found using Eq. 9–1.

$$\bar{x} = \frac{\Sigma \tilde{x}_i W_i}{\Sigma W_i} = \frac{-2 \text{ ft}(1 \text{ lb}) + 0 \text{ ft}(3 \text{ lb}) + 3 \text{ ft}(2 \text{ lb}) + 5 \text{ ft}(3 \text{ lb})}{1 \text{ lb} + 3 \text{ lb} + 2 \text{ lb} + 3 \text{ lb}}$$

$$= \frac{19 \text{ lb-ft}}{9 \text{ lb}} = 2.11 \text{ ft} \qquad \qquad Ans.$$

The four particles shown in Fig. 9–2a may, therefore, be replaced by a 9-lb particle acting at G, 2.11 ft to the right of B. See Fig. 9–2b.

Example 9–2

Locate the center of gravity G of the three particles shown in Fig. 9–3a.

(a)   (b)

Fig. 9–3

Solution

Since the particle weights all act parallel to the z axis, Fig. 9–3a, they create no moment about this axis; hence, the calculations for the position $\bar{x}$ and $\bar{y}$ of the center of gravity for the system of particles can be visualized when the coordinate axis is oriented as shown. Using Eq. 9–3, we obtain

$$\bar{x} = \frac{\Sigma \tilde{x}_i W_i}{\Sigma W_i} = \frac{3 \text{ ft}(1 \text{ lb}) + 1 \text{ ft}(2 \text{ lb}) + 0 \text{ ft}(3 \text{ lb})}{1 \text{ lb} + 2 \text{ lb} + 3 \text{ lb}} = 0.833 \text{ ft} \quad Ans.$$

$$\bar{y} = \frac{\Sigma \tilde{y}_i W_i}{\Sigma W_i} = \frac{2 \text{ ft}(1 \text{ lb}) + 3 \text{ ft}(2 \text{ lb}) + 0 \text{ ft}(3 \text{ lb})}{1 \text{ lb} + 2 \text{ lb} + 3 \text{ lb}} = 1.33 \text{ ft} \quad Ans.$$

To determine the location $\bar{z}$ it may be helpful to imagine the coordinate axis as being rotated 90° about the x axis as shown in Fig. 9–3b. (Although it is not necessary, this is done here to "visualize" how the moments of forces are developed.) For this orientation of the axes, it is possible to see how both $\bar{x}$ and $\bar{z}$ are determined. In particular,

$$\bar{z} = \frac{\Sigma \tilde{z}_i W_i}{\Sigma W_i} = \frac{0 \text{ ft}(1 \text{ lb}) + 3 \text{ ft}(2 \text{ lb}) + 2 \text{ ft}(3 \text{ lb})}{1 \text{ lb} + 2 \text{ lb} + 3 \text{ lb}} = 2.00 \text{ ft} \quad Ans.$$

$$\bar{y} = \frac{\int y \, dw}{W}$$

centroid means geometry only

9-1. Locate the center of gravity G of the five particles with respect to the origin O.

Prob. 9–1

Sett 4
Nov. 5

9-2. Locate the center of gravity $(\bar{x}, \bar{y}, \bar{z})$ of the three particles.

Prob. 9–2

9-3. Locate the center of gravity $(\bar{x}, \bar{y}, \bar{z})$ of the four particles.

Prob. 9–3

9-4. Determine the location (x, y) of a 7-lb particle so that the three particles have a center of gravity located at the origin of coordinates.

Prob. 9–4

9-5. If the four particles can be replaced by a single 10-lb particle acting at a distance 2 ft to the left of the origin, determine the position $\tilde{x}_p$ and the weight W_p of particle P.

Prob. 9–5

9-6. Locate the coordinates $(\bar{x}, \bar{y})$ of the center of gravity for the system of four particles lying on the xy plane.

Prob. 9–6

9-3. Center of Gravity and Centroid of a Body

The principles used to determine the center of gravity for a system of discrete particles may also be used to determine the center of gravity for continuous systems which essentially contain an infinite number of particles. For example, consider the problem of determining the center of gravity G of a thin wire of length L placed along the x axis, Fig. 9–4a. The location of the center of gravity of the wire may be *approximated* by dividing the wire into segments and considering each segment to be a separate particle. For example, the ith particle, shown in the figure, has a length Δx_i and weight ΔW_i. Provided the cross-sectional area of the wire is constant, the density of the wire may be represented as a *linear density* ρ_L which has units of weight per unit length. In general, ρ_L will vary along the length of the wire, i.e., $\rho_L = \rho_L(x)$. The weight of the ith segment of wire therefore becomes $\Delta W_i = \rho_L \Delta x_i$.

When the weight ΔW_i is replaced at the origin O, it is also necessary to include a moment $\Delta M_i = \tilde{x}_i \Delta W_i$, where $\tilde{x}_i$ represents the *algebraic distance* from the origin of coordinates to the ith segment of wire. When the weight of all the n segments of wire are transferred to the origin, Fig. 9–4b, the total force becomes $W = \Sigma_{i=1}^{n} \Delta W_i = \Sigma_{i=1}^{n} \rho_L \Delta x_i$, and the resultant moment becomes $M_O = \Sigma_{i=1}^{n} \tilde{x}_i \Delta W_i = \Sigma_{i=1}^{n} \tilde{x}_i \rho_L \Delta x_i$. Since $\mathbf{M}_O$ is perpendicular to the total weight $\mathbf{W}$, the weight may be transferred to a point, $\bar{x}$, which locates the center of gravity of the wire. This requires

$$\bar{x} = \frac{M_O}{W} = \frac{\sum\limits_{i=1}^{n} \tilde{x}_i \rho_L \Delta x_i}{\sum\limits_{i=1}^{n} \rho_L \Delta x_i}$$

Taking smaller and smaller Δx_i segments until $\Delta x_i \rightarrow 0$, and hence $n \rightarrow \infty$, we obtain an integral relation which defines the exact position of the center of gravity, i.e.,

$$\bar{x} = \frac{\lim\limits_{\Delta x_i \rightarrow 0} \sum\limits_{i=1}^{n} \tilde{x}_i \rho_L \Delta x_i}{\lim\limits_{\Delta x_i \rightarrow 0} \sum\limits_{i=1}^{n} \rho_L \Delta x_i} = \frac{\int_0^L \tilde{x}\rho_L \, dx}{\int_0^L \rho_L \, dx}$$

See Fig. 9–4c.

In particular, a differential element of length dx, is located at an arbitrary distance $\tilde{x} = x$ from the origin as shown in Fig. 9–4d. If the linear density ρ_L is constant—that is, if the wire is made of homogeneous material, then, using the above formulation, we have

$$\bar{x} = \frac{\rho_L \int_0^L x \, dx}{\rho_L \int_0^L dx} = \frac{\rho_L \dfrac{L^2}{2}}{\rho_L L} = \frac{L}{2}$$

(a)

(b)

(c)

(d)

Fig. 9–4

The total weight of the wire $W = \rho_L L$ may therefore be represented as a concentrated force acting at the midpoint of the wire, Fig. 9–4d.

In general, the geometry for a slender rod or wire may be described in three directions by a *curve* such as shown in Fig. 9–5. If the wire has a total length L and linear density ρ_L, the center of gravity G is determined using the formulas

Fig. 9–5

$$\bar{x} = \frac{\int_L \tilde{x}\rho_L\,dL}{\int_L \rho_L\,dL}, \qquad \bar{y} = \frac{\int_L \tilde{y}\rho_L\,dL}{\int_L \rho_L\,dL}, \qquad \bar{z} = \frac{\int_L \tilde{z}\rho_L\,dL}{\int_L \rho_L\,dL} \qquad (9\text{-}4)$$

The derivation of these formulas is based upon the moment principle used for finding the center of gravity of the straight wire just described. As in the case of a system of particles, the center of gravity G may lie at a point in space, not located on the body, Fig. 9–5.

When a continuous body assumes the shape of a flat or curved surface having an area A and constant thickness, such as a plate or shell, the center of gravity may be found by dividing the surface into a number of area elements. The differential element shown in Fig. 9–6 has an area

Fig. 9–6

dA and corresponding weight $dW = \rho_A\, dA$, where ρ_A represents a *surface density* or weight per unit area of the material composing the surface. In general, ρ_A may vary across the surface. In a manner similar to using the moment principle just described for linear elements, it may be shown that the center of gravity G for a surface has coordinates

$$\overline{x} = \frac{\int_A \widetilde{x}\rho_A\, dA}{\int_A \rho_A\, dA}, \qquad \overline{y} = \frac{\int_A \widetilde{y}\rho_A\, dA}{\int_A \rho_A\, dA}, \qquad \overline{z} = \frac{\int_A \widetilde{z}\rho_A\, dA}{\int_A \rho_A\, dA} \qquad (9\text{--}5)$$

A body of volume V, having a *density* or weight per unit volume ρ_V, may be divided into volume segments, having volume dV and weight $dW = \rho_V\, dV$, Fig. 9–7. It may then be shown that the center of gravity G for the body has coordinates

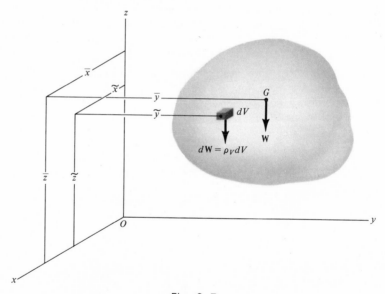

Fig. 9–7

$$\overline{x} = \frac{\int_V \widetilde{x}\rho_V\, dV}{\int_V \rho_V\, dV}, \qquad \overline{y} = \frac{\int_V \widetilde{y}\rho_V\, dV}{\int_V \rho_V\, dV}, \qquad \overline{z} = \frac{\int_V \widetilde{z}\rho_V\, dV}{\int_V \rho_V\, dV} \qquad (9\text{--}6)$$

If the material composing the body is *homogeneous*, the density ρ will be *constant* throughout, and therefore, this term will factor out of the

integrals and hence *cancel* from both the numerator and denominator in Eqs. 9-4, 9-5, and 9-6. The resulting formulas will then *depend* purely on the *size* and *shape* of the body. As mentioned earlier, when the calculations depend entirely upon the geometrical properties of the solid, the point $(\bar{x}, \bar{y}, \bar{z})$ is called the *centroid* of the body.*

For uniform rods or wires having a *constant linear density* ρ_L,

$$\bar{x} = \frac{\int_L \tilde{x}\, dL}{\int_L dL}, \qquad \bar{y} = \frac{\int_L \tilde{y}\, dL}{\int_L dL}, \qquad \bar{z} = \frac{\int_L \tilde{z}\, dL}{\int_L dL} \qquad (9\text{-}7)$$

For thin homogeneous plates and shells having a uniform thickness and *constant surface density* ρ_A,

$$\bar{x} = \frac{\int_A \tilde{x}\, dA}{\int_A dA}, \qquad \bar{y} = \frac{\int_A \tilde{y}\, dA}{\int_A dA}, \qquad \bar{z} = \frac{\int_A \tilde{z}\, dA}{\int_A dA} \qquad (9\text{-}8)$$

For volume shapes of constant density ρ_V,

$$\bar{x} = \frac{\int_V \tilde{x}\, dV}{\int_V dV}, \qquad \bar{y} = \frac{\int_V \tilde{y}\, dV}{\int_V dV}, \qquad \bar{z} = \frac{\int_V \tilde{z}\, dV}{\int_V dV} \qquad (9\text{-}9)$$

The *centroids* of some shapes may be partially or completely specified by using *symmetry conditions*. In cases where the shape has an axis of symmetry, the centroid of the shape will lie along that axis. For example, the centroid C for the homogeneous wire shown in Fig. 9-8 must lie along the y axis, since for every elemental length dL at a distance $+x$ to the right of the y axis, there is an identical element at a distance $-x$ to the left of the y axis. The total moment for all of the elements about the axis of symmetry will therefore cancel, that is, $\int_L \tilde{x}\, dL = 0$ so that $\bar{x} = 0$. In cases where a shape has two or three axes of symmetry, it follows that the centroid lies at the intersection of these axes. See Fig. 9-9.

---

*Of course, a nonhomogeneous body also has a centroid. In this case, however, the centroid and center of gravity will *not* coincide.

Fig. 9–8

Fig. 9–9

(a)

(b)

(c)

In general, when integrating Eqs. 9–4 through 9–9 it is best to choose a coordinate system which simplifies the equations of the boundary as much as possible. For example, polar coordinates are generally appropriate in solving problems having circular boundaries. The differential element should be chosen so that only one integration is required to cover the entire region. This can be done provided the element is of differential size or thickness in one direction only. Such an element is called a *first-order element*. If it is not possible to use first-order elements, a second-order element (having a differential size in two directions) should be chosen in preference to a third-order element. For second-order elements two integrations are required.

The example problems given at the end of this section illustrate the techniques for computing the center of gravity and centroids for lines, surfaces, and volumes using first-order elements. It is important that you study these examples carefully in order to obtain a clear understanding of how to apply the necessary formulas. In general, the following step-by-step procedure may be applied for computing the center of gravity or centroid of a solid using a first-order element for integration.

1. Specify the coordinate axes, and choose a differential element for the solid. For lines (wires, rods) this element is represented as a differential line segment; for areas (plates, shells) the element is generally a rectangle, having a finite height and differential width; and for volumes (bodies) the element is either a circular disk, having a finite radius and differential thickness, or a shell having a finite length and radius, and differential thickness.

2. Construct the element so that it intersects the surface boundary of the line, area, or volume at *some arbitrary point*. This fixes the dimensions of the element.

3. Express the length, area, or volume of the element in terms of the coordinates used to define the boundary of the solid.

4. Determine the perpendicular distance from the coordinate axes to the *centroid or center of gravity of the element*. For an x, y, z coordinate system, these dimensions are expressed as $\tilde{x}$, $\tilde{y}$, and $\tilde{z}$.

5. Substitute the data computed in steps 3 and 4 into the appropriate equations (Eqs. 9–4 through 9–9) and perform the integration. (For computing the center of gravity of nonhomogeneous solids, using Eqs. 9–4, 9–5, or 9–6, the material density must be expressed in terms of the coordinates of the boundary.) In particular, integration using a first-order element is accomplished when the function in the integrand is expressed in terms of the *same variable as the differential thickness of the element*. The limits of the integral are then defined from the two extreme locations of the element's differential thickness, so that when the elements are summed (or the integration performed), the entire region is covered.

This step-by-step procedure is used in solving the following examples.

Example 9–3

Locate the centroid of the circular wire segment shown in Fig. 9–10.

Solution
Step 1: Polar coordinates will be used to solve this problem since the arc is circular. The differential element is shown in Fig. 9–10.
Step 2: The element intersects the curve at (R, θ).
Step 3: The element length is $dL = R\, d\theta$.
Step 4: The element centroid is located at $\tilde{x} = R\cos\theta$, $\tilde{y} = R\sin\theta$.
Step 5: Using Eqs. 9–7, and integrating with respect to θ, we obtain

$$\bar{x} = \frac{\displaystyle\int_L \tilde{x}\, dL}{\displaystyle\int_L dL} = \frac{\displaystyle\int_0^{\pi/2} (R\cos\theta)R\, d\theta}{\displaystyle\int_0^{\pi/2} R\, d\theta} = \frac{R^2\displaystyle\int_0^{\pi/2}\cos\theta\, d\theta}{R\displaystyle\int_0^{\pi/2} d\theta} = \frac{2R}{\pi} \quad Ans.$$

Fig. 9–10

Fig. 9–11

$$\bar{y} = \frac{\int_L \tilde{y}\, dL}{\int_L dL} = \frac{\int_0^{\pi/2} (R \sin \theta) R\, d\theta}{\int_0^{\pi/2} R\, d\theta} = \frac{R^2 \int_0^{\pi/2} \sin d\theta}{R \int_0^{\pi/2} d\theta} = \frac{2R}{\pi} \qquad Ans.$$

Example 9–4

Locate the centroid of the wire bent into the shape of a parabolic arc, as shown in Fig. 9–11.

Solution

Step 1: The differential element is shown in Fig. 9–11.

Step 2: The element intersects the curve at (x, y).

Step 3: The element length is determined using the Pythagorean Theorem.

$$dL = \sqrt{(dx)^2 + (dy)^2} = \sqrt{\left(\frac{dx}{dy}\right)^2 + 1}\, dy$$

Since $x = y^2$, then $dx/dy = 2y$. Therefore

$$dL = \sqrt{(2y)^2 + 1}\, dy$$

Step 4: The centroid of the element is located at $\tilde{x} = x$, $\tilde{y} = y$.

Step 5: Using Eqs. 9–7 and integrating with respect to y, we obtain

$$\bar{x} = \frac{\int_L \tilde{x}\, dL}{\int_L dL} = \frac{\int_0^1 x\sqrt{4y^2 + 1}\, dy}{\int_0^1 \sqrt{4y^2 + 1}\, dy} = \frac{\int_0^1 y^2 \sqrt{4y^2 + 1}\, dy}{\int_0^1 \sqrt{4y^2 + 1}\, dy}$$

$$= \frac{0.739}{1.412} = 0.412 \text{ ft} \qquad\qquad Ans.$$

$$\bar{y} = \frac{\int_L \tilde{y}\, dL}{\int_L dL} = \frac{\int_0^1 y\sqrt{4y^2 + 1}\, dy}{\int_0^1 \sqrt{4y^2 + 1}\, dy} = \frac{0.848}{1.470} = 0.577 \text{ ft} \qquad Ans.$$

Example 9–5

Locate the centroid of the shaded area shown in Fig. 9–12a.

(a)

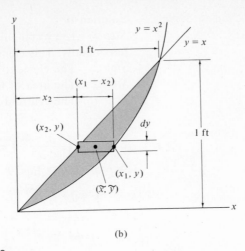

(b)

Fig. 9-12

Solution I

Step 1: The differential element is shown in Fig. 9–12a.

Step 2: The element intersects the curves at points (x, y_1) and (x, y_2).

Step 3: The element area is $dA = (y_2 - y_1)\, dx$.

Step 4: The centroid of the element is located at $\tilde{x} = x$,

$$\tilde{y} = y_1 + \left(\frac{y_2 - y_1}{2}\right) = \frac{y_1 + y_2}{2}.$$

Step 5: Using Eqs. 9–8, and integrating with respect to y, we obtain

$$\bar{x} = \frac{\int_A \tilde{x}\, dA}{\int_A dA} = \frac{\int_0^1 x(y_2 - y_1)\, dx}{\int_0^1 (y_2 - y_1)\, dx} = \frac{\int_0^1 x(x - x^2)\, dx}{\int_0^1 (x - x^2)\, dx}$$

$$= \frac{\int_0^1 (x^2 - x^3)\, dx}{\int_0^1 (x - x^2)\, dx} = \frac{1/12}{1/6} = 0.5 \text{ ft} \qquad\qquad Ans.$$

$$\bar{y} = \frac{\int_A \tilde{y}\, dA}{\int_A dA} = \frac{\int_0^1 \left(\frac{y_1 + y_2}{2}\right)(y_2 - y_1)\, dx}{\int_0^1 (y_2 - y_1)\, dx} = \frac{\int_0^1 \left(\frac{x^2 + x}{2}\right)(x - x^2)\, dx}{\int_0^1 (x - x^2)\, dx}$$

$$= \frac{\frac{1}{2}\int_0^1 (x^2 - x^4)\, dx}{\int_0^1 (x - x^2)\, dx} = \frac{1/15}{1/6} = 0.4 \text{ ft} \qquad\qquad Ans.$$

319

Solution II

Step 1: The rectangular element may be chosen so that it is orientated horizontally. This element is shown in Fig. 9–12b.

Step 2: The element intersects the curves at points (x_2, y) and (x_1, y).

Step 3: The area of the element is $dA = (x_1 - x_2)\,dy$.

Step 4: The centroid of the element is located at

$$\tilde{x} = x_2 + \left(\frac{x_1 - x_2}{2}\right) = \frac{x_1 + x_2}{2}, \quad \tilde{y} = y.$$

Step 5: Using Eqs. 9–8, and integrating with respect to y, we obtain

$$\bar{x} = \frac{\displaystyle\int_A \tilde{x}\,dA}{\displaystyle\int_A dA} = \frac{\displaystyle\int_0^1 \left(\frac{x_1 + x_2}{2}\right)(x_1 - x_2)\,dy}{\displaystyle\int_0^1 (x_1 - x_2)\,dy} = \frac{\displaystyle\int_0^1 \left(\frac{\sqrt{y} + y}{2}\right)(\sqrt{y} - y)\,dy}{\displaystyle\int_0^1 (\sqrt{y} - y)\,dy}$$

$$= \frac{\dfrac{1}{2}\displaystyle\int_0^1 (y - y^2)\,dy}{\displaystyle\int_0^1 (\sqrt{y} - y)\,dy} = \frac{1/12}{1/6} = 0.5 \text{ ft} \qquad\qquad Ans.$$

$$\bar{y} = \frac{\displaystyle\int_A \tilde{y}\,dA}{\displaystyle\int_A dA} = \frac{\displaystyle\int_0^1 y(x_1 - x_2)\,dy}{\displaystyle\int_0^1 (x_1 - x_2)\,dy} = \frac{\displaystyle\int_0^1 y(\sqrt{y} - y)\,dy}{\displaystyle\int_0^1 (\sqrt{y} - y)\,dy}$$

$$= \frac{\displaystyle\int_0^1 (y^{3/2} - y^2)\,dy}{\displaystyle\int_0^1 (\sqrt{y} - y)\,dy} = \frac{1/15}{1/6} = 0.4 \text{ ft} \qquad\qquad Ans.$$

Example 9–6

Determine the distance $\bar{y}$ to the centroid of the area of the triangle shown in Fig. 9–13.

Solution

Step 1: Consider a rectangular element having thickness dy and *variable length* x', Fig. 9–13. By similar triangles, $b/h = x'/(h - y)$ or $x' = \dfrac{b}{h}(h - y)$.

Step 2: The element intersects the sides of the triangle at a height y above the x axis.

Fig. 9–13

Step 3: The area of the element is $dA = x'\,dy = \dfrac{b}{h}(h - y)\,dy.$

Step 4: The centroid of the element is located a distance $\tilde{y} = y$ from the x axis.

Step 5: Using the second of Eqs. 9–8, and integrating with respect to y, we obtain

$$\bar{y} = \frac{\displaystyle\int_A \tilde{y}\,dA}{\displaystyle\int_A dA} = \frac{\displaystyle\int_0^h y\,\frac{b}{h}(h - y)\,dy}{\displaystyle\int_0^h \frac{b}{h}(h - y)\,dy}$$

$$= \frac{\dfrac{b}{h}\displaystyle\int_0^h (hy - y^2)\,dy}{\dfrac{b}{h}\displaystyle\int_0^h (h - y)\,dy} = \frac{h}{3} \qquad\qquad Ans.$$

Example 9–7

Locate the centroid for the area of a quarter circle shown in Fig. 9–14a.

Solution I

Step 1: Using polar coordinates, we choose the element in the shape of a triangle, Fig. 9–14a. (Actually the shape is a circular sector; however, neglecting higher-order differentials, the element becomes triangular.)

Step 2: The element intersects the curve at point (R, θ).

Step 3: The area of the element is $dA = \frac{1}{2}(R)(R\,d\theta) = \dfrac{R^2}{2}\,d\theta.$

Step 4: Using the results in Example 9–6, the centroid of the (triangular) element is located at $\tilde{x} = \frac{2}{3}R \cos\theta,\ \tilde{y} = \frac{2}{3}R \sin\theta.$

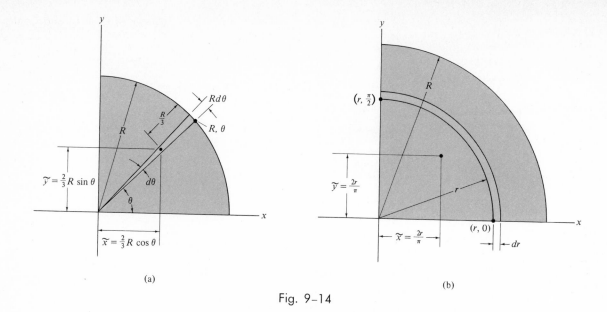

(a)

(b)

Fig. 9–14

Step 5: Using Eqs. 9–8, and integrating with respect to θ, we obtain

$$\bar{x} = \frac{\int_A \tilde{x}\, dA}{\int_A dA} = \frac{\int_0^{\pi/2} \left(\frac{2}{3}R \cos\theta\right)\frac{R^2}{2}\, d\theta}{\int_0^{\pi/2} \frac{R^2}{2}\, d\theta}$$

$$= \frac{\frac{R^3}{3}\int_0^{\pi/2} \cos\theta\, d\theta}{\frac{R^2}{2}\int_0^{\pi/2} d\theta} = \frac{4}{3}\frac{R}{\pi} \qquad\qquad Ans.$$

$$\bar{y} = \frac{\int_A \tilde{y}\, dA}{\int_A dA} = \frac{\int_0^{\pi/2} \left(\frac{2}{3}R \sin\theta\right)\frac{R^2}{2}\, d\theta}{\int_0^{\pi/2} \frac{R^2}{2}\, d\theta}$$

$$= \frac{\frac{R^3}{3}\int_0^{\pi/2} \sin\theta\, d\theta}{\frac{R^2}{2}\int_0^{\pi/2} d\theta} = \frac{4}{3}\frac{R}{\pi} \qquad\qquad Ans.$$

Solution II

Step 1: The differential element may be chosen in the form of a circular arc having a thickness dr as shown in Fig. 9–14*b*.

Step 2: The element intersects the axes at points $(r, 0)$ and $\left(r, \dfrac{\pi}{2}\right)$.

Step 3: The area of the element is $dA = \dfrac{2\pi r}{4}\, dr$.

Step 4: Since the centroid of a 90° circular arc was determined in Example 9–3, then for the element $\tilde{x} = \dfrac{2r}{\pi}, \tilde{y} = \dfrac{2r}{\pi}$.

Step 5: Using Eqs. 9–8, and integrating with respect to r, we obtain

$$\bar{x} = \frac{\displaystyle\int_A \tilde{x}\, dA}{\displaystyle\int_A dA} = \frac{\displaystyle\int_0^R \frac{2r}{\pi}\left(\frac{2\pi r}{4}\right) dr}{\displaystyle\int_0^R \frac{2\pi r}{4}\, dr} = \frac{\displaystyle\int_0^R r^2\, dr}{\dfrac{\pi}{2}\displaystyle\int_0^R r\, dr} = \frac{4}{3}\frac{R}{\pi} \qquad Ans.$$

$$\bar{y} = \frac{\displaystyle\int_A \tilde{y}\, dA}{\displaystyle\int_A dA} = \frac{\displaystyle\int_0^R \frac{2r}{\pi}\left(\frac{2\pi r}{4}\right) dr}{\displaystyle\int_0^R \frac{2\pi r}{4}\, dr} = \frac{\displaystyle\int_0^R r^2\, dr}{\dfrac{\pi}{2}\displaystyle\int_0^R r\, dr} = \frac{4}{3}\frac{R}{\pi} \qquad Ans.$$

Example 9–8

Locate the centroid for the paraboloid of revolution, which is generated by revolving the shaded area shown in Fig. 9–15a about the y axis.

Solution I

Since the volume is *symmetric* with respect to the y axis,

$$\bar{x} = \bar{z} = 0 \qquad Ans.$$

Step 1: To simplify the integration process, a first-order element having the shape of a *thin disk* is chosen. As shown in Fig. 9–15a, this element has a radius of $r = z$ and a thickness of dy.

Step 2: The element intersects the generating curve at point $(0, y, z)$.

Step 3: The volume of the element is $dV = (\pi z^2)\, dy$.

Step 4: The centroid of the element is located at $\tilde{y} = y$.

Step 5: Using the second of Eqs. 9–9, and integrating with respect to y, we have

$$\bar{y} = \frac{\displaystyle\int_V \tilde{y}\, dV}{\displaystyle\int_V dV} = \frac{\displaystyle\int_0^1 y(\pi z^2)\, dy}{\displaystyle\int_0^1 (\pi z^2)\, dy} = \frac{\displaystyle\int_0^1 \pi y^2\, dy}{\displaystyle\int_0^1 \pi y\, dy} = 0.667 \text{ ft} \qquad Ans.$$

Solution II

Step 1:   As shown in Fig. 9–15b, the volume element can be chosen in the form of a *thin cylindrical shell,* where the shell radius is $r = z$ and the thickness is dz.

Step 2:   The element intersects the generating curve at point $(0, y, z)$.

Step 3:   The volume of the element is $dV = 2\pi r \, dA = 2\pi z (1 - y) \, dz$.

Step 4:   The centroid of the element is located at

$$\tilde{y} = y + \frac{1 - y}{2} = \frac{1 + y}{2}$$

Step 5:   Using the second of Eqs. 9–9, and integrating with respect to z, we obtain

$$\bar{y} = \frac{\displaystyle\int_V \tilde{y} \, dV}{\displaystyle\int_V dV} = \frac{\displaystyle\int_0^1 \left(\frac{1 + y}{2}\right) 2\pi z (1 - y) \, dz}{\displaystyle\int_0^1 2\pi z (1 - y) \, dz}$$

$$= \frac{\pi \displaystyle\int_0^1 z(1 - z^4) \, dz}{2\pi \displaystyle\int_0^1 z(1 - z^2) \, dz} = 0.667 \text{ ft} \qquad\qquad Ans.$$

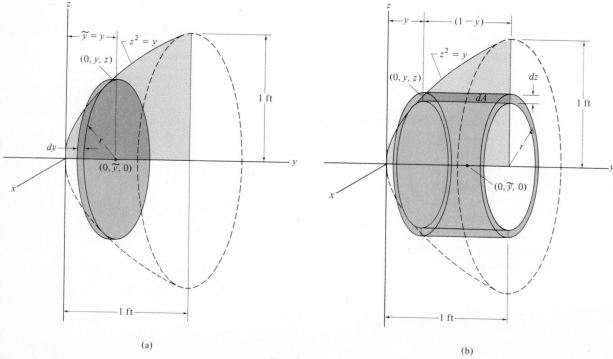

(a)   (b)

Fig. 9–15

Example 9-9

Locate the center of gravity of the right circular cylinder shown in Fig. 9-16a if the density of the cylinder varies directly with its distance from the base, such that $\rho_V = 2z$ lb/ft^3.

Solution

For reasons of material symmetry

$$\bar{x} = \bar{y} = 0 \qquad\qquad Ans.$$

Step 1: A disk element is chosen for integration, Fig. 9-16a, since the *density of this entire element is constant* for a given value of z. This element has a constant radius of $\frac{1}{2}$ ft and a thickness of dz.

Step 2: The element is located along the z axis at point $(0, 0, z)$.

Step 3: The volume of the element is $dV = \pi \left(\dfrac{1}{2}\right)^2 dz$.

Step 4: The centroid of the element is located at $\tilde{z} = z$.

Step 5: Using the third of Eqs. 9-6, and integrating with respect to z, noting that $\rho_V = 2z$, we obtain

$$\bar{z} = \frac{\displaystyle\int_V \rho_V \tilde{z}\, dV}{\displaystyle\int_V \rho_V\, dV} = \frac{\displaystyle\int_0^1 (2z)(z)\pi \left(\frac{1}{2}\right)^2 dz}{\displaystyle\int_0^1 (2z)\pi \left(\frac{1}{2}\right)^2 dz}$$

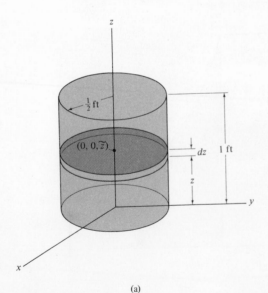

(a)

(b)

Fig. 9-16

$$\bar{z} = \frac{\dfrac{\pi}{2}\displaystyle\int_0^1 z^2 \, dz}{\dfrac{\pi}{2}\displaystyle\int_0^1 z \, dz} = 0.667 \text{ ft}$$

Ans.

If a shell element were chosen for integration, as shown in Fig. 9–16b, the density of the material composing the shell would *vary* along the height of the shell; hence the location of $\bar{z}$ for the element is not easily determined. It would, therefore, be necessary to further subdivide the shell into second- or third-order differential elements to locate $\bar{z}$. Doing this simply complicates the integration process, since use of second- and third-order differential elements requires two and three integrations, respectively, to obtain the location of the center of gravity.

Problems

15 /Nov. 10

9-7. Locate the centroid of the thin wire bent in the form of the parabola.

9-8. Locate the centroid of the shaded area.

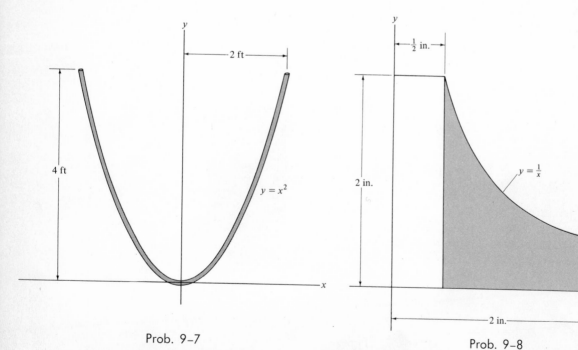

Prob. 9–7

Prob. 9–8

9-9. Locate the centroid of the wire bent into the shape of the circular arc.

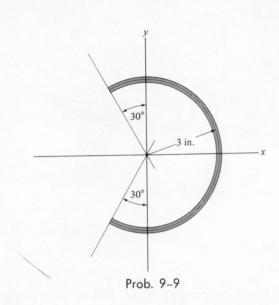

Prob. 9-9

9-10. Locate the centroid of the shaded area.

Prob. 9-10

9-11. Locate the centroid of the area under the curve $y = 3 \sin 2x$ from $x = 0$ to $x = \pi/2$.

9-12. Locate the centroid of the shaded area.

Prob. 9-12

$15/11$

$\bar{X} = \dfrac{49}{3\pi}$

9-13. Locate the centroid of the shaded elliptical sector.

$\dfrac{\pi a b}{4}$

$\dfrac{\pi}{4} \, \dfrac{23}{4}$

Prob. 9-13

$15/11 \cdot 10$

9-14. Locate the centroid of the wire bent into the shape shown.

Prob. 9-14

327

9-15. Locate the centroid of the shaded area.

Prob. 9-15

9-16. Locate the centroid of the shaded area.

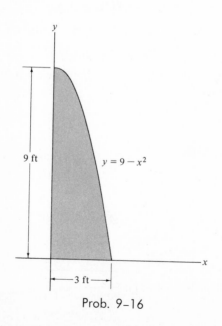

Prob. 9-16

9-17. Locate the centroid of the thin cylindrical shell.

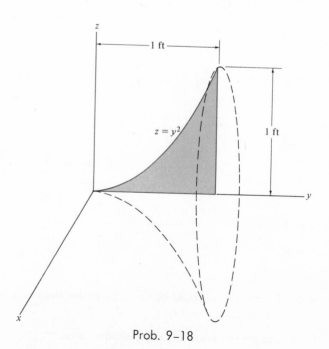

Prob. 9-17

11-12
sol 16

9-18. Locate the centroid of the "bell-shaped" volume formed by revolving the shaded area shown about the y axis.

Prob. 9-18

9-19. Locate the centroid of the volume generated by revolving the shaded area shown about the z axis.

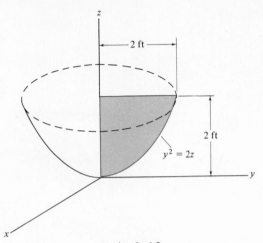

Prob. 9-19

9-20. Locate the center of gravity of the hemisphere. The density of the material varies linearly from zero at the origin O to ρ_0 at the surface. (*Hint:* Establish a differential element using the results of Prob. 9-23.)

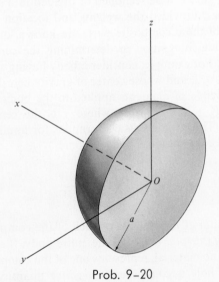

Prob. 9-20

9-21. Locate the centroid of the cone formed by rotating the shaded area about the y axis.

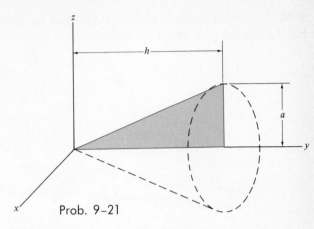

Prob. 9-21

9-22. Locate the centroid of the volume generated by revolving the shaded area shown about the y axis.

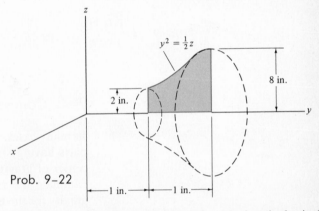

Prob. 9-22

9-23. Show that the centroid of the thin hemispherical shell is located at $(a/2, 0, 0)$.

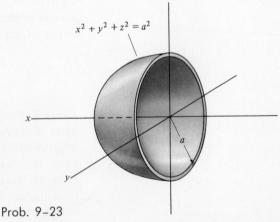

Prob. 9-23

329

9-24. Locate the centroid of the solid formed by rotating the shaded area about the aa axis.

9-25. Locate the centroid for the half cone.

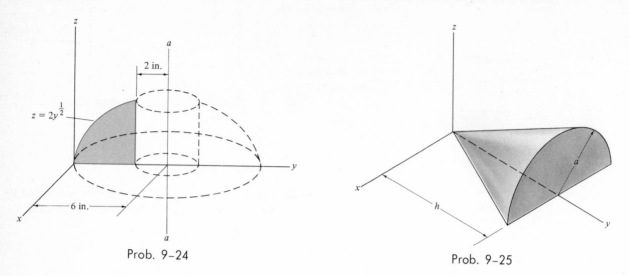

Prob. 9-24

Prob. 9-25

9-4. Composite Bodies

In many cases, a body or figure can be sectioned or divided into several parts having simpler shapes. Provided the weights and location of the center of gravity of each of these composite parts are known, one can eliminate the need for using integration in determining the center of gravity for the entire body. For example, consider a body having several parts which weigh W_1, W_2, . . ., and whose center of gravity coordinates $\tilde{x}$ are $\tilde{x}_1$, $\tilde{x}_2$, ... The principle of moments applied to the body yields $\bar{x}(W_1 + W_2 + . . .) = \tilde{x}_1 W_1 + \tilde{x}_2 W_2 + . . .$, where $\bar{x}$ is the center of gravity of the entire body. Thus, the necessary formulas for finding the center of gravity for a three-dimensional body become

$$\bar{x} = \frac{\Sigma \tilde{x} W}{\Sigma W}, \qquad \bar{y} = \frac{\Sigma \tilde{y} W}{\Sigma W}, \qquad \bar{z} = \frac{\Sigma \tilde{z} W}{\Sigma W} \qquad (9\text{-}10)$$

where ΣW represents the sum of the weights of each of the component parts of the body, or simply the *total weight* of the entire body. If a *hole,* or geometric region having no material, represents one of the component parts, the "weight" of the hole is considered a *negative* quantity. See Example 9–11. $\tilde{x}$, $\tilde{y}$, and $\tilde{z}$ in Eq. 9–10 represent the *algebraic distances* from the center of gravity of each component part to the origin of the coordinates.

When the entire body has a *constant density,* the center of gravity coincides with the geometric center or centroid of the body. The centroid for composite lines, areas, and volumes can be found using relations analogous to Eq. 9–10; however, the W's are replaced by L's, A's, and V's, respectively. Centroids for common shapes of lines, areas, shells, and volumes are given in Appendix B.

Computation errors for composite shapes can be kept to a minimum by labeling each component part on a sketch of the body and arranging the calculations in a tabular form. This procedure is used in the following two examples.

Example 9–10

Locate the centroid of the composite wire shown in Fig. 9–17a.

(a)

(b)

Fig. 9–17

Solution

The wire is divided into three segments, as shown in Fig. 9–17b. The centroid of segment 1 is determined either by integration or using the formula in Appendix B. The calculations are listed in the following table:

Segment	L (in.)	$\tilde{x}$ (in.)	$\tilde{y}$ (in.)	$\tilde{z}$ (in.)	$\tilde{x}L$ (in.²)	$\tilde{y}L$ (in.²)	$\tilde{z}L$ (in.²)
1	$\pi(6) = 18.85$	6	−3.82	0	113.1	−72.0	0
2	4	0	2	0	0	8	0
3	2	0	0	−1	0	8	−2
	$\Sigma L = 24.85$				$\Sigma \tilde{x}L = 113.1$	$\Sigma \tilde{y}L = -56.0$	$\Sigma \tilde{z}L = -2.0$

Thus,

$$\bar{x} = \frac{\Sigma \tilde{x} L}{\Sigma L} = \frac{113.1}{24.85} = 4.55 \text{ in.} \qquad Ans.$$

$$\bar{y} = \frac{\Sigma \tilde{y} L}{\Sigma L} = \frac{-56.0}{24.85} = -2.26 \text{ in.} \qquad Ans.$$

$$\bar{z} = \frac{\Sigma \tilde{z} L}{\Sigma L} = \frac{-2.0}{24.85} = -0.08 \text{ in.} \qquad Ans.$$

Example 9-11

Locate the center of gravity of the composite shape shown in Fig. 9–18a. The surface density of the vertical and horizontal plates are 6 lb/ft² and 4 lb/ft², respectively. The 2-ft-diameter cylinder has a density of 160 lb/ft³.

Solution

The assembly is divided into four segments, as shown in Fig. 9–18b. Segment 4 must be taken as a "negative" shape in order that the four pieces, when added together, yield the total composite shape shown in Fig. 9–18a. The location of the centroid for segment 4 can be found using the formula for a semi-circular area given in Appendix B. Because of *symmetry* of the composite shape, $\bar{y} = 0$. The computations for $\bar{x}$ and $\bar{z}$ are given in the following table:

Segment	W (lb)	$\tilde{x}$ (ft)	$\tilde{z}$ (ft)	$\tilde{x}W$ (ft-lb)	$\tilde{z}W$ (ft-lb)
1	4(4)4 = 64.0	2	0	128.0	0
2	$\pi(1)^2(2)160 = 1{,}005.3$	2	1	2,010.6	1,005.3
3	5(4)6 = 120.0	0	2.5	0	300.0
4	$-\frac{1}{2}\pi(2)^2 6 = -37.7$	0	5 − 0.85 = 4.15	0	−156.5
	$\Sigma W = 1{,}151.6$			$\Sigma \tilde{x}W = 2{,}138.6$	$\Sigma \tilde{z}W = 1{,}148.8$

Applying Eqs. 9–10 yields

$$\bar{x} = \frac{\Sigma \tilde{x} W}{\Sigma W} = \frac{2{,}138.6}{1{,}151.6} = 1.857 \text{ ft} \qquad Ans.$$

$$\bar{y} = 0 \qquad Ans.$$

$$\bar{z} = \frac{\Sigma \tilde{z} W}{\Sigma W} = \frac{1{,}148.8}{1{,}151.6} = 0.998 \text{ ft} \qquad Ans.$$

(a)

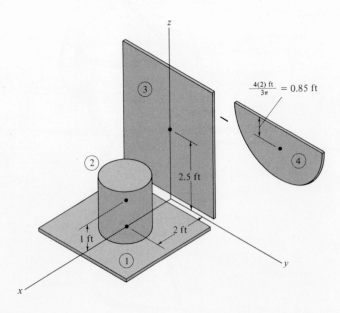

(b)

Fig. 9–18

Problems

9-26. Locate the centroid of the wire.

Prob. 9-26

9-27. Locate the center of gravity of the homogeneous rod.

Prob. 9-27

9-28. Determine the length of the wire segment AB such that the centroid for the entire composite is located at point C.

Prob. 9-28

9-29. Locate the centroid of the wire.

Prob. 9-29

9-30. Locate the centroid of the wire.

Prob. 9-30

9-31. Locate the centroid for the cross-sectional area of the beam.

Prob. 9-31

9-32. Locate the centroid of the shaded area.

Prob. 9-32

9-33. Locate the centroid of the shaded area.

Prob. 9-33

9-34. Locate the centroid of the shaded area.

Prob. 9-34

9-35. Locate the centroid of the shaded area.

Prob. 9-35

335

9-36. Locate the centroid for the cross-sectional area of the beam.

Prob. 9–36

9-37. Locate the centroid of the solid.

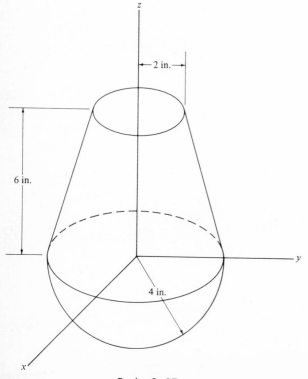

Prob. 9–37

9-38. Locate the centroid for the area of the built-up structural section.

Prob. 9–38

9-39. Locate the centroid of the solid.

Prob. 9–39

9-40. Locate the centroid of the thin plate.

Prob. 9-40

9-42. The frustum has a hole drilled through its center. Locate the centroid for this piece.

Prob. 9-42

9-41. Locate the center of gravity of the composite part. The density of materials A and B are 150 lb/ft³ and 400 lb/ft³, respectively.

Prob. 9-41

9-43. Determine the center of gravity of the airplane. The weights of the various items are shown in the table.

	Item	Weight (lb)
1	Tail assembly	150
2	Fuselage assembly	650
3	Wing assembly	700
4	Engine section	170
5	Fixed equipment	800

Prob. 9-43

9-44. Locate the center of gravity of the assembly. The density for the materials is as follows: rod R, $\rho_L = 0.6$ lb/ft; thin plate B, $\rho_A = 5$ lb/ft^2; and block C, $\rho_V = 450$ lb/ft^3.

9-46. Locate the centroid of the sheet-metal bracket.

Prob. 9–44

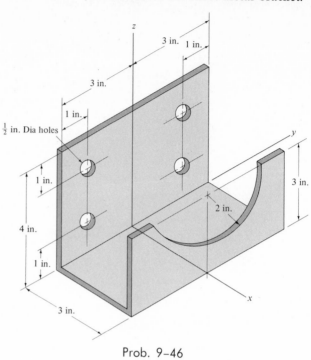

Prob. 9–46

9-45. By approximation, locate the centroid of the plate.

Prob. 9–45

338

The *theorems of Pappus and Guldinus,* which were first developed by Pappus of Alexandria during the third century A.D. and somewhat later by the Swiss mathematician Paul Guldin or Guldinus (1577–1643), are used to find the surface area or volume of any solid of revolution.

A *surface area of revolution* is generated by revolving a *curve* about a fixed axis. For example, the surface area of a cone is generated by rotating the line segment *AB*, shown in Fig. 9–19, about the *x* axis. In

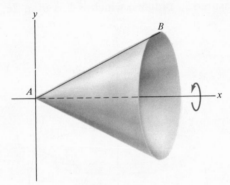

Fig. 9–19

a similar manner, a *volume of revolution* is generated by revolving an *area* about a fixed axis. Thus, the volume of a cone is formed by rotating the triangular area *ABC*, shown in Fig. 9–20, about the *x* axis.

The statements and proofs of the theorems of Pappus and Guldinus follow. The proofs require that the generating curves and areas do *not* cross the axis about which they are generated, although they may touch it.

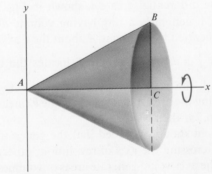

Fig. 9–20

Theorem I.   *The area of a surface of revolution equals the product of the length of the generating curve and the distance traveled by the centroid of the curve in generating the surface area.*

   Proof.   When a differential length dL of a curve is revolved about the y axis, as shown in Fig. 9–21, it generates a ring having a surface area $dA = 2\pi x\, dL$. The entire surface area generated by revolving L about the y axis is therefore $A = 2\pi \int_L x\, dL$. Since $\int_L x\, dL = \bar{x}L$, where $\bar{x}$ locates the centroid C of the generating curve L, the area becomes $A = 2\pi\bar{x}L$. If the surface area is formed by revolving the line L through an angle θ (measured in radians) which is less than 2π, then $A = \theta\bar{x}L$.

Fig. 9–21

Theorem II.   *The volume of a surface of revolution equals the product of the generating area and the distance traveled by the centroid of the area in generating the volume.*

Proof.   When the differential area dA, shown in Fig. 9–22, is revolved about the y axis, it generates a ring having volume $dV = 2\pi x\, dA$. The entire volume generated by revolving A about the y axis is therefore $V = 2\pi \int_A x\, dA$. Since $\int_A x\, dA = \bar{x}A$, where $\bar{x}$ locates the centroid C of the generating area A, the volume becomes $V = 2\pi\bar{x}A$. If the volume is formed by revolving the area A through an angle θ (measured in radians), which is less than 2π, then $V = \theta\bar{x}A$.

   In these proofs, it should be noted that the generating curve or area is restricted from crossing the axis of revolution; otherwise two sections on either side of the axis would generate areas or volumes having opposite signs and hence would cancel each other.

Fig. 9-22

Using Theorem I of Pappus-Guldinus, prove that the surface area of a sphere is $4\pi R^2$.

Solution

From Theorem I of Pappus-Guldinus, we know that the surface area generated by a curve equals the product of the curve length and the distance "traveled" by the centroid of the curve. A sphere is generated by rotating the semicircular arc shown in Fig. 9-23 about the x axis. Using

Fig. 9-23

Appendix B, it is seen that the centroid of this arc is located at a distance of $2R/\pi$ from the axis of rotation. The centroid moves through a distance of $2\pi \, (2R/\pi) = 4R$ in generating the sphere shown in Fig. 9–23. Hence, the surface area of the sphere becomes

$$A = 4R(\pi R) = 4\pi R^2 \qquad \textit{Ans.}$$

Example 9–13

Determine the weight of concrete needed to construct the arched beam shown in Fig. 9–24. The density of concrete is $\rho_c = 150 \ \text{lb/ft}^3$.

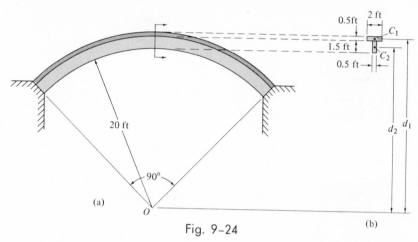

Fig. 9–24

Solution

The weight of the arch may be determined by using the density ρ_c provided we know the volume of concrete. The cross-sectional area of the beam is a "T-section" composed of two rectangles having centroids at points C_1 and C_2, as shown in Fig. 9–24b. Using Theorem II of Pappus-Guldinus, we know that the volume generated by each of these areas is equal to the product of the given cross-sectional area and the distance traveled by its centroid. In generating the arch, the centroid passes through an angle of $\theta = \pi(90°/180°) = 1.57$ radians. Point C_1 acts at a distance of $d_1 = 20 + (1.5 + 0.25) = 21.75$ ft from the center of rotation, point O. C_2 acts at $d_2 = 20 + 0.75 = 20.75$ ft from O. The total volume is thus

$$V_T = \Sigma\theta\bar{x}A$$

$$= (1.57 \ \text{rad})(21.75 \ \text{ft})(2 \ \text{ft})(0.5 \ \text{ft}) + (1.57 \ \text{rad})(20.75 \ \text{ft})(1.5 \ \text{ft})(0.5 \ \text{ft})$$

$$= 58.6 \ \text{ft}^3$$

The required weight of concrete is the product of the density and the volume.

$$W = \rho_c(V_T) = (150 \ \text{lb/ft}^3)(58.6 \ \text{ft}^3) = 8,790 \ \text{lb} \qquad \textit{Ans.}$$

Problems

9–47. Using the theorem of Pappus-Guldinus, determine the surface area of a cone, using an inclined line segment to generate the cone. See Fig. 9–19.

9–48. Determine the surface area and volume of the torus.

8 in.

4 in.

Prob. 9–48

9–49. A steel wheel has a sectional view shown. Determine the total weight of the wheel if $\rho_V = 0.25$ lb/in^3.

2 in.

$\frac{1}{2}$ in.

$\frac{1}{2}$ in.

5 in.

$\frac{1}{2}$ in.

Prob. 9–49

9–50. Determine the weight of the wedge which is generated by rotating a right triangle of base 6 in. and height 3 in. through an angle of 30°. The density of the material is $\rho_V = 0.22$ lb/in^3.

30°

3 in.

6 in.

Prob. 9–50

9–51. A V-belt has an inner radius of 6 in., and a cross-sectional area as shown. Determine the volume of material used in making the V-belt.

30°

$\frac{1}{2}$ in.

$\frac{1}{4}$ in.

$\frac{1}{4}$ in.

Prob. 9–51

343

9-52. Using integration, determine the area and the centroidal distance $\bar{y}$ of the shaded area. Then, using Theorem II of Pappus-Guldinus, determine the volume of a paraboloid formed by revolving the area about the x axis.

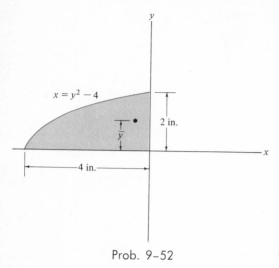

Prob. 9-52

9-53. Compute both the area A and the centroidal distance $\bar{x}$ for the shaded region. Using Theorem II of Pappus-Guldinus, compute the volume of the solid generated by revolving the shaded area about the aa axis.

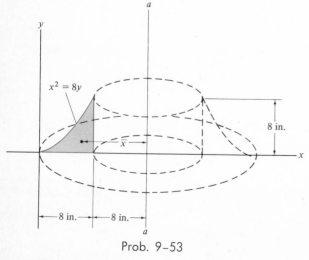

Prob. 9-53

9-54. Determine both the surface area and the volume of material for the casting shown.

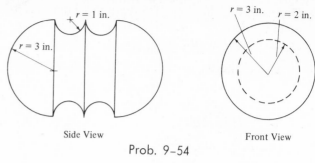

Side View   Front View

Prob. 9-54

9-55. The surface of the water tank is to be painted with two coats of noncorrosive paint. If a gallon of paint covers 300 ft², determine by approximate means (a) the volume of water which may be contained in the tank, and (b) the number of gallons of paint required to paint the surface of the tank.

Prob. 9-55

344

9-56. A circular sea wall is made of concrete. Determine the total weight of the wall if the concrete has a density of 150 lb/ft³.

Prob. 9-56

*9-6. Pressure Loadings on Submerged Surfaces

An important application of force distributed over a surface area is that caused by the action of pressure acting on a surface submerged in a fluid. Provided the fluid is at rest, the pressure force will always act *normal* to the submerged surface.

According to Pascal's law, the pressure p, measured as a force per unit area and acting at any point in a fluid is the same in all directions. If ρ represents the density (weight/volume) of a fluid, the magnitude of pressure p, acting at a point located a depth z from the surface of the fluid is

$$p = \rho z \quad \text{(force/area)} \qquad (9\text{-}11)$$

This equation is valid only for fluids which are incompressible, as is the case of most liquids. Gases on the other hand are compressible fluids, and since their density changes significantly with both vertical height and temperature, Eq. 9-11 cannot be used. All matter lying on the earth's surface is subjected to an atmospheric pressure of approximately 14.7 lb/in.² at sea level. Hence, Eq. 9-11 is actually a measure of the gauge pressure exerted by the liquid at some point. The *gauge pressure* is that pressure greater than the atmospheric pressure. In studying the action of pressures acting on a surface, we will not be concerned with atmospheric pressure, since it acts over the total surface and yields a zero force resultant.

Let us now consider what effect a liquid having a density ρ_l has at points A, B, and C, located on the top of a submerged plate, shown in Fig. 9–25. Since both points A and B are at the same depth z_2 from the surface of the liquid, the *pressure* at these points (and all other points along the line containing A and B) has a magnitude of $p_2 = \rho_l z_2$. Point C acts at a depth z_1 from the liquid surface; hence, $p_1 = \rho_l z_1$. In all cases, the pressure acts *normal* to the submerged plate. The *force* acting on an element of surface area dA, located at points A and B, has a magnitude of $dF_2 = p_2\,dA = \rho_l z_2\,dA$. Similarly, $dF_1 = \rho_l z_1\,dA$ for point C. (See Fig. 9–25.)

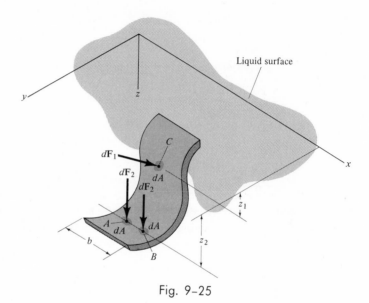

Fig. 9–25

To find the entire pressure force distribution acting across the top of a submerged plate, consider the *flat plate of variable width x,* shown in Fig. 9–26. The plane of this plate makes an angle θ with the horizontal. A strip of area $dA = x\,dy'$, located a distance z below the surface of liquid, has a pressure distribution $p = \rho_l z$ (force/area) acting on it. The magnitude of the differential force $d\mathbf{F}$, acting on the area dA located at the top surface of the plate, is equal to the *volume* of the shaded element. Hence, $dF = dV = p\,dA = \rho_l z(x\,dy')$. The magnitude of the resultant force acting over the plate is found by integration, and is therefore,

$$F = \int_V dV = V$$

This total volume is described by the plate area as its base and linear varying pressure distribution as its altitude, Fig. 9–26. To account for any

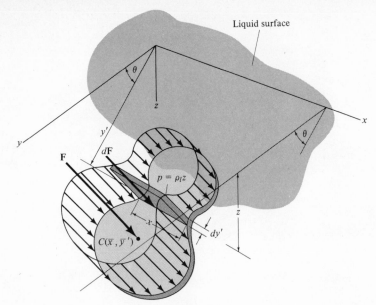

Fig. 9–26

moments created by the pressure loading, the resultant force **F** produces an equivalent moment, provided it acts through the *centroid of the volume* V; that is, **F** must act through coordinate point $C(\overline{x}, \overline{y}')$ located on the surface of the plate defined by

$$\overline{x} = \frac{\displaystyle\int_V \widetilde{x}\, dV}{\displaystyle\int_V dV}, \qquad \overline{y}' = \frac{\displaystyle\int_V \widetilde{y}'\, dV}{\displaystyle\int_V dV}$$

This point should *not* be mistaken for the centroid of the plate area.

Consider now the submerged *rectangular flat plate having a constant width b,* shown in Fig. 9-27a. In this case a strip of area $dA = b\, dy'$, located a distance z below the surface of the liquid, is subjected to a pressure distribution $p = \rho_l z$ (force/area). The magnitude of the differential force acting across this strip of area is therefore $dF = \rho_l z\, dA = \rho_l z(b\, dy')$. As stated previously, in the case of the plate having a *variable width,* the resultant pressure force is equal to the volume under the loading diagram and the force acts through the centroid of this volume. For this rectangular plate, however, the pressure distribution along the *width* of the entire plate is uniform. Hence, the loading function can easily be projected as an area, shown in two dimensions, rather than using the loading function represented geometrically as a volume (see

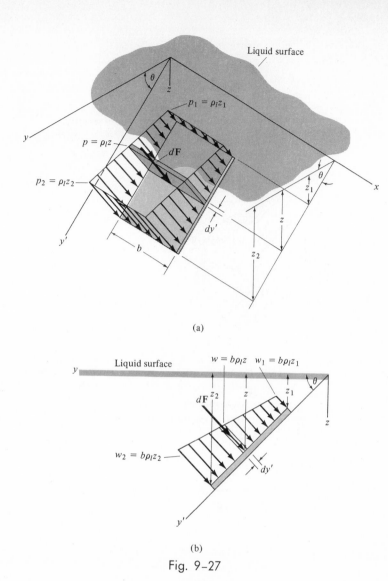

(a)

(b)

Fig. 9–27

Sec. 4–11). On this diagram, Fig. 9–27b, the loading function $w = bp = b\rho_l z$(force/length) represents a trapezoidal area. The intensity of loading varies linearly from $w_1 = b\rho_l z_1$ to $w_2 = b\rho_l z_2$ across the surface. From the theory of Sec. 4–11, the *magnitude of the resultant force* **F**, created by this distributed load, is equal to the *area* under this trapezoid. The *line of action* of this force passes through the *centroid* of the trapezoidal area.

In the case of a submerged *curved surface,* the resultant force of a distributed pressure loading involves considerably more calculation than

for a flat plate, because the pressure acting normal to the plate continually changes direction. For example, if the rectangular plate in Fig. 9–27 is curved, Fig. 9–28a, the pressure distribution acting along the length of the plate is as shown in Fig. 9–28b. The y and z components of force $\mathbf{F}$ acting on the surface can be found by integration, noting from the figure that $p/dL = p_y/dy = p_z/dz$. Hence,

$$F_y = \int_L w\, dy = \int_L bp\, dy = b \int_0^L p_y\, dL \qquad (9\text{–}12)$$

(a)

(b)

Fig. 9–28

$$F_z = \int_L w \, dz = \int_L bp \, dz = b \int_0^L p_z dL \qquad (9\text{--}13)$$

The magnitude of **F** then becomes $F = \sqrt{(F_y)^2 + (F_z)^2}$. This magnitude represents the *area* under the loading diagram in Fig. 9–28b, or the *volume* under the loading diagram in Fig. 9–28a. **F** acts through the centroid of this loading diagram normal to the surface of the plate.

Carrying out the integration needed to find both the magnitude of **F** and its location on a plate having a curved surface is, in many cases, not easy since the direction of the pressure forces changes. For this reason, a simpler method will now be discussed for finding this resultant force. This method is applicable *only* to *flat and curved plates having a constant width*. To illustrate the method, consider again the curved plate shown in Fig. 9–28a. Because the width of the plate is constant, the pressure loadings acting on the top surface of the plate can be graphed in two dimensions. Using the methods previously discussed, we can calculate the forces acting along the separate *horizontal and vertical projections* of the plate, Fig. 9–29. The force $\mathbf{F}_{AD}$ acting on the vertical projection AD has a magnitude which equals the area under the trapezoidal pressure diagram, and acts through the centroid C_{AD} of this area. (The magnitude of this force is equivalent to F_y defined by Eq. 9–12.) The pressure distribution along the horizontal plane AB is constant, since all points lying in this plane are at equal distances from the surface of the liquid. The magnitude of force $\mathbf{F}_{AB}$ is then simply the area under the rectangle. This force acts through the centroid C_{AB} (or midpoint) of AB. This would be the only vertical force acting if the surface shape were BAD. However, the curved surface BD must support the *weight of fluid* $\mathbf{W}_f$ contained within the area BAD in addition to the forces $\mathbf{F}_{AD}$ and $\mathbf{F}_{AB}$. The magnitude

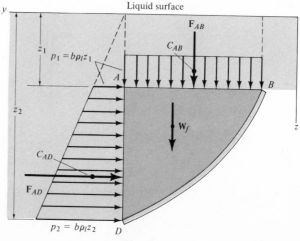

Fig. 9–29

of $\mathbf{W}_f$ is $W_f = \rho_l(b)(\text{area } BAD)$. This force acts through the centroid of the area BAD. Why? The magnitude of the resultant of $\mathbf{F}_{AB}$ and $\mathbf{W}_f$ is then equivalent to F_z defined by Eq. 9–13. Summing the three coplanar forces shown in Fig. 9–29, one obtains the single force resultant $\mathbf{F}$ acting normal to the plate and passing through the centroid of the loading diagram, Fig. 9–28a.

Example 9–14

The rectangular plate $ABCD$ shown in Fig. 9–30a serves as an access opening to a water tank. Determine the magnitude and location P of the resultant force of the hydrostatic load acting on the plate. The density of water is $\rho_w = 62.4 \text{ lb/ft}^3$.

Solution I

A three-dimensional view of the pressure loading acting on the plate is shown in Fig. 9–30b. The pressure acting along AB is

$$p_{AB} = \rho_w z_{AB} = 62.4 \text{ lb/ft}^3(14 \sin 45° \text{ ft}) = 617.6 \text{ lb/ft}^2$$

and the pressure acting along CD is

$$p_{CD} = \rho_w z_{CD} = 62.4 \text{ lb/ft}^3(12 \sin 45° \text{ ft}) = 529.4 \text{ lb/ft}^2$$

The resultant force acting on the plate equals the volume under this pressure diagram. This force acts through the centroid of the volume. Because of the symmetry of the volume shape, the centroid must lie along the y' axis shown in Fig. 9–30b. Hence,

$$\bar{x} = 0 \qquad\qquad\qquad Ans.$$

Since the plate has a *constant width* of 3 ft, the pressure loading may be projected in two dimensions, as shown in Fig. 9–30c. The loading per unit length along AB becomes

$$w_{AB} = b p_{AB} = (3 \text{ ft})(617.6 \text{ lb/ft}^2) = 1{,}853 \text{ lb/ft}$$

Similarly, the loading per unit length along CD is

$$w_{CD} = b p_{CD} = (3 \text{ ft})(529.4 \text{ lb/ft}^2) = 1{,}588 \text{ lb/ft}$$

Since the loading varies linearly with depth, the distributed loading along the length of the plate is trapezoidal, as shown. The resultant force $\mathbf{F}_R$ of the trapezoidal loading, and its location, may be obtained using the data listed in Appendix B.

$$F = (\text{area of trapezoid})$$
$$= \tfrac{1}{2}(2 \text{ ft})(1{,}853 \text{ lb/ft} + 1{,}588 \text{ lb/ft}) = 3{,}441 \text{ lb} \qquad Ans.$$

The force acts through the centroid P, where

$$d = \frac{1}{3}\left(\frac{2(1{,}588 \text{ lb/ft}) + 1{,}853 \text{ lb/ft}}{1{,}588 \text{ lb/ft} + 1{,}853 \text{ lb/ft}}\right) 2 \text{ ft} = 0.97 \text{ ft} \qquad Ans.$$

measured to the right of *AB,* Fig. 9–30*d.*

Solution II

Since the plate has a *constant width,* the projection method may be used to solve this problem. As shown in Fig. 9–30*e,* the *vertical projection* of the plate, *EA,* is subjected to a distributed trapezoidal loading. The loading at *E* is

$$w_E = b\rho_w z_E = 62.4 \text{ lb/ft}^3(3 \text{ ft})(12 \sin 45° \text{ ft}) = 1{,}588 \text{ lb/ft}$$

and the loading at *A* is

$$w_A = b\rho_w z_A = 62.4 \text{ lb/ft}^3(3 \text{ ft})(14 \sin 45° \text{ ft}) = 1{,}853 \text{ lb/ft}$$

Using Appendix B, the magnitude of the resultant force $\mathbf{F}_T$ created by this distribution is

$$F_T = (\text{area of trapezoid})$$
$$= \tfrac{1}{2}(2 \sin 45° \text{ ft})(1{,}853 \text{ lb/ft} + 1{,}588 \text{ lb/ft}) = 2{,}433 \text{ lb}$$

This resultant acts through the centroid of the area,

$$t = \frac{1}{3}\left(\frac{2(1{,}588 \text{ lb/ft}) + 1{,}853 \text{ lb/ft}}{1{,}588 \text{ lb/ft} + 1{,}853 \text{ lb/ft}}\right)(2 \sin 45° \text{ ft}) = 0.69 \text{ ft}$$

measured upward from point *A.* See Fig. 9–30*f.*

In a similar manner, the magnitude of the resultant force F_R acting on the *horizontal projection EC* in Fig. 9–30*e,* is

$$F_R = (\text{area of rectangle})$$
$$= (2 \cos 45° \text{ ft})1{,}588 \text{ lb/ft} = 2{,}246 \text{ lb}$$

This force passes through the centroid of the rectangular area located at

$$r = \tfrac{1}{2}(2 \cos 45° \text{ ft}) = 0.71 \text{ ft}$$

from either end *A* or *C.* See Fig. 9–30*f.*

The force **W** equals the weight of water contained within the triangular shaped block *AEC.* Hence,

$$W = 62.4 \text{ lb/ft}^3[\tfrac{1}{2}(2 \sin 45° \text{ ft})(2 \cos 45° \text{ ft})3 \text{ ft}]$$
$$= 187 \text{ lb}$$

The centroid for this loading is

$$s = \tfrac{1}{3}(2 \sin 45° \text{ ft}) = 0.47 \text{ ft}$$

measured to the right of *A.*

(a)

(b)

(c)

(d)

(e)

(f)

Fig. 9–30

The three forces F_T, F_R, and W are shown acting on the plate in Fig. 9–30f. The resultant force is therefore the vector sum, i.e.,

$$F = \sqrt{(2{,}246 \text{ lb} + 187 \text{ lb})^2 + (2{,}433 \text{ lb})^2} = 3{,}441 \text{ lb} \qquad Ans.$$

where

$$\theta = \tan^{-1} \frac{2{,}246 \text{ lb} + 187 \text{ lb}}{2{,}433 \text{ lb}} = 45° \qquad \searrow^{\theta}{}_{F} \qquad Ans.$$

Since θ is measured from the horizontal, the resultant force acts *perpendicular* to the plate. (This result is expected since the entire distributed load acts perpendicular to the plate, Fig. 9–30b.)

The location, d, of the resultant force, Fig. 9–30d, can be determined from a balance of moments about point A. It is required that the moment produced by F about point A (Fig. 9–30d) be equivalent to the moment of each of the three forces F_T, F_R, and W about this point, Fig. 9–30f. Hence, choosing clockwise moments positive,

$\langle M_A = (3{,}441 \text{ lb})d = (2{,}433 \text{ lb})0.69 \text{ ft} + (187 \text{ lb})0.47 \text{ ft}$

$+ (2{,}246 \text{ lb})0.71 \text{ ft}$

or

$$(3{,}441 \text{ lb})d = 3{,}340 \text{ lb-ft}$$

so that,

$$d = \frac{3{,}340 \text{ lb-ft}}{3{,}441 \text{ lb}} = 0.97 \text{ ft} \qquad Ans.$$

Example 9–15

Determine the magnitude and location of the resultant force acting on the triangular end plates of the water trough shown in Fig. 9–31a. The density of water is $\rho_w = 62.4 \text{ lb/ft}^3$.

Solution

The pressure loading acting on the end plate E is shown in Fig. 9–31b. The magnitude of the resultant force F is equal to the volume of this diagram. Choosing the differential volume element shown in the figure yields

$$dF = dV = p\, dA = \rho_w z(2x\, dz) = 2(62.4 \text{ lb/ft}^3)zx\, dz$$

The equation of line AB is

$$x = \frac{2 - z}{2}$$

Hence, substituting and integrating with respect to z from 0 to 2 ft yields,

$$F = V = \int_V dV = \int_0^2 (124.8)z \left(\frac{2 - z}{2} \right) dz$$

$$= 62.4 \int_0^2 (2z - z^2)dz = 83.2 \text{ lb} \qquad Ans.$$

(a)

(b)

Fig. 9–31

This resultant passes through the *centroid of the volume*. Because of symmetry,

$$\bar{x} = 0 \qquad \qquad Ans.$$

Since $\tilde{z} = z$ for the volume element, then

$$\bar{z} = \frac{\int_V \tilde{z}\, dV}{\int_V dV} = \frac{\int_0^2 z\, 124.8z\left(\frac{2-z}{2}\right) dz}{83.2\ \text{lb}} = \frac{62.4 \int_0^2 (2z^2 - z^3)\, dz}{83.2\ \text{lb}}$$

or

$$\bar{z} = 1.0\ \text{ft} \qquad \qquad Ans.$$

Problems

9-57. Determine the magnitude and location of the resultant hydrostatic force acting on the dam. The width of the dam is 20 ft. The density of water is $\rho_w = 62.4$ lb/ft³.

12 ft

Water

Prob. 9–57

9-58. The wind pressure acting on a triangular sign is approximately uniform. Determine the resultant force and couple due to this loading at point O.

3 ft

$\frac{1}{8}$ ft

.3 ft/lb²

4 ft

4 ft

z

O

y

x

Prob. 9–58

9-59. The wind loading acting on a square plate has a pressure distribution which is parabolic. Determine the magnitude and location of the resultant force.

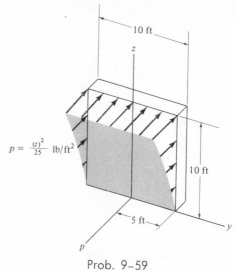

10 ft

z

$p = \frac{(z)^2}{25}$ lb/ft²

10 ft

5 ft

y

p

Prob. 9–59

9-60. Wind blowing across a roof creates a suction pressure on the leeward side of the roof. Determine the magnitude and direction of the resultant force of this pressure distribution.

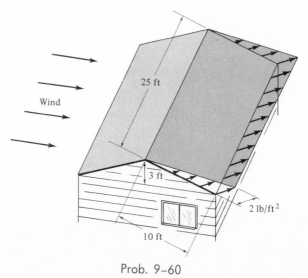

25 ft

Wind

3 ft

10 ft

2 lb/ft²

Prob. 9–60

9-61. Determine the resultant hydrostatic force acting per unit width on the sea wall; $\rho_w = 62.4 \text{ lb/ft}^3$.

$y = \frac{3}{4} x^2$

4 ft

12 ft

Water

Prob. 9–61

9-62. The rectangular plate is subjected to a distributed load over the surface of the plate given by the expression $p = p_o \sin(\pi x/a) \sin(\pi y/b)$, where p_o represents the intensity of the load acting at the center of the plate. Determine the magnitude and location of the resultant force acting on the plate.

p_o

Prob. 9–62

9-63. The storage tank contains oil having a density of $\rho_o = 40 \text{ lb/ft}^3$. If the tank is 6 ft wide, calculate the resultant force acting on the inclined side AB of the tank caused by the oil pressure.

6 ft

2 ft

B

6 ft

30°

Oil

A

4 ft

Prob. 9–63

9-64. The form is used to cast columns. Determine the resultant force acting on the plates A and B when the form is filled with a liquid having a density of $\rho_c = 150 \text{ lb/ft}^3$.

2 ft

1 ft

3 ft

B

8 ft

A

1 ft

1 ft

Prob. 9–64

357

9-65. A 4-ft-square tank, 10 ft high, is filled to a depth of 3 ft with water and 7 ft with oil. Determine the magnitude and location of the resultant force created by these two fluids along the side of the tank; $\rho_o = 50 \text{ lb/ft}^3$, and $\rho_w = 62.4 \text{ lb/ft}^3$.

Prob. 9-65

9-67. Determine the magnitude of the resultant force acting on the gate ABC due to hydrostatic pressure. The gate is 4 ft wide; $\rho_w = 62.4 \text{ lb/ft}^3$.

Prob. 9-67

9-66. Determine the magnitude and direction of the resultant force acting on each of the cover plates A and B. The density of liquid is $\rho_l = 50 \text{ lb/ft}^3$.

Prob. 9-66

358

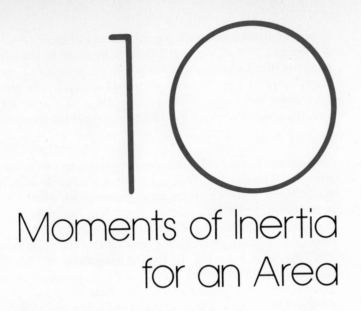

Moments of Inertia for an Area

10-1. Definition of Moments of Inertia for Areas

In Chapter 4 it was shown that a distributed load acting along a beam may be replaced by a single concentrated force equal in magnitude to the area of the loading diagram and having a line of action which passes through the *centroid* of this area. You will recall that the *centroid of an area* is determined by considering the *first moment of the area* about an axis; that is, it is necessary to evaluate integrals of the form $M = \int_A x \, dA$.

There are many important topics in mechanics which require evaluation of an integral of the *second moment of an area* about an axis. These integrals are of the form $\int_A x^2 \, dA$, where x is the shortest distance from the element to an axis which is either perpendicular to, or lying in the plane of, the area. Often in engineering practice such integrals are referred to as the *moment of inertia for the area*. This phrase is actually a misnomer; however, it has been adopted because of the similarity with integrals of the same form related to mass.*

The moment of inertia of an area originates when one computes the *moment* of a linearly varying distributed force system about a given axis. A common example of this type of loading occurs in the study of fluid mechanics. It was pointed out in Sec. 9–6 that the pressure p, or

*The moment of inertia for mass is discussed in *Engineering Mechanics: Dynamics*.

$$\sqrt{x} = \int_A^B x \, dv$$

force per unit area, exerted at a point located a distance z below the surface of a liquid is $p = \rho z$, where ρ is the density of the liquid. Thus, the magnitude of force exerted by a liquid on the area dA of the submerged plate shown in Fig. 10–1a is $dF = p \, dA = \rho z \, dA$. The moment of this force about the x axis of the plate is $dM = dF z = \rho z^2 \, dA$ and, therefore, the moment created by the entire distributed (pressure) loading is

$$M = \rho \int_A z^2 \, dA.$$

The area moment of inertia also appears when one relates the *normal stress,* or force per unit area, acting on a transverse cross section of an elastic beam to the applied external moment **M,** which causes bending of the beam. From the theory of strength of materials, it can be shown that the normal stress σ within the beam varies linearly with its distance from an axis passing through the centroid C of the beam's cross-sectional area; that is, $\sigma = kz$, Fig. 10–1b. The magnitude of force $d\mathbf{F}$ acting on the area element dA, shown in the figure, is therefore $dF = \sigma \, dA = kz \, dA$. Since this force is located a distance z from the y axis, the moment of $d\mathbf{F}$ about the y axis is $dM = dF z = kz^2 \, dA$. The resulting moment of the entire stress distribution is caused by the applied moment **M.** Therefore, for equilibrium it is necessary that $M = k \int_A z^2 \, dA$.

As a third example to illustrate the formulation of area moments of inertia, consider the elastic twisting of a circular shift about its longitudinal axis, as shown in Fig. 10–1c. The applied external torsional moment **T** causes a *shearing stress* distribution at a transverse cross-section which varies linearly along any radial line segment; i.e., $\sigma = kr$. The magnitude of the applied torque **T** may be related to the internal stress distribution by summing moments about the x axis of the shaft, which results in

$$T = \int_A r \, dF = k \int_A r\sigma \, dA = k \int_A r^2 \, dA.$$

Since the moments of inertia for areas frequently appear in design formulas used in fluid mechanics, strength of materials, and structural mechanics, it is important that the student become familiar with the methods used to compute these quantities.

In general, the moment of inertia for an element of area dA, about an axis either lying in the plane of this area or normal to it and lying at a perpendicular distance s away from the area element, is defined by the expression

$$dI = s^2 \, dA \qquad\qquad (10\text{–}1)$$

If dA represents a segment of the area A, the moment of inertia for the entire area A is found by integrating over the entire area; hence,

$$I = \int_A s^2 \, dA$$

(a)

(b)

(c)

Fig. 10–1

Since both dA and s^2 are always positive quantities, the moment of inertia for an area element is always a *positive* quantity. Furthermore, the moment of inertia of the finite area, represented as the sum or integral of these differential moments, is also positive.

From Eq. 10–1, it is seen that the dimensions for the moment of inertia of an area are in terms of length raised to the fourth power, e.g., in.^{4}, ft^4, or m^4.

Consider the area A shown in Fig. 10–2 which lies in the xy plane. In accordance with Eq. 10–1, the moment of inertia of the differential planar area, $dA = dx\,dy$, about the x, y, and z axes, respectively, is defined as

$$dI_x = y^2\,dA$$
$$dI_y = x^2\,dA$$
$$dJ_O = r^2\,dA$$

For the entire planar area A, the moments of inertia become

Fig. 10–2

$$I_x = \int_A y^2\,dA \qquad (10\text{--}2a)$$

$$I_y = \int_A x^2\,dA \qquad (10\text{--}2b)$$

$$J_O = \int_A r^2\,dA = I_x + I_y \qquad (10\text{--}2c)$$

The terms I_x and I_y are called the *rectangular moments of inertia* for the area, and J_O is referred to as the *polar moment of inertia*. From Fig. 10–2, it is seen that $r^2 = x^2 + y^2$, so that the polar moment of inertia represents the *sum* of the rectangular moments of inertia for any two perpendicular axes which intersect the polar (or z) axis, Eq. 10–2c. In particular, the rectangular moments of inertia are required for the fluid pressure and bending problems mentioned earlier, whereas the torsional computation requires use of the polar moment of inertia. The techniques for the integration of Eqs. 10–2 are given in Sec. 10–4.

10–2. Parallel-Axis Theorem for an Area

If the moment of inertia for an area is given about an axis, it is sometimes desirable to calculate this quantity about a parallel axis. The *parallel-axis theorem* may be used for this calculation, provided the perpendicular distances from the centroid of the area to both axes are known. (This theorem relates the moment of inertia about an axis to the moment of inertia about a parallel axis passing through the centroid of the area.)

To derive this theorem, let us consider finding the moment of inertia of the shaded area shown in Fig. 10–3 about the bb axis. Using Eq. 10–1, we have

$$dI = (x + d)^2 \, dA$$

where d represents the perpendicular distance between the centroidal $\bar{y}$ axis and the parallel bb axis. Expanding this expression and integrating each term to obtain the moment of inertia for the entire area yields

$$I = \int_A (x + d)^2 \, dA$$

$$= \int_A x^2 \, dA + 2d \int_A x \, dA + d^2 \int_A dA$$

or

$$I = \bar{I} + 2d \int_A x \, dA + Ad^2 \tag{10–3}$$

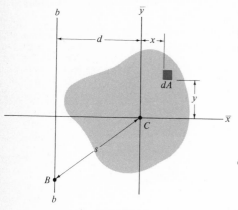

Fig. 10–3

The second term on the right-hand side of Eq. 10–3 represents the first moment of the area about the $\bar{y}$ axis. Since this axis passes through the *centroid* C of the area, this term is zero, in which case Eq. 10–3 becomes

$$I = \bar{I} + Ad^2 \tag{10–4}$$

In words, the moment of inertia I about an axis is equal to the moment of inertia $\bar{I}$ of the area about a parallel axis passing through the centroid plus the product of the area A and the square of the perpendicular distance d between the axes.

In a similar manner, it can be shown that the parallel-axis theorem for determining the polar moment of inertia J_B about an axis perpendicular to the $\bar{x}\bar{y}$ plane and passing through point B is

$$J_B = \bar{J} + As^2 \tag{10–5}$$

Here $\bar{J}$ represents the polar moment of inertia of the area A about the $\bar{z}$ axis which passes through the centroid C and is perpendicular to the $\bar{x}\bar{y}$ plane, and s is the distance between points B and C, that is, the perpendicular distance between the parallel axes, Fig. 10–3.

10–3. Radius of Gyration of an Area

The *radius of gyration* of an area is a useful term in structural mechanics that serves to determine the strength of columns. For an area A having a moment of inertia I_x about the x axis, we define the radius of gyration

k_x, having units of length, using a formula similar to Eq. 10–1, namely,

$$I_x = k_x^2 A$$

or

$$k_x = \sqrt{\frac{I_x}{A}} \qquad (10\text{–}6a)$$

In a similar manner, the radii of gyration k_y and k_O may be defined as

$$k_y = \sqrt{\frac{I_y}{A}} \qquad (10\text{–}6b)$$

$$k_O = \sqrt{\frac{J_O}{A}} \qquad (10\text{–}6c)$$

Substituting Eqs. 10–6 into Eq. 10–2c, we see that

$$k_O^2 = k_x^2 + k_y^2$$

Hence, the square of the radius of gyration of a planar area about the polar (or z) axis is equal to the sum of the squares of its radii of gyration about the x and y axes.

10–4 Moments of Inertia for a Planar Area by Integration

When all the boundaries for a planar area can be expressed by mathematical functions, Eqs. 10–2 may be used to determine the moments of inertia for the area. The integration is simplified, provided the coordinate system chosen is compatible with the boundaries. For example, it is generally convenient to use polar coordinates for areas having circular boundaries.

When a planar area element is chosen which has a differential size in two directions (second-order element) as in Fig. 10–2, a double integration must be performed to evaluate the moment of inertia. Provided the limits of integration can be determined, it is often easier, and hence more convenient, to choose a first-order element, that is an element having a differential size in only one direction, since it simplifies the integration process. In accordance with Eq. 10–1, *the entire differential element must lie at the same distance from the moment axis.* This will always be the case when a second-order element $dA = dx\,dy$ is chosen for integration (see Fig. 10–2). If a first-order element is chosen, however, this condition

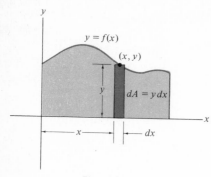

$y = f(x)$

(x, y)

$dA = y\, dx$

Fig. 10–4

is satisfied only when the element is *properly orientated* with respect to the axis. For example, the rectangular element $dA = y\, dx$, shown in Fig. 10–4, satisfies this requirement when computing the moment of inertia of the shaded area about the y axis, since all parts of the element lie at the *same distance* x from the y axis. (The element is infinitesimally "thin" in the x direction.) However, when the moment of inertia for the area is required about the x axis, one cannot directly apply Eq. 10–1 using the same element shown in the figure. The element is not infinitesimally thin in the y direction, and thus parts of its entirety lie at different distances from the x axis. It is possible, however, to use this same element to calculate I_x for the area. To do this, it is *first necessary to obtain the moment of inertia for the (rectangular) element about its centroidal axis, and then determine the moment of inertia for this element about the x axis using the parallel-axis theorem.* This technique is further explained in Examples 10–2 and 10–3.

Example 10–1

Determine the radius of gyration for the rectangular area shown in Fig. 10–5a with respect to (a) the centroidal $\bar{x}$ axis, (b) an axis perpendicular to the $\bar{x}\bar{y}$ plane and passing through the centroid O, and (c) an axis parallel to the $\bar{x}$ axis and passing through the base of the rectangle.

Solution

Part (a). The radius of gyration $k_{\bar{x}}$ may be found after first obtaining the moment of inertia $\bar{I}_{\bar{x}}$. The second-order differential element shown in Fig. 10–5a is chosen for integration. The element is located at the point (x, y). Note that the entire element is located at a distance y from the x axis. The entire region is covered if we integrate from $x = -b/2$ to $x = b/2$, and from $y = -h/2$ to $y = h/2$. Hence,

(a)

Fig. 10–5a

$$\bar{I}_{\bar{x}} = \int_A y^2\, dA = \int_{-h/2}^{h/2} \int_{-b/2}^{b/2} y^2\, dx\, dy = b \int_{-h/2}^{h/2} y^2\, dy = \tfrac{1}{12}bh^3 \quad (1)$$

Applying Eq. 10–6a, we obtain

$$k_{\bar{x}} = \sqrt{\frac{\bar{I}_{\bar{x}}}{A}} = \sqrt{\frac{\tfrac{1}{12}bh^3}{bh}} = \frac{h}{\sqrt{12}} \qquad\qquad Ans.$$

Part (b). The rectangular moment of inertia $\bar{I}_{\bar{y}}$ may be found by interchanging the dimensions b and h in Eq. (1), in which case

$$\bar{I}_{\bar{y}} = \tfrac{1}{12}hb^3$$

Using Eq. 10–2c, we obtain

$$\bar{J}_O = \bar{I}_{\bar{x}} + \bar{I}_{\bar{y}} = \tfrac{1}{12}bh(h^2 + b^2)$$

From Eq. 10–6c,

$$k_O = \sqrt{\frac{\overline{J_O}}{A}} = \sqrt{\frac{\frac{1}{12}bh(h^2 + b^2)}{bh}} = \sqrt{\frac{h^2 + b^2}{12}} \qquad Ans.$$

Part (c). We can obtain the moment of inertia I_{x_b}, computed with respect to an axis passing through the base of the rectangle by applying Eq. 10–2a, using a first-order element. As shown in Fig. 10–5b, a rectangular element having a width b, and thickness dy has been chosen for integration. We have selected this element because all points on the element act at the *same distance y* from the x_b axis. Hence,

$$I_{x_b} = \int_A y^2 \, dA = \int_0^h y^2 b \, dy = \tfrac{1}{3}bh^3$$

This same result can be obtained by applying the parallel-axis theorem, Eq. 10–4 in conjunction with Eq. (1) in Part (a):

$$I_{x_b} = I_{\bar{x}} + Ad^2$$
$$= \tfrac{1}{12}bh^3 + bh\left(\frac{h}{2}\right)^2 = \tfrac{1}{3}bh^3$$

The radius of gyration is therefore

$$k_{x_b} = \sqrt{\frac{I_{x_b}}{A}} = \sqrt{\frac{\tfrac{1}{3}bh^3}{bh}} = \frac{h}{\sqrt{3}} \qquad Ans.$$

(b)

Fig. 10–5b

Example 10–2

Compute the polar moment of inertia for the shaded area shown in Fig. 10–6a with respect to a polar axis passing through the origin O of the coordinate system.

Solution

The polar moment of inertia J_O can easily be computed provided I_x and I_y are known. The moment of inertia I_x can be computed by choosing elements of area *parallel* to the x axis. When this is done, *all parts of the element* lie at an *equal distance y* from the axis. An element intersecting the arbitrary point (x, y) on the curve is shown in Fig. 10–6a. Hence, integrating with respect to y from $y = 0$ to $y = 2$ in. yields

$$I_x = \int_A y^2 \, dA = \int_A y^2(4 - x) \, dy$$
$$= \int_0^2 y^2\left[4 - \left(\frac{y^2}{2} + 2\right)\right] dy = 2.13 \text{ in.}^4$$

(a)

Fig. 10–6a

This same result can be obtained by using elements parallel to the y axis, such as shown in Fig. 10–6b. In this case, however, the entire element does *not* lie at a fixed distance from the x axis, and therefore the parallel-axis theorem must be used to determine the *moment of inertia of the element* with respect to the axis. For a rectangle having a base b and height h, the moment of inertia about its centroidal $\bar{x}$ axis has been computed in Example 10–1. There it was found that $\bar{I}_{\bar{x}} = \frac{1}{12}bh^3$. For the differential element shown in Fig. 10–6b, $b = dx$ and $h = y$, and thus $d\bar{I}_{\bar{x}} = \frac{1}{12}dx\,y^3$. The centroid of the element is located at coordinates $(\tilde{x}, \tilde{y})$, where

$$\tilde{x} = x, \qquad \tilde{y} = \frac{y}{2}$$

Hence, the moment of inertia of the element about the x axis is

$$dI_x = d\bar{I}_{\bar{x}} + dA\,(\tilde{y})^2$$
$$= \frac{1}{12}\,dx\,y^3 + y\,dx\left(\frac{y}{2}\right)^2 = \frac{1}{3}y^3\,dx$$

Therefore, integrating this result with respect to x from $x = 2$ in. to $x = 4$ in. yields the same result as before,

$$I_x = \int_A dI_x = \int_A \frac{1}{3}y^3\,dx = \int_2^4 \frac{1}{3}[2(x-2)]^{3/2}\,dx = 2.13 \text{ in.}^4$$

When I_y is obtained using the rectangular element shown in Fig. 10–6b, the *entire element* lies at a distance x from the y axis; therefore, the moment of inertia is*

$$I_y = \int_A x^2\,dA = \int_A x^2 y\,dx = \int_2^4 x^2[2(x-2)]^{1/2}\,dx = 28.04 \text{ in.}^4$$

If rectangular elements orientated as shown in Fig. 10–6a are used for computing I_y, the parallel-axis theorem must be used to compute the moment of inertia of the element about the y axis. Why? From Fig. 10–6a, the centroid for the element is located at $(\tilde{x}, \tilde{y})$, where

$$\tilde{x} = x + \left(\frac{4-x}{2}\right) = \frac{4+x}{2}, \qquad \tilde{y} = y$$

Hence, in this case,

$y^2 = 2(x-2)$

2 in.

2 in. 2 in.

(b)

Fig. 10–6b

---

*If the parallel-axis theorem were used to determine this result, the moment of inertia of the element chosen about its centroidal axis becomes $d\bar{I}_{\bar{y}} = \frac{1}{12}y(dx)^3$. Thus $dI_y = \frac{1}{12}y(dx)^3 + x^2\,dA$. The first term on the right goes to zero, since it is a third-order differential, whereas dA is of the second order. The resulting integration is thus $I_y = \int_A x^2\,dA$, which is the same as that used above.

$$dI_y = d\bar{I}_{\bar{y}} + dA\, d^2$$

$$= \frac{1}{12}\, dy\, (4 - x)^3 + (4 - x)\, dy \left(\frac{4 + x}{2}\right)^2$$

or

$$dI_y = \frac{1}{3}(64 - x^3)\, dy$$

Integrating with respect to y from $y = 0$ to $y = 2$ in. yields the same result as before,

$$I_y = \int_A dI_y = \frac{1}{3}\int_0^2 (64 - x^3)\, dy$$

$$= \frac{1}{3}\int_0^2 \left[64 - \left(\frac{y^2 + 4}{2}\right)^3\right] dy = 28.04 \text{ in.}^4$$

The polar moment of inertia is therefore

$$J_O = I_x + I_y = 2.13 \text{ in.}^4 + 28.04 \text{ in.}^4$$

$$= 30.17 \text{ in.}^4 \qquad\qquad\qquad Ans.$$

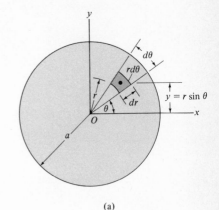

(a)

Example 10-3

For the circular area shown in Fig. 10–7a, determine the moment of inertia with respect to the x axis.

Solution I

This problem is most conveniently solved by using double integration and polar coordinates. For an arbitrary element having an area $dA = (r\, d\theta)\, dr$ and located at point (r, θ), as shown in Fig. 10–7a, the moment of inertia about the x axis is

$$I_x = \int_A y^2\, dA = \int_0^a \int_0^{2\pi} (r\sin\theta)^2\, r\, d\theta\, dr = \int_0^a r^3\, dr \int_0^{2\pi} \sin^2\theta\, d\theta$$

$$= \left.\frac{r^4}{4}\right]_0^a \left.\left(\frac{\theta}{2} - \frac{\sin 2\theta}{r}\right)\right]_0^{2\pi} = \frac{\pi a^4}{4} \qquad Ans.$$

(b)

Solution II

The result may be obtained by using a rectangular element and single integration. For the element chosen as shown in Fig. 10–7b,

$$I_x = \int_A y^2\, dA = \int_A y^2(2x)\, dy$$

Integrating with respect to y,

$$I_x = \int_{-a}^a y^2(2\sqrt{a^2 - y^2})\, dy = \frac{\pi a^4}{4} \qquad Ans.$$

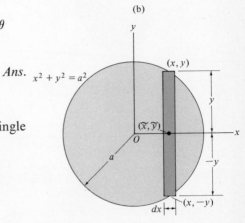

(c)

Fig. 10–7

Solution III

When rectangular elements are chosen as shown in Fig. 10–7c, the parallel-axis theorem must be used. Why? The centroid for the element has coordinates $\tilde{x} = x$, and $\tilde{y} = 0$. Hence, applying Eq. 10–4, we have

$$d\bar{I}_x = dI_{\tilde{x}} + dA\,(d)^2$$

$$= \frac{1}{12}\,dx\,(2y)^3 + 2y\,dx\,(0)$$

or

$$dI_x = \frac{2}{3}(y)^3\,dx$$

Therefore, integrating with respect to x,

$$I_x = \int_{-a}^{a} \frac{2}{3}(a^2 - x^2)^{3/2}\,dx = \frac{\pi a^4}{4} \qquad\qquad Ans.$$

Problems

10-1. The irregular area has a moment of inertia about the AA axis of 500 in.⁴ If the total area is 12 in.², determine the moment of inertia of the area about the BB axis. The CC axis passes through the centroid of the area.

10-3. Determine the moments of inertia of the shaded elliptical area about the x and about the y axis, and the polar moment of inertia about the origin O and about the point P (1 in., 0).

Prob. 10–1

Prob. 10–3

10-2. Determine the moment of inertia about the x axis of an area which is bounded by the lines $x = 0$, $y = 1$ in., and the parabola $y = 3x^2$.

10-4. Determine the moments of inertia of an area, bounded by the curve $y^2 = 2x$, and the lines $x = 2$ in. and $y = 0$, about the x and about the y axis.

10-5. Determine the moment of inertia of the shaded area about the y axis. Solve this problem two ways, using rectangular differential elements having a thickness of dx and dy, respectively.

$4y = 4 - x^2$

1 ft

2 ft

Prob. 10-5

10-6. Determine the moment of inertia I_x and I_y of the shaded portion of the cosine curve about the x and about the y axis.

2 in.

$y = 2 \cos x$

Prob. 10-6

10-7. The polar moment of inertia of the irregular area is $J_C = 900$ in.⁴ at the centroid point C. If the moment of inertia about the y axis is 600 in.⁴, and the moment of inertia about the x' axis is 800 in.⁴, determine the area A.

C

A

6 in.

x'

Prob. 10-7

10-8. Determine the polar moment of inertia of the shaded area (a) with respect to an axis perpendicular to the xy plane and passing through the origin of coordinates O, and (b) with respect to an axis perpendicular to the xy plane and passing through the point P (2 in., 0).

2 in.

P

O

2 in. — 2 in.

$y^2 = x$

Prob. 10-8

10-9. Determine the moment of inertia of the shaded area about the x axis.

2 in.

1 in.

$4y = x^2$

Prob. 10-9

10-10. Determine the moment of inertia of the area of the circular sector about the x axis.

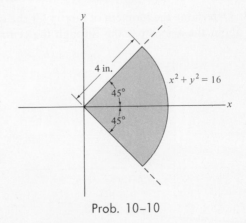

4 in.

$x^2 + y^2 = 16$

45°

45°

Prob. 10-10

369

10-11. For the shaded area, determine (a) the radius of gyration k_y about the y axis, and (b) the radius of gyration k_O about an axis perpendicular to the xy plane and passing through the origin O.

Prob. 10-11

10-12. Compute the moment of inertia of the triangular area about (a) an axis passing through the base of the triangle, and (b) the x axis (the centroidal axis of the triangle).

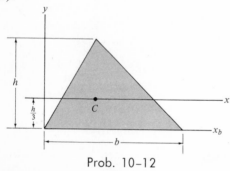

Prob. 10-12

10-13. Determine the moment of inertia for the quarter circle about the $\bar{x}$ axis passing through the centroid C.

Prob. 10-13

10-14. Determine the moment of inertia of the shaded area about the x and about the y axis.

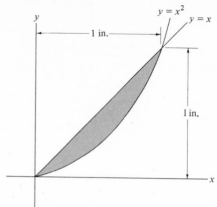

Prob. 10-14

10-15. Determine the radius of gyration k_x of the shaded area with respect to the x axis.

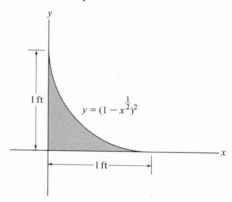

Prob. 10-15

10-5. Moments of Inertia of Composite Areas

A composite area generally consists of a series of connected "simpler" area shapes, such as semicircles, rectangles, and triangles. The moments of inertia for each of these simpler shapes can be determined by using the methods of integration discussed in Sec. 10-4. Since the integral used to compute the moment of inertia for a composite area may be subdivided into integrals computed over the simpler parts, the total moment of inertia for the composite area with respect to a given axis is obtained by "adding together" the moments of inertia of the composite parts about the same axis. The parallel-axis theorem should be used to transfer the moment of inertia of each composite area to the desired axis. When the region of an area has a hole, the moment of inertia for the region is found by "subtracting" the moment of inertia for the hole from the moment of inertia for the entire area, including the hole.

The moments of inertia for some common area shapes are given in Table 10-1. Furthermore, in many cases beams and columns are composed of two or more structural elements, such as angles and channels, which are welded together. The moment of inertia or radius of gyration for such a composite cross-sectional area is frequently required when using structural design formulas. The moments of inertia for standard

Table 10-1 Moments of Inertia for Common Geometric Area Shapes

1. Rectangle $\qquad A = bh \qquad \overline{I}_{\overline{x}} = \frac{1}{12}bh^3$

2. Triangle $\qquad A = \frac{bh}{2} \qquad \overline{I}_{\overline{x}} = \frac{1}{36}bh^3$

3. Quarter Circle $\qquad A = \frac{\pi a^2}{4} \qquad \overline{I}_{\overline{x}} = 0.0549a^4$

4. Semi circle $\qquad A = \frac{\pi a^2}{2} \qquad \overline{I}_{\overline{x}} = 0.1098a^4$

5. Circle $\qquad A = \pi a^2 \qquad \overline{I}_{\overline{x}} = \frac{1}{4}\pi a^4$

6. Ellipse $\qquad A = \pi ab \qquad \overline{I}_{\overline{x}} = \frac{1}{4}\pi ab^3$

structural elements are tabulated in engineering handbooks and design codes. For purposes here, a representative tabulation, taken from the American Institute of Steel Construction manual, is given in Table 10–2.

The following two example problems illustrate the method for finding the moment of inertia for composite areas using the data in Tables 10–1 and 10–2.

Table 10–2   Properties of Rolled Steel Shapes

Type	Nominal Size	wt/Length (lb/ft)	Area (in.²)	$\bar{I}_{\bar{x}}$ (in.⁴)	$\bar{y}$ (in.)	$\bar{I}_{\bar{y}}$ (in.⁴)	$\bar{x}$ (in.)
Wide Flange (W)	30 × 15*	210	61.9	9,890	—	757	—
	18 × 17	60	17.7	986	—	50.1	—
	8 × 8	67	19.7	272	—	88.6	—
Standard Channel (C)	15 × 3⅜	40	11.8	349	—	9.23	0.778
	12 × 3	30	8.82	162	—	5.14	0.674
	10 × 2⅝	25	7.35	91.2	—	3.36	0.690
Angle (L)	4 × 3 × ⅜†	8.5	2.48	3.96	1.28	1.92	0.782
	3 × 3 × ¼	4.9	1.44	1.24	0.842	1.24	0.842
	2 × 2 × ⅜	4.7	1.36	0.479	0.636	0.479	0.636

*First figure is depth; second is width.   †Third figure is thickness.

Example 10-4

Compute the moment of inertia of the composite area shown in Fig. 10-8a about the x axis.

(a)

(b)

Fig. 10-8

Solution

For convenience, the composite area may be obtained by *subtracting* the semicircle from the rectangle and triangle as shown in Fig. 10-8b. The moments of inertia for each of these composite parts about the x axis may be computed using the parallel-axis theorem and the data contained in Table 10-1.

Semicircle:

$$d = 8 - \frac{4(3)}{3\pi} = 6.73 \text{ in.}$$

$$I_x = \bar{I}_x + Ad^2$$

$$= 0.1098(3 \text{ in.})^4 + \frac{\pi(3 \text{ in.})^2}{2}(6.73 \text{ in.})^2 = 649 \text{ in.}^4$$

Rectangle:

$$I_x = \bar{I}_x + Ad^2$$

$$= \tfrac{1}{12}(6 \text{ in.})(8 \text{ in.})^3 + (6 \text{ in.})(8 \text{ in.})(4 \text{ in.})^2 = 1,024 \text{ in.}^4$$

Triangle:

$$I_x = \bar{I}_x + Ad^2$$

$$= \tfrac{1}{36}(6 \text{ in.})(6 \text{ in.})^3 + \tfrac{1}{2}(6 \text{ in.})(6 \text{ in.})(2 \text{ in.})^2 = 108 \text{ in.}^4$$

The moment of inertia for the entire composite area is thus

$$I_x = -649 \text{ in.}^4 + 1,024 \text{ in.}^4 + 108 \text{ in.}^4 = 483 \text{ in.}^4 \qquad \textit{Ans.}$$

Example 10–5

Compute the moment of inertia about the centroidal axis of the composite beam cross section shown in Fig. 10–9a. Neglect the inertia of the rivets.

Fig. 10–9

Solution

The necessary geometrical properties for the wide-flange beam and channel, which are listed in Table 10–2, are shown in Fig. 10–9b.

It is first necessary to determine the centroidal distance $\bar{y}$ for the *entire section*, Fig. 10–9a. This is done by using the techniques discussed in Sec. 9–4. With reference to the base axis x_b, the distance $\bar{y}$ is

$$\bar{y} = \frac{\Sigma \bar{y}A}{\Sigma A}$$

$$= \frac{(9 \text{ in.})17.7 \text{ in.}^2 + (18 \text{ in.} + 0.51 \text{ in.} - 0.674 \text{ in.})8.82 \text{ in.}^2}{17.7 \text{ in.}^2 + 8.82 \text{ in.}^2}$$

$$= 11.94 \text{ in.}$$

Using the parallel-axis theorem, we can now compute the moment of inertia for the wide-flange beam and channel about the $\bar{x}$ axis.

Wide-flange beam:

$$I_{\bar{x}} = \bar{I}_x + Ad^2$$

$$= 986 \text{ in.}^4 + 17.7 \text{ in.}^2(11.94 \text{ in.} - 9 \text{ in.})^2 = 1{,}139 \text{ in.}^4$$

Channel:

$$I_{\bar{x}} = \bar{I}_x + Ad^2$$

$$= 5.14 \text{ in.}^4 + 8.82 \text{ in.}^2 (18 \text{ in.} + 0.51 \text{ in.} - 0.674 - 11.94 \text{ in.})^2 = 312 \text{ in.}^4$$

Thus, the moment of inertia for the entire section becomes

$$I_{\bar{x}} = 1,139 \text{ in.}^4 + 312 \text{ in.}^4 = 1,451 \text{ in.}^4 \qquad \textit{Ans.}$$

Problems

10-16. Determine the moment of inertia of the shaded section with respect to a horizontal axis passing through the centroid.

Prob. 10–16

10-18. Determine the moments of inertia of the shaded area with respect to the x and with respect to the y centroidal axes.

Prob. 10–18

10-17. Determine the radius of gyration k_x of the shaded area about the x centroidal axis.

Prob. 10–17

10-19. Determine the moments of inertia of the shaded area with respect to the x and with respect to the y axis.

Prob. 10–19

10-20. Determine the moment of inertia of the shaded area with respect to a horizontal axis passing through the centroid of the section.

6 in.

1 in.

1 in.

6 in.

1 in.

Prob. 10-20

10-21. Determine the moment of inertia of the composite section with respect to a horizontal axis passing through the centroid of the cross section.

3 in.

8 in.

1 in.

4 in.

Prob. 10-21

10-22. Determine the moment of inertia of the composite beam cross section with respect to the x centroidal axis. Neglect the inertia of the rivets.

$\frac{1}{2}$ in. $\times$ $6\frac{1}{4}$ in.

L3 $\times$ 3 $\times$ $\frac{1}{4}$

$\frac{1}{4}$ in.

29 in.

x ——————————— x

Prob. 10-22

10-23. Determine the moment of inertia of the beam cross section with respect to the x centroidal axis. Neglect the inertia of the rivets.

C 10 $\times$ 25

0.526 in.

L3 $\times$ 3 $\times$ $\frac{1}{4}$

x ——————————— x

15 in. $\times$ $\frac{1}{2}$ in. Plate

L2 $\times$ 2 $\times$ $\frac{3}{8}$

Prob. 10-23

10-24. Determine the polar moment of inertia of the shaded area about the origin of coordinates, O, located at the centroid.

Prob. 10-24

10-25. The cross section of the beam consists of two angles, a wide-flange beam, and a cover plate, all welded together as shown. Determine the moment of inertia of this section with respect to the x centroidal axis.

Prob. 10-25

10-26. The composite cross section for the column consists of two cover plates riveted to two channels. Determine the radii of gyration k_x and k_y with respect to the centroidal x and y axes.

Prob. 10-26

10-27. A column has a builtup cross section as shown. Neglecting the inertia of the lacing and rivets, determine the radius of gyration of the column about the x centroidal axis.

Prob. 10-27

377

10-28. Determine the moment of inertia of the composite area with respect to the x' axis. The area is symmetric with respect to the y axis.

Prob. 10–28

10-29. Determine the moment of inertia of the composite area with respect to the x axis.

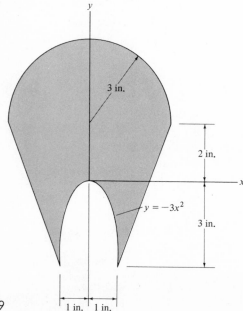

Prob. 10–29

10-30. Determine the moments of inertia of the "Z" section with respect to the x and with respect to the y centroidal axes.

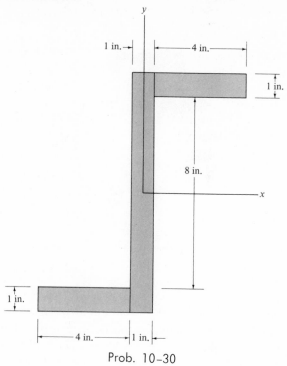

Prob. 10–30

10-31. Determine the moment of inertia of the composite area with respect to the x axis.

Prob. 10–31

378

*10–6. Products of Inertia of Areas

The rectangular moment of inertia for a planar area about an axis will usually vary, depending upon how the axis is orientated with respect to the area. In some applications of structural design it is necessary to know the orientation of those axes which give, respectively, the maximum and minimum moment of inertia for the area. Techniques which are used to find these axes will be discussed in Sec. 10–7. To use these techniques it is required that one know the product of inertia for the area as well as its rectangular moments of inertia for a given orientation of the axes.

The *product of inertia* dI_{xy} for an elemental area dA, shown in Fig. 10–10, with respect to the x and y axis, is defined as

$$dI_{xy} = xy \, dA$$

For the total area A, shown in Fig. 10–10, this equation becomes

$$I_{xy} = \int_A xy \, dA \qquad (10\text{–}7)$$

Fig. 10–10

The integration of Eq. 10–7 may be performed by using double or single integration. Although the element of area is more easily established with double integration, there is less confusion in establishing the limits for single integration.

Like the moment of inertia, the product of inertia has units of length raised to the fourth power, e.g., in.^{4} or ft^4. Since x or y may be a negative quantity, while the element of area is always positive, the product of inertia may be positive, negative, or zero, depending upon the location and orientation of the coordinate axes.

The product of inertia I_{xy} for the area will be *zero* if the x or y axis, or both axes are *axes of symmetry* for the area. To show this consider the area given in Fig. 10–11, which happens to be symmetrical with respect to the x axis. As shown in the figure, for every element dA located at the point (x, y) there is a corresponding element dA located at $(x, -y)$. The products of inertia for these elements are respectively $xy \, dA$ and $-xy \, dA$. Since one product of inertia is the negative of the other, the sum of the products of inertia for all of the elements of area which are chosen in this way will cancel each other. Hence, the product of inertia for the total area becomes zero.

The *parallel-axis theorem* for products of inertia of planar areas may be developed in a manner similar to that for rectangular moments of inertia. Consider the shaded area shown in Fig. 10–12, where $\bar{x}$ and $\bar{y}$ represent a set of axes passing through the *centroid C* of the area, and x and y represent a corresponding set of parallel axes. The product of inertia for the elemental area dA, with respect to the x and y axes, is

Fig. 10–11

$$dI_{xy} = xy\,dA = (x_C + \bar{x})(y_C + \bar{y})\,dA$$

Expanding this expression and integrating each term to obtain the total product of inertia yields

$$I_{xy} = \int_A \bar{x}\bar{y}\,dA + x_C \int_A \bar{y}\,dA + y_C \int_A \bar{x}\,dA + x_C y_C \int_A dA$$

or

$$I_{xy} = \bar{I}_{\bar{x}\bar{y}} + x_C \int_A \bar{y}\,dA + y_C \int_A \bar{x}\,dA + x_C y_C A$$

In this expression, $\bar{I}_{\bar{x}\bar{y}}$ represents the product of inertia for the area with respect to the centroidal axes. The integrals in the second and third terms on the right vanish, since the moments of the area are taken about the centroidal axis. Therefore, the parallel-axis theorem for the product of inertia becomes

$$I_{xy} = \bar{I}_{\bar{x}\bar{y}} + A\,x_C y_C$$

In words, the product of inertia I_{xy} is equal to the product of inertia $\bar{I}_{\bar{x}\bar{y}}$, plus the product of the area A, times the perpendicular distances x_C and y_C which lie between the $\bar{x}\bar{y}$ and xy axes. In particular, it is important that the algebraic signs for x_C and y_C be maintained when applying this formula.

Fig. 10–12

*10–7. Principal Moments of Inertia

Consider the planar area A and the coordinate axes xy and uv, shown in Fig. 10–13. In structural mechanics, it is sometimes necessary to calculate the moments and product of inertia I_u, I_v, and I_{uv} for the area with respect to the inclined u and v axes, when the values for θ, I_x, I_y, and I_{xy} are *known*. Referring to Fig. 10–13, we may relate the perpendicular distance from the area element dA to the axes of the two coordinate systems by using the transformation

$$u = x\cos\theta + y\sin\theta$$
$$v = y\cos\theta - x\sin\theta$$

Using these equations, the moments and product of inertia of dA about the u and v axes become

$$dI_u = v^2\,dA = (y\cos\theta - x\sin\theta)^2\,dA$$
$$dI_v = u^2\,dA = (x\cos\theta + y\sin\theta)^2\,dA$$

Fig. 10–13

and

$$dI_{uv} = uv \, dA = (x \cos \theta + y \sin \theta)(y \cos \theta - x \sin \theta) \, dA$$

Expanding each of these expressions and integrating, realizing that

$$I_x = \int_A y^2 \, dA, \; I_y = \int_A x^2 \, dA, \text{ and } I_{xy} = \int_A xy \, dA, \text{ we obtain}$$

$$I_u = I_x \cos^2 \theta + I_y \sin^2 \theta - 2I_{xy} \sin \theta \cos \theta$$
$$I_v = I_x \sin^2 \theta + I_y \cos^2 \theta + 2I_{xy} \sin \theta \cos \theta$$

and

$$I_{uv} = I_x \sin \theta \cos \theta - I_y \sin \theta \cos \theta + I_{xy}(\cos^2 \theta - \sin^2 \theta)$$

Using the trigonometric identities, $\sin 2\theta = 2 \sin \theta \cos \theta$ and $\cos 2\theta = \cos^2 \theta - \sin^2 \theta$, we can reduce the above equations to the form

$$I_u = \frac{I_x + I_y}{2} + \frac{I_x - I_y}{2} \cos 2\theta - I_{xy} \sin 2\theta \qquad (10\text{–}8a)$$

$$I_v = \frac{I_x + I_y}{2} - \frac{I_x - I_y}{2} \cos 2\theta + I_{xy} \sin 2\theta \qquad (10\text{–}8b)$$

and

$$I_{uv} = \frac{I_x - I_y}{2} \sin 2\theta + I_{xy} \cos 2\theta \qquad (10\text{-}8c)$$

If Eqs. 10-8a and 10-8b are added together, we see that the polar moment inertia about the z axis passing through point O is *independent* of the orientation of the u and v axes, i.e.,

$$J_O = I_u + I_v = I_x + I_y$$

From Eqs. 10-8, it may be seen that the moments and product of inertia, I_u, I_v, and I_{uv}, will, in general, have a different value for each value of θ. It is sometimes important to determine the orientation of the axis about which the moments of inertia for the area are maximum and minimum. This particular set of axes is called the *principal axes* of the area through the point O. The corresponding moments of inertia with respect to these axes are called the *principal moments of inertia* at point O. In general, *every point* has its own set of principal axes. However, in structural mechanics, the centroid is important.

The angle $\theta = \theta_p$, which gives the principal moments of inertia for the area, may be found by differentiating Eq. 10-8a with respect to θ, and setting the result equal to zero. Thus,

$$\frac{dI_u}{d\theta} = -2\left(\frac{I_x - I_y}{2}\right) \sin 2\theta - 2I_{xy} \cos 2\theta = 0$$

and therefore,

$$\tan 2\theta_p = \frac{-I_{xy}}{\dfrac{I_x - I_y}{2}} \qquad (10\text{-}9)$$

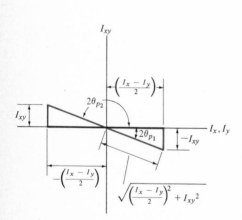

Fig. 10-14

Equation 10-9 has two roots θ_{p_1} and θ_{p_2} which locate the axes for maximum and minimum moments of inertia. Due to the nature of the tangent, the values of $2\theta_{p_1}$ and $2\theta_{p_2}$ are 180° apart, so that θ_{p_1} and θ_{p_2} are 90° apart. Assuming that I_{xy} and $(I_x - I_y)$ are both positive quantities, we can obtain the sine and cosine of $2\theta_{p_1}$ and $2\theta_{p_2}$ from the triangles shown in Fig. 10-14, which are based upon Eq. 10-9.

$$\text{For } \theta_{p_1} = \begin{cases} \sin 2\theta_{p_1} = -I_{xy} \left/ \sqrt{\left(\dfrac{I_x - I_y}{2}\right)^2 + I_{xy}^2} \right. \\[4ex] \cos 2\theta_{p_1} = \left(\dfrac{I_x - I_y}{2}\right) \left/ \sqrt{\left(\dfrac{I_x - I_y}{2}\right)^2 + I_{xy}^2} \right. \end{cases}$$

$$\text{For } \theta_{p_2} = \begin{cases} \sin 2\theta_{p_2} = I_{xy} \left/ \sqrt{\left(\dfrac{I_x - I_y}{2}\right)^2 + I_{xy}^2} \right. \\[4ex] \cos 2\theta_{p_2} = -\left(\dfrac{I_x - I_y}{2}\right) \left/ \sqrt{\left(\dfrac{I_x - I_y}{2}\right)^2 + I_{xy}^2} \right. \end{cases}$$

If we substitute these two sets of trigonometric relations into Eq. 10–8a, (or Eq. 10–8b) and simplify, the result reduces to

$$I_{\substack{max \\ min}} = \frac{I_x + I_y}{2} \pm \sqrt{\left(\frac{I_x - I_y}{2}\right)^2 + I_{xy}^2} \qquad (10\text{--}10)$$

Depending upon the sign chosen, this result gives the maximum or minimum moment of inertia for the area. Furthermore, if the above trigonometric relations for θ_{p_1} and θ_{p_2} are substituted into Eq. 10–8c, it may be shown that $I_{uv} = 0$; that is, the *product of inertia with respect to the principal axes is zero.* As shown in Sec. 10–6 the product of inertia is zero with respect to a symmetrical axis; therefore, it follows that *any symmetrical axis represents a principal axis of inertia for the area.*

*10–8. Mohr's Circle for Moments of Inertia

Equations 10–8 through 10–10 have a graphical solution which is convenient to use and generally easy to remember. Squaring Eqs. 10–8a and 10–8c, and adding, we find that

$$\left(I_u - \frac{I_x + I_y}{2}\right)^2 + I_{uv}^2 = \left(\frac{I_x - I_y}{2}\right)^2 + I_{xy}^2 \qquad (10\text{--}11)$$

In a given problem, I_u and I_{uv} are *variables,* and I_x, I_y, and I_{xy} are *known constants.* Thus, Eq. 10–11 may be written in compact form as

$$(I_u - a)^2 + I_{uv}^2 = R^2$$

When this equation is plotted on the $I_u I_{uv}$ plane, the resulting graph represents a circle of radius

$$R = \sqrt{\left(\frac{I_x - I_y}{2}\right)^2 + I_{xy}^2}$$

having its center located at point (a, O), where $a = (I_x + I_y)/2$. The circle so constructed is called *Mohr's circle,* named after the German engineer Otto Mohr (1835–1918).

There are several methods for plotting Mohr's circle as defined by Eq. 10–11. The main purpose in using the circle here is to have a convenient procedure for transforming the moments of inertia I_x, I_y, and I_{xy} into the principal moments of inertia for a given planar area. The following step-by-step procedure provides a method for doing this.

1. Establish an xy axis for the area having the origin located at the point P of interest, and determine I_x, I_y, and I_{xy} (see Fig. 10–15a).

2. Construct a rectangular coordinate system of axes such that the abscissa represents the rectangular moment of inertia, I, and the ordinate represents the product of inertia I_{xy} (see Fig. 10–15b).

3. Determine the location for the center of the circle, O, which is located at a distance $(I_x + I_y)/2$ from the origin (see Fig. 10–15b).

4. Plot the "controlling point" A having coordinates (I_x, I_{xy}). By definition, I_x is always positive, whereas I_{xy} will be either positive or negative (see Fig. 10–15b).

5. Connect the controlling point A found in step 4 with the center of the circle, and determine the distance OA by trigonometry. This distance represents the radius of the circle (see Fig. 10–15b).

6. Draw the circle having the radius found in step 5. The points where the circle intersects the abscissa give the values of the principal moments of inertia $(I_u)_{min}$ and $(I_u)_{max}$. Notice that the *product of inertia will be zero at these points* (see Fig. 10–15b).

7. To find the direction of the major principal axis, determine by trigonometry the angle 2α, *measured from the radius OA to the direction line of the positive abscissa*. This angle represents twice the angle from the x axis of the area in question to the axis of maximum moment of inertia, $(I_u)_{max}$. Both the angle on the circle, 2α, and the angle on the area, α, *must be measured in the same sense,* as shown in Fig. 10–15. The axis for minimum moment of inertia $(I_u)_{min}$ is perpendicular to the axis for $(I_u)_{max}$.

Using trigonometry, we may verify these statements as being in accordance with the equations developed in Sec. 10–7.

Example 10–6

Determine the product of inertia of (a) the rectangle and (b) the triangle, shown in Fig. 10–16a, with respect to both the $\bar{x}$ and $\bar{y}$ centroidal axes, and also with respect to the x and y axes.

Solution

Part (a). Since $\bar{x}$ and $\bar{y}$ are *axes of symmetry* for the rectangle, the product of inertia for the rectangle with respect to these axes is zero. Hence,

$$I_{\bar{x}\bar{y}} = 0 \qquad \text{Ans.}$$

Using the parallel-axis theorem, we find that the product of inertia with respect to the x and y axes is

$$I_{xy} = \bar{I}_{\bar{x}\bar{y}} + A\, x_C y_C$$

$$= 0 + bh\,\frac{b}{2}\,\frac{h}{2}$$

$$= \frac{b^2 h^2}{4} \qquad \text{Ans.}$$

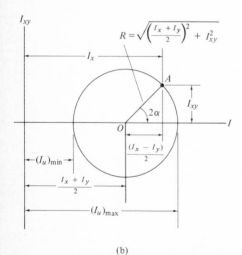

Axis for major principal moment of inertia, $(I_u)_{max}$

(a)

$$R = \sqrt{\left(\frac{I_x + I_y}{2}\right)^2 + I_{xy}^2}$$

(b)

Fig. 10–15

(a)

(b)

(c)

(d)

Fig. 10–16

Part (b). The product of inertia of the triangle with respect to the x and y axes can be determined by using integration. If a second-order element located at point (x, y) is chosen, as shown in Fig. 10–16b, then

$$dI_{xy} = xy\, dA = xy\, dy\, dx$$

Integrating on y from 0 to $y = (h/b)x$ and then on x from 0 to b we obtain,

$$I_{xy} = \int_A dI_{xy} = \int_A xy\, dy\, dx = \int_0^b \int_0^{(h/b)x} xy\, dy\, dx$$

$$= \int_0^b x\, dx \left[\frac{1}{2} y^2 \right]_0^{(h/b)x}$$

$$= \frac{1}{2} \frac{h^2}{b^2} \int_0^b x^3\, dx = \frac{b^2 h^2}{8} \qquad\qquad Ans.$$

As a second method, if first-order elements are used for integration, the parallel-axis theorem must be used. The *product of inertia for the differential element* shown in Fig. 10–16c is

$$dI_{xy} = d\bar{I}_{\tilde{x}\tilde{y}} + dA\, \tilde{x}\tilde{y}$$

where $(\tilde{x}, \tilde{y})$ represents the *centroid for the element*. Since $d\bar{I}_{\tilde{x}\tilde{y}} = 0$, (due to symmetry) and $\tilde{x} = x$, $\tilde{y} = y/2$, then

$$dI_{xy} = 0 + (y\, dx)\, x\, \frac{y}{2} = \left(\frac{h}{b} x\, dx \right) x\, \frac{h}{2b} x$$

or

$$dI_{xy} = \frac{h^2}{2b^2} x^3\, dx$$

Integrating with respect to x from 0 to b,

$$I_{xy} = \frac{h^2}{2b^2} \int_0^b x^3\, dx = \frac{b^2 h^2}{8} \qquad\qquad Ans.$$

As a third method of solution, consider elements having thickness dy and area $dA = (b - x)dy$, as shown in Fig. 10–16d. The centroid is located at point $\tilde{x} = x + (b - x)/2 = (b + x)/2$, $\tilde{y} = y$. Hence,

$$dI_{xy} = d\bar{I}_{\tilde{x}\tilde{y}} + dA\, \tilde{x}\tilde{y}$$

$$= 0 + (b - x)\, dy \left(\frac{b + x}{2} \right) y$$

$$= \left(b - \frac{b}{h} y \right) dy \left(\frac{b + \frac{b}{h} y}{2} \right) y$$

Integrating with respect to y from 0 to h,

$$I_{xy} = \frac{1}{2} \int_0^h y \left(b^2 - \frac{b^2}{h^2} y^2 \right) dy = \frac{b^2 h^2}{8} \qquad \textit{Ans.}$$

The product of inertia for the triangle with respect to its centroidal axis, Fig. 10–16a, may be computed using the above result and the parallel-axis theorem.

$$I_{xy} = \bar{I}_{\bar{x}\bar{y}} + A x_C y_C$$

$$\frac{b^2 h^2}{8} = \bar{I}_{\bar{x}\bar{y}} + \frac{1}{2} bh \left(\frac{1}{3} h \right) \left(\frac{2}{3} b \right)$$

Hence,

$$I_{\bar{x}\bar{y}} = \frac{1}{72} b^2 h^2 \qquad \textit{Ans.}$$

Example 10–7

Determine the principal moments of inertia for the "Z" section shown in Fig. 10–17a with respect to an axis passing through the centroid C.

Solution

To compute the principal moments of inertia, it is first necessary to determine the moments and product of inertia of the area with respect to the x and y axes shown. This can be done by assuming that the cross-sectional area consists of three rectangles, A, B, and D, as shown in Fig. 10–17b. Using the parallel-axis theorem for rectangles A and D, the computations are as follows:

$$I_x = [\tfrac{1}{12}(1)(3)^3 + (1)(3)(2)^2] + [\tfrac{1}{12}(6)(1)^3] + [\tfrac{1}{12}(1)(3)^3$$
$$+ (1)(3)(2)^2] = 29 \text{ in.}^4$$

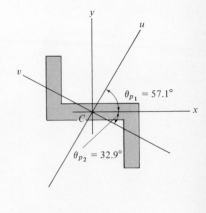

(a)   (b)   (c)

Fig. 10–17

$$I_y = [\tfrac{1}{12}(3)(1)^3 + (3)(1)(-2.5)^2] + [\tfrac{1}{12}(1)(6)^3]$$
$$+ [\tfrac{1}{12}(3)(1)^3 + (3)(1)(2.5)^2] = 56 \text{ in.}^4$$

$$I_{xy} = [0 + (3)(1)(-2.5)(2)] + [0] + [0 + (3)(1)(2.5)(-2)] = -30 \text{ in.}^4$$

Using Eq. 10-9, the principal axes are orientated as follows:

$$\tan 2\theta_p = \frac{-I_{xy}}{\dfrac{I_x - I_y}{2}} = \frac{30}{\dfrac{29 - 56}{2}} = -2.22$$

$$2\theta_{p_1} = 114.2° \quad \text{and} \quad 2\theta_{p_2} = -65.8°$$

Thus,

$$\theta_{p_1} = 57.1° \quad \text{and} \quad \theta_{p_2} = -32.9°$$

The u and v (principal) axes are therefore orientated as shown in Fig. 10-17c.

The principal moments of inertia with respect to the u and v axes are determined using Eq. 10-10. Hence,

$$(I_u)^{\max}_{\min} = \frac{I_x + I_y}{2} \pm \sqrt{\left(\frac{I_x - I_y}{2}\right)^2 + I_{xy}^2}$$

$$= \frac{29 + 56}{2} \pm \sqrt{\left(\frac{29 - 56}{2}\right)^2 + (-30)^2}$$

$$= 42.5 \text{ in.}^4 \pm 32.9 \text{ in.}^4$$

or

$$(I_u)_{\max} = 75.4 \text{ in.}^4, \qquad (I_u)_{\min} = 9.6 \text{ in.}^4 \qquad \qquad Ans.$$

Specifically, the *maximum moment of inertia* $I_{\max} = 75.4$ in.^{4} occurs with respect to the u axis *since, by inspection, most of the cross-section area is farthest away from this axis.* (To show this mathematically, substitute $\theta = 57.1°$ into Eq. 10-8a.)

Example 10-8

Using Mohr's circle, determine the principal moments of inertia for the "Z" section shown in Fig. 10-18a with respect to an axis passing through the centroid C.

Solution

The problem will be solved using the step-by-step procedure outlined in Sec. 10-8.

Step 1: The moments of inertia and the product of inertia have been determined in Example 10-7 with respect to the xy axis shown in Fig. 10-18a as $I_x = 29$ in.^{4}, $I_y = 56$ in.^{4}, and $I_{xy} = -30$ in.^{4}.

Steps 2, 3, 4, and 5: The I and I_{xy} axes are shown in Fig. 10–18b. The center of the circle, O, lies at a distance $(I_x + I_y)/2 = (29 + 56)/2 = 42.5$ in.^{4} from the origin. When the controlling point $A(29$ in.^{4}, -30 in.$^4)$ is connected to point O, the radius OA is determined from the triangle OBA using the Pythagorean theorem.

$$R = \sqrt{(13.5)^2 + (-30)^2} = 32.9 \text{ in.}^4$$

Step 6: The circle is constructed in Fig. 10–18c and intersects the I axes at points (75.4 in.^{4}, 0) and (9.6 in.^{4}, 0). Hence,

$$(I_u)_{\max} = 75.4 \text{ in.}^4 \qquad \textit{Ans.}$$
$$(I_u)_{\min} = 9.6 \text{ in.}^4 \qquad \textit{Ans.}$$

(a)

(b)

(c)

Fig. 10–18

(d)

Step 7: From Fig. 10–18c, the angle 2α as determined from the circle (*measured counterclockwise* from OA to the direction of the *positive I* axis) is

$$2\alpha = 180° - \sin^{-1}\frac{BA}{OA} = 180° - \sin^{-1}\frac{30}{32.9} = 114.2°$$

The principal axis for $(I_u)_{\max} = 75.4$ in.^{4} is therefore orientated at an angle $\alpha = 57.1°$ (measured *counterclockwise*) from the *positive x* axis to the *positive u* axis. The *v* axis is perpendicular to this axis. The results are shown in Fig. 10–18d.

Problems

10-32. Determine the product of inertia of the shaded area with respect to the x and y axes.

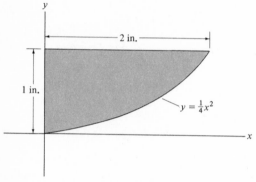

Prob. 10–32

10-33. Determine the product of inertia of the shaded area with respect to the x and y axes.

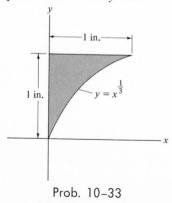

Prob. 10–33

10-34. Determine the product of inertia of the shaded section of the ellipse with respect to the x and y axes.

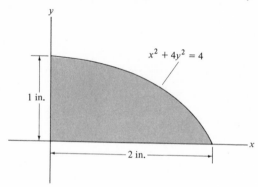

Prob. 10–34

10-35. Determine the product of inertia of the shaded area with respect to the x and y axes.

Prob. 10–35

10-36. Determine the product of inertia of the quarter circle with respect to the x and y axes.

$x^2 + y^2 = 4$

2 in.

2 in.

Prob. 10-36

10-37. Determine the product of inertia of the shaded portion of the parabola with respect to the x and y axes.

1 in.

2 in.

$y = 2x^2$

Prob. 10-37

10-38. Determine the product of inertia I_{xy} of half the shaded portion of the parabola in Prob. 10-37, bounded by the lines $y = 2$ in. and $x = 0$.

10-39. Determine the product of inertia of the shaded area with respect to the x and y axes passing through the centroid C.

4 in.

1 in.

$\frac{1}{2}$ in.

5 in. C

3.5 in.

1 in.

4 in.

Prob. 10-39

10-40. Determine the product of inertia for the area with respect to the x and y axes passing through the centroid C.

$\frac{1}{2}$ in. y

5 in.

C

$\frac{1}{2}$ in.

3 in.

Prob. 10-40

391

10–41. Determine the product of inertia of the "Z" section with respect to the x and y axes passing through the centroid C.

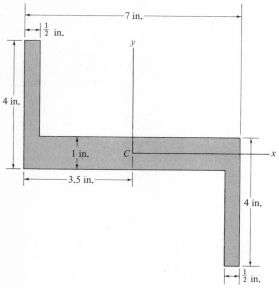

Prob. 10–41

10–42. Compute the moments of inertia of the shaded area with respect to the x' and y' axes.

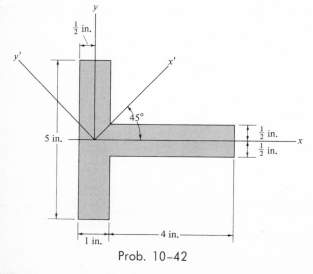

Prob. 10–42

10–43. Determine the orientation of the principal centroidal axis for the composite area. What are the moments of inertia with respect to these axes?

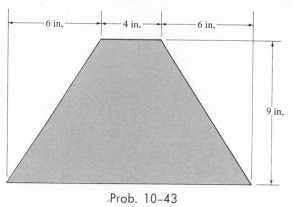

Prob. 10–43

10–44. Determine the moments of inertia and the product of inertia for the rectangular area with respect to the x' and y' axes passing through the centroid C.

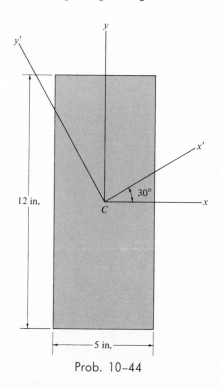

Prob. 10–44

10-45. Compute the moments of inertia for the C 12 × 30 channel section with respect to the inclined x' and y' axes passing through the centroid C.

Prob. 10-45

10-46. Determine the principal moments of inertia of the composite area with respect to a pair of axes passing through the centroid C. Use the equations developed in Sec. 10-7. $I_{xy} = -81$ in.^{4}

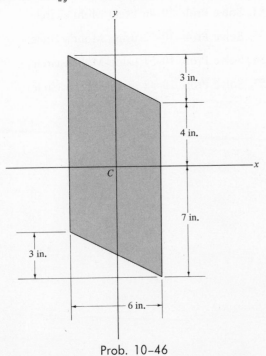

Prob. 10-46

10-47. Determine the product of inertia for the area cross section of the W 18 × 60 beam with respect to the x' and y' axes.

Prob. 10-47

10-48. The "Z" section has an $I_x = 55.53$ in.^{4} and $I_y = 13.87$ in.^{4}. Determine the principal moments of inertia for axes passing through the centroid C. Use the equations developed in Sec. 10-7. Assume all corners to be square.

Prob. 10-48

393

10-49. Determine the moments of inertia and the product of inertia for the semi-circular area with respect to the x' and y' axes.

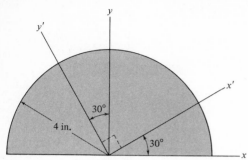

Prob. 10-49

10-50. Determine the product of inertia for the angle with respect to the x and y axes, passing through the centroid C. Assume all corners to be square.

Prob. 10-50

10-51. Compute the principal moments of inertia for the $\llcorner 2 \times 2 \times \frac{3}{8}$ angle with respect to the x and y axes passing through the centroid C. Use the equations developed in Sec. 10-7. Assume all corners to be square.

Prob. 10-51

10-52. Determine the directions of the principal axes acting through point O, and the principal moments of inertia for the rectangular area about this point.

Prob. 10-52

10-53. Construct Mohr's circle for the shaded area in Prob. 10-43.

10-54. Solve Prob. 10-46 using Mohr's circle.

10-55. Solve Prob. 10-52 using Mohr's circle.

10-56. Solve Prob. 10-51 using Mohr's circle.

10-57. Solve Prob. 10-48 using Mohr's circle.

11

Virtual Work

*11-1. Definition of Work and Virtual Work

Up to this point, all equilibrium problems have been solved utilizing the equations of equilibrium and the concept of a free-body diagram. There is, however, another method for solving equilibrium problems which is based on the principles of work and energy. This method, often referred to as *the principle of virtual work,* is found to be very useful in solving certain types of equilibrium problems, particularly those involving a series of connected rigid bodies.

To understand the principle of virtual work, it is first necessary to define the *work done by a force* **F** when the force undergoes an infinitesimal displacement $d\mathbf{s}$ along a path s. This situation is shown in Fig. 11-1a.* If β is the angle made between the tails of the force and displacement vectors, the differential amount of work done by the force **F** is a *scalar quantity* defined by the dot product,

$$dW = \mathbf{F} \cdot d\mathbf{s} = F \, ds \cos \beta \tag{11-1}$$

This work may be interpreted in one of two ways, either as the product of the force magnitude F and the magnitude of the component of the displacement vector in the direction of **F**, that is, $ds \cos \beta$, Fig. 11-1b, or as the product of ds times the component of force magnitude in the direction of displacement, $F \cos \beta$, Fig. 11-1c. If $F \cos \beta$ and ds are in the same direction, the work is "positive"; if they are in opposite directions, the work is "negative." As seen from Eq. 11-1, if **F** is *perpendicular* to the differential displacement $d\mathbf{s}$, the work done by **F** is zero. For example, the component $F \sin \beta$, shown in Fig. 11-1c, does no work when

---

*The magnitude of the force is assumed to be constant during the infinitesimal displacement $d\mathbf{s}$. Accounting for a change made in the magnitude of force during the displacement would only result in higher-order terms which would go to zero in the limit.

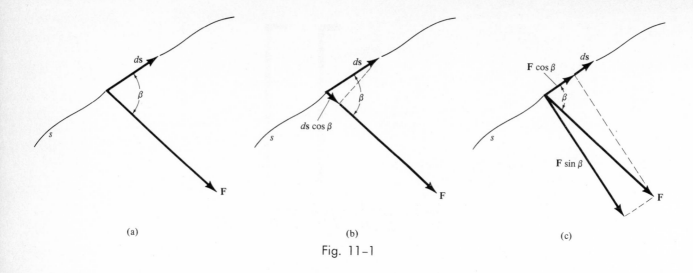

(a)　　　　　　　(b)　　　　　　　(c)

Fig. 11-1

the force is displaced a distance $d\mathbf{s}$. The units of work are force times displacement, e.g., ft-lb or m-kg$_f$, the same as those for a moment; however, the two concepts are in no way related. A moment is a vector quantity, whereas work is a scalar.

The two forces of a couple do work only when the couple rotates through an angle in the plane of the couple. Consider, for example, the couple moment $M = Fr$ acting on the rigid body shown in Fig. 11-2a. When the body undergoes a small displacement (shown dashed) the force $-\mathbf{F}$ acting at A moves through a displacement $d\mathbf{s}_A$ to point A', while the force $\mathbf{F}$ at B moves through a displacement $d\mathbf{s}_B$ to B'. To analyze the motion, the displacement of the body is separated into two *independent motions*, consisting of a *translation* and a *rotation*. For example, assume that the entire body is first translated by an amount $d\mathbf{s}_A$, Fig. 11-2b, and then rotated about point A' through an angle $d\theta$. When the motion is separated in this manner, the displacement $d\mathbf{s}_B$ is divided into two components to represent each motion, that is, $d\mathbf{s}_B = d\mathbf{s}_A + d\mathbf{s}_\theta$. Here the force $\mathbf{F}$ acting at B undergoes a *differential displacement,* such that the work done when $\mathbf{F}$ is displaced is independent of the path taken by the force in moving from the beginning to the end of the displacement. Therefore, the work done in displacing $-\mathbf{F}$ along the path $d\mathbf{s}_A + d\mathbf{s}_\theta$ is equivalent to the work done in moving $\mathbf{F}$ along $d\mathbf{s}_B$. The total work done by the couple is the sum of the work done by the two forces during the translation plus the work done during the rotation. During the *translation $d\mathbf{s}_A$*, the work done by force $\mathbf{F}$ is *canceled* by the work of $-\mathbf{F}$ since the forces are equal in magnitude but opposite in direction. When the body is *rotated* about point A', $-\mathbf{F}$ does no work since this point does not move. Force $\mathbf{F}$ moves through a displacement $d\mathbf{s}_\theta$, consequently, the total work is $dW = F\,d\mathbf{s}_\theta = Fr\,d\theta$. Since the product Fr is equal to the

(a)

(b)

Fig. 11-2

magnitude M of the couple, then

$$dW = M \, d\theta \qquad (11\text{-}2)$$

Clearly, this result indicates that a *couple does work only* during its *rotation* and *not* during a *translation*.

The definitions of work done by both a force and a couple have been presented in terms of infinitesimal displacements ds or rotations $d\theta$. These differentials represent *actual movements*. On the other hand, a *virtual movement* indicates displacements or rotations which are *assumed* and do not actually exist. These displacements and rotations are first-order differential quantities and will be denoted by the symbols δs and $\delta\theta$, respectively. If a system of forces acts on a body originally at rest, the forces will do work when the body undergoes a virtual movement. The *virtual work* done by a force $\mathbf{F}$ undergoing a virtual displacement $\delta \mathbf{s}$ is

$$\delta W = \mathbf{F} \cdot \delta \mathbf{s} = F \cos \theta \, \delta s$$

When a couple moment $\mathbf{M}$ undergoes a virtual rotation $\delta\theta$, the virtual work is

$$\delta W = M \, \delta\theta$$

where the rotation $\delta\theta$ occurs in the plane of the couple.

*11-2. Principle of Virtual Work for a Particle

The principle of virtual work for a particle initially at rest may be stated as follows: *If the virtual work done by the system of external forces acting on a particle is zero for every independent virtual displacement of the*

particle, then the particle is in equilibrium. Mathematically, this principle may be expressed as

$$\delta W = 0$$

where δW represents the virtual work done by *all the external forces* acting on the particle during a virtual displacement. In three dimensions, an unconstrained particle may have three independent virtual displacements, and therefore three independent virtual work equations may be written for the particle. For example, consider the particle P, shown in Fig. 11–3, which has a weight W and is subjected to the action of three forces $\mathbf{F}_1$, $\mathbf{F}_2$, and $\mathbf{F}_3$. Expressing each force in Cartesian vector notation, we have

$$\mathbf{F}_1 = (F_x)_1 \mathbf{i} + (F_y)_1 \mathbf{j} + (F_z)_1 \mathbf{k}$$
$$\mathbf{F}_2 = (F_x)_2 \mathbf{i} + (F_y)_2 \mathbf{j} + (F_z)_2 \mathbf{k}$$
$$\mathbf{F}_3 = (F_x)_3 \mathbf{i} + (F_y)_3 \mathbf{j} + (F_z)_3 \mathbf{k}$$
$$\mathbf{W} = W\mathbf{k}$$

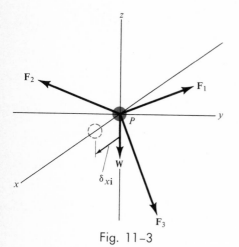

Fig. 11–3

If the particle is given a virtual displacement $\delta x\,\mathbf{i}$, as shown in Fig. 11–3, the virtual work equation for this displacement is

$$\delta W = 0; \qquad \mathbf{F}_1 \cdot \delta x\,\mathbf{i} + \mathbf{F}_2 \cdot \delta x\,\mathbf{i} + \mathbf{F}_3 \cdot \delta x\,\mathbf{i} + \mathbf{W} \cdot \delta x\,\mathbf{i} = 0$$

Substituting the vector components of each force into this equation, and carrying out the dot product operation yields

$$(F_x)_1\,\delta x + (F_x)_2\,\delta x + (F_x)_3\,\delta x = 0$$

or

$$\left(\sum_{i=1}^{3} (F_x)_i \right) \delta x = 0$$

Since δx has been imposed, this factor cannot be zero; therefore, the sum of the force components in the x direction must equal zero, i.e.,

$$\sum_{i=1}^{3} (F_x)_i = 0$$

This condition is equivalent to writing the *equilibrium equation* $\Sigma F_x = 0$ for the particle. In a similar manner, two other virtual work equations may be written for the particle by assuming virtual displacements $\delta y\,\mathbf{j}$ and $\delta z\,\mathbf{k}$, acting in the y and z directions respectively. Doing this amounts to satisfying the equilibrium equations $\Sigma F_y = 0$ and $\Sigma F_z = 0$ for the particle.

Writing the virtual work equations is therefore another way of stating both the necessary and sufficient conditions of equilibrium for the particle. No real advantage is gained, however, in using this method for the solution of particle equilibrium problems. This is because the virtual displacement in a given direction appears directly in each term of the

virtual work equation; hence, when factored, it leaves an equation which is the same as that which would be obtained by using an equilibrium equation written for the particle in the same direction. The only real purpose of applying the method of virtual work to the equilibrium of a single particle as stated here is simply to provide an introduction to the concepts involved.

*11–3. Principle of Virtual Work for a Rigid Body

Since a rigid body does not deform when subjected to a given loading, the total work done by the *internal forces* which hold the particles of a rigid body in equilibrium is *zero*. If the body moves through an arbitrary displacement, each of the internal forces occur in equal and opposite collinear pairs; consequently, the corresponding work done by each pair of internal forces cancels. For equilibrium of the body, therefore, it is necessary that the resultant external force and moment acting on the body be equal to zero. Thus, in a manner similar to that for a particle, the principle of virtual work for a rigid body may be stated as follows: *If the virtual work done by a system of external forces and couple moments acting on the rigid body is zero for every independent virtual displacement of the rigid body, then the rigid body is in equilibrium.*

When applying the principle of virtual work to equilibrium problems involving rigid bodies, it is necessary to take into account *both* translational and rotational movements of the body. A rigid body subjected to a general force system can have *six* independent virtual motions, namely, three coordinate translations δx, δy, and δz, and three rotations $\delta \alpha$, $\delta \beta$, and $\delta \gamma$, about the *x, y,* and *z* axes, respectively. An independent virtual work equation may be written for each of these six motions. These equations will then correspond to the *six* scalar equilibrium equations for a body. When the rigid body is subjected to a planar system of forces it has two independent translations on the surface of the plane and one rotation about an axis perpendicular to the plane. These *three* motions will yield *three* independent virtual work equations which correspond to the *three* force and moment equilibrium equations for the body.

Consider, for example, the beam shown in Fig. 11–4a, which has negligible weight. The reactions at supports *A* and *B* can be found by *removing the constraining action at each support and allowing the beam to displace in the direction of each reaction.* Thus, to determine the horizontal reaction at *B*, the roller at *B* and the pin at *A* are removed, and the beam is allowed to rotate through an angle $\delta \theta$ about the pin at *A*. This causes a virtual displacement $\delta \mathbf{b}$ along the direction of $\mathbf{F}_B$ and a movement $\delta \mathbf{c}$ under the 10-lb load, as shown on the free-body diagram (Fig. 11–4b). The support forces at *A* do no work since the pin at *A* does

(a)

(b)

(c)

Fig. 11–4

not translate during the motion. The virtual work equation for this movement becomes

$$\delta W = 0; \qquad 10 \text{ lb } \delta c = F_B \, \delta b = 0$$

To solve this equation for the unknown F_B, it is first necessary to relate δc to δb. This can easily be done by relating both of these displacements to $\delta\theta$. From Fig. 11–4b, $\delta c = (5 \text{ ft}) \delta\theta$ and $\delta b = (4 \text{ ft}) \delta\theta$.* Substituting these expressions into the previous equation, we obtain

$$10 \text{ lb } (5 \text{ ft}) \, \delta\theta - F_B(4 \text{ ft}) \, \delta\theta = 0$$

Factoring out the virtual displacement term yields

$$(50 \text{ lb-ft} - 4 \text{ ft } F_B) \, \delta\theta = 0$$

According to this virtual work equation, the rotation about point A corresponds to computing moments of the forces about point A, as indicated by the term in parentheses. Since $\delta\theta \neq 0$, solution of the above equation requires that

*The displacement acting along the direction of $\mathbf{F}_B$ is $\delta\mathbf{b}$, as shown in Fig. 11-4b. Note, since $\delta\theta$ is a differential of the first order, the actual displacement at B, $\delta\mathbf{b}'$, may be approximated by $\delta\mathbf{b}$.

$$F_B = \frac{50 \text{ lb-ft}}{4 \text{ ft}} = 12.5 \text{ lb}$$

i.e., $\Sigma M_A = 0$.

The vertical reactive force $\mathbf{A}_y$ at A may be determined by *removing the constraints and allowing the entire beam to be translated by an amount* $\delta \mathbf{y}$ *in the direction of* A_y. This virtual displacement is shown in Fig. 11-4c. The virtual work equation becomes

$$\delta W = 0; \qquad A_y \, \delta y - 10 \text{ lb } \delta y = 0$$
$$(A_y - 10 \text{ lb}) \, \delta y = 0$$

The term in parentheses is equivalent to vertical summation of forces. Since $\delta y \neq 0$, solution requires that

$$A_y = 10 \text{ lb}$$

i.e., $\Sigma F_y = 0$. How would you determine $\mathbf{A}_x$?

As in the case of a particle, no added advantage is gained by solving rigid body equilibrium problems using the principle of virtual work. The above example simply illustrates how the method is applied; and from the results, it can be seen that the solution may be obtained in a more direct manner by using the equations of equilibrium.

*11–4. Degrees of Freedom for a System

The method of virtual work is particularly suited for solving equilibrium problems that involve a series of several *connected* rigid bodies. Since the virtual work done by a force depends upon the direction of displacement, the number of *independent virtual work equations* which can be written for a system of connected rigid bodies will depend upon the number of *independent virtual displacements* which can be made by the system. *For a system of connected bodies the number of independent virtual displacements equals the number of independent coordinates needed to specify completely the location of all members of the system* (provided none of the constraints or supports for the system are removed). Each independent coordinate gives the system a *degree of freedom* which must be consistent with the constraining action of the supports. Thus, an *n*-degree-of-freedom system requires *n* independent coordinates to specify its location with respect to a fixed reference point. Furthermore, for this system we may write *n* independent virtual work equations.

Several examples of connected systems are shown in Fig. 11-5. It is assumed that the *geometry of each system is known*. The piston and crank arrangement shown in Fig. 11-5a is an example of a one-degree-of-freedom system. The independent coordinate θ may be used to specify

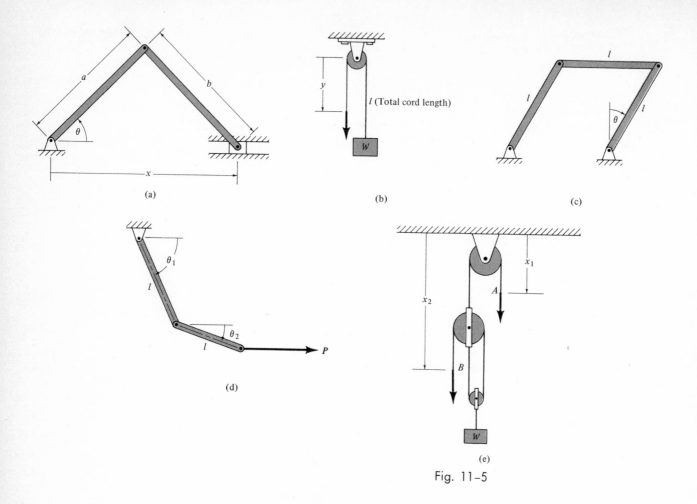

(a)

(b)

(c)

(d)

(e)

Fig. 11–5

the location of the two connecting links and the piston head. The coordinate x could also be used as the independent coordinate. However, since the piston is constrained to move between the walls, the distance x is *not* independent of θ and can be related to θ using the dimensions a and b of the connecting links. For example, if we use the cosine law, $b^2 = a^2 + x^2 - 2ax \cos \theta$. Other examples of one-degree-of-freedom systems are given in Fig. 11–5b and Fig. 11–5c.

The double pendulum and compound pulley arrangement shown in Figs. 11–5d and 11–5e, respectively, are examples of two-degree-of-freedom systems. To specify the location of each member of the double pendulum, both coordinate angles θ_1 and θ_2 must be known, since a rotation of one link is independent of the rotation of the other. To determine the location of the weight W attached to the rope and pulley arrangement in Fig. 11–5e, it is necessary to know the location of both end points A and B of the ropes. Provided the ropes are inextensible,

have known lengths, and do not slip on the pulleys, the independent coordinates x_1 and x_2, Fig. 11-5e, may be used to locate these end points.

As discussed in previous sections, a particle unrestrained to move in space has three degrees of freedom which may be specified by the three coordinate directions x, y, and z. A rigid body free to move in space has six degrees of freedom; three coordinate directions x, y, and z are needed to specify its translation, and three coordinate angles α, β, and γ specify its rotation. If a rigid body is constrained to move on a plane, it has three degrees of freedom—two translations on the surface of the plane, and one rotation about an axis perpendicular to the plane.

*11-5. Equilibrium of Multiconnected Rigid Bodies and the Principle of Virtual Work

The principle of virtual work when applied to problems involving the equilibrium of a particle or a rigid body, as discussed in Secs. 11-2 and 11-3, has no real advantage in obtaining the problem solution. This is because after the virtual work equation is applied, the virtual displacement, common to every term, factors out of the equation, leaving an equation which may have been obtained from direct application of the equilibrium equations. The purpose of the previous discussion was to provide an introduction to the concepts used in solving problems by the method of virtual work. The distinct advantage in using the principle of virtual work is realized, however, when solving equilibrium problems involving a series of *ideally connected rigid bodies*. By definition, two bodies are ideally connected provided the connections are frictionless. When an ideally connected system of rigid bodies is given a virtual displacement consistent with the constraints of the system, only the active forces do work during the displacement. *Active forces* are defined as those forces which tend to set the system in motion. Externally applied forces (and couples), weight, and other body forces are examples of active forces. *Reactive forces* tend to prevent motion of the system. These forces occur at the supports and connections. Assuming that the supports do not move, and since the connective forces occur in equal and opposite pairs, the reactive forces do no work during the displacement. In particular, a reactive force becomes an active force when the constraint causing the reaction is *removed*, and the system of connected bodies is given a displacement in the *direction of the force*.

The principle of virtual work may be stated in a form appropriate for ideally connected systems of rigid bodies as follows: *An ideally connected system of rigid bodies is in equilibrium provided the virtual work done by all the active forces acting on the system is zero for all virtual displacements consistent with the constraints imposed by the connections and supports.*

Symbolically, this principle may be stated as

$$\delta W = 0 \qquad (11\text{--}3)$$

where δW represents the total work done by the active forces (and couples) acting on the system during a virtual displacement.

It has been pointed out that if a system has n degrees of freedom it takes n independent coordinates to specify the location of the system. It is thus possible to write a virtual work equation if only *one* of these coordinates undergoes a virtual displacement while the remaining $(n - 1)$ coordinates are held fixed. Hence, for the entire system it is possible to write n independent virtual work equations, one for each independent displacement. In many cases, a system of connected bodies has only one degree of freedom, and therefore only one virtual work equation needs to be written for the (one) virtual displacement.

The method of virtual work, applied to the solution of equilibrium problems of multiconnected bodies, requires more mathematical sophistication (using calculus and trigonometry) than the conventional vector approach (using the equations of equilibrium). However, once the equations of virtual work are established, the equilibrium position for the system may be obtained directly, *without* the necessity for having to dismember the system to obtain relationships between forces occurring at the connections.

When multiconnected rigid bodies are subjected to the effects of *friction* at their connections, or at points of contact, any advantage of using the method of virtual work will generally be lost. The work done by the frictional forces must be included in any equation of virtual work written for the body having a displacement tangent to a surface of contact. This work is negative, since the frictional force always acts in the opposite direction of motion. In any case, if the frictional force is related to the normal force $(F = \mu N)$, the system of connected bodies will have to be *dismembered* in order to determine the magnitude of the frictional force.* Since the process of dismembering the system results in analyzing a series of *rigid bodies,* the most expedient way of obtaining the equilibrium forces is by using the equations of equilibrium *directly,* rather than applying the principle of virtual work.

In applying the principle of virtual work to the solution of equilibrium problems, it is *first necessary* to draw a free-body diagram of the entire system of connected bodies. From this diagram one can then determine which forces (or couples) do work (active forces) when the system undergoes a virtual displacement. The virtual work equation is then written

---

*The bodies must be dismembered in order to allow a displacement in the direction of **N**. By writing a virtual work equation in this direction, it is possible to compute the magnitude of **N**, and thereby obtain the value of $F = \mu N$.

(a)

(b)

Fig. 11–6

for a *displacement* of the system which is made in the *direction of an unknown force*. Using mathematics, the virtual work of all the active forces is computed during this displacement. By relating the virtual displacements of each active force to a common virtual displacement of the system, this common displacement can then be factored out of the virtual work equation. The resulting factor can then generally be solved for the unknown force. The following examples give a detailed treatment of this procedure.

Example 11–1

Using the principle of virtual work, determine the angle θ for equilibrium of the two-member linkage shown in Fig. 11–6a. Each member weighs 10 lb. The spring has a stiffness of $k = 1$ lb/in. and is unstretched when $\theta = 0°$.

Solution

The system has only one degree of freedom, since the location of both links may be specified by the single coordinate θ. As shown on the free-body diagram in Fig. 11–6b, when θ undergoes a virtual rotation $\delta\theta$, only the active forces $\mathbf{F}_{BA}$ and the two 10-lb weights do work. Why? If the origin of an xy coordinate system is established at the *fixed* pin support D, the location of these active forces may be specified by the variables x_B and y_w, as shown in the figure. Expressing these dimensions in terms of the angle θ, we obtain

$$x_B = 2(10 \cos \theta \text{ in.})$$
$$y_w = \tfrac{1}{2}(10 \sin \theta \text{ in.})$$

405

Differentiating to obtain the virtual displacements yields

$$\delta x_B = -20 \sin \theta \, \delta\theta \text{ in.} \tag{1}$$

$$\delta y_w = 5 \cos \theta \, \delta\theta \text{ in.} \tag{2}$$

It is seen by the signs of Eqs. (1) and (2), and indicated in Fig. 11–6b, that an *increase* in $\delta\theta$ causes a *decrease* in x_B and an *increase* in y_w.

For the arbitrary angle θ, the spring has been stretched a distance of $\Delta x_B = (20 \text{ in.} - x_B)$, and therefore $F_{BA} = k \, \Delta x_B = (1 \text{ lb/in.})(20 \text{ in.} - x_B) = 20(1 - \cos \theta) \text{ lb}$. Since $\mathbf{F}_{BA}$ and $\mathbf{w}$ act in the same direction as $\delta\mathbf{x}_B$ and $\delta\mathbf{y}_w$, respectively, the virtual work equation for the displacement is

$$\delta W = 0; \qquad w \, \delta y_w + w \, \delta y_w + F_{BA} \, \delta x_B = 0$$

Substituting Eqs. (1) and (2), to relate the virtual displacements to the common virtual displacement $\delta\theta$, and expressing F_{BA} in terms of θ,

$$10 \text{ lb } (5 \cos \theta \, \delta\theta \text{ in.}) + 10 \text{ lb } (5 \cos \theta \, \delta\theta \text{ in.})$$
$$+ 20(1 - \cos \theta) \text{ lb } (-20 \sin \theta \, \delta\theta \text{ in.}) = 0$$

Factoring out the *common displacement,*

$$[100 \cos \theta - 400(1 - \cos \theta) \sin \theta] \, \delta\theta \text{ in.-lb} = 0$$

Since $\delta\theta \neq 0$, therefore,

$$100 \cos \theta - 400(1 - \cos \theta) \sin \theta = 0$$

or

$$\tan \theta \, (1 - \cos \theta) = 0.25$$

Solving for θ by trial and error, we have

$$\theta = 42.9° \qquad\qquad\qquad\qquad\qquad \textit{Ans.}$$

If this problem had been solved by using the equations of equilibrium, it would have been necessary to dismember the links and apply three scalar equations to each link. Using the principles of virtual work via calculus, we have eliminated the need for doing this and have arrived at the answer in a very direct manner.

Example 11-2

Using the principle of virtual work, determine the horizontal force **P** required to maintain the equilibrium position of the mechanism shown in Fig. 11–7a, when $\theta = 30°$. Neglect the weight of the links.

Solution

The mechanism has one degree of freedom and therefore the location of each member may be specified using the variable θ. When θ undergoes

(a)

(b)

Fig. 11-7

a virtual displacement $\delta\theta$, as shown on the free-body diagram in Fig. 11-7b, links AB and EC rotate by the same amount since they have the same length, and link BC only translates. Since a couple does work *only* when it rotates, the work done by $\mathbf{M}_2$ is zero. The reactive forces at A and E do no work since the supports do not translate.

The coordinates x_B and x_D locate the (horizontal) position of the forces $\mathbf{F}$ and $\mathbf{P}$ with respect to the *fixed points* A and E. From Fig. 11-7b,

$$x_B = 4 \sin \theta \text{ ft}$$
$$x_D = 2 \sin \theta \text{ ft}$$

Thus,

$$\delta x_B = 4 \cos \theta \; \delta\theta \text{ ft}$$
$$\delta x_D = 2 \cos \theta \; \delta\theta \text{ ft}$$

Applying the equation of virtual work, noting that $\mathbf{P}$ is opposite to positive δx_D displacement, and hence does negative work, we obtain

$$\delta W = 0; \qquad M_1 \, \delta\theta + F \, \delta x_B - P \, \delta x_D = 0$$

Relating each of the virtual displacements to the *common* virtual displacement $\delta\theta$, yields

$$5 \text{ lb-ft } \delta\theta + 2 \text{ lb } (4 \cos \theta \; \delta\theta \text{ ft}) - P(2 \cos \theta \; \delta\theta \text{ ft}) = 0$$
$$(5 \text{ ft-lb} + 8 \cos \theta \text{ ft-lb} - 2P \cos \theta \text{ ft}) \, \delta\theta = 0$$

Since $\delta\theta \neq 0$,

$$P = \frac{5 + 8 \cos \theta}{2 \cos \theta} \text{lb}$$

At the required equilibrium position $\theta = 30°$, therefore,

$$P = \frac{5 + 8 \cos 30°}{2 \cos 30°} \text{ lb}$$
$$= 6.89 \text{ lb} \qquad \qquad Ans.$$

Example 11–3

Using the principle of virtual work, determine the force which the spring must exert in order to hold the mechanism shown in Fig. 11–8a in equilibrium when $\theta = 45°$. Neglect the weight of the members.

Solution

The xy coordinate system has its origin located at the *fixed point A,* as shown on the free-body diagram in Fig. 11–8b. The system has one degree of freedom, defined by the angle θ. Why? The spring stretches when link AB undergoes a counterclockwise virtual rotation $\delta\theta$. If the spring is removed, $\mathbf{F}_s$ becomes an active force. Noting that the 20-lb force is directed in the negative δy_B direction, the virtual work equation for the displacement becomes

$$\delta W = 0; \qquad -20 \text{ lb } \delta y_B + F_s \, \delta x_C = 0 \qquad (1)$$

Forces $\mathbf{A}_x$, $\mathbf{A}_y$, and $\mathbf{C}_y$ do no work during this movement. Why? From Fig. 11–8b, using the "law of cosines," we have

$$(7 \text{ ft})^2 = (6 \text{ ft})^2 + (x_C)^2 - 2(6 \text{ ft})x_C \cos \theta \qquad (2)$$

The variation or differential of the above equation is then

$$0 = 0 + 2x_C \, \delta x_C - 12 \text{ ft } \delta x_C \cos \theta + 12 \text{ ft } x_C \sin \theta \; \delta\theta$$

(a)

(b)

Fig. 11-8

Thus,

$$\delta x_C = \frac{12x_C \sin \theta}{12 \cos \theta - 2x_C} \delta\theta \text{ ft} \tag{3}$$

Also,

$$y_B = 6 \sin \theta \text{ ft}$$

Therefore,

$$\delta y_B = 6 \cos \theta \, \delta\theta \text{ ft} \tag{4}$$

When Eqs. (3) and (4) are substituted into Eq. (1) this yields,

$$-20 \text{ lb } (6 \cos \theta \, \delta\theta \text{ ft}) + F_s \frac{12x_C \sin \theta \text{ ft}}{12 \cos \theta \text{ ft} - 2x_C} \delta\theta = 0$$

or

$$\left[-120 \cos \theta + \frac{F_s(12x_C \sin \theta)}{12 \cos \theta - 2x_C} \right] \delta\theta \text{ ft-lb} = 0$$

Hence,

$$F_s = \frac{120 \cos \theta (12 \cos \theta - 2x_C)}{12x_C \sin \theta} \text{ lb}$$

At the required equilibrium position $\theta = 45°$, the corresponding value of x_C can be found by using Eq. (2), in which case

$$x_C^2 - 12 \cos 45° \text{ ft } x_C - 13 \text{ ft}^2 = 0$$

Solving for the positive root, we obtain,

$$x_C = 9.81 \text{ ft}$$

409

Thus,

$$F_s = \frac{120 \cos 45° \,[12 \cos 45° - 2(9.81)]}{12(9.81) \sin 45°} \text{lb}$$

$$= -11.35 \text{ lb}$$ *Ans.*

The negative sign indicates that $\mathbf{F}_s$ was shown acting in the wrong direction in Fig. 11–8b. Hence, the spring must be subjected to a *compressive force* in order to maintain equilibrium of the mechanism.

Example 11–4

Using the principle of virtual work, determine the equilibrium position of the "double" pendulum shown in Fig. 11–9a. Neglect the weight of the links.

Solution

The position for equilibrium may be determined by specifying the angles θ_1 and θ_2 in terms of the applied loading and the dimensions of the pendulum. The system has two degrees of freedom, since both θ_1 and θ_2 must be known to locate the position of both links. Measured from the *fixed point O*, the coordinate x_B may be related to θ_1 and θ_2 by the equation

$$x_B = l \sin \theta_1 + l \sin \theta_2 \qquad (1)$$

If we hold θ_1 *fixed* and vary θ_2 by an amount $\delta\theta_2$, as shown in Fig. 11–9b, the virtual work equation becomes

$$[\delta W = 0]_{\theta_2}; \qquad P(\delta x_B)_{\theta_2} - M \,\delta\theta_2 = 0 \qquad (2)$$

P and M represent the magnitudes of the applied force and couple acting on link AB. The work term for the couple is negative, since $\mathbf{M}$ is directed opposite to the rotation of $\delta\theta_2$. To obtain the variation of δx_B in terms of $\delta\theta_2$ it is necessary to take the *partial derivative* of x_B with respect to θ_2 in Eq. (1) since x_B is a function of θ_1 and θ_2. Hence,

$$\frac{\partial x_B}{\partial \theta_2} = l \cos \theta_2$$

Then

$$(\delta x_B)_{\theta_2} = l \cos \theta_2 \,\delta\theta_2$$

Substituting into Eq. (2), we have

$$(Pl \cos \theta_2 - M) \,\delta\theta_2 = 0$$

Since $\delta\theta_2 \neq 0$, then

$$\theta_2 = \cos^{-1} \frac{M}{Pl} \qquad \text{*Ans.*}$$

Fig. 11-9

When θ_2 is held *fixed*, and θ_1 is allowed to vary by an amount $\delta\theta_1$, as shown in Fig. 11-9c, the virtual work equation becomes

$$[\delta W = 0]_{\theta_1}; \qquad -M\,\delta\theta_1 - M\,\delta\theta_1 + P(\delta x_B)_{\theta_1} = 0 \qquad (3)$$

The first two terms are negative since the couples **M** act opposite to the rotation of $\delta\theta_1$. Using Eq. (1), we obtain the variation of x_B with θ_1:

$$\frac{\partial x_B}{\partial \theta_1} = l\cos\theta_1$$

or

$$(\delta x_B)_{\theta_1} = l\cos\theta_1\,\delta\theta_1$$

Substituting into Eq. (3) yields

$$(-2M + Pl\cos\theta_1)\,\delta\theta_1 = 0$$

Since $\delta\theta_1 \neq 0$, then

$$\theta_1 = \cos^{-1}\frac{2M}{Pl} \qquad\qquad Ans.$$

Problems

Neglect the weights of the members in the following problems unless specified otherwise.

11-1. Use the method of virtual work to compute the tensions in cables AB and AC. The lamp weighs 10 lb.

Prob. 11-1

11-2. Block A weighs 50 lb and rests on the smooth inclined plane. Determine the maximum weight w for equilibrium. The pulley at B is frictionless.

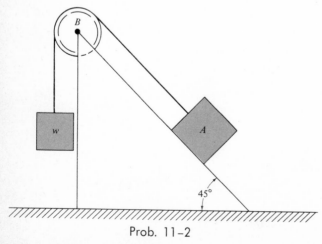

Prob. 11-2

11-3. Determine the vertical force **P** needed to maintain equilibrium of the lever.

Prob. 11-3

11-4. Determine the angle θ required for equilibrium of the bent lever. The lever weighs 2 lb/ft.

Prob. 11-4

11-5. The uniform rod AB weighs 10 lb and has a 100-lb weight attached at its end. If the spring is unstretched when $\theta = 60°$, determine the angle θ for equilibrium.

Prob. 11-5

11-6. The uniform links AB and BC each weigh 2 lb. Determine the horizontal force **P** required to hold the mechanism in the position shown. The spring has an unstretched length of 6 in.

Prob. 11-6

11-7. Calculate the equilibrium position x for the 15-lb block. The spring is unstretched when $x = 2$ in. The horizontal plane and pulleys are frictionless.

Prob. 11-7

11-8. Compute the force developed in the spring required to keep beam AB in equilibrium when $\theta = 35°$.

Prob. 11-8

11-9. If small weights w_A and w_B are added to points A and B of the desk lamp, the lamp will balance at any angle θ. Determine these required weights if the lamp weighs w lb.

Prob. 11-9

11-10. The pin-connected mechanism is constrained at A by a pin and at B by a smooth sliding block. If $\theta = 45°$, determine the vertical force **P** which must be applied to block B to hold the mechanism in equilibrium.

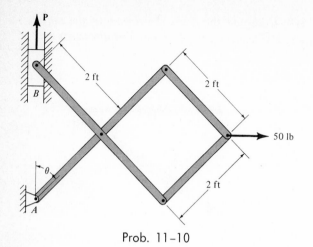

Prob. 11-10

11-11. Each of the three links of the mechanism has a weight of 10 lb. Determine the angle θ for equilibrium. The springs are unstretched when $\theta = 0°$.

Prob. 11-11

11-12. Determine the vertical reaction at B for the compound beam. The support at C is smooth.

Prob. 11-12

11-13. The uniform rod OA weighs 2 lb, and when the rod is in the vertical position, the spring is unstretched. Determine the position θ for equilibrium.

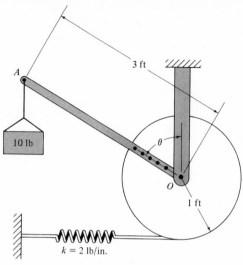

Prob. 11-13

11-14. A force **P** is applied at the handle of the toggle press. Determine the compressive force developed at C for an arbitrary angle θ of the handle.

Prob. 11-14

414

11-15. The piston C moves vertically between the two smooth walls. If the spring has a stiffness of $k = 15$ lb/in. and is unstretched when $\theta = 0°$, determine the couple moment $\mathbf{M}$ which must be applied to the pin at A to hold the mechanism in equilibrium when $\theta = 20°$.

Prob. 11-15

11-16. Determine the vertical force $\mathbf{P}$ needed to hold the 100-lb block in equilibrium. Pulley A weighs 5 lb. Neglect the weight of the ropes and assume that the inclined plane is smooth.

Prob. 11-16

11-17. The members of the mechanism are pin connected. Determine the angle θ necessary to hold the 50-lb horizontal force in equilibrium. The spring is unstretched when $\theta = 90°$.

Prob. 11-17

11-18. When $\theta = 20°$, the 50-lb block compresses the two vertical springs 4 in. If the uniform links AB and CD each weigh 10 lb, determine the magnitude of the applied couple moments $\mathbf{M}$ needed to maintain equilibrium when $\theta = 20°$.

Prob. 11-18

11-19. Each of the uniform beams AB and BC has a center of gravity located at their midpoints. If all contacting surfaces are smooth and beam BC weighs 20 lb, determine the weight of beam AB required for equilibrium in the position shown.

Prob. 11-19

11-20. Determine the force in member BC of the truss using the principle of virtual work. (*Hint:* Cut member BC, exposing the internal force acting in this member.)

Prob. 11-20

11-21. Determine the vertical force P required to maintain equilibrium of the mechanism. The spring has a stiffness $k = 3$ lb/in. and is compressed 2 in. when in the position shown.

Prob. 11-21

11-22. For the press shown, it takes n turns of the worm shaft at A to produce one turn of gear B. Determine the compression developed at C if a load P is applied to the handle. The pitch of the screw s is p.

Prob. 11-22

11-23. Determine the load P required to lift the 100-lb load using the differential hoist. The lever arm is fixed to the upper pulley and turns with it.

Prob. 11-23

11-24. Determine the horizontal force **F** required to maintain equilibrium of the slider mechanism when $\theta = 60°$.

Prob. 11-24

11-25. Determine the angles θ_1 and θ_2 for equilibrium of the two uniform links. Each link has a weight of w lb.

Prob. 11-25

11-26. Replace the 50-lb weight in Prob. 11-28 by a 50-lb *horizontal* force acting to the right, and determine the angle θ for equilibrium. Use the data in Prob. 11-28.

11-27. Determine the horizontal force **P** and vertical force **Q** necessary to maintain the equilibrium position of the two uniform links. Each link weighs 20 lb.

Prob. 11-27

11-28. The spring acting on the mechanism has an unstretched length of 1 ft. Determine the angle θ for equilibrium.

Prob. 11-28

11-29. Determine the reaction at B for the compound beam.

Prob. 11-29

417

*11-6. Definition of Potential Energy

Equation 11-1 defines the work done by a force **F** when the force undergoes an infinitesimal displacement d**s**. If the displacement path has a finite length s, Fig. 11-1a, the total work done is

$$W = \int_s \mathbf{F} \cdot d\mathbf{s} = \int_s F \cos \beta \, ds \qquad (11-4)$$

The integration is carried out over the entire length of the path, and as a result, in order to evaluate the integral it is necessary to obtain a relation between the force component $F \cos \beta$ and the displacement path s.

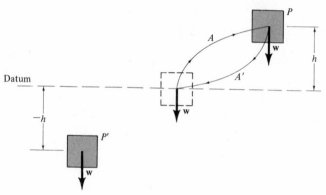

Fig. 11-10

In some instances, however, the work done by a force will be *independent of the path* followed and will depend only upon the initial and final points of the path. A force which has this property is called a *conservative force*. For example, a weight w, initially at some fixed horizontal reference or *datum*, as shown in Fig. 11-10, may be moved to position P, a distance h *above* the datum, along any arbitrary path A. Regardless of this path, the total work W done by the gravitational force w is independent of the path and depends only on the vertical height h. Thus,

$$W = -wh$$

The work is negative since **w** acts in the opposite direction of upward displacement. In a similar manner, the work done by lowering the weight w a distance h back to the datum along the arbitrary path A' is

$$W = wh$$

Why is the work positive?

Since the work done by a gravitational force is independent of the path and depends only on the initial and final points of the path, *the gravitational force is called a conservative force.*

The force developed by an elastic spring is also a conservative force. As shown in Fig. 11–11, the work done by a spring force *acting on a body,* such that the spring is elongated or compressed a distance d away from its equilibrium position, is

Fig. 11–11

$$W = \int F\,dx = -\int_0^{-d} kx\,dx = -\int_0^d kx\,dx = -\tfrac{1}{2}kd^2$$

The amount of work done *on the body* is thus dependent only upon the amount of extension *or* compression of the spring and the spring stiffness k. In *both* cases, the work is negative, since the spring exerts a force *on the body* which is opposite to its displacement.

In contrast to a conservative force, consider a *friction force* exerted on a moving object by a fixed surface. The work done by the friction force *depends upon the path*; the longer the path, the greater the work. When the object is slid across a fixed surface to some point and then returned to its original position, the friction force reverses direction when the block is returned, so that instead of recovering the work done during the first displacement, we must again do work on returning the body back to its original position. Thus, *a friction force is a nonconservative force.* The work done by this force is dissipated from the body in the form of heat.

When work is done on a body by a conservative force,* it gives the body *energy*. The energy stored in a body because of its elastic state, or energy measured due to the vertical position of a body, is known as *potential energy*. The total potential energy of a body or system of bodies is therefore the *sum* of its elastic potential energy V_e and its potential energy due to gravitation V_g, i.e.,

$$V = V_e + V_g \qquad (11\text{–}5)$$

Provided the weight w, shown in Fig. 11–10, is *above* the datum at position P, it has a *positive* potential energy,

$$V_g = wh$$

In this position, the gravitational force has a capacity of doing positive work on the body in moving it back to the datum. When the body is at position P', *below* the datum, the potential energy is *negative*, $V_g = -wh$, since work must be done on the body to move the weight back up to the datum level. In the next section it will be shown that the datum can be chosen at any elevation. This is so since the solution of

---

*A detailed discussion of conservative forces is given in *Engineering Mechanics: Dynamics*.

mechanics problems depends only upon a *change* in the potential energy, not the potential energy itself.

The potential energy stored in an elastic spring when the spring is *elongated or compressed* a distance d is

$$V_e = \tfrac{1}{2}kd^2$$

where k is the spring stiffness. In the deformed position, the spring has the capacity of doing work in returning an attached block to its original position. Hence, the elastic potential energy is always positive.

From this discussion, it may be concluded that the *work* done by a weight (gravitational) force is *opposed* to the *potential energy* gained or lost by the weight—for example, a weight force which acts *on the body* does *negative work* when the body is moved above a datum; this, however, causes a corresponding *increase* in the potential energy of the body. In a similar manner, a spring force does *negative work* when it acts *on a body* which compresses or elongates it; however, this causes an *increase* in the potential energy of the body. As a general statement, the total work done by all the conservative forces acting on a system of ideally connected rigid bodies is equal to a corresponding decrease in the potential energy of the system. This statement may be expressed mathematically as

$$W = -V + \text{const} \tag{11-6}$$

The constant is needed because the choice of a datum used to measure the potential energy is *arbitrary*. From this equation it may be seen that when the work is a minimum, the potential energy is a maximum and vice versa.

*11-7. Potential Energy Criterion for Equilibrium

When an ideally-connected system of bodies is acted upon by a series of conservative forces, the total potential energy stored in the system will depend upon the positions of each of the bodies relative to some datum or fixed reference position of the bodies. For example, a block of weight w resting on a spring, as shown in Fig. 11–12a, has a total potential energy relative to a datum plane located at the *unstretched length* of the spring of

$$
\begin{aligned}
V &= V_e + V_g \\
&= \tfrac{1}{2}kx^2 - wx
\end{aligned}
\tag{11-7}
$$

Using this equation, we may compute the total potential energy V for any deformed position x of the spring measured from the chosen datum plane.

The potential energy criterion for equilibrium may be formulated by using Eq. 11–6. Taking the derivative of this equation, we obtain

Fig. 11–12

$dW = -dV$. If the independent coordinates undergo *virtual displacements* rather than actual displacements, this equation becomes $\delta W = -\delta V$. For equilibrium, the principle of virtual work requires $\delta W = 0$, and therefore, knowing the potential-energy function for the system, we can also require

$$\delta V = 0 \qquad\qquad (11\text{-}8)$$

Therefore, when *an ideal system of multiply connected bodies is in equilibrium, the first variation or change in the potential energy function is zero.* Provided the system has only one degree of freedom, this change is computed by taking the *first derivative* of the potential-energy function and setting it equal to zero. For example, to determine the equilibrium position for the spring and block shown in Fig. 11-12*a*, a differential movement δx of the system causes a change in potential energy δV. Applying the equilibrium criterion to Eq. 11-7, we have

$$\delta V = \frac{dV}{dx}\,\delta x = (kx - w)\,\delta x = 0$$

Since δx is a virtual movement of the system, which is not equal to zero, we require that the term in the parentheses be zero. Hence, the equilibrium position $x = x_{\text{eq.}}$ becomes

$$x_{\text{eq.}} = \frac{w}{k}$$

We may obtain this *same result* by applying the equation of equilibrium $\Sigma F_y = 0$ to the forces acting on the free-body diagram of the block, Fig. 11-12*b*.

When the ideally-connected system has n degrees of freedom, the total potential energy stored in the system will be a function of n independent coordinates q_n; i.e., $V = V(q_1, q_2, \ldots, q_n)$. Applying the equilibrium criterion $\delta V = 0$ we must compute the total change in potential energy δV by using the "chain rule" of differential calculus; that is, the total variation or total differential of V must be equal to zero. Thus,

$$\delta V = \frac{\partial V}{\partial q_1}\,\delta q_1 + \frac{\partial V}{\partial q_2}\,\delta q_2 + \cdots + \frac{\partial V}{\partial q_n}\,\delta q_n = 0$$

Since the virtual displacements $\delta q_1, \delta q_2, \ldots, \delta q_n$ are independent of one another, it is necessary that

$$\frac{\partial V}{\partial q_1} = 0, \qquad \frac{\partial V}{\partial q_2} = 0, \qquad \cdots, \qquad \frac{\partial V}{\partial q_n} = 0 \qquad (11\text{-}9)$$

As in the case of virtual work, *we can therefore write n independent potential energy equations for a system having n degrees of freedom.*

Provided the potential-energy function for a system is known, the method of obtaining the solution is considerably simplified. All that is

required for equilibrium is that the derivative of the total potential energy be equal to zero. It will be shown in the next section that there is an added feature in knowing the potential-energy function for an equilibrium configuration, and that is, the *stability* of the equilibrium position may be studied.

*11-8. Stability of Equilibrium

Fig. 11-13

Consider the position of a small sphere resting at one point on each of the paths shown in Fig. 11-13. Each point represents an equilibrium state for the sphere. When the sphere is at point *A,* it is said to be in *stable equilibrium.* The sphere, when given a small displacement up the hill, will always *return* to its original equilibrium position. At point *A,* its total potential energy is a *minimum.* When the sphere is at point *B,* it is in *neutral equilibrium.* A small displacement of the sphere either to the left or right of point *B* will not alter the equilibrium condition of the sphere. It will *remain* in equilibrium in the displaced position. In this case, the potential energy of the sphere is *constant.* When the sphere is at point *C,* it is in *unstable equilibrium.* A small displacement of the sphere will cause its potential energy to be *decreased,* and therefore the sphere will roll further *away* from its equilibrium position. At point *C,* the potential energy of the sphere is a *maximum.*

To summarize, this simple example illustrates three *types of equilibrium* which can occur. *Stable equilibrium* occurs when a small displacement of the system returns the system to its original position. In this case the potential energy stored in the system is a minimum. In *neutral equilibrium,* the system may be displaced by a small amount, in which case the system remains in the displaced state; that is, the potential energy

Stable Equilibrium

(a)

Neutral Equilibrium

(b)

Unstable Equilibrium

(c)

Fig. 11-14

of the system remains constant during a displacement. *Unstable equilibrium* exists when the system is displaced by a small amount and not only does not return to its original equilibrium position but continues to go further away from the equilibrium position. In this equilibrium position, the potential energy is a maximum.

In particular, for *equilibrium* of a system having a single degree of freedom, it has been shown that the *first derivative* of the potential-energy function for the system must be equal to zero. The function V may have zero slope at points where the function is maximum, minimum, or at an inflection point. As shown in Fig. 11-14, each of these points represents an equilibrium position for the system.

Often one is interested in investigating the stability of a particular equilibrium configuration. The *type of equilibrium* may be investigated by computing the *second derivative* of the potential-energy function and substituting in the value of $x = x_{eq.}$ at the equilibrium position. When $d^2V/dx^2 > 0$, Fig. 11-14a, the equilibrium position is stable. When $d^2V/dx^2 = 0$, Fig. 11-14b, the system is in neutral equilibrium. And finally, when $d^2V/dx^2 < 0$, Fig. 11-14c, the equilibrium position is termed unstable.

The criterion for investigating stability becomes increasingly complex as the number of degrees of freedom for the system increases. For a system having two degrees of freedom, with independent coordinates (x, y), it may be verified (using the calculus of functions of two variables) that equilibrium and stability occur at a point $(x_{eq.}, y_{eq.})$ when

$$\frac{\partial V}{\partial x} = \frac{\partial V}{\partial y} = 0$$

$$\left[\left(\frac{\partial^2 V}{\partial x \partial y}\right)^2 - \left(\frac{\partial^2 V}{\partial x^2}\right)\left(\frac{\partial^2 V}{\partial y^2}\right)\right] < 0$$

$$\left[\frac{\partial^2 V}{\partial x^2} + \frac{\partial^2 V}{\partial y^2}\right] > 0$$

at the point. Both equilibrium and instability occur when

$$\frac{\partial V}{\partial x} = \frac{\partial V}{\partial y} = 0$$

$$\left[\left(\frac{\partial^2 V}{\partial x \partial y}\right)^2 - \left(\frac{\partial^2 V}{\partial x^2}\right)\left(\frac{\partial^2 V}{\partial y^2}\right)\right] < 0$$

$$\left[\frac{\partial^2 V}{\partial x^2} + \frac{\partial^2 V}{\partial y^2}\right] < 0$$

at the point.

The following examples illustrate both the procedures for solving equilibrium problems using the method of potential energy and determining the type of equilibrium.

Example 11-5

The uniform link shown in Fig. 11–15a has a weight of $w = 20$ lb and length of 6 ft. The spring is unstretched when $\theta = 0°$. Using the method of potential energy, determine the angle θ for equilibrium and investigate the stability at the equilibrium position.

Solution

The datum is established at the top of the link when the *spring is unstretched*, Fig. 11–15b. (Location of the datum is arbitrary. Why?) The system has one degree of freedom. When the link is located at the *equilibrium position*, defined by the angle θ, the spring increases its potential energy by stretching, and the weight decreases its potential energy since it moves further below the datum plane. The potential-energy function for the system may therefore be written as

$$V = V_e + V_g$$
$$= \frac{1}{2} kx^2 - w\left(x + \frac{l}{2}\cos\theta - \frac{l}{2}\right)$$

The potential-energy function will be expressed in terms of the independent variable θ. Since $l = x + l\cos\theta$ or $x = l(1 - \cos\theta)$, then

$$V = \frac{1}{2} kl^2(1 - \cos\theta)^2 - \frac{wl}{2}(1 - \cos\theta)$$

Computing the first derivative to determine the equilibrium position, we obtain,

$$\delta V = 0; \quad \frac{dV}{d\theta}\delta\theta = \left[kl^2(1 - \cos\theta)\sin\theta - \frac{wl}{2}\sin\theta\right]\delta\theta = 0$$

Since $\delta\theta \neq 0$, then

$$l\left[kl(1 - \cos\theta) - \frac{w}{2}\right]\sin\theta = 0$$

This equation is satisfied provided

$$\sin\theta = 0$$

for which

$$\theta = 0° \qquad\qquad Ans.$$

or

$$kl(1 - \cos\theta) - \frac{w}{2} = 0$$

$$\theta = \cos^{-1}\left(1 - \frac{w}{2kl}\right)$$

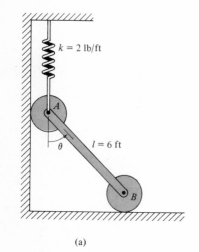

$k = 2$ lb/ft

$l = 6$ ft

(a)

Datum

$F = kx$

$\frac{1}{2}\cos\theta$

(b)

Fig. 11–15

Substituting in the numerical data for k, l, and w, we have

$$\theta = \cos^{-1}\left[1 - \frac{20 \text{ lb}}{2(2 \text{ lb/ft})(6 \text{ ft})}\right]$$

$$= 80.4° \qquad \qquad \qquad \qquad Ans.$$

Computing the second derivative of V to determine the type of equilibrium gives

$$\frac{d^2V}{d\theta^2} = kl^2(1 - \cos\theta)\cos\theta + kl^2\sin\theta\sin\theta - \frac{wl}{2}\cos\theta$$

or

$$\frac{d^2V}{d\theta^2} = kl^2(\cos\theta - \cos 2\theta) - \frac{wl}{2}\cos\theta$$

Substituting the values $\theta = 0°$ and $\theta = 80.4°$ and the numerical values of the constants yields

$$\left.\frac{d^2V}{dx^2}\right|_{\theta=0°} = 2 \text{ lb/ft } (6 \text{ ft})^2(\cos 0° - \cos 0°) - \frac{20 \text{ lb } (6 \text{ ft})}{2}\cos 0°$$

$$\frac{d^2V}{dx^2} = -60 \text{ ft-lb} < 0 \qquad \text{(unstable at } \theta = 0°) \qquad \qquad Ans.$$

$$\left.\frac{d^2V}{dx^2}\right|_{\theta=80.4°} = 2 \text{ lb/ft } (6 \text{ ft})^2(\cos 80.4° - \cos 160.8°) - \frac{20 \text{ lb } (6 \text{ ft})}{2}\cos 80.4°$$

$$\frac{d^2V}{dx^2} = 70.0 \text{ ft-lb} > 0 \qquad \text{(stable at } \theta = 80.4°) \qquad \qquad Ans.$$

Example 11-6

Determine the weight w required for equilibrium of the uniform 10-lb rod shown in Fig. 11-16a when $\theta = 30°$. The spring is unstretched when $\theta = 0°$. Investigate the stability at the equilibrium position.

Solution

The potential-energy function for the system may be determined by establishing the datum through point A and measuring the displaced position θ, as shown in Fig. 11-16b. When $\theta = 0°$, the weight is assumed to be suspended $(y_w)_1$ below the datum. In the final position the weight is $(y_w)_2$ below the datum. Thus, when the spring is stretched, the weight moves downward a distance $[(y_w)_2 - (y_w)_1]$ *below* the datum. The potential energy function is therefore

$$V = V_e + V_g$$

$$= \tfrac{1}{2}(30 \text{ lb/in.})(15 \sin\theta \text{ in.})^2 + 10 \text{ lb}\left(\frac{15 \sin\theta \text{ in.}}{2}\right) - w[(y_w)_2 - (y_w)_1] \quad (1)$$

(a)

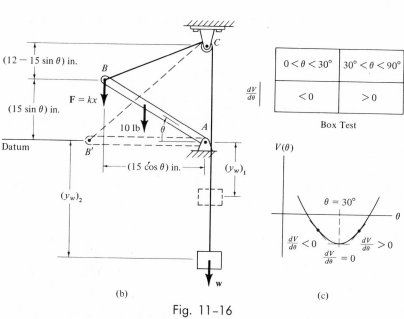

(b)

	$0 < \theta < 30°$	$30° < \theta < 90°$
$\dfrac{dV}{d\theta}$	< 0	> 0

Box Test

(c)

Fig. 11–16

The distance $[(y_w)_2 - (y_w)_1]$ may be related to the variable θ by measuring the difference in cord length $B'C$ and BC. Since

$$B'C = \sqrt{(15)^2 + (12)^2} = 19.21 \text{ in.}$$

and

$$BC = \sqrt{(15 \cos \theta)^2 + (12 - 15 \sin \theta)^2} \text{ in.}$$
$$= \sqrt{369 - 360 \sin \theta} \text{ in.}$$

then

$$[(y_w)_2 - (y_w)_1] = B'C - BC$$
$$= 19.21 \text{ in.} - \sqrt{369 - 360 \sin \theta} \text{ in.}$$

Substituting this relation into Eq. (1), we have

$$V = \tfrac{1}{2}(30 \text{ lb/in.})(15 \sin \theta \text{ in.})^2 + 10 \text{ lb}\left(\frac{15 \sin \theta \text{ in.}}{2}\right)$$
$$- w(19.21 \text{ in.} - \sqrt{369 - 360 \sin \theta} \text{ in.}) \qquad (2)$$

For equilibrium we require $\delta V = 0$; hence,

$$\frac{dV}{d\theta} \, \delta\theta = \left(6{,}750 \sin \theta \cos \theta + 75 \cos \theta - \left(\frac{w}{2}\right)\frac{360 \cos \theta}{\sqrt{369 - 360 \sin \theta}}\right)\delta\theta = 0$$

or

$$\frac{dV}{d\theta} \, \delta\theta = \frac{15 \cos \theta}{\sqrt{369 - 360 \sin \theta}} [(450 \sin \theta$$
$$+ 5) \sqrt{369 - 360 \sin \theta} - 12 \, w] \, \delta\theta = 0$$

Substituting the value of $\theta = 30°$ into this expression and simplifying, we have

$$\frac{dV}{d\theta}\bigg|_{\theta=30°} \delta\theta = 0.945\,(3{,}162 - 12w)\,\delta\theta = 0$$

Thus,

$$w = \frac{3{,}162}{12} = 264 \text{ lb} \qquad\qquad Ans.$$

Taking the second derivative of Eq. (2) to determine the type of equilibrium, we obtain,

$$\frac{d^2V}{d\theta^2} = 6{,}750 \cos 2\theta - 75 \sin \theta - \frac{w}{4}(369 - 360 \sin \theta)^{-3/2} (360 \cos \theta)^2$$
$$+ 180w \sin \theta \, (369 - 360 \sin \theta)^{-1/2}$$

For the equilibrium position $\theta = 30°$, $w = 264$ lb, hence,

$$\frac{d^2V}{d\theta^2} = 2{,}600 \text{ in.-lb} > 0 \qquad (\text{stable equilibrium at } \theta = 30°) \; Ans.$$

When the first derivative of the potential-energy function becomes complicated, as in this case, the "box test" provides a simplified way for determining whether the *second derivative* is positive, negative, or zero at the equilibrium position. After obtaining $w = 264$ lb, compute $\dfrac{dV}{d\theta}$ using values of θ just to the right and just to the left of $\theta = 30°$, as shown in Fig. 11–16c. From the calculations we can then reason from the behavior of the slope that the curve is concave upward as shown by the corresponding graph. Therefore, $\dfrac{d^2V}{d\theta^2}\bigg|_{\substack{w=264 \\ \theta=30°}} > 0$ (stable equilibrium).

(a)

(b)

(c)

Fig. 11–17

Example 11-7

A homogeneous block having a weight w rests on the top surface of a smooth cylinder, as shown in Fig. 11–17a. Show that this is a condition of unstable equilibrium.

Solution

The datum is established through the center of gravity, at the equilibrium position of the block. If the block is displaced by an amount θ from the equilibrium position as shown in Fig. 11–17b, the potential-energy function may be written in the form

$$V = V_e + V_g$$
$$= 0 - wy$$

From Fig. 11–17c,

$$y = \left(R + \frac{h}{2}\right)(1 - \cos\theta)$$

Thus,

$$V = -w\left(R + \frac{h}{2}\right)(1 - \cos\theta)$$

For equilibrium,

$$\delta V = \frac{dV}{d\theta}\,\delta\theta = -w\left(R + \frac{h}{2}\right)(\sin\theta)\,\delta\theta = 0$$

Obviously $\theta = 0°$ is the equilibrium position which satisfies this equation.

Taking the second derivative of V, we have

$$\frac{d^2V}{d\theta^2} = -w\left(R + \frac{h}{2}\right)\cos\theta$$

At $\theta = 0°$,

$$\left.\frac{d^2V}{d\theta^2}\right|_{\theta=0°} = -w\left(R + \frac{h}{2}\right) < 0$$

Since all the constants are positive, the block is in unstable equilibrium.

428

Problems

11-30. If the potential energy for a conservative one-degree-of-freedom system is expressed by the relation $V = 4x^3 - x^2 - 3x + 10$, determine the positions for equilibrium and investigate the stability at each of these positions.

11-31. Work Prob. 11-30 using the potential-energy function $V = 10 \cos 2\theta + 24 \sin \theta$, $0 \le \theta \le 180°$.

11-32. If the potential energy for a conservative two-degree-of-freedom system is expressed by the relation $V = 3y^2 + 2x^2$, determine the position of equilibrium and investigate the stability at each of these positions.

11-33. The spring acting on the mechanism has an unstretched length when $\theta = 30°$. Determine the position θ for equilibrium using the principle of potential energy, and investigate the stability of the mechanism at this position.

Prob. 11-33

11-34. Using the method of potential energy, determine the equilibrium position x for the 5-lb block. The spring is unstretched when $x = 2$ in. Investigate the stability at the equilibrium position. The inclined plane is smooth.

Prob. 11-34

11-35. Solve Prob. 11-7 using the principle of potential energy, and investigate the stability at the equilibrium position.

11-36. Solve Prob. 11-13 using the principle of potential energy, and investigate the stability at the equilibrium position.

11-37. Solve Prob. 11-28 using the principle of potential energy, and investigate the stability at the equilibrium position.

11-38. Using the method of potential energy, determine the angle θ for equilibrium of the 10-lb weight. The spring is unstretched when $\theta = 0°$. Investigate the stability in the equilibrium position.

Prob. 11-38

429

11-39. The uniform beam weighs 100 lb. If the contacting surfaces are smooth, determine the angle θ for equilibrium using the principle of potential energy. Investigate the stability of the beam when it is in the equilibrium position. The spring has an unstretched length of 4 ft.

Prob. 11-39

11-41. The two blocks weigh 10 lb each and are suspended from the ends of springs as shown. Using the method of potential energy, determine the deflection of the springs, x_1 and x_2, for both blocks. The springs are unstretched when $x_1 = x_2 = 0$.

Prob. 11-41

11-40. The uniform rod AB weighs 10 lb. If the spring has an unstretched length when $\theta = 60°$, determine the angle θ for equilibrium using the principle of potential energy. Investigate the stability at the equilibrium position.

Prob. 11-40

11-42. The uniform beam AB weighs 100 lb. If both springs DE and BC are unstretched when $\theta = 90°$, determine the angle θ for equilibrium using the principle of potential energy. Investigate the stability at the equilibrium position. Both springs always act in the horizontal position because of the roller guides at C and E.

Prob. 11-42

430

11–43. The uniform beam OA weighs 150 lb. Determine the weight w which will hold the beam in equilibrium when $\theta = 30°$. Neglect the size of the pulley at B.

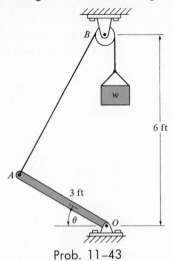

Prob. 11–43

11–44. A block made of homogeneous material rests on top of the cylindrical surface. Derive the relationship between the radius of the cylinder, r, and the dimension of the block, b, for stable, unstable, and neutral equilibrium. (*Hint:* Establish the potential energy function for *small* angle θ; i.e., approximate $\sin \theta \approx 0$, and $\cos \theta \approx 1 - \dfrac{\theta^2}{2}$.)

Prob. 11–44

11–45. Using the method of potential energy, determine the weight w required to deform the mechanism so that $\theta = 45°$. The springs are unstretched when $\theta = 60°$. Neglect the weights of the members.

Prob. 11–45

11–46. The uniform 50-lb beam AB is pin-connected at both of its end points. The rod BC, having negligible weight, passes through a swivel block at C. If the spring has a stiffness of $k = 5$ lb/ft and is unstretched when $\theta = 0°$, determine the angle θ for equilibrium of the beam. Investigate the stability at the equilibrium position.

Prob. 11–46

11-47. Determine the spring constant k so that the 50-lb beam in Prob. 11–46 is in equilibrium when $\theta = 45°$. The spring is unstretched when $\theta = 0°$.

11-48. The 60-lb hemisphere supports a cylinder having a density of 311 lb/ft³. If the radii of the cylinder and hemisphere are both 5 in., determine the maximum height h of the cylinder which will produce neutral equilibrium in the position shown.

Prob. 11–48

11-49. The cylinder is made of two materials and has a center of gravity at point G. Show that when G lies directly above the centroid C of the cylinder, the equilibrium position is unstable.

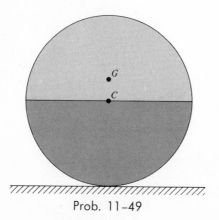

Prob. 11–49

11-50. Determine the position of stable equilibrium for the cylinder in Prob. 11–49.

11-51. A right circular cone is attached to the hemisphere. If both pieces have the same density, $\rho = 50$ lb/ft³, determine the maximum height h of the cone if the configuration is to be in neutral equilibrium at the position shown.

Prob. 11–51

11-52. Using the method of potential energy, establish two equilibrium equations which relate the two angles θ_1 and θ_2. The spring is unstretched when $\theta_1 = \theta_2 = 0°$. The pulley support at P is movable such that it keeps the cord BP horizontal.

Prob. 11–52

432

Computer Program for the Simultaneous Solution of Linear Algebraic Equations

The application of the equations of static equilibrium or the equations of dynamic motion to the solution of problems involving three-dimensional force systems generally results in the necessity of solving a set of linear algebraic equations. The computer program listed in Fig. A–1 may be used to solve a series of n linear, nonhomogeneous, algebraic equations for n unknowns ($n \leq 21$). These equations are solved numerically using a Gauss elimination procedure. (Refer to a textbook on numerical analysis.)

The program is written in "Fortran IV" computer language, which is commonly used in most computer systems today. In using this program one must supply the *job control language cards (JCL)*, the *program statement cards,* and the *input data cards*. Personnel at the computer center should be able to lend assistance both in obtaining the proper *JCL* cards and in arranging these cards in the proper order in the program. *In general, the cards are arranged in the following manner: JCL cards, program statement cards, JCL cards, input data cards, JCL cards.* The *program statement cards* are composed of the program statements listed in Fig. A–1. Each of these statements must be punched on a separate computer card in the proper card columns, as indicated in the figure. The input data is recorded on cards which are contained at the end of the program.

```
|1|2|3|4|5|6|7| Card column

|C| | | | | |   PROGRAM FOR SOLVING SIMULTANEOUS EQUATIONS
| | | | | | |   DIMENSION A(21,21),B(21),X(21),C(21,3)
| | | | | | |   READ(5,1)RCH
| | | | |1|   FORMAT(F10.4)
| | | | | |   N=RCH
              READ(5,2)((A(I,J),J=1,N),I=1,N)
         2    FORMAT(8F10.4)
              READ(5,2)(B(I),I=1,N)
              WRITE(6,3)
         3    FORMAT(35H1SOLUTION OF SIMULTANEOUS EQUATIONS//)
              WRITE(6,4)
         4    FORMAT(//13H THE A MATRIX/)
              DO 5 I=1,N
         5    WRITE(6,6)(A(I,J),J=1,N)
         6    FORMAT(1X,1PE16.7,6(2X,1PE16.7))
              WRITE(6,8)
         7    FORMAT(4X,2HX(,I2,2H)=,1PE16.7/)
         8    FORMAT(//13H THE B MATRIX/)
              DO 9 I=1,N
         9    WRITE(6,6)B(I)
              DO 10 I=1,N
        10    C(I,1)=0.0
              II=0
        11    AMAX=-1.0
              DO 16 I=1,N
              IF(C(I,1))16,12,16
        12    DO 15 J=1,N
              IF(C(J,1))15,13,15
        13    T=ABS(A(I,J))
              IF(T-AMAX)15,15,14
        14    IR=I
              IC=J
              AMAX=T
        15    CONTINUE
        16    CONTINUE
              IF(AMAX)27,32,17
        17    C(IC,1)=IR
              IF(IR-IC)18,20,18
        18    DO 19 J=1,N
              T=A(IR,J)
              A(IR,J)=A(IC,J)
        19    A(IC,J)=T
              II=II+1
              C(II,2)=IC
        20    P=A(IC,IC)
              A(IC,IC)=1.0
              P=1.0/P
              DO 21 J=1,N
        21    A(IC,J)=A(IC,J)*P
              DO 24 I=1,N
              IF(I-IC)22,24,22
        22    T=A(I,IC)
              A(I,IC)=0.0
              DO 23 J=1,N

        23    A(I,J)=A(I,J)-A(IC,J)*T
```

```
24 CONTINUE
   GO TO 11
25 IC=C(II,2)
   IR=C(IC,1)
   DO 26 I=1,N
   T=A(I,IR)
   A(I,IR)=A(I,IC)
26 A(I,IC)=T
   II=II-1
27 IF(II)25,28,25
28 DO 29 I=1,N
   X(I)=0.0
   DO 29 K=1,N
29 X(I)=X(I)+A(I,K)*B(K)
   WRITE(6,30)
30 FORMAT(///10H SOLUTIONS//)
   DO 31 I=1,N
31 WRITE(6,7)I,X(I)
   GO TO 34
32 WRITE(6,33)
33 FORMAT(///27H EQUATIONS CANNOT BE SOLVED)
34 STOP
   END
```

Fig. A-1

Each numerical element of data is punched anywhere within a field of 10 spaces (card columns) on the cards. Since a standard computer card contains 80 spaces, one can put at most *eight elements of input data on a card*. Each element *must* contain a decimal point, and these elements must be rounded off to, at most, four numeric characters after the decimal point.

Consider, for example, solving the following set of four linear algebraic equations for the unknowns x_1, x_2, x_3, and x_4.

$$A_{11}x_1 + A_{12}x_2 + A_{13}x_3 + A_{14}x_4 = B_1$$
$$A_{21}x_1 + A_{22}x_2 + A_{23}x_3 + A_{24}x_4 = B_2$$
$$A_{31}x_1 + A_{32}x_2 + A_{33}x_3 + A_{34}x_4 = B_3$$
$$A_{41}x_1 + A_{42}x_2 + A_{43}x_3 + A_{44}x_4 = B_4$$

The coefficients of these equations may be arranged in the following matrix form

$$[A][x] = [B]$$

or

$$\begin{bmatrix} A_{11} & A_{12} & A_{13} & A_{14} \\ A_{21} & A_{22} & A_{23} & A_{24} \\ A_{31} & A_{32} & A_{33} & A_{34} \\ A_{41} & A_{42} & A_{43} & A_{44} \end{bmatrix} \begin{bmatrix} x_1 \\ x_2 \\ x_3 \\ x_4 \end{bmatrix} = \begin{bmatrix} B_1 \\ B_2 \\ B_3 \\ B_4 \end{bmatrix}$$

To use the computer program for the solution of these equations, the input data (cards) must consist of the following:

1. Card 1 always contains the number of equations. In this case there are four equations, hence, the number 4.0 is punched anywhere within spaces 1 through 10 on the first card.
2. Cards 2 and 3 contain the A-matrix coefficients. (In all there are 16 coefficients.) Since there are 80 columns per card, and each coefficient is to be punched anywhere within a field of 10 spaces, the eight elements $A_{11}, A_{12}, A_{13}, A_{14}, A_{21}, A_{22}, A_{23}$, and A_{24} are punched *in this order* on the second card. The third card contains the remaining eight elements $A_{31}, A_{32}, A_{33}, A_{34}, A_{41}, A_{42}, A_{43}$, and A_{44}. (If there were more than 16 elements in the A-matrix, one would continue listing the data on cards 4, 5, etc.)
3. Card 4 contains the B-matrix coefficients. Each of these four elements is punched anywhere within a 10-space field on the card. They are entered in the following order: B_1, B_2, B_3, B_4.

When the program is received back from the computer center, the input data along with the answers (*output*) will be printed on paper. The answers are printed using "E-field" notation. This notation is similar to scientific notation. For example,

$$3.02 \text{ E } 02 = 3.02 \times 10^2 = 302.0$$
$$3.02 \text{ E } 00 = 3.02 \times 10^0 = 3.02$$
$$3.02 \text{ E } - 02 = 3.02 \times 10^{-2} = 0.0302$$

The following example illustrates the use of the program.

Example A-1

Solve the three equations

$$-x_1 - 2x_2 + 3x_3 = 2$$
$$2x_1 + x_2 + x_3 = -3$$
$$x_1 + 3x_2 + 4x_3 = 1$$

Solution:

Arranging the coefficients of three equations in the matrix form $[A][x] = [B]$ yields

$$\begin{bmatrix} -1 & -2 & 3 \\ 2 & 1 & 1 \\ 1 & 3 & 4 \end{bmatrix} \begin{bmatrix} x_1 \\ x_2 \\ x_3 \end{bmatrix} = \begin{bmatrix} 2 \\ -3 \\ 1 \end{bmatrix}$$

The input data is punched on four cards, as shown in Fig. A-2a. The first card contains the number of equations to be solved. The second and third cards contain the coefficients of the A-matrix, and the fourth card contains the B-matrix coefficients.

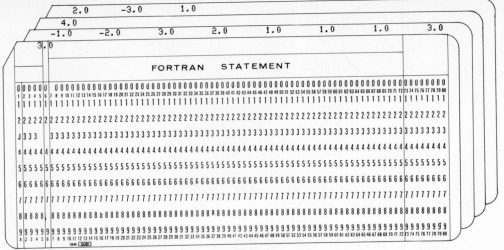

Fig. A–2a

SOLUTION OF SIMULTANEOUS EQUATIONS

THE A MATRIX

  -1.0000000E 00   -2.0000000E 00   3.0000000E 00
   2.0000000E 00   1.0000000E 00   1.0000000E 00
   1.0000000E 00   3.0000000E 00   4.0000000E 00

THE B MATRIX

   2.0000000E 00
  -3.0000000E 00
   1.0000000E 00

SOLUTIONS

  X(1)= -1.9285698E 00
  X(2)=   5.0000006E-01
  X(3)=   3.5714293E-01

Fig. A–2b

The printed output is shown in Fig. A-2*b*. Thus, the answers are

$$x_1 = -1.93 \qquad\qquad Ans.$$
$$x_2 = 0.50 \qquad\qquad Ans.$$
$$x_3 = 0.36 \qquad\qquad Ans.$$

B

Centroids of Line, Area, and Volume Elements

Circular arc segment $$\bar{x} = \frac{r \sin \alpha}{\alpha}$$	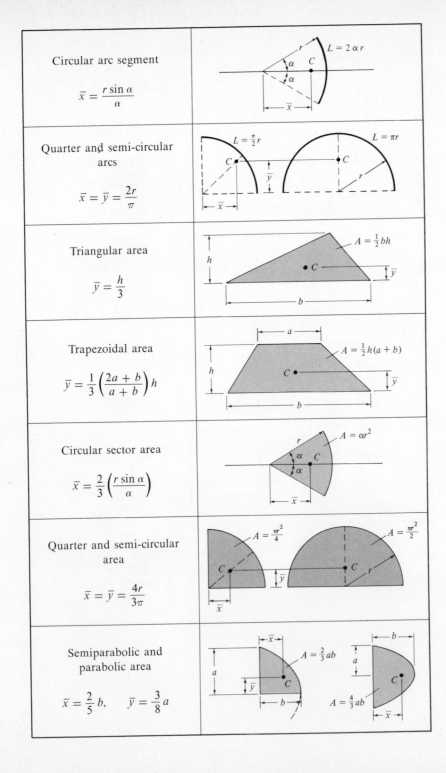
Quarter and semi-circular arcs $$\bar{x} = \bar{y} = \frac{2r}{\pi}$$	
Triangular area $$\bar{y} = \frac{h}{3}$$	
Trapezoidal area $$\bar{y} = \frac{1}{3}\left(\frac{2a + b}{a + b}\right)h$$	
Circular sector area $$\bar{x} = \frac{2}{3}\left(\frac{r \sin \alpha}{\alpha}\right)$$	
Quarter and semi-circular area $$\bar{x} = \bar{y} = \frac{4r}{3\pi}$$	
Semiparabolic and parabolic area $$\bar{x} = \frac{2}{5}b, \quad \bar{y} = \frac{3}{8}a$$	

Semiparabolic area $$\bar{x} = \frac{3}{4}b, \qquad \bar{y} = \frac{3}{10}a$$	$A = \frac{ab}{3}$
Hemispherical surface $$\bar{y} = \frac{r}{2}$$	$A = 2\pi r^2$
Conical surface $$\bar{y} = \frac{h}{3}$$	$A = 2\pi rh$
Hemispherical volume $$\bar{y} = \frac{3}{8}r$$	$V = \frac{2}{3}\pi r^3$
Conical volume $$\bar{y} = \frac{h}{4}$$	$V = \frac{1}{3}\pi r^2 h$
Paraboloid of revolution volume $$\bar{y} = \frac{h}{3}$$	$V = \frac{1}{2}\pi r^2 h$

Table of Conversion Factors (FPS)→(MKGFS)

Fundamental Quantities:

Length 1 foot (ft) = 12 inches (in.) = 0.3048 meter (m) = 30.48 centimeters (cm). The centimeter scale shown here approximates its actual size, and represents a comparison of length with the inch scale.

Force 1 pound (lb) = 0.4536 kilogram-force (kg_f). A weight equivalent to 1 kg_f is shown here in comparison to 1 lb of the same material.

Time The second (sec) is the standard unit of time in both systems of units.

Derived Quantity	Unit of Measurement (FPS)	To Convert (FPS) → (MKGFS) Multiply by	Unit of Measurement (MKGFS)
Moment of a Force Couple Torque	lb-ft	0.1383	kg_f-m
Linear load intensity Linear density	lb/ft	1.4882	kg_f/m
Surface load intensity Surface density Pressure (stress)	lb/ft^2	4.8825	kg_f/m^2
Volume load intensity Density	lb/ft^3	16.0187	kg_f/m^3
Area Moment of inertia	ft^4	0.008631	m^4
Mass Moment of inertia	ft-lb-sec^2	0.1383	m-kg_f-sec^2
Velocity	ft/sec	0.3048	m/sec
Acceleration	ft/sec^2	0.3048	m/sec^2
Work (energy)	ft-lb	0.1383	m-kg_f
Power	ft-lb/sec	0.1383	m-kg_f/sec
Impulse Momentum	lb-sec	0.4536	kg_f-sec
Frequency	Hz	1.0000	Hz
Angular velocity	rad/sec	1.0000	rad/sec
Angular acceleration	rad/sec^2	1.0000	rad/sec^2
Angular impulse Angular momentum	lb-ft-sec	0.1383	kg_f-m-sec

Answers to Even-Numbered Problems

2-2. 4.8 lb. $^{11°}$⟋

2-4. 84.6°.

2-6. 400 lb →, 300 lb.↑

2-8. 374 lb. ⟍$^{49°}$

2-10. 82.3°.

2-12. 21.6 ft/sec. $^{86°}$⟋

2-14. (a) $F_{AC} = 226$ lb, $F_{AB} = 298$ lb.
(b) 70°.

2-16. 133.1 lb, 52.1°.

2-18. 458 lb, 23.6°.

2-20. (a) 40.3 lb, $\alpha = 75.6°$, $\beta = 51.7°$, $\gamma = 138.1°$.
(b) 26 in., $\alpha = 112.6°$, $\beta = 22.6°$, $\gamma = 90°$.

2-22. $\cos \alpha = 0.226$, $\cos \beta = -0.793$, $\cos \gamma = -0.566$, 8.83 ft.

2-24. (a) 2.24 ft, $\alpha = 153.2°$, $\beta = 63.5°$, $\gamma = 90°$.
(b) 4.82 ft, $\alpha = 78.0°$, $\beta = 58.8°$, $\gamma = 33.9°$.

2-26. $\theta = 38.7°$, $\phi = 59.1°$, $\psi = 36.9°$.

2-28. 374 lb, $\alpha = 48.7°$, $\beta = 138.7°$.

2-30. $\alpha = 69.0°$, $\beta = 74.5°$, $\gamma = 26.7°$.

2-32. 202.4 lb, $\alpha = 106.2°$, $\beta = 78.1°$, $\gamma = 159.7°$.

2-34. $\beta = 115.3°$, $\mathbf{r} = \{8.66\mathbf{i} - 4.27\mathbf{j} + 2.59\mathbf{k}\}$ ft.

2-36. 5.48 in., $\alpha = 68.6°$, $\beta = 79.5°$, $\gamma = 24.1°$.

2-38. $\mathbf{F}_{BA} = \{14.3\mathbf{j} - 4.43\mathbf{k}\}$ lb, $\mathbf{F}_{DC} = \{13.27\mathbf{i} + 14.29\mathbf{j} - 4.42\mathbf{k}\}$ lb.

2-40. 32.4 lb, 2.98 ft.

3-2. $F_D = 1,596$ lb, $F_B = 2,042$.

3-4. $F_A = 13.08$ lb, $F_B = 9.07$ lb, 36.9°.

3-6. 152.1 lb, 48.2°.

3-8. 11.16 lb.

3-10. $T = 3,030$ lb, $F = 2,260$ lb.

3-12. 12.25 lb.

3-14. 6.88 in.

3-16. 24.9 in.

3-18. 3.40 ft.

3-20. $N = 2.50$ lb, $W = 4.48$ lb.

3-22. 419 lb.

3-24. 96.6 lb, $\alpha = 138.4°$, $\beta = 131.5°$, $\gamma = 91.9°$.

3-26. 1.79×10^{-3} lb.

3-28. $F_{DE} = 125$ lb, $F_{DF} = 707$ lb.

3-30. (5.05 in., 1.05 in., 0).

3-32. $T_{BD} = 47.1$ lb, $T_{CD} = 60.0$ lb, $W = 67.8$ lb.

3-34. $F_{AB} = 210$ lb, $F_{CB} = 201$ lb, $F_{OB} = 24.3$ lb.

3-36. $P = F = 0$, $R = 200$ lb.

3-38. $F_{OA} = 202$ lb, $F_{OC} = 216$ lb, $F_{OD} = 58.7$ lb.

3-40. 639 lb, $\alpha = 77.2°$, $\beta = 148.2°$, $\gamma = 118.6°$.

4-2. (a) $-48i - 48j - 32k$.
(b) $-38i + 52j - 47k$.

4-6. $\{-176.8k\}$ lb-in.$\}$

4-8. $\{119i + 148.8j\}$ lb-ft.

4-10. 310 lb-in., $\alpha = 93.8°$, $\beta = 104.4°$, $\gamma = 14.9°$.

4-12. 31.6 lb.

4-14. $M_A = \{-144.3i - 289k\}$ lb-in.

4-16. (a) $M_D = \{-13.6i - 174.2j + 96.6k\}$ lb-ft.
(b) $M_D = \{-13.6i - 174.2j + 96.6k\}$ lb-ft.

4-18. $F_{AB} = 16.75$ lb C; $F_{AC} = 34.5$ lb T; $F_{AD} = 26.0$ lb C;
no: $M_{P'} = \{10i\}$ ft-lb.

4-20. $\alpha = 77.8°$, $\beta = 29.6°$, $\gamma = 63.4°$.

4-22. 909 lb.

4-24. 58.4 lb-in., $\alpha = 142°$, $\beta = 90°$, $\gamma = 128°$, any point.

4-26. $F = 4.70$ lb, $P = 6.26$ lb.

4-28. 40 lb.

4-30. 100 lb.

4-32. 60 lb-in., 10 lb.

4-34. 8.40 ft.

4-36. $M_A = 1,628$ lb-in., $\alpha = 169.4°$, $\beta = 100.6°$, $\gamma = 90°$;
$M_O = 2,474$ lb-in., $\alpha = 166°$, $\beta = 104°$, $\gamma = 90°$.

4-38. $F_O = \{25j - 60k\}$ lb,
$M_O = \{-487i - 312j - 103.9k\}$ lb-in.

4-40. 200 lb $\rightarrow$, 1,000 lb-in.$\}$

4-44. $3x + 5y + 6z = 20$.

4-46. 160.0 lb-in.

4-48. 80.3°.

4-50. $M_O = \{152.8i - 40j - 38.2k\}$ lb-ft; $\alpha = 19.9°$,
$\beta = 104.2°$, $\gamma = 103.6°$; $M_{OD} = \{152.8i\}$ lb-ft.

4-52. $M_x = \{44.2i\}$ lb-ft.

4-54. (a) $F_x = 10$ lb, $F_y = 7.05$ lb, $F_z = 10$ lb.
(b) Projections are along the same orthogonal
coordinate system.

4-56. $F_R = \{2.18i + 3.77j\}$ lb, point O.

4-58. $F_R = \{193i + 100j + 248k\}$ lb,
$M_O = \{1,104i - 2,072k\}$ lb-in.

4-60. $F_R = \{-1.94i + 3.24j - 3.20k\}$ lb,
$M_A = \{9.71i + 5.83j\}$ lb-in.

4-62. $(-1.39$ ft, -1.24 ft, $0)$.

4-64. 667 lb$\downarrow$, 1,667 lb-ft.$\}$

4-66. 162.5 lb$\downarrow$, 6.8 ft from A.

4-68. $F_R = 0$, $M_R = 100$ kip-ft.$($

4-70. 8,013 lb$\uparrow$, 187,600 lb-ft.$($ Use Appendix B.

4-72. $F_R = \{-31.4k\}$ kips, $M_A = \{200i + 114.2j\}$ kip-ft.

5-2. 3 unknowns, A_x, A_y, T.

5-4. 3 unknowns, F_A, F_B, F_C.

5-6. 2 unknowns, T_A, T_B.

5-8. 3 unknowns, A_x, A_y, T.

5-10. 3 unknowns, B_x, B_y, N_A.

5-12. 3 unknowns, A_x, A_y, T_B.

5-14. $T_A = 40$ lb, $T_B = 20$ lb.

5-16. $N_A = N_B = N_C = 47.0$ lb $^{70°}\searrow$, $N_R = 51.3$ lb. $\measuredangle^{20°}$

5-18. $A_x = 17.4$ lb $\rightarrow$, $A_y = 197.7$ lb$\uparrow$, $F_B = 50.8$ lb. $^{70°}\nearrow$

5-20. $T = 112.5$ lb, $F = 187.5$ lb, $\theta = 0°$.

5-22. $T = 74.6$ lb, $A_x = 33.4$ lb $\rightarrow$, $A_y = 61.3$ lb.$\downarrow$

5-24. $A_y = 137.5$ lb$\uparrow$, $B_x = 0$, $B_y = 137.5.\uparrow$

5-26. $A_x = 52.7$ lb $\rightarrow$, $A_y = 49.6$ lb$\uparrow$, $F_B = 60.8$ lb. $^{30°}\searrow$

5-28. $A_x = 69.3$ lb $\rightarrow$, $A_y = B_y = 74.8$ lb.$\downarrow$

5-30. $T = 20.4$ lb, $N = 4.10$ lb.

5-32. $A_x = 760$ lb$\rightarrow$, $A_y = 400$ lb$\downarrow$, $F_B = 990$ lb $^{45°}\searrow$

5-34. $\alpha = 43.5°$, $\beta = 51.5°$, 87.3 lb.

5-36. 1,564 lb.

5-38. $N_A = N_B = W/(2 \cos \alpha)$.

5-40. 195.4 lb, 51.4°.

5-42. $W = 3,000$ lb.

5-44. $F_R = 53.9$ lb, $\alpha = 90°$, $\beta = 68.2°$, $\gamma = 21.9°$;
$M_R = 416.7$ lb-ft, $\alpha = 24.2°$, $\beta = 112.6°$, $\gamma = 81.7°$.

5-46. $P = 75$ lb, $A_x = 37.5$ lb, $A_y = 0$, $A_z = 75$ lb;
$B_x = 112.5$ lb, $B_z = 75$ lb.

5-48. $A_x = A_y = 0$, $A_z = 2,000$ lb; $M_A = 13,730$ lb-ft, $\alpha = 79.5°$, $\beta = 10.5°$, $\gamma = 90°$.

5-50. $E_x = E_y = 0$, $E_z = 2$ lb, $F_A = F_B = 7$ lb.

5-52. $A_x = 90$ lb, $A_z = 80$ lb; $B_x = 30$ lb, $B_y = 0$, $B_z = 125$ lb, $\Sigma M_z \neq 0$.

5-54. $T_{AB} = 31.4$ lb, $C_x = 13.9$ lb, $C_y = 3.73$ lb, $C_z = 32.2$ lb, $M_{Cy} = 20.9$ lb-ft, $M_{Cz} = 36.2$ lb-ft.

5-56. $A_x = 0$, $A_y = 0$, $A_z = 600$ lb, $M_{Ax} = 0$, $M_{Az} = 0$.

5-58. $A_x = 37.5$ lb, $A_y = 0$, $A_z = 37.5$ lb, $T_{BD} = T_{CD} = 70.2$ lb.

5-60. $C_x = C_y = 0$, $A_z = B_z = C_z = 5.33$ lb.

5-62. $P = 6.38$ lb, $A_x = 4.50$ lb, $A_y = 11.9$ lb, $B_x = 4.50$ lb, $B_y = 15.69$ lb, $B_z = 40.0$ lb.

6-2. $F_{AB} = 1,179$ lb, C; $F_{AC} = 833$ lb, T; $F_{CB} = 373$ lb, C.

6-4. $F_{AB} = 978$ lb, C; $F_{AF} = 437$ lb, T; $F_{BC} = 437$ lb, C; $F_{BF} = 875$ lb, T; $F_{CF} = 177$ lb, T; $F_{EF} = 312$ lb, T; $F_{ED} = 312$ lb, T; $F_{CE} = 500$ lb, T; $F_{CD} = 699$ lb, C.

6-6. $F_{CL} = 0.50$ kip, T; $F_{MC} = 1.12$ kips, C; $F_{CD} = 5.00$ kips, T.

6-8. $F_{AG} = 12.26$ kips, C; $F_{GF} = 8.66$ kips, C; $F_{BC} = 7.99$ kips, T.

6-10. $F_{AG} = 10.8$ kips, C; $F_{BF} = 1.20$ kips, T; $F_{CD} = 0$.

6-12. $F_{BC} = F_{BA} = 224$ lb, C; $F_{DB} = F_{EB} = 0$; $F_{DC} = F_{DE} = F_{AE} = 100$ lb, T.

6-14. $F_{AB} = F_{BC} = 0.687P$, T; $F_{AD} = 0.471P$, T; $F_{BD} = 0.333P$, T; $F_{DC} = 0.943P$, C.

6-16. $F_{EB} = 1,077$ lb, C; $F_{ED} = 77.7$ lb, C.

6-18. $F_{GF} = 2,167$ lb, C; $F_{BC} = 1,010$ lb, T.

6-20. $F_{GF} = 6$ kips, C; $F_{FC} = 1$ kip, C; $F_{BC} = 7.33$ kips, T.

6-22. $F_{FE} = 1,000$ lb, T; $F_{FB} = 1,500$ lb, C; $F_{BC} = 2,240$ lb, C.

6-24. $F_{GA} = 100$ lb, T; $F_{FC} = F_{FB} = 0$.

6-26. $F_{HG} = 1,146$ lb, C; $F_{BG} = 105.2$ lb, T.

6-28. $F_{DE} = 7.38$ kips, T; $F_{LE} = 0$.

6-30. $F_{JI} = 12.5$ kips, C; $F_{DM} = 0$.

6-32. $F_{AB} = 220$ lb, C; $F_{CB} = F_{DB} = 312$ lb, T.

6-34. $F_{BC} = F_{AC} = 163$ lb, C; $F_{AB} = 0$; $F_{DC} = 231$ lb, T; $F_{DB} = F_{AD} = 116$ lb, T.

6-36. $F_{AD} = 161.2$ lb, T.

6-38. $F_A = 434$ lb $\angle 64°$ $F_C = 220$ lb. $30°$

6-40. $F_{AC} = 2.36$ kips, $\angle 45°$; $C_x = C_y = A_x = A_y = 1.67$ kips.

6-42. $F_{BC} = 297$ lb, $D_x = 60.0$ lb, $D_y = 210$ lb.

6-44. $A_x = 138$ lb, $A_y = 185$ lb; $C_x = 138$ lb, $C_y = 585$ lb.

6-46. 5,600 lb.

6-48. $F_A = 45.8$ lb, $F_B = 15.7$ lb, $F_C = 29.2$ lb, $F_D = 13.5$ lb.

6-50. $F_{BC} = 350$ lb, $A_x = 303$ lb, $A_y = 25$ lb.

6-52. $F_B = 83.3$ lb, $\angle 36.9°$; $F_C = 83.3$ lb. $36.9°$

6-54. $T = 20$ lb, $R_{BA} = 10$ lb.

6-56. $B_x = 303$ lb, $B_y = 363$ lb, $F_C = 303$ lb.

6-58. 69.3 lb.

6-60. $A_x = 285$ lb, $A_y = 236$ lb, $B_x = 111.6$ lb, $B_y = 336$ lb, $C_x = 285$ lb, $C_y = 436$ lb.

6-62. $A_x = 137.5$ lb, $A_y = 37.5$ lb, $B_x = 275$ lb, $B_y = 75$ lb, $D_x = 275$ lb, $D_y = 175$ lb, $F_E = 265$ lb. $45°$

6-64. $A_x = 12.11$ k, $A_y = 2.61$ k, $D_x = 12.11$ k, $D_y = 6.38$ k.

6-66. $A_x = 1,455$ lb, $A_y = 40.0$ lb; $C_x = 1,455$ lb, $C_y = 760$ lb.

6-68. $F_{AD} = 17,690$ lb, $F_{EF} = 1,000$ lb.

7-2. $A_D = 1.14$ kips, $V_D = 1.00$ kip, $M_D = 1.50$ kip-ft.

7-4. $A_F = 0$, $V_F = 3.33$ kips, $M_F = 10$ kip-ft, $A_E = 0$, $V_E = -0.42$ kip, $M_E = 11.67$ kip-ft.

7-6. $A_D = -300$ lb, $V_D = 100$ lb, $M_D = 200$ lb-ft, $A_E = 0$, $V_E = 50$ lb, $M_E = -25$ lb-ft.

7-8. $A_D = -70.7$ lb, $V_D = 0$, $M_D = -82.8$ lb-ft; $A_E = -50.0$ lb, $V_E = 50.0$ lb, $M_E = -50$ lb-ft.

7-10. $A_x = 0$, $A_y = A_z = 60$ lb, $M_x = -60$ lb-ft, $M_y = 0$, $M_z = -180$ lb-ft.

7-12. $A_E = -12.11$ k, $V_E = 1.61$ k, $M_E = 12.11$ k-ft.

7-14. $A_x = -100$ lb, $A_y = 0$, $A_z = 50$ lb; $M_x = 100$ lb-ft, $M_y = 50$ lb-ft, $M_z = 200$ lb-ft.

7-16. $A_x = 0$, $A_y = 0$, $A_z = 80$ lb; $M_x = 288$ lb-ft, $M_y = -36$ lb-ft, $M_z = 0$.

7-18. $0 \le x < 5$, $V = 9.5 - 2x$, $M = 9.5x - x^2$;
$5 < x < 8$, $V = -0.5$, $M = 25 - 0.5x$;
$8 < x \le 10$, $V = -10.5$, $M = 105 - 10.5x$.

7-20. $0 \le x < 4$, $V = 4.43$, $M = 4.43x$;
$4 < x < 10$, $V = -0.57$, $M = 20 - 0.57x$;
$10 < x < 14$, $V = -5.57$, $M = -5.57x + 70$;
$14 < x \le 18$, $V = 18 - x$, $M = -x^2/2 + 18x - 162$.

7-22. $0 \le x < 2$, $V = 3$, $M = 3x - 36$;
$2 < x < 3$, $V = 5$, $M = 5x - 40$;
$3 < x \le 6$, $V = 5$, $M = 5x - 30$.

7-24. Segment CD:
$0 \le x < 2$, $V = 0.919$, $M = 0.919x$;
$2 < x \le 8$, $V = 0.306$, $M = -0.306x + 2.45$.
Segment AB:
$0 \le x < 12$, $V = -0.15x + 0.875$,
$M = -0.075x^2 + 0.875x$;
$12 < x \le 14$, $V = -0.15x + 2.1$,
$M = -0.075x^2 + 2.1x - 14.7$.

7-26. $0 \le x < 4$, $V = 5$, $M = 15 + 5x$;
$4 < x < 34$, $V = 0$, $M = 35$;
$34 < x < 38$, $V = -5$, $M = 205 - 5x$.

7-28. $0 \le x < 6$, $V = 50x - 83$, $M = 83x - 25x^2$;
$6 < x < 8$, $V = -50x + 500$,
$M = -25x^2 + 500x - 2,500$;
$8 < x \le 9$, $V = 100$, $M = 100x - 900$.

7-30. $0 \le x < 3$, $V = \frac{1}{2}x^2 - 3x$, $M = \frac{1}{6}x^3 - \frac{3}{2}x^2$;
$3 < x < 10$, $V = -6.5$, $M = 10.5 - 6.5x$;
$10 < x \le 13$, $V = 29.7 - x$,
$M = 29.7x - x^2/2 - 301.5$.

7-32. $0 \le x < 8$, $V = 166.7$, $M = 166.7x$;
$8 < x < 10$, $V = -833.3$, $M = 8,000 - 833.3x$;
$10 < x < 12$, $V = 166.7$, $M = 166.7x - 2,000$;
$12 < x \le 15$, $V = -333.3$, $M = 5,000 - 333.3x$.

7-34. See Prob. 7–18.

7-36. $0 \le x \le 15$, $V = 1506.7 - 200x$,
$M = 1506.7x - 100x^2 - 100$.

7-38. $0 \le x < 5$, $V = 66.7$, $M = 66.7x$;
$5 < x < 10$, $V = -33.3$, $M = 500 - 33.3x$;
$10 < x \le 18$, $V = -33.3$, $M = 600 - 33.3x$.

7-40. $x = 0$, $V = 633.3$, $M = 0$; $x = 4^{\mp}$, $V = 633.3, 433.3$,
$M = 2,533.3$; $x = 7^{\mp}$, $V = 433.3, 233.3$,
$M = 3,833.3$; $x = 13$, $V = 0$, $M = 4,633.3$; $x = 24$,
$V = -966.7$, $M = 0$.

7-42. $x = 0$, $V = 0$, $M = 0$; $x = 7.5$, $V = 3.25$.
$M = 18.75$; $x = 15$, $V = 0$, $M = 37.5$.

7-44. $x = 0$, $V = -0.45$, $M = 0$; $x = 6$, $V = -0.75$,
$M = -3.30$; $x = 12$, $V = -1.05$, $M = -9$; $x = 16$,
$V = 0$, $M = -9$.

7-46. $x = 0$, $V = 6$, $M = -19$; $x = 6$, $V = 0$, $M = -1$;
$x = 10$, $V = 0$, $M = -1$.

7-48. 2,320 lb.

7-50. $y = (h/L^3)x^3$, $T_{\max} = (w_o L/2)\sqrt{1 + (L/3h)^2}$.

7-52. 50.2 ft, 184.9 lb.

7-54. 546 ft, 506 lb, 82.1 ft.

7-56. 75.2 lb, 46.3 ft.

7-58. 162.0 ft

7-60. $h/l = 0.141$.

8-2. $22.6°$.

8-4. 17.14 lb.

8-6. 17.3 lb, no.

8-8. 4.5 in.

8-10. 8 in.

8-12. 20.3 lb.

8-14. 547 lb.

8-16. 59.7 lb.

8-18. $P = 60.0$ lb, $T = 15.0$ lb.

8-20. 2.31 ft.

8-22. 73.8 lb $\le P \le 233.8$ lb.

8-24. $s = b/\mu_s$

8-26. 306 lb.

8-28. $P = 15.63$ lb, $N_A = 14.66$ lb, $N_B = 89.5$ lb.

8-30. $20.4°$.

8-32. Motion of crate impends, $P = 30$ lb.

8-34. $85.4°$.

8-36. (a) 574 lb, (b) 1,648 lb.

8-38. $2 \tan^{-1}\mu$.

8-40. $28.1°$.

8-42. $33.4°$.

8–46. 3,060 lb.

8–48. 98.2 lb-ft.

8–50. 33.6 lb-ft.

8–52. (a) 208 lb, (b) 42.4 lb.

8–54. 1,615 lb.

8–56. $\frac{1}{2}\mu PR$.

8–58. $N = 481$ lb, $F = 101.1$ lb.

8–60. 62.8 ft-lb.

8–62. 3.90 lb.

8–64. 98.0 lb.

8–66. (a) 27.2 lb, (b) 3.68 lb.

8–68. 101.8 lb.

8–70. 16.79 lb.

8–72. 5.34 ft.

8–76. 0.00781 in.

8–78. 78.8 lb.

8–80. 187.5 lb.

9–2. (1.44 ft, 2.22 ft, 2.67 ft).

9–4. (1.57 ft, 2.29 ft).

9–6. (1.30 ft, 2.30 ft, 0).

9–8. (1.08 in., 0.540 in.).

9–10. (2.4 in., 1.5 in.).

9–12. (1.61 in., 2.66 in.).

9–14. (0.587 ft, 0.413 ft).

9–16. (1.13 in., 3.60 in.).

9–18. (0, 0.833 ft, 0).

9–20. $(\frac{4}{5}a, 0, 0)$.

9–22. (0, 1.64 in., 0).

9–24. (0, 6 in., 1.39 in.).

9–26. (1.54 in., 1.72 in., 0.5 in.).

9–28. 1.55 in.

9–30. (1.12 in., 1.12 in., 1.36 in.).

9–32. (2.11 in., 2.11 in.).

9–34. (1.95 in., 0.904 in.).

9–36. (0, 1.64 in.).

9–38. (0, 8.99 in.).

9–40. (-0.262 in., 0.262 in.).

9–42. (0, 0, 2.34 in.).

9–44. (2.43 in., -0.075 in., 2.36 in.).

9–46. (1.174 in., 0, 1.132 in.).

9–48. 118.4 in.^{2}, 59.2 in.^{3}

9–50. 1.04 lb.

9–52. $A = 5.33$ in.^{2}, $\bar{y} = 0.75$ in., $V = 25.1$ in.^{3}

9–54. 156.1 in.^{2}, 118.8 in.^{3}

9–56. 3,120,000 lb.

9–58. $\mathbf{M}_o = \{-120\mathbf{j} - 20.25\mathbf{k}\}$ lb-ft.

9–60. 250 lb. $\measuredangle^{72.5°}$

9–62. $(4/\pi^2)(p_0\,a\,b)$, $(a/2, b/2, 0)$.

9–64. $F_A = 671$ lb. $\searrow^{26.6°}$, $F_B = 8,400$ lb. $\rightarrow$

9–66. $F_A = 1,600$ lb, $F_B = 4,240$ lb.

10–2. 0.1650 in.^{4}

10–4. $I_x = 2.13$ in.^{4}, $I_y = 4.57$ in.^{4}

10–6. $I_x = 3.56$ in^4, $I_y = 1.870$ in.^{4}

10–8. $J_o = 81.6$ in.^{4}, $J_{(2,0)} = 21.9$ in.^{4}

10–10. 18.30 in.^{4}

10–12. (a) $I_{xb} = bh^3/12$, (b) $I_{\bar{x}} = bh^3/36$.

10–14. $I_x = 0.0357$ in.^{4} $I_y = 0.050$ in.^{4}

10–16. $\bar{y} = 1.75$ in. (from top), 36.3 in.^{4}

10–18. $I_x = 141.3$ in.^{4}, $I_y = 21.3$ in.^{4}

10–20. $\bar{y} = 2.33$ in. (from top), 86.0 in.^{4}

10–22. 2,947 in.^{4}

10–24. 92.7 in.^{4}

10–26. $k_x = 6.22$ in., $k_y = 2.66$ in.

10–28. 180.6 in.^{4}

10–30. $I_x = 246$ in.^{4}, $I_y = 61.5$ in.^{4}

10–32. 0.667 in.^{4}

10–34. 0.5 in.^{4}

10–36. 2 in.^{4}

10-38. 0.667 in.^{4}

10-40. $\bar{x} = 0.75$ in., $\bar{y} = 1.75$ in., $I_{xy} = -2.81$ in.^{4}

10-42. $I_{x'} = 20.8$ in.^{4}, $I_{y'} = 20.8$ in.^{4}

10-44. $I_{x'} = 571$ in.^{4}, $I_{y'} = 274$ in.^{4}

10-46. $\theta = 8.70°$, $I_{max} = 727$ in.^{4}, $I_{min} = 186$ in.^{4}

10-48. $I_{max} = 64.1$ in.^{4}, $I_{min} = 5.33$ in.^{4}

10-50. 0.740 in.^{4}

10-52. $\theta = -27.3°$, $62.7°$, $I_{max} = 154.0$ in.^{4}, $I_{min} = 15.98$ in.^{4}

10-54. Center (456.5 in.^{4}, 0), $R = 271$ in.^{4}, $I_{min} = 185.5$ in.^{4}, $I_{max} = 727.5$ in.^{4}

10-56. Center (0.479 in.^{4}, 0), $R = 0.273$ in.^{4}, $I_{min} = 0.206$ in.^{4}, $I_{max} = 0.752$ in.^{4}

11-2. 35.4 lb.

11-4. 65.8°.

11-6. 5.28 lb.

11-8. 142.8 lb.

11-10. 75 lb.

11-12. 23.3 lb.

11-14. 2.5 $P/\sin \theta$.

11-16. 27.5 lb.

11-18. 130.0 lb-ft.

11-20. $(\frac{2}{3})P$

11-22. $2\pi lP/pn$.

11-24. 3.46 lb.

11-26. 90°, 16.6°.

11-28. 6.0°.

11-30. $x = 0.590$, stable; $x = -0.424$, unstable.

11-32. (0, 0), stable.

11-34. 2.83 in., stable.

11-36. $\theta = 0°$, unstable; $\theta = 76.6°$, stable.

11-38. $\theta = 90°$, unstable; $\theta = 5.90°$, stable.

11-40. $\theta = 90°$, unstable; $\theta = 57.6°$, stable.

11-42. $\theta = 90°$, stable; $\theta = 9.47°$, unstable.

11-44. $r > b/2$, stable; $r = b/2$, neutral; $r < b/2$, unstable.

11-46. 39.8°, stable.

11-48. 3.99 in.

11-50. $\theta = 180°$, i.e., G is directly below C.

11-52. $36 \sin \theta_1 + 135 \cos \theta_1 (\sin \theta_2 - \sin \theta_1) = 0$.
$36 \sin \theta_2 + 135 \cos \theta_2 (\sin \theta_2 - \sin \theta_1)$
$-60 \cos \theta_2 = 0$.

Index

Engineering Mechanics: DYNAMICS

R.C. Hibbeler

Department of Engineering Mechanics
Illinois Institute of Technology

Macmillan Publishing Co., Inc.
New York
Collier Macmillan Publishers
London

This book is the second part of *Engineering Mechanics: Statics and Dynamics,* copyright © 1974 by R. C. Hibbeler.

Macmillan Publishing Co., Inc.
866 Third Avenue, New York, New York 10022

Collier-Macmillan Canada, Ltd.

Library of Congress Cataloging in Publication Data

Hibbeler, R. C.
  Engineering mechanics: dynamics.

  Companion volume to Engineering mechanics: statics. These two works are also published together as Engineering mechanics: statics and dynamics.
  1. Dynamics. I. Title.
TA352.H5   620.1'04   73-17146
ISBN 0-02-354080-X

Printing: 1 2 3 4 5 6 7 8   Year: 4 5 6 7 8 9 0

Preface

The purpose of this book is to provide the student with a clear and thorough presentation of the theory and application of engineering mechanics as it is currently required for the engineering curriculum. One of the major aims of the book is to help develop the student's ability to analyze problems—a most important skill for any engineer.

Although engineering mechanics is based upon relatively few principles, it is these principles which provide a necessary means for the solution of many problems relating to present-day engineering design and analysis. In this book, emphasis is given to both the understanding and the application of these principles, so that the student will have a firm basis for understanding how these principles are utilized both in more advanced courses of study and in engineering practice.

For teaching purposes, as each new principle is introduced, it is first applied to simple situations. Specifically, each principle is applied first to a particle, then to a rigid body having planar motion, and finally to the most general case of a body having spatial motion. This presentation allows the instructor some flexibility for teaching the material, since each of the three cases is elaborated upon in separate chapters.

The contents of each chapter are separated into well-defined units. Each unit contains the development and explanation of a specific topic, illustrative example problems, and a set of problems designed to test the student's ability to apply the theory. In any set, the problems are generally arranged in order of increasing difficulty, and the answers to all even-numbered problems are given in the back of the book. Many of the problems depict realistic situations encountered in engineering practice. It is hoped that this realism will stimulate student interest in engineering mechanics and furthermore will help the student to develop his skills in reducing any such problem from its physical description to a model or symbolic representation whereby the principles of mechanics may be applied. All numerical problems are stated in terms of the currently used British system of units; however, if it is deemed necessary, some of the problems may be worked with metric units, using the table of conversion factors given in Appendix C.

Diagrams are used extensively throughout the book, since they provide an important means for obtaining a complete understanding of both the theory and its applications. By emphasizing the use of diagrams in the examples, the student is given a logical and orderly procedure for the solution of problems. Furthermore, when diagrams are included in the analysis, physical insight relating to the problem is maintained. For example, the free-body diagram is introduced early in the book, and it is used whenever a force analysis is required for a problem solution.

Mathematics provides a systematic means of applying the principles of mechanics. The student is expected to have prior knowledge of algebra, geometry, trigo-

nometry, and some calculus. Vector analysis is introduced in this book at points where it is most applicable. Its use often provides a convenient means for presenting concise derivations of the theory, and it makes possible a simple and systematic solution of many complicated three-dimensional problems. Occasionally, the example problems are solved in several ways in an effort to compare the use of a vector analysis with other mathematical techniques based upon a conventional scalar approach. In this way, the student develops the ability to use mathematics as a tool whereby the solution of any problem may be carried out in the most direct and effective manner.

The contents of the book are divided into 11 chapters.* Due to the nature of this subject, the book provides the instructor with some versatility in presenting the material. Two suggested teaching sequences follow.

Sequence I   *Chapter*
Kinematics of Particles and Rigid
Bodies   12, 13
Kinetics of Particles and Planar Kinetics
of Rigid Bodies
    Equations of Motion   14, 15
    Work and Energy   16, 17
    Impulse and Momentum   18, 19
Sequence II
Kinematics of Particles   12
Kinetics of Particles
    Equations of Motion   14
    Work and Energy   16
    Impulse and Momentum   18
Kinematics of Rigid Bodies   13
Planar Kinetics of Rigid Bodies
    Equations of Motion   15
    Work and Energy   17
    Impulse and Momentum   19

Time permitting, some of the material involving three-dimensional rigid-body motions may be included

*This book is also available in a combined volume, *Engineering Mechanics: Statics and Dynamics*.

in the course. The kinematics of this motion are discussed in Sections 13–10 through 13–13. Detailed treatment of mass moment of inertia of a body is given in Chapter 20. (The mass moment of inertia concept is briefly introduced in Section 15–4 and forms a suitable basis for solving planar motion problems.) The kinetics of the spatial motions of a rigid body are given in Chapter 21. Chapter 22 ("Vibrations") may be included, provided the student has had the necessary mathematical background. Sections of the book which are beyond the basic mechanics course are indicated by a star and may be omitted. (Note that this more advanced material also provides a suitable reference for basic principles when it is covered in more advanced courses.)

The author has endeavored to write a textbook that will appeal to both student and instructor alike. In doing so, it must be admitted that the development of this book was not accomplished singlehandedly. Through the years, many people have given both their support and their help. Although I cannot list them all, I wish to acknowledge the valuable suggestions and comments made by Professors W. G. Plumtree, California State University at Los Angeles; M. E. Raville, Georgia Institute of Technology; A. Pytel, Pennsylvania State University; G. W. Washa, University of Wisconsin; R. Schaefer, University of Missouri at Rolla; and P. K. Mallick, Illinois Institute of Technology. Gratitude is extended to the engineering students at Youngstown State University and Illinois Institute of Technology for their assistance both in the preparation and classroom testing of the manuscript. I wish also to thank my secretaries, Joanne and Leslie, who tirelessly typed and retyped the manuscript. Lastly, I would like to praise the publishing staff for presenting the manuscript in such fine artistic form.

It is hoped that all instructors and students using this textbook will find it most suitable for the purpose for which it is intended. Since they can provide the most critical review of this work, the author welcomes and will gladly acknowledge any comments they make regarding its contents.

Russell C. Hibbeler

Contents

Engineering Mechanics:
DYNAMICS

12

Kinematics of a Particle

12-1. Introductory Remarks, Kinematics of Particles

Engineering mechanics consists of a study of both statics and dynamics. *Statics* deals with the equilibrium of bodies at rest or moving with constant velocity, whereas *dynamics* deals entirely with bodies in motion. In general, dynamics is more complicated than statics, since the laws of motion relate the forces acting on the body to the geometry of the motion. Because statics consists primarily of the analysis of forces acting on bodies, a prior knowledge of statics is generally a prerequisite to dynamics.

Dynamics is subdivided into two parts: (1) *Kinematics,* which is concerned only with the *geometry of motion,* and (2) *Kinetics,* which is concerned with the *force analysis* of bodies in motion. Since the study of kinetics requires a knowledge of kinematics in order to relate the motion of the body to the forces which cause the motion, the principles of kinematics will be discussed first.

For simplicity in developing the methods of equilibrium used in statics,* the equilibrium of a particle was treated first, followed by a discussion of rigid-body equilibrium. In a similar manner, the dynamics of a particle will be discussed before encountering topics in rigid body dynamics. Specifically, rigid-body dynamics is generally more complicated than particle dynamics, since rotational motion of the body must be accounted for.

Recall that a *particle* is defined as a small portion of matter such that

---

*See *Engineering Mechanics: Statics.*

its dimension or size is of no consequence in the analysis of a physical problem. In most problems encountered, one is interested in bodies of a finite size, such as rockets, projectiles, or vehicles. Such objects may be considered as particles, provided the distance over which they travel is large compared to their size, and any rotation of the body can be neglected. In general, motion of a particle may be classified in one of two ways: motion along a straight-line path, which is called *rectilinear motion,* and motion along a curved path, which is referred to as *curvilinear motion.* In both cases, the motion is characterized by specifying the particle's displacement, velocity, and acceleration.

The motion of a particle measured with respect to a *fixed coordinate system,* will be studied first. This type of motion is referred to as *absolute motion.* Afterwards we shall study the *relative motion* between two particles using a *translating coordinate system.* In some problems the relative motion between two particles may be conveniently analyzed when the motion of one of the particles is described with respect to a *translating and rotating coordinate system.* This more general study of relative particle motion will be presented in Chapter 13, since it depends upon prior knowledge of the kinematics of line segments.

12-2. Rectilinear Velocity and Acceleration of a Particle

The simplest motion of a particle is motion occurring along a straight-line path, called *rectilinear motion.* Consider the particle at point P shown in Fig. 12–1. The scalar *position* (or *coordinate*) s is used to *locate* the particle at any given instant relative to the fixed origin O. If s is positive the particle is located to the right of the origin, whereas if s is negative, it is located to the left.

The *displacement* of the particle is defined as the *change* in its *position.* This is represented in Fig. 12–1 by the symbol Δs. When the particle's final position is to the right of its initial position, Δs is positive, Fig. 12–1; when the displacement is to the left, Δs is negative.

The displacement of a particle must be distinguished from the distance the particle travels. Specifically, the *distance* traveled is defined as the *total length of path* traversed by the particle—which is always positive.

Let us now consider that the particle moves through a positive displacement Δs from P to P' during the time interval Δt, Fig. 12–1. The *average velocity* of the particle during this time interval is defined by the displacement of the particle divided by the elapsed time, i.e.,

$$v_{avg} = \frac{\Delta s}{\Delta t}$$

By taking smaller values of Δt and consequently smaller values of Δs,

Fig. 12–1

we obtain the *instantaneous velocity* of the particle, which is defined as

$$v = \lim_{\Delta t \to 0} \frac{\Delta s}{\Delta t}$$

or

$$v = \frac{ds}{dt} \qquad (12\text{--}1)$$

For both the average and instantaneous velocities, the *direction* of the velocity is positive or negative depending upon whether the displacement is positive or negative. For example, if the particle is moving to the right as shown in Fig. 12–1, the velocity is positive. The *magnitude* of the velocity is known as the *speed* of the particle. If the displacement is expressed in feet or meters, and the time in seconds, the speed is expressed as ft/sec or m/sec. Occasionally the term average speed is used. The *average speed* is defined as the total distance of path traveled by a particle divided by the elapsed time. This term should not be confused with the average velocity. See Example 12–3.

Provided the instantaneous velocities for the particle are known at the two points P and P', then the *average acceleration* for the particle during the time interval Δt is defined as

$$a_{avg} = \frac{\Delta v}{\Delta t}$$

where Δv represents the difference in the velocities during the time interval Δt.

The *instantaneous acceleration* of the particle at time t is found by taking smaller values of Δt giving smaller values of Δv until

$$a = \lim_{\Delta t \to 0} \frac{\Delta v}{\Delta t}$$

or

$$a = \frac{dv}{dt} \qquad (12\text{--}2)$$

Using Eq. 12–1, we also have

$$a = \frac{d^2 s}{dt^2} \qquad (12\text{--}3)$$

Both the average and instantaneous accelerations will be positive or negative, depending upon whether the change in velocity is positive or negative. In particular, when the velocity of the particle is *decreasing*,

the particle is said to be *decelerating*. If this occurs, the acceleration acts in the *opposite direction* of the *velocity* since the particle is slowing down. Also, note that when the *velocity* is *constant,* the *acceleration* is *zero.* Units commonly used to express the magnitude of acceleration are ft/sec^2 or m/sec^2.

A differential relation involving the displacement, velocity, and acceleration along the path may be obtained by eliminating the time differential dt in Eqs. 12–1 and 12–2, i.e.,

$$dt = \frac{ds}{v} = \frac{dv}{a}$$

so that

$$a\,ds = v\,dv \tag{12–4}$$

When a functional relationship describing either the displacement, velocity, or acceleration of a particle is known, the functional relations describing the other kinematic quantities may be obtained by either the proper differentiation or integration of Eqs. 12–1 through 12–4. If integration is necessary, data regarding the initial or final conditions of the particle are used to determine either the constant of integration if an indefinite integral is used, or the limits of integration if a definite integral is used. See Example 12–1.

Four common types of problems involving the differential equations just described are as follows:

1. *Acceleration given as a function of time, $a = f(t)$.* Substitution into Eq. 12–2 yields $dv = f(t)\,dt$, which may then be integrated directly to obtain $v = g(t)$. The displacement is obtained by substituting for v into Eq. 12–1, which gives $ds = g(t)\,dt$. Integrating yields $s = h(t)$.
2. *Acceleration given as a function of velocity, $a = f(v)$.* Substituting into Eq. 12–2 yields $dv = f(v)\,dt$, or $dv/f(v) = dt$. Integration gives $v = g(t)$. The displacement is obtained by using Eq. 12–1, as in Case 1. Substituting for v and integrating gives $s = h(t)$.
3. *Acceleration given as a function of displacement, $a = f(s)$.* Substitution into Eq. 12–4 yields $f(s)\,ds = v\,dv$. Integrating yields $v = g(s)$. Substituting for v into Eq. 12–1 we have $g(s) = ds/dt$ or $ds/g(s) = dt$. Integrating gives $s = h(t)$.
4. *Acceleration is constant, $a = a_c$.* Substitution into Eq. 12–2 yields

$$\frac{dv}{dt} = a_c$$

The velocity is obtained by integration, assuming the initial velocity is v_1, at $t = 0$. Using a definite integral,

$$\int_{v_1}^{v} dv = a_c \int_{0}^{t} dt$$

$$v - v_1 = a_c t$$

or

$$v = v_1 + a_c t \qquad\qquad (12\text{-}5a)$$

Substituting this value of v into Eq. 12-1, we have

$$\frac{ds}{dt} = v_1 + a_c t$$

Integrating this equation, assuming the initial position of the particle is at s_1 when $t = 0$, we obtain

$$\int_{s_1}^{s} ds = \int_{0}^{t} (v_1 + a_c t)\, dt$$

$$s - s_1 = v_1 t + \tfrac{1}{2} a_c t^2$$

or

$$s = s_1 + v_1 t + \tfrac{1}{2} a_c t^2 \qquad\qquad (12\text{-}5b)$$

Equation 12-4 may also be integrated directly, with the condition that $v = v_1$ when $s = s_1$.

$$\int_{s_1}^{s} a_c\, ds = \int_{v_1}^{v} v\, dv$$

$$a_c (s - s_1) = \tfrac{1}{2}(v^2 - v_1^2)$$

or

$$v^2 = v_1^2 + 2a_c(s - s_1) \qquad\qquad (12\text{-}5c)$$

The correct magnitude and signs of s_1, v_1, and a_c, used in Eqs. 12-5, are determined from the chosen origin and positive direction of the s axis. It is important to remember that these equations are useful *only when the acceleration is either constant or zero*. A common example of constant accelerated motion occurs when a body falls freely and is subjected only to a gravitational force. Specifically, this force creates a *constant downward acceleration* on the body of approximately 32.2 ft/sec^2 or 9.80 m/sec^2.

The following examples illustrate some practical applications of this theory.

Example 12-1

A small projectile is fired into a fluid medium with an initial velocity of 200 ft/sec. If the fluid resistance causes a deceleration on the projectile which is equal to $a = -0.4v^3$ ft/sec^2, where v is measured in ft/sec, determine both the velocity v and position s as functions of time after the projectile is fired.

Solution

This problem is equivalent to the type in Case 2, mentioned previously.* Since Eq. 12-2 relates velocity to time, we can obtain the velocity as a function of time by integrating this equation, with the initial condition that $v = 200$ ft/sec at $t = 0$.

$$\frac{dv}{dt} = a = -0.4v^3$$

$$\int_{200}^{v} \frac{dv}{-0.4v^3} = \int_{0}^{t} dt$$

$$\frac{1}{0.8}\left[\frac{1}{v^2} - \frac{1}{(200)^2}\right] = t - 0†$$

$$v = \left\{\frac{1}{\left[\frac{1}{(200)^2} + 0.8t\right]^{1/2}}\right\} \text{ ft/sec} \qquad Ans.$$

In a similar manner, the displacement s as a function of time is obtained by using Eq. 12-1, with the initial condition $s = 0$ at $t = 0$.

$$\frac{ds}{dt} = v = \frac{1}{\left[\frac{1}{(200)^2} + 0.8t\right]^{1/2}}$$

$$\int_{0}^{s} ds = \int_{0}^{t} \frac{dt}{\left[\frac{1}{(200)^2} + 0.8t\right]^{1/2}}$$

$$s = \left\{\frac{1}{0.4}\left[\frac{1}{(200)^2} + 0.8t\right]^{1/2} - \frac{1}{80}\right\} \text{ft} \qquad Ans.$$

Example 12-2

A man throws a ball in the vertical direction off the side of a 200 ft cliff, as shown in Fig. 12-2. If the initial velocity of the ball is 25 ft/sec,

Fig. 12-2

*Why can't we use Eqs. 12-5 to solve this problem?

†We can obtain the *same result* by evaluating a constant of integration rather than using definite limits on the integral. For example, integrating $dt = dv/(-0.4v^3)$ yields $t = 1/0.8[1/v^2] + c$. Using the condition that at $t = 0$, $v = 200$ ft/sec, we obtain upon substituting, $c = -1/0.8 \, [1/(200)^2]$.

determine (a) the maximum height h reached by the ball with reference to the ground level, and (b) the speed of the ball just before it hits the ground. During the entire time the ball is in motion, it is subjected to a constant downward acceleration of 32.2 ft/sec^2 due to gravity. Neglect the effect of air resistance.

Solution

Part (a).   The coordinate axis for position $s = 0$ is taken at the ground level as shown in the figure. Thus, the ball is thrown from an initial height of $s_A = 200$ ft. Since the ball is thrown upward at $t = 0$, it is subjected to a velocity of $v_A = 25$ ft/sec (positive since it is in the same direction as positive displacement). For the entire motion, the acceleration is *constant* such that $a_c = -32.2$ ft/sec^2. (Note that a_c is negative since it acts in a direction *opposite* to positive velocity or positive displacement. Hence, the ball is actually being decelerated from A to B, Fig. 12–2a.) Since a_c is *constant,* throughout the entire motion we may relate displacement to velocity using Eq. 12–5c. When the ball reaches its maximum height s_B its final velocity is $v_B = 0$. Thus,

$$v_B^2 = v_A^2 + 2a_c(s_B - s_A)$$
$$0 = (25)^2 + 2(-32.2)(s_B - 200)$$

so that,

$$s_B = 209.7 \text{ ft} \qquad\qquad Ans.$$

Part (b).   To obtain the velocity v_C of the ball just before it hits the ground, apply Eq. 12–5c between points B and C, Fig. 12–2,

$$v_C^2 = v_B^2 + 2a_c(s_C - s_B)$$
$$v_C^2 = 0 + 2(-32.2)(0 - 209.7)$$
$$v_C = \pm 116.2 \text{ ft/sec}$$

From the physical aspects of the problem, the ball is moving *downward,* hence

$$v_C = -116.2 \text{ ft/sec} \qquad\qquad Ans.$$

Example 12–3

A particle moves along a horizontal path such that the speed v is given by $v = 10t^2 - 4t$, where v is measured in ft/sec and t is the time in seconds. Initially the particle is located 4 ft to the *left* of the origin 0, as shown in Fig. 12–3a. Determine (a) the distance traveled by the particle during the time interval $t = 0$ to $t = 1$ sec, (b) the average velocity and the average speed of the particle during this time interval, and (c) the acceleration at $t = 1$ sec.

Solution

Part (a). Since the velocity is related to time, a function which relates displacement to time may be found by integrating Eq. 12-1, with the condition that at $t_1 = 0$, $s_1 = -4$ ft (negative since the particle is located to the left of the origin).*

$$ds = v\, dt$$

$$ds = (10t^2 - 4t)\, dt$$

$$\int_{-4}^{s} ds = 10 \int_{0}^{t} t^2\, dt - 4 \int_{0}^{t} t\, dt$$

$$s = (3.33t^3 - 2t^2 - 4)\text{ft} \tag{1}$$

The graph of the velocity function of the particle is shown in Fig. 12-3b. It can be seen from this graph that for $0 < t < 0.4$ sec the velocity is negative, which means the particle is traveling to the left, and for $t > 0.4$ sec the velocity is positive and hence the particle is traveling to the right. Specifically, since the velocity of the particle changes sign when $t = 0.4$ sec, the particle reverses its direction at this time. The location of the particle when $t = 0$, $t = 0.4$ sec, and $t = 1$ sec can be computed from Eq. (1). This yields

$$s|_{t=0} = -4 \text{ ft}$$

$$s|_{t=0.4 \text{ sec}} = -4.11 \text{ ft}$$

$$s|_{t=1 \text{ sec}} = -2.67 \text{ ft}$$

The negative signs indicate that the particle is located at the respective points located to the left of the origin as shown in Fig. 12-3c. Thus, when the particle is traveling to the *left*, i.e., $0 < t < 0.4$ sec, the displacement is $\Delta s_1 = (-4.11 \text{ ft} - (-4 \text{ ft})) = -0.11$ ft, Fig. 12-3c. When the particle is traveling to the *right*, i.e., for $0.4 \text{ sec} < t < 1$ sec, $\Delta s_2 = (-2.67 \text{ ft} - (-4.11 \text{ ft})) = 1.44$ ft. The *total distance traveled* for $0 < t < 1$ sec is therefore

$$\Delta s_T = |\Delta s_1| + |\Delta s_2| = 0.11 + 1.44 = 1.55 \text{ ft} \qquad Ans.$$

Part (b). The *displacement* from $t = 0$ to $t = 1$ sec is

$$\Delta s = \Delta s_1 + \Delta s_2 = -0.11 \text{ ft} + 1.44 \text{ ft} = 1.33 \text{ ft}$$

See Fig. 12-3c. Hence, the average velocity is

$$v_{avg} = \frac{\Delta s}{\Delta t} = \frac{1.33 \text{ ft}}{1 \text{ sec}} = 1.33 \text{ ft/sec} \qquad Ans.$$

By definition, the average speed is defined in terms of the total distance traveled. Hence

*Why can't Eqs. 12-5 be used for the solution of this problem?

(a)

(b)

(c)

Fig. 12-3

$$(v_{sp})_{avg} = \frac{\Delta s_T}{\Delta t} = \frac{1.55 \text{ ft}}{1 \text{ sec}} = 1.55 \text{ ft/sec} \qquad \textit{Ans.}$$

Part (c). The acceleration, expressed as a function of time, is obtained from Eq. 12–2, namely,

$$a = \frac{dv}{dt} = \frac{d}{dt}(10t^2 - 4t)$$

$$a = 20t - 4$$

Thus, when $t = 1$ sec,

$$a|_{t=1\,\text{sec}} = 16 \text{ ft/sec}^2 \qquad \textit{Ans.}$$

Problems

12-1. A small metal particle passes through a fluid medium. The particle is subjected to the attraction of a magnet. The displacement of the particle is observed to be $s = 0.5t^3 + 4t$, where s is measured in inches and t in seconds. Determine the changes in the displacement, the velocity, and the acceleration of the particle from $t = 1$ sec to $t = 3$ sec.

12-2. After traveling a distance of 100 ft, a particle reaches a velocity of 30 ft/sec, starting from rest. Determine the constant linear acceleration given to the particle.

12-3. A particle moves along a 200-ft straight-line path. Starting from rest, it attains a velocity of 60 ft/sec in 175 ft with constant acceleration. Then after being subjected to another constant acceleration it attains a final velocity of 80 ft/sec. Determine the average velocity and acceleration of the particle for the entire displacement.

12-4. A particle moves in a straight line in such a manner that at any instant the displacement of the particle, measured from a fixed point O on the path, is given by the equation $s = 12t^3 + 2t^2 + 3t$, where s is in feet and t is in seconds. Determine the displacement, velocity, and acceleration of the particle at $t = 2$ sec.

12-5. A particle is moving along a straight-line path such that its displacement is given by $s = 10t^2 + 2$, where s is given in feet and t in seconds. Determine

(a) the displacement of the particle during the time interval from $t = 1$ sec to $t = 5$ sec, (b) the average velocity of the particle during this time interval, and (c) the acceleration at $t = 1$ sec.

12-6. A particle moves along a straight-line path with an acceleration described by the function $a = 4t^2 - 2$, where a is measured in ft/sec^2 and t is measured in seconds. When $t = 0$, the particle is located 2 ft to the left of the origin, and when $t = 2$ sec, it is 20 ft to the left of the origin. Determine the position of the particle when $t = 4$ sec.

12-7. A particle moves along a straight-line path such that the displacement is defined by the equation $s = 4t^3 - 16t^2 + 3$, where s is expressed in feet and t in seconds. From time $t = 0$, determine (a) the total distance traveled in order to reduce the particle velocity to zero, and (b) the time at which the acceleration is zero.

12-8. Starting from rest, a car traveled 40 ft during a 3-sec time interval and 30 ft during the next 2 sec. Determine the average velocity of the car during the entire 5-sec interval.

12-9. A particle moves along a straight-line path with an acceleration defined by the equation $a = kt^3 + 4$, where t is in seconds and a is measured in ft/sec^2. Determine the constant k, knowing that $v = 12$ ft/sec when $t = 1$ sec, and that $v = -2$ ft/sec when $t = 2$ sec.

12–10. A particle moves along a straight-line path with an acceleration defined by the equation $a = 5/s$, where s is measured in feet and a is in ft/sec^2. Determine the velocity of the particle when $s = 2$ ft. The particle starts from rest when $s = 1$ ft.

12–11. An object freely falling through the atmosphere has an acceleration defined by the equation $a = 32.2[1 - v^2/(16 \times 10^4)]$ ft/sec^2. If the object is released from rest, determine (a) the velocity at the end of time $t = 15$ sec, and (b) the terminal or maximum attainable velocity of the object.

12–12. An automobile is traveling at a speed of 60 mi/hr, when the brakes are suddenly applied. If the deceleration rate of the automobile is 10 ft/sec^2, determine the time required to stop the car. How far will the car travel before stopping?

12–13. The velocity of a particle falling downward through a fluid medium is directly proportional to the distance the particle travels from its initial position. When $t = 1$ sec, $s = 1$ ft downward, $v = 2$ ft/sec, and $a = 0.6$ ft/sec^2. Determine the distance traveled by the particle when $t = 5$ sec.

12–14. A particle is traveling to the right, past a given point with a velocity of 80 ft/sec and receiving a constant acceleration of 12 ft/sec^2 to the left. Determine the velocity and displacement of the particle 4 sec after passing the given point.

12–15. An automobile, traveling along a straight road, has an initial velocity of 30 mph to the right and changes its speed uniformly to 60 mph to the left. During the entire motion the time rate of change of velocity for the automobile is maintained at 8 ft/sec^2. Determine the total distance the automobile must travel in order to change its velocity.

12–16. A race car uniformly accelerates at 10 ft/sec^2 from rest, reaches a maximum speed of 60 mph, and then decelerates uniformly to a stop. Determine the total elapsed time of motion if the distance traveled was 1,500 ft.

12–17. A ball is released from rest from the top of a 500-ft building at the same time that a second ball is thrown upward from the ground level. If the balls pass one another at a height of 150 ft, determine the speed at which the second ball was thrown upward. Assume that the acceleration of each ball is constant, acting downward with a magnitude of 32.2 ft/sec^2.

*12–3. Graphical Solutions

In some cases it is difficult to obtain mathematical functions which describe the position, velocity, and acceleration of a particle. When this occurs, a graphical procedure utilizing Eqs. 12–1 and 12–2 or Eq. 12–4 may be used to obtain curves which can be plotted to describe the motion.

Provided the linear displacements of a particle can be *experimentally determined* for several periods of time, an *st curve* (displacement versus time) for the particle can be plotted. An example of such a curve is shown in Fig. 12–4a. In this case the curve is represented with a curved segment *OA* and straight line *AB*. The ordinates of this curve give the linear displacement of the particle from the origin at any instant of time. Since the velocity is $v = ds/dt$ (Eq. 12–1), the velocity of the particle v can be determined at any given instant t by measuring the *slope* of the tangent of the *st* curve at time t. (In a graphical sense, the slope is measured using a ruler or protractor.) Hence, the *vt curve* (velocity versus time) can be established from measurements of the slope at various points along the *st* curve. For example, measurements of the slope v_1, v_2, v_3 at the

Fig. 12-4

intermediate points (t_1, s_1), (t_2, s_2), (t_3, s_3) on the *st* curve in Fig. 12-4*a* yield the construction shown in Fig. 12-4*b*. In a similar manner, the *at* curve (acceleration versus time), Fig. 12-4*c*, may be constructed given the fact that $a = dv/dt$ (Eq. 12-2). In this case, the slopes a_1, a_2, a_3 are measured from the *vt* curve at points (t_1, v_1), (t_2, v_2), (t_3, v_3), as shown in Fig. 12-4*b*.

If the *at* curve is given, Fig. 12-5*a*, using integration, we can construct the *vt* curve and *st* curve. For example, since $a = dv/dt$, then $dv = a\,dt$. Hence, the change in speed dv is equal to the differential area $a\,dt$, shown colored in Fig. 12-5*a*. For a very small finite time, $t_2 - t_1$, we have

$$\int_{v_1}^{v_2} dv = \int_{t_1}^{t_2} a \, dt$$

or

$$v_2 - v_1 = \int_{t_1}^{t_2} a \, dt$$

(a)

(b)

(c)

Fig. 12-5

The *area* under the *at* curve during the time interval $t_2 - t_1$, shown shaded in Fig. 12–5a, is therefore equal to the change in speed $(v_2 - v_1)$ which takes place while the particle moves during this same time interval, Fig. 12–5b. (In a graphical sense, the area may be approximated by a trapezoid.) In a similar manner, $v = ds/dt$, or $ds = v\,dt$, so that

$$\int_{s_1}^{s_2} ds = \int_{t_1}^{t_2} v\,dt$$

or

$$s_2 - s_1 = \int_{t_1}^{t_2} v\,dt$$

Graphically, this equation indicates that the *area* under the *vt* curve between times t_1 and t_2 (shown shaded in Fig. 12–5b) represents the *change in displacement* $(s_2 - s_1)$ of the particle during the same time interval, Fig. 12–5c.

Since the *vt* graph is constructed on the basis of *integrating* the *at* graph, and the *st* graph is found by *integrating* the *vt* graph, it might be noted that integrating an *at* graph which is *constant* (zero-degree curve) yields a *vt* graph which is *linear* (first-degree curve) and consequently an *st* graph which is *parabolic* (second-degree curve). In general, if the *at* graph is a polynomial of degree n, then the *vt* and *st* graphs are polynomials of degrees $n + 1$ and $n + 2$, respectively. See Example 12–4.

Generally, it is better to construct the *vt* curve and *st* curve given the *at* curve rather than starting with the *st* curve and constructing the *vt* and *at* curves. This is because increments of areas can be graphically measured more accurately than slope. When measuring increments of areas, it should be remembered that areas lying *above* the *t* axis correspond to an *increase* in *v* or *s*, while those lying *below* the *t* axis indicate a *decrease* in *v* or *s*. See Example 12–5.

The two graphical procedures may be summarized by the following four statements:

1. The instantaneous value of the speed at any time t is equal to the *slope* of the *st* curve at time t.
2. The magnitude of the instantaneous acceleration at time t is equal to the *slope* of the *vt* curve at time t.
3. The *area* under the *vt* curve during the time interval $(t_2 - t_1)$ represents the change in displacement $(s_2 - s_1)$ of the particle during this time interval.
4. The *area* under the *at* curve during the time interval $(t_2 - t_1)$ represents the change in velocity $(v_2 - v_1)$ of the particle during this time interval.

In some cases two other types of motion curves are used. They are

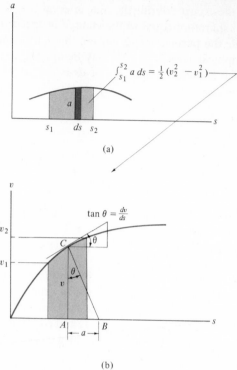

Fig. 12–6

called the *vs curve* (velocity versus displacement) and the *as curve* (acceleration versus displacement), Fig. 12–6. Knowing the *as* curve, the *vs* curve can be constructed using Eq. 12–4, since

$$\int_{v_1}^{v_2} v\, dv = \int_{s_1}^{s_2} a\, ds$$

Then,

$$\tfrac{1}{2}(v_2^2 - v_1^2) = \int_{s_1}^{s_2} a\, ds$$

Thus, small segments of area under the *as* curve equal one-half the difference in the squares of the speed, Fig. 12–6a. Hence, by approximation of the area $\int_{s_1}^{s_2} a\, ds$ using the *as* graph, it is possible to compute a value of v_2 knowing an initial value of v_1, Fig. 12–6b.

If the *vs* curve is known, the *as* curve is determined using the construction procedure shown in Fig. 12–6b. At any point C, the slope of the curve is constructed. The normal to this slope intersects the s axis at point B. The distance AB represents the magnitude of acceleration a of the

particle (Fig. 12–6a) since the two triangles shown in Fig. 12–6b are similar, and because $\tan \theta = dv/ds = a/v$. (Recall $a\,ds = v\,dv$.) Of course, one must work with a consistent set of units when using this procedure; for example, if v is measured in ft/sec and s in feet, then a will be measured in ft/sec².

Example 12–4

An experiment is performed in order to determine the acceleration due to gravity g that is acting on a small sphere which is released from rest. The downward displacements of the sphere are measured along the path at different time intervals. The results are shown in the table at right. Graph the motion curves for the falling sphere, and show that $g = 32.2$ ft/sec².

time t (sec)	displacement s (ft)
0	0
1	16.1
2	64.4
4	257.6
6	579.6

Solution

When the data listed in the table are plotted, the resulting st graph is parabolic, as shown in Fig. 12–7a. The speed v is equivalent to the slope measured at each of the data points on this graph, i.e., $v = ds/dt$. Obviously, for such measurements one must approximate using small finite measurements $v = \Delta s/\Delta t$. Measuring the slope is a graphical process and generally involves the use of a protractor or ruler. The results of the measurements are shown in Fig. 12–7a. From this data, the vt graph plots as a straight sloping line, Fig. 12–7b. The slope of this line is easily computed from the graph. For example, $a = \Delta v/\Delta t = (193.2 - 0)/(6 - 0) = 32.2$ ft/sec². The at graph is therefore a horizontal line, Fig. 12–7c, such that $a_c = g$ 32.2 ft/sec² at all times.

Since the acceleration is *constant,* one can use Eqs. 12–5 to obtain the *equations* which describe the graphs shown in Fig. 12–7. The initial conditions of motion are: $s_1 = 0$, $v_1 = 0$ at $t = 0$. Therefore, we obtain,

$$v = v_1 + a_c t$$
$$v = 32.2t \quad \text{(Fig. 12–7b)} \tag{1}$$

and

$$s = s_1 + v_1 t + \tfrac{1}{2} a_c t$$
$$s = 16.1 t^2 \quad \text{(Fig. 12–7a)} \tag{2}$$

Note how the at graph, which is *constant,* yields a vt graph which is linear, Eq. (1), and a st graph which is parabolic, Eq. (2).

Example 12–5

A particle moves along a straight-line path such that the velocity of the particle is described by the graph shown in Fig. 12–8a. If $s = 20$ ft

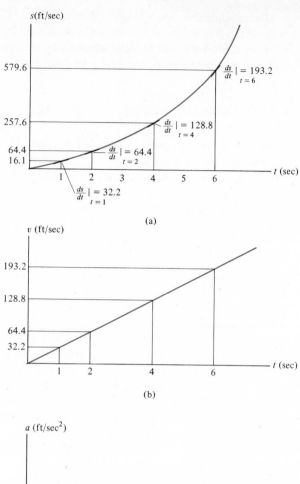

Fig. 12-7

when $t = 0$, construct the at and st graphs for $0 \leq t \leq 25$ sec, and determine the distance traveled by the particle during this time interval.

Solution

The at graph is plotted by using the data measured from the vt graph in accordance with the equation $a = dv/dt$, Fig. 12-8b. For the time

(a)

(b)

(c)

Fig. 12-8

interval $0 \leq t < 5$ sec, the slope of the vt graph is *constant*. Thus, $a = \Delta v/\Delta t = (50 \text{ ft/sec} - 0)/(5 \text{ sec} - 0) = 10 \text{ ft/sec}^2$. When $5 \text{ sec} < t < 15$ sec, $a = 0$, since the slope of the vt curve is zero during this interval of time. Finally, for $15 \text{ sec} < t < 25$ sec, $a = \Delta v/\Delta t = (-50 \text{ ft/sec} - (50 \text{ ft/sec}))/(25 \text{ sec} - 15 \text{ sec}) = -10 \text{ ft/sec}^2$.

The st graph is plotted by using the data of the vt graph and the equation $(s_2 - s_1) = \int_{t_1}^{t_2} v\,dt$, Fig. 12–8c. Initially the position of the particle is $s_1 = 20$ ft when $t = 0$. During the time interval $0 < t < 5$ sec, the change in displacement of the particle is equal to the *area* under the vt curve, i.e., $s_2 - s_1 = \frac{1}{2}(5 \text{ sec})(50 \text{ ft/sec}) = 125$ ft. Thus, at $t = 5$ sec, the position of the particle is $s_2 = 20$ ft $+ 125$ ft $= 145$ ft from the origin, Fig. 12–8c. Since the velocity curve is *integrated* (theoretically) to obtain the displacement curve, and the velocity curve is *linear* from $0 < t < 5$ sec, Fig. 12–8a, the displacement curve is *parabolic* within this region.* In a similar manner, from 5 sec $< t <$ 15 sec, the particle undergoes a displacement of $s_3 - s_2 = 50$ ft/sec (15 sec $-$ 5 sec) $= 500$ ft. Thus, at $t = 15$ sec, the particle is located $s_3 = 145$ ft $+ 500$ ft $= 645$ ft from the origin. Since the velocity curve is constant for 5 sec $< t <$ 15 sec, by integration, the displacement graph is linear during this time interval. Continuing in this manner, we can construct the entire st curve as shown in Fig. 12–8c; it consists of parabolic and straight-line segments.

The correctness of this diagram may be checked by noting that the slope of the st diagram at any instant of time t is equivalent to the magnitude of velocity v at that instant of time $(ds/dt = v)$. For example, since the velocity is linearly *increasing* during the time interval $0 < t < 5$ sec, the parabolic line segment of the st curve within this region is concave upward, that is, the slope measured from one point to the next is *increasing*. Likewise, the vt curve is linearly *decreasing* from 15 sec $< t \le$ 25 sec. Therefore, the *slope* measured from one point to the next along the parabolic sector of the st curve within this region *decreases*. Hence, the parabola is concave downward for 15 sec $< t <$ 25 sec. Note that the *displacement* of the particle is the same at $t = 15$ sec and $t = 25$ sec. This is due to the change in direction of velocity at $t = 20$ sec, Fig. 12–8a.

Since the positive and negative signs for v on the vt graph indicate the sense of direction (displacement) of the particle, the *distance traveled* by the particle during the time interval $0 \le t \le 25$ sec is computed as the *absolute value* of the shaded area under the vt curve in Fig. 12–8a. This area is computed in segments as

$$d = \int_0^{25} |v|\,dt$$

$$d = \tfrac{1}{2}(5)(50) + (15 - 5)\,50 + \tfrac{1}{2}(20 - 15)\,50 + \tfrac{1}{2}(25 - 20)(50)$$
$$d = 125 \text{ ft} + 500 \text{ ft} + 125 \text{ ft} + 125 \text{ ft} = 875 \text{ ft} \qquad \textit{Ans.}$$

---

*To show this mathematically, for $0 < t < $ sec, $v = 10t$ (Fig. 12–8a). Thus, $ds = 10t\,dt$, so that $s = 5t^2 + c$. At $t = 0$, $s = 20$ ft, so that $c = 20$ and $s = 5t^2 + 20$. Fig. 12–8c. Also note that $a = dv/dt = 10$, Fig. 12–8b.

Problems

12-18. Measurements of the velocity of a particle moving along a straight line have been recorded in the table. Plot the vt and the at graphs for the motion, and determine the acceleration when $t = 7$ sec.

v ft/sec	t sec
0	0
5	1
20	4
40	6
68	8
104	10
148	12

Prob. 12-18

12-19. A small race car starts from rest and moves along a straight track with an acceleration as shown. Construct the vt and the st graphs for the motion. What is the speed and displacement at $t = 60$ sec?

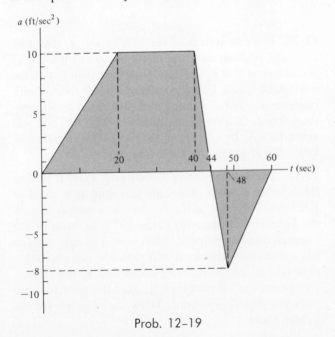

Prob. 12-19

12-20. A man riding a freight elevator accidentally drops a package off the elevator when the elevator is 200 ft from the ground level. If the elevator maintains a constant upward velocity of 5 ft/sec, determine how high the elevator is from the ground the instant the package hits the ground. Draw the vt curve for the package. Assume that the acceleration of the package is constant, acting downward with a magnitude of 32.2 ft/sec^2.

12-21. A toy rocket A is launched vertically with an initial velocity of 100 ft/sec from a height of 150 ft. At the same instant another toy rocket B is launched vertically from the ground with an initial velocity of 150 ft/sec. Determine the time at which they pass one another. Assume that both rockets are in free flight at the instant they are launched. The acceleration due to gravity is constant and acts downward with a magnitude of 32.2 ft/sec^2. Draw the vt curve for each rocket.

Prob. 12-21

453

12-22. The rocket shown is subjected to an engine thrust giving the rocket an acceleration which varies linearly with time, as shown. If the rocket is launched vertically, compute the height h traveled by the rocket and the velocity of the rocket at the moment the fuel burns out ($t = 20$ sec).

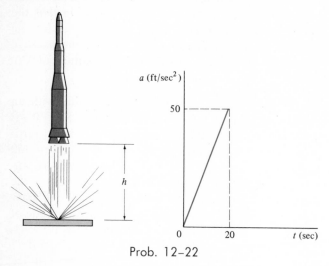

Prob. 12-22

12-23. A particle starts from rest and is subjected to an acceleration in accordance with the graph. Construct the vt graph for the motion, and determine the distance traveled during the time interval $2 \text{ sec} \le t \le 6 \text{ sec}$.

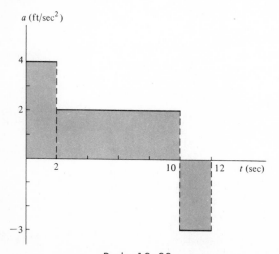

Prob. 12-23

12-24. The st graph for a small race car has been experimentally determined. From the data, construct the vt and at graphs for the car, and determine the magnitude of the maximum deceleration the car has attained.

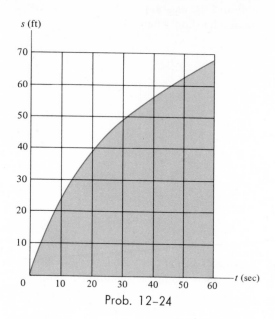

Prob. 12-24

12-25. Two cars start from rest side by side at the same time and position and race along a straight track. Car A accelerates at 4 ft/sec² for 35 sec and then maintains a constant speed. Car B accelerates at 10 ft/sec² until reaching a speed of 45 mph and then maintains a constant speed. Determine the time at which the cars will again be side by side. How far has each car traveled? Draw the vt graphs for each car.

12-26. The fighter plane F is following 1,000 ft behind the bomber B. Both planes are traveling at a speed of 200 mph. The pilot in the fighter plane wishes to pass the bomber and place his plane 500 ft in front of the bomber, resuming a speed of 200 mph. The fighter plane has a maximum velocity of 600 mph and can accelerate at a rate of 300 ft/sec² and decelerate at 100 ft/sec². Determine the shortest time it takes the pilot to complete the passing operation. Draw the vt graph for the fighter plane.

Prob. 12-26

12-27. The *vs* graph was determined experimentally to describe the straight-line motion of a rocket sled. Using the data of this graph, determine the acceleration of the sled when $s = 100$ ft, and when $s = 200$ ft.

Prob. 12-27

12-4. Curvilinear Velocity and Acceleration of a Particle

The motion of a particle along a curved path is called *curvilinear motion*. Because this path may be represented in three dimensions, *vector analysis* will be used to formulate the position, velocity, and acceleration of the particle. In this section we will discuss some general aspects of curvilinear

motion. There are three types of orthogonal coordinate systems commonly used to describe the motion. Each of these will be discussed in detail in subsequent articles of this chapter.

In the discussion of particle rectilinear motion a *scalar analysis* was used. This was convenient, since magnitudes are determined by the value of the scalar and the sense of direction along the (straight-line) path by the sign of the scalar. From the analysis it was shown that particle rectilinear motion is related by derivatives of scalar functions (see Eqs. 12–1 to 12–4). In elementary calculus it is shown that the derivative of a scalar function $f(s)$, where s is a scalar, is given as

$$\frac{df(s)}{ds} = \lim_{\Delta s \to 0} \frac{f(s + \Delta s) - f(s)}{\Delta s}$$

That is, the derivative is defined as the ratio of *change* made in the value of $f(s)$ to a corresponding *change* made in the variable s. Higher-order derivatives follow naturally from this definition.

Fig. 12-9

A *vector* on the other hand possesses both a magnitude *and* a direction. In defining the derivative of a vector, it is therefore necessary to account for *changes* made in *both* the *magnitude and direction* of the vector. Consider, for example, the *position vector* $\mathbf{r} = \mathbf{r}(t)$, Fig. 12–9a, which locates the position $P(x, y, z)$ of a particle at any instant of time as the particle moves along the curved path *s*. Suppose that the particle undergoes a small *displacement* $\Delta\mathbf{r}$ along the path to point $P'(x + \Delta x, y + \Delta y, z + \Delta z)$ during the time interval Δt. As a consequence of this, the vector $\mathbf{r}(t)$ changes in both magnitude and direction and becomes vector $\mathbf{r}(t + \Delta t)$. From the definition of the derivative of a vector,

$$\mathbf{v} = \frac{d\mathbf{r}(t)}{dt} = \lim_{\Delta t \to 0} \frac{\mathbf{r}(t + \Delta t) - \mathbf{r}(t)}{\Delta t}$$

Due to vector subtraction, the displacement $\Delta\mathbf{r} = \mathbf{r}(t + \Delta t) - \mathbf{r}(t)$, Fig. 12–9a. As $\Delta t \to 0$, $\Delta\mathbf{r}$ approaches the *tangent line* to the curve at point *P*. Thus,

$$\mathbf{v} = \lim_{\Delta t \to 0} \frac{\Delta\mathbf{r}}{\Delta t}$$

$$\mathbf{v} = \frac{d\mathbf{r}}{dt} \qquad (12\text{–}6)$$

Vector $\mathbf{v}$ is the *instantaneous velocity*. As shown in Fig. 12–9b, the *direction* of this vector is *always tangent to the path of motion*. The speed *v* or *magnitude* of $\mathbf{v}$ may be obtained by noting that the magnitude of displacement is the length of the straight-line segment from *P* to *P'*, Fig. 12–9a. Realizing that the length of this segment approaches the arc length Δs as $\Delta t \to 0$, we have

$$v = \lim_{\Delta t \to 0} \frac{PP'}{\Delta t} = \lim_{\Delta t \to 0} \frac{\Delta s}{\Delta t}$$

or

$$v = \frac{ds}{dt} \qquad (12\text{–}7)$$

Thus, the speed may be obtained by differentiating the path function *s* with respect to time.

The *instantaneous acceleration* of the particle at time *t* is the time rate of change of velocity. In general, this vector is *not* directed along the tangent line to the path; instead, its direction depends upon the relative changes made in the velocity vector. To illustrate this, consider the movement of the particle from point *P* to point *P'* during the instant of time Δt, Fig. 12–9c. During this movement the velocity vector $\mathbf{v}$ for the particle, which *always* remains tangent to the path, experiences a

change in *both* its magnitude and direction. To study these changes, consider an x', y', z' coordinate system measured in units of speed, Fig. 12–9*d*. The two velocity vectors in Fig. 12–9*c* are plotted in Fig. 12–9*d* with their tails located at the fixed origin O'. The tip of the velocity vectors reach points on the dashed curve. This curve is called a *hodograph,* and when constructed, it describes the locus of points of the tip of the velocity vector.* Hence, as $\Delta t \to 0$, $\Delta v \to 0$, and in the limit $\Delta \mathbf{v}$ becomes tangent to the hodograph at point A. We define the acceleration of the particle during this instant of time as

$$\mathbf{a} = \lim_{\Delta t \to 0} \frac{\Delta \mathbf{v}}{\Delta t}$$

$$\mathbf{a} = \frac{d\mathbf{v}}{dt} = \frac{d^2\mathbf{r}}{dt^2} \qquad (12\text{–}8)$$

By nature of the definition, the acceleration acts in the same direction as the limiting value of $\Delta \mathbf{v}$, that is, *tangent* to the *hodograph* at point A, Fig. 12–9*e*. Note, in general however, $\mathbf{a}$ is *not tangent* to the *path of motion,* Fig. 12–9*f*.

If the velocity vector $\mathbf{v}$ for a particle is given as a suitable function of time, $\mathbf{v} = \mathbf{v}(t)$, it is possible to specify the position vector $\mathbf{r}(t)$ for the particle by integration. Since

$$\mathbf{v}(t) = \frac{d\mathbf{r}(t)}{dt}$$

then

$$\mathbf{r}(t) = \int \mathbf{v}(t) \, dt$$

In a similar manner, provided the acceleration is known, the velocity vector may be obtained from

$$\mathbf{v}(t) = \int \mathbf{a}(t) \, dt$$

The vector constant of integration resulting from the above two equations may be obtained using the initial conditions of the problem.

12–5. Differentiation and Integration of Vector Functions

The rules for differentiation and integration of the sums and products of scalar functions apply as well to vector functions. Consider, for exam-

---

*In a similar manner, the path s, Fig. 12–9*a*, describes the locus of points for the tip of the position vector $\mathbf{r}$.

ple, the two vector functions $\mathbf{A}(s)$ and $\mathbf{B}(s)$. Provided these functions are smooth and continuous for all s,

$$\frac{d}{ds}(\mathbf{A} + \mathbf{B}) = \lim_{\Delta s \to 0} \frac{\Delta(\mathbf{A} + \mathbf{B})}{\Delta s} = \lim_{\Delta s \to 0} \frac{\mathbf{A}(s + \Delta s) - \mathbf{A}(s)}{\Delta s}$$

$$+ \lim_{\Delta s \to 0} \frac{\mathbf{B}(s + \Delta s) - \mathbf{B}(s)}{\Delta s}$$

or

$$\frac{d}{ds}(\mathbf{A} + \mathbf{B}) = \frac{d\mathbf{A}}{ds} + \frac{d\mathbf{B}}{ds} \qquad (12\text{-}9)$$

In a similar manner, from Eq. 12-9, it may be shown that

$$\int (\mathbf{A} + \mathbf{B}) \, ds = \int \mathbf{A} \, ds + \int \mathbf{B} \, ds \qquad (12\text{-}10)$$

Using the fundamental definition of the derivative, we can show that the derivative of a product of a scalar function $f(s)$ and a vector function $\mathbf{A}(s)$ is

$$\frac{d}{ds}[f(s)\mathbf{A}(s)] = \frac{df}{ds}\mathbf{A} + f\frac{d\mathbf{A}}{ds} \qquad (12\text{-}11)$$

If vector $\mathbf{A}$ is constant in both magnitude and direction for all values of s, then $\mathbf{A}(s) = \mathbf{C}$ and Eq. 12-11 becomes,

$$\frac{d}{ds}[f(s)\mathbf{C}] = \frac{df}{ds}\mathbf{C} = g(s)\mathbf{C} \qquad (12\text{-}12)$$

When integrating Eq. 12-12, we can factor the constant vector $\mathbf{C}$ out of the integral and obtain

$$\int \mathbf{C}g(s) \, ds = \mathbf{C} \int g(s) \, ds \qquad (12\text{-}13)$$

To obtain the derivative for the cross product of two vectors $\mathbf{A}$ and $\mathbf{B}$, write

$$\frac{d}{ds}(\mathbf{A} \times \mathbf{B}) = \lim_{\Delta s \to 0} \frac{\mathbf{A}(s + \Delta s) \times \mathbf{B}(s + \Delta s) - \mathbf{A}(s) \times \mathbf{B}(s)}{\Delta s}$$

Adding and subtracting the cross-product term $\mathbf{A}(s) \times \mathbf{B}(s + \Delta s)$ with the terms in the numerator of the above equation gives

$$\frac{d}{ds}(\mathbf{A} \times \mathbf{B}) = \lim_{\Delta s \to 0} \frac{\mathbf{A}(s + \Delta s) \times \mathbf{B}(s + \Delta s) - \mathbf{A}(s) \times \mathbf{B}(s + \Delta s)}{\Delta s}$$

$$+ \lim_{\Delta s \to 0} \frac{\mathbf{A}(s) \times \mathbf{B}(s + \Delta s) - \mathbf{A}(s) \times \mathbf{B}(s)}{\Delta s}$$

Using the distributive property of the cross product, we have

$$\frac{d}{ds}(\mathbf{A} \times \mathbf{B}) = \lim_{\Delta s \to 0} \frac{[\mathbf{A}(s + \Delta s) - \mathbf{A}(s)] \times \mathbf{B}(s + \Delta s)}{\Delta s}$$

$$+ \lim_{\Delta s \to 0} \frac{\mathbf{A}(s) \times [\mathbf{B}(s + \Delta s) - \mathbf{B}(s)]}{\Delta s}$$

In the limit,

$$\frac{d}{ds}(\mathbf{A} \times \mathbf{B}) = \left(\frac{d\mathbf{A}}{ds} \times \mathbf{B}\right) + \left(\mathbf{A} \times \frac{d\mathbf{B}}{ds}\right) \tag{12-14}$$

The order of cross-product multiplication is important in this formula, since the cross-product operation is noncommutative, e.g., $\mathbf{A} \times \mathbf{B} = -\mathbf{B} \times \mathbf{A}$.

In a similar manner, it may be shown that for the dot product

$$\frac{d}{ds}(\mathbf{A} \cdot \mathbf{B}) = \frac{d\mathbf{A}}{ds} \cdot \mathbf{B} + \mathbf{A} \cdot \frac{d\mathbf{B}}{ds} \tag{12-15}$$

For a constant vector $\mathbf{C}$, the dot and cross-product operations involving vector integration may be written as

$$\int \mathbf{C} \cdot \mathbf{A}(s) \, ds = \mathbf{C} \cdot \int \mathbf{A}(s) \, ds \tag{12-16}$$

and

$$\int \mathbf{C} \times \mathbf{A}(s) \, ds = \mathbf{C} \times \int \mathbf{A}(s) \, ds \tag{12-17}$$

The above formulations involving vector differentiation and integration will be useful in establishing the necessary relations in both kinematics and kinetics.

12-6. Curvilinear Motion of a Particle: Rectangular Components

If the position vector $\mathbf{r}$ of a particle is defined at each instant of time in terms of its rectangular components, it becomes convenient to also express the motion (velocity and acceleration) of the particle in terms of rectangular components. For example, let

$$\mathbf{r} = x\mathbf{i} + y\mathbf{j} + z\mathbf{k} \tag{12-18}$$

denote a position vector which defines the location of a particle in space at point $P(x, y, z)$ as the particle is moving along the curved path s, Fig. 12-10a. It is to be understood that as the particle moves along the curve,

(a)

(b)

(c)

Fig. 12–10

the x, y, z components of $\mathbf{r}$ are generally functions of time. By definition, the first and second time derivatives of $\mathbf{r}$ yield the velocity and acceleration vectors, respectively, for the particle. Recall, however, that when a vector is differentiated it is necessary to account for changes in *both* magnitude and direction of the vector. Using the distributive property of vector differentiation (Eq. 12–9), we obtain for the velocity,

$$\mathbf{v} = \frac{d\mathbf{r}}{dt} = \frac{d}{dt}(x\mathbf{i}) + \frac{d}{dt}(y\mathbf{j}) + \frac{d}{dt}(z\mathbf{k})$$

From Eq. 12–11 the derivative of the first component of $\mathbf{v}$ is

$$\frac{d}{dt}(x\mathbf{i}) = \frac{dx}{dt}\mathbf{i} + x\frac{d\mathbf{i}}{dt} \qquad (12\text{–}19)$$

The second term on the right side of this equation is zero since the *xyz* reference frame is *fixed,* and therefore, the unit vector direction **i** does not change with time. Differentiation of the **j** and **k** components may be carried out in a similar manner. The velocity vector then becomes

$$\mathbf{v} = v_x\mathbf{i} + v_y\mathbf{j} + v_z\mathbf{k}$$
$$= \frac{dx}{dt}\mathbf{i} + \frac{dy}{dt}\mathbf{j} + \frac{dz}{dt}\mathbf{k} \tag{12-20}$$

Recall from Sec. 12–4 that the velocity **v** is at all times *tangent to the path* traveled by the particle, Fig. 12–10*b*.

In a similar manner, the acceleration vector may be obtained by taking the first time derivative of Eq. 12–20 (or the second time derivative of Eq. 12–18). This yields,

$$\mathbf{a} = a_x\mathbf{i} + a_y\mathbf{j} + a_z\mathbf{k}$$
$$= \frac{dv_x}{dt}\mathbf{i} + \frac{dv_y}{dt}\mathbf{j} + \frac{dv_z}{dt}\mathbf{k} \tag{12-21}$$
$$= \frac{d^2x}{dt^2}\mathbf{i} + \frac{d^2y}{dt^2}\mathbf{j} + \frac{d^2z}{dt^2}\mathbf{k}$$

Hence, the acceleration components a_x, a_y, and a_z represent, respectively, the first time derivatives of velocity, v_x, v_y, and v_z, or the second time derivatives of displacement, x, y, and z. Since the acceleration vector represents the change made in the velocity vector, in general the acceleration will *not* be tangent to the path traveled by the particle, Fig. 12–10*c*.

12-7. Motion of a Projectile

The motion of a projectile is an important application of curvilinear particle motion which we can describe using rectangular components.* Provided the projectile is in *free flight* and air resistance can be neglected, the only force acting on the projectile is its weight. Because of gravitation, this force creates a *constant downward acceleration* on the particle of approximately $g = 32.2$ ft/sec² or 9.80 m/sec².

As an example to illustrate particle motion consider a projectile fired from a gun located at point (x_1, y_1), shown in Fig. 12–11. For simplicity, the path of the projectile is located in the *xy* plane. The projectile has an initial velocity $\mathbf{v}_1$, with components $(\mathbf{v}_x)_1$ and $(\mathbf{v}_y)_1$. Since the *y* axis is directed *upward,* the components of acceleration are always $\mathbf{a}_x = 0$, and

*The force analysis of a projectile will be explained when the kinetics of a particle is studied. (Chapter 14.)

Fig. 12–11

$\mathbf{a}_y = -g\mathbf{j}$. The motion of the projectile in the x direction is always *independent* of its motion in the y direction. These two *component* motions however yield the *resultant* motion when summed vectorially using Eqs. 12–20 and 12–21. Since the acceleration components $\mathbf{a}_x$ and $\mathbf{a}_y$ are always *constant,* we can use Eqs. 12–5 to obtain the velocity and displacement of the projectile. In particular, substituting $(a_c)_x = 0$, $v_1 = (v_x)_1$, and $s_1 = x_1$ into Eqs. 12–5, we obtain,

$$v_x = (v_x)_1$$
$$x = x_1 + (v_x)_1 t \qquad\qquad (12\text{–}22)$$

In a similar manner, if $(a_c)_y = -g$, $v_1 = (v_y)_1$, and $s_1 = y_1$ Eqs. 12–5 become,

$$v_y = (v_y)_1 - gt$$
$$y = y_1 + (v_y)_1 t - \tfrac{1}{2} g t^2 \qquad\qquad (12\text{–}23)$$
$$v_y^2 = (v_y)_1^2 - 2g(y - y_1)$$

The resultant motion of the particle is therefore described by the vectors

$$\mathbf{a} = -g\mathbf{j}$$
$$\mathbf{v} = (v_x)_1\mathbf{i} + [(v_y)_1 - gt]\mathbf{j}$$

or

$$\mathbf{v} = (v_x)_1\mathbf{i} + \sqrt{[(v_y)_1^2 - 2g(y - y_1)]}\,\mathbf{j}$$

and

$$\mathbf{r} = [x_1 + (v_x)_1 t]\mathbf{i} + [y_1 + (v_y)_1 t - \tfrac{1}{2} g t^2]\mathbf{j}$$

None of the above results should be memorized, instead, the derivation

should be understood, based upon Eqs. 12–5. In particular, the scalar functions representing the **i** and **j** components of the position vector **r** are time-parametric equations of a parabola. The path or *trajectory* of the projectile is therefore *parabolic,* as shown in Fig. 12–11.

Example 12–6

A ball is thrown from a position 5 ft above the ground to the roof of a 100-ft building, as shown in Fig. 12–12*a*. If the initial velocity of the ball is 110 ft/sec, inclined at an angle of 60° from the horizontal, determine (a) the maximum height *h* attained by the ball, and (b) the horizontal distance *d* from the point where the ball was thrown to where it strikes the roof.

Solution
Part (a). If we establish the origin of coordinates at the point where the ball is given its initial velocity $\mathbf{v}_0$ this vector may be written in rectangular component form as

$$\mathbf{v}_O = (v_x)_0\mathbf{i} + (v_y)_0\mathbf{j} = \{110\cos 60°\mathbf{i} + 110\sin 60°\mathbf{j}\}\ \text{ft/sec}$$
$$= \{55\mathbf{i} + 95.3\mathbf{j}\}\ \text{ft/sec} \tag{1}$$

The (independent) vertical motion of the ball is shown in Fig. 12–12*b*. When the ball reaches its highest point *A*, its vertical component of velocity is zero. We can therefore compute the height *h* using the third of Eqs. 12–23, since velocity is related to displacement by this equation.

$$(v_y)_A^2 = (v_y)_0^2 - 2g(y - y_0)$$
$$0 = (95.3)^2 - 2(32.2)(h - 5)$$
$$h = 146\ \text{ft} \qquad\qquad Ans.$$

Part (b). The distance *d* is computed by analyzing the motion in the *x* direction. Before doing this, however, it is first necessary to determine the total time during which the ball travels. We can find the time needed for the ball to reach its highest point *A* by using the first of Eqs. 12–23, with the data given in Fig. 12–12*b*:

$$(v_y)_A = (v_y)_0 - gt_{OA}$$
$$0 = 95.3 - (32.2)t_{OA}$$
$$t_{OA} = \frac{95.3}{32.2} = 2.96\ \text{sec}$$

With the initial conditions shown in Fig. 12–12*c*, we can find the time required for the ball to fall from point *A* to *B* by using the second of Eqs. 12–23:

(a)

(b)

(c)

Fig. 12-12

$$y_B = y_A + (v_y)_A t_{AB} - \tfrac{1}{2}g(t_{AB})^2$$
$$-46 = 0 + 0 - \tfrac{1}{2}(32.2)(t_{AB})^2$$
$$t_{AB} = 1.69 \text{ sec}$$

Thus, the ball is in flight for

$$t_{OB} = t_{OA} + t_{AB} = 2.96 \text{ sec} + 1.69 \text{ sec} = 4.65 \text{ sec.}$$

Since the ball is not accelerated in the x direction, it moves with *constant velocity* in this direction, namely $(v_x)_O = 55$ ft/sec.

We can find the distance d traveled during the time interval t_{OB} by using Eq. 12–22.

$$(x_B) = x_0 + (v_x)_0 t_{OB}$$
$$d = 0 + 55(4.65)$$
$$= 255 \text{ ft} \qquad \qquad Ans.$$

Example 12–7

Two cannon balls are fired simultaneously from points A and B, as shown in Fig. 12–13. The cannon ball fired at B rolls freely on the ground and follows the straight-line path BC. If the cannon ball at A is fired with a muzzle velocity of $\mathbf{v}_1$, determine (a) the equation which describes the path AC, (b) the elevation h of the cannon at A so that the cannon ball strikes the tree at C, and (c) neglecting the resistance effects of the ground, the required muzzle velocity $\mathbf{v}'$ of the cannon ball at B such that both cannon balls reach point C at the same time.

Solution

Part (a).   To obtain the equation of the path, $y = f(x)$, the x and y motions of the cannon ball can each be expressed in terms of the time parameter t as the cannon ball moves along the path. This parameter may then be eliminated to yield $y = f(x)$. Proceeding in this manner, the time required for the cannon ball to reach the arbitrary point $(x, -y)$ along the path can be found by using the second of Eqs. 12–23, namely,

$$y = y_1 + (v_y)_A t - \tfrac{1}{2}gt^2$$
$$-y = 0 + 0 - \tfrac{1}{2}g(t)^2$$
$$t = \sqrt{\frac{2y}{g}}$$

Similarly, in the x direction we have from Eq. 12–22,

$$s_x = (s_x)_A + (v_x)_A t$$
$$x = 0 + v_1 t$$

Substituting for t,

$$x = v_1 \sqrt{\frac{2y}{g}}$$

Solving this equation for y, choosing the negative sign since the path is downward, yields

$$y = -\frac{g}{2v_1^2} x^2 \qquad \qquad (1) \; Ans.$$

This *parabolic path AC* is shown in Fig. 12–13.

Fig. 12-13

Part (b). The elevation h may be determined by substituting $y = -h$ when $x = s$ into Eq. (1):

$$h = \frac{g}{2v_1^2} s^2 \qquad\qquad Ans.$$

Part (c). Since the cannon ball at A travels with a constant *horizontal* velocity v_1 throughout the length of its path and this motion is independent of its vertical motion, the required muzzle velocity of the cannon ball at B must be

$$v' = v_1 \qquad\qquad Ans.$$

When this occurs, both cannon balls arrive at C at the same time.

Problems

12-28. If the velocity vector for a particle is given as

$$v(t) = \{8t^2\mathbf{i} + 12t\mathbf{j} + 5\mathbf{k}\} \text{ ft/sec}$$

determine the derivative (acceleration) with respect to time.

12-29. If

$$\mathbf{A} = 12t^2\mathbf{i} + 2\mathbf{j}$$
$$\mathbf{B} = 3t^3\mathbf{j} + 2t^2\mathbf{k}$$

determine (a) $\dfrac{d}{dt}(\mathbf{A} \cdot \mathbf{B})$, and (b) $\dfrac{d}{dt}(\mathbf{A} \times \mathbf{B})$.

12-30. For vectors $\mathbf{A}$ and $\mathbf{B}$ given in Prob. 12-29, (a) determine $\dfrac{d}{dt}(\mathbf{A} + \mathbf{B})$, and (b) integrate $\int (\mathbf{A} + \mathbf{B})\, dt$ from $t = 0$ to $t = 2$ sec.

12-31. If

$$\mathbf{A} = 4t^2\mathbf{i} + 3t^3\mathbf{j} + 2\mathbf{k}$$
$$\mathbf{B} = 6t^2\mathbf{i} + \mathbf{j} + t^3\mathbf{k}$$

determine (a) $\displaystyle\int_{t=0}^{t=1} \mathbf{B} \cdot \mathbf{A}\, dt$, and (b) $\displaystyle\int_{t=0}^{t=1} \mathbf{A} \times \mathbf{B}\, dt$

12-32. If

$$A = 4t^2\mathbf{i} + 6t\mathbf{j} + 2t^3\mathbf{k},$$

determine (a) $\dfrac{d}{dt}(t^2\mathbf{A})$, and (b) $\displaystyle\int_{t=1}^{t=2} \dfrac{t}{2}\mathbf{A}\,dt$.

12-33. The curvilinear motion of a particle is defined by $x = 3t^2$, $y = 4t + 2$, and $z = 6t^3 - 8$, where the displacement is given in feet and the time in seconds. Determine the magnitude and direction of the velocity and acceleration of the particle when $t = 3$ sec.

12-34. A particle moves with curvilinear motion in the xy plane such that the y component of motion is described by the equation $y = 7t^3$, where y is in feet and t is in seconds. When $t = 1$ sec, the speed is 60 ft/sec. If the acceleration of the particle in the x direction is zero and the particle starts from rest at the origin when $t = 0$, determine the velocity of the particle when $t = 2$ sec.

12-35. The velocity of a particle released from rest is given by $\mathbf{v} = 16t^2\mathbf{i} + 4t^3\mathbf{j} + (5t + 2)\mathbf{k}$, where v is given in in./sec and t in seconds. Determine the magnitude of acceleration and the location of the particle when $t = 2$ sec.

12-36. The nozzle of a garden hose discharges water at the rate of 45 ft/sec. If the hose is held at ground level, determine the maximum height reached by the water, and the corresponding angle α at which the nozzle must be held from the horizontal.

12-37. Show that if a projectile is fired with an initial velocity of v_1 ft/sec, the maximum range (farthest distance) the projectile will travel is given by $R_{max} = v_1^2/g$, where $g = 32.2$ ft/sec^2.

12-38. A bird, flying horizontally with a velocity of 18 ft/sec, is carrying a worm in its mouth. If the worm accidentally drops out of the bird's mouth, determine the height h of the bird so that the worm drops into a bucket of water. The bird is 100 ft, horizontally, from the bucket when the worm is released.

12-39. A golf ball lands at a distance of 150 yd away from the point at which it was struck. If the initial velocity of the ball is directed at 20° with the horizontal,

determine the initial speed of the ball and the highest elevation of the ball above the ground.

12-40. The ball is launched at an angle of 90° from a 20° slope. If the initial velocity of the ball is 40 ft/sec, determine the range R.

$v_B = 40$ ft/sec

20°

R

Prob. 12-40

12-41. Small packages resting on the conveyor belt fall into the loading car. If the conveyor is running at a speed of 7 ft/sec, determine the range of distance s which the loading car may be placed from the wall so that the packages enter the car.

$v_C = 7$ ft/sec

30°

10 ft

s 3 ft

Prob. 12-41

12-42. The man throws a ball with an initial velocity of 50 ft/sec. Determine the two angles α_1 and α_2 at which he can throw the ball such that it lands in the bucket.

Prob. 12-42

12-43. Determine the minimum speed v_t which the toboggan must have when it approaches the jump at point A so that it reaches the other side of the gorge.

Prob. 12-43

12-44. The boy standing at A throws a snowball 4 ft from the ground with an initial velocity of 50 ft/sec. The snowball strikes the foot of a skier when the skier arrives at point B. Determine the horizontal distance R from the boy to the skier. (Experimental verification is not recommended.)

Prob. 12-44

12-45. The toy rocket is launched from a 30° slope with an initial velocity of 60 ft/sec. Determine the maximum height h the rocket attains, and the distance R downrange where the rocket strikes the ground. After launching, the rocket is in free flight.

Prob. 12-45

12-46. A boy throws a ball horizontally with a velocity of 66 ft/sec off the top of a bridge in an effort to hit the top surface AB of a passing truck traveling at a constant velocity of 50 ft/sec. If the top of the truck at point B is at point C at time $t = 0$, determine the range of time the boy can wait before throwing the ball and still hit the truck. (Experimental verification is not recommended.)

Prob. 12-46

12-47. Work Prob. 12-46 assuming that the truck has a constant acceleration of $a_t = 1$ ft/sec² and that its initial velocity when point B is at point C is 50 ft/sec.

12–8. Curvilinear Motion of a Particle: Cylindrical Components

In some engineering problems it is often convenient to express particle motion using a system of fixed *cylindrical coordinates*. These coordinates are defined by the *radial component r*, the *transverse component θ*, and the *axial component z*. For example, consider the position of the particle located at point $P(r, \theta, z)$ along the space curve *s* shown in Fig. 12–14*a*. The axial component *z* is identical to that used for rectangular coordinates. The radial component *r* is directed normal to the *z* axis, with the positive direction extending away from the origin. Using the right-hand rule, with the thumb pointing in the positive *z* direction, the fingers curl in the direction of positive *θ*, measured from the *x* axis. The magnitude of *θ* is generally measured in radians, where 1 rad $= 360°/2\pi$. The set of cylindrical coordinates so defined are *mutually orthogonal*. A set of unit vectors acting in the direction of positive *r*, *θ*, and *z*, will be designated as $\mathbf{u}_r$, $\mathbf{u}_\theta$, and $\mathbf{u}_z$, respectively, Fig. 12–14*a*. In particular, $\mathbf{u}_z$ is equivalent to **k** used for rectangular coordinates.

From Fig. 12–14*a*, the cylindrical coordinates (r, θ, z) may be related to rectangular coordinates (x, y, z) by using the transformation equations

$$x = r \cos \theta$$
$$y = r \sin \theta \qquad (12\text{–}24)$$
$$z = z$$

or

$$r = \sqrt{x^2 + y^2}$$
$$\theta = \tan^{-1}\frac{y}{x} \qquad (12\text{–}25)$$

The position vector $\mathbf{r}_P$, which locates the particle, Fig. 12–14*a*, may be expressed in terms of components in the $\mathbf{u}_r$ and $\mathbf{u}_z$ directions, i.e.,

$$\mathbf{r}_P = r\mathbf{u}_r + z\mathbf{u}_z \qquad (12\text{–}26)$$

Taking the time derivative of this equation in accordance with Eq. 12–11 to obtain the velocity, we obtain

$$\mathbf{v} = \frac{d\mathbf{r}_P}{dt} = \frac{dr}{dt}\mathbf{u}_r + r\frac{d\mathbf{u}_r}{dt} + \frac{dz}{dt}\mathbf{u}_z + z\frac{d\mathbf{u}_z}{dt} \qquad (12\text{–}27)$$

Since $\mathbf{u}_z$ *does not change with time* for a fixed reference, the last term in the above equation is zero. The term $d\mathbf{u}_r/dt$ *changes only in direction* with respect to time, since by definition the magnitude of this vector is always unity. As shown in Fig. 12–14*b*, the *direction* of $\mathbf{u}_r$ changes *only* when *θ* changes. (If *r* or *z* changes do you see why $\mathbf{u}_r$ does *not* change?

Fig. 12–14

Refer to Fig. 12–14*a*.) Hence, when θ changes by an amount $\Delta\theta$ because of a change in time Δt, $\mathbf{u}_r$ becomes *unit vector* $\mathbf{u}_r'$, Fig. 12–14*b*. When both vectors are extended back to point A and subtracted vectorially, an isosceles triangle is formed, as shown in Fig. 12–14*c*. The time change in $\mathbf{u}_r$ is thus $\Delta\mathbf{u}_r$. The *magnitude* of $\Delta\mathbf{u}_r$ is $\Delta u_r = 2(u_r \sin \Delta\theta/2)$. By definition $u_r = 1$, and for small angles, $\sin \Delta\theta/2 \approx \Delta\theta/2$. Therefore $\Delta u_r \approx \Delta\theta$. As $\Delta t \to 0$, (or $\Delta\theta \to 0$), vector $\Delta\mathbf{u}_r$ acts in the *direction* of $\mathbf{u}_\theta$ (Fig. 12–14*c*). Thus, in the limit $\Delta\mathbf{u}_r \to \Delta\theta\mathbf{u}_\theta$, so that the time derivative of $\mathbf{u}_r$ becomes

$$= \lim_{\Delta t \to 0} \frac{\Delta\mathbf{u}_r}{\Delta t} = \left(\lim_{\Delta t \to 0} \frac{\Delta\theta}{\Delta t} \right)\mathbf{u}_\theta$$

$$\frac{d\mathbf{u}_r}{dt} = \frac{d\theta}{dt}\mathbf{u}_\theta \tag{12-28}$$

Using the above results, we may therefore write Eq. 12–27 in component form as

$$\mathbf{v} = \mathbf{v}_r + \mathbf{v}_\theta + \mathbf{v}_z$$
$$= \frac{dr}{dt}\mathbf{u}_r + r\frac{d\theta}{dt}\mathbf{u}_\theta + \frac{dz}{dt}\mathbf{u}_z \qquad (12\text{--}29)$$

In particular, note that the term $d\theta/dt$ provides a measure of the rate of change of the angle θ with time. Common units used for the measurement are rad/sec.

Taking the time derivative of Eq. 12–29, and recalling that $d\mathbf{u}_z/dt = 0$, we obtain the acceleration $\mathbf{a}$ for the particle.

$$\mathbf{a} = \frac{d\mathbf{v}}{dt} = \frac{d^2 r}{dt^2}\mathbf{u}_r + \frac{dr}{dt}\frac{d\mathbf{u}_r}{dt} + \frac{dr}{dt}\frac{d\theta}{dt}\mathbf{u}_\theta +$$
$$r\frac{d^2\theta}{dt^2}\mathbf{u}_\theta + r\frac{d\theta}{dt}\frac{d\mathbf{u}_\theta}{dt} + \frac{d^2 z}{dt^2}\mathbf{u}_z \quad (12\text{--}30)$$

To evaluate the term involving $d\mathbf{u}_\theta/dt$, it is necessary to consider a change made in the direction of $\mathbf{u}_\theta$. As in the case of $\mathbf{u}_r$, for a fixed frame of reference, only a change in θ causes a change in $\mathbf{u}_\theta$, Fig. 12–14d. (Do you understand why $\mathbf{u}_\theta$ does not change with a change in coordinates r and z? Refer to Fig. 12–14a.) The new (or changed) position of $\mathbf{u}_\theta$ is designated as $\mathbf{u}'_\theta$. When these two unit vectors are drawn from a common point, the vector subtraction forms an isosceles triangle, as shown in Fig. 12–14e. The time change in $\mathbf{u}_\theta$ is thus $\Delta\mathbf{u}_\theta$. From the triangle the *magnitude* of $\Delta\mathbf{u}_\theta$ is $\Delta u_\theta = 2(u_\theta \sin \Delta\theta/2) \approx 2(1)\Delta\theta/2 = \Delta\theta$. The *direction* of $\Delta\mathbf{u}_\theta$ approaches $-\mathbf{u}_r$ as $\Delta t \to 0$ (Fig. 12–14e). Thus in the limit $\Delta\mathbf{u}_\theta \to -\Delta\theta\mathbf{u}_r$ and so

$$= \lim_{\Delta t \to 0}\frac{\Delta\mathbf{u}_\theta}{\Delta t} = \left(\lim_{\Delta t \to 0}\frac{-\Delta\theta}{\Delta t}\mathbf{u}_r\right)$$
$$\frac{d\mathbf{u}_\theta}{dt} = -\frac{d\theta}{dt}\mathbf{u}_r \qquad (12\text{--}31)$$

Using the above result and Eq. 12–28, we may write the acceleration of the particle as expressed by Eq. 12–30 in cylindrical component form as

$$\mathbf{a} = \mathbf{a}_r + \mathbf{a}_\theta + \mathbf{a}_z$$
$$= \left[\frac{d^2 r}{dt^2} - r\left(\frac{d\theta}{dt}\right)^2\right]\mathbf{u}_r + \left(r\frac{d^2\theta}{dt^2} + 2\frac{dr}{dt}\frac{d\theta}{dt}\right)\mathbf{u}_\theta + \frac{d^2 z}{dt^2}\mathbf{u}_z \quad (12\text{--}32)$$

In particular, the term $d^2\theta/dt^2 = d/dt(d\theta/dt)$ is a measure of the change made in the rate of change of θ during an instant of time. Common units of measurement are rad/sec².

When motion of the particle is restricted to the xy plane, $z = 0$ and

polar coordinates r and *θ* are used to describe the motion. Equations 12–29 and 12–32 reduce to

$$\mathbf{v} = \mathbf{v}_r + \mathbf{v}_\theta$$
$$= \frac{dr}{dt}\mathbf{u}_r + r\frac{d\theta}{dt}\mathbf{u}_\theta \qquad (12\text{–}33)$$

and

$$\mathbf{a} = \mathbf{a}_r + \mathbf{a}_\theta$$
$$= \left[\frac{d^2r}{dt^2} - r\left(\frac{d\theta}{dt}\right)^2\right]\mathbf{u}_r + \left(r\frac{d^2\theta}{dt^2} + 2\frac{dr}{dt}\frac{d\theta}{dt}\right)\mathbf{u}_\theta \qquad (12\text{–}34)$$

Application of these equations is given in Examples 12–8 and 12–9.

12–9. Curvilinear Motion of a Particle: Tangential and Normal Components

When the path along which the particle is moving is *known,* it is often convenient to describe the motion of the particle by using *n* and *t* coordinates which act normal and tangent to the path, respectively, and have their *origin located at the particle.* A set of such coordinates is shown in Fig. 12–15*a*. The positive tangent axis *t* is directed along the space curve from the particle in the direction of *increasing s,* as shown in the figure. This direction is always uniquely specified. There are, however, an infinite number of straight lines which can be constructed normal to the tangent axis at the given point *P.* We will choose the normal axis *n* directed from the particle *toward* the *center of curvature O'* of the curve at the point.* This normal is called the *principle normal* to the curve at *P.* The plane containing the *nt* axis is often referred to as the *osculating plane.* The orientation of this plane depends upon the location of the particle *P* along the space curve. The unit vectors which act in the direction of the positive *n* and *t* axes will be called $\mathbf{u}_n$ and $\mathbf{u}_t$, respectively.

---

*Geometrically, each *differential segment* of a curve is constructed by using a circle having a unique radius (radius of curvature) and a center (center of curvature). At each point along the curve both the radius of curvature and the center of curvature are different. In particular, if the curve lies in the *xy* plane, the radius of curvature ρ can be determined from the equation

$$\frac{1}{\rho} = \left|\frac{\dfrac{d^2y}{dx^2}}{\left[1 + \left(\dfrac{dy}{dx}\right)^2\right]^{3/2}}\right|$$

The derivation of this result is given in any standard calculus text.

The velocity of the particle traveling over the curved path, Fig. 12–15a, may be expressed as

$$\mathbf{v} = \frac{d\mathbf{r}}{dt}$$

Since the velocity vector always acts tangent to the path of motion, the velocity of the particle may be expressed in terms of the unit vector $\mathbf{u}_t$ by rewriting this equation in the form

$$\mathbf{v} = \frac{d\mathbf{r}}{dt} = \lim_{\Delta t \to 0} \frac{\Delta \mathbf{r}}{\Delta t} = \lim_{\Delta t \to 0} \frac{\Delta \mathbf{r}}{\Delta s} \frac{\Delta s}{\Delta t}$$

The term $\Delta \mathbf{r}$ represents the vector displacement of the particle in moving from point P to P' during the time increment Δt, Fig. 12–15b. As $\Delta t \to 0$

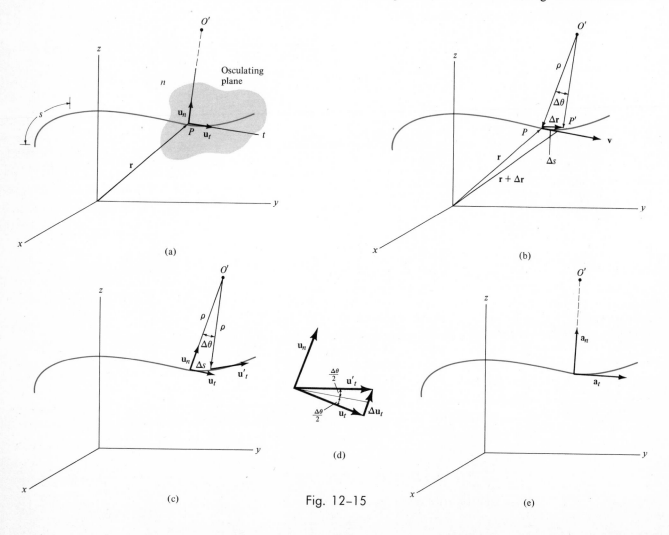

Fig. 12–15

(and consequently $\Delta\theta \to 0$) the *direction* of $\Delta\mathbf{r}$ approaches $\mathbf{u}_t$ shown in Fig. 12–15a. The *magnitude* $\Delta r \to \Delta s$, Fig. 12–15b. Thus, in the limit $\Delta\mathbf{r} \to \Delta s\,\mathbf{u}_t$, so that

$$\mathbf{v} = \lim_{\Delta t \to 0} \frac{\Delta s}{\Delta s}\mathbf{u}_t\frac{\Delta s}{\Delta t} = \frac{ds}{dt}\mathbf{u}_t$$

Specifically, the velocity acts in the tangent direction, so that

$$\mathbf{v} = \mathbf{v}_t$$
$$= \frac{ds}{dt}\mathbf{u}_t = v\mathbf{u}_t \qquad (12\text{–}35)$$

where v represents the *magnitude* of the instantaneous velocity of the particle, ds/dt, at point P, Fig. 12–15b.

The acceleration of the particle is the time rate of change of the velocity vector; thus, using Eq. 12–35, we have

$$\mathbf{a} = \frac{d\mathbf{v}}{dt} = \frac{dv}{dt}\mathbf{u}_t + v\frac{d\mathbf{u}_t}{dt} \qquad (12\text{–}36)$$

The time derivative of the unit vector $\mathbf{u}_t$ may be evaluated with reference to Fig. 12–15c. When the particle moves through a displacement Δs during the time Δt, the unit vector $\mathbf{u}_t$ becomes unit vector $\mathbf{u}_t'$. To obtain this change the two vectors are subtracted vectorially at a common point, forming the isosceles triangle shown in Fig. 12–15d. Since the angle subtended by the radius of curvature ρ across the arc Δs is $\Delta\theta$, this same angle is formed between the unit vectors $\mathbf{u}_t$ and $\mathbf{u}_t'$. From Fig. 12–15d, the *magnitude* of $\Delta\mathbf{u}_t$ is $\Delta u_t = 2(u_t \sin \Delta\theta/2) \approx \Delta\theta$. (The $\sin \Delta\theta/2 \approx \Delta\theta/2$ for small angles and $u_t = 1$.) As $\Delta t \to 0$, vector $\Delta\mathbf{u}_t$ acts in the *direction* of $\mathbf{u}_n$. Thus, in the limit $\Delta\mathbf{u}_t \to \Delta\theta\mathbf{u}_n$, so that the time derivative of $\mathbf{u}_t$ is

$$\frac{d\mathbf{u}_t}{dt} = \lim_{\Delta t \to 0}\frac{\Delta\mathbf{u}_t}{\Delta t} = \lim_{\Delta t \to 0}\frac{\Delta\theta}{\Delta t}\mathbf{u}_n = \frac{d\theta}{dt}\mathbf{u}_n = \frac{d\theta}{ds}\frac{ds}{dt}\mathbf{u}_n$$

From Fig. 12–15c, $\Delta s = \rho\,\Delta\theta$, so that $d\theta/ds = 1/\rho$ when $\Delta t \to 0$. Also $v = ds/dt$. Thus,

$$\frac{d\mathbf{u}_t}{dt} = \frac{v}{\rho}\mathbf{u}_n$$

If we substitute this result into Eq. 12–36, the two components of the acceleration vector become

$$\mathbf{a} = \mathbf{a}_t + \mathbf{a}_n$$
$$= \frac{dv}{dt}\mathbf{u}_t + \frac{v^2}{\rho}\mathbf{u}_n \qquad (12\text{–}37)$$
$$= v\frac{dv}{ds}\mathbf{u}_t + \frac{v^2}{\rho}\mathbf{u}_n$$

These two components of acceleration are shown in Fig. 12–15e. The *tangential component* $\mathbf{a}_t$ represents a *change* made in the *magnitude* of the *velocity* as denoted by dv/dt or $v\,dv/ds$ (Eq. 12–4). The sense of direction of this component is either in the direction of motion or opposed to the direction of motion, depending upon whether the speed of the particle is increasing or decreasing. The *normal component* $\mathbf{a}_n$ is *always directed toward the center of curvature of the path*. From the derivation, this term represents the *change* made in the *direction* of the *velocity*. Its magnitude is equal to the square of the speed of the particle divided by the radius of curvature of the path at point P.

The acceleration of the particle is zero only when *both* of its components are equal to zero. If the particle moves along the path at a *constant speed*, only the normal component exists. When motion of the particle is along a straight-line path or at an inflection point along a curved path, the radius of curvature becomes infinite, so that only the tangential component exists. Thus, unlike the velocity, the acceleration is generally *not* tangent to the path of motion.

Fig. 12–16

Engineering structures and machinery can be designed to reduce the acceleration of particles moving on or across their surfaces by slowly changing the curvature of the path taken by the particle. For example, the surface of a high-speed cam is designed without sudden changes in curvature in order to produce smooth transitions in acceleration to contacting parts. In a similar manner, circular highway curves most often contain transitions, called spirals, so that cars leaving a straight portion of road can enter the curve without causing discomfort to the passengers or extra stress on the car.

Example 12–8

The 4-in. long rod *AB* shown in Fig. 12–16 is threaded to a bolt. The rod has an angular rotation of 5 rad/sec and is advancing upward along the axis of the bolt at a rate of 1 in./sec. Determine the cylindrical components of velocity and acceleration of point *A* on the rod with respect to the fixed coordinate system shown.

Solution

Since the angular rate of rotation is constant, i.e., $d\theta/dt = 5$ rad/sec (positive since rotation is *increasing* with time in the positive θ direction), then $d^2\theta/dt^2 = 0$. The velocity of point *A* along the *z* axis is also positive, having a constant magnitude of $dz/dt = 1$ in./sec. Thus, $d^2z/dt^2 = 0$. The radius $r = 2$ in. is constant, thus $dr/dt = 0$. Substituting this data into Eqs. 12–29 and 12–32, we have

$$\mathbf{v}_A = \frac{dr}{dt}\mathbf{u}_r + r\frac{d\theta}{dt}\mathbf{u}_\theta + \frac{dz}{dt}\mathbf{u}_z$$

$$= 0 + 2(5)\mathbf{u}_\theta + (1)\mathbf{u}_z$$

$$= \{10\mathbf{u}_\theta + \mathbf{u}_z\} \text{ in./sec} \qquad \textit{Ans.}$$

$$\mathbf{a}_A = \left[\frac{d^2r}{dt^2} - r\left(\frac{d\theta}{dt}\right)^2\right]\mathbf{u}_r + \left(r\frac{d^2\theta}{dt^2} + 2\frac{dr}{dt}\frac{d\theta}{dt}\right)\mathbf{u}_\theta + \frac{d^2z}{dt^2}\mathbf{u}_z$$

$$= [0 - 2(5)^2]\mathbf{u}_r + [2(0) - 2(0)5]\mathbf{u}_\theta + 0$$

$$= \{-50\mathbf{u}_r\} \text{ in./sec}^2 \qquad \textit{Ans.}$$

The path of motion for point *A* is a spiral. From the results the velocity $\mathbf{v}_A$ is tangent to this path and the acceleration $\mathbf{a}_A$ is directed toward the *z* axis, Fig. 12–16.

Example 12–9

A thin rod *AO*, shown in Fig. 12–17*a*, is rotating in the *xy* plane with an angular motion of $d\theta/dt = 3t^{1/3}$ rad/sec, where *t* is measured in seconds. During this time a small collar *B* is sliding outward along *OA*

with a radial velocity of $dr/dt = (7 - 2t)$ in./sec, where t is measured in seconds. Determine the velocity and acceleration of the collar relative to the fixed xy coordinate system when $t = 1$ sec. At this instant $r = 3$ in. and $\theta = 60°$.

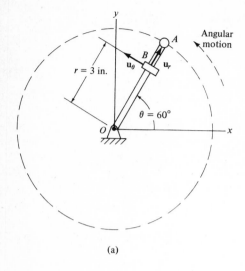

(a)

Solution

Since motion occurs in the xy plane (or $r\theta$ plane) Eqs. 12–33 and 12–34 will be used for the solution. From the problem statement, the terms used in these equations for $t = 1$ sec are

$$r = 3 \text{ in.}$$

$$\left.\frac{dr}{dt}\right|_{t=1\,\text{sec}} = (7 - 2t)\Big|_{t=1\,\text{sec}} = 5 \text{ in./sec}$$

$$\left.\frac{d^2r}{dt^2}\right|_{t=1\,\text{sec}} = -2 \text{ in./sec}^2$$

$$\theta = 60°$$

$$\left.\frac{d\theta}{dt}\right|_{t=1\,\text{sec}} = 3t^{1/3}\Big|_{t=1\,\text{sec}} = 3 \text{ rad/sec}$$

$$\left.\frac{d^2\theta}{dt^2}\right|_{t=1\,\text{sec}} = t^{-2/3} = 1 \text{ rad/sec}^2$$

Hence the velocity of the collar is

$$\mathbf{v} = \frac{dr}{dt}\mathbf{u}_r + r\frac{d\theta}{dt}\mathbf{u}_\theta$$

$$= 5\mathbf{u}_r + 3(3)\mathbf{u}_\theta$$

$$= \{5\mathbf{u}_r + 9\,\mathbf{u}_\theta\} \text{ in./sec}$$

(b)

The magnitude of $\mathbf{v}$ is therefore

$$v = \sqrt{(5)^2 + (9)^2} = 10.3 \text{ in./sec} \qquad \textit{Ans.}$$

With reference to Fig. 12–17b,

$$\beta = \tan^{-1}\tfrac{9}{5} = 60.9°$$

Thus, the angle which $\mathbf{v}$ makes with the x axis is

$$\beta + 60° = 120.9° \qquad \textit{Ans.}$$

(c)

Fig. 12–17

The acceleration of the collar is computed from Eq. 12–34:

$$\mathbf{a} = \left[\frac{d^2r}{dt^2} - r\left(\frac{d\theta}{dt}\right)^2\right]\mathbf{u}_r + \left(r\frac{d^2\theta}{dt^2} + 2\frac{dr}{dt}\frac{d\theta}{dt}\right)\mathbf{u}_\theta$$

$$= [-2 - 3(3)^2]\mathbf{u}_r + [3(1) + 2(5)3]\mathbf{u}_\theta$$

$$= \{-29\mathbf{u}_r + 33\mathbf{u}_\theta\} \text{ in./sec}^2$$

Thus,

$$a = \sqrt{(-29)^2 + (33)^2} = 43.9 \text{ in./sec}^2 \qquad \textit{Ans.}$$

From Fig. 12–17c,

$$\phi = \tan^{-1} \tfrac{29}{33} = 41.3°$$

The angle which **a** makes with the x axis is therefore,

$$\phi + 90° + 60° = 191.3° \qquad \textit{Ans.}$$

Example 12–10

A small bead is forced to move along the parabolic path AO, shown in Fig. 12–18, with a constant speed of 6 ft/sec. Determine the velocity and acceleration of the bead when the bead arrives at B.

Solution

The velocity vector is always directed tangent to the path. Since $dy/dx = 2x$; then $\dfrac{dy}{dx}\bigg|_{x=1} = 2$. Hence at B, the velocity vector makes an angle of

$$\beta = \tan^{-1} 2 = 63.4° \qquad \textit{Ans.}$$

with the x axis as shown in Fig. 12–18.

The acceleration is computed using Eq. 12–37. It is first necessary, however, to determine the radius of curvature of the path at point B (1 ft, 1 ft).* Since $d^2y/dx^2 = 2$, then

$$\frac{1}{\rho} = \left| \frac{\dfrac{d^2y}{dx^2}}{\left[1 + \left(\dfrac{dy}{dx}\right)^2\right]^{3/2}} \right| = \frac{2}{[1 + (2x)^2]^{3/2}} \text{ ft}^{-1}$$

At point B (1 ft, 1 ft),

$$\frac{1}{\rho} = \frac{2}{[1 + 4]^{3/2}} = 0.179 \text{ ft}^{-1}$$

Thus,

$$\rho = 5.59 \text{ ft}$$

Using Eq. 12–37, the acceleration becomes

$$\mathbf{a}_B = \frac{dv}{dt}\mathbf{u}_t + \frac{v^2}{\rho}\mathbf{u}_n$$

Fig. 12–18

*See the footnote on p. 473.

$$= 0\mathbf{u}_t + \frac{(6)^2}{5.59}\mathbf{u}_n$$

$$= \{6.44\mathbf{u}_n\} \ \text{ft/sec}^2$$

In particular, note that the *acceleration* has no component in the *tangential direction* since the *magnitude of velocity* does not change. (The component of acceleration in the *normal direction* accounts for the *change* in the *direction of velocity*.) Since $\mathbf{a}_B$ acts in the direction of the positive n axis, it makes an angle of $\beta + 90° = 153.4°$ with the positive x axis. Hence,

$$a_B = 6.44 \ \text{ft/sec}^2 \qquad\qquad Ans.$$

Example 12-11

A small particle starts from rest at point A and is forced to travel along the *horizontal path ABCD* shown in Fig. 12-19a. During the entire motion, the magnitude of *tangential acceleration* is given by the equation $a_t = 0.2t$ ft/sec², where t is in seconds. Determine the acceleration of the particle when it is located at point B and at point C.

Solution

The acceleration is computed from Eq. 12-37. In order to use this equation, however, it is first necessary to compute the velocity and tangential component of acceleration at points B and C. Hence, from Eq. 12-37,

$$a_t = \frac{dv}{dt} = 0.2t$$

Since the particle starts from rest,

$$\int_0^v dv = \int_0^t 0.2t \ dt$$

$$v = 0.1t^2 \qquad (1)$$

From Eq. 12-35, we have

$$v = \frac{ds}{dt} = 0.1t^2$$

$$\int_0^s ds = \int_0^t 0.1t^2 \ dt$$

$$s = 0.033t^3 \qquad (2)$$

The parameter t may be eliminated between Eqs. (1) and (2) yielding velocity as a function of displacement.

(a)

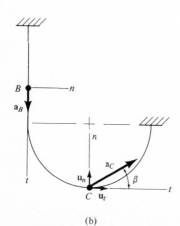

(b)

Fig. 12-19

$$v = 0.965s^{2/3}$$

By substitution of the appropriate value of s, we obtain the speed at points B and C.

$$v_B = 0.965(2 \text{ ft})^{2/3} = 1.53 \text{ ft/sec}$$

$$v_C = 0.965\left[3 \text{ ft} + \frac{2\pi(2 \text{ ft})}{4}\right]^{2/3} = 3.24 \text{ ft/sec}$$

In a similar manner, combining Eq. (2) and $a_t = 0.2t$, we have

$$a_t = 0.621s^{1/3}$$

Thus,

$$(a_t)_B = \left(\frac{dv}{dt}\right)_B = 0.621(2 \text{ ft})^{1/3} = 0.783 \text{ ft/sec}^2$$

$$(a_t)_C = \left(\frac{dv}{dt}\right)_C = 0.621\left[3 \text{ ft} + \frac{2\pi(2 \text{ ft})}{4}\right]^{1/3} = 1.14 \text{ ft/sec}^2$$

At point B, $\rho_B \to \infty$, so that the normal component of acceleration in Eq. 12–37 approaches zero (v_B does not change *direction*). Hence,

$$\mathbf{a}_B = \left(\frac{dv}{dt}\right)_B \mathbf{u}_t + \frac{v_B^2}{\rho_B}\mathbf{u}_n$$

$$= \{0.783\mathbf{u}_t\} \text{ ft/sec}^2$$

Thus,

$$a_B = 0.783 \text{ ft/sec}^2 \qquad \downarrow \mathbf{a}_B \qquad\qquad Ans.$$

The result is shown in Fig. 12–19b.

At point C, $\rho_C = 2$ ft; therefore,

$$\mathbf{a}_C = \left(\frac{dv}{dt}\right)_C \mathbf{u}_t + \frac{v_C^2}{\rho_C}\mathbf{u}_n$$

$$= 1.14 \text{ ft/sec}^2 \, \mathbf{u}_t + \frac{(3.24 \text{ ft/sec})^2}{2 \text{ ft}}\mathbf{u}_n$$

$$= \{1.14\mathbf{u}_t + 5.25\mathbf{u}_n\} \text{ ft/sec}^2$$

The magnitude of $\mathbf{a}_C$ is

$$a_C = \sqrt{(1.14)^2 + (5.25)^2} = 5.37 \text{ ft/sec}^2$$

The direction is determined as shown in Fig. 12–19b, where

$$\beta = \tan^{-1}\frac{5.25}{1.14} = 77.8°$$

Thus,

$$a_C = 5.37 \text{ ft/sec}^2 \qquad \angle\beta \qquad \mathbf{a}_C \qquad\qquad Ans.$$

12-48. The motion of a particle moving along a space curve may be expressed in terms of its rectangular components in the form $x = 10t$, $y = -4t$, and $z = 3t^2$, where x, y, and z are measured in inches and t is measured in seconds. Express the velocity and acceleration of the particle in terms of cylindrical components, and determine the magnitudes of velocity and acceleration when $t = 1$ sec.

12-49. An airplane is flying horizontally with a velocity of 200 mi/hr and an acceleration of 3 mi/hr². If the propeller has a diameter of 6 ft and is rotating at an angular rate of 120 rad/sec, determine the magnitude of velocity and acceleration of a particle located on the tip of the propeller.

12-50. The rod OA rotates counterclockwise in a circle with a constant angular rate of 5 rad/sec. A slider block located at B moves freely both on the straight rod OA and the curved rod whose shape is a limaçon described by the equation $r = 2 - \cos \theta$, where θ is given in radians and r is in feet. Determine the speed of the slider block when $\theta = 30°$.

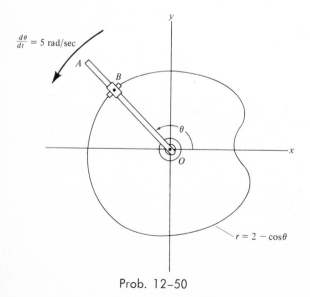

Prob. 12-50

12-51. Determine the magnitude of acceleration of the slider block in Prob. 12-50 when $\theta = 30°$.

12-52. The motion of the rod B is controlled by the rotation of link OA which has a groove cut into it. At the instant shown, the link is rotating with a constant angular rate of 6 rad/sec. Determine the magnitude of velocity and acceleration of rod B when $\theta = 90°$. The spiral path is defined by the equation $r = 3\theta$.

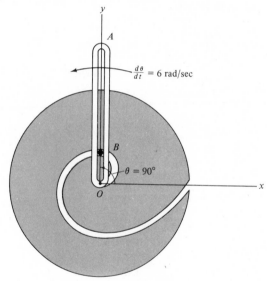

Prob. 12-52

12-53. The rim of a 26-in.-diameter bicycle wheel has a tangential speed of 10 ft/sec. Determine the normal component of acceleration of a point on the rim of the wheel.

12-54. A particle is moving along a circular path having a radius of 4 in. such that its position as a function of time is given by $\theta = \cos 2t$, where θ is in radians and t is in seconds. Determine the acceleration of the particle when $\theta = 30°$. The particle starts from rest when $\theta = 0°$.

12-55. The small washer is sliding down the cord OA. When the washer is half way down the cord, its speed is 12 in./sec, and its acceleration (directed towards point O) has a magnitude of 1 in./sec². Determine the velocity and acceleration vectors of the washer at this point in terms of its cylindrical components.

Prob. 12-55

12-56. Calculate the time derivative of **a** as expressed by Eq. 12-32. The quantity represents the time rate of change of the acceleration and is often referred to as a *jerk*. This term is useful for measuring passenger discomfort on elevators, trains, etc.

12-57. A particle is moving along a curved path at a constant speed of 60 ft/sec. The radii of curvature of the path at points P and P' are 20 and 50 ft, respectively. If it takes the particle 20 sec to go from P to P', determine the normal acceleration of the particle at points P and P'.

12-58. The jet plane is traveling along a horizontal path with a velocity of 500 mph. If the plane is 10,000 ft from the ground, determine the rate of rotation $d\theta/dt$ of the radar tracking instrument when $\theta = 40°$.

Prob. 12-58

12-59. An old-fashioned train is traveling with a constant velocity of 60 ft/sec along the curved path shown. Determine the velocity and acceleration of the train when it reaches the obstacle located at point A. The coordinates are measured in feet.

Prob. 12-59

483

12-60. The automobile has a velocity of 80 ft/sec at point A and an acceleration $\mathbf{a}$ of 10 ft/sec^2, acting in the direction shown. Determine the radius of curvature of the path at point A and the tangential component of acceleration.

Prob. 12-60

12-61. An automobile is on a highway curve having a radius of 800 ft. If the acceleration of the automobile is 5 ft/sec^2, determine the constant speed at which the automobile is traveling.

12-62. The satellite S shown travels around the earth with a constant speed of 13,000 mph in a circular path. If the acceleration of the satellite is 2 ft/sec^2, determine the altitude h. Assume the earth's diameter to be 7,950 mi.

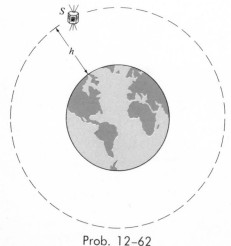

Prob. 12-62

12-63. A train travels along a circular curve having a radius of 2,000 ft. If the speed of the train is uniformly increased from 6 to 8 mph in 1 min, determine the magnitude of the acceleration when the speed of the train is 7 mph.

12-64. The small bead B travels along the curve having a constant speed of 10 ft/sec. Determine the acceleration of the bead when it is located at point (1 ft, 1 ft), and sketch this vector on the curve. (*Hint:* See footnote p. 473.)

Prob. 12-64

12-65. A ball is kicked from A along the path shown with an initial velocity of 60 ft/sec. Determine the radius of curvature of the path at points B and C.

Prob. 12-65

12-66. A particle moves along a curve $y = 5x^2 - 2$ with a constant velocity of 5 ft/sec. Determine the point along the curve where the maximum acceleration occurs, and compute its value. (*Hint:* See footnote p. 473.)

12-67. The motion of the rod B is controlled by the movements of the slider mechanisms A and C. At the instant shown, the slider A is traveling at a constant

velocity of 6 in./sec in the direction shown, while C is traveling with a velocity of 10 in./sec and has an acceleration of 3 in./sec² in the direction shown. Determine the speed of rod B and the radius of curvature of the path of the rod at this instant.

12-68. The rod B is confined to move in the circular groove and is forced into motion by the slotted guide A. When the rod is in the position shown, the magnitudes of its velocity and acceleration are 12 in./sec and 8 in./sec², respectively. Knowing that the slotted guide A is moving up the plane, determine the velocity and acceleration of A at the instant shown.

Prob. 12-68

Prob. 12-67

12-10. Relative Motions of Two Particles

Up to this point only the methods for determining the *absolute* displacement, velocity, and acceleration of a particle have been considered. This has been done using a single *fixed* reference frame for measurement. Such an approach to a solution is called an *absolute motion analysis*. There are many cases, however, where the path of motion for a particle may

be rather complicated, so that it may be feasible instead to analyze the motion *in parts* by using two or more frames of reference rather than a single frame. For example, the motion of a particle located at the tip of an airplane propeller, while the plane is in flight, is more easily described if one observes first the motion of the airplane from a fixed reference and then superimposes (vectorially) the motion of the particle from a reference attached to the airplane. Any type of coordinates—rectangular (Sec. 12–6), cylindrical (Sec. 12–8), etc.—may be chosen to describe these two different motions. This type of analysis is called a *relative motion analysis*. In this section only *translating frames of reference* will be considered for the analysis. Relative motion analysis of particles using rotating frames of reference will be treated in Secs. 13–12 and 13–13, since such analysis depends upon prior knowledge of the kinematics of line segments.

In general, the relative motion between two particles may occur in one of two ways: The motions may be independent or dependent. We will presently study each of these cases separately.

Case I: Independent Relative Motion. An example of independent relative motion is shown in Fig. 12–20, where particles A and B move arbitrarily along the paths aa and bb, respectively. The *absolute position* of both particles, $\mathbf{r}_A$ and $\mathbf{r}_B$, is measured from the common origin O of the *fixed x, y,* and z reference frame. The origin of a second frame of reference x', y', and z' is attached to and moves with particle A. The axes of this frame are *only permitted to translate* relative to the fixed frame. The *relative position* of point B with respect to point A can be observed from this moving frame and is designated by the vector $\mathbf{r}_{B/A}$, called a *relative position vector*. Using vector addition, we can relate the three vectors shown in the figure by the equation

$$\mathbf{r}_B = \mathbf{r}_A + \mathbf{r}_{B/A}$$

The equation for the velocity is determined by taking the time derivative of this equation:

$$\mathbf{v}_B = \mathbf{v}_A + \frac{d\mathbf{r}_{B/A}}{dt}$$

Fig. 12–20

Differentiation of the last term is straightforward and yields the *relative velocity vector* $\mathbf{v}_{B/A}$, that is, the velocity of B with respect to A. To show this operation in a simplified way, the x', y', and z' coordinate axes are oriented such that they remain parallel at all times to the x, y, and z coordinate axes during the motion, Fig. 12–20. Hence, the $\mathbf{i}$, $\mathbf{j}$, and $\mathbf{k}$ unit vectors are *always* parallel to the respective $\mathbf{i}'$, $\mathbf{j}'$, and $\mathbf{k}'$ unit vectors. (This is *not* the case if the $x'y'z'$ coordinate system is rotating. Refer to Secs. 13–12 and 13–13.) Vector components in both reference systems will therefore *change only in magnitude* and not in direction during a given

time interval, so that $d\mathbf{r}_{B/A}/dt = \mathbf{v}_{B/A}$. In the more general case, the x', y', and z' axes may be oriented in any arbitrary direction, in which case the components of a vector in both reference frames will be different. However, again the components will experience only changes in magnitude since the axes will not change direction. The relative velocity equation may therefore be written as

$$\mathbf{v}_B = \mathbf{v}_A + \mathbf{v}_{B/A}$$

Since $\mathbf{v}_A$ and $\mathbf{v}_B$ are both measured by a *fixed* observer, these velocities are termed *absolute velocities*.

A second time derivative yields a similar vector relationship between the *absolute* and *relative accelerations* of particles A and B:

$$\mathbf{a}_B = \mathbf{a}_A + \mathbf{a}_{B/A}$$

The *origin* of the translating x', y', and z' axes may also be located at particle B. Provided this is the case, the relative motion equations for position, velocity, and acceleration become

$$\mathbf{r}_A = \mathbf{r}_B + \mathbf{r}_{A/B}$$
$$\mathbf{v}_A = \mathbf{v}_B + \mathbf{v}_{A/B}$$
$$\mathbf{a}_A = \mathbf{a}_B + \mathbf{a}_{A/B}$$

where $\mathbf{r}_{A/B}$ represents the relative position of particle A with respect to B, etc. Notice that $\mathbf{r}_{A/B} = -\mathbf{r}_{B/A}$, so that $\mathbf{v}_{A/B} = -\mathbf{v}_{B/A}$ and $\mathbf{a}_{A/B} = -\mathbf{a}_{B/A}$. It is necessary to use *vector addition* when applying these equations. For simplification in representing the motion along the paths, the vectors may be expressed by using a suitable type of reference: rectangular, cylindrical, etc. See Examples 12–12 and 12–14.

Case II: Dependent Relative Motion. This type of motion generally occurs if particles are interconnected. A common example consists of blocks attached to inextensible cords which are wrapped around pulleys. For instance, the movement of block A downward *along the inclined plane* in Fig. 12–21a will cause a corresponding movement of block B up to the inclined plane. To show this in another way, the inclined plane may be removed and the motions of both blocks restricted only to vertical displacements, Fig. 12–21b. The coordinates s_A and s_B (positive direction downward) are used to locate the position of blocks A and B, respectively, from the horizontal datum containing the fixed point E. These two vertical displacements represent the actual movements of the blocks along the inclined planes, *not* the vertical displacements of the blocks. For the pulley system in Fig. 12–21b, s_A and s_B are related by

$$s_A + s_B = l$$

(a)

(b)

Fig. 12–21

where l is a constant and represents the length of cord *except* the *constant sector* of cord *CD*. Knowing the displaced position s_A of block *A*, we can therefore find the distance s_B of block *B* using the equation

$$s_B = l - s_A$$

Taking the time derivative of this equation,

$$\frac{ds_B}{dt} = 0 - \frac{ds_A}{dt}$$

These two terms represent the magnitude of the velocity of each block, i.e.,

$$v_B = -v_A$$

The negative sign indicates that positive motion of one block (downward) causes a corresponding negative motion (upward) of the other block.

In a similar manner, time differentiation of the velocities yields the acceleration, namely

$$a_B = -a_A$$

A more complicated example involving the *dependent motions* of blocks and pulleys is shown in Fig. 12–22a. Since there are *two independent cables* in this system, the motions of blocks *A* and *B* may be related by first *separating* the system of the pulleys and the two cables, as shown in Figs. 12–22b and 12–22c, and then studying the motions of each cable. From Fig. 12–22b, the displaced positions of block *A* and the center of pulley *C* may be related by the equation

$$s_A + 2s_C = l_1 \tag{12–38}$$

This equation requires a cable length of $2s_C$ for pulley *C*, since the length of cable segment supporting this pulley is effectively $2s_C$. (As shown, the measurement is made with respect to a horizontal datum passing through the fixed point *D*.) l_1 represents the constant cable length *except* the constant circular sectors *EF* and *GH,* and sector *IJ*.

In a similar manner, the motion of pulley *C* may be related to the motion of block *B* from the same fixed datum, Fig. 12–22c. This requires,

$$s_B + (s_B - s_C) = l_2$$

or

$$2s_B - s_C = l_2 \tag{12–39}$$

Here l_2 represents the cable length excluding sectors *KL* and *MN*. If we eliminate the common pulley displacement s_C between Eqs. 12–38 and 12–39, the displacement equation for the two blocks becomes

$$s_A + 4s_B = 2l_2 + l_1$$

(a)

(b)

(c)

Fig. 12-22

Taking successive time derivatives, realizing that l_1 and l_2 are constants, we have,

$$v_A = -4v_B$$
$$a_A = -4a_B$$

Thus, motion of block A is four times the opposite motion of block B.

The following example problems further illustrate the use of this theory for solving relative motion problems.

Example 12-12

Water drips from a faucet at the rate of ten drops per second, as shown in Fig. 12-23. Determine the vertical separation between two consecutive drops after the lower drop has attained a velocity of 8 ft/sec.

Solution

The relative motion that occurs is *independent* rectilinear motion. If we denote the first and second drops of water as B and A, respectively, Fig. 12-23, the separation distance $s_{B/A}$ (displacement of B with respect to A) can be found by the equation

$$s_{B/A} = s_B - s_A \qquad (1)$$

Here the origin of the fixed frame of reference is located at the head of the faucet (datum) with positive displacement s, downward. The origin of the moving frame (not shown) is at drop A.

Fig. 12-23

489

The equations developed in Sec. 12–2 will be used for the solution, since each drop is subjected to rectilinear motion, having a *constant downward acceleration* of $a_c = g = 32.2$ ft/sec^2. Applying Eq. 12–5a with the initial condition $v_1 = 0$ (at $t = 0$), the velocity of each drop is

$$v = v_1 + a_c t$$
$$v = 32.2t$$

Thus, the time required for drop B to attain a velocity of 8 ft/sec is

$$t_B = \frac{8}{32.2} = 0.248 \text{ sec}$$

The displacement of each drop as a function of time can be found using Eq. 12–5b with the initial condition $s_1 = 0$, $v_1 = 0$.

$$s = s_1 + v_1 t + \tfrac{1}{2} a_c t^2$$
$$s = 16.1t^2 \tag{2}$$

Thus, the displacement of drop B in 0.248 sec is,

$$s_B = 16.1(0.248)^2 = 0.990 \text{ ft}$$

Since 10 drops fall per second (one drop each tenth second), drop A falls for a time $t_A = 0.248$ sec $- 0.1$ sec $= 0.148$ sec before drop B attains a velocity of 8 ft/sec. Using Eq. (2) we see that the displacement of drop A is,

$$s_A = 16.1(0.148)^2 = 0.352 \text{ ft}$$

Applying Eq. (1) yields

$$s_{B/A} = s_B - s_A = 0.990 \text{ ft} - 0.352 \text{ ft} = 0.638 \text{ ft} \qquad \textit{Ans.}$$

Example 12–13

The freight elevator shown in Fig. 12–24a is operated by an electric motor located at A. If the motor winds up its attached cable at a speed of 15 ft/sec, determine (a) the speed at which the elevator rises, and (b) the relative velocity at which the counterweight W moves with respect to the elevator.

Solution
Part (a). Because of the cable and pulley system, the motion is *dependent*. To determine the speed at which the elevator rises, an *equivalent pulley system* representing cable $ABCDE$ is shown in Fig. 12–24b. The coordinate s_A is used to locate a point *on the cable* which is moving upward with a velocity of 15 ft/sec, whereas, s_F determines the position of the elevator. Both coordinates are measured with respect to the datum which passes through the fixed point D. The length of cable, l—except the constant

(a)

(b)   (c)

Fig. 12-24

sections of cable which pass around the two pulleys and sector JI—may be related to the displacements s_A and s_F by the equation

$$2s_F + (s_F - s_A) = l$$

or

$$3s_F - s_A = l$$

Since l is a constant, taking the time derivative of this equation, we have

$$3v_F - v_A = 0$$

Therefore, the velocity of the elevator is

$$3v_F - (-15 \text{ ft/sec}) = 0$$
$$v_F = -5 \text{ ft/sec} \quad (\text{upward}) \qquad \textit{Ans.}$$

Part (b). An equivalent pulley system for cable GHW is shown in Fig. 12–24c. The length of cable—except for constant sectors of the cable which pass around the two pulleys and sector KL—may be related to the coordinates s_F and s_W by the equation

$$2s_F + s_W = l'$$

(By proper choice of the constant l', s_F here represents an equivalent elevator position to that shown in Fig. 12–24b. Actually this is of no consequence since time rates of change of the coordinates are needed for the solution.) Taking the time derivative, we have,

$$2v_F + v_W = 0$$

Thus,

$$2(-5 \text{ ft/sec}) + v_W = 0$$

or

$$v_W = 10 \text{ ft/sec} \quad (\text{downward})$$

To determine the velocity of the weight with respect to the elevator, $\mathbf{v}_{W/F}$, the relative velocity equation may be used in scalar form, since the paths of motion are parallel:

$$
\begin{aligned}
v_{W/F} &= v_W - v_F \\
&= 10 \text{ ft/sec} - (-5 \text{ ft/sec}) \\
&= 15 \text{ ft/sec} \qquad \textit{Ans.}
\end{aligned}
$$

Since the answer is positive, it appears to an observer standing in the elevator that the counterweight is moving downward or approaching at a speed of 15 ft/sec.

Example 12–14

A 900-ft train traveling on a bridge at a constant speed of 60 mph crosses over the intersection of the right-hand lane of a road, as shown in Fig. 12–25a. If an automobile A is 1,200 ft from the crossing at the time that the train reaches the crossing, determine (a) the constant speed v_A which the automobile must maintain in order to meet the crossing

$s_A = 1,200$ ft

v_A

$v_T = 60$ mph

E

45°

900 ft

(a)

$v_A = 80$ mph

$v_{E/A}$

45°

$v_E = 60$ mph

(b)

s'_A

y'

y

$s''_A = 600$ ft

x'

A

$s_{E/A}$

0 45°

x

$s_E = 450$ ft

E

45°

(c)

Fig. 12-25

just as the end of the train, E, crosses over the bridge, (b) the relative
velocity at which the driver of the automobile sees the end of the train

traveling, and (c) the distance from the automobile to the end of the last car of the train at the instant the train is halfway over the lane.

Solution

Part (a). The motion of the train and the automobile are *independent* of one another. Both vehicles, however, have rectilinear motion with constant velocity. For the solution it is first necessary to convert the speed of the train to the proper set of units.*

$$v_T = \left(60 \; \frac{\text{mi}}{\text{hr}}\right) \frac{1 \; \text{hr}}{3,600 \; \text{sec}} \; \frac{5,280 \; \text{ft}}{1 \; \text{mi}} = 88.0 \; \text{ft/sec}$$

The time it takes for the end of the train, E, to clear the crossing can be found by integrating Eq. 12–1, or using Eq. 12–5b with $s_1 = 0$, $a_c = 0$. It is assumed that at $t = 0$, the initial positions of the train and automobile are as shown in Fig. 12–25a. Using Eq. 12–5b, we have

$$s = v_1 t$$

For the train, $v_1 = 88$ ft/sec. When $s = 900$ ft, therefore

$$t = \frac{900 \; \text{ft}}{88 \; \text{ft/sec}} = 10.23 \; \text{sec}$$

During this time the automobile must travel 1,200 ft. Hence, the required constant speed of the automobile is determined from,

$$s_A = v_A t \tag{1}$$

or

$$v_A = \frac{1,200 \; \text{ft}}{10.23 \; \text{sec}} = 117.3 \; \text{ft/sec}$$

$$= (117.3 \; \text{ft/sec}) \frac{60 \; \text{mph}}{88 \; \text{ft/sec}}$$

$$= 80 \; \text{mph} \qquad\qquad\qquad Ans.$$

Part (b). We can find the relative velocity of the end of the train with respect to the automobile, $v_{E/A}$, by using the vector equation

$$\mathbf{v}_E = \mathbf{v}_A + \mathbf{v}_{E/A}$$

The vector addition is shown in Fig. 12–25b. Note that $v_E = v_T = 60$ mph. Using the law of cosines to determine the speed $v_{E/A}$, we have

$$v_{E/A} = \sqrt{(v_E)^2 + (v_A)^2 - 2(v_A)(v_E) \cos 45°}$$
$$= \sqrt{(60)^2 + (80)^2 - 2(60)80(0.707)} \; \text{mph}$$
$$= 56.7 \; \text{mph} \qquad\qquad\qquad Ans.$$

---

*It may be convenient to remember that 60 mph = 88 ft/sec when converting mph to fps.

Hence, as the motorist approaches the bridge, he sees the end of the train approaching at a speed of 56.7 mph. Once he has crossed under the bridge, the end of the train is seen to move away from him at this speed.

Part (c). When the train passes halfway over the center of the right lane, Fig. 12–25c, the distance s'_A the automobile has traveled can be determined by using Eq. (1):

$$s'_A = (117.2 \text{ ft/sec})\frac{10.23 \text{ sec}}{2} = 600 \text{ ft}$$

Hence, the automobile is $s''_A = s_A - s'_A = (1,200 \text{ ft} - 600 \text{ ft}) = 600$ ft from the bridge at this instant. With the origin of the fixed xy coordinate system at the crossing, and the origin of the translating $x'y'$ coordinate system at the automobile, Fig. 12–25c, the required distance $s_{E/A}$ can be found using the law of cosines applied to the vector triangle ($\mathbf{s}_E = \mathbf{s}''_A + \mathbf{s}_{E/A}$), as shown in the figure.

$$
\begin{aligned}
s_{E/A} &= \sqrt{(s''_A)^2 + (s_E)^2 - 2(s''_A)(s_E) \cos 45°} \\
&= \sqrt{(600)^2 + (450)^2 - 2(600)450(0.707)} \text{ ft} \\
&= 425 \text{ ft} \hspace{4cm} \textit{Ans.}
\end{aligned}
$$

Problems

12-69. Two billiard balls A and B are moving at constant velocity on the surface of a pool table along the paths shown. If the balls are traveling with velocities of $v_A = 2$ ft/sec and $v_B = 3.5$ ft/sec, determine the relative speed of ball A with respect to ball B at the instant shown, and the distance between the two balls when $t = 1$ sec.

Prob. 12-69

12-70. The two blocks A and B rest on a smooth inclined plane. If block B moves down the plane at a speed of 12 ft/sec, determine the speed at which block A moves up the inclined plane.

Prob. 12-70

12-71. Determine the speed at which the block B is raised if the end of the cable at C is pulled downward with a speed of 10 ft/sec. What is the relative velocity of the cable with respect to block B?

Prob. 12-71

12-72. Two planes are flying at the same altitude. If the velocities of each plane are as shown in the figure and the pilot in plane A sees plane B approaching him at a relative speed of 120 mph, determine the angle α made between their straight-line courses.

Prob. 12-72

12-73. The small mine car is being lifted up the inclined plane using the motor M and the rope and pulley arrangement shown. Determine the speed at which the cable must be taken up by the motor in order to move the mine car up the plane with a constant speed of 4 ft/sec.

Prob. 12-73

12-74. Three blocks A, B, and C move in a straight line with constant velocities. If the relative velocity of block A with respect to block B is 10 ft/sec (moving to the right), and the relative velocity of block B with respect to block C is observed to be -5 ft/sec (moving to the left), determine the absolute velocities of blocks A and B. The velocity of block C is 6 ft/sec to the right.

12-75. Sand falls vertically from a bin onto a chute, as shown. If the sand is moving at a speed of $v_C = 6$ ft/sec down the chute, determine the relative speed of the sand falling on the chute to the sand sliding down the chute. Perform the calculation at a time just before the sand hits the chute.

496

Prob. 12-75

12-76. A toy car C is traveling along a straight path toward A with a constant velocity of $v_C = 2.5$ ft/sec. If

Prob. 12-76

a second car B, starting from rest, is directed toward point A, determine the constant acceleration it must have in order to cause a collision when it reaches point A. What is the relative velocity and acceleration of car B with respect to car C when the collision occurs?

12-77. If blocks A, B, and C move downward with a velocity of 1 ft/sec, 2 ft/sec, and 3 ft/sec, respectively, at the instant shown, determine the velocity of block D and the relative velocity of block A with respect to block D.

Prob. 12-77

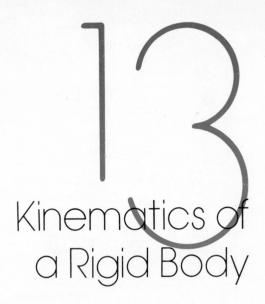

13

Kinematics of a Rigid Body

13-1. Types of Rigid-Body Motion

A rigid body is considered as a combination of a large number of particles in which all the particles remain at a fixed distance from one another. In this chapter we shall discuss the kinematics, or the geometry of motion, for a rigid body. Once you thoroughly understand these principles we will then be able to apply the equations of motion which relate the forces acting on the body to the motion of the body.

In general, a rigid body may be subjected to three types of *displacement:* translation, rotation about a fixed axis, and rotation about a fixed point. *Translation* occurs when any straight line on the body remains parallel to its original direction during the motion. When the paths of motion for all the particles of the body are along parallel straight lines, Fig. 13–1a, the motion is called *rectilinear translation*. When the path of motion is along curved paths which are all parallel, Fig. 13–1b, the translation is said to be *curvilinear translation*. When a rigid body *rotates about a fixed axis,* all the particles of the rigid body, except those which lie on the axis of rotation, have a velocity and acceleration. The path taken by each of the moving particles is that of a circle, Fig. 13–1c. Rigid body *rotation about a fixed point O* requires that the distance r from the fixed point to any particle P of the body remain invariant. Hence, the path of motion for the particle P lies on the surface of a sphere having a radius r and centered at the fixed point. Motion about a fixed point applies to spinning tops and gyroscopes, Fig. 13–1d.

During an infinitesimal period of time (hereafter referred to as an instant of time) dt, the general motion of a rigid body can be considered

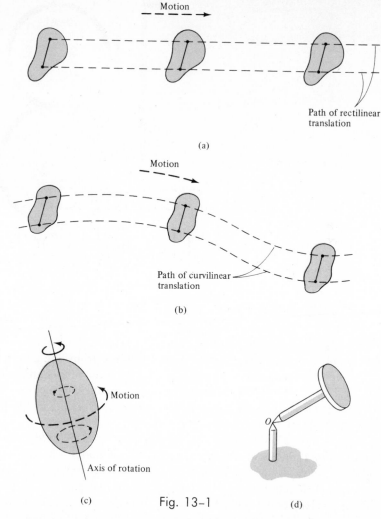

Motion

Path of rectilinear translation

(a)

Motion

Path of curvilinear translation

(b)

Motion

Axis of rotation

(c)

O

(d)

Fig. 13–1

as a combination of two of the three types of motion just discussed. The first part of the motion consists of a *pure translation* of an arbitrary *base point* of the body, and the second part represents a *pure rotation* of the body about that point. As an example, consider the general motion of the body shown in Fig. 13–2a during the time interval dt.* Motion is made in reference to the arbitrary base point A located in the body. To obtain the final position at time $t + dt$, the *entire body* is first *translated* by an amount $d\mathbf{s}_A$, Fig. 13–2b, which brings point A to its *final position;* the body is then *rotated about point A* by an amount $d\boldsymbol{\theta}$ until the body coincides with its final position, Fig. 13–2c. Superimposing both of these

*The body shown in Fig. 13–2 is an arbitrarily shaped object which, in reality, can represent a stone, hammer, automobile, etc.

(a)   (b)   (c)

Fig. 13-2

effects, we obtain the required infinitesimal displacement of the body. The base point A, which is located on the axis of rotation, undergoes only translation. All the points which do not lie on this axis of rotation experience a displacement caused by the angular displacement $d\boldsymbol{\theta}$ in addition to the translational displacement $d\mathbf{s}_A$.

In this discussion, reference has been made only to infinitesimal translations and rotations, that is, those that occur during an *instant* of time dt. Both infinitesimal translations and infinitesimal rotations are vectors; therefore, when their effects are added vectorially, they yield the correct

(a)

(b)

Fig. 13-3

displacement vector for any point in the body. Had the motions been formulated during a *finite* time interval, it would have been necessary to superimpose finite translations and finite rotations. Although a finite translation is a vector, a *finite rotation is not a vector*. To show this, consider the three finite rotations given to the block shown in Fig. 13–3*a*. Each rotation has a specified magnitude and direction as shown by the vector arrow. The resultant orientation of the block is shown at the right. When these *same* three rotations are *added* in *another manner*, Fig. 13–3*b*, the resultant direction of the block is *not* the same as it was in Fig. 13–3*a*. Therefore, since finite rotations are *not commutative*, they cannot be classified as vectors. A proof that *infinitesimal rotations are vectors* is given in Appendix E.

When each of the particles of a rigid body move on a path equidistant to a fixed plane, the rigid body is said to undergo *plane motion*. The entire motion of the body can then be represented by the motion of points of the body which lie in the same plane. Thus, the case of *general plane motion* can be thought of as a translation which occurs in the plane *and* a rotation about an axis perpendicular to the plane. In particular, translation can be represented as plane motion having no rotation, and rotation about a fixed axis is plane motion without translation.* In the case of *general motion* (spatial motion) of a rigid body, each point of the body moves along a path which is not contained in a plane; hence, all points in the body will have different velocities and accelerations at a given instant of time.

In this chapter all five types of rigid-body motion will be considered. The analysis of these motions will be presented in order of increasing complexity:

1. Translation
2. Rotation about a fixed axis
3. General plane motion
4. Rotation about a fixed point
5. General rigid body motion

It will be shown that for *any* of these types of motion, provided the motions of any two points of the body are known, the motion of any other point in the body can be determined. In the following discussion, the kinematic analysis is based on the use of coordinate systems which *translate* relative to one another. In the last two sections of this chapter however, we will consider a more general study of this analysis by making use of coordinate systems which *both* translate and rotate.

---

*Consider the crankshaft *AB*, connecting rod *BC*, and the piston at *C* shown in Fig. 13–27*a*. All three parts undergo plane motion. The piston is subjected to translation, the crankshaft undergoes rotation about a fixed axis, and the connecting rod has general plane motion.

Consider a rigid body which is subjected to either rectilinear or curvilinear translation, Fig. 13-4. The location of points A and B in the rigid body is defined relative to the fixed $x\,y\,z$ reference frame by using position vectors $\mathbf{r}_A$ and $\mathbf{r}_B$. The $x'\,y'\,z'$ coordinate system is *fixed in the body,* and therefore its origin translates with the base point A. The position of point B with respect to point A is denoted by the relative position vector $\mathbf{r}_{B/A}$. This vector is measured from the origin of the translating coordinate system. The three vectors shown in the figure are related by the equation

$$\mathbf{r}_B = \mathbf{r}_A + \mathbf{r}_{B/A}$$

Taking the time derivative, we have

$$\mathbf{v}_B = \mathbf{v}_A + \frac{d\mathbf{r}_{B/A}}{dt}$$

where $\mathbf{v}_A$ and $\mathbf{v}_B$ denote the absolute velocity of points A and B, respectively, since these vectors are measured from the x, y, and z coordinate axes. The *magnitude* of $\mathbf{r}_{B/A}$ is constant by definition of a rigid body. Also, from the definition of rigid-body translation, the *direction* of $\mathbf{r}_{B/A}$ does not change with time. This is because at all times during the motion the axes of both coordinate systems remain *parallel* to one another, and therefore the components of $\mathbf{r}_{B/A}$ remain the *same* in both coordinate systems. Hence, $d\mathbf{r}_{B/A}/dt = 0$, and therefore,

$$\mathbf{v}_B = \mathbf{v}_A$$

Taking the time derivative of this equation yields

$$\mathbf{a}_B = \mathbf{a}_A$$

Fig. 13-4

These two equations indicate that *all points* in a rigid body subjected to either curvilinear or rectilinear *translation* move with the *same velocity and acceleration*. Since all points of the body have the same motion, the kinematics of particle motion discussed in Chapter 12 may be applied to determine the kinematics of rigid bodies in translation.

13–3. Angular Velocity and Angular Acceleration of a Line

Like a rigid body, the general displacement of a line during an *instant* of time may be represented first by a translation of the line to some point and then a rotation of the line about this point until the line coincides with its final position. Let us consider these separate motions when the line segment *ABC*, Fig. 13–5a, is moved to position *A'B'C'* during the time interval *dt*. In Fig. 13–5b, the line is first translated from the base point *B* by an infinitesimal amount $d\mathbf{s}_B$ so that point *B* moves to its final position *B'*. Then the line undergoes an infinitesimal rotation *dθ* until it coincides with its final position *A'B'C'*. Figure 13–5c shows the translational and rotational movements relative to base point *A*. During the translation, in both cases (Figs. 13–5b and 13–5c), the vectors $d\mathbf{s}_B$ and $d\mathbf{s}_A$ act in the *same direction* but have *different magnitudes*. Since the dashed lines shown in Figs. 13–5b and 13–5c are parallel to one another, the amount of angular displacement *dθ* is the same for both cases. From this discussion we may conclude that the *magnitude* of translation is *different* for every base point chosen on the line, yet the amount of *rotation* of the line is the *same* about the selected base point.

Since a line segment may be thought of as a succession of particles of a rigid body, the kinematics for translation of line segments has essentially been discussed in the previous section regarding rigid-body translation. Hence, we will be concerned only with the rotational motion of the line. Consider the case where the line *ABC* moves in the plane of the page when it is rotated from *ABC* to *A'B'C'*, Figs. 13–5a and 13–5b. As shown in Appendix E, *angular displacement dθ* is a *vector* quantity.

(a)

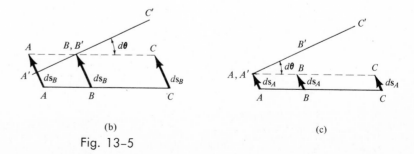

(b)   (c)

Fig. 13–5

The *magnitude* of this vector, $d\boldsymbol{\theta}$, is commonly measured in radians (or revolutions, where 1 rev = 2π rad). The *directional sense* is determined by using the right-hand rule. With the fingers of the right-hand curled from ABC to $A'B'C'$, Fig. 13–5, the thumb points in the directional sense of $d\boldsymbol{\theta}$. Hence, $d\boldsymbol{\theta}$ is *perpendicular* to the plane containing ABC and $A'B'C'$ such that the vector arrowhead is directed out of this plane. Since the rotation is the *same* about every point on the line, $d\boldsymbol{\theta}$ is a *free vector*.

The time rate of change of angular displacement is called the *angular velocity*. Thus,

$$\boldsymbol{\omega} = \frac{d\boldsymbol{\theta}}{dt}$$

This vector has the same direction and sense as the angular displacement $d\boldsymbol{\theta}$, that is, perpendicular to the plane of rotation. The magnitude of $\boldsymbol{\omega}$ is

$$\omega = \frac{d\theta}{dt} \qquad (13\text{--}1)$$

Common units for measurement are rad/sec.

In a similar manner, the *angular acceleration* $\boldsymbol{\alpha}$ is the time rate of change of the angular velocity vector, i.e.,

$$\boldsymbol{\alpha} = \frac{d\boldsymbol{\omega}}{dt} = \frac{d^2\boldsymbol{\theta}}{dt^2}$$

The magnitude of $\boldsymbol{\alpha}$ is therefore

$$\alpha = \frac{d\omega}{dt} = \frac{d^2\theta}{dt^2} \qquad (13\text{--}2)$$

This term is commonly measured in units of rad/sec². The line of direction of $\boldsymbol{\alpha}$ is the same as that for $\boldsymbol{\omega}$; however, its *sense* depends upon whether $\boldsymbol{\omega}$ is increasing or decreasing with time. In particular, if $\boldsymbol{\omega}$ is decreasing $\boldsymbol{\alpha}$ is called an *angular deceleration*. Since $d\boldsymbol{\theta}$ is a free vector, $\boldsymbol{\omega}$ and $\boldsymbol{\alpha}$ are also free vectors. You may have already recognized that Eqs. 13–1 and 13–2 have been used in Sec. 12–8 when we discussed the curvilinear motion of a particle, using cylindrical components.

By eliminating the time differential from Eqs. 13–1 and 13–2, it is possible to obtain a relation between the angular acceleration, angular velocity, and the angular displacement:

$$dt = \frac{d\theta}{\omega} = \frac{d\omega}{\alpha}$$

or

$$\alpha \, d\theta = \omega \, d\omega \qquad (13\text{--}3)$$

The similarity between the differential relations developed here and those developed for rectilinear motion, Sec. 12–2, should become apparent. If a rigid body is subjected to *plane motion,* these equations may be integrated to obtain a set of equations which describe the angular motion of a point located on the rigid body. For example, if the *angular acceleration is constant,* $\alpha = \alpha_c$, the above differential equations, when integrated, yield a set of formulas which relate the angular velocity, angular displacement, and time, which are similar to Eqs. 12–5. The results are,

$$\omega = \omega_1 + \alpha_c t$$

$$\theta = \theta_1 + \omega_1 t + \tfrac{1}{2}\alpha_c t^2 \qquad (13\text{–}4)$$

$$\omega^2 = \omega_1^2 + 2\alpha_c(\theta - \theta_1)$$

where θ_1 and ω_1 are the initial values of angular displacement and angular velocity, respectively. It is important to remember that *these equations are useful only when the angular acceleration is either constant or zero.*

13–4. Rotation of a Rigid Body About a Fixed Axis

When a rigid body rotates about a fixed axis, all points of the body not lying on the axis of rotation travel in circular paths having centers on the axis of rotation. Let us consider the rigid body in Fig. 13–6a which is rotating about the fixed z axis. An arbitrary point $P(r, \theta, z)$, located either on the surface or within the body, travels along the dotted circular path having its center at point A. As shown in the figure, the origin of

(a)

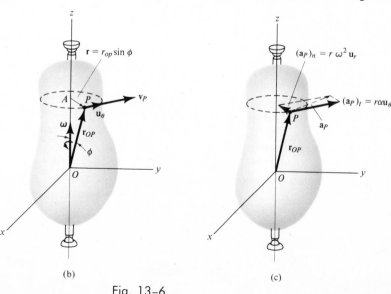

(b)   (c)

Fig. 13–6

the fixed $x\,y\,z$ coordinate system is located at the *arbitrarily chosen point* O on the z axis of rotation. The axes form a right-hand coordinate system, in which the fingers curl in the direction of rotation and the thumb points along the positive z axis. With the coordinate axes so established, the positive direction of the angular velocity $\boldsymbol{\omega}$ is along the positive z axis in accordance with the right-hand rule. The angular-acceleration vector $\boldsymbol{\alpha}$ acts in the positive z direction when ω is increasing and in the opposite sense when ω is decreasing.

The velocity and acceleration of point P may be determined in terms of the angular motion of the body. Using the cylindrical coordinate unit vectors $\mathbf{u}_r$, $\mathbf{u}_\theta$, and $\mathbf{u}_z$, Fig. 13–6a, the position vector which locates point P can be written as

$$\mathbf{r}_{OP} = r\mathbf{u}_r + z\mathbf{u}_z$$

The velocity is determined by taking the time derivative of this vector. Using the chain rule for vector differentiation, Eq. 12–11, we have,

$$\mathbf{v}_P = \frac{d\mathbf{r}_{OP}}{dt} = \frac{dr}{dt}\mathbf{u}_r + r\frac{d\mathbf{u}_r}{dt} + \frac{dz}{dt}\mathbf{u}_z + z\frac{d\mathbf{u}_z}{dt}$$

Since r, z, and $\mathbf{u}_z$ do not change with time as P moves along the dotted path, this equation reduces to

$$\mathbf{v}_P = r\frac{d\mathbf{u}_r}{dt}$$

Using Eq. 13–1 and Eq. 12–28, i.e., $d\mathbf{u}_r/dt = (d\theta/dt)\mathbf{u}_\theta$, yields

$$\mathbf{v}_P = r\frac{d\theta}{dt}\mathbf{u}_\theta = r\omega\mathbf{u}_\theta \tag{13–5}$$

The *magnitude* of $\mathbf{v}_P$ is therefore

$$v_P = r\omega \tag{13–6}$$

The *direction* of the velocity is tangent to the circular path and has the same sense as $\mathbf{u}_\theta$, Fig. 13–6b. Since $r = r_{OP}\sin\phi$, Eq. 13–5 may be rewritten in the form

$$\mathbf{v}_P = (r_{OP}\sin\phi)\omega\mathbf{u}_\theta = r_{OP}\omega\sin\phi\,\mathbf{u}_\theta$$

This equation defines the cross product* between $\boldsymbol{\omega}$ and $\mathbf{r}_{OP}$, i.e.,

---

*By definition of the cross-product, $\mathbf{C} = \mathbf{A} \times \mathbf{B}$, the resultant vector $\mathbf{C}$ has a *magnitude* of $C = AB\sin\theta$, where θ is the angle made between the tails of both vectors; and a *direction* defined by the right-hand rule, curling the fingers of the right hand from $\mathbf{A}$ to $\mathbf{B}$ ("A cross B"), the thumb points in the direction of $\mathbf{C}$. For example, $\mathbf{i} \times \mathbf{j} = \mathbf{k}$ since the magnitude $k = (1)(1)\sin 90° = 1$ and the direction "i cross j" is $\mathbf{k}$. Similarly, $\mathbf{k} \times \mathbf{i} = \mathbf{j}$, etc. If $\mathbf{A}$ and $\mathbf{B}$ are expressed in Cartesian component form, the resultant vector $\mathbf{C}$ is determined by evaluating the determinant

$$\mathbf{C} = \mathbf{A} \times \mathbf{B} = \begin{vmatrix} \mathbf{i} & \mathbf{j} & \mathbf{k} \\ A_x & A_y & A_z \\ B_x & B_y & B_z \end{vmatrix}$$

$$\mathbf{v}_P = \boldsymbol{\omega} \times \mathbf{r}_{OP} \qquad (13\text{–}7)$$

Here $\mathbf{r}_{OP}$ is a vector drawn from *any point O* on the axis of rotation *to* point P. Notice how the correct direction of $\mathbf{v}_P$ is established by using the right-hand rule. The fingers of the right hand are curled from $\boldsymbol{\omega}$ to $\mathbf{r}_{OP}$. The thumb indicates the correct directional sense of $\mathbf{v}_P$, Fig. 13-6*b*.

The acceleration of P is determined from the time derivative of Eq. 13-5. Again by the chain rule, we have,

$$\mathbf{a}_P = \frac{dr}{dt}\omega\mathbf{u}_\theta + r\frac{d\omega}{dt}\mathbf{u}_\theta + r\omega\frac{d\mathbf{u}_\theta}{dt}$$

Since r is constant, the first term on the right-hand side is zero. Hence,

$$\mathbf{a}_P = r\frac{d\omega}{dt}\mathbf{u}_\theta + r\omega\frac{d\mathbf{u}_\theta}{dt}$$

Using Eq. 13-2 and Eq. 12-31, i.e., $d\mathbf{u}_\theta/dt = -(d\theta/dt)\mathbf{u}_r$, yields

$$\mathbf{a}_P = (\mathbf{a}_P)_t + (\mathbf{a}_P)_n = r\alpha\mathbf{u}_\theta - r\omega^2\mathbf{u}_r$$

As shown in Fig. 13-6*c*, these two terms represent the *tangential and normal components of the acceleration*. The *tangential component* of acceleration has a *magnitude* of

$$(a_P)_t = r\alpha \qquad (13\text{–}8)$$

As in the case of motion of a particle moving along a curved path—see Sec. 12-9—this component represents the *change* made in the *magnitude* of the *velocity* of P. If the speed of P is *increasing*, then $(\mathbf{a}_P)_t$ acts in the *same direction* as $\mathbf{v}_P$. On the other hand, if the speed is *decreasing*, $(\mathbf{a}_P)_t$ acts in the *opposite direction* of $\mathbf{v}_P$. The normal component of acceleration, $(\mathbf{a}_P)_n$, represents the *change* made in the *direction* of the *velocity* vector during the instant of time. The *magnitude* of this component is

$$(a_P)_n = r\omega^2 = \frac{v_P^2}{r} \qquad (13\text{–}9)$$

This component is always directed toward the *center* of the circular path.

Like the velocity, the acceleration of point P may be expressed in terms of the vector cross-product. Taking the derivative of Eq. 13-7 we obtain,

$$\mathbf{a}_P = \frac{d\mathbf{v}_P}{dt} = \frac{d\boldsymbol{\omega}}{dt} \times \mathbf{r}_{OP} + \boldsymbol{\omega} \times \frac{d\mathbf{r}_{OP}}{dt}$$

Recalling that $\boldsymbol{\alpha} = d\boldsymbol{\omega}/dt$, and using Eq. 13-7 yields

(a)

$v_A = 1$ ft/sec

(b)

$(a_A)_n = 1$ ft/sec^2

$(a_A)_t = 2$ ft/sec^2

ω, α

a_A

(c)

Fig. 13-7

$$\mathbf{a}_P = (\mathbf{a}_P)_t + (\mathbf{a}_P)_n = \boldsymbol{\alpha} \times \mathbf{r}_{OP} + \boldsymbol{\omega} \times (\boldsymbol{\omega} \times \mathbf{r}_{OP}) \qquad (13\text{-}10)$$

Again, the position vector $\mathbf{r}_{OP}$ extends from an *arbitrary point O*, located *on* the axis of rotation, to point P. The reader should verify the direction of each of the two component accelerations in Eq. 13-10, using the definition of the cross product for the vectors $\mathbf{r}_{OP}$, $\boldsymbol{\omega}$, and $\boldsymbol{\alpha}$ shown in Fig. 13-6a. Also, from the cross-product definition, the magnitude of the acceleration components are defined by Eqs. 13-8 and 13-9.

Example 13-1

A cord is wrapped around a wheel which is initially at rest as shown in Fig. 13-7a. If a force **F** is applied to the cord and gives it an acceleration of $a = 4t$ ft/sec^2, where t is in seconds, determine (a) the angular velocity of the wheel at $t = 1$ sec, (b) the magnitude of the velocity and acceleration of point A at $t = 1$ sec, and (c) the number of turns the wheel makes during the first second.

Solution

Part (a). The wheel is subjected to rotation about a fixed axis passing through point O. Thus, point P on the wheel is subjected to motion about a curved path, and therefore the acceleration of this point can be represented by both tangential and normal components. In particular, the tangential component of acceleration is equivalent to the acceleration of the cord since the cord is connected to the wheel and tangent to it at point P. Hence, using Eq. 13-8, we have

$$(a_P)_t = r\alpha$$

or

$$4t \text{ ft/sec}^2 = (2 \text{ ft})\alpha$$
$$\alpha = 2t \text{ rad/sec}^2 \qquad (1)$$

509

We can find the angular velocity of the wheel by using this result and integrating Eq. 13–2 with the initial condition that $\omega_1 = 0$ at $t_1 = 0$. (Why can't we use Eqs. 13–4 to obtain our results?) Thus,

$$\alpha = \frac{d\omega}{dt} = 2t$$

$$\int_0^\omega d\omega = \int_0^t 2t \; dt$$

$$\omega = t^2 \, \text{rad/sec} \tag{2}$$

In this formula, when t is measured in seconds, ω is given in rad/sec. Thus, for $t = 1$ sec,

$$\omega = (1)^2 = 1 \; \text{rad/sec} \qquad \rangle \omega \qquad\qquad Ans.$$

Part (b). We can determine the velocity of point A at $t = 1$ sec by using Eq. 13–6 with $\omega = 1$ rad/sec. Hence,

$$v_A = r\omega$$

$$v_A = (1 \text{ ft}) \, (1 \text{ rad/sec}) = 1 \text{ ft/sec} \qquad\qquad Ans.$$

The velocity vector is shown in Fig. 13–7b. It is, of course, *tangent to the path of motion.*

The acceleration of point A at $t = 1$ sec has both normal and tangential components. Why? These two components may be obtained using Eq. 13–8, Eq. (1), and Eq. 13–9:

$$(a_A)_t = r\alpha = (1 \text{ ft})[2(1)] \text{ rad/sec}^2 = 2 \text{ ft/sec}^2$$

and

$$(a_A)_n = r\omega^2 = (1 \text{ ft})(1 \text{ rad/sec})^2 = 1 \text{ ft/sec}^2$$

These vector components are shown in Fig. 13–7c. In particular, note that $\mathbf{a}_A$ is *not* tangent to the path. The magnitude of the acceleration of point A is thus

$$a_A = \sqrt{(a_A)_t^2 + (a_A)_n^2}$$

$$a_A = \sqrt{(2 \text{ ft/sec}^2)^2 + (1 \text{ ft/sec}^2)^2} = 2.24 \text{ ft/sec}^2 \qquad\qquad Ans.$$

Part (c). The number of turns the wheel makes in 1 sec can be computed using Eq. 13–1, since this equation relates the magnitude of the angular velocity ω to the rotation θ and time t. (Why can't we use Eqs. 13–4?) The angular velocity is related to time by Eq. (2), thus

$$\frac{d\theta}{dt} = \omega = t^2$$

or

$$d\theta = t^2 \; dt$$

Using the initial condition $\theta = 0°$ at $t = 0$, we have

$$\int_{0°}^{\theta} d\theta = \int_{0}^{t} t^2 \, dt$$

or

$$\theta = \tfrac{1}{3}t^3 \text{ rad}$$

When t is measured in seconds, θ is given in radians. Hence, for $t = 1$ sec,

$$\theta = \tfrac{1}{3}(1)^3 = \tfrac{1}{3} \text{ rad}$$

Since there are 2π radians to one revolution,

$$\theta = \tfrac{1}{3} \text{ rad} \left(\frac{1 \text{ rev}}{2\pi \text{ rad}}\right) = 0.053 \text{ rev} \qquad \textit{Ans.}$$

Example 13–2

The bent rod shown in Fig. 13–8 rotates about the fixed axis AB with an angular deceleration of $\alpha = 2$ rad/sec^2. Determine the velocity and acceleration of point C at the instant shown, when $\omega = 10$ rad/sec. Express the answer in vector form with respect to the fixed $x\,y\,z$ coordinate system.

Solution

Vector Eqs. 13–7 and 13–10 will be used for the solution. It is first necessary to express ω and α in Cartesian vector form. Since both of these vectors lie along the fixed axis AB, their line of action can be determined using the unit vector $\mathbf{u}_{AB}$, Fig. 13–8.

$$\mathbf{u}_{AB} = \frac{\mathbf{r}_{AB}}{r_{AB}} = \frac{2\mathbf{i} + 2\mathbf{j} - 2\mathbf{k}}{\sqrt{(2)^2 + (2)^2 + (-2)^2}} = 0.577\mathbf{i} + 0.577\mathbf{j} - 0.577\mathbf{k}$$

Since $\omega = 10$ rad/sec, $\alpha = 2$ rad/sec^2, and α acts in the direction $-\mathbf{u}_{AB}$,

$$\omega = (10 \text{ rad/sec})\mathbf{u}_{AB} = \{5.77\mathbf{i} + 5.77\mathbf{j} - 5.77\mathbf{k}\} \text{ rad/sec}$$
$$\alpha = (2 \text{ rad/sec}^2)(-\mathbf{u}_{AB}) = \{-1.15\mathbf{i} - 1.15\mathbf{j} + 1.15\mathbf{k}\} \text{ rad/sec}^2$$

The velocity of point C is determined by using Eq. 13–7. Suitable position vectors in this equation are represented by $\mathbf{r}_{AC} = \{1\mathbf{j}\}$ ft or $\mathbf{r}_{BC} = \{-2\mathbf{i} - 1\mathbf{j} + 2\mathbf{k}\}$ ft, Fig. 13–8. Note that *both* of these vectors are drawn from a point which lies on the axis of rotation to point C. Hence,

$$\mathbf{v}_C = \omega \times \mathbf{r}_{AC} = \begin{vmatrix} \mathbf{i} & \mathbf{j} & \mathbf{k} \\ 5.77 & 5.77 & -5.77 \\ 0 & 1 & 0 \end{vmatrix} = \{5.77\mathbf{i} + 5.77\mathbf{k}\} \text{ ft/sec} \quad \textit{Ans.}$$

or

$$\mathbf{v}_C = \omega \times \mathbf{r}_{BC} = \begin{vmatrix} \mathbf{i} & \mathbf{j} & \mathbf{k} \\ 5.77 & 5.77 & -5.77 \\ -2 & -1 & 2 \end{vmatrix} = \{5.77\mathbf{i} + 5.77\mathbf{k}\} \text{ ft/sec} \quad \textit{Ans.}$$

Fig. 13–8

We obtain the acceleration by using Eq. 13–10. Again, the vector $\mathbf{r}_{OP}$ used in this equation must be a vector drawn from any point lying on the axis of rotation to point C, e.g., $\mathbf{r}_{AC}$ or $\mathbf{r}_{BC}$. Using vector $\mathbf{r}_{AC}$, we have

$$\mathbf{a}_C = \boldsymbol{\alpha} \times \mathbf{r}_{AC} + \boldsymbol{\omega} \times (\boldsymbol{\omega} \times \mathbf{r}_{AC})$$

$$\mathbf{a}_C = \begin{vmatrix} \mathbf{i} & \mathbf{j} & \mathbf{k} \\ -1.15 & -1.15 & 1.15 \\ 0 & 1 & 0 \end{vmatrix} + (5.77\mathbf{i} + 5.77\mathbf{j} - 5.77\mathbf{k})$$

or

$$\mathbf{a}_C = -1.15\mathbf{i} - 1.15\mathbf{k} + \begin{vmatrix} \mathbf{i} & \mathbf{j} & \mathbf{k} \\ 5.77 & 5.77 & -5.77 \\ 5.77 & 0 & 5.77 \end{vmatrix} \qquad \times \begin{vmatrix} \mathbf{i} & \mathbf{j} & \mathbf{k} \\ 5.77 & 5.77 & -5.77 \\ 0 & 1 & 0 \end{vmatrix}$$

Evaluating the last determinant and combining the respective $\mathbf{i}$, $\mathbf{j}$, and $\mathbf{k}$ components yields

$$\mathbf{a}_C = \{32.1\mathbf{i} - 66.6\mathbf{j} - 34.4\mathbf{k}\} \text{ ft/sec}^2 \qquad \textit{Ans.}$$

Problems

13–1. A wheel has an initial angular velocity of 10 rev/sec and a constant acceleration of 3 rad/sec². Determine the number of turns required for the wheel to acquire an angular speed of 20 rev/sec. What time is required?

13–2. The link is pinned at O and moves because of the action of the rod R. If the link starts from rest when $\theta = 0°$ and the rod R has a constant acceleration of 2 ft/sec² to the right, determine the angular velocity and angular acceleration of the link after $t = 1$ sec.

13–3. The spin drier of a washing machine has a constant angular acceleration of 2 rev/sec², starting from rest. Determine how many turns it makes in (a) 10 sec, and (b) from the fifth to the sixth second.

13–4. The hoisting gear A has an initial angular velocity of 60 rad/sec and a deceleration of 1 rad/sec². Determine the velocity and acceleration of the weight W which is being hoisted by gear B after $t = 3$ sec.

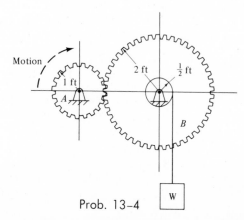

Prob. 13–4

13–5. Work Prob. 13–4 assuming that the angular deceleration of gear A is defined by the relation $\alpha = e^{-t}$, when t is given in seconds and α in rad/sec².

Prob. 13–2

13-6. The motor M is used via the gear arrangement to lift the weight W. If the motor shaft is turning at a constant rate of 6 rad/sec, determine the velocity and acceleration of the weight.

Prob. 13-6

Prob. 13-7

13-8. The torsional pendulum (disk) undergoes oscillations in the horizontal plane, such that the angle of rotation, measured from the equilibrium position, is given by $\theta = \frac{1}{2}\sin 3t$, where θ is measured in radians and t is in seconds. Determine the maximum velocity of point A located at the periphery of the disk while the pendulum is oscillating. What is the acceleration of point A in terms of t?

Prob. 13-8

13-7. The disk is partially submerged in a heavy liquid and is driven by a motor in the counterclockwise direction. It is found that the angular motion of the disk can be described by the relation $\theta = t + 4t^2$, where θ is in radians and t is measured in minutes. Determine the number of revolutions, the angular velocity, and angular acceleration of the disk when $t = 90$ sec.

13-9. The rod is bent into the shape of a sine curve and is rotated about the y axis by connecting the spindle S to a drive motor M. If the rod starts in the position shown, and a motor drives it with an angular acceleration $\alpha = 0.01\,e^t$, where t is measured in seconds and α is given in rad/sec^2, determine the magnitude of the angular velocity and the angular displacement of the rod when $t = 3$ sec. Locate the point on the rod which

has the greatest velocity and acceleration. What is the magnitude of the velocity and acceleration of this point when $t = 3$ sec?

Prob. 13–9

13–10. The three gears are meshed together. If gear A is rotating clockwise with an angular velocity of $\omega_A = 3$ rad/sec, determine the angular velocity of gears B and C. Gear B is one unit having radii of 2 and 5 in.

Prob. 13–10

13–11. The board rests on the surface of two drums. At the instant shown, it has an acceleration of 1 ft/sec², while points located on the periphery of the drums have a constant acceleration of 3 ft/sec². If the board does not slip on the surface of the drums, compute the velocity of the board.

Prob. 13–11

13–12. The link l moves from rest with an acceleration of 20 ft/sec². If it is attached to an inextensible cord which is wound around the disk shown in the figure, determine the angular acceleration of the disk as the link moves and the angular velocity after the disk has completed 10 revolutions. How many revolutions will the disk turn after it has first completed 10 revolutions and the link continues to move for 4 sec longer?

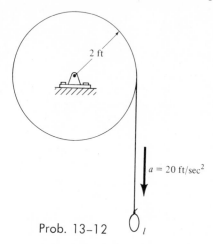

Prob. 13–12

13–13. The drive wheel A of the tape recorder has a constant angular velocity of 10 rad/sec. At a particular instant, the radius of tape wound on each wheel is as shown. If the tape has a thickness of 1.5×10^{-6} in., determine the angular acceleration of wheel B.

Prob. 13–13

514

13-5. General Plane Motion of a Rigid Body: Velocity

It was pointed out in Sec. 13-1 that the general motion of a rigid body during an instant of time may be considered in two steps: the *translation* of an arbitrary base point or reference point in the body to its final position, and then a *rotation* of the body about this base point, so that the body coincides with its final position. In order to conveniently analyze these two separate motions, two coordinate systems will be used, Fig. 13-9. The first coordinate system $x\,y\,z$ is *fixed*, and therefore it will be used to measure *absolute* displacements, velocities, and accelerations of points in the rigid body. The origin of the second coordinate system $x'\,y'\,z'$ will be fixed to the base point A in the body. The axes of this coordinate system will only be allowed to *translate* with respect to the fixed frame. Motion of the body will therefore be observed relative to the base point A, considered first as a translation of point A as observed from x, y, and z, and then a rotation of the body about point A as observed from x', y', and z'.

Fig. 13-9

The position vector $\mathbf{r}_A$ specifies the location of the origin A of the translating coordinate system. The relative position vector $\mathbf{r}_{B/A}$ locates the position of an arbitrary point B in the body with respect to an observer stationed at A. We can determine the location of point B with respect to the origin of the fixed frame of reference by using the vector equation

$$\mathbf{r}_B = \mathbf{r}_A + \mathbf{r}_{B/A} \qquad (13\text{-}11)$$

Taking the derivative of this equation with respect to time, we have

$$v_B = v_A + \frac{d\mathbf{r}_{B/A}}{dt}$$

The terms $\mathbf{v}_A$ and $\mathbf{v}_B$ represent the absolute velocity of points A and B, respectively. The *magnitude* of $\mathbf{r}_{B/A}$ does *not* change with respect to time because the body is rigid; however, the *direction* of this vector does change because of the rotational motion of the body about point A. Since the axes of the translating coordinate system remain *parallel* to the fixed axes at all times, the *components* of $\mathbf{r}_{B/A}$ projected on each coordinate system will be the *same* at any given instant of time. The time derivative of $\mathbf{r}_{B/A}$ is therefore the same, using either x, y, and z or x', y', and z' coordinates to measure the motion. The time rate of change of $\mathbf{r}_{B/A}$ is denoted as the relative velocity $\mathbf{v}_{B/A}$. This vector represents the velocity of point B, measured with respect to an observer stationed at point A. Thus,

$$\mathbf{v}_B = \mathbf{v}_A + \mathbf{v}_{B/A} \tag{13–12}$$

Consider now the rigid body shown in Fig. 13–10a. This body undergoes *general plane motion,* which means that the path of motion of any particle of the body is contained in a plane which remains parallel to a fixed reference plane. The motion will be studied in the xy plane during the instant of time dt. When this occurs, points A and B undergo infinitesimal translations $d\mathbf{s}_A$ and $d\mathbf{s}_B$, respectively. In Fig. 13–10b this motion is analyzed with respect to the base point A. Under these conditions, point B first moves in the plane through an infinitesimal displacement of $d\mathbf{s}_A$ to point B'. The body is then rotated by an amount $d\boldsymbol{\theta}$ about an axis perpendicular to the plane and passing through point A until point B' coincides with B. In other words, the body is first considered to translate, then to rotate about the origin of the $x'y'$ axis passing through the base point. *Kinematic diagrams* of the *velocity* which correspond to these displacements are shown in Fig. 13–10c. Notice that the *translation* of the body yields the velocity $\mathbf{v}_A = d\mathbf{s}_A/dt$ for both points A and B, and the *rotation* about point A yields the relative velocity of B with respect to A, $\mathbf{v}_{B/A}$.

Since the body appears *only* to *rotate* relative to an *observer located at A,* point B follows the dashed *circular path* having a radius of curvature $r_{B/A}$ as shown. $\mathbf{v}_{B/A}$ is tangent to this path and is therefore *perpendicular* to $\mathbf{r}_{B/A}$, such that $\mathbf{v}_{B/A} = \boldsymbol{\omega} \times \mathbf{r}_{B/A}$ (Eq. 13–7). In this equation $\boldsymbol{\omega}$ is the instantaneous angular velocity of the body, and $\mathbf{r}_{B/A}$ is the relative position vector drawn *from A to B,* i.e., "B with respect to A." Knowing the relative velocity $\mathbf{v}_{B/A}$, the absolute velocity of point B can be determined by using Eq. 13–12, written in the form

$$\mathbf{v}_B = \mathbf{v}_A + (\boldsymbol{\omega} \times \mathbf{r}_{B/A}) \tag{13–13}$$

Fig. 13-10

The three velocity terms in this equation are located at point B on the three kinematic diagrams in Fig. 13-10c. In accordance with Eq. 13-13, the vector addition is shown in the right-hand figure.

The final position of the body may also be obtained by choosing point B as a base point, Fig. 13–10d. Here the body is first translated by an amount equal to the infinitesimal displacement $d\mathbf{s}_B$ of point B (the origin of the $x'y'$ axis) and then rotated about this point. As discussed in Sec. 13–3, the angular rotation $d\boldsymbol{\theta}$ of line segment AB is the *same* as shown in both Fig. 13–10b and Fig. 13–10d. The kinematic diagrams of velocity are shown in Fig. 13–10e. Adding the three velocity vectors at point A as shown yields

$$\mathbf{v}_A = \mathbf{v}_B + (\boldsymbol{\omega} \times \mathbf{r}_{A/B}) \tag{13–14}$$

In comparing Eqs. 13–13 and 13–14 and noting Figs. 13–10c and 13–10e, it is seen that $\mathbf{v}_{B/A} = -\mathbf{v}_{A/B}$.

The above analysis may be used in a practical manner to study the motions of several members which are pin-connected or in contact with one another. It should be realized, however, that Eq. 13–13 can be applied *only between two points located on the same rigid body*. Since motion occurs in the plane, Eq. 13–13 may be written in the form of *two scalar equations*. The terms in these equations then represent the vector components of the motion for any two points on the body in the x and y directions. Using these equations, we may solve for at most *two unknown* scalar quantities. For example, if both the angular velocity and the velocity of one point on the body are known, the velocity (magnitude and direction) of any other point on the body can be determined. When applying Eq. 13–13 in *scalar form*, mistakes are generally avoided by accompanying the problem solution with a set of three kinematic diagrams of velocity, such as those shown in Fig. 13–10c. With these drawings, one can establish the correct direction for each of the velocity vectors. Once the motion of one member has been analyzed, it is possible to obtain the motion of *other connected members* by realizing that *points of contact, or joints which are pin-connected,* have the *same absolute velocity* (see Examples 13–4 and 13–6).

The following example problems illustrate the procedure used in applying Eq. 13–13 to determine the velocity of points on rigid bodies subjected to plane motion. All the problems are solved by first establishing the kinematical diagrams of velocity and then determining the scalar velocity component equations with the aid of these diagrams. This procedure for solution often gives "insight" into the physical aspects of the kinematics. In Sec. 13–6 we will solve some of these example problems using Eq. 13–13 in Cartesian vector form.

Example 13–3

The link shown in Fig. 13–11a is guided by the two blocks at A and B, which move in the fixed slots. At the instant shown, the velocity of

(a)

(c)

Fig. 13-11

block A is 20 ft/sec downward. Determine the velocities of points B and C located on the link at this same instant.

Solution

Since the velocity $\mathbf{v}_A$ is directed downward, the velocity $\mathbf{v}_B$ must be directed *horizontally* to the right. (Both points A and B are restricted to move along the fixed slots.) This motion causes the link to rotate *counter-clockwise;* that is, the angular velocity $\boldsymbol{\omega}$ acts outward, perpendicular to the plane of motion. Knowing the magnitude and direction of $\mathbf{v}_A$ and the line of direction of $\mathbf{v}_B$ and $\boldsymbol{\omega}$, we can apply Eq. 13–13 between points A and B in order to solve for the *two unknown* magnitudes v_B and ω. The three kinematic velocity diagrams are shown in Fig. 13–11b, where A has been chosen as the base point. (How do you know A is the base point?) In accordance with the cross product, $\mathbf{v}_{B/A} = \boldsymbol{\omega} \times \mathbf{r}_{B/A}$, or by nature of the rotation, $\mathbf{v}_{B/A}$ acts up to the right, perpendicular to $\mathbf{r}_{B/A}$. The magnitude of this velocity is

$$v_{B/A} = \omega r_{B/A} = \omega(1 \text{ ft})$$

Referring to the three velocity components *acting at point B* in each of the three figures, we can write

$$\mathbf{v}_B = \mathbf{v}_A + (\boldsymbol{\omega} \times \mathbf{r}_{B/A}) \tag{1}$$

or

$$v_B = 20 \text{ ft/sec} + \omega(1 \text{ ft})$$
$$\rightarrow \qquad\quad \downarrow \qquad\quad \angle^{45°}$$

Equating the horizontal and vertical components of this equation, with the assumed positive directions to the right and upward, we have,

$(\xrightarrow{+})$ $\qquad\qquad\qquad\qquad v_B = \omega \cos 45°$

$(+\uparrow)$ $\qquad\qquad\qquad\qquad 0 = -20 + \omega \sin 45°$

Thus,

$$\omega = 28.3 \text{ rad/sec}$$

and

$$v_B = 20 \text{ ft/sec} \qquad\qquad\qquad\qquad Ans.$$

Since both results are *positive*, the *sense of direction* of vectors $\mathbf{v}_B$ and $\boldsymbol{\omega}$ are indeed *correct* as shown in Fig. 13–11b.

Knowing the angular velocity of the link, we can obtain the magnitude and direction of the velocity of point C (the two unknowns) by applying Eq. 13–13 between points A and C. The kinematic diagrams, with the base point at A, are shown in Fig. 13–11c. The magnitude of the relative velocity term $\mathbf{v}_{C/A}$ is

$$v_{C/A} = \omega r_{C/A} = 28.3 \text{ rad/sec} (1.5 \text{ ft}) = 42.4 \text{ ft/sec}$$

Do you understand why $\mathbf{v}_{C/A}$ is directed as shown in the figure? Referring

to the three velocity components acting at point C, Fig. 13–11c, and applying Eq. 13–13, we have

$$\mathbf{v}_C = \mathbf{v}_A + \mathbf{v}_{C/A} \qquad (2)$$

or

$$v_C = 20 \text{ ft/sec} + 42.4 \text{ ft/sec}$$
$$\measuredangle^{\theta} \qquad\qquad \downarrow \qquad\qquad \measuredangle^{45°}$$

Equating the horizontal and vertical components of this equation, we have

$(\overset{+}{\rightarrow})$ $\qquad\qquad v_C \cos \theta = 42.4(\cos 45°)$

$(+\uparrow)$ $\qquad\qquad v_C \sin \theta = -20 + 42.4(\sin 45°)$

Dividing one equation into the other eliminates v_C and allows a solution for θ. v_C is obtained by resubstitution of θ into one of the equations. The results are

$$v_C = 31.6 \text{ ft/sec} \qquad\qquad Ans.$$
$$\theta = 18.4° \qquad \measuredangle^{\theta} \mathbf{v}_C$$

Using the results for v_B and ω, obtained previously, it is suggested that you apply Eq. 13–13 between points B and C to obtain these same results for $\mathbf{v}_C$.

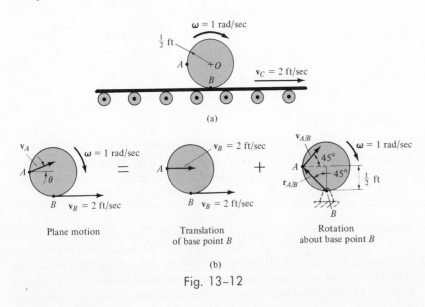

(a)

Plane motion $\qquad$ Translation of base point B $\qquad$ Rotation about base point B

(b)

Fig. 13–12

Example 13–4

The cylinder shown in Fig. 13–12a rolls freely on the surface of a conveyor belt which is moving at 2 ft/sec. Assuming that no slipping occurs between the cylinder and the belt, determine the velocity of point A located on the periphery of the cylinder. The cylinder has a clockwise angular velocity of $\omega = 1$ rad/sec at the instant shown.

Solution

Since no slipping occurs, point B on the cylinder has the same velocity as the conveyor. Knowing the angular velocity of the cylinder, Eq. 13–13 may be applied between points B and A to determine the magnitude and direction of $\mathbf{v}_A$. Choosing point B as a base point, the three kinematic diagrams for velocity are shown in Fig. 13–12b. The relative velocity $\mathbf{v}_{A/B}$ has a magnitude of

$$v_{A/B} = \omega r_{A/B} = (1 \text{ rad/sec})\left(\frac{\frac{1}{2} \text{ ft}}{\cos 45°}\right) = 0.707 \text{ ft/sec}$$

This vector acts perpendicular to $\mathbf{r}_{A/B}$. Why? Applying Eq. 13–13, we have

$$\mathbf{v}_A = \mathbf{v}_B + (\boldsymbol{\omega} \times \mathbf{r}_{A/B})$$
$$v_A = 2 \text{ ft/sec} + 0.707 \text{ ft/sec} \qquad (1)$$
$$\measuredangle^{\theta} \qquad \rightarrow \qquad \measuredangle^{45°}$$

Equating components in the horizontal and vertical direction yields

$(\xrightarrow{+})$ $\qquad\qquad v_A \cos \theta = 2 + 0.707(\cos 45°)$

$(+\uparrow)$ $\qquad\qquad v_A \sin \theta = 0.707(\sin 45°)$

Which give

$$v_A = 2.55 \text{ ft/sec} \qquad\qquad Ans.$$
$$\theta = 11.3° \qquad \measuredangle^{\theta}\, \mathbf{v}_A$$

Example 13–5

The link AB, shown in Fig. 13–13a, is attached to the rod CD by the smooth collar at B. If AB has a counterclockwise angular velocity of 2 rad/sec at the instant shown, determine the angular velocity of rod CD at this instant.

Solution

To obtain the rotational velocity of rod CD, it is first necessary to determine the velocity of the smooth collar which is connected to link AB. The pin at A is fixed, and the link AB is subjected only to a *pure rotation about a fixed axis* through point A. Hence, $\mathbf{v}_B$ acts in a direction *perpendicular* to $\mathbf{r}_{B/A}$, Fig. 13–13b. Why? The magnitude of $\mathbf{v}_B$ is determined by using Eq. 13–6, i.e.,

$$v_B = \omega_{AB} r_{B/A} = 2 \text{ rad/sec}(\tfrac{1}{2} \text{ ft}) = 1 \text{ ft/sec}$$

This velocity is separated into two components, shown in Fig. 13–13c. The component of $\mathbf{v}_B$ which is parallel to the rod, $(\mathbf{v}_B)_y$, lies along the

Fig. 13-13

axis of the rod and represents the velocity at which the collar is *slipping over* the rod. Only the perpendicular component of velocity $(v_B)_x$ gives motion to rod CD since any displacement of the collar in the horizontal direction displaces the rod.

Since rod CD is in *pure rotation about a fixed axis* passing through C, we may obtain the angular velocity of the rod using the data in Fig. 13-13*d*.

$$(v_B)_x = \omega_{CD} r_{B/C}$$
$$0.5 \text{ ft/sec} = \omega_{CD}(\tfrac{1}{4} \text{ ft})$$
$$\omega_{CD} = 2 \text{ rad/sec} \qquad \curvearrowleft \omega_{CD} \qquad\qquad \textit{Ans.}$$

Example 13-6

The bar AB of the linkage shown in Fig. 13-14*a* has a clockwise angular velocity of 30 rad/sec when $\theta = 60°$. Compute the angular velocity of members BC and DC at this instant.

Solution

From the figure both bars AB and DC are subjected to rotation about a fixed axis, and bar BC is subjected to general plane motion. The magnitude and direction of $\mathbf{v}_B$ may be determined from the kinematic diagram shown in Fig. 13-14*b*. Due to the rotation, $\mathbf{v}_B$ *always* acts perpendicular to $\mathbf{r}_{B/A}$. The magnitude of $\mathbf{v}_B$ is

$$v_B = \omega_{AB} r_{B/A} = 30 \text{ rad/sec}(2 \text{ ft}) = 60 \text{ ft/sec}$$

From the kinematic diagram of bar DC, Fig. 13-14*c*, $\mathbf{v}_C$ acts perpendicular to $\mathbf{r}_{C/D}$ and has a magnitude of

$$v_C = \omega_{DC} r_{C/D}$$
$$= \omega_{DC}(1 \text{ ft}) \qquad\qquad (1)$$

523

(a)

Plane motion

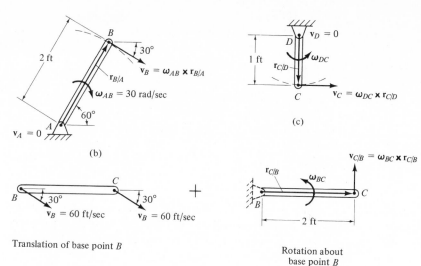

(b)

(c)

(d)

Fig. 13-14

Knowing $\mathbf{v}_B$ and the direction of $\mathbf{v}_C$, we can determine the two unknown magnitudes v_C and ω_{BC} (angular velocity of rod BC) by applying Eq. 13-13 between points B and C lying on rod BC. Choosing point B as a base point, the kinematic diagrams for velocity on BC are shown in Fig. 13-14d. The angular velocity ω_{BC} acts perpendicular to the plane of motion with an *assumed* directional sense that is counterclockwise, that is, out of the plane. Thus,

$$\mathbf{v}_C = \mathbf{v}_B + (\omega_{BC} \times \mathbf{r}_{C/B})$$
$$v_C = 60 \text{ ft/sec} + \omega_{BC}(2 \text{ ft})$$
$$\rightarrow \qquad \searrow^{30°} \qquad \uparrow$$

Equating the horizontal and vertical components yields

$(\xrightarrow{+})$ $\qquad\qquad\qquad v_C = 60 \cos 30° = 52.0 \text{ ft/sec}$

$(+\uparrow)$ $\qquad\qquad\qquad 0 = -60 \sin 30° + \omega_{BC} 2$

$\qquad\qquad \omega_{BC} = 15 \text{ rad/sec} \qquad \curvearrowright \omega_{BC}$ $\qquad\qquad$ *Ans.*

The angular velocity of rod DC is computed by using Eq. (1):

$$52.0 \text{ ft/sec} = \omega_{DC}(1 \text{ ft})$$
$$\omega_{DC} = 52 \text{ rad/sec} \qquad \curvearrowright \omega_{DC} \qquad\qquad \textit{Ans.}$$

Both answers are positive, and hence, the assumed directions of ω_{BC} and ω_{DC} are correct.

13–6. Method for Determining Velocity by Using Cartesian Unit Vectors

The example problems in Sec. 13–5 were all solved with the aid of three kinematic diagrams for velocity. These diagrams are useful, because (1) they provide a physical "insight" as to the meaning of each of the terms in Eq. 13–13, and (2) the directional components of the velocities can be quickly established with the aid of these diagrams, thus keeping computations to a minimum. It is possible, however, to apply Eq. 13–13 without drawing the three kinematic diagrams; instead we can express each of the vectors in this equation in terms of Cartesian unit vectors. This vector approach is very methodical in application and thus has an advantage for solving problems involving mechanisms with *several con-nected members*. In applying this method, however, the physical aspects of the kinematics are often lost. Because of the added vector algebra which must be included in the solution, in general more computations must be made to obtain the complete solution.

The following two example problems illustrate this method for solving problems involving the general plane motion of rigid bodies.

Example 13–7

The link shown in Fig. 13–15a is guided by the two blocks at A and B which move in the fixed slots. At the instant shown, the velocity of the block at A is 20 ft/sec downward. Determine the velocities of points B and C located on the link during this same instant.

Solution

This problem is the same as Example 13–3. A kinematic diagram showing the velocity of points A, B, and C and the angular velocity of

(a)   (b)

Fig. 13–15

the link is given in Fig. 13–15*b*. Each of these vectors may be expressed in terms of **i**, **j**, and **k** components:

$$\mathbf{v}_A = \{-20\mathbf{j}\} \text{ ft/sec}, \qquad \mathbf{v}_B = v_B\mathbf{i}, \qquad \boldsymbol{\omega} = \omega\mathbf{k}, \qquad \mathbf{v}_C = v_C \cos\theta\mathbf{i} + v_C \sin\theta\mathbf{j}$$

Applying Eq. 13–13 between points *A* and *B* on the link in order to solve for the two unknowns ω and v_B, we have

$$\mathbf{v}_B = \mathbf{v}_A + (\boldsymbol{\omega} \times \mathbf{r}_{B/A})$$
$$v_B\mathbf{i} = -20\mathbf{j} + [\omega\mathbf{k} \times (1\sin 45°\mathbf{i} - 1\cos 45°\mathbf{j})]$$

or

$$v_B\mathbf{i} = -20\mathbf{j} + \omega\sin 45°\mathbf{j} + \omega\cos 45°\mathbf{i}$$

Equating the **i** and **j** components gives

$$v_B = \omega\cos 45°, \qquad 0 = -20 + \omega\sin 45°$$

Thus,

$$\omega = 28.3 \text{ rad/sec}$$
$$v_B = 20 \text{ ft/sec} \qquad\qquad\qquad Ans.$$

Knowing the velocity $\mathbf{v}_B$ and the angular velocity $\boldsymbol{\omega}$, we can obtain the magnitude and direction of $\mathbf{v}_C$ by applying Eq. 13–13 between points *B* and *C*, i.e.,

$$\mathbf{v}_C = \mathbf{v}_B + (\boldsymbol{\omega} \times \mathbf{r}_{C/B})$$
$$v_C \cos\theta\mathbf{i} + v_C \sin\theta\mathbf{j} = 20\mathbf{i} + [28.3\mathbf{k} \times (\tfrac{1}{2}\sin 45°\mathbf{i} - \tfrac{1}{2}\cos 45°\mathbf{j})]$$

Expanding and equating the **i** and **j** components yields

$$v_C \cos\theta = 20 + 14.2\cos 45°$$
$$v_C \sin\theta = 14.2\sin 45°$$

Solving, we have

$$v_C = 31.6 \text{ ft/sec} \qquad\qquad\qquad Ans.$$
$$\theta = 18.4° \qquad \angle^{\theta\ \mathbf{v}_C} \qquad\qquad Ans.$$

It is suggested that you obtain this same solution for $\mathbf{v}_C$ by applying Eq. 13–13 between points *A* and *C*.

Example 13–8

The bar *AB* of the linkage shown in Fig. 13–16*a* has a clockwise angular velocity of 30 rad/sec when $\theta = 60°$. Compute the angular velocity of members *BC* and *DC* at this instant.

Solution

This problem is the same as Example 13–6. Using Cartesian unit

(a)   Fig. 13–16   (b)

vectors, we may apply Eq. 13–13 between the end points of link BC. (Again it is repeated, that Eq. 13–13 may only be applied between any two points lying on the *same rigid body*.) Note that links AB and DC each undergo motion about a fixed axis and in particular $\omega_{AB} = \{-30\mathbf{k}\}$ rad/sec, Fig. 13–16b. Assuming that $\omega_{BC} = \omega_{BC}\mathbf{k}$ and $\omega_{DC} = \omega_{DC}\mathbf{k}$, we have for link AB,

$$\mathbf{v}_B = \omega_{AB} \times \mathbf{r}_{B/A}$$
$$= (-30\mathbf{k}) \times (2\cos 60°\mathbf{i} + 2\sin 60°\mathbf{j})$$

or

$$\mathbf{v}_B = \{52\mathbf{i} - 30\mathbf{j}\} \text{ ft/sec} \qquad (1)$$

and for link BC,

$$\mathbf{v}_C = \mathbf{v}_B + (\omega_{BC} \times \mathbf{r}_{C/B})$$

Using Eq. (1), we have

$$\mathbf{v}_C = 52\mathbf{i} - 30\mathbf{j} + [(\omega_{BC}\mathbf{k}) \times (2\mathbf{i})]$$
$$= 52\mathbf{i} + (2\omega_{BC} - 30)\mathbf{j} \qquad (2)$$

For link DC,

$$\mathbf{v}_C = \omega_{DC} \times \mathbf{r}_{C/D}$$
$$= (\omega_{DC}\mathbf{k}) \times (-1\mathbf{j})$$
$$= \omega_{DC}\mathbf{i} \qquad (3)$$

Substituting Eq. (3) into Eq. (2), we have

$$\omega_{DC}\mathbf{i} = 52\mathbf{i} + (2\omega_{BC} - 30)\mathbf{j}$$

Equating the respective $\mathbf{i}$ and $\mathbf{j}$ components and solving, we have

$$\omega_{DC} = 52 \text{ rad/sec} \qquad \zeta\omega_{DC} \qquad \qquad Ans.$$

$$\omega_{BC} = 15 \text{ rad/sec} \qquad \zeta\omega_{BC} \qquad \qquad Ans.$$

Problems

13-14. The rod *AB* is constrained to move in the grooved slots. If the velocity of the slider block at *B* is 2 ft/sec to the left, compute the velocity of the block at *A* and the angular velocity of the rod at the instant shown.

Prob. 13-14

13-15. The rigid body of arbitrary shape moves in the plane. If *h* and θ are known, and $v_A = v_B = v$, compute the angular velocity of the body.

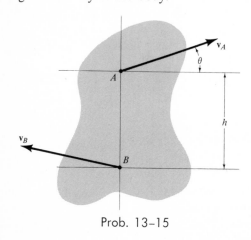

Prob. 13-15

13-16. The center of the wheel is moving to the right with a speed of 2 ft/sec. If no slipping occurs at the ground, *A*, determine the velocity of points *B* and *C*.

Prob. 13-16

13-17. When the slider block *C* is in the position shown, the link *AB* has a clockwise angular velocity of 2 rad/sec. Determine the velocity of block *C* at this instant.

Prob. 13-17

13-18. The gear rests in a fixed horizontal rack. A cord is wrapped around the inner core of the gear so that it remains horizontally tangent to the inner core at *A*. If the cord is pulled with a constant velocity of 2 ft/sec, determine the velocity of the center of the gear, *C*.

528

Prob. 13-18

Prob. 13-21

13-19. Work Prob. 13-18, assuming that the cord is wrapped around the gear in the opposite sense, so that the end of the cord remains horizontally tangent to the inner core at B, and is pulled to the right at 2 ft/sec.

13-20. The link AB moves with a clockwise angular velocity of 5 rad/sec. A pin passes through the link BD at C in order to guide its motion. When the mechanism is in the position shown, determine the velocity of point D and the angular velocity of rod BD.

13-22. The wheel slips on the surface of the smooth horizontal plane. At a given instant, point B has a downward velocity of 2 ft/sec. Determine the velocity of point A at this instant.

Prob. 13-20

Prob. 13-22

13-21. The clockwise angular velocity of the link AB is 2 rad/sec. Determine the angular velocity of the connecting links BC and CD at the instant shown.

13-23. If the wheel in Prob. 13-22 is rotating clockwise with an angular velocity of 3 rad/sec, determine the required velocity of point B so that the velocity of point A is equal to zero.

13-24. Compute the velocity of the rod R for each of the three cam arrangements for any angle ϕ of the cam. The cam rotates at ω.

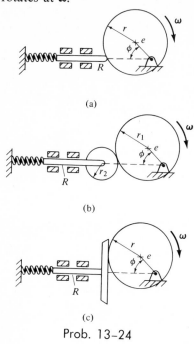

(a)

(b)

(c)

Prob. 13-24

13-25. Link CD has a counterclockwise angular velocity of 3 rad/sec at the instant shown. Determine the angular velocity of link AB at this instant.

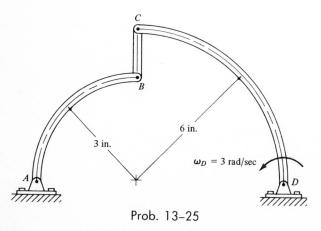

Prob. 13-25

13-26. Link AB has an angular velocity of $\omega_{AB} = 3$ rad/sec. Determine the angular velocity of the wheel at the instant shown. The collar at C is free to rotate.

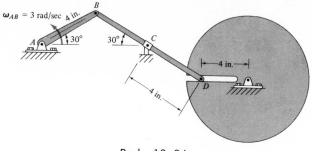

Prob. 13-26

13-27. The piston P is moving upward with a velocity of 300 in./sec at the instant shown. Determine the angular velocity of the crankshaft AB at this instant.

Prob. 13-27

13-28. Determine the velocity of the center of gravity G of the connecting rod in Prob. 13-27 at the instant shown. The piston is moving upward with a velocity of 300 in./sec.

13-29. The link AB has a constant clockwise angular velocity of 3 rad/sec. The collar at B slides freely over the rod CD. Determine the velocity of point D on the rod at the instant shown and when $\theta = 0°$.

Prob. 13-29

13-31. The link AB rotates clockwise about the fixed point A with an angular velocity of 3 rad/sec. If the frame F is stationary, determine the angular velocity of gear C at the instant shown.

Prob. 13-31

13-30. The rod CD has a downward velocity of 6 ft/sec. The rod is pinned at C to gear B. Determine the velocity of the gear rack A at the instant shown.

Prob. 13-30

531

13–7. Instantaneous Center of Zero Velocity: Plane Motion

Provided the angular velocity of a body and absolute motion of a base point A on the body are known, we can determine the absolute motion of any other point B on the body by using Eq. 13–12, i.e., $\mathbf{v}_B = \mathbf{v}_A + \mathbf{v}_{B/A}$. Specifically, the velocity of point B is determined by assuming that point B has a velocity $\mathbf{v}_A$, plus (vectorially) a relative velocity $\mathbf{v}_{B/A}$. The relative velocity term is determined by assuming that the body is instantaneously rotating about the base point A (see Fig. 13–10c).

A simple way to compute $\mathbf{v}_B$ is to choose a base point IC which has *zero velocity*, i.e., $\mathbf{v}_{IC} = 0$, and about which the body is rotating at a *given instant*. This point is called the *instantaneous center of zero velocity* (IC) and it lies on the *instantaneous axis of zero velocity* (IA). For plane motion, the IA is always perpendicular to the plane used to represent the motion. The intersection of the IA with this plane defines the IC. Once the IC is determined (the method for which will be explained later) the velocity of point B on the body may be computed at the given instant by applying Eq. 13–13 with the IC as a base point; i.e.,

$$\mathbf{v}_B = \mathbf{v}_{IC} + (\boldsymbol{\omega} \times \mathbf{r}_{B/IC})$$

Since $\mathbf{v}_{IC} = 0$,

$$\mathbf{v}_B = \boldsymbol{\omega} \times \mathbf{r}_{B/IC} \qquad (13\text{--}15)$$

For *general plane motion* of the body, $\boldsymbol{\omega}$ will *always be perpendicular* to $\mathbf{r}_{B/IC}$—which is contained in the plane of motion. Consequently, from the cross product, the magnitude of $\mathbf{v}_B$ is*

$$v_B = \omega(r_{B/IC}) \qquad (13\text{--}16)$$

In general, the location of the instantaneous center of zero velocity for a rigid body subjected to general plane motion can be easily determined, provided one knows the lines of action for the velocities of any two points A and B on the body, Fig. 13–7a. In this regard, construct at points A and B line segments $r_{A/IC}$ and $r_{B/IC}$, which are *perpendicular* to the lines of action of $\mathbf{v}_A$ and $\mathbf{v}_B$, respectively. Extending these perpendiculars to their *point of intersection* locates the IC at the instant considered. In accordance with Eq. 13–15, if the instantaneous angular velocity $\boldsymbol{\omega}$ of the body is known, we may determine the velocities of points A and B

---

*Equations 13–15 and 13–16 are identical in form to Eqs. 13–7 and 13–6, which represent *rotation of a body about a fixed axis*. A fixed axis of rotation has zero velocity *and* zero acceleration. In general, the instantaneous axis of zero velocity *has* an acceleration since this point follows a curved path.

Fig. 13–17

by using Eq. 13–16, as shown in the figure. In this particular case, the *IC* is located *off* the body. As a special case, if the lines of action of the velocities $\mathbf{v}_A$ and $\mathbf{v}_B$ are *parallel* to one another, provided one knows the magnitude of $\mathbf{v}_A$ and $\mathbf{v}_B$, the *IC* can be located by similar triangles, Figs. 13–17*b* and 13–17*c*.

In general, the instantaneous center of zero velocity is *not a fixed point;* rather *it moves* as the body changes its position. The locus of points which describes the path of motion for the *IC* is called a *centrode,* Fig. 13–17*a*. A point on the centrode acts as the *IC* for the body only for an *instant*. Thus, this point is *not* a point of zero acceleration, and as a result this point, in general, cannot be used to obtain accelerations. The *IC* simply provides a convenient means of establishing the velocities of various points located on a rigid body at a given *instant* of time. At the given *instant,* we may think of the body as *"extended and pinned"* at the *IC* when computing the velocity of any point on the body, Fig. 13–18. The

Fig. 13–18

533

body then rotates about this pin with an instantaneous angular velocity ω. (Recall that ω is a free vector.) Provided the location of the *IC* is *known*, the *magnitude* of velocity for the arbitrary points A, B, and C on the body can thus be determined as the product of the magnitude of the angular velocity ω of the body and the radial line drawn from the *IC* to the point, Eq. 13–16. The line of action of each velocity vector is perpendicular to the radial line and the velocity has a *directional sense* which tends to direct the point in a manner consistent with the angular rotation of the radial line, Fig. 13–18.

Example 13–9

Determine the location of the instantaneous center of zero velocity for (a) the wheel shown in Fig. 13–19a which is rolling without slipping along the ground, (b) the link BC shown in Fig. 13–19b, and (c) the crankshaft BC shown in Fig. 13–19c.

Solution

Part (a). The wheel rolls without slipping, and therefore the point of *contact* of the wheel with the ground has *zero velocity*. Hence, this point represents the instantaneous center of zero velocity for the wheel, Fig. 13–19d. If it is imagined that the wheel is momentarily pinned at this

Fig. 13–19

point, the velocity of points A, B, O, etc., can then be found using Eq. 13–16, as shown in the figure.

Part (b). The center of zero velocity of rod CD is always at point D since the rod is subjected to rotation about a fixed axis which passes through this point, Fig. 13–19b. In a similar manner, the center of zero velocity of rod AB is at point A. Points B and C therefore follow circular paths of motion as shown. Since the velocity is always *tangent* to the *path,* at the instant considered, the velocities of point C on rod CD and point B on rod AB are directed vertically downward along the axis of the rod, Fig. 13–19e. Also, since the link CB is *rigid,* no vertical displacement occurs between points B and C so that $\mathbf{v}_C = \mathbf{v}_B$. Radial lines drawn normal to these two velocities form parallel lines which intersect at infinity. Thus, to satisfy Eq. 13–15, rod CB must have zero rotation. When the rod is in the position shown, it therefore *translates momentarily* with a speed $v_C = \omega_{CD}b$. An instant later, however, the rod will move to a new position, causing the instantaneous center for the rod to also *change* its position.

Part (c). As shown in Fig. 13–19c, link AB rotates about the fixed point A. Thus, point B has a speed of $v_B = \omega d$, caused by the clockwise rotation of the link AB. This velocity is perpendicular to AB, Fig. 13–19f. The motion of point B causes the piston to move forward *horizontally* with a velocity $\mathbf{v}_C$. Point C on the rod moves horizontally with this same velocity. When lines are drawn perpendicular to $\mathbf{v}_B$ and $\mathbf{v}_C$, Fig. 13–19f, they intersect at the *IC*. The magnitudes of $\mathbf{r}_{B/IC}$ and $\mathbf{r}_{C/IC}$ are determined strictly from geometry.

Example 13–10

Rod DE, shown in Fig. 13–20a, moves with a speed of 3 ft/sec. Determine the angular velocity of arm AB and the velocity of point C on link BD at the instant shown.

Solution

Since rod DE is subjected to translational motion, due to the constraint, the speed of point D (and any other point on DE) is 3 ft/sec directed upward along the axis of rod DE. This movement causes arm AB to rotate about point A in a clockwise direction, and hence, the velocity of point B is directed as shown in Figs. 13–20b and 13–20c, that is, perpendicular to line AB. The instantaneous center of zero velocity for rod BD is located at the intersection of the line segments drawn perpendicular to the velocities $\mathbf{v}_B$ and $\mathbf{v}_D$, Fig. 13–20b. From the geometry,

$$r_{B/IC} = 4 \tan 30° \text{ ft} = 2.31 \text{ ft}$$

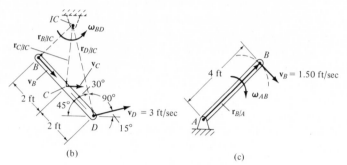

Fig. 13–20

$$r_{D/IC} = \frac{4 \text{ ft}}{\cos 30°} = 4.62 \text{ ft}$$

and

$$r_{C/IC} = \sqrt{(r_{B/IC})^2 + (2 \text{ ft})^2}$$

or

$$r_{C/IC} = \sqrt{(2.31 \text{ ft})^2 + (2 \text{ ft})^2} = 3.06 \text{ ft}$$

Assuming that link BD is momentarily pinned at the IC, as shown, the angular velocity of the link is

$$\omega_{BD} = \frac{v_D}{r_{D/IC}} = \frac{3 \text{ ft/sec}}{4.62 \text{ ft}} = 0.649 \text{ rad/sec} \qquad \jmath\omega_{BD}$$

The velocities of points B and C are therefore

$$v_B = \omega_{BD}(r_{B/IC}) = 0.649 \text{ rad/sec} (2.31 \text{ ft}) = 1.50 \text{ ft/sec}$$

and

$$v_C = \omega_{BD}(r_{C/IC}) = 0.649 \text{ rad/sec} (3.06 \text{ ft}) = 1.99 \text{ ft/sec} \qquad Ans.$$

Link AB is subjected to rotation about a fixed axis passing through A. Since the velocity of point B has been computed, the angular velocity of the link can easily be determined. With reference to Fig. 13–20c,

$$\omega_{AB} = \frac{v_B}{r_{B/A}} = \frac{1.50 \text{ ft/sec}}{4 \text{ ft}} = 0.375 \text{ rad/sec} \qquad \jmath\omega_{AB} \qquad Ans.$$

Problems

13-32. The instantaneous center of zero velocity for the body is located at the point (1 ft, 4 ft). If the body has an angular velocity of 3 rad/sec, as shown, determine the relative velocity of B with respect to A.

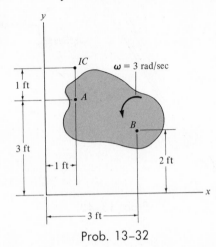

Prob. 13-32

13-33. The center of the wheel has a velocity of 3 ft/sec up the inclined plane. At the same time, it has a clockwise angular velocity of $\omega = 3$ rad/sec which causes the wheel to slip at its contact point A. Determine the velocity of point A.

Prob. 13-33

13-34. Solve Prob. 13-14 using the method of instantaneous center of zero velocity.

13-35. Solve Prob. 13-17 using the method of instantaneous center of zero velocity.

13-36. Solve Prob. 13-21 using the method of instantaneous center of zero velocity.

13-37. The velocity of the slider block C is 4 ft/sec up the inclined groove. Determine the angular velocity of links AB and BC and the velocity of point B at the instant shown.

Prob. 13-37

13-38. The planetary gear A is pinned at B. Link BC rotates clockwise at an angular velocity of 8 rad/sec, while the outer gear rack rotates counterclockwise at an angular velocity of 2 rad/sec. Determine the angular velocity of gear A.

Prob. 13-38

13-39. Solve Prob. 13-30 using the method of instantaneous center of zero velocity.

13-40. The wheel rolls on its hub without slipping on the horizontal surface. If the velocity of the center of the wheel is 2 ft/sec to the right, determine the velocity of points A and B at the instant shown.

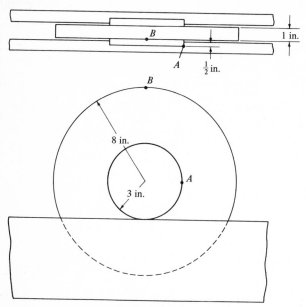

Prob. 13-40

13-41. Knowing that the angular velocity of link CD is 4 rad/sec for the mechanism, determine the velocity of point E on link BC and the angular velocity of link AB at the instant shown.

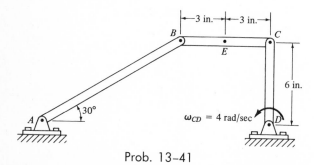

Prob. 13-41

13-42. The slider block C is fixed to the wheel which has a counterclockwise rotation of 6 rad/sec. Determine the angular rotation of the slotted arm AB and the velocity of point B at the instant shown.

Prob. 13-42

13-43. Gear C turns clockwise with an angular velocity of 2 rad/sec, while the connecting link AB rotates counterclockwise at 4 rad/sec. Determine the angular velocity of gear D.

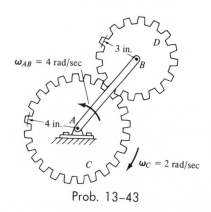

Prob. 13-43

538

13–8. General Plane Motion of a Rigid Body: Acceleration

The equation which relates the acceleration of two points on a rigid body subjected to plane motion may be determined by differentiating Eq. 13–13 with respect to time. Thus,

$$\frac{d\mathbf{v}_B}{dt} = \frac{d\mathbf{v}_A}{dt} + \frac{d}{dt}(\boldsymbol{\omega} \times \mathbf{r}_{B/A})$$

The derivative of the cross product is obtained by using Eq. 12–14. Hence,

$$\mathbf{a}_B = \mathbf{a}_A + \frac{d\boldsymbol{\omega}}{dt} \times \mathbf{r}_{B/A} + \boldsymbol{\omega} \times \frac{d\mathbf{r}_{B/A}}{dt}$$

Recalling that $d\boldsymbol{\omega}/dt = \boldsymbol{\alpha}$ and $d\mathbf{r}_{B/A}/dt = \mathbf{v}_{B/A} = (\boldsymbol{\omega} \times \mathbf{r}_{B/A})$, we have

$$\mathbf{a}_B = \mathbf{a}_A + \boldsymbol{\alpha} \times \mathbf{r}_{B/A} + \boldsymbol{\omega} \times (\boldsymbol{\omega} \times \mathbf{r}_{B/A}) \qquad (13\text{--}17)$$

From Eq. 13–12, we may also write,

$$\mathbf{a}_B = \mathbf{a}_A + \mathbf{a}_{B/A} \qquad (13\text{--}18)$$

where

$$\begin{aligned} \mathbf{a}_{B/A} &= (\mathbf{a}_{B/A})_t + (\mathbf{a}_{B/A})_n \\ &= \boldsymbol{\alpha} \times \mathbf{r}_{B/A} + \boldsymbol{\omega} \times (\boldsymbol{\omega} \times \mathbf{r}_{B/A}) \end{aligned} \qquad (13\text{--}19)$$

Since $\mathbf{r}_{B/A}$ is perpendicular to $\boldsymbol{\alpha}$ and $\boldsymbol{\omega}$, the magnitudes of the two relative acceleration components are

$$(a_{B/A})_t = \alpha r_{B/A} \qquad (13\text{--}20)$$

and

$$(a_{B/A})_n = \omega^2 r_{B/A} \qquad (13\text{--}21)$$

Each of the terms in Eq. 13–17 or Eq. 13–18 may be defined with reference to Fig. 13–21. As shown in this figure, the accelerated plane motion of the rigid body, Fig. 13–21a, is analyzed as the sum of a pure translation of a chosen base point A and a rotation about the base point, in the same manner as was done for the velocity. For example, knowing the acceleration of point A, $\mathbf{a}_A$, the angular velocity $\boldsymbol{\omega}$, and the angular acceleration $\boldsymbol{\alpha}$ of the body, we can determine the acceleration of point B, $\mathbf{a}_B$, by first considering point B to have a known (instantaneous) acceleration of $\mathbf{a}_A$, Fig. 13–21b. The body is then considered to rotate

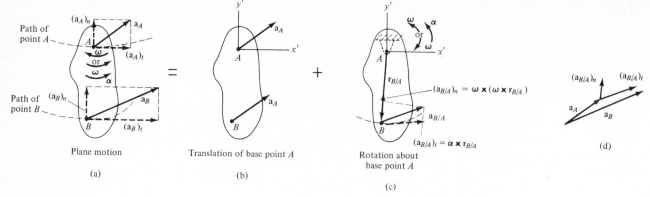

Plane motion

(a)

$=$

Translation of base point A

(b)

$+$

Rotation about
base point A

(c)

(d)

Fig. 13–21

(instantaneously) *about point A* with the known angular velocity $\boldsymbol{\omega}$ and acceleration $\boldsymbol{\alpha}$, Fig. 13–21c. This *rotation* is observed with respect to the origin of the $x'y'$ frame of reference, which is fixed at the base point A and *translates* with an acceleration of $\mathbf{a}_A$. Since the moving frame of reference $x'y'$ does *not* rotate, the values of $\boldsymbol{\alpha}$ and $\boldsymbol{\omega}$ are used for calculating the relative acceleration of point B with respect to A. Because the path is a *curve,* having a radius of curvature of $r_{B/A}$, two components are used to express the relative motion. The tangential component, $\boldsymbol{\alpha} \times \mathbf{r}_{B/A}$, always acts tangent to the path of motion and has a magnitude of $\alpha r_{B/A}$ (Eq. 13–20). The normal component, $\boldsymbol{\omega} \times (\boldsymbol{\omega} \times \mathbf{r}_{B/A})$, is *always directed toward the center of rotation,* in this case toward point A, and has a magnitude of $\omega^2 r_{B/A}$ (Eq. 13–21).

The acceleration vector $\mathbf{a}_B$ is determined by adding (vectorially) each of the component acceleration vectors at point B, shown in Figs. 13–21b and 13–21c. This addition is shown in Fig. 13–21d. The resultant is expressed by Eq. 13–17.

It is important to remember that unlike the velocity, the accelerations of points A and B, in general, are *not* tangent to their path of motion. Why? It was shown in Secs. 12–9 and 13–4 that when a point moves along a *curved path* the acceleration of the point has two components directed normal and tangent to the path. In particular, the component acting *tangent* to the path, $\mathbf{a}_t$, accounts for the time rate of *change* made in the *magnitude of the velocity* of the point. The normal component $\mathbf{a}_n$ *always acts toward the center of curvature of the path* and represents the time rate of *change* made in the *velocity direction*. These two components are shown in Fig. 13–21a for the acceleration of points A and B, since these points are traveling on *curved paths* at the instant considered.

Equation 13–17 applies between any two points located on the *same* rigid body which is subjected to general plane motion. When two bodies are pin-connected, the points coincident at the pin move with the same

540

Fig. 13-22

acceleration, since the *path of motion* over which they act is the *same*. For example, point *B* lying on either connecting rod *AB* or *CB* of the mechanism shown in Fig. 13–22 has the same acceleration, since the rods are *pin connected* at *B*. The accelerations, however, are not equal when a common point of contact between two bodies moves along *different* paths. Consider the two gears in contact with one another in Fig. 13–23. Point *A* is located on gear *B*, and point *A'* is located on gear *C*. As shown in the figure, when these two points are coincident, the *tangential* components of acceleration of both points are the *same*. However, because both points follow *different curved paths*, their normal components are *not* equal. Therefore, $\mathbf{a}_A \neq \mathbf{a}_{A'}$. Do you see that $\mathbf{v}_A = \mathbf{v}_{A'}$ at the instant considered? Recall that the velocity is always *tangent* to the path of motion.

Since Eq. 13–17 is a vector equation, it is equivalent to *two* scalar equations in the plane. Hence, with only two equations it is possible to solve for at most *two* unknowns. In any given problem, these unknowns may be represented by the magnitude and/or direction of $\mathbf{a}_A$ or $\mathbf{a}_B$, and the magnitude of $\boldsymbol{\alpha}$ ($\boldsymbol{\alpha}$ always acts perpendicular to the plane of motion). The vector $\mathbf{r}_{B/A}$ is determined from the geometry of the problem, and $\boldsymbol{\omega}$ is determined by performing a kinematic analysis of velocity *before* applying Eq. 13–17.

When using Eq. 13–17 or Eq. 13–18 in *scalar form*, errors can usually be avoided provided the problem solution is accompanied with a set of three kinematic diagrams, such as those shown in Fig. 13–21. With these

Fig. 13-23

drawings we can establish the correct sense of direction for each of the acceleration vector components. Since at most six vector components must be added together for each application of Eq. 13–18, the solution is often facilitated using the polygon method of vector addition. After adding all the known components together first, we can generally solve for the remaining two unknowns by using trigonometry. (In Sec. 13–9, we will discuss a method of solution using Cartesian unit vectors.)

Example 13–11

The rod AB shown in Fig. 13–24a is confined to move along the inclined plane at A and the circular arc at B. If point A has an acceleration of 3 ft/sec² and a velocity of 2 ft/sec, both directed down the plane when the rod becomes horizontal, determine the angular acceleration of the rod at this instant.

Solution

Velocity Analysis. Before applying the acceleration equation, Eq. 13–17, to the rod, it is first necessary to determine the angular velocity ω of the rod and the velocity of point B. Since the velocity of point A is known and the velocity of point B is *tangent* to the *curved path* of motion at point B, the instantaneous center of zero velocity for the rod is located as shown in Fig. 13–24b. From the geometry,

$$r_{A/IC} = r_{B/IC} = \frac{\frac{10}{2}\ \text{ft}}{\cos 45°} = 7.07\ \text{ft}$$

Thus,

$$\omega = \frac{v_A}{r_{A/IC}} = \frac{2\ \text{ft/sec}}{7.07\ \text{ft}} = 0.283\ \text{rad/sec}$$

Also,

$$v_B = \omega(r_{B/IC}) = 0.283\ \text{rad/sec}\ (7.07\ \text{ft}) = 2\ \text{ft/sec}$$

Show that one obtains these same results by applying Eq. 13–13 to the rod.

Acceleration Analysis. Equation 13–17 may now be applied between points A and B on the rod. Choosing point A as a base point, the kinematic diagrams of acceleration are shown in Fig. 13–24c. Notice that the path of motion of point B is along a *curve* having a constant radius of curvature, $\rho = 4$ ft. Because of this *curved motion,* the acceleration of point B is expressed in terms of its tangential and normal components, $(\mathbf{a}_B)_t$ and $(\mathbf{a}_B)_n$. In particular, the magnitude of the *normal component* of acceleration of any point following a curved path is determined by using Eq. 12–37. In the case of point B,

$$(a_B)_n = v_B^2/\rho = (2\ \text{ft/sec})^2/4\ \text{ft} = 1\ \text{ft/sec}^2$$

(a)

(b)

(c)

Fig. 13-24

Equation 13-17 applied to the acceleration vectors acting at point B in Fig. 13-24c yields:

$$(\mathbf{a}_B)_t + (\mathbf{a}_B)_n = \mathbf{a}_A + \boldsymbol{\alpha} \times \mathbf{r}_{B/A} + \boldsymbol{\omega} \times (\boldsymbol{\omega} \times \mathbf{r}_{B/A})$$

Expressing each of the terms in this equation in scalar form, which requires the use of Eqs. 13-20 and 13-21, we have

$$(a_B)_t + 1 \text{ ft/sec}^2 = 3 \text{ ft/sec}^2 + \alpha(10 \text{ ft}) + (0.283 \text{ rad/sec})^2(10 \text{ ft})$$

Equating components in the horizontal and vertical direction yields:

$(\xrightarrow{+})$ $(a_B)_t \cos 45° - 1 \cos 45° = 3 \cos 45° + 0 - (0.283)^2(10)$

Hence,

$$(a_B)_t = 2.87 \text{ ft/sec}^2$$

$(+\uparrow)$ $(a_B)_t \sin 45° + 1 \sin 45° = -3 \sin 45° + \alpha(10) + 0$

Substituting for $(a_B)_t$ and solving for α, we get

$$\alpha = 0.486 \text{ rad/sec}^2   \quad \zeta\alpha \qquad\qquad Ans.$$

Example 13–12

The collar C, shown in Fig. 13–25a, is moving downward with an acceleration of 1 ft/sec², and at the instant shown it has a speed of 12 ft/sec. Determine the angular acceleration of link AB at this instant.

Fig. 13–25

Solution

Velocity Analysis. Before applying Eq. 13–17 it is first necessary to determine the angular velocities of links AB and CB. The velocity of point C is directed downward, and due to the rotational motion of link AB about the fixed point A, the velocity of point B is directed horizontally (to the right), Fig. 13–25b. Knowing this, the instantaneous center of zero velocity for link CB is located as shown in Fig. 13–25c. From the figure, $r_{C/IC} = r_{B/IC} = 2$ ft and ω_{CB} is assumed to act counterclockwise. Hence,

$$v_C = \omega_{CB}(r_{C/IC})$$
$$12 \text{ ft/sec} = \omega_{CB}(2 \text{ ft})$$
$$\omega_{CB} = 6 \text{ rad/sec}$$

Therefore,

$$v_B = \omega_{CB}(r_{B/IC}) = 6 \text{ rad/sec } (2 \text{ ft}) = 12 \text{ ft/sec}$$

From Fig. 13–25b,

$$\omega_{AB} = \frac{v_B}{r_{B/A}} = \frac{12 \text{ ft/sec}}{2 \text{ ft}} = 6 \text{ rad/sec}$$

Acceleration Analysis. Since the collar moves along a *straight-line path* $a_C = 1$ ft/sec,^{2} directed downward. (Would this direction be downward if the path for C was *curved?*) Eq. 13–17 may be applied between points B and C since both of these points lie on the *same* rigid body. Choosing point C as a base point, the kinematic diagrams for acceleration are shown in Fig. 13–25d. At the instant shown, the link has an angular velocity of $\omega_{CB} = 6$ rad/sec, as computed above, and an angular acceleration of α_{CB}. In accordance with the direction of these vectors, do you understand why $(\mathbf{a}_{B/C})_t$ and $(\mathbf{a}_{B/C})_n$ are *directed* as shown in Fig. 13–25d? Since point B moves along a *curved path* having a radius of curvature of $r_{B/A} = 2$ ft, the acceleration of point B is resolved into two components, $(\mathbf{a}_B)_x$ and $(\mathbf{a}_B)_y$. The magnitudes of these two components can be expressed in terms of the angular motion of link AB since links CB and AB are *pin connected.* Assuming link AB has an angular acceleration of $\boldsymbol{\alpha}_{AB}$, as shown in Fig. 13–25e, the magnitudes of the acceleration components $(\mathbf{a}_B)_x$ and $(\mathbf{a}_B)_y$ are therefore:

$$(a_B)_x = \alpha_{AB}(r_{B/A}) = \alpha_{AB}(2 \text{ ft}) \qquad \rightarrow \qquad (1)$$
$$(a_B)_y = (\omega_{AB})^2(r_{B/A}) = (6 \text{ rad/sec})^2(2 \text{ ft}) = 72 \text{ ft/sec}^2 \qquad \uparrow$$

Applying Eq. 13–17 in reference to the vectors shown at point B in Fig. 13–25d we have,

$$\mathbf{a}_B = (\mathbf{a}_B)_x + (\mathbf{a}_B)_y = \mathbf{a}_C + (\boldsymbol{\alpha}_{CB} \times \mathbf{r}_{B/C}) + \boldsymbol{\omega}_{CB} \times (\boldsymbol{\omega}_{CB} \times \mathbf{r}_{B/C})$$

Substituting in the necessary data,

$\alpha_{AB}(2 \text{ ft}) + 72 \text{ ft/sec}^2 = 1 \text{ ft/sec}^2 + \alpha_{CB}(2\sqrt{2} \text{ ft}) + (6 \text{ rad/sec})^2(2\sqrt{2} \text{ ft})$

$\quad\quad\quad\rightarrow \quad\quad\quad\quad \uparrow \quad\quad\quad\quad \downarrow \quad\quad\quad \angle\!\!\!\!^{\nearrow\,45°} \quad\quad\quad {}^{45°}\!\!\searrow$

Equating components in the horizontal and vertical directions,

$(\xrightarrow{+}) \quad \alpha_{AB}(2) + 0 = 0 + \alpha_{CB}(2\sqrt{2}) \cos 45° - (6)^2(2\sqrt{2}) \cos 45°$

or

$$2\alpha_{AB} = 2\alpha_{CB} - 72$$

$(+\uparrow) \quad 0 + 72 = -1 + \alpha_{CB}(2\sqrt{2}) \sin 45° + (6)^2(2\sqrt{2}) \sin 45°$

or

$$1 = 2\alpha_{CB}$$

Solving,

$$\alpha_{CB} = 0.5 \text{ rad/sec}^2$$
$$\alpha_{AB} = -35.5 \text{ rad/sec}^2$$

The negative sign for α_{AB} indicates that $\boldsymbol{\alpha}_{AB}$ acts in the opposite direction to that shown in Fig. 13–25e. (By Eq. (1), this also indicates that the direction of $(\mathbf{a}_B)_x$ is opposite to that shown in Figs. 13–25d and 13–25e.) Hence,

$$\alpha_{AB} = 35.5 \text{ rad/sec}^2 \quad \searrow \boldsymbol{\alpha}_{AB} \quad\quad\quad\quad Ans.$$

13–9. Method for Determining Acceleration by Using Cartesian Unit Vectors

The three kinematic diagrams for acceleration used in the solution of the example problems in Sec. 13–8 provide a useful means of accounting for each of the component terms in Eq. 13–17. Furthermore, these diagrams provide a means for obtaining a physical "insight" regarding the kinematics of the problem. As in the case of velocity (Sec. 13–6), it is possible to apply Eq. 13–17 directly without drawing the three kinematic diagrams. This is done by expressing each of the vectors in this equation in terms of Cartesian unit vectors. Since this approach is very methodical in application, it has a distinct advantage for solving problems involving mechanisms with several connected members. Although geometry is less of a problem when using a vector approach for the solution, this method is not as direct as the one discussed previously. In general, more computations will have to be made because of the added vector algebra.

The method is illustrated in the solution of the following two example problems.

Fig. 13–26

Example 13–13

The collar C shown in Fig. 13–26a is moving downward with an acceleration of 1 ft/sec^2, and at the instant shown, it has a speed of 12 ft/sec. Determine the angular acceleration of link AB at this instant.

Solution

This problem is the same as Example 13–12. Using Cartesian unit vectors, we may express the position vectors $\mathbf{r}_{B/A}$ and $\mathbf{r}_{B/C}$, located on AB and CB, and the angular motions shown in Fig. 13–26b as

$$\mathbf{r}_{B/A} = \{-2\mathbf{j}\} \text{ ft} \qquad \mathbf{r}_{B/C} = \{2\mathbf{i} - 2\mathbf{j}\} \text{ ft}$$
$$\boldsymbol{\omega}_{AB} = \omega_{AB}\mathbf{k} \qquad \boldsymbol{\alpha}_{AB} = \alpha_{AB}\mathbf{k}$$
$$\boldsymbol{\omega}_{CB} = \omega_{CB}\mathbf{k} \qquad \boldsymbol{\alpha}_{CB} = \alpha_{CB}\mathbf{k}$$
$$\mathbf{v}_C = \{-12\mathbf{j}\} \text{ ft/sec} \qquad \mathbf{a}_C = \{-1\mathbf{j}\} \text{ ft/sec}^2$$

Velocity Analysis. Since link AB is subjected to rotational motion about a *fixed axis* passing through point A, using Eq. 13–7 we have:

$$\mathbf{v}_B = \boldsymbol{\omega}_{AB} \times \mathbf{r}_{B/A}$$
$$\mathbf{v}_B = (\omega_{AB}\mathbf{k}) \times (-2\mathbf{j}) = 2\omega_{AB}\mathbf{i}$$

Link CB undergoes general plane motion, thus using Eq. 13–13 and the above result:

$$\mathbf{v}_B = \mathbf{v}_C + \boldsymbol{\omega}_{CB} \times \mathbf{r}_{B/C}$$
$$2\omega_{AB}\mathbf{i} = -12\mathbf{j} + (\omega_{CB}\mathbf{k}) \times (2\mathbf{i} - 2\mathbf{j})$$

or

$$2\omega_{AB}\mathbf{i} = -12\mathbf{j} + 2\omega_{CB}\mathbf{j} + 2\omega_{CB}\mathbf{i}$$

Equating the respective $\mathbf{i}$ and $\mathbf{j}$ components gives

$$2\omega_{AB} = 2\omega_{CB}$$
$$0 = -12 + 2\omega_{CB}$$

Thus,

$$\omega_{AB} = \omega_{CB} = 6 \text{ rad/sec}$$

Acceleration Analysis. Applying Eq. 13–10 between points *A* and *B* on link *AB*, we have

$$\mathbf{a}_B = \boldsymbol{\alpha}_{AB} \times \mathbf{r}_{B/A} + \boldsymbol{\omega}_{AB} \times (\boldsymbol{\omega}_{AB} \times \mathbf{r}_{B/A})$$
$$\mathbf{a}_B = (\alpha_{AB}\mathbf{k}) \times (-2\mathbf{j}) + [6\mathbf{k}] \times [(6\mathbf{k}) \times (-2\mathbf{j})]$$

or

$$\mathbf{a}_B = 2\alpha_{AB}\mathbf{i} + 72\mathbf{j}$$

Applying Eq. 13–17 between points *B* and *C* on rod *BC*, using the above value for $\mathbf{a}_B$, yields

$$\mathbf{a}_B = \mathbf{a}_C + \boldsymbol{\alpha}_{CB} \times \mathbf{r}_{B/C} + \boldsymbol{\omega}_{CB} \times (\boldsymbol{\omega}_{CB} \times \mathbf{r}_{B/C})$$
$$2\alpha_{AB}\mathbf{i} + 72\mathbf{j} = -\mathbf{j} + (\alpha_{CB}\mathbf{k}) \times (2\mathbf{i} - 2\mathbf{j}) + [6\mathbf{k}] \times [(6\mathbf{k}) \times (2\mathbf{i} - 2\mathbf{j})]$$

or

$$2\alpha_{AB}\mathbf{i} + 72\mathbf{j} = -\mathbf{j} + 2\alpha_{CB}\mathbf{j} + 2\alpha_{CB}\mathbf{i} - 72\mathbf{i} + 72\mathbf{j}$$

Equating the respective **i** and **j** components:

$$2\alpha_{AB} = 2\alpha_{CB} - 72 \tag{1}$$
$$72 = -1 + 2\alpha_{CB} + 72 \tag{2}$$

Hence,

$$\alpha_{CB} = 0.5 \text{ rad/sec}^2$$
$$\alpha_{AB} = -35.5 \text{ rad/sec}^2$$

Thus,

$$\boldsymbol{\alpha}_{AB} = \{-35.5\mathbf{k}\} \text{ rad/sec}^2 \qquad Ans.$$

Example 13–14

The crankshaft *AB* of an automobile engine turns with a clockwise angular acceleration of 20 rad/sec², Fig. 13–27a. Determine the acceleration of the piston at the instant when *AB* is in the position shown and has a clockwise angular velocity of 10 rad/sec.

Solution

Points *B* and *C* can be located using position vectors $\mathbf{r}_{B/A}$ and $\mathbf{r}_{C/B}$ on *AB* and *BC*, Fig. 13–27b. These vectors can be written as

$$\mathbf{r}_{B/A} = \{-\tfrac{1}{4}\sin 45°\mathbf{i} + \tfrac{1}{4}\cos 45°\mathbf{j}\} \text{ ft} = \{-0.177\mathbf{i} + 0.177\mathbf{j}\} \text{ ft}$$
$$\mathbf{r}_{C/B} = \{\tfrac{3}{4}\sin 13.6°\mathbf{i} + \tfrac{3}{4}\cos 13.6°\mathbf{j}\} \text{ ft} = \{0.177\mathbf{i} + 0.729\mathbf{j}\} \text{ ft}$$

Since the piston is subjected to translation, and the line of action of the angular motion is perpendicular to the plane of motion, we have

$$\boldsymbol{\omega}_{AB} = \{-10\mathbf{k}\} \text{ rad/sec} \qquad \boldsymbol{\alpha}_{AB} = \{-20\mathbf{k}\} \text{ rad/sec}^2$$
$$\boldsymbol{\omega}_{BC} = \omega_{BC}\mathbf{k} \qquad \boldsymbol{\alpha}_{BC} = \alpha_{BC}\mathbf{k}$$
$$\mathbf{v}_C = v_C\mathbf{j} \qquad \mathbf{a}_C = a_C\mathbf{j}$$

Velocity Analysis. Since the crankshaft AB is subjected to pure rotation about a fixed axis, using Eq. 13–7 we get

$$\mathbf{v}_B = \boldsymbol{\omega}_{AB} \times \mathbf{r}_{B/A}$$
$$= (-10\mathbf{k}) \times (-0.177\mathbf{i} + 0.177\mathbf{j})$$
$$= 1.77\mathbf{j} + 1.77\mathbf{i}$$

The angular velocity of rod BC may be determined by applying Eq. 13–13 to points B and C on the rod and using the above result.

$$\mathbf{v}_C = \mathbf{v}_B + \boldsymbol{\omega}_{BC} \times \mathbf{r}_{C/B}$$
$$v_C\mathbf{j} = 1.77\mathbf{j} + 1.77\mathbf{i} + (\omega_{BC}\mathbf{k}) \times (0.177\mathbf{i} + 0.729\mathbf{j})$$
$$v_C\mathbf{j} = 1.77\mathbf{j} + 1.77\mathbf{i} + 0.177\omega_{BC}\mathbf{j} - 0.729\omega_{BC}\mathbf{i}$$

Equating the respective $\mathbf{i}$ and $\mathbf{j}$ components,

$$0 = 1.77 - 0.729\omega_{BC}$$
$$v_C = 1.77 + 0.177\omega_{BC}$$

Hence,

$$\omega_{BC} = 2.42 \text{ rad/sec}, \qquad v_C = 2.20 \text{ ft/sec}$$

Acceleration Analysis. Point B moves in a *circular* path because of the rotation of rod AB about a fixed axis passing through point A. Using Eq. 13–10, we find the acceleration of this point to be

$$\mathbf{a}_B = \boldsymbol{\alpha}_{AB} \times \mathbf{r}_{B/A} + \boldsymbol{\omega}_{AB} \times (\boldsymbol{\omega}_{AB} \times \mathbf{r}_{B/A})$$
$$= (-20\mathbf{k}) \times (-0.177\mathbf{i} + 0.177\mathbf{j}) + [-10\mathbf{k}]$$
$$\times [(-10\mathbf{k}) \times (-0.177\mathbf{i} + 0.177\mathbf{j})]$$
$$= \{21.2\mathbf{i} - 14.14\mathbf{j}\} \text{ ft/sec}^2$$

Since point C moves along a *straight-line path*, $\mathbf{a}_C = a_C\mathbf{j}$, Fig. 13–27b. Applying Eq. 13–17 to rod BC and using the value of $\boldsymbol{\omega}_{BC} = \{2.42\mathbf{k}\}$ rad/sec, we have

$$\mathbf{a}_C = \mathbf{a}_B + \boldsymbol{\alpha}_{BC} \times \mathbf{r}_{C/B} + \boldsymbol{\omega}_{BC} \times (\boldsymbol{\omega}_{BC} \times \mathbf{r}_{C/B})$$
$$a_C\mathbf{j} = 21.2\mathbf{i} - 14.14\mathbf{j} + (\alpha_{BC}\mathbf{k}) \times (0.177\mathbf{i} + 0.729\mathbf{j})$$
$$+ [2.42\mathbf{k}] \times [(2.42\mathbf{k}) \times (0.177\mathbf{i} + 0.729\mathbf{j})]$$
$$= 21.2\mathbf{i} - 14.14\mathbf{j} + 0.177\alpha_{BC}\mathbf{j} - 0.729\alpha_{BC}\mathbf{i} - 1.04\mathbf{i} - 4.27\mathbf{j}$$

(a)

(b)

Fig. 13–27

Equating the respective **i** and **j** components,

$$0 = 20.16 - 0.729\alpha_{BC}$$
$$a_C = 0.177\alpha_{BC} - 18.41$$

Solving, we have

$$\alpha_{BC} = 27.7 \text{ rad/sec}^2$$
$$a_C = -13.5 \text{ ft/sec}^2 \qquad\qquad Ans.$$

The negative sign indicates that the piston is decelerating; i.e., $\mathbf{a}_C = \{-13.5\mathbf{j}\}$ ft/sec^2. This causes the speed v_C of the piston to decrease until the crankshaft AB and the connecting rod BC become vertical, at which time the piston is momentarily at rest.

Problems

13-44. The center of the wheel moves to the right with a velocity of 20 ft/sec and has an acceleration of 10 ft/sec^2 at the instant shown. Determine the acceleration of points A and B at this instant. Assume that the wheel does not slip at A.

Prob. 13-45

Prob. 13-44

13-45. The center of the pulley is being lifted vertically with an acceleration of 3 ft/sec^2. At the instant shown the velocity is 6 ft/sec. Determine the acceleration of points A and B. Assume that the rope does not slip on the pulley's surface.

13-46. Determine the acceleration of points B and D located on the planetary gear A in Prob. 13-38. Use the data given in Prob. 13-38.

13-47. The link AB is turning clockwise with a constant angular velocity of 10 rad/sec. Determine the angular acceleration of link DC at the instant shown and the acceleration of point C.

Prob. 13-47

Prob. 13-49

13-48. At a given instant, the slider block *A* has the velocity and deceleration shown. Determine the acceleration of block *B* and the angular acceleration of the link at this same instant.

Prob. 13-48

13-49. At a certain instant link *AB* has a clockwise angular velocity of 2 rad/sec and a counterclockwise angular deceleration of 2 rad/sec². Determine the acceleration of point *C*, and the angular acceleration of link *CD* at the instant shown.

13-50. At a given instant, the slider block *B* is traveling to the right with the velocity and acceleration as shown. Determine the angular acceleration of the wheel at this instant. Assume the pin connection at *A* is located at the periphery of the wheel.

Prob. 13-50

13-51. The slider block *B* is moving to the right with an acceleration of 2 ft/sec². At the instant shown, its velocity is 6 ft/sec. Determine the angular acceleration of link *AB* and the acceleration of point *A* at this instant.

Prob. 13-51

551

13–52. The wheel is rotating counterclockwise with an angular acceleration of 3 rad/sec², and at the instant shown it has an angular velocity of 2 rad/sec. Determine the acceleration of the slider block A at this instant.

Prob. 13–52

13–53. Determine the angular acceleration of link AB if link DE has the instantaneous angular velocity and angular deceleration shown.

Prob. 13–53

13–54. Determine the instantaneous acceleration of point C located on the mechanism in Prob. 13–53.

13–55. Assume that the block C in Prob. 13–37 has an acceleration of 1 ft/sec² and an instantaneous velocity of 4 ft/sec, both acting up the inclined slot. Determine the angular acceleration of link AB at the instant shown.

13–56. The collar C has an instantaneous velocity and deceleration as shown. Determine the angular acceleration of links AB and BC at the instant shown.

Prob. 13–56

13–57. The disk A has an instantaneous angular velocity of 2 rad/sec, and an angular acceleration of 4 rad/sec², both acting clockwise. Determine the acceleration of point C on wheel B. The wheels are in contact with one another and *do not slip*. The extension rod DE is pin-connected at points located on the rim of each wheel.

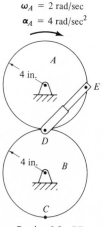

Prob. 13–57

13–58. At a given instant the gear racks have the velocities and accelerations shown. Determine the acceleration of points A and B.

552

Prob. 13–58

13-60. Determine the acceleration of points A and B when $\theta = 0°$ for the link in Prob. 13–59.

13-61. The roller A is confined to move in the semi-circular groove. If the disk turns with a constant angular velocity of 3 rad/sec, determine the acceleration of roller A at the instant shown.

13-59. Gear C is turning at a constant angular velocity of 6 rad/sec. Determine the acceleration of points A and B at the instant when $\theta = 90°$.

Prob. 13–59

Prob. 13–61

*13-10. Rotation of a Rigid Body About a Fixed Point

When a rigid body rotates about a fixed point the distance r from the point to any particle P of the body is the *same* for *any position* of the body. Thus, the path of motion for the particle P lies on the *surface of a sphere* having a radius r and centered at the fixed point. In general, therefore, the path taken by the particle is described in three dimensions. Since motion along this path occurs only as a series of angular displacements made during a finite time interval, it is perhaps wise to first develop a familiarity with some of the properties of angular displacements. *Euler's theorem* is important in this regard. This theorem states that *any number of rotational displacements about different axes through a point is equivalent to a single rotational displacement.* A procedure for obtaining this single equivalent rotation for a pair of arbitrary rotations will presently be demonstrated. Any number of rotations can, in steps, be combined into pairs, and each pair further reduced to combine into one rotation.

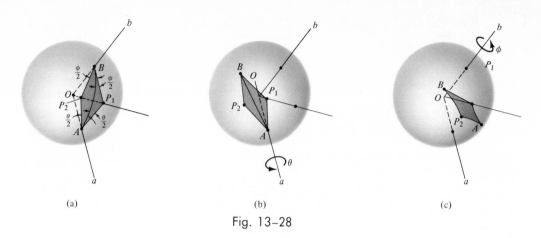

Fig. 13-28

Since a sphere represents the locus of points traced by all the possible paths of a particle P, located in a rigid body, let us, for purposes of simplicity, consider the rigid body itself to be a sphere which is allowed to rotate about its central fixed point O. This situation is shown in Fig. 13-28a. The sphere is to be rotated about the fixed axes Oa and Ob by amounts θ and ϕ, respectively. Before the sphere is rotated, draw on the sphere (Fig. 13-28a) the arc joining points A and B. Also, construct arcs from A and B which lie at angles of $\theta/2$ and $\phi/2$, respectively, on both sides of AB. These arcs intersect at points P_1 and P_2, as shown in the figure. From the construction, the corresponding sides and angles of spherical triangles AP_1B and AP_2B are equal.

When the sphere is first rotated about the *fixed axis Oa* by an amount θ, the graphical construction moves to the position shown in Fig. 13-28b. Because of the nature of the construction, point P_1 coincides with the position of where P_2 was located before the rotation. Next, rotate the sphere about the *fixed axis Ob* by an amount ϕ. This returns only the point P_1 back to its original position. The other points move off to some other location, as shown in Fig. 13-28c. Thus, the two rotations of θ and ϕ about the fixed axes Oa and Ob are equivalent to a single rotation about an axis passing through O and P_1.

If the rotational displacements had taken place first by an amount ϕ about axis Ob and then θ about axis Oa, the result would have been equivalent to a single rotation about a fixed axis passing through OP_2. Since the *order of addition* of the rotations is important, this example again shows that *finite rotations* are *not* vectors.* If successive *infinitesimal rotations* of the sphere were considered, Euler's theorem would again be valid. Furthermore, the resultant single rotation would be *unique* for both rotations, since infinitesimal rotations *are* vector quantities and thus, the order of addition is immaterial (see Appendix E).

*Refer also to the discussion which relates to Fig. 13-3.

From this discussion we may conclude that motion of a rigid body about a fixed point during a very small time interval Δt is actually a rotation $\Delta \boldsymbol{\theta}$ of the body about an *instantaneous axis of rotation*. Consequently, as $\Delta t \to 0$, the limit $\Delta \boldsymbol{\theta}/\Delta t$ approaches the angular velocity $\boldsymbol{\omega}$ of the body. As in the case of rotation about a fixed axis, the line of action of $\boldsymbol{\omega}$ is in the *same direction* as the instantaneous axis of rotation. This is illustrated in Fig. 13–29 for the body rotating about the fixed point O. If the position vector $\mathbf{r}_{OP}$ defines the location of a particle P of the body at the instant considered, the velocity $\mathbf{v}_P$ of the particle is thereby determined by the cross product.

Fig. 13–29

$$\mathbf{v}_P = \boldsymbol{\omega} \times \mathbf{r}_{OP} \qquad (13\text{–}22)$$

Differentiating this expression with respect to time yields the acceleration

$$\mathbf{a}_P = (\mathbf{a}_P)_t + (\mathbf{a}_P)_n = \boldsymbol{\alpha} \times \mathbf{r}_{OP} + \boldsymbol{\omega} \times (\boldsymbol{\omega} \times \mathbf{r}_{OP}) \qquad (13\text{–}23)$$

These equations are identical to Eqs. 13–7 and 13–10, which were determined for a rigid body rotating about a fixed axis. Specifically, in the case of *fixed-axis rotation,* the angular acceleration $\boldsymbol{\alpha}$ of the body defines *only* a *change* in the *magnitude* of $\boldsymbol{\omega}$ since the line of action of $\boldsymbol{\alpha}$ is always the *same* as $\boldsymbol{\omega}$. For motion about a *fixed point,* however, the body's angular acceleration $\boldsymbol{\alpha}$ accounts for a change in *both* the *magnitude and direction* of $\boldsymbol{\omega}$. (Recall that $\boldsymbol{\omega}$ always acts in the direction of the instantaneous axis of rotation which, in general, is changing during each instant.) Therefore, $\boldsymbol{\alpha}$ is *not,* in general, directed along the axis of rotation. Fig. 13–29.

The direction of $\boldsymbol{\alpha}$ can be determined by drawing a curve which describes the path taken by the tip of the $\boldsymbol{\omega}$ vector. Vector $\boldsymbol{\alpha}$ must act tangent to this path at any given instant, since the time rate of change of $\boldsymbol{\omega}$ is equal to $\boldsymbol{\alpha}$. As the instantaneous axis of rotation (or the line of action of $\boldsymbol{\omega}$) changes in space, the locus of points generates a fixed *space cone*. If the change in this axis is viewed with respect to the body, the axis generates a *body cone*, Fig. 13–30. At a given instant these cones are tangent along the instantaneous axis of rotation and, when the body is in motion, the body cone appears to roll on the fixed space cone. See Example 13–15.

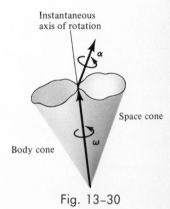

Fig. 13–30

*13–11. General Motion of a Rigid Body

The most general motion of a rigid body moving in space occurs when any point of the body has a specified velocity $\mathbf{v}$ and acceleration $\mathbf{a}$, and the body is rotating with an angular velocity $\boldsymbol{\omega}$ and angular acceleration $\boldsymbol{\alpha}$. In general, none of these four vectors will be collinear with one another,

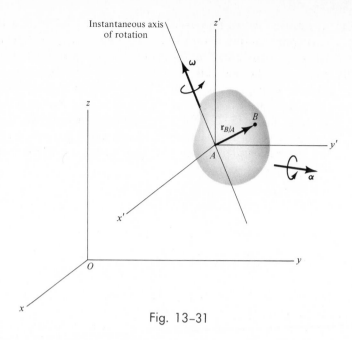

Fig. 13–31

since **a** and **α** measure the change made in *both* the *magnitude and direction* of **v** and **ω**, respectively.

The general motion of the body may be analyzed, at a given instant, as the sum of a translation in which all points of the body have the same instantaneous velocity and acceleration as a selected base point, and a single rotation of the body about an *instantaneous axis of rotation* passing through the base point. The rotational aspect of the motion is, of course, the consequence of Euler's theorem presented in Sec. 13–10. Shown in Fig. 13–31 is a rigid body subjected to general motion. The translation and rotation are conveniently analyzed using two coordinate systems to study each motion separately. The absolute motion of points A and B are measured from the fixed xyz frame of reference. The origin of the $x'y'z'$ coordinate system is fixed to the base point A in the body. As in the case of plane motion, the axes of this coordinate system are only allowed to *translate* with respect to the fixed frame. Thus, since the body is rigid, the motion of point B relative to an observer stationed at A is the same as motion of the body about a fixed point. This relative rotational motion occurs about an instantaneous axis of rotation and is defined by Eqs. 13–22 and 13–23. Recalling the discussion pertaining to the relative velocity and acceleration Eqs. 13–13 and 13–17, the absolute velocity and acceleration of point B can be determined by the equations

$$\mathbf{v}_B = \mathbf{v}_A + (\boldsymbol{\omega} \times \mathbf{r}_{B/A}) \tag{13–24}$$

and

$$a_B = a_A + (\alpha \times r_{B/A}) + \omega \times (\omega \times r_{B/A}) \qquad (13\text{--}25)$$

Here ω and α represent the angular velocity and angular acceleration of the body at the instant of time considered. These two equations are identical to those describing the general plane motion of a rigid body. However, difficulty in application arises in general motion because all the vectors used in these equations are not collinear. Recall that for plane motion, α and ω are always collinear, because the instantaneous axis of rotation is always *perpendicular* to the plane of motion. Hence, for plane motion, α measures only the change in magnitude of ω. For general motion, however, the line of action of ω is always directed along the instantaneous axis of rotation and α is *not* always along this axis because it measures both the change in magnitude and direction of ω.

Example 13–15

The gyro top shown in Fig. 13–32a is spinning about a horizontal axis with a constant angular velocity of $\omega_s = 3$ rad/sec, while the horizontal shaft of the top is precessing (rotating) about the vertical shaft at a constant rate of $\omega_p = 1$ rad/sec. Determine the velocity and acceleration of point A on the top when the top is in the position shown.

Solution

The top is rotating about the *fixed point O*. To determine the velocity and acceleration of point A it is first necessary to determine the resultant angular velocity ω and angular acceleration α of the top. (These vectors are used in Eqs. 13–22 and 13–23.) The resultant angular velocity is simply the vector addition of the spin and precession. Thus,

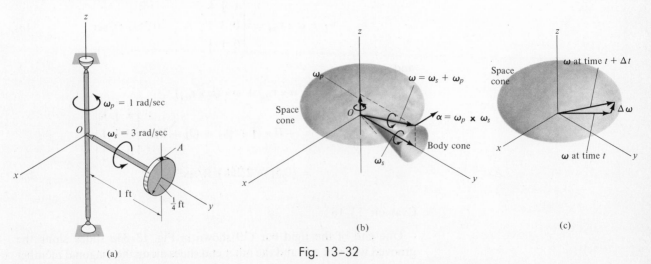

(a)   (b)   (c)

Fig. 13-32

$$\boldsymbol{\omega} = \boldsymbol{\omega}_s + \boldsymbol{\omega}_p = \{3\mathbf{j} + 1\mathbf{k}\} \text{ rad/sec}$$

At first glance it may appear that the top is not actually rotating with this angular velocity since it is generally more difficult to imagine the resultant of angular motions in comparison with linear motions. To further understand the angular motion, the appearance of the problem may be changed without changing its characteristics. This is done by replacing the top with a body cone rolling over an inverted stationary space cone, Fig. 13–32*b*. The instantaneous axis of rotation is along the line of contact of the cones. This axis defines the direction of the resultant $\boldsymbol{\omega}$, having components $\boldsymbol{\omega}_s$ and $\boldsymbol{\omega}_p$.

Although the magnitude of $\boldsymbol{\omega}$ is constant, its direction is changing. This change in direction of $\boldsymbol{\omega}$ gives rise to the angular acceleration $\boldsymbol{\alpha}$ of the top. As shown in Fig. 13–32*c*, the change in direction of $\boldsymbol{\omega}$ (or the instantaneous axis of rotation) during the time interval Δt is $\Delta \boldsymbol{\omega}$. By definition,

$$\boldsymbol{\alpha} = \lim_{\Delta t \to 0} \frac{\Delta \boldsymbol{\omega}}{\Delta t} = \frac{d\boldsymbol{\omega}}{dt}$$

Thus, $\boldsymbol{\alpha}$ acts in a *direction* which is *tangent* to the head of $\boldsymbol{\omega}$, that is, in the limiting direction of $\Delta \boldsymbol{\omega}$. The *magnitude* of $\boldsymbol{\alpha}$ is $\omega_p(\omega_s) = 3$ rad/sec². Both the direction and magnitude of $\boldsymbol{\alpha}$ can be accounted for by the cross product:

$$\boldsymbol{\alpha} = \boldsymbol{\omega}_p \times \boldsymbol{\omega}_s = 1\mathbf{k} \times 3\mathbf{j} = \{-3\mathbf{i}\} \text{ rad/sec}^2$$

See Fig. 13–32*b*.

Having determined $\boldsymbol{\omega}$ and $\boldsymbol{\alpha}$, we can compute the velocity and acceleration of point A using Eqs. 13–22 and 13–23, in which case

$$\mathbf{v}_A = \boldsymbol{\omega} \times \mathbf{r}_{OA} = \begin{vmatrix} \mathbf{i} & \mathbf{j} & \mathbf{k} \\ 0 & 3 & 1 \\ 0 & 1 & \frac{1}{4} \end{vmatrix} = \{-0.25\mathbf{i}\} \text{ ft/sec} \qquad \textit{Ans.}$$

and

$$\mathbf{a}_A = \boldsymbol{\alpha} \times \mathbf{r}_{OA} + \boldsymbol{\omega} \times (\boldsymbol{\omega} \times \mathbf{r}_{OA})$$

$$= -3\mathbf{i} \times (\mathbf{j} + \tfrac{1}{4}\mathbf{k}) + (3\mathbf{j} + \mathbf{k}) \times \begin{vmatrix} \mathbf{i} & \mathbf{j} & \mathbf{k} \\ 0 & 3 & 1 \\ 0 & 1 & \frac{1}{4} \end{vmatrix}$$

$$= \{0.5\mathbf{j} - 2.25\mathbf{k}\} \text{ ft/sec}^2 \qquad \textit{Ans.}$$

Example 13–16

One end of the rigid bar CD shown in Fig. 13–33*a* slides along the grooved wall slot AB, and the other end slides along the diagonal member

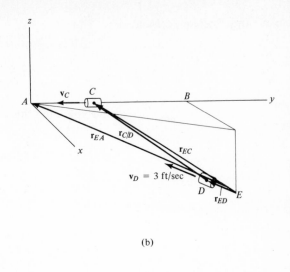

(a)

(b)

Fig. 13–33

AE of the fixed triangular frame. When the collar at D is 2 ft from E, it is moving toward A at a speed of 3 ft/sec. Determine the velocity of the slider at C and the angular velocity of the rigid bar. The bar is connected to sliders at its end points by means of ball-and-socket joints. Assume that the angular velocity of the bar is directed normal to the axis of the bar.

Solution

Referring to Fig. 13–33b, we may express vectors $\mathbf{v}_C$, $\mathbf{v}_D$, $\mathbf{r}_{C/D}$, and $\boldsymbol{\omega}_{CD}$ using Cartesian unit vectors defined from the fixed xyz reference frame:

$$\mathbf{v}_C = -v_C\mathbf{j}$$

$$\mathbf{v}_D = (3 \text{ ft/sec})\frac{\mathbf{r}_{EA}}{r_{EA}} = (3 \text{ ft/sec})\left(\frac{-3}{\sqrt{125}}\mathbf{i} + \frac{-10}{\sqrt{125}}\mathbf{j} + \frac{4}{\sqrt{125}}\mathbf{k}\right)$$
$$= \{-0.805\mathbf{i} - 2.68\mathbf{j} + 1.07\mathbf{k}\} \text{ ft/sec}$$

$$\mathbf{r}_{C/D} = \mathbf{r}_{EC} - \mathbf{r}_{ED} = \{-3\mathbf{i} - 6\mathbf{j} + 4\mathbf{k}\} \text{ ft} - (2 \text{ ft})\frac{\mathbf{r}_{EA}}{r_{EA}}$$

$$= \{-3\mathbf{i} - 6\mathbf{j} + 4\mathbf{k} + \frac{6}{\sqrt{125}}\mathbf{i} + \frac{20}{\sqrt{125}}\mathbf{j} - \frac{8}{\sqrt{125}}\mathbf{k}\} \text{ ft}$$

$$= \{-2.46\mathbf{i} - 4.21\mathbf{j} + 3.28\mathbf{k}\} \text{ ft}$$

$$\boldsymbol{\omega}_{CD} = \omega_x\mathbf{i} + \omega_y\mathbf{j} + \omega_z\mathbf{k}$$

Rod CD is subjected to general motion. Why? The velocity of point C on the rod may be related to the velocity of point D by the equation

$$\mathbf{v}_C = \mathbf{v}_D + (\boldsymbol{\omega}_{CD} \times \mathbf{r}_{C/D})$$

Thus,

$$-v_C\mathbf{j} = -0.805\mathbf{i} - 2.68\mathbf{j} + 1.07\mathbf{k} + \begin{vmatrix} \mathbf{i} & \mathbf{j} & \mathbf{k} \\ \omega_x & \omega_y & \omega_z \\ -2.46 & -4.21 & 3.28 \end{vmatrix}$$

Expanding this expression and equating the respective $\mathbf{i}$, $\mathbf{j}$, and $\mathbf{k}$ components yields

$$3.28\omega_y + 4.21\omega_z - 0.805 = 0 \tag{1}$$
$$-3.28\omega_x - 2.46\omega_z + v_C - 2.68 = 0 \tag{2}$$
$$-4.21\omega_x + 2.46\omega_y + 1.07 = 0 \tag{3}$$

There are four unknowns in these three equations,* namely, ω_x, ω_y, ω_z, and v_C. To determine a fourth equation it is necessary to specify the *direction* of the angular velocity $\boldsymbol{\omega}_{CD}$. In particular, since the bar is connected at its end points by means of ball-and-socket joints, the component of $\boldsymbol{\omega}_{CD}$ acting *along* the axis of the rod has no effect in changing the velocities $\mathbf{v}_C$ and $\mathbf{v}_D$ of the collars. This is because the rod is *free to rotate* about its axis. Therefore, any *arbitrary value* of $\boldsymbol{\omega}_{CD}$ can be assumed in this direction without changing the solution of Eqs. (1), (2), and (3). However, if for example, $\boldsymbol{\omega}_{CD}$ acts *perpendicular* to the axis of the rod, as stated in the problem, it must have a *unique magnitude* to satisfy the solution. Perpendicularity is guaranteed provided the dot product†

$$\boldsymbol{\omega}_{CD} \cdot \mathbf{r}_{C/D} = 0$$
$$(\omega_x\mathbf{i} + \omega_y\mathbf{j} + \omega_z\mathbf{k}) \cdot (-2.46\mathbf{i} - 4.21\mathbf{j} + 3.28\mathbf{k}) = 0$$

or

$$-2.46\omega_x - 4.21\omega_y + 3.28\omega_z = 0 \tag{4}$$

Equations (1) through (4) may be solved simultaneously by using, for example, the computer program listed in Appendix A. The solution yields

$$\omega_x = 0.255 \text{ rad/sec}$$
$$\omega_y = 0.0 \text{ rad/sec}$$
$$\omega_z = 0.191 \text{ rad/sec}$$
$$v_C = 3.99 \text{ ft/sec}$$

---

*Although this is the case, we may solve these three equations for the magnitude of v_C. For example, solve Eq. (1) for ω_y in terms of ω_z. Substitute this into Eq. (3) and thereby solve for ω_x in terms of ω_z. Substitute this result in Eq. (2), and you will find that ω_z will cancel out, which will allow a solution for v_C.

†By definition of the dot product $\mathbf{A} \cdot \mathbf{B} = AB \cos\theta$, where θ is the angle made between the tails of both vectors. For example, $\mathbf{i} \cdot \mathbf{i} = (1)(1) \cos 0° = 1$. Similarly, $\mathbf{i} \cdot \mathbf{j} = 0$, etc. If $\mathbf{A}$ and $\mathbf{B}$ are expressed in Cartesian component form, then $\mathbf{A} \cdot \mathbf{B} = A_x B_x + A_y B_y + A_z B_z$.

Hence,

$$\boldsymbol{\omega}_{CD} = \{0.255\mathbf{i} + 0.191\mathbf{k}\} \text{ rad/sec} \qquad \textit{Ans.}$$

and

$$\mathbf{v}_C = \{-3.99\mathbf{j}\} \text{ ft/sec} \qquad \textit{Ans.}$$

Problems

13-62. Gear B rotates in the xy plane with an angular velocity of $\omega = 6$ rad/sec. Gear A is fixed to the shaft C which is connected to a ball-and-socket joint at O. If the shaft C remains in the yz plane while gear A turns on gear B, determine the angular velocity of gear A and the speed of one of the gear teeth on A.

Prob. 13-62

13-63. Gear B in Prob. 13-62 is stationary, while the shaft C, connected to gear A, rotates about the z axis with an angular velocity of $\omega = 3$ rad/sec. Determine the angular acceleration of gear A and the rate of rotation of gear A about shaft C.

13-64. Show that the location of the instantaneous axis of rotation for gear A in Prob. 13-63 lies along the y axis.

13-65. The gyro rotor is attached to a ball-and-socket joint at O. At the instant shown, point A on the rotor has a velocity of $\mathbf{v}_A = \{-21\mathbf{i} - 8\mathbf{j} + 18\mathbf{k}\}$ ft/sec. Determine the direction of the instantaneous axis of rotation for the rotor if $\omega_x = 2$ rad/sec.

Prob. 13-65

13-66. An automobile maintains a constant speed of 40 ft/sec. Assuming that the 2.2-ft-diameter tires do not slip on the pavement, determine the magnitude of the angular velocity and angular acceleration of the tires if the automobile enters a horizontal curve having a 100 ft radius.

13-67. The electric fan is mounted on a swivel support such that the fan rotates about the z axis at a constant rate of $\omega_z = 1$ rad/sec. The fan blade is spinning at $\omega_s = 60$ rad/sec. If $\phi = 45°$ for the motion, determine the angular velocity and the angular acceleration of the blade.

Prob. 13-67

Prob. 13-69

13-68. The propeller of the airplane is rotating at a speed of $\omega_s\mathbf{i}$, while the plane is undergoing a turn at a constant rate of ω_t. Compute the angular acceleration of the propeller if (a) the turn is horizontal, i.e., $\omega_t\mathbf{k}$, and (b) the turn is vertical, downward, i.e., $\omega_t\mathbf{j}$.

13-70. The construction boom OA is rotating about its vertical axis with an angular velocity of $\omega_z = 0.7$ rad/sec while it is rotating downward with an angular velocity of $\omega_B = 0.4$ rad/sec. Compute the velocity and acceleration of point A located at the tip of the boom.

Prob. 13-68

Prob. 13-70

13-69. The anemometer located on the ship is spinning about its own axis at a rate ω_s, while the ship is rolling about the x axis at the rate ω_x and about the y axis at the rate of ω_y. Compute the angular velocity and angular acceleration of the anemometer at the instant when the ship is level as shown. Assume that all components of angular velocity are constant and that the rolling motion caused by the sea is independent in the x and y directions.

13-71. The radar-tracking antenna is following a jet plane. At the instant $\theta = 0°$ and $\phi = 60°$, the angular rate of change is $d\theta/dt = 0.2$ rad/sec and $d\phi/dt = 0.5$ rad/sec. Compute the velocity and acceleration of the signal horn A at this instant. The distance OA is 4 ft.

Prob. 13–71

13-72. The rod is attached to smooth collars at its end-points by means of ball-and-socket joints. Determine the velocity of collar B if collar A is moving downward at a constant speed of 2 ft/sec. Can the angular velocity of the rod be determined? Why or why not?

$v_A = 2$ ft/sec

3 ft

5 ft

2 ft

Prob. 13–72

13-73. Gear A is fixed to the crankshaft S, while gears C and B are free to rotate. The crankshaft is rotating at 50 rad/sec about its axis. Determine the magnitude of the angular velocity of the propeller and the angular acceleration of gear B.

1 in.

4 in.

Prob. 13–73

13-74. Gears A and B are fixed while gears C and D are free to rotate on the shaft S. The shaft is turning about the z axis at a constant rate ω_z. Determine the magnitude of the angular velocity and angular acceleration of gear C.

ω_z

4 in.

2 in.

4 in.

4 in.

8 in.

Prob. 13–74

13–75. The corner C of the triangular plate rests on the horizontal plane, while end points A and B are restricted to move along the grooved slots. At the instant shown, A is moving downward with a constant velocity of 2 in./sec. Determine the angular velocity of the plate and the velocity of B and C.

Prob. 13–75

13–76. Rod AB is attached to two collars by means of ball-and-socket joints. The joint at B is attached to a disk which is rotating at a constant angular velocity of 2 rad/sec. Determine the velocity and acceleration of the collar at A at the instant shown. Assume that the angular velocity of the rod is directed perpendicular to the axis of the rod.

Prob. 13–76

13–77. Determine the acceleration of collar B in Prob. 13–72. Assume that the angular velocity of rod AB is directed perpendicular to the axis of the rod.

13–78. Determine the angular acceleration of the plate in Prob. 13–75 and the acceleration of points B and C.

*13–12. Relative Velocity with Respect to a Rotating Axis

Up to this point, the equations which relate the position, velocity, and acceleration of one point on a rigid body relative to another have been established with respect to a frame of reference which *translates* relative to a fixed frame of reference. When angular motions of the body become complicated, as in the case of rigid-body motion about a fixed point and general motion of a rigid body, it is often convenient, and sometimes easier, to employ a frame of reference having axes which are fixed in, and move with, the body. These axes will, in general, *both rotate and translate* relative to the fixed frame of reference. With this set of axes, an analysis of the *components* of the rotational motions (ω and α) may be considered rather than the resultants. (Refer to the difficulty encountered in studying the motion of the gyro top of Example 13–15.) Frames of reference which both translate and rotate are also useful for analyzing the motion of two points on a mechanism which are not located on the *same* rigid body. Furthermore, we can conveniently analyze the kinematics of relative particle motion using these reference frames when one of the particles is moving along a path which is subjected to rotation.

We will presently develop an equation which relates the absolute velocities of two points relative to a frame of reference subjected to both a translation and a rotation. Because of the generality, these points may represent two particles moving independently of one another or two points located on the same (or different) rigid bodies. Consider the two points A and B shown in Fig. 13–34a. The location of these points is completely determined using the position vectors $\mathbf{r}_A$ and $\mathbf{r}_B$ which are measured with respect to the *fixed XYZ* coordinate system. As shown in the figure, the base point A represents the origin of the *xyz* coordinate system, which is assumed to be both *translating* and *rotating* with respect to the fixed frame. The velocity of translation of point A is $\mathbf{v}_A$, and the rate of rotation of the *xyz* axis will be designated as $\boldsymbol{\Omega}$. *Both of these vectors ($\mathbf{v}_A$, $\boldsymbol{\Omega}$) are absolute quantities measured relative to the fixed frame of reference.* The position of point B with respect to point A is determined by means of the position vector $\mathbf{r}_{B/A}$. Thus,

$$\mathbf{r}_B = \mathbf{r}_A + \mathbf{r}_{B/A}$$

The absolute velocity of point B is determined by taking the time derivative of this equation which yields

$$\mathbf{v}_B = \mathbf{v}_A + \frac{d\mathbf{r}_{B/A}}{dt} \tag{13–26}$$

The relative position vector $\mathbf{r}_{B/A}$ may be expressed in terms of either $\mathbf{I}$, $\mathbf{J}$, and $\mathbf{K}$ unit vectors, or in terms of $\mathbf{i}$, $\mathbf{j}$, and $\mathbf{k}$ unit vectors. Although

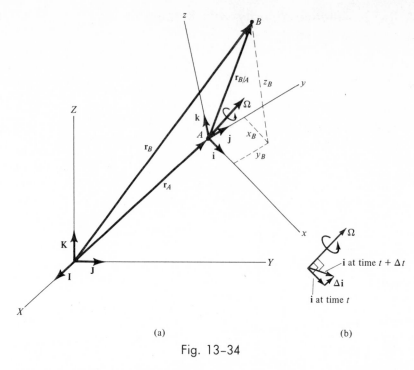

(a)

(b)

Fig. 13-34

the *magnitude* of $\mathbf{r}_{B/A}$ is the *same* when measured in *both* coordinate systems, the *direction* of this vector is different. For our purposes $\mathbf{r}_{B/A}$ will be *measured relative to the moving frame of reference,* since motion of point B will be measured relative to this reference. If point B is located at coordinates (x_B, y_B, z_B), the relative position vector directed from point A to B may then be expressed as

$$\mathbf{r}_{B/A} = x_B\mathbf{i} + y_B\mathbf{j} + z_B\mathbf{k} \tag{13-27}$$

From this equation, the last term in Eq. 13-26 may be computed.

$$\frac{d\mathbf{r}_{B/A}}{dt} = \frac{dx_B}{dt}\mathbf{i} + x_B\frac{d\mathbf{i}}{dt} + \frac{dy_B}{dt}\mathbf{j} + y_B\frac{d\mathbf{j}}{dt} + \frac{dz_B}{dt}\mathbf{k} + z_B\frac{d\mathbf{k}}{dt}$$

or

$$\frac{d\mathbf{r}_{B/A}}{dt} = \left(\frac{dx_B}{dt}\mathbf{i} + \frac{dy_B}{dt}\mathbf{j} + \frac{dz_B}{dt}\mathbf{k}\right) + \left(x_B\frac{d\mathbf{i}}{dt} + y_B\frac{d\mathbf{j}}{dt} + z_B\frac{d\mathbf{k}}{dt}\right) \tag{13-28}$$

The three terms in the first set of parentheses on the right side of this equation represent the components of velocity of point B as seen by an observer located at the origin A and attached to the moving coordinate system. These three terms will be denoted by the vector $\mathbf{v}_{B/A}$. The three terms in the second set of parentheses represent the instantaneous time rate of change of the unit vectors $\mathbf{i}$, $\mathbf{j}$, and $\mathbf{k}$ as measured by an observer

located in the fixed coordinate system. By definition, the magnitude of each of these vectors is one unit, and therefore, only their direction changes. This *directional change* is due to the *rotation* of the moving reference system. To obtain these time derivatives, consider, for example, the change in orientation of the **i** vector during an instant of time Δt, Fig. 13–34b. It is seen that this *change,* $\Delta \mathbf{i}$, occurs tangent to the path described by the head of the vector (in the limiting direction of $\Delta \mathbf{i}$ as $\Delta t \to 0$). The rate of change is simply the product of the magnitude of **i** and the instantaneous angular velocity Ω of the rotating coordinate system, i.e., $(1)(\Omega)$. Thus, accounting for *both* the *magnitude* Ω and *direction* $\Delta \mathbf{i}$, we have

$$\frac{d\mathbf{i}}{dt} = \lim_{\Delta t \to 0} \frac{\Delta \mathbf{i}}{\Delta t} = \Omega \times \mathbf{i}$$

Similarly,

$$\frac{d\mathbf{j}}{dt} = \Omega \times \mathbf{j} \quad \text{and} \quad \frac{d\mathbf{k}}{dt} = \Omega \times \mathbf{k}$$

Substituting these results into Eq. 13–28 we obtain

$$\frac{d\mathbf{r}_{B/A}}{dt} = \mathbf{v}_{B/A} + [x_B(\Omega \times \mathbf{i}) + y_B(\Omega \times \mathbf{j}) + z_B(\Omega \times \mathbf{k})]$$

Using the distributive property of the vector cross-product,

$$\frac{d(\mathbf{r}_{B/A})}{dt} = \mathbf{v}_{B/A} + \Omega \times (x_B\mathbf{i} + y_B\mathbf{j} + z_B\mathbf{k})$$

From Eq. 13–27,

$$\frac{d(\mathbf{r}_{B/A})}{dt} = \mathbf{v}_{B/A} + \Omega \times \mathbf{r}_{B/A} \qquad (13\text{–}29)$$

The velocity of B is obtained by substituting Eq. 13–29 into Eq. 13–26:

$$\mathbf{v}_B = \mathbf{v}_A + \Omega \times \mathbf{r}_{B/A} + \mathbf{v}_{B/A} \qquad (13\text{–}30)$$

For convenience, the definitions of each of the variables contained in Eq. 13–30 are repeated as follows:

$\mathbf{v}_B$ = absolute velocity of point B

$\mathbf{v}_A$ = absolute velocity of the origin A of the moving frame of reference

$\mathbf{v}_{B/A}$ = relative velocity of point B measured with respect to the origin A of the moving frame of reference

Ω = instantaneous angular velocity of the moving frame of reference

$\mathbf{r}_{B/A}$ = position vector drawn from point A to point B and measured with respect to the moving frame of reference

Comparing Eq. 13–30 with Eq. 13–13, which is valid for a translating frame of reference, it is observed that the only difference between the equations is represented by the term $\mathbf{v}_{B/A}$. When applying Eq. 13–30 it is often useful to understand physically what each of the terms contained in this equation represents. In order of appearance these terms are defined as follows:

$\mathbf{v}_B$ $\begin{cases} \text{absolute velocity of } B \end{cases}$

(equals)

$\mathbf{v}_A$ $\begin{cases} \text{absolute velocity of origin} \\ \text{of } xyz \text{ frame} \end{cases}$

(plus) motion of xyz frame

$\mathbf{\Omega} \times \mathbf{r}_{B/A}$ $\begin{cases} \text{angular velocity effect caused} \\ \text{by rotation of } xyz \\ \text{frame} \end{cases}$

(plus)

$\mathbf{v}_{B/A}$ $\begin{cases} \text{relative velocity of } B \text{ with} \\ \text{respect to origin } A \end{cases}$ motion of particle B within xyz frame

Equation 13–29 may be generalized by defining the time derivative of any vector **A** which is described using the coordinates of a rotating reference frame. In comparison with the terms defined for $\mathbf{r}_{B/A}$ in Eq. 13–29, we may write

$$\frac{d\mathbf{A}}{dt} = \left(\frac{d\mathbf{A}}{dt}\right)_{x,y,z} + \mathbf{\Omega} \times \mathbf{A} \qquad (13\text{–}31)$$

The time derivative $d\mathbf{A}/dt$ of **A** as observed from the fixed XYZ frame of reference consists, therefore, of two parts: $(d\mathbf{A}/dt)_{x,y,z}$ represents the time rate of change of **A** as observed from the xyz rotating frame of reference, and $\mathbf{\Omega} \times \mathbf{A}$ represents the change of **A** caused by the rotation of the xyz frame.*

*13–13. Relative Acceleration with Respect to a Rotating Axis

The absolute acceleration of a point B which is observed from a fixed coordinate system may be expressed in terms of its motion with respect to a rotating or moving system of coordinates. This is done by taking the time derivative of Eq. 13–30:

$$\frac{d\mathbf{v}_B}{dt} = \frac{d\mathbf{v}_A}{dt} + \frac{d\mathbf{\Omega}}{dt} \times \mathbf{r}_{B/A} + \mathbf{\Omega} \times \frac{d\mathbf{r}_{B/A}}{dt} + \frac{d\mathbf{v}_{B/A}}{dt}$$

*In analyzing motions in the previous sections, $\mathbf{\Omega} = 0$; i.e., the xyz axes *translate* relative to the XYZ axes, so that by Eq. 13–31 the time derivatives are the *same* in both references.

$$a_B = a_A + \alpha \times r_{B/A} + \Omega \times \frac{dr_{B/A}}{dt} + \frac{dv_{B/A}}{dt} \qquad (13\text{--}32)$$

Since

$$v_{B/A} = \frac{dx_B}{dt}i + \frac{dy_B}{dt}j + \frac{dz_B}{dt}k$$

the derivative $dv_{B/A}/dt$ may be obtained by using a procedure similar to that followed in going from Eq. 13–27 to Eq. 13–29, or directly by using Eq. 13–31 with $A = v_{B/A}$ and realizing that $(dv_{B/A}/dt)_{x,y,z} = a_{B/A}$. In any case,

$$\frac{dv_{B/A}}{dt} = a_{B/A} + \Omega \times v_{B/A} \qquad (13\text{--}33)$$

Substituting Eqs. 13–29 and 13–33 into Eq. 13–32 and rearranging terms yields

$$a_B = a_A + \alpha \times r_{B/A} + \Omega \times (\Omega \times r_{B/A}) + 2\Omega \times v_{B/A} + a_{B/A} \qquad (13\text{--}34)$$

The definition of each of the vectors used in this equation is given as follows:

a_B = absolute acceleration of point B

a_A = absolute acceleration of the origin A of the moving frame of reference

$a_{B/A}, v_{B/A}$ = relative acceleration and velocity of point B measured with respect to the origin of the moving frame of reference

α, Ω = instantaneous angular acceleration and angular velocity of the moving frame of reference

$r_{B/A}$ = position vector drawn from point A to point B, and measured with respect to the moving frame of reference

If motions of points A and B are along *curved paths,* it is convenient to compute the acceleration terms a_B, a_A, and $a_{B/A}$ on the basis of their normal and tangential components.

The first two terms and the last term of Eq. 13–34 have been defined above. The third term, $\alpha \times r_{B/A}$, represents the contribution of acceleration introduced by the angular acceleration of the moving coordinate system with respect to the fixed coordinate system. The fourth term, $\Omega \times (\Omega \times r_{B/A})$, is called the *centripetal acceleration* and represents the acceleration component introduced by the angular velocity of the moving system with respect to the fixed system. The fifth in Eq. 13–34, $2\Omega \times v_{B/A}$, is called the *Coriolis acceleration.* This term represents the difference between the acceleration of point B measured with respect to a point located at the

origin of a nonrotating axis which is translating with the *xyz* frame, and the acceleration of *B* measured with respect to the rotating *xyz* frame of reference. The Coriolis acceleration is an important component of the acceleration which must be considered whenever rotating reference frames are used—even in plane motion. This term must be considered, for example, when studying the accelerations and forces which act on rockets, long-range projectiles, or other bodies having motions which are largely affected by the rotation of the earth.

Summarizing these results, we have

$\mathbf{a}_B$ $\begin{cases} \text{absolute acceleration of } B \end{cases}$

(equals)

$\mathbf{a}_A$ $\begin{cases} \text{absolute acceleration} \\ \text{of origin } A \\ \text{of } xyz \text{ frame} \end{cases}$

(plus)

$\boldsymbol{\alpha} \times \mathbf{r}_{B/A}$ $\begin{cases} \text{angular acceleration} \\ \text{effect caused by} \\ \text{rotation of } xyz \\ \text{frame} \end{cases}$ motion of *xyz* frame

(plus)

$\boldsymbol{\Omega} \times (\boldsymbol{\Omega} \times \mathbf{r}_{B/A})$ $\begin{cases} \text{angular velocity} \\ \text{effect caused by} \\ \text{rotation of } xyz \text{ frame} \end{cases}$

(plus)

$2\boldsymbol{\Omega} \times \mathbf{v}_{B/A}$ $\begin{cases} \text{combined effect of particle} \\ B \text{ moving} \\ \text{relative to } xyz \\ \text{coordinates and rotation} \\ \text{of } xyz \text{ frame} \end{cases}$ interacting motion

(plus)

$\mathbf{a}_{B/A}$ $\begin{cases} \text{relative acceleration of } B \\ \text{with respect to origin } A \end{cases}$ motion of particle *B* within *xyz* frame

The following example problems illustrate the application of both Eqs. 13–30 and 13–34. The vectors in these equations may be expressed in terms of components projected either along the *X*, *Y*, and *Z* axes, or along the *x*, *y*, and *z* axes. Selection of a given set of axes is arbitrary as long as all the vector components used in the problem solution are expressed in terms of the coordinates chosen (see Example 13–17). In general, several choices may be made for orienting the frames of reference. *Most often, solutions are easily obtained if the axes are parallel or collinear and the origins are coincident at the instant considered.*

Example 13-17

The rod AB shown in Fig. 13-35 rotates counterclockwise with a constant angular velocity of 2 rad/sec. Determine the velocity and acceleration of point C located on the double collar when $\theta = 45°$. The collar consists of two slider blocks which are constrained to move along the circular shaft and the rod AB.

Solution

The origin of both coordinate systems is located at point A. The xy frame is fixed in, and rotates with rod AB. This coordinate system rotates with an angular velocity $\Omega = \{2\mathbf{K}\}$ rad/sec. Applying Eq. 13-30 gives

$$\mathbf{v}_C = \mathbf{v}_A + \Omega \times \mathbf{r}_{C/A} + \mathbf{v}_{C/A} \qquad (1)$$

We shall express all vector quantities using $\mathbf{I}, \mathbf{J}, \mathbf{K}$ unit vectors, appropriate to the fixed coordinate system. Since the block is restricted to move along the *fixed* circular rod, its velocity when $\theta = 45°$ will be

$$\mathbf{v}_C = -v_C \mathbf{I} \qquad (2)$$

Since the origins of both coordinate systems are fixed, $\mathbf{v}_A = 0$. Point C has a velocity

$$\mathbf{v}_{C/A} = v_{C/A}(-0.707\mathbf{I} + 0.707\mathbf{J})$$

relative to an observer fixed at A and rotating with the xy coordinate system. Also,

$$\mathbf{r}_{C/A} = \{-2\mathbf{I} + 2\mathbf{J}\} \text{ ft}$$

Substituting these quantities into Eq. (1), we have

$$-v_C\mathbf{I} = 0 + 2\mathbf{K} \times (-2\mathbf{I} + 2\mathbf{J}) - 0.707v_{C/A}\mathbf{I} + 0.707v_{C/A}\mathbf{J}$$

or

$$-v_C\mathbf{I} = -4\mathbf{J} - 4\mathbf{I} - 0.707v_{C/A}\mathbf{I} + 0.707v_{C/A}\mathbf{J}$$

Equating the respective $\mathbf{I}$ and $\mathbf{J}$ components yields

$$-v_C = -4 - 0.707v_{C/A}$$
$$0 = -4 + 0.707v_{C/A}$$

Hence,

$$v_C = 8.0 \text{ ft/sec}$$
$$v_{C/A} = 5.66 \text{ ft/sec}$$

So that from Eq. (2),

$$\mathbf{v}_C = \{-8\mathbf{I}\} \text{ ft/sec} \qquad \textit{Ans.}$$

This *same result* could have been obtained if the calculations had been

Fig. 13-35

performed using the x and y coordinates. For this case the various terms to be used in Eq. (1) are

$$\mathbf{v}_C = -0.707v_C\mathbf{i} + 0.707v_C\mathbf{j}$$
$$\mathbf{v}_A = 0$$
$$\mathbf{v}_{C/A} = v_{C/A}\mathbf{j}$$
$$\boldsymbol{\Omega} = \{2\mathbf{k}\}\ \text{rad/sec}$$
$$\mathbf{r}_{C/A} = \{2\sqrt{2}\mathbf{j}\}\ \text{ft}$$

Substituting into Eq. (1), we have

$$-0.707v_C\mathbf{i} + 0.707v_C\mathbf{j} = 0 + (2\mathbf{k}) \times (2\sqrt{2}\mathbf{j}) + v_{C/A}\mathbf{j}$$

Expanding and equating the $\mathbf{i}$ and $\mathbf{j}$ components, we write

$$-0.707v_C = -5.66$$
$$0.707v_C = v_{C/A}$$

Solving, we again obtain

$$v_C = 8.0\ \text{ft/sec} \qquad\qquad Ans.$$
$$v_{C/A} = 5.66\ \text{ft/sec}$$

Applying Eq. 13–34, we may write the acceleration of the collar C as

$$\mathbf{a}_C = \mathbf{a}_A + \boldsymbol{\alpha} \times \mathbf{r}_{C/A} + \boldsymbol{\Omega} \times (\boldsymbol{\Omega} \times \mathbf{r}_{C/A}) + 2\boldsymbol{\Omega} \times \mathbf{v}_{C/A} + \mathbf{a}_{C/A} \qquad (3)$$

The acceleration $\mathbf{a}_A = 0$, since the origins of both references coincide at all times during the motion. Also, $\boldsymbol{\alpha} = 0$, since AB (or the xy reference) is rotating with constant angular velocity. The collar slides along the *straight* rod AB; therefore, $\mathbf{a}_{C/A} = a_{C/A}(-0.707\mathbf{I} + 0.707\mathbf{J})$. The path of C is along the fixed *curved* shaft; thus the acceleration $\mathbf{a}_C$ has *both* normal and tangential components. In particular, the normal component of acceleration has a magnitude of $v_C^2/\rho = (8\ \text{ft/sec})^2/2\ \text{ft} = 32\ \text{ft/sec}^2$. Thus,

$$\mathbf{a}_C = -(a_C)_t\mathbf{I} - \frac{v_C^2}{\rho}\mathbf{J} = -(a_C)_t\mathbf{I} - 32\mathbf{J} \qquad (4)$$

Substituting this, and the other computed data into Eq. (3), we have

$$-(a_C)_t\mathbf{I} - 32\mathbf{J} = 0 + 0 + [2\mathbf{K}] \times [2\mathbf{K} \times (-2\mathbf{I} + 2\mathbf{J})]$$
$$+ 2[(2\mathbf{K}) \times 5.66(-0.707\mathbf{I} + 0.707\mathbf{J})] - 0.707a_{C/A}\mathbf{I} + 0.707a_{C/A}\mathbf{J}$$

Expanding and equating the respective $\mathbf{I}$ and $\mathbf{J}$ components, we write

$$-(a_C)_t = -8 - 0.707a_{C/A}$$
$$0 = +8 + 0.707a_{C/A}$$

Solving yields

$$(a_C)_t = 0$$
$$a_{C/A} = -11.32\ \text{ft/sec}^2$$

Hence, from Eq. (4),

$$\mathbf{a}_C = \{-32\mathbf{J}\} \text{ ft/sec}^2 \qquad\qquad Ans.$$

Since the computed magnitude of $\mathbf{a}_{C/A}$ is negative, the sense of direction of this vector was assumed incorrectly. Thus, at the instant when $\theta = 45°$, an observer stationed at A and rotating with the rod AB sees the collar C moving *away* from him with a velocity $v_{C/A} = 5.66$ ft/sec, and having a *deceleration* of $a_{C/A} = 11.32$ ft/sec^2.

Computations for the accelerations may also be performed using $\mathbf{i}, \mathbf{j}, \mathbf{k}$ vector components in the same manner as was shown for the velocities. The computations will, however, be rather cumbersome, since the xy coordinates are not compatible with the curved-path motion of the collar.

Example 13–18

A motor M is fixed to the surface of a platform which has an angular motion, as shown in Fig. 13–36. A shaft AC is attached to the motor and rotates about the axis of the motor at an angular speed of $\omega_M = 3$ rad/sec and has an angular acceleration of $\alpha_M = 1$ rad/sec^2, as shown. A smooth collar B is attached to the rod. At the instant the rod is in the vertical position, the collar has a downward velocity of 10 ft/sec and a downward acceleration of 3 ft/sec^2, along the rod. Determine the absolute velocity and acceleration of the collar at this instant.

Fig. 13–36

Solution

As shown in Fig. 13–36, the fixed X, Y, Z reference axis is located at the center of the platform, and the moving x, y, z frame of reference is fixed to the rotating shaft at point A. The relative velocity and acceleration equations for the collar are

$$\mathbf{v}_B = \mathbf{v}_A + \boldsymbol{\Omega} \times \mathbf{r}_{B/A} + \mathbf{v}_{B/A} \qquad (1)$$

and

$$\mathbf{a}_B = \mathbf{a}_A + \boldsymbol{\alpha} \times \mathbf{r}_{B/A} + \boldsymbol{\Omega} \times (\boldsymbol{\Omega} \times \mathbf{r}_{B/A}) + 2\boldsymbol{\Omega} \times \mathbf{v}_{B/A} + \mathbf{a}_{B/A} \qquad (2)$$

The terms used in these equations may be computed from the problem data. With reference to the coordinate axis shown in Fig. 13–36, and noting that point A travels along a *curved path*, we have

$$\mathbf{a}_A = \boldsymbol{\alpha}_P \times \mathbf{r}_{OA} + \boldsymbol{\omega}_P \times (\boldsymbol{\omega}_P \times \mathbf{r}_{OA})$$
$$= 2\mathbf{k} \times 4\mathbf{i} + 5\mathbf{k} \times (5\mathbf{k} \times 4\mathbf{i})$$
$$= \{8\mathbf{j} - 100\mathbf{i}\} \ \text{ft/sec}^2$$

$$\mathbf{a}_{B/A} = \{-3\mathbf{k}\} \ \text{ft/sec}^2$$
$$\boldsymbol{\alpha} = \boldsymbol{\alpha}_M + \boldsymbol{\alpha}_P = \{\mathbf{i} + 2\mathbf{k}\} \ \text{rad/sec}^2$$
$$\boldsymbol{\Omega} = \boldsymbol{\omega}_M + \boldsymbol{\omega}_P = \{3\mathbf{i} + 5\mathbf{k}\} \ \text{rad/sec}^2$$
$$\mathbf{v}_A = \boldsymbol{\omega}_P \times \mathbf{r}_{OA} = 5\mathbf{k} \times 4\mathbf{i} = \{20\mathbf{j}\} \ \text{ft/sec}$$
$$\mathbf{v}_{B/A} = \{-10\mathbf{k}\} \ \text{ft/sec}$$
$$\mathbf{r}_{B/A} = \{-\mathbf{k}\} \ \text{ft}$$

Substituting these terms into Eqs. (1) and (2), expanding, and simplifying yields

$$\mathbf{v}_B = 20\mathbf{j} + (3\mathbf{i} + 5\mathbf{k}) \times (-\mathbf{k}) - 10\mathbf{k}$$
$$= \{23\mathbf{j} - 10\mathbf{k}\} \ \text{ft/sec} \qquad \qquad Ans.$$

and

$$\mathbf{a}_B = 8\mathbf{j} - 100\mathbf{i} + (\mathbf{i} + 2\mathbf{k}) \times (-\mathbf{k}) + (3\mathbf{i} + 5\mathbf{k}) \times [(3\mathbf{i} + 5\mathbf{k}) \times (-\mathbf{k})]$$
$$+ 2[(3\mathbf{i} + 5\mathbf{k}) \times (-10\mathbf{k})] - 3\mathbf{k}$$
$$= \{-115\mathbf{i} + 69\mathbf{j} + 6\mathbf{k}\} \ \text{ft/sec}^2 \qquad \qquad Ans.$$

Example 13–19

The pendulum shown in Fig. 13–37 is attached to a rolling support at B which is restricted to move down the inclined plane at a constant speed of 10 ft/sec. An electric motor M is attached to the center of the pendulum and drives an armature AD at a speed of $\omega_M = 8$ rad/sec clockwise relative to the pendulum bob. Determine the velocity and acceleration of point A on the armature when the pendulum is in the position shown. The pendulum has an angular velocity of 2 rad/sec, perpendicular to the plane of the page and is spinning about the axis of rod BC with an angular velocity of 3 rad/sec.

Solution

The relative velocity and acceleration equations relate the motion of the origin (or base point) of the rotating frame of reference to the motion of a point relative to this reference. For this problem, *three* frames of reference will be considered for the solution. As shown in Fig. 13–37, the origin of the *fixed XY* frame of reference coincides with point B on the roller at the instant considered. Both *rotating frames* of reference, xy and $x'y'$, are fixed in and move with the pendulum; and have origins located at points B and D, respectively. The analysis will proceed as follows: From the fixed $X\,Y$ and rotating $x\,y$ frames, the motion of point D (origin of x' and y') will be determined by using the known motion of the origin (base point) B of x and y. Then, using the fixed XY and rotating $x'y'$ frames, we shall determine the motion of point A in terms of the (previously determined) motion of the new origin (base point) D. Proceeding in this manner, we write

Fig. 13–37

$$v_D = v_B + \Omega \times r_{D/B} + v_{D/B} \qquad (1)$$

$$a_D = a_B + \alpha \times r_{D/B} + \Omega \times (\Omega \times r_{D/B}) + 2\Omega \times v_{D/B} + a_{D/B} \qquad (2)$$

The terms in these equations are determined from the problem data, appropriate to the XY and xy frames of reference. Thus,

$$v_B = 10 \text{ ft/sec } (-\cos 45°i - \sin 45°j) = \{-7.07i - 7.07j\} \text{ ft/sec}$$

$$v_{D/B} = 0$$

$$\Omega = \{3j - 2k\} \text{ rad/sec}$$

$$r_{D/B} = \{-4j\} \text{ ft}$$

$$a_B = 0$$

$$a_{D/B} = 0$$

$$\alpha = 0$$

Substituting into Eqs. (1) and (2), and simplifying, we obtain

$$v_D = -7.07i - 7.07j + (3j - 2k) \times (-4j) + 0$$
$$= \{-15.07i - 7.07j\} \text{ ft/sec}$$

and

$$a_D = 0 + 0 + [3j - 2k] \times [(3j - 2k) \times (-4j)] + 0 + 0$$
$$= \{16j + 24k\} \text{ ft/sec}^2$$

Using these results, the motion of A may be related to D by the equations

$$v_A = v_D + \Omega \times r_{A/D} + v_{A/D} \qquad (3)$$

$$a_A = a_D + \alpha \times r_{A/D} + \Omega \times (\Omega \times r_{A/D}) + 2\Omega \times v_{A/D} + a_{A/D} \qquad (4)$$

Besides v_D and a_D, which have just been computed, the terms used in these equations appropriate to the XY and $x'y'$ frames of reference are

$$\mathbf{r}_{A/D} = \{-\tfrac{1}{2}\mathbf{i}\} \text{ ft}$$
$$\mathbf{v}_{A/D} = \boldsymbol{\omega}_M \times \mathbf{r}_{A/D} = (-8\mathbf{k}) \times (-\tfrac{1}{2}\mathbf{i}) = \{4\mathbf{j}\} \text{ ft/sec}$$
$$\boldsymbol{\Omega} = \{3\mathbf{j} - 2\mathbf{k}\} \text{ rad/sec}$$
$$\boldsymbol{\alpha} = 0$$
$$\mathbf{a}_{A/D} = \boldsymbol{\omega}_M \times (\boldsymbol{\omega}_M \times \mathbf{r}_{A/D}) = [-8\mathbf{k}] \times [(-8\mathbf{k}) \times (-\tfrac{1}{2}\mathbf{i})] = \{32\mathbf{i}\} \text{ ft/sec}^2$$

(Note that $\mathbf{a}_{A/D}$ has no tangential component of acceleration since $\boldsymbol{\alpha} = 0$.)
Substituting into Eq. (3) and (4), and solving yields

$$\mathbf{v}_A = -15.07\mathbf{i} - 7.07\mathbf{j} + (3\mathbf{j} - 2\mathbf{k}) \times (-\tfrac{1}{2}\mathbf{i}) + 4\mathbf{j}$$
$$= \{-15.07\mathbf{i} - 2.07\mathbf{j} + 1.5\mathbf{k}\} \text{ ft/sec} \qquad\qquad Ans.$$
$$\mathbf{a}_A = 16\mathbf{j} + 24\mathbf{k} + 0 + [3\mathbf{j} - 2\mathbf{k}] \times [(3\mathbf{j} - 2\mathbf{k}) \times (-\tfrac{1}{2}\mathbf{i})]$$
$$\qquad\qquad\qquad + 2[(3\mathbf{j} - 2\mathbf{k}) \times (4\mathbf{j})] + 32\mathbf{i}$$
$$= \{54.5\mathbf{i} + 16\mathbf{j} + 24\mathbf{k}\} \text{ ft/sec}^2 \qquad\qquad Ans.$$

Problems

13-79. Determine the *maximum* velocity and acceleration of a body resting on the surface of the earth caused by the rotation of the earth about its own axis and the axis of the sun. For simplicity, assume that both the sun and the earth's polar axes are parallel as shown. The earth's orbit is circular of radius 92,500,000 mi, and the earth is a sphere of radius 4,000 mi.

92,500,000 mi

Prob. 13-79

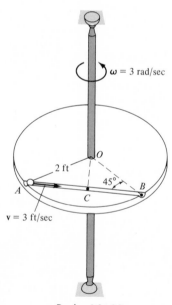

$\omega = 3$ rad/sec

2 ft

45°

A C B

O

$v = 3$ ft/sec

Prob. 13-80

13-80. The small sphere is moving along the grooved slot of a rotating platform with a velocity of 3 ft/sec relative to the platform. What is the magnitude of the Coriolis acceleration of the sphere when it is at points C and B?

13-81. The satellite is traveling with a velocity of 600 mph in the Northern Hemisphere at a latitude of 30° from south to north. Determine the Coriolis acceleration of the satellite with respect to the center of the earth. The radius of the earth is 3,960 mi, and the satellite is 70 mi from the surface of the earth.

Prob. 13-81

Prob. 13-83

13-82. At a given instant the 4-ft rod AC is rotating about the vertical axis BC with an angular velocity of $\omega_{BC} = 5$ rad/sec, having an angular acceleration of $\alpha_{BC} = 2$ rad/sec^2. At this same instant $\theta = 60°$, and link AC is rotating downward with an angular velocity of $\omega_{AC} = 1$ rad/sec, having an angular acceleration of $\alpha_{AC} = 2$ rad/sec^2. Determine the magnitude of velocity and acceleration of point A on the link at this instant.

13-84. The disk is fastened to link AB, which has the motion shown. If the *disk* rotates clockwise with an angular velocity of $\omega = 1$ rad/sec and has an angular acceleration of $\alpha = 2$ rad/sec^2, both *measured relative to the link*, determine the velocity and acceleration of point C on the disk, at the instant shown.

Prob. 13-82

Prob. 13-84

13-83. The two jet planes are flying at the same elevation. Plane A is flying along a straight-line path at 300 mph, and plane B is flying along a circular path at 400 mph. Compute the relative velocity at which the pilot of B sees plane A at the instant shown.

13-85. A particle P rolls along the slot with a velocity of 3 ft/sec and an acceleration of 1 ft/sec^2, both *measured relative to the disk*, and directed away from the center. If the disk has an angular velocity and angular acceleration as shown, determine the velocity and acceleration of the particle at the instant shown.

577

Prob. 13-87

13-86. At the instant shown, link CD has an angular velocity of 3 rad/sec and an angular acceleration of 1 rad/sec². Determine the angular velocity and angular acceleration of rod AB at this instant.

13-88. Link AB is confined to move in the vertical plane by means of the slider blocks at ends A and B. The smooth collar at C has a relative velocity of 4 ft/sec and a relative acceleration of 4 ft/sec², both directed towards A and *measured with respect to the link*. Determine the velocity and acceleration of the collar C at the instant shown.

Prob. 13-86

13-87. The double collar E is attached to rod AB and slides freely over rod CD. At the instant shown, rod AB has an angular velocity of 5 rad/sec, and an angular acceleration of 3 rad/sec², both measured clockwise. Determine the angular velocity and angular acceleration of rod CD at this instant.

Prob. 13-88

13-89. The collar E is attached to, and pivots about, rod AB while it slides on rod CD. If rod AB has an angular velocity of 6 rad/sec and an angular acceleration of 1 rad/sec², both acting clockwise, determine the angular velocity and the angular acceleration of rod CD at the instant shown.

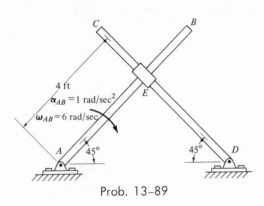

Prob. 13-89

13-90. The double pendulum consists of two rods. Rod AB has an absolute angular velocity of 3 rad/sec, and rod BC has an absolute angular velocity of 2 rad/sec., both measured counterclockwise. Determine the velocity and acceleration of point C at the instant shown.

Prob. 13-90

13-91. If rod AB in Prob. 13-90 has a counterclockwise angular acceleration of 3 rad/sec², determine the acceleration of point C on the pendulum. Use the angular velocity data given in Prob. 13-90.

13-92. The particle P slides around the circular hoop with a constant angular velocity of $d\theta/dt = 5$ rad/sec. The hoop rotates about the x axis at a constant rate of $\omega = 3$ rad/sec. At the instant shown, the hoop is in the xy plane and the angle $\theta = 45°$. Determine the velocity and acceleration of the particle at this instant.

Prob. 13-92

13-93. The frame $ABCD$ is rotating about the x axis with an angular velocity of $\omega = \{-6i\}$ rad/sec and has an angular acceleration of $\alpha = \{-2i\}$ rad/sec². The rotating rod EGH has an angular motion *relative to the frame*, as shown in the figure. If a small bead P is moving along the rod from G to H with a velocity of 1 ft/sec and has an acceleration of 1 ft/sec², both measured relative to rod EGH, determine the velocity and acceleration of the bead at the instant shown.

Prob. 13-93

13-94. The wheel is rolling on the horizontal surface without slipping. The constant velocity of its midpoint C is $\mathbf{v}_C = \{12\mathbf{j}\}$ ft/sec. While the wheel is rolling, a disk portion A of the wheel is rotating relative to the wheel with an angular velocity of $\boldsymbol{\omega}_A = \{4\mathbf{k}\}$ rad/sec. Determine the velocity and acceleration of point B on the disk at the instant shown. The x axis is directed out of the plane of the page.

13-96. The tower crane is turning, while the trolley T is moving outward. At the instant shown, the concrete bucket is also swinging toward the vertical such that $d\theta/dt = -6$ rad/sec and $d^2\theta/dt^2 = 2$ rad/sec^2, measured with respect to the trolley. If the cable AB is being shortened at a constant rate of 3 ft/sec, compute the velocity and acceleration of the bucket B at the instant shown.

Prob. 13–94

13-95. Determine the velocity and acceleration of point B on the disk in Prob. 13-94 if the center of the wheel is accelerating at $a_C = 3$ ft/sec^2 to the right and the angular acceleration of the disk A is $\boldsymbol{\alpha}_A = \{2\mathbf{k}\}$ rad/sec^2. Use the velocity data given in Prob. 13-94, and assume that the wheel does not slip as it rolls.

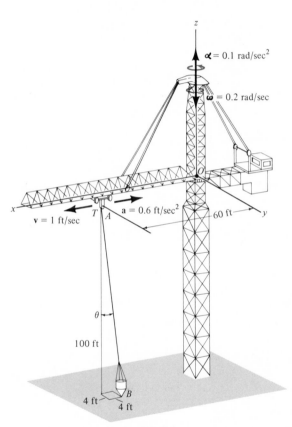

Prob. 13–96

14

Kinetics of a Particle: Forces and Accelerations

14-1. Newton's Laws of Motion

Many of the earlier notions about dynamics were dispelled after Galileo performed experiments in 1590 to study the motions of pendulums and falling bodies. The conclusions drawn from these experiments gave some insight into the effects of forces acting on bodies in motion. The general motion of a body subjected to forces was not known, however, until 1687, when Sir Isaac Newton first stated three basic laws governing the motion of a particle. These laws were formulated on the basis of his studies made in regards to the motion of the planets. (In considering the dynamics of a planet about the sun, Newton neglected the rotational effects of the planet about its own axis and assumed the planet to be a particle. This, of course, is justified since the distance a planet travels is very large in comparison to its size.) Although Newton's three laws as initially stated pertain only to the motion of a particle, in later portions of this book we will see how these laws may be extended and applied to the motion of rigid bodies. In a slightly reworded form, Newton's three laws of motion are as follows:

First Law.　A particle originally at rest, or moving at a constant velocity, will continue to remain at rest, or move with a constant velocity along a straight line, provided there is no unbalanced force acting on the particle.

Second Law.　A particle acted upon by an unbalanced force $\mathbf{F}$ receives

an acceleration **a** that is in the direction of the force and has a magnitude which is directly proportional to the force.

Third Law. For every force acting on a particle, the particle exerts an equal, opposite, and collinear reactive force.

The first and third laws were used extensively in developing the concepts of statics and are also considered in dynamics. Newton's second law of motion, however, is the most important principle used in the study of dynamics. This law forms the basis for the study of kinetics, since it *relates* the *motion* of a particle to the *forces* which act on it. In this regard, measurements of forces and accelerations can be performed in a laboratory. And, hence, in accordance with the second law of motion, if a known *unbalanced force* F_1 is applied to a particle, the acceleration a_1 of the particle may be measured. Since the force and acceleration are directly proportional, the constant of proportionality, m, may be determined from the ratio $m = F_1/a_1$. Provided the units of measurement are consistent, a different unbalanced force F_2 applied to the particle will create an acceleration a_2, such that $F_2/a_2 = m$. In both cases the ratio will be the same and the acceleration and the force, thought of as vectors, will have the *same* sense of direction. The constant of proportionality, m, is called the *mass* of the particle. (The concept of mass is discussed fully in Sec. 14–2.) When more than one force acts on the particle, the resultant force F_R is determined by vector summation, i.e., $F_R = \Sigma F$. For this more general case we may write Newton's second law of motion in mathematical form as

$$\Sigma F = m a \qquad (14\text{–}1)$$

This vector equation is often referred to as the *equation of motion* for the particle. As an example of the use of this equation, consider the particle P, shown in Fig. 14–1, which has a mass m and is subjected to the action of two forces F_1 and F_2. The resultant of these two forces yields an *unbalanced force* $F_R = F_1 + F_2$. According to Eq. 14–1, the particle therefore moves in the *direction* of F_R with an acceleration having a *magnitude* of $a = F_R/m$. In particular, if the unbalanced force acting on the particle is zero, i.e., $F_R = 0$, the particle acceleration will also be zero. As a result, the particle will either remain at rest or move in a straight-line path with a constant velocity. Such are the conditions of *static equilibrium*.

Both fixed and moving coordinate systems have been employed in the study of kinematics. If we use Eq. 14–1, however, it is required that measurements of the acceleration take place from an *inertial frame of reference* (sometimes referred to as a *Newtonian frame of reference*). *Such a system of axes does not rotate and is either fixed or translates in a given direction with a constant velocity (zero acceleration).* This definition ensures that the particle *acceleration* measured by observers in two different inertial frames of reference will always be the *same.* When studying the

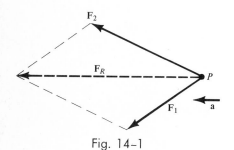

Fig. 14–1

motions of rockets and satellites, it is justifiable to consider the inertial reference frame as fixed to the stars. For most engineering applications, dynamics problems concerned with motions on or near the surface of the earth may be solved by using an inertial frame of reference which is assumed fixed to the earth. Even though the earth *rotates* both about its own axis and about the sun, the normal components of acceleration created by these rotations can be neglected in most computations.

The conclusions drawn from many past *experiments* have verified the use of Eq. 14–1 to describe the general motion of bodies. In 1905, however, Albert Einstein placed limitations on the use of this equation. In developing the theory of relativity, he had discovered that *time* was not an absolute quantity as assumed by Newton. As a consequence, it has been shown that Newton's second law fails to accurately predict the behavior of a particle, especially when the particle approaches the speed of light (186,000 mi/sec = 3.0×10^8 m/sec). Developments of the theory of quantum mechanics by Schrödinger and his colleagues indicate further that conclusions drawn from Eq. 14–1 are also invalid when particles move within an atomic distance of one another. For the most part, however, these requirements regarding particle speed and size are generally not encountered in engineering problems; therefore, these effects will not be considered in this book.

14–2. Units, Mass, and Newton's Law of Gravitation

Four fundamental quantities commonly used as a basis for measurement in mechanics are force, mass, length, and time. In general, each of these quantities is defined on the basis of an arbitrarily chosen *unit* or "standard." By applying a simple experimental process to compare quantities of the same kind to the standard unit, we can then form a basis for defining the standard quantities. Since *all* four quantities are *related* by Newton's second law of motion, Eq. 14–1, we *cannot* select units for measuring *all* these quantities arbitrarily. Instead, the equality $\mathbf{F} = m\mathbf{a}$ is maintained if three of the four units (called *primary units*) are arbitrarily defined and the fourth unit is derived from the equation.

Four important systems of units used to describe the four fundamental quantities are listed in Table 14–1. Note that the two *absolute systems of units* are defined using length, mass and time as primary units. The system is called absolute since the defined measurements of the primary units can be made at *any location*. For example, the *mass* of a body may be regarded as a property which is used to determine the resistance of matter to a change in its motion. This property is more fundamental (or absolute) than specifying the *weight* of the body because the weight or gravitational force which one body exerts upon another changes in

Table 14-1 System of Units

Type of System	Name of System	Length	Mass	Force	Time
Absolute	Metric (MKS)	meter (m)	kilogram (kg)	newton* $\left(\dfrac{\text{kg-m}}{\text{sec}^2}\right)$	second (sec)
Absolute	(CGS)	centimeter (cm)	gram (g)	dyne* $\left(\dfrac{\text{g-cm}}{\text{sec}^2}\right)$	second (sec)
Gravitational	Metric Gravitational (MKGFS)	meter (m)	metric slug* $\left(\dfrac{\text{kg}_f\text{-sec}^2}{\text{m}}\right)$	kilogram force (kg_f)	second (sec)
Gravitational	British Gravitational (FPS)	foot (ft)	slug* $\left(\dfrac{\text{lb-sec}^2}{\text{ft}}\right)$	pound (lb)	second (sec)

*Derived unit

magnitude, depending upon where the measurements for weight are made. Thus a body having a weight of 1 lb at the surface of the earth, as measured by a *spring scale,* would weigh 1/4 lb at a height of 4,000 mi from the earth's surface, or approximately 1/6 lb on the surface of the moon. However, the mass of the body will remain *constant* regardless of where the body is located—comparison of masses is usually made using a lever-arm balance.

The two absolute systems of units listed in Table 14-1 are the MKS and CGS systems. In the MKS system, the unit of length is defined as the *meter* (m) which is taken to be 1,650,763.73 wavelengths in a vacuum of the orange-red line of the spectrum of krypton-86. The unit of mass is the *kilogram* (kg), defined by a bar of platinum alloy kept at the International Bureau of Weights and Measures located in Sèvres, France. And, the standard unit of time is the *second* (sec), defined by the duration of 9,192,631,770 cycles of radiation associated with a specified transition of the isotope cesium-133. In the CGS system, length, mass, and time are defined by the *centimeter* (cm), *gram* (g), and *second*. Specifically, 1 g = 0.001 kg and 1 cm = 0.01 m. The unit of force in these systems is a *derived unit*. Hence, in the MKS system, a *newton* is the amount of force needed to give 1 kg mass an acceleration of 1 m/sec² ($\mathbf{F} = m\mathbf{a}$); whereas in the CGS system, a *dyne* is defined as the amount of force which gives a mass of 1 g an acceleration of 1 cm/sec². These two absolute systems of units are used almost universally by scientists throughout the world.

Since most experiments involve a direct measurement of force, engi-

neers prefer to use a *gravitational system of units* in preference to an absolute system. In a gravitational system the three primary units are length, force and time, Table 14–1. Mass, therefore, becomes a derived unit. In formulating the MKGFS system, the units of length (m) and time (sec) are the same as in the MKS system. The MKGFS system, however, is often confused with the MKS system since the unit of mass in the MKS system and the unit of force in the MKGFS system have the same name—the kilogram. Due to frequent misunderstanding in solving problems, it should therefore be kept firmly in mind that the kilogram (kg) is a unit of mass in the MKS system and the kilogram-force (kg_f) is a unit of force in the MKGFS system. Specifically, the unit of mass in the MKGFS system is the *metric slug* defined as the amount of mass to which a force of 1 kg_f imparts an acceleration of 1 m/sec². On the other hand, in the FPS system of units, the unit of mass is called a *slug*, which is equal to the amount of mass which is accelerated at 1 ft/sec² when acted upon by a force of 1 lb.* (By comparison, 1 lb = 0.4536 kg_f.) Hence, as shown in Fig. 14–2a, when a body weighs 1 lb, applying Eq. 14–1, we have

$$\Sigma \mathbf{F} = m\mathbf{a}$$
$$1 \text{ lb} = (1 \text{ slug})(1 \text{ ft/sec}^2)$$
$$1 \text{ slug} = 1 \text{ lb-sec}^2/\text{ft}$$

Likewise, when a body weighing W lb is located at a point where the acceleration of the body due to gravity is g ft/sec², Fig. 14–2b, the mass m of the body is then

$$\Sigma \mathbf{F} = m\mathbf{a}$$
$$W \text{ lb} = (m \text{ slug})(g \text{ ft/sec}^2)$$

or

$$m \text{ (slug)} = \frac{W}{g}\left(\frac{\text{lb-sec}^2}{\text{ft}}\right) \tag{14–2}$$

In particular, when a freely falling body is located at sea level and at a latitude of 45° (considered the standard location), the acceleration due to gravity is approximately $g = 32.2$ ft/sec² (9.8 m/sec²). If the weight of the body is measured at this standard location the mass is then

*Currently the FPS system of units is predominantly used by American engineers; however, by act of Congress, the MKGFS units of measurement are intended to replace the FPS units. Although this change in units will come in due time, in this book we will continue to use the FPS system of units as a means of applying the principles of mechanics. The conversion factors given in Appendix C should enable the reader to directly convert all the physical quantities in mechanics from FPS to MKGFS units if this becomes desirable.

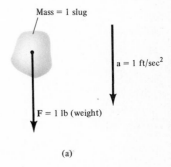

Mass = 1 slug

$a = 1$ ft/sec²

$\mathbf{F} = 1$ lb (weight)

(a)

Mass = m(slug) = $\frac{W}{g}(\frac{\text{lb/sec}^2}{\text{ft}})$

$a = g$ ft/sec²

$\mathbf{F} = W$ lb (weight)

(b)

Fig. 14–2

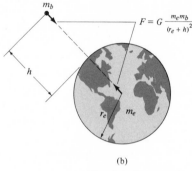

(b)

Fig. 14–3

computed by using Eq. 14–2.* For most engineering problems we can compute the mass of a body using a value of $g = 32.2$ ft/sec², associated with a weight W lb measured at or near the earth's surface. Actually, as stated previously, the magnitude of weight force will vary with its location on the earth's surface and distance above it. In the general case, the mass must be determined by using the acceleration of gravity at the location *where* the measurement has taken place. When this is done, the *mass* of the body will always remain the *same*, even though the *weight* and the *acceleration of gravity change*.

The fact that a body has a weight force which depends upon the position of the body relative to the earth or its position relative to any other object is an example of Newton's law of gravitation. This law states that between any two particles having a mass of m_1 and m_2, there is a force of attraction $\mathbf{F}$ which acts along the line joining the particles and has a magnitude directly proportional to the product of the two masses

*In the MKGFS system Eq. 14–2 becomes

$$m(\text{metric slug}) = \frac{W}{g}\left(\frac{\text{kg}_f\text{-sec}^2}{\text{m}}\right)$$

Here W is measured in kg_f and $g = 9.80$ m/sec².

and inversely proportional to the square of the distance r between the masses. In mathematical form,

$$F = G\frac{m_1 m_2}{r^2} \qquad (14\text{-}3)$$

The constant G is called the *universal gravitational constant,* and experiments have shown that it has a magnitude of 3.44×10^{-8} lb-ft^2/slug^2(6.67×10^{-11}m^3/kg-sec^2).

From Eq. 14-3, it is now possible to see why the weight of a body will vary, depending upon where it is measured. In particular, when a body is located on the *surface* of the earth, the earth's mass is considered to be concentrated at its center, and the gravitational force (weight) is computed as shown in Fig. 14-3a. When the weight is measured at an altitude h above the earth's surface the force is computed in a similar manner, as shown in Fig. 14-3b. The gravitational force expressed by Eq. 14-3 is valid for *every pair* of bodies. However, for bodies near the surface of the earth, the earth's gravitational force is the only one of any appreciable magnitude, so that gravitational forces from surrounding bodies may be neglected in the calculations. (See Prob. 14-1).

14-3. Method of Problem Solution

When applying Eq. 14-1 to the solution of kinetic problems, the solution should *always* include a free-body diagram and an inertia-vector diagram. The *free-body diagram* provides a convenient means of accounting for all the forces ($\Sigma \mathbf{F}$) which act on the particle.* The diagram which shows the *inertia-force vector m*$\mathbf{a}$ acting on the particle is termed an *inertia-vector diagram*. The word "inertia" is incorporated in this terminology because mass m is related to the *inertia* or resistance of a body to a change in its motion. The change in motion is a consequence of the acceleration $\mathbf{a}$ given to the mass. Hence, the inertia-force vector has a *magnitude* of ma and a *sense of direction* which is the same as $\mathbf{a}$.†

It is important to realize that the inertia-force vector is actually *not* a force. The inertia of a body manifests itself as a force whenever a change in motion (acceleration) of the body is produced. Specifically, this change in motion is caused by the applied forces. To understand this further, consider a passenger riding in a car which is accelerating. The forward motion of the car creates a horizontal force which the *seat* exerts on the back of the passenger. By Newton's second law of motion it is this

---

*See Chapter 3 of *Engineering Mechanics: Statics* for a detailed discussion of free-body diagrams for particles.

†Later in our study of rigid-body kinetics, we will introduce the concept of an inertia couple.

unbalanced force which gives the passenger a forward acceleration. No force exists which pushes his back toward the seat, although this is the sensation he receives.

Application of Eq. 14–1 becomes very direct and methodical once the correct free-body and inertia-vector diagrams for the particle are drawn and "equated." Consider, for example, the block, shown in Fig. 14–4*a*, resting on a smooth surface. The block has a weight of W lb and is subjected to a cable tension of T lb. With this in mind, the correct free-body and inertia-vector diagrams for the block are shown in Fig. 14–4*b*. Having constructed these two diagrams, all of the force vectors and the inertia-force vector used in Eq. 14–1 can now easily be accounted for.

A second method of applying Eq. 14–1 exists which is a simple extension of the method just described. If Eq. 14–1 is rewritten in the form $\Sigma \mathbf{F} - m\mathbf{a} = 0$, we only need *one diagram* to accompany the problem solution. This diagram should show all the forces acting on the particle, including the inertia-force vector acting in this reverse sense $(-m\mathbf{a})$. When this is done, it appears as if the inertia-force vector balances the resultant of the external forces. The state of equilibrium, so produced, is called *dynamic equilibrium,* and as a result, the principles of statics may be applied to the solution of the problem. For the block shown in Fig. 14–4*a* the required diagram is shown in Fig. 14–4*c*. This method of applying Eq. 14–1 is often referred to as the *D'Alembert principle,* named after the French mathematician Jean le Rond d'Alembert (1717–1783). Historically, the D'Alembert principle was developed as an aid to understand the concepts of dynamics in terms of statics, which at the time were more fully understood. The advantages of this approach, however, are mainly incorporated in the first method, where the analysis of the forces (using the free-body diagram) is kept separated from the motion analysis (using the inertia-vector diagram).

If application of the equations of motion does not yield enough information to obtain a complete solution to a given problem other equations must be written. These equations might include frictional equations which relate the coefficient of kinetic friction to the frictional and normal forces acting at surfaces of contact, and kinematic equations imposed by the constraints of the problem.

14–4. Equations of Motion for a System of Particles

Newton's second law of motion will now be extended to include a system of n particles isolated within an enclosed region in space as shown in Fig. 14–5*a*. In particular, there is no restriction in the way the particles are connected, and as a result, the following analysis will apply equally well to the motion of a solid, liquid, or gas system. At the instant con-

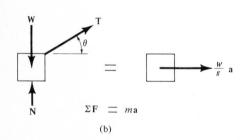

(a)

$$\Sigma \mathbf{F} \;=\; m\mathbf{a}$$

(b)

$$\Sigma \mathbf{F} - m\mathbf{a} = 0$$

(c)

Fig. 14–4

sidered, there are two types of forces acting on the ith particle of the system: the *resultant force* $\mathbf{F}_i$ which is caused by interactions *external* to the system of particles (e.g., contact forces or gravitation with external bodies and magnetic or electrical forces), and the *internal reactive forces* $\mathbf{f}_{ij}$ which act between the ith particle and each of the other j particles within the system. The free-body and inertia-vector diagrams for the ith particle are shown in Fig. 14–5b. Applying Newton's law of motion, Eq. 14–1, to the ith particle, yields,

$$\Sigma \mathbf{F} = m\mathbf{a}; \quad \mathbf{F}_i + \mathbf{f}_{i1} + \mathbf{f}_{i2} + \mathbf{f}_{i3} + \cdots + \mathbf{f}_{ij} + \cdots + \mathbf{f}_{in} = m_i \mathbf{a}_i$$

In a similar manner, if Newton's law of motion is applied to each of the other particles of the system, similar equations will result. If these equations are written for all n particles and added together vectorially, we have

$$\Sigma \mathbf{F} + \Sigma \mathbf{f}_{ij} = \Sigma m_i \mathbf{a}_i$$

Here $\Sigma \mathbf{F}$ represents the vector sum of *all the external forces* acting on the system of particles and $\Sigma \mathbf{f}_{ij}$ is the vector sum of *all the internal forces* acting within the system. This internal summation is equivalent to *zero* since, by Newton's third law of motion, all the internal forces occur in equal and opposite collinear pairs. Hence, the above equation reduces to

$$\Sigma \mathbf{F} = \Sigma m_i \mathbf{a}_i \qquad (14\text{–}4)$$

If $\mathbf{r}_G$ is a position vector which locates the *center of mass G* for the particles, Fig. 14–5a, then by definition of the center of mass,

$$m\mathbf{r}_G = \Sigma m_i \mathbf{r}_i$$

where $m = \Sigma m_i$ is the total mass of all n particles. Differentiating this equation twice with respect to time, assuming no mass is entering or leaving the system*, yields

$$m\frac{d^2\mathbf{r}_G}{dt^2} = \Sigma m_i \frac{d^2\mathbf{r}_i}{dt^2}$$

or

$$m\mathbf{a}_G = \Sigma m_i \mathbf{a}_i$$

where $\mathbf{a}_G$ is the acceleration of the center of mass of the system. Substituting this result into Eq. 14–4, yields

$$\Sigma \mathbf{F} = m\mathbf{a}_G \qquad (14\text{–}5)$$

Thus, the acceleration of the mass center, $\mathbf{a}_G$, of the system of n particles

*A case in which m is a function of time (variable mass) is discussed in Sec. 18–7.

(a)

(b)

Fig. 14–5

can be determined by applying the sum of the external forces $\Sigma\mathbf{F}$, acting on the system of particles, to a single "fictitious" particle having a mass $m = \Sigma m_i$. This fictitious particle is located at the center of mass G of all the particles.

Since, in reality, all particles must have finite size to possess mass, Eq. 14–5 justifies the use of Newton's law of motion for a body which is represented as a single particle.

14–5. Equations of Motion for a Particle: Rectangular Coordinates

When the particle is moving relative to an inertial xyz frame of reference both the forces acting on the particle, as well as its acceleration, may be expressed in terms of rectangular components. Applying Eq. 14–1, we have

$$\Sigma\mathbf{F} = m\mathbf{a}$$
$$\Sigma F_x\mathbf{i} + \Sigma F_y\mathbf{j} + \Sigma F_z\mathbf{k} = m(a_x\mathbf{i} + a_y\mathbf{j} + a_z\mathbf{k})$$

Equating the respective $\mathbf{i}$, $\mathbf{j}$, and $\mathbf{k}$ components satisfies the solution of this equation and yields the following *three scalar* equations:

$$
\begin{aligned}
\Sigma F_x &= ma_x \\
\Sigma F_y &= ma_y \\
\Sigma F_z &= ma_z
\end{aligned}
\tag{14–6}
$$

After the free-body and inertia-vector diagrams for the particle are drawn, the equations of motion may *then* be applied to solve for the unknowns. In particular, if the force system acting on the particle is *three-dimensional,* the *vector* Eq. 14–1 should be used for the solution. This equation reduces to the three scalar equations (Eqs. 14–6) after each of the vectors on the free-body and inertia-vector diagrams are expressed in Cartesian component form. On the other hand, if all the forces acting on the particle lie in the xy plane, the acceleration of the particle will also occur in this plane. The last of Eqs. 14–6 is not applicable and there are, at most, *two scalar equations* which may be used for the solution.

(a)

(b)

Fig. 14–6

Example 14–1

A small 5-lb sphere is fired vertically upward from the ground, with an initial velocity of 100 ft/sec, Fig. 14–6a. If it is assumed that the atmospheric drag force can be neglected, determine the maximum height to which the sphere will travel.

Solution

Neglecting the drag force, the free-body and inertia-vector diagrams are shown in Fig. 14–6b. Applying Eq. 14–6,

$$+\uparrow \Sigma F_z = ma_z; \qquad -5 = \frac{5}{32.2} a$$

$$a = -32.2 \text{ ft/sec}^2$$

The result indicates that the sphere, like every projectile subjected to free-flight motion near the earth's surface, is subjected to an acceleration of 32.2 ft/sec². (Refer to Sec. 12–7.) The negative sign indicates that the sphere is being decelerated as it moves upward, that is, the inertia vector in Fig. 14–6b acts in the opposite direction.

Since the acceleration is *constant*, we may use Eq. 12–5c (kinematics) to obtain the maximum height h. Initially $s_1 = 0$ and $v_1 = 100$ ft/sec, and at the maximum height $v_2 = 0$, therefore

$$v_2^2 = v_1^2 + 2a_c(s_2 - s_1)$$
$$0^2 = (100)^2 + 2(-32.2)(h - 0)$$

Thus,

$$h = 155.3 \text{ ft} \qquad\qquad Ans.$$

Example 14–2

The particle P, shown in Fig. 14–7a, weighs 7 lb and is acted upon by two magnetic forces $\mathbf{F}_1 = \{4\mathbf{i} - 3\mathbf{j} + 4\mathbf{k}\}$ lb and $\mathbf{F}_2 = \{2\mathbf{i} + 2\mathbf{j} + 2\mathbf{k}\}$ lb. Provided these forces remain constant while the particle is in motion, determine the location and velocity of P two seconds after it is released from rest.

Solution

The free-body and inertia-vector diagrams for the particle are shown in Fig. 14–7b. In particular, both the magnitude and direction of the inertia vector are unknown since $\mathbf{a}$ is unknown. Because the force system is three dimensional, a vector approach will be used for the solution. Applying Eq. 14–1 gives

$$\Sigma \mathbf{F} = m\mathbf{a}; \qquad \mathbf{F}_1 + \mathbf{F}_2 + \mathbf{W} = m\mathbf{a}$$

Since the weight $\mathbf{W} = \{-7\mathbf{k}\}$ lb, we have

$$4\mathbf{i} - 3\mathbf{j} + 4\mathbf{k} + 2\mathbf{i} + 2\mathbf{j} + 2\mathbf{k} - 7\mathbf{k} = \frac{7}{32.2} \mathbf{a}$$

$$6\mathbf{i} - \mathbf{j} - \mathbf{k} = \frac{7}{32.2} \mathbf{a}$$

Letting $\mathbf{a} = d^2\mathbf{r}/dt^2$ (kinematics), where $\mathbf{r}$ denotes a position vector directed from the origin of coordinates to the particle, yields

(a)

(b)

Fig. 14–7

$$\frac{d^2\mathbf{r}}{dt^2} = \frac{32.2}{7}(6\mathbf{i} - \mathbf{j} - \mathbf{k})$$

Integrating this equation,

$$\mathbf{v} = \frac{d\mathbf{r}}{dt} = \frac{32.2}{7}(6\mathbf{i} - \mathbf{j} - \mathbf{k})t + \mathbf{C}_1$$

and

$$\mathbf{r} = \frac{32.2}{14}(6\mathbf{i} - \mathbf{j} - \mathbf{k})t^2 + \mathbf{C}_1 t + \mathbf{C}_2$$

The vector constants of integration $\mathbf{C}_1$ and $\mathbf{C}_2$ may be obtained using the initial conditions, i.e., $t = 0$, $\mathbf{r} = 0$, and $\mathbf{v} = 0$. Substituting into the above equations, $\mathbf{C}_1 = \mathbf{C}_2 = 0$. Thus,

$$\mathbf{v} = \{4.6(6\mathbf{i} - \mathbf{j} - \mathbf{k})t\}\ \text{ft/sec}$$
$$\mathbf{r} = \{2.3(6\mathbf{i} - \mathbf{j} - \mathbf{k})t^2\}\ \text{ft}$$

At $t = 2$ sec,

$$\mathbf{v}\bigg|_{t=2\ \text{sec}} = \{55.2\mathbf{i} - 9.2\mathbf{j} - 9.2\mathbf{k}\}\ \text{ft/sec} \qquad \textit{Ans.}$$

$$\mathbf{r}\bigg|_{t=2\ \text{sec}} = \{55.2\mathbf{i} - 9.2\mathbf{j} - 9.2\mathbf{k}\}\ \text{ft} \qquad \textit{Ans.}$$

Example 14-3

The 50-lb block, resting on the inclined plane shown in Fig. 14–8a, is subjected to a 40-lb horizontal force. If the coefficient of kinetic friction between the block and plane is $\mu_k = 0.3$, determine the acceleration of the block if (a) the block is initially at rest and (b) the block is initially moving up the plane with a velocity of 60 ft/sec.

Solution
Part (a). Using Eq. 14–2, the mass of the block is

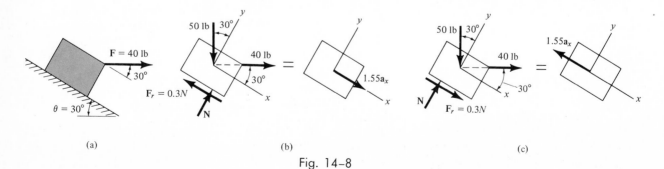

(a)   (b)   (c)

Fig. 14-8

$$m = \frac{W}{g} = \frac{50 \text{ lb}}{32.2 \text{ ft/sec}^2} = 1.55 \text{ slug}$$

The free-body and inertia-vector diagrams are shown in Fig. 14-8b. It is *assumed* that the acceleration of the block is down along the surface of the plane. If this assumption proves to be incorrect, either the magnitude of normal force **N** will be negative (which indicates that the block will tend to lift off the surface of the plane), or the acceleration $\mathbf{a}_x$ will be negative. The frictional force $\mathbf{F}_r$ opposes the motion of the block and is related to the normal force **N** by the equation $F_r = \mu_k N$.

Applying Eqs. 14-6, in reference to Fig. 14-8b, we have

$$+\searrow \Sigma F_x = ma_x; \quad -0.3N + 40 \cos 30° + 50 \sin 30° = 1.55a_x$$
$$+\nearrow \Sigma F_y = ma_y; \quad N - 50 \cos 30° + 40 \sin 30° = 0$$

Solving, we obtain

$$N = 23.3 \text{ lb}$$
$$a_x = 34.0 \text{ ft/sec}^2 \qquad \qquad Ans.$$

Since both N and a_x are positive our assumption was correct. Hence, the block has a *constant* acceleration as it moves *down* the plane.

Part (b). As the block moves *up the plane,* the frictional force acts to oppose this motion. The free-body and inertia-vector diagrams are shown in Fig. 14-8c. It is *assumed* that the block accelerates up the plane as shown. Applying the equations of motion, we have

$$+\searrow \Sigma F = ma_x; \quad 0.3 \ N + 40 \cos 30° + 50 \sin 30° = -(1.55)a_x$$
$$+\nearrow \Sigma F = ma_y; \quad N - 50 \cos 30° + 40 \sin 30° = 0$$

Solving,

$$N = 23.3 \text{ lb}$$
$$a_x = -43.0 \text{ ft/sec}^2 \qquad \qquad Ans.$$

Since the magnitude of $\mathbf{a}_x$ is negative its direction is opposite to that shown (assumed) on the inertia-vector diagram. As a result, the block is *decelerating* as it moves up the plane.

Example 14-4

The 400-lb block *H* shown in Fig. 14-9a is released from rest and travels down the smooth inclined plane. If the weights of the pulleys and the cables are neglected and the pulleys are frictionless, determine the speed which block *E* attains in 1 sec. What is the tension developed in each of the cables?

Fig. 14-9

Solution

Since Eqs. 14-6 apply to the motion of a single particle, the motion of blocks H and E will be discussed separately. The necessary free-body and inertia-vector diagrams for each of the blocks are shown in Figs. 14-9b and 14-9c, respectively. Block H moves down the plane, causing block E to move upward. Applying Eqs. 14-6 to the vectors listed on these diagrams, we obtain for block H,

$$+ \nearrow \Sigma F_x = ma_x; \qquad T_H - 400 \sin 45° = -\frac{400}{32.2} a_H \qquad (1)$$

$$+ \nwarrow \Sigma F_y = ma_y; \qquad N - 400 \cos 45° = 0 \qquad (2)$$

and for block E,

$$+ \uparrow \Sigma F_y = ma_y; \qquad T_E - 20 = \frac{20}{32.2} a_E \qquad (3)$$

There are 5 unknowns (T_H, N, T_E, a_H, and a_E) in Eqs. (1) through (3). Since the mass of the pulley is neglected and the pulley is frictionless, a relation between T_H and T_E may be obtained from the free-body diagram of the pulley at B, as shown in Fig. 14–9d.

$$2T_E = T_H \tag{4}$$

A fifth equation may be obtained by studying the kinematics of the pulley arrangement thereby relating the accelerations of both blocks. Using the technique developed in Sec. 12–10, an equivalent pulley system is established for the blocks as shown in Fig. 14–9e. It is seen that

$$2s_H + s_E = l$$

where l is constant and represents the total length of the vertical segments of cable. Differentiating this expression twice with respect to time yields

$$2a_H = a_E \tag{5}$$

The *signs* in this equation are consistent with the assumed accelerations shown by the directions given in Figs. 14–9b and 14–9c; that is, a negative x acceleration in Fig. 14–9b gives the positive y acceleration in Fig. 14–9c. This is important since the set of Equations (1) through (5) must be solved simultaneously. The results of this solution are,

$$T_H = 80.4 \text{ lb} \qquad Ans.$$
$$T_E = 40.2 \text{ lb} \qquad Ans.$$
$$N = 282.8 \text{ lb}$$
$$a_H = 16.3 \text{ ft/sec}^2$$
$$a_E = 32.6 \text{ ft/sec}^2$$

Notice that the accelerations are *constant* since all forces acting on the blocks are constant. Because of this, the velocity of block E may be obtained using Eq. 12–5a. Since the block starts from rest, $(v_E)_1 = 0$, we have

$$v_E = (v_E)_1 + a_E t$$
$$= 0 + (32.6)t$$

At $t = 1$ sec,

$$v_E = 32.6 \text{ ft/sec} \qquad Ans.$$

Problems

14-1. Determine the gravitational attraction between two 10-in. spheres which are just touching each other. The weight of each sphere (gravitational attraction toward the earth) is 50 lb each.

14-2. The planets Mars and Earth have diameters of 4,200 and 7,900 mi, respectively. The mass of Mars is 0.108 that of Earth. If a body weighs 200 lb on the surface of the Earth, what would its weight be on Mars? Determine the acceleration due to gravity on Mars.

14-3. Neglecting the effects of friction and the mass of the pulley and cord in the system, determine the acceleration at which the 30-lb block will descend. What is the tension in the cord?

Prob. 14-3

14-4. The 15 and 25-lb weights are suspended over a pulley by a rope. Neglecting the mass of the rope and the pulley, determine the acceleration of both weights.

Prob. 14-4

14-5. Determine the force which a boy weighing 75 lb exerts on the floor of an elevator when the elevator is (a) at rest, (b) descending with a constant velocity of

6 ft/sec, and (c) ascending with a constant acceleration of 2 ft/sec².

14-6. The weight of an elevator operator varies between 130 and 170 lb while he is riding his elevator. When the elevator is at rest the operator weighs 153 lb. Determine how fast his elevator car can accelerate, going up and going down.

14-7. Work Prob. 14-6 assuming the elevator operator weighs 140 lb when the elevator is at rest.

14-8. A parachutist weighing 170 lb falls freely for 4 sec before opening his chute. If the chute opens in 3 sec and reduces the man's speed to 30 ft/sec, compute the total force acting on the ropes of the parachute during this time. Neglect the weight of the parachute.

14-9. The 20-lb block B rests on the surface of a table for which the coefficient of friction is $\mu = 0.1$. Determine the speed at which the 10-lb block A will move after falling 2 ft from rest.

Prob. 14-9

14-10. The double inclined plane supports two blocks A and B, each having a weight of 10 lb. If the coefficient of friction between the blocks and the plane is $\mu_s = 0.35$, show that the blocks are in equilibrium. If block A is given an initial velocity of 4 ft/sec down the plane, compute the distance it travels before coming to rest.

Prob. 14-10

14-11. Determine the weight of block A, required to move block B 10 ft up the inclined plane in 4 sec starting from rest. Neglect the weights of the pulleys and cable.

Prob. 14-11

14-12. A smooth 2-lb collar C fits loosely on the horizontal shaft. If the spring is unextended when the collar is in the dotted position, determine the velocity of the collar when $s = 1$ ft. The collar is given an initial horizontal velocity of 15 ft/sec when $s = 0$.

Prob. 14-12

14-13. Determine the tension developed in the cables and the acceleration of each weight. Neglect the weight of the pulleys and cables.

Prob. 14-13

14-14. The train shown consists of three cars, each car weighing 20 tons. If a constant frictional force of 150 lb acts at the wheels of each car while the train moves forward with an acceleration of 2 ft/sec², determine the normal force which the engine A exerts on the tracks. What is the tension force exerted on the couplings which hold each of the cars together?

Prob. 14-14

14-15. Two blocks A and B are connected by separate cords fastened to the single pulley C. The cord connecting block A is wound around the outer surface of the

597

pulley, while the cord from block *B* unwinds from the pulley's inner surface. Determine the acceleration of block *A* and the tension in the cords. Neglect the weight of the cords and pulley *C*. The horizontal surface supporting block *A* is smooth.

Prob. 14–15

14-16. Work Prob. 14–15 by replacing block *B* by a 25-lb vertical force.

14-17. Determine the towing force created by the motor *M* which is required to move the 500-lb crate up the inclined plane with a constant acceleration of 1 ft/sec². The coefficient of friction between the crate and the plane is $\mu = 0.3$. Neglect the weight of the pulleys and cable.

Prob. 14–17

14-18. Block *B* rests upon a smooth surface. If the coefficient of friction between *A* and *B* is $\mu = 0.4$, determine the acceleration of each block if (a) $F = 6$ lb, and (b) $F = 50$ lb.

Prob. 14–18

14-19. A motor boat has a total weight of 1,200 lb and is traveling at a speed of 40 mph. When the engine is turned off, the speed suddenly drops to 25 mph in 60 sec. If the water developes a drag force on the boat having a magnitude of Cv, where v is the speed of the boat, determine the value of the constant *C*.

14-20. The chain is 15 ft long and weighs 2 lb/ft. If $\mu = 0.22$, determine the velocity at which the end *A* will pass point *B* when the chain is released from rest.

Prob. 14–20

14-21. The 10-lb block is traveling with a velocity of 10 ft/sec over a smooth plane. If a force $F = 2.5t$, where t is in seconds and F is in pounds, suddenly acts on the block for 3 sec, determine the final velocity of the block and the distance the block travels during the time the force was acting.

Prob. 14–21

14-22. A 150-lb parachutist jumps from a plane and attains a speed of 50 ft/sec before his parachute opens. If the air resistance acting on the parachute is $0.67v^2$, where v is the speed of the parachute given in ft/sec, determine the terminal or maximum speed of the man as he falls.

14-23. The 6-lb ball is also subjected to the action of forces: $F_1 = 2i + 6j - 2tk$, $F_2 = t^2i - 4tj - k$, and $F_3 = -2ti$, where the force is given in pounds and time in seconds. Determine the magnitude of displacement of the ball with respect to its original position 2 sec after being released from rest. Assume that the directions of the forces acting on the ball are constant during the time interval.

Prob. 14–23

14-24. The two particles A and B weigh 2 and 5 lb, respectively. If they are also acted upon by the same force system, namely, $F_1 = 2i + 6j - 2k$, $F_2 = 3i - k$, and $F_3 = i - t^2j - 2k$, where the force is given in pounds and t is in seconds, determine the distance between them 3 sec after they have been released from rest.

Prob. 14–24

14-25. A chain hangs across the smooth peg. Determine the time required for it to slide off the peg when end A is released from rest.

Prob. 14–25

14-26. Work Prob. 14–25, assuming that friction between the chain and the peg is equal to the weight of 1 ft of the chain. Neglect the size of the peg.

14–6. Equations of Motion for a Particle: Cylindrical Coordinates

The acceleration of a particle written in terms of its cylindrical components (r, θ, and z) has been developed in Sec. 12–8 (Eq. 12–32). When the forces acting on the particle are resolved into components along the unit-vector directions $\mathbf{u}_r$, $\mathbf{u}_\theta$, and $\mathbf{u}_z$, Fig. 14–10, Newton's second law of motion may be expressed as

$$\Sigma \mathbf{F} = m\mathbf{a}$$

$$\Sigma F_r \mathbf{u}_r + \Sigma F_\theta \mathbf{u}_\theta + \Sigma F_z \mathbf{u}_z = m\left[\frac{d^2r}{dt^2} - r\left(\frac{d\theta}{dt}\right)^2\right]\mathbf{u}_r$$

$$+ m\left(r\frac{d^2\theta}{dt^2} + 2\frac{dr}{dt}\frac{d\theta}{dt}\right)\mathbf{u}_\theta + m\frac{d^2z}{dt^2}\mathbf{u}_z$$

Equating the respective $\mathbf{u}_r$, $\mathbf{u}_\theta$, and $\mathbf{u}_z$ components yields the following three scalar equations:

$$\Sigma F_r = m\left[\frac{d^2r}{dt^2} - r\left(\frac{d\theta}{dt}\right)^2\right]$$

$$\Sigma F_\theta = m\left(r\frac{d^2\theta}{dt^2} + 2\frac{dr}{dt}\frac{d\theta}{dt}\right) \qquad (14\text{--}7)$$

$$\Sigma F_z = m\frac{d^2z}{dt^2}$$

These equations form a useful means for solving certain types of problems involving rockets, satellites, and planetary bodies subjected to gravitational forces. In this regard see Sec. 14–8.

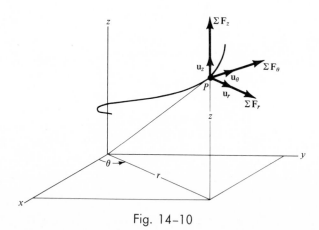

Fig. 14–10

Fig. 14–11

14–7. Equations of Motion for a Particle: Normal and Tangential Components

When a particle moves over a curved path which is known, the vector equation of motion for the particle may be expressed in terms of its normal and tangential components. Substituting Eq. 12–37 into Eq. 14–1, we have

$$\Sigma \mathbf{F} = m\mathbf{a}$$

$$\Sigma F_t \mathbf{u}_t + \Sigma F_n \mathbf{u}_n = m\mathbf{a}_t + m\mathbf{a}_n$$

or

$$\Sigma F_t \mathbf{u}_t + \Sigma F_n \mathbf{u}_n = m\frac{dv}{dt}\mathbf{u}_t + m\frac{v^2}{\rho}\mathbf{u}_n$$

ΣF_n and ΣF_t represent the sum of all the force components acting on the particle in the normal and tangential directions, respectively, Fig. 14–11. Equating the respective $\mathbf{u}_n$ and $\mathbf{u}_t$ components, we have

$$\Sigma F_t = ma_t = m\frac{dv}{dt}$$

$$\Sigma F_n = ma_n = m\frac{v^2}{\rho}$$

(14–8)

It was shown in Sec. 12–9 that the tangential component of acceleration

$\mathbf{a}_t$ represents a *change* in the *magnitude* of the *velocity*. From the first of Eqs. 14–8, this change is caused by the sum of the *tangential force components* ΣF_t acting on the particle. If the speed of the particle is increasing, the resultant of this tangential force summation acts in the same direction as the velocity. If the speed is decreasing, it acts in the opposite sense. The *normal force components* ΣF_n create a *change* in the *direction* of the *velocity* of the particle. Like the normal component of acceleration, $\mathbf{a}_n$, the resultant of this force summation *always acts toward the center of curvature, O,* of the path, Fig. 14–11. In particular, when the particle is traveling on a constrained *circular path* with a *constant* speed, there is a normal force exerted on the particle by the constraint. This force is termed a *centripetal force*. The equal but opposite force exerted by the particle on the constraint is called a *centrifugal force*.

Example 14–5

From experimental measurements, the motion of a 5-lb particle is defined in terms of cylindrical coordinates by the relations $r = t^2 + 2t$, $\theta = 3t + 2$, and $z = t^3 + 4$, where r and z are expressed in feet, θ in radians, and t in seconds. Determine the magnitude of the resultant force acting on the particle at the instant when $t = 1$ sec.

Solution

The orthogonal components of force F_r, F_θ, and F_z can be found by applying Eqs. 14–7. Computing the required derivatives yields

$$\frac{dr}{dt} = 2t + 2; \qquad \frac{d^2r}{dt^2} = 2$$

$$\frac{d\theta}{dt} = 3; \qquad \frac{d^2\theta}{dt^2} = 0$$

$$\frac{d^2z}{dt^2} = 6t$$

Therefore,

$$\Sigma F_r = m\left[\frac{d^2r}{dt^2} - r\left(\frac{d\theta}{dt}\right)^2\right]; \qquad F_r = \frac{5}{32.2}[2 - (t^2 + 2t)3^2]$$

$$\Sigma F_\theta = m\left(r\frac{d^2\theta}{dt^2} + 2\frac{dr}{dt}\frac{d\theta}{dt}\right); \qquad F_\theta = \frac{5}{32.2}[(t^2 + 2t)0 + 2(2t + 2)3]$$

$$\Sigma F_z = m\frac{d^2z}{dt^2}; \qquad F_z = \frac{5}{32.2}6t$$

When $t = 1$ sec,

$$F_r\bigg|_{t=1\,\text{sec}} = \frac{5}{32.2}(-25) = -3.88 \text{ lb}$$

$$F_\theta\bigg|_{t=1\,\text{sec}} = \frac{5}{32.2}(24) = 3.73 \text{ lb}$$

$$F_z\bigg|_{t=1\,\text{sec}} = \frac{5}{32.2}(6) = 0.93 \text{ lb}$$

The magnitude of force acting on the particle is then

$$F = \sqrt{(F_r)^2 + (F_\theta)^2 + (F_z)^2}$$
$$= \sqrt{(-3.88)^2 + (3.73)^2 + (0.93)^2} \text{ lb}$$
$$= 5.46 \text{ lb} \qquad\qquad\qquad\qquad Ans.$$

Example 14–6

The ball B is fastened to the end of a 2-ft-long string, Fig. 14–12a. As the ball moves with a constant speed, it describes a horizontal circular path in which the string OB generates the surface of a cone. Determine the speed of the ball along its circular path if $\theta = 45°$.

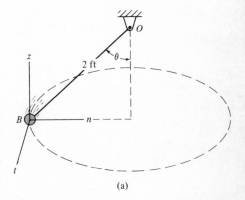

(a)

Solution

The free-body and inertia-vector diagrams for the ball are shown in Fig. 14–12b. The mass of the ball may be represented as $m = W/g$, where $g = 32.2 \text{ ft/sec}^2$. Since the ball moves around the circle with *constant speed*, only a normal component of acceleration $\mathbf{a}_n$ (directed toward the center of curvature) is present. Why? This component has a magnitude of $a_n = v^2/\rho = v^2/(2 \sin 45° \text{ ft})$.

Using the orthogonal coordinate axis shown in Fig. 14–12a and applying the equations of motion, we obtain

$$+\uparrow \Sigma F_z = ma_z; \qquad T\sin 45° - W = 0 \qquad (1)$$

$$\overset{+}{\rightarrow} \Sigma F_n = ma_n; \qquad T\cos 45° = \frac{W}{g}\frac{v^2}{2\sin 45°} \qquad (2)$$

Rearranging Eq. (1) and dividing Eq. (2) by Eq. (1), we obtain the speed v, i.e.,

$$v = \sqrt{2(32.2)\cos 45°}$$
$$= 6.75 \text{ ft/sec} \qquad\qquad Ans.$$

(b)

Fig. 14–12

Example 14–7

Determine the slope angle θ which a circular track must have so that the wheels of the sports car, shown in Fig. 14–13a, will not have to depend

(a)

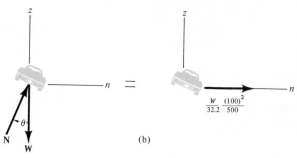

(b)

Fig. 14-13

upon friction to prevent the car from sliding either up or down the banked curve. The car travels at a constant speed of 100 ft/sec. The radius of the track is 500 ft.

Solution

Since the problem statement stipulates no friction, the free-body and inertia-vector diagrams for the car are shown in Fig. 14-13b. The car is assumed to weigh W lb. Using the nz axis shown, and applying the equations of motion, realizing that $a_n = v^2/\rho$, we have

$$+\uparrow \Sigma F_z = ma_z; \qquad\qquad N \cos \theta - W = 0$$

$$\overset{+}{\rightarrow} \Sigma F_n = ma_n; \qquad\qquad N \sin \theta = \frac{W}{32.2} \frac{(100)^2}{500}$$

Eliminating N and W from these two equations by division, and solving for θ, we obtain

$$\tan \theta = \frac{(100)^2}{32.2(500)}$$

$$\theta = \tan^{-1}(0.621) = 31.8° \qquad\qquad Ans.$$

A force summation in the tangential direction of motion is of no consequence to the solution of this problem. If it were considered note that $a_t = dv/dt = 0$ since the car moves with *constant speed*.

Example 14-8

A small satellite *s* revolves around the earth in a circular orbit, Fig. 14–14*a*. Determine the constant speed in which the satellite circles the earth if the radius of its orbit is $r = 9,000$ miles. (The mass of the earth is 4.09×10^{23} slugs.)

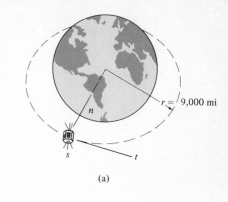

(a)

Solution

The only force acting on the satellite is the gravitational attraction of the earth. If m_s and m_e denote the mass of the satellite and earth, respectively, this gravitational force is determined by using Eq. 14–3. The free-body and inertia-vector diagrams are shown in Fig. 14–14*b*. Applying the second of Eqs. 14–8, we have

$$\Sigma F_n = ma_n; \qquad G\frac{m_e m_s}{r^2} = m_s \frac{v^2}{r}$$

Therefore,

$$v = \sqrt{\frac{Gm_e}{r}}$$

$$= \sqrt{\frac{(3.44 \times 10^{-8} \text{ ft}^3/\text{slug-sec}^2)(4.09 \times 10^{23} \text{ slug})}{(9,000 \text{ mi})(5,280 \text{ ft/mi})}}$$

(b)

Fig. 14–14

or

$$v = 17,200 \text{ ft/sec} = 11,730 \text{ mi/hr} \qquad \textit{Ans.}$$

Example 14-9

A point on the periphery of a small 3-lb disk is attached to a long cord as shown in Fig. 14–15*a*. The other end of the cord is attached to a ball-and-socket joint located at the center of a rotating platform. The maximum tension which the cord can sustain is 6 lb. If the disk is placed on the platform and released from rest, as shown, determine the time it takes before the cord breaks. The coefficient of kinetic friction between the disk and the platform is $\mu_k = 0.1$. Assume the platform rotates at a very high rate of speed so that the disk always slides on the platform surface before the cord breaks.

Solution

The free-body and inertia-vector diagrams for the disk are shown in Fig. 14–15*b*. Note that the disk has *both* normal and tangential components of acceleration due to the unbalanced forces **T** and **F** which act on the disk. In particular, since sliding is occurring, the frictional force has a magnitude of $F = \mu_k N = 0.1 N$ and a direction which opposes the

605

(a)

(b)

Fig. 14-15

relative motion between the contacting surfaces, Fig. 14–15b. Applying the equations of motion,

$$\Sigma F_z = ma_z; \qquad N - 3 = 0 \qquad (1)$$

$$\Sigma F_t = ma_t; \qquad 0.1\,N = \frac{3}{32.2}a_t \qquad (2)$$

$$\Sigma F_n = ma_n; \qquad T = \frac{3}{32.2}\left(\frac{v^2}{2}\right) \qquad (3)$$

Since the cord breaks when $T = 6$ lb, we can solve Eq. (3) for the critical speed v_{cr}, needed to break the cord. Solving all the equations,

$$N = 3 \text{ lb}$$
$$a_t = 3.22 \text{ ft/sec}^2$$
$$v_{cr} = 11.35 \text{ ft/sec}$$

From kinematics, since $a_t = dv/dt$, and a_t is *constant*, we have

$$\int dv = \int a_t\,dt$$
$$v = v_1 + a_t\,t$$

where v_1 is a constant of integration. (This result is identical to Eq. 12–5a.) Initially, $v_1 = 0$, so that the time required to break the cord is

$$11.35 = 0 + (3.22)t$$
$$t = 3.52 \text{ sec} \qquad \qquad Ans.$$

Example 14-10

A block having a weight w is given an initial velocity of 1 ft/sec when it is at the top of the smooth cylinder shown in Fig. 14–16a. If the cylinder has a radius of 10 ft, determine the angle $\theta_{\max}$ at which the block leaves the surface of the cylinder.

Solution

For convenience, the angle θ, Fig. 14–16a, increases in the direction of motion of the block. The free-body and inertia-vector diagrams for the block, when the block is located at the *general position* θ, are shown in Fig. 14–16b. The block must have a tangential acceleration $\mathbf{a}_t$ since the *speed* is always *increasing* as the block falls. Applying the equations of motion yields

$$+\swarrow \Sigma F_n = ma_n; \qquad -N + w\cos\theta = \frac{w}{g}\frac{v^2}{10} \qquad (1)$$

$$\searrow + \Sigma F_t = ma_t; \qquad w\sin\theta = \frac{w}{g}a_t \qquad (2)$$

These two equations contain four unknowns, N, v, a_t, and θ. At the instant when $\theta = \theta_{\max}$, however, the block leaves the surface of the cylinder so that $N = 0$.

A third equation for the solution may be obtained using kinematics. In particular, the magnitude of tangential acceleration $\mathbf{a}_t$ may be related to the speed of the block v, and the angle θ. From Eq. 12–37,

$$a_t = v\frac{dv}{ds}$$

or,

$$a_t\,ds = v\,dv$$

But $ds = 10d\theta$, Fig. 14–16a; thus,

$$a_t = \frac{v\,dv}{10d\theta} \qquad (3)$$

Substituting Eq. (3) into Eq. (2) and separating the variables, we have

$$10g\sin\theta\,d\theta = v\,dv$$

Integrating both sides of this equation yields

$$-10g\cos\theta + C = \frac{v^2}{2}$$

where C is a constant of integration. Hence,

$$v^2 = 2(C - 10g\cos\theta) \qquad (4)$$

(a)

(b)

Fig. 14–16

When $\theta = 0$, $v_1 = 1$ ft/sec; so that

$$1^2 = 2(C - 10g)$$

or

$$C = (\tfrac{1}{2} + 10g)$$

Equation (4) becomes

$$v^2 = 20g(1 - \cos\theta) + 1$$

Substituting into Eq. (1) with $N = 0$, and solving for $\cos\theta_{max}$, we have

$$\cos\theta_{max} = \frac{20g + 1}{30g} = 0.668$$

$$\theta_{max} = 48.1° \qquad\qquad Ans.$$

Problems

14-27. The motion of a 5-lb particle in two dimensions is described in terms of polar coordinates as $r = 2t + 10$ and $\theta = 5t^2 - 6t$, where t is measured in seconds, r is in feet, and θ is in radians. Determine the magnitude and direction of the unbalanced force acting on the particle when $t = 2$ sec.

14-28. The 10-lb metal spool rides freely along the rotating rod. The angular rate of rotation of the rod is 6 rad/sec. This rate of rotation is increasing at 2 rad/sec², as shown in the figure. Relative to the rod, the spool has a velocity of 3 ft/sec and an acceleration of 1 ft/sec², both directed away from the center O when $r = 3$ ft. Determine the magnitude of unbalanced force which acts on the spool at this instant.

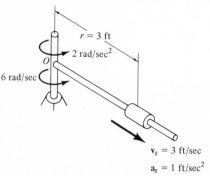

Prob. 14-28

14-29. A 5-lb particle P moves along the Archemedian spiral path. The motion of P is controlled by the tangential force **F** such that the angular rate of rotation of line segment OP is $d\theta/dt = -6$ rad/sec. Determine the total force which the particle exerts on the path when it arrives at point A. The distance r is measured in feet and θ is measured in radians.

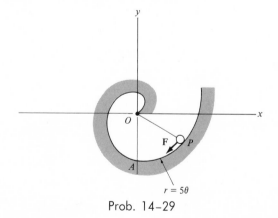

Prob. 14-29

14-30. The tube rotates in the horizontal xy plane at a constant angular rate of $d\theta/dt = 4$ rad/sec. Ball B starts at the origin with an initial radial velocity of 1 ft/sec, and moves outward through the tube. Deter-

mine the speed of the ball at the instant that it leaves the outer end of the tube.

Prob. 14–30

14–31. The boy, weighing 100 lb, is sliding down the spiral chute at a constant speed. Determine the magnitude of force he exerts on the chute if the spiral has a radius of $r = 6$ ft and $\theta = 2t$, $z = -t$.

Prob. 14–31

14–32. A 10-lb metal spool rides along a smooth rod. Provided the rod has a constant angular rate of rotation of $d\theta/dt = 2$ rad/sec, write the equations of motion for the spool. If r and θ for the spool are zero when $\theta = 0°$, determine r as a function of θ.

Prob. 14–32

14–33. Compute the mass of the sun knowing that the distance from the earth to the sun is 92×10^6 mi.

14–34. Prove that if the block is released from rest at the top point B of the smooth path, the velocity it attains when it reaches point A is equal to the same velocity it attains when it falls freely through a distance h.

Prob. 14–34

14–35. A bucket filled with water weighs 40 lb and is whirled in a verticle circle having a radius of 5 ft. Determine the minimum speed of the bucket when it is at the top of the circle so that the cord does not slacken. Compute the tension in the cord when the bucket returns to the bottom of the circle after rounding the top.

609

14-36. A pilot weighs 150 lb and is traveling at a constant speed of 100 mph. He wishes to turn an inside vertical loop, as shown, so that when he is upside down at A, he exerts an upward force of 30 lb against the seat of the plane. Determine the radius r of the required loop.

Prob. 14-36

14-37. The 5-lb pendulum bob B is released from rest in the position shown by cutting the string AB. Determine the tension in string BC immediately after AB is cut and when the pendulum reaches point D.

Prob. 14-37

14-38. The earth's rotation has an effect on the weight of bodies lying on its surface. If this rotation could be altered, determine the required velocity of a man caused by the rotation of the earth such that he would weigh one half as much as he does under normal earth rotation. Assume that the man is standing at the equator. The mass of the earth is 4.09×10^{23} slugs, and the equatorial radius of the earth is 4,000 mi.

610

14-39. One of the rides at an amusement park is called a "Rotor." For this ride passengers stand on the floor of the circular room as shown. The room is then allowed to rotate at a constant angular rate about the z axis. When this happens, the riders are forced up against the circular wall of the room because of centrifugal force. The floor of the room is then lowered. If the coefficient of friction between each passenger and the wall is approximately $\mu = 0.4$, determine the translational velocity of the wall so that no person can slip down the wall.

Prob. 14-39

14-40. The sports car is traveling along a 20° banked road having a curvature of $\rho = 500$ ft. If the coefficient of friction between the tires and the road is $\mu = 0.2$, determine the maximum and minimum safe speeds for travel with respect to skidding up or down the slope.

Prob. 14-40

14-41. The block rests at a distance of 3 ft from the center of the horizontal disk. The coefficient of friction between the block and the platform is $\mu = 0.3$. If the platform rotates about the z axis with a constant angular speed, determine the maximum speed which the block can attain before slipping off the platform.

Prob. 14-41

14-42. Work Prob. 14-41 assuming that the platform starts rotating from rest so that the block is given a constant tangential acceleration of 2 ft/sec².

14-43. The automobile is going over the crest of a vertical hill in which $\rho = 250$ ft. Determine the maximum speed at which the car may travel and not move tangentially off the road.

Prob. 14-43

14-44. Small 5-lb packages ride on the surface of the conveyor belt. If the belt starts from rest and increases to a constant speed of 5 ft/sec in 2 sec, determine the angle θ so that none of the packages slip on the inclined surface AB of the belt. The coefficient of friction between the belt and a package is $\mu = 0.3$. At what angle ϕ do the packages first begin to slip off the surface of the belt after the belt is moving at its constant speed of 5 ft/sec?

Prob. 14-44

14-45. Electrons having a mass m are emitted from a source S with an initial velocity $\mathbf{v}_0$ between the space of two oppositely charged plates. When an electric field $\mathbf{V}$ or voltage gradient is applied across the plates, the electrons are subjected to a constant force acting in the direction of the field and having a magnitude of eV/w, where e is the charge of the electron and w is the distance between the plates. Determine the maximum voltage which may be applied so that the electrons do not strike the positively charged plate.

Prob. 14-45

14-46. In the cathode-ray tube, electrons having a mass m are emitted from a source point S and travel horizontally with an initial velocity $\mathbf{v}$. After passing between the grid plates, they receive a vertical force having a magnitude of eV/w, where e is the charge of an electron, V is the applied voltage acting across the plates, and w is the distance between the plates. After passing clear of the plates, the electrons travel in straight lines and strike the screen at A. Determine the deflection s of the electrons in terms of the dimension of the voltage plate and tube.

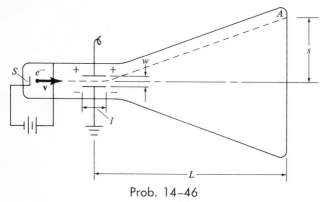

Prob. 14-46

14-47. A small 3-lb block, initially at rest at point A, slides along the smooth parabolic surface. Determine the normal force acting on the block when it is at point B.

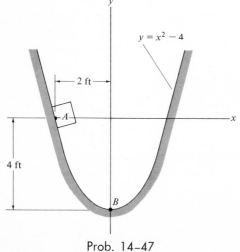

Prob. 14-47

14-48. A small 6-lb ball rolls along the vertical circular slot. If the ball is released from rest when $\theta = 10°$, determine the force which it exerts on the slot when it arrives at points A and B.

Prob. 14-48

14-49. A small 2-lb weight is attached to the vertex of the right-circular cone using a light 1-ft-long cord AB. The cone is then rotated at a constant angular rate about the z axis such that the weight attains a speed of 1 ft/sec. At this speed, determine the tension in the cord AB and the reaction which the cone exerts on the weight.

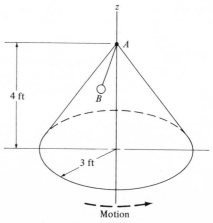

Prob. 14-49

*14–8. Central-Force Motion and Space Mechanics

If a particle is moving under the influence of a force having a line of action which is always directed toward a fixed point the motion is called *central-force motion*. This type of motion is commonly caused by electrostatic and gravitational forces. Before we consider the analysis of central forces, we will first discuss some general characteristics of all central-force motions.

In Fig. 14–17a, the particle P having a mass m is acted upon only by the central force $\mathbf{F}$. The free-body and inertia-vector diagrams for the particle are given in Fig. 14–17b. Using polar coordinates (r, θ), the equations of motion (Eq. 14–7) become

$$-F = m\left[\frac{d^2r}{dt^2} - r\left(\frac{d\theta}{dt}\right)^2\right]$$
$$0 = m\left(r\frac{d^2\theta}{dt^2} + 2\frac{dr}{dt}\frac{d\theta}{dt}\right)$$
(14–9)

The second of these equations may be written in the form

$$\frac{1}{r}\left[\frac{d}{dt}\left(r^2\frac{d\theta}{dt}\right)\right] = 0$$

so that integrating yields

$$r^2\frac{d\theta}{dt} = C_1$$
(14–10)

where C_1 is a constant of integration. From Fig. 14–17a it can be seen that the shaded area inscribed by the radius vector $\mathbf{r}$, as $\mathbf{r}$ moves through an angle $d\theta$, is $dA = \frac{1}{2}r^2\,d\theta$. If we define the *areal velocity* as

$$\frac{dA}{dt} = \frac{1}{2}r^2\frac{d\theta}{dt}$$
(14–11)

then, by comparison with Eq. 14–10, it may be seen that the areal velocity for a particle moving under central force motion is *constant*. To obtain the *path of motion*, the independent variable t must be eliminated from Eqs. 14–9 in order to obtain a differential equation in terms of r and θ. Using the chain rule of calculus and Eq. 14–10 the time derivatives of Eq. 14–9 may be replaced by

$$\frac{dr}{dt} = \frac{dr}{d\theta}\frac{d\theta}{dt} = \frac{C_1}{r^2}\frac{dr}{d\theta}$$

$$\frac{d^2r}{dt^2} = \frac{d}{dt}\left(\frac{C_1}{r^2}\frac{dr}{d\theta}\right) = \frac{d}{d\theta}\left(\frac{C_1}{r^2}\frac{dr}{d\theta}\right)\frac{d\theta}{dt} = \left[\frac{d}{d\theta}\left(\frac{C_1}{r^2}\frac{dr}{d\theta}\right)\right]\frac{C_1}{r^2}$$

Substituting a new dependent variable $\xi = 1/r$ into the second equation it is seen that

(a)

(b)

Fig. 14–17

$$\frac{d^2r}{dt^2} = -C_1^2\xi^2\frac{d^2\xi}{d\theta^2}$$

Also, the square of Eq. 14–10 becomes

$$\left(\frac{d\theta}{dt}\right)^2 = C_1^2\xi^4$$

Substituting these last two equations into the first of Eqs. 14–9, we have

$$-C_1^2\xi^2\frac{d^2\xi}{d\theta^2} - C_1^2\xi^3 = -\frac{F}{m}$$

or

$$\frac{d^2\xi}{d\theta^2} + \xi = \frac{F}{mC_1^2\xi^2} \tag{14–12}$$

Equation 14–12 defines the path over which the particle travels when it is subjected to the central force* **F**.

In order to show how to apply this equation to the solution of mechanics problems, we will now consider the force of gravitational attraction. The magnitude of this force is given by Eq. 14–3 as discussed in Sec. 14–2. Some common examples of central-force systems which depend upon gravitation include the motion of the moon and artificial satellites about the earth, and the motion of the planets about the sun.

As a typical problem in space mechanics, let us determine the trajectory of a space satellite or space vehicle launched into orbit with an initial velocity v_o, Fig. 14–18. It will be assumed that this velocity is initially *parallel* to the tangent at the surface of the earth, as shown in the figure. (If v_o acts at some initial angle θ_o to the earth's surface tangent, the initial velocity will depend upon the characteristics of the powered-flight trajectory.) Just after the satellite is released into free flight the only force acting upon it is the gravitational force of the earth. (Gravitational attractions involving other bodies such as the moon or sun will be neglected. Actually for orbits close to the earth, their effect is small in comparison with the gravitational force of the earth.) According to Newton's law of gravitation, force **F** will always act between the mass centers of the earth and the satellite, Fig. 14–18. If we apply Eq. 14–3, this force of attraction has a magnitude of

$$F = G\frac{M_e m}{r^2}$$

where M_e and m represent the mass of the earth and the satellite, respectively, G is the gravitational constant, and r is the distance between the

Free flight trajectory

v_o

F Satellite

Power flight trajectory

$r = r_o$

Launching

F

Fig. 14–18

*In the derivation, **F** is considered positive when it is directed toward point O. If **F** is oppositely directed, the right-hand side of Eq. 14–12 should be negative.

mass centers. Setting $\xi = 1/r$ in the previous equation and substituting the result into Eq. 14–12, we obtain

$$\frac{d^2\xi}{d\theta^2} + \xi = \frac{GM_e}{C_1^2} \qquad (14\text{--}13)$$

This second-order ordinary differential equation has constant coefficients and is nonhomogeneous. The total solution is represented as the sum of the complementary and particular solutions. The complementary or homogeneous solution is the solution of Eq. 14–13 when the term on the right is equal to zero. It is,

$$\xi_c = A \sin \theta + B \cos \theta = C_2 \cos (\theta - \alpha)$$

where A and B, or C_2 and α, are constants of integration. The particular solution is

$$\xi_p = \frac{GM_e}{C_1^2}$$

Thus, the total solution to Eq. 14–13 is

$$\xi = \xi_c + \xi_p$$
$$= \frac{1}{r} = C_2 \cos (\theta - \alpha) + \frac{GM_e}{C_1^2} \qquad (14\text{--}14)$$

The validity of this result may be checked by substitution into Eq. 14–13.

Equation 14–14 represents the *free-flight trajectory* of the satellite. It is the equation of a conic section expressed in terms of polar coordinates. A *conic section* is defined as the locus of a point which moves in a plane in such a way that the ratio of its distance from a fixed point to its distance from a fixed line is constant. The fixed point is called the *focus*, and the fixed line is called the *directrix*. The constant ratio is called the *eccentricity* of the conic and is denoted by e. In Fig. 14–19, the focus is point F and the directrix is line DD. The eccentricity is

$$e = \frac{FP}{PA}$$

which may be written in the form

$$FP = r = e(PA) = e[p - r \cos(\theta - \alpha)]$$

or

$$\frac{1}{r} = \frac{1}{p} \cos(\theta - \alpha) + \frac{1}{ep}$$

Comparing this equation with Eq. 14–14, we see that the eccentricity of the conic section for the trajectory is

Fig. 14–19

$$e = \frac{C_2 C_1^2}{GM_e} \qquad (14\text{--}15)$$

and the fixed distance from the focus to the directrix is

$$p = \frac{1}{C_2} \qquad (14\text{--}16)$$

Provided the polar angle θ is measured about the x axis (an axis of symmetry since it is perpendicular to the directrix DD), the angle α is zero, Fig. 14–19. The measured trajectory of the satellite actually begins with respect to the x axis (axis of symmetry), and therefore Eq. 14–14 reduces to

$$\frac{1}{r} = C_2 \cos\theta + \frac{GM_e}{C_1^2} \qquad (14\text{--}17)$$

The constants C_1 and C_2 are determined from the data obtained for the position and velocity of the satellite at the end of the *power-flight trajectory*. For example, if the initial height or radius vector to the space vehicle is $\mathbf{r}_o$ (measured from the center of the earth) and its initial speed is $\mathbf{v}_o$ at the beginning of its free flight, Fig. 14–20, then the constant C_1 may be obtained from Eq. 14–10. When $\theta = \alpha = 0°$, the velocity $\mathbf{v}_o$ has no radial component; therefore, from Eq. 12–33, $v_o = r_o(d\theta/dt)$, so that

$$C_1 = r_o^2 \frac{d\theta}{dt}$$

or

$$C_1 = r_o v_o \qquad (14\text{--}18)$$

To determine C_2, use Eq. 14–14 with $\theta - \alpha = 0°$, $r = r_o$, and substitute Eq. 14–18 for C_1:

$$C_2 = \frac{1}{r_o}\left(1 - \frac{GM_e}{r_o v_o^2}\right) \qquad (14\text{--}19)$$

The equation for the free-flight trajectory therefore becomes

$$\frac{1}{r} = \frac{1}{r_o}\left(1 - \frac{GM_e}{r_o v_o^2}\right)\cos\theta + \frac{GM_e}{r_o^2 v_o^2} \qquad (14\text{--}20)$$

The type of path taken by the satellite is determined from the value of the eccentricity of the conic section as given by Eq. 14–15. If

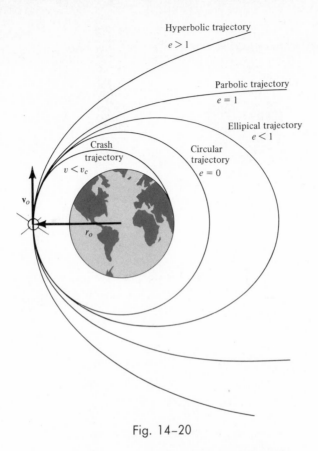

Hyperbolic trajectory
$e > 1$

Parbolic trajectory
$e = 1$

Ellipical trajectory
$e < 1$

Crash
trajectory
$v < v_c$

Circular
trajectory
$e = 0$

v_o

r_o

Fig. 14–20

$e = 0$,	free-flight trajectory is a circle
$e = 1$,	free-flight trajectory is a parabola
$e < 1$,	free-flight trajectory is an ellipse
$e > 1$,	free-flight trajectory is a hyperbola

(14-21)

A graph of each of these trajectories is shown in Fig. 14–20. From the curves it is seen that when the satellite follows a parabolic path, it is "on the border" of not returning to its initial starting point. The initial velocity at launching required for the satellite to follow a parabolic path is called the *escape velocity*. This velocity v_e can be determined by using the second of Eqs. 14–21 with Eqs. 14–15, 14–18, and 14–19. It is left as an exercise to show that

$$v_e = \sqrt{\frac{2GM_e}{r_o}}$$

(14-22)

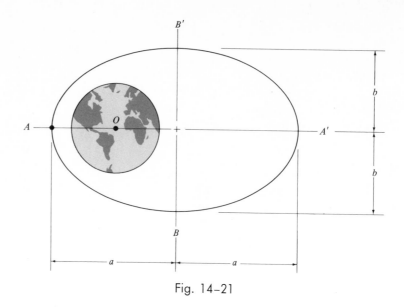

Fig. 14–21

We can find the velocity v_c required to launch a satellite into a *circular orbit* using the first of Eqs. 14–21. Since e is related to C_1 and C_2, Eq. 14–15, C_2 must be zero to satisfy this equation (from Eq. 14–18 C_1 cannot be zero); and, therefore, using Eq. 14–19, we have

$$v_c = \sqrt{\frac{GM_e}{r_o}} \qquad (14\text{–}23)$$

Provided r_o represents a minimum height for launching, in which frictional resistance from the atmosphere is neglected, then velocities of launching which are less than v_c will cause the satellite to reenter the earth's atmosphere and either burn up or crash, Fig. 14–20.

All the trajectories attained by planets and most satellites are elliptical, Fig. 14–21. For an earth orbit, the *minimum distance* from the orbit to the center of the earth (which is located at one of the foci of the ellipse) is OA and can be found using Eq. 14–20 with $\theta = 0°$. Therefore,

$$OA = r_o \qquad (14\text{–}24)$$

This distance is called the *perigee* of the orbit. The *apogee* or maximum distance OA' can be found using Eq. 14–20 with $\theta = 180°$. Thus,

$$OA' = \frac{r_o}{\left(\dfrac{2GM_e}{r_o v_o^2} - 1\right)} \qquad (14\text{–}25)$$

With reference to Fig. 14–21, the semi-major axis a of the ellipse is

$$a = \frac{OA + OA'}{2} \qquad (14\text{–}26)$$

Using analytical geometry it can be shown that the minor axis b is determined from the equation

$$b = \sqrt{(OA)(OA')} \qquad (14\text{–}27)$$

Furthermore, by direct integration, the area of an ellipse is

$$A = \pi ab = \frac{\pi}{2}(OA + OA')\sqrt{(OA)(OA')} \qquad (14\text{–}28)$$

The areal velocity has been defined by Eq. 14–11. Substituting Eq. 14–11 into Eq. 14–10, we have

$$2\frac{dA}{dt} = C_1$$

Integrating yields

$$A = \frac{C_1}{2}T$$

where T is the *period* of time required to make one orbital revolution. Using Eq. 14–28, we have the period

$$T = \frac{\pi}{C_1}(OA + OA')\sqrt{(OA)(OA')} \qquad (14\text{–}29)$$

In addition to predicting the orbital trajectory of earth satellites, the theory developed in this section is valid, as a surprisingly close approximation, in predicting the actual motion of the planets traveling around the sun. In this case, the mass of the sun M_s should be substituted for M_e when using the appropriate formulas.

The fact that the planets do indeed follow elliptic orbits about the sun was first discovered by the German astronomer Johannes Kepler (1571–1630). His discovery was made *before* Newton had developed his laws of motion and the law of gravitation. Kepler's laws, after 20 years of labor, are summarized by the following three statements:

1. Every planet moves in its orbit such that the line joining it to the sun sweeps over equal areas in equal intervals of time whatever the line's length.
2. The orbit of every planet is an ellipse with the sun placed at one of its foci.
3. The square of the period of any planet is directly proportional to the cube of the semi-major axis of its orbit.

The mathematical derivation of these laws can easily be shown from the theory developed in this section.

Example 14–11

A satellite is launched 400 miles from the surface of the earth with an initial velocity of 20,000 mph, acting parallel to the surface of the earth, Fig. 14–22. Assuming that the radius of the earth is 3,960 miles and that its mass is 4.09×10^{23} slug, determine (a) the eccentricity of the orbital path, (b) the velocity of the satellite at apogee, and (c) the period of revolution.

Solution
Part (a).   We may obtain the eccentricity of the orbit using Eq. 14–15. The constants C_1 and C_2 are first determined by using Eqs. 14–18 and 14–19. Since

$$r_o = (3{,}960 \text{ mi} + 400 \text{ mi})5{,}280 \text{ ft/mi} = 2.30 \times 10^7 \text{ ft}$$

and

$$v_o = 20{,}000 \text{ mi/hr} = 2.93 \times 10^4 \text{ ft/sec}$$

then

$$C_1 = r_o v_o = (2.30 \times 10^7)(2.93 \times 10^4) = 6.74 \times 10^{11} \text{ ft}^2/\text{sec}$$

$$C_2 = \frac{1}{r_o}\left(1 - \frac{GM_e}{r_o v_o^2}\right) = \frac{1}{2.30 \times 10^7}\left[1 - \frac{(3.44 \times 10^{-8})(4.09 \times 10^{23})}{(2.30 \times 10^7)(2.93 \times 10^4)^2}\right]$$

or

$$C_2 = 1.25 \times 10^{-8} \text{ ft}^{-1}$$

Hence,

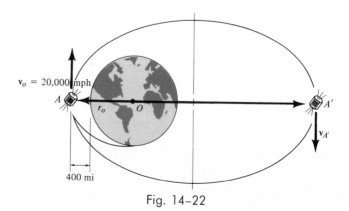

$v_o = 20{,}000$ mph

r_o

O

A'

$v_{A'}$

400 mi

Fig. 14–22

$$e = \frac{C_2 C_1^2}{GM_e} = \frac{(1.25 \times 10^{-8})(6.74 \times 10^{11})^2}{(3.44 \times 10^{-8})(4.09 \times 10^{23})} = 0.404 < 1 \qquad Ans.$$

Compared with Eq. 14–21, the orbit is an *ellipse*.

Part (b). From Eq. 14–18 the product $r_o v_o$ is constant. If the satellite were launched at the apogee A', shown in Fig. 14–22, with a velocity $v_{A'}$, the same orbit would be maintained provided

$$C_1 = r_o v_o = OA' v_{A'} = 6.74 \times 10^{11} \text{ ft}^2/\text{sec}$$

so that

$$v_{A'} = \frac{6.74 \times 10^{11}}{OA'}$$

Using Eq. 14–25, we have

$$OA' = \frac{r_o}{\left(\dfrac{2GM_e}{r_o v_o^2} - 1\right)} = \frac{2.30 \times 10^7}{\left[\dfrac{2(3.44 \times 10^{-8})(4.09 \times 10^{23})}{(2.30 \times 10^7)(2.93 \times 10^4)^2} - 1\right]}$$

$$OA' = 5.41 \times 10^7 \text{ ft}$$

Thus,

$$v_{A'} = \frac{6.74 \times 10^{11}}{5.41 \times 10^7}$$

$$= 12,460 \text{ ft/sec} = 8,500 \text{ mi/hr} \qquad Ans.$$

Part (c). The time for one revolution is determined by using Eq. 14–29. Since $OA = r_o$,

$$T = \frac{\pi}{C_1}(OA + OA')\sqrt{(OA)(OA')}$$

$$= \frac{\pi}{6.74 \times 10^{11}}(2.30 \times 10^7 + 5.41 \times 10^7)\sqrt{(2.30 \times 10^7)(5.41 \times 10^7)}$$

$$= 1.268 \times 10^4 \text{ sec} = 3.52 \text{ hr} \qquad Ans.$$

Problems

In the following problems, assume that the radius of the earth is 3,960 mi, the earth's mass is 4.09×10^{23} slugs, the mass of the sun is 1.97×10^{23} slugs, and the gravitational constant is $G = 3.44 \times 10^{-8}$ (lb-ft^2)/slug^2.

14-50. Show that the minimum escape velocity for a satellite launched into orbit about the earth is given by Eq. 14–23. Determine the escape velocity of a satellite located 250 mi from the surface of the earth.

14-51. A satellite is launched with an initial velocity of 2,500 mph, parallel to the surface of the earth. Determine the required altitude (or range of altitudes) for

launching if the free-flight trajectory is to be (a) circular, (b) parabolic, (c) elliptical, and (d) hyperbolic.

14-52. The earth has an eccentricity ratio of $e = 0.077$ in its orbit around the sun. Knowing that its farthest distance from the sun is 93,800,000 mi, compute the velocity at which it is traveling at this point. Determine the equation in polar coordinates which describes the orbit of the earth.

14-53. A rocket is in free-flight elliptical orbit around the planet Venus as shown. Knowing that the perigee and apogee of the orbit are 5,000 and 15,000 mi, respectively, determine (a) the velocity of the vehicle at point A', (b) the required velocity which the vehicle must attain at A so that it goes into a 5,000-mi free-flight circular orbit around Venus, and (c) compute the periods of both the circular and elliptical orbits. The mass of Venus is 0.82 times the mass of the earth. The planet's radius is 3,850 mi.

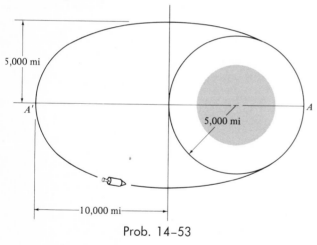

5,000 mi

A'

5,000 mi

A

10,000 mi

Prob. 14-53

14-54. With what free-flight velocity must the rocket in Prob. 14-53 be traveling so that it can leave its elliptical orbit at A and travel along a hyperbolic trajectory having an eccentricity of $e = 1.5$?

14-55. A satellite is placed into orbit at a speed of 20,000 ft/sec, parallel to the surface of the earth. Determine the proper altitude of the satellite such that its orbit remains circular. What will happen to the satellite

if its initial speed is 10,000 ft/sec when placed into orbit at the calculated altitude?

14-56. If the orbit of an asteroid has an eccentricity of $e = 0.056$ about the sun, determine the perigee of the orbit. The period of revolution is 29.5 yr. The orbit's apogee is 1.24×10^9 mi.

14-57. The rocket is docked next to a satellite located 8,000 mi above the surface of the earth. If the satellite is traveling in a circular orbit, determine the tangential velocity which must suddenly be given to the rocket, relative to the satellite such that it travels in free flight away from the satellite on a parabolic trajectory as shown.

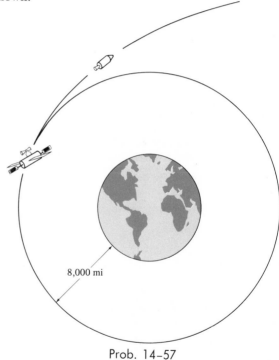

8,000 mi

Prob. 14-57

14-58. The planet Jupiter travels around the sun with an elliptical orbit such that the eccentricity is $e = 0.048$. If the perigee between Jupiter and the sun is 4.40×10^8 mi, determine (a) the velocity at perigee, and (b) the apogee.

14-59. The rocket shown is originally in a circular orbit 4,000 mi above the surface of the earth. It is to be placed into a second circular orbit having an altitude of 10,000 mi. After a short pulse of power at A the rocket travels in free flight from the first orbit to the second orbit. Determine the time required to get to the outer orbit along the path AB.

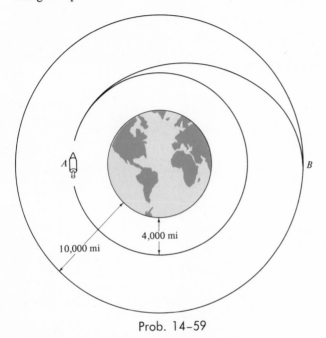

4,000 mi

10,000 mi

Prob. 14-59

14-60. The rocket is traveling in free flight along an elliptical trajectory $A'A$. The planet has no atmosphere, and its mass is 0.60 times that of the earth's. If the rocket has an apogee and perigee as shown in the figure, determine the velocity of the rocket when it is at point A.

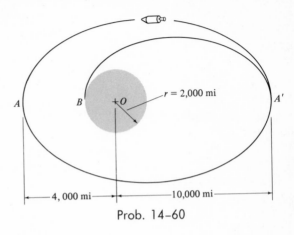

$r = 2,000$ mi

4,000 mi

10,000 mi

Prob. 14-60

14-61. If the rocket in Prob. 14-60 is to land on the surface of the planet, determine the required free-flight speed it must have at A' so that the landing occurs at B. How long does it take for the rocket to land, in going from A' to B?

15

Planar Kinetics of Rigid Bodies: Forces and Accelerations

15–1. Introductory Remarks

As shown in Chapter 14, the study of kinetics is based upon Newton's second law of motion, which may be written in mathematical form as $\Sigma\mathbf{F} = m\mathbf{a}$. This equation relates the resultant force $\Sigma\mathbf{F}$ acting on a particle to the mass m and acceleration $\mathbf{a}$ of the particle. Since a rigid body has a definite size and shape, Newton's second law of motion may be applied to *each* of the particles contained in the body. When this is done, a set of equations (the equations of motion) may then be derived which relate the forces acting on the body to the path, mass, and the acceleration of the *mass center* of the body.

In Chapter 13 the kinematics of rigid-body motion were presented in order of increasing complexity, that is, translation, rotation about a fixed axis, plane motion, rotation about a fixed point, and general motion. Our study of rigid-body kinetics will be presented in somewhat the same order. In this chapter we will begin by deriving the equations of motion for rigid bodies subjected to general plane motion. Then we will proceed to discuss application of these equations to specific problems of rigid-body translation, rotation about a fixed axis, and finally general plane motion. The motion of a rigid body subjected to any of these three types of motions may be analyzed in a fixed reference plane, because the path of motion of each particle of the body is a plane curve parallel to the reference plane. A kinetic study of these motions may be referred to as

the *kinetics of planar motions* or simply *planar kinetics*. In the analysis we will limit our study of planar kinetics to rigid bodies considered to be *symmetrical with respect to the fixed reference plane*. Thus, the motion of the body may be represented by a *slab* having the same mass as the body. All the forces acting on the body may then be projected into the plane of the slab. When this is done, the rotational equations of motion may be derived in a rather simplified manner, resulting in equations which reduce to a very concise form.

The more complex study of the kinetics of rigid body motions, which includes planar motion of unsymmetrical rigid bodies, motion about a fixed point, and the general motion of a rigid body, will be presented in Chapter 21. Analysis of this type of motion requires prior knowledge of the principles of angular momentum.

15–2. Planar Kinetic Equations of Motion

The derivation of the equations of planar rotational motion will be given in this section. Before doing this, however, we will first discuss some aspects of the translational equation of motion.

Consider the rigid body (slab) shown in Fig. 15–1*a* which has a mass *m* and is subjected to motion viewed in the *xy* plane. An *inertial frame of reference x, y, z* has its origin at point *O*. By definition, these axes do not rotate and are either fixed or translate with a constant velocity. A second x', y', z' frame of reference is located at the center of mass, *G*, of the body. This reference frame *translates* relative to the inertial frame. At the *instant* considered, the applied *external force system* causes the slab to have an angular acceleration $\boldsymbol{\alpha}$ and angular velocity $\boldsymbol{\omega}$; the center of mass of the slab has an acceleration $\mathbf{a}_G$.

Consider the arbitrary *i*th particle of the body having a mass Δm_i. The free-body and inertia-vector diagrams for this particle are shown in Fig. 15–1*b*. There are two types of forces which act on this particle: the *resultant force* $\mathbf{F}_i$ which is caused by the interactions *external* to the slab (e.g., contact forces or gravitation with external bodies, and magnetic or electrical forces) and *internal* reactive forces $\mathbf{f}_{ij}$ which act between the *i*th particle and each of the other *j* particles within the system. In Sec. 14–4 we considered the force analysis of any closed system of *n* particles by applying Newton's second law of motion to an arbitrary *i*th particle of the system and summed the results to include all the other *n* particles. In the case considered here, the closed system of *n* particles is contained within the boundary of the body (or slab). From the analysis given in Sec. 14–4, we may therefore conclude that

$$\Sigma \mathbf{F} = m\mathbf{a}_G \qquad (15\text{–}1)$$

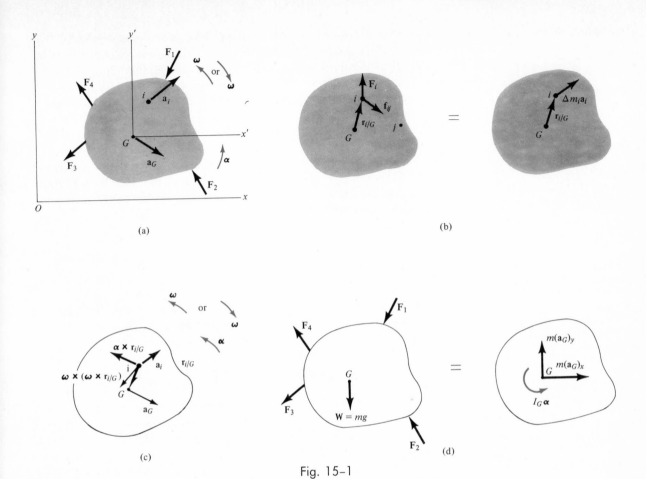

Fig. 15-1

which is identical to Eq. 14-5. Equation 15-1 represents Newton's second law of motion for a rigid body. It states that *the resultant force of all the external forces ΣF acting on the body is equal to the total mass m of the body times the acceleration of the mass center a_G of the body.* If $\mathbf{F}_R$ represents the resultant external force $\mathbf{F}_R = \Sigma \mathbf{F}$ which acts on the body, we may conclude from Eq. 15-1 that $\mathbf{F}_R$ has a *magnitude* ma_G and a *line of action* which is *always* collinear with the inertia-force vector $m\mathbf{a}_G$.

For motion of the body (or slab) in the xy plane, the external force system acting on the body may be expressed in terms of x and y components. Also, the acceleration of the body's mass center may be written as $\mathbf{a}_G = (a_G)_x\mathbf{i} + (a_G)_y\mathbf{j}$. Vector Eq. 15-4 may then be written in the form of two independent scalar equations:

$$\Sigma F_x = m(a_G)_x$$
$$\Sigma F_y = m(a_G)_y \tag{15-2}$$

Each of these equations states that the sum of the components of the external force system acting on the body in a given direction (x or y) is equal to the mass of the body times the acceleration component of the body's mass center in the given direction (x or y). Specifically, these two scalar equations represent the effects caused by the *translational accelerated motion* of the body.

We will now study what effects are caused by the *rotational accelerated motion* of the body by *summing* the *moments* created by the forces acting on all the particles of the slab about the mass center G. As shown in Fig. 15–1b, vector $\mathbf{r}_{i/G}$ locates the ith particle of the body with respect to the point G. Hence, the moment of the external force resultant $\mathbf{F}_i$ and the internal force $\mathbf{f}_{ij}$ about point G (free-body diagram) must be equivalent to the moment created by the inertia force $\Delta m_i \mathbf{a}_i$ about point G (inertia-vector diagram), i.e.,

$$(\mathbf{r}_{i/G} \times \mathbf{F}_i) + (\mathbf{r}_{i/G} \times \mathbf{f}_{ij}) = (\mathbf{r}_{i/G} \times \Delta m_i \mathbf{a}_i) \qquad (15\text{–}3)$$

As discussed in Sec. 13–8, during an instant of time the acceleration $\mathbf{a}_i$ of the ith particle may be analyzed by first considering a translation of point G (the base point) and then a rotation of the body about point G. Since the body has an instantaneous angular acceleration $\boldsymbol{\alpha}$ and angular velocity $\boldsymbol{\omega}$, Fig. 15–1c, then

$$\mathbf{a}_i = \mathbf{a}_G + \boldsymbol{\alpha} \times \mathbf{r}_{i/G} + \boldsymbol{\omega} \times (\boldsymbol{\omega} \times \mathbf{r}_{i/G})$$

Since the direction of the relative normal acceleration component always acts in the direction of $-\mathbf{r}_{i/G}$, Fig. 15–1c, we may write this equation as:

$$\mathbf{a}_i = \mathbf{a}_G + \boldsymbol{\alpha} \times \mathbf{r}_{i/G} - \omega^2 \mathbf{r}_{i/G}$$

Substituting into Eq. 15–3, using the distributive property of the cross-product, and noting, by definition of the cross product, that $-\Delta m_i \omega^2(\mathbf{r}_{i/G} \times \mathbf{r}_{i/G}) = 0$, we get

$$(\mathbf{r}_{i/G} \times \mathbf{F}_i) + (\mathbf{r}_{i/G} \times \mathbf{f}_{ij}) = (\Delta m_i\, \mathbf{r}_{i/G} \times \mathbf{a}_G) + \Delta m_i\, \mathbf{r}_{i/G} \times (\boldsymbol{\alpha} \times \mathbf{r}_{i/G})$$

If equations like these are written for all n particles of the slab and added together, the result becomes:

$$\sum (\mathbf{r}_{i/G} \times \mathbf{F}_i) + \sum (\mathbf{r}_{i/G} \times \mathbf{f}_{ij}) =$$
$$\left(\sum \Delta m_i \mathbf{r}_{i/G} \right) \times \mathbf{a}_G + \sum \Delta m_i \mathbf{r}_{i/G} \times (\boldsymbol{\alpha} \times \mathbf{r}_{i/G}) \quad (15\text{–}4)$$

The first term on the left represents the sum of the moments of all the external forces acting on the particle about point G, $\Sigma \mathbf{M}_G$. The second term is zero since the moment created by the internal force $\mathbf{f}_{ij}$ is equal, opposite, and collinear with the moment created by $\mathbf{f}_{ji}$ (the internal force which particle i exerts on particle j). Furthermore, the first term on the right side of Eq. 15–4 is also zero. This is because $\Sigma\, \Delta m_i \mathbf{r}_{i/G} = (\Sigma m_i)\overline{\mathbf{r}}$,

and the position vector $\bar{r}$ relative to G is equal to zero by definition of the center of mass. Hence, our result reduces to the form

$$\Sigma \mathbf{M}_G = \Sigma \, \Delta m_i \mathbf{r}_{i/G} \times (\boldsymbol{\alpha} \times \mathbf{r}_{i/G}) \qquad (15\text{--}5)$$

The term on the right side may be further simplified by noting that $\mathbf{r}_{i/G} \times (\boldsymbol{\alpha} \times \mathbf{r}_{i/G})$ is a vector having a magnitude of $r_{i/G}^2 \, \alpha$ and acting perpendicular to the plane of the slab. Realizing that $\Sigma \mathbf{M}_G$ also acts in this direction, we can write Eq. 15–5 in the scalar form

$$\Sigma M_G = (\Sigma \, \Delta m_i \, r_{i/G}^2) \alpha$$

Letting the particle size $\Delta m_i \rightarrow dm$, the summation sign becomes an integral since the number of particles $n \rightarrow \infty$. Representing $r_{i/G}$ by a more generalized dimension r, we have

$$\Sigma M_G = \left(\int_m r^2 \, dm \right) \alpha$$

The term in the integrand represents the "second moment" of the mass element dm about the mass center. The integral of this is called the *mass moment of inertia* of the body about an axis (the z' axis) passing through point G and directed perpendicular to the plane of the slab. (Further discussion of the mass moment of inertia is given in Sec. 15–4.) For simplicity, we will represent this term by I_G. Our final result is therefore

$$\Sigma M_G = I_G \alpha$$

This scalar equation states that the *summation of the moments of all the external forces about the mass center G is equal to the product of the mass moment of inertia of the body about point G and the magnitude of the angular acceleration of the body.* Specifically, this moment equation may be applied only at point G.

Using Eqs. 15–2, the *three scalar equations* describing the general plane motion of a rigid body may therefore be written as

$$\begin{aligned}
\Sigma F_x &= m(a_G)_x \\
\Sigma F_y &= m(a_G)_y \qquad (15\text{--}6) \\
\Sigma M_G &= I_G \alpha
\end{aligned}$$

A general procedure should be followed when applying these equations to the solution of rigid-body planar kinetic problems. As in the case of particle kinetic problems, free-body and inertia-vector diagrams for the body should be drawn *before* applying the equations of motion.* In Chapter 14, we saw that the inertia-vector diagram for a particle was

*See Chapter 6 of *Engineering Mechanics: Statics* for a discussion of free-body diagrams for rigid bodies.

simply an outlined shape of the particle, which included the inertia-force vector $m\mathbf{a}$. Since accelerated rigid bodies subjected to general plane motion undergo *both* a translation *and* a rotation, *two types* of inertia vectors must be shown on the inertia-vector diagram for the body. In particular, the *inertia-force vector* has a magnitude of ma_G and acts in the same direction as $\mathbf{a}_G$. Likewise, the *inertia-couple vector* has a magnitude of $I_G\alpha$ and a direction defined by $\boldsymbol{\alpha}$.

The free-body and inertia-vector diagrams for the slab in Fig. 15–1a are shown in Fig. 15–1d. Note that the weight $W = mg$ of the body passes through the mass center and has been included on the free-body diagram since it is an external force. When both the free-body and the inertia-vector diagrams are correctly drawn, accounting for all the terms in the equations of motion is greatly simplified. Actually, this graphical representation permits a complete freedom of choice in selecting a point for summing moments. (This advantage is discussed further in the remaining sections of this chapter.)

If the three equations of motion do not provide a complete solution for the unknown quantities in a problem, additional equations may be obtained by using *kinematic relationships* between the motion of different points in the body. If the body rests on a rough contacting surface and provided *slipping is impending or motion is occurring,* the existing frictional force may be related to the normal force using the frictional equation $F = \mu N$.*

A *scalar solution* for *planar kinetic problems* is generally recommended. A vector approach to the solution is mathematically more sophisticated than a scalar approach; however, a vector solution becomes advantageous only when the geometry of the problem becomes complicated, which occurs most often in three dimensions. A scalar solution is more direct and lends greater insight regarding the physical effects caused by each of the applied forces, moments, and inertia vectors.

15–3. Equations of Motion: Translation of a Rigid Body

When a rigid body undergoes a translation, all the particles contained in the body have the *same acceleration*. There are two types of translation defined in Chapter 13: rectilinear translation, whereby motion of each of the particles is directed along the same straight-line path, and curvilinear translation, in which all the particles of the body follow curved paths which are all parallel.

---

*Since many of the problems in rigid body kinetics involve friction, it is suggested that the reader review the material on friction, covered in Secs. 8–1 and 8–2 of *Engineering Mechanics: Statics.*

By definition, *translational motion* requires that $\boldsymbol{\alpha} = \boldsymbol{\omega} = \mathbf{0}$. The equations of motion then become

$$\Sigma F_x = m(a_G)_x$$
$$\Sigma F_y = m(a_G)_y \qquad (15\text{--}7)$$
$$\Sigma M_G = 0$$

The last equation requires that the summation of the *moments* of all the *external forces* acting on the body *about the body's center of mass G* be equal to *zero*.

As stated in Sec. 15–2, the resultant $\mathbf{F}_R = \Sigma \mathbf{F}_x + \Sigma \mathbf{F}_y$ of the external forces acting on the body is equal in magnitude and direction to the inertia-force vector $m\mathbf{a}_G = m(\mathbf{a}_G)_x + m(\mathbf{a}_G)_y$. Since the inertia vectors represent the resultant effect of the accelerated motion, the line of action of $\mathbf{F}_R$ can be determined using tne moment equation. To show this consider the rigid body (slab) in Fig. 15–2a, having a translational acceleration $\mathbf{a} = \mathbf{a}_G$. This motion is caused by the applied external force system shown. If the body undergoes *rectilinear translation,* the free-body and inertia-vector diagrams of the body are as shown in Fig. 15–2b. The weight $W = mg$ must be included on the free-body diagram since it is an external force. At the *instant* considered, the origin of the inertial reference is located at the center of mass G. Since $I_G\boldsymbol{\alpha} = 0$, only the inertia-force vector $m\mathbf{a}_G$ is shown on the inertia-vector diagram. When moments are summed on the inertia-vector diagram about point G, *zero moment* is created by this inertia-force vector since it is applied at point G. Thus, the line of action of the *resultant* of the external forces acting on the body must also pass through point G in order to satisfy $\Sigma M_G = 0$. As a consequence, moments may be summed about *any point* other than the mass center G, *provided* one accounts for the moment created by the inertia vector $m\mathbf{a}_G$ about the point. For example, if moments are summed about point A, Fig. 15–2b, it is required that the moment of the external forces acting on the free-body diagram about point A equal the moment of the inertia-force on the inertia-vector diagram about point A. Thus, $\Sigma \mathbf{M}_A = \mathbf{r}_{AG} \times m\mathbf{a}_G$. Because A lies on the line of action of $\mathbf{a}_G$, the moment of $m\mathbf{a}_G$ about point A is zero. Thus, a suitable set of three equations of motion is

$$\Sigma F_x = m(a_G)_x$$
$$\Sigma F_y = m(a_G)_y$$
$$\Sigma M_A = 0$$

If point B, Fig. 15–2b, is chosen, which lies at a perpendicular distance d from the line of action of the inertia-force vector, we have

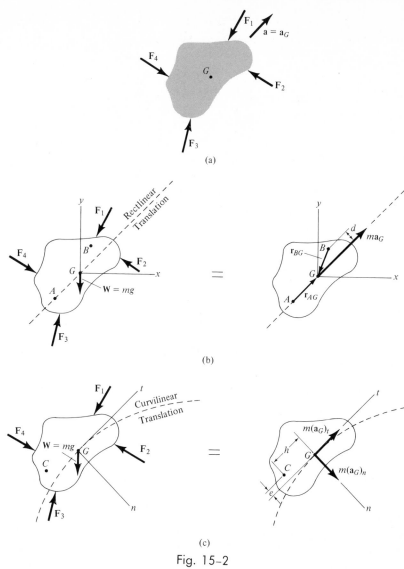

Fig. 15–2

$$\Sigma\mathbf{M}_B = \mathbf{r}_{BG} \times m\mathbf{a}_G$$

which reduces to the scalar form

$$\zeta + \Sigma M_B = (d)ma_G$$

Here, the moment of the external forces about B (free-body diagram) equals the moment of the inertia-force vector about point B (inertia-vector diagram).

When the rigid body is subjected to *curvilinear translation*, it is often

convenient to use coordinate axes which are oriented in the normal and tangential directions of motion, Fig. 15–2c. The three scalar equations of motion then become

$$\Sigma F_n = m(a_G)_n = m\frac{v_G^2}{\rho}$$

$$\Sigma F_t = m(a_G)_t = m\frac{dv_G}{dt} \qquad (15\text{–}8)$$

$$\Sigma M_G = 0$$

The last equation may be replaced by a moment summation about the arbitrary point C, Fig. 15–2c, *provided* this equation accounts for the moments, $\Sigma(M_C)_{IV}$, of each of the two inertia-force vector components about this point. From the inertia-vector diagram, if for example, e and h represent the perpendicular distance from point C to the line of action of the components, the required moment equation becomes

$$\zeta + \Sigma M_C = \Sigma(M_C)_{IV} = h(m(a_G)_n) - e(m(a_G)_t)$$

When computing moments, the inertia-force vector is seen to have the same properties as a force. The moment created by this vector when choosing a moment point other than the center of mass G, is easily obtained from the geometry of the body associated with the inertia-vector diagram. This is done in the same way in which moments of the external force system about a point are computed using the free-body diagram.

Example 15–1

A uniform 250-lb crate rests on the inclined surface for which $\mu = 0.2$. If a force of 400 lb is applied to the crate as shown in Fig. 15–3a, compute the acceleration of the crate and the normal force which the plane exerts on the crate. Assume the crate is originally at rest.

Solution

It is first necessary to draw the free-body and inertia-vector diagrams of the crate, Fig. 15–3b. Why? It will be assumed that the applied force of 400 lb causes the crate to slide *up* the plane. Under these conditions the normal force $\mathbf{N}$ acts at a distance of $0 < x < 1.5$ ft from the center line of the crate. If the crate slides as assumed, it is subjected to rectilinear acceleration. Establishing the x and y axes at the mass center G, and applying the equation of motion, we have

$$+\nearrow \Sigma F_x = m(a_G)_x; \qquad 400 - F - 250 \sin 20° = \frac{250}{32.2}a \qquad (1)$$

$$+\nwarrow \Sigma F_y = m(a_G)_y; \qquad N - 250 \cos 20° = 0 \qquad (2)$$

(a)

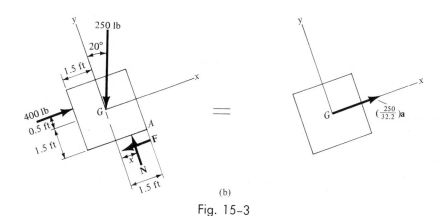

(b)

Fig. 15-3

$$\zeta + \Sigma M_G = 0; \quad -400(0.5) + N(x) - F(1.5) = 0 \tag{3}$$

Since sliding is assumed to occur,

$$F = 0.2N \tag{4}$$

Solving Eqs. (1) through (4) for the four unknowns yields

$$F = 47.0 \text{ lb}$$
$$N = 234.9 \text{ lb} \qquad \qquad Ans.$$
$$x = 1.15 \text{ ft}$$
$$a = 34.5 \text{ ft/sec}^2 \qquad \qquad Ans.$$

Since $x = 1.15 < 1.5$ ft and the magnitude of acceleration is positive, the crate *slides up the plane* as originally assumed. If $x > 1.5$ ft, the normal force would act at the corner point A of the crate and the problem would have to be reworked, assuming that tipping occurred. Under these conditions, Eq. (4) would not be valid since the crate may not be on the verge of sliding at the instant it begins to tip.

Example 15-2

The 50-lb beam *BD* shown in Fig. 15-4*a* is supported by two rods having negligible weight. When the beam is in the position shown, it has a speed of 10 ft/sec because of clockwise rotation of the rods. Determine the tensile force created in each rod and the angular acceleration of the rods at this instant.

Solution I

The beam moves with *curvilinear translation* since points *B, D*, and

Fig. 15-4

the center of mass G, move along circular paths, each path having the same radius of 3 ft. Using normal and tangential coordinates, the free-body and inertia-vector diagrams for the beam are shown in Fig. 15–4b. Because of the translation, point G has the *same* acceleration and velocity as point B, which is connected to both the rod and the beam. By studying the angular motion of rod AB, Fig. 15–4c, we note that the tangential component of acceleration $\mathbf{a}_t$ acts downward to the left due to the clockwise direction of $\boldsymbol{\alpha}$. Furthermore, the normal component of acceleration $\mathbf{a}_n$ is *always* directed toward the center of curvature (toward point A for rod AB). Since the speed of the beam is stated in the problem, this component of acceleration has a magnitude of

$$a_n = \frac{v^2}{\rho} = \frac{(10 \text{ ft/sec})^2}{3 \text{ ft}} = 33.3 \text{ ft/sec}^2$$

Note that the direction of the inertia-force vectors $m(\mathbf{a}_G)_t$ and $m(\mathbf{a}_G)_n$ in Fig. 15–4b are the same as $\mathbf{a}_t$ and $\mathbf{a}_n$ in Fig. 15–4c since the mass m is a scalar. Applying the equations of motion, we have

$$+\nwarrow \Sigma F_n = m(a_G)_n; \qquad T_{BA} + T_{DC} - 50 \cos 30° = \frac{50}{32.2}(33.3) \qquad (1)$$

$$+\nearrow \Sigma F_t = m(a_G)_t; \qquad -50 \sin 30° = \frac{-50}{32.2}(a_G)_t \qquad (2)$$

$$\zeta+\Sigma M_G = 0; \qquad -(T_{BA} \cos 30°)2 + (T_{DC} \cos 30°)2 = 0 \qquad (3)$$

Simultaneous solution of these three equations yields

$$T_{BA} = T_{DC} = 47.5 \text{ lb} \qquad \qquad Ans.$$
$$(a_G)_t = 16.1 \text{ ft/sec}^2$$

Note: These same results may also be obtained by applying the translational equations of motion (Eqs. 15–2) in the *horizontal* and *vertical* directions.

The angular acceleration of each rod may now be found with reference to Fig. 15–4c, for which

$$a_t = (a_G)_t = r_{AB}\alpha$$

or

$$\alpha = \frac{(a_G)_t}{r_{AB}} = \frac{16.1 \text{ ft/sec}^2}{3 \text{ ft}} = 5.37 \text{ rad/sec}^2 \qquad \qquad Ans.$$

Solution II

This problem may be solved by using a vector approach. At the instant when $\theta = 30°$ the origin of the xy axis is established at B, Fig. 15–4d. For convenience, these axes are oriented parallel to the n and t axis used in Fig. 15–4b. Applying Eq. 15–1, we have

$$\Sigma F = ma_G; \qquad T_{BA}\mathbf{j} + T_{DC}\mathbf{j} - 50 \sin 30°\mathbf{i} - 50 \cos 30°\mathbf{j}$$

$$= -\frac{50}{32.2}(a_G)_t\mathbf{i} + \frac{50}{32.2}(a_G)_n\mathbf{j}$$

Equating the respective $\mathbf{i}$ and $\mathbf{j}$ components and noting that $(a_G)_n = 33.3$ ft/sec² as in Solution I, we obtain the two scalar equations

$$\Sigma F_x = m(a_G)_x; \qquad\qquad 50 \sin 30° = \frac{50}{32.2}(a_G)_t \qquad\qquad (4)$$

$$\Sigma F_y = m(a_G)_y; \quad T_{BA} + T_{DC} - 50 \cos 30° = \frac{50}{32.2}(33.3) \qquad\qquad (5)$$

The above equations are identical to Eqs. (2) and (1). Moments will be summed about point B. Since this point is not the mass center of the beam, the moment of the two inertia-force vector components shown on the inertia-vector diagram must be included as terms on the right side of the moment equation. Thus,

$$\Sigma M_G = \Sigma (M_G)_{IV}; \quad \mathbf{r}_{BG} \times (-50 \sin 30°\mathbf{i} - 50 \cos 30°\mathbf{j})$$

$$+ \mathbf{r}_{BD} \times T_{DC}\mathbf{j} = \mathbf{r}_{BG} \times m\mathbf{a}_G$$

The vectors $\mathbf{r}_{BG}$ and $\mathbf{r}_{BD}$ represent position vectors directed from point B to points G and D, respectively. With reference to the x, y coordinate system shown in Fig. 15-4d, these vectors may be expressed in $\mathbf{i}$, $\mathbf{j}$ component notation as

$$\mathbf{r}_{BG} = \{2 \cos 30°\mathbf{i} - 2 \sin 30°\mathbf{j}\} \text{ ft}$$
$$\mathbf{r}_{BD} = \{4 \cos 30°\mathbf{i} - 4 \sin 30°\mathbf{j}\} \text{ ft}$$

Substituting into the moment equation yields

$$(2 \cos 30°\mathbf{i} - \sin 30°\mathbf{j}) \times (-50 \sin 30°\mathbf{i} - 50 \cos 30°\mathbf{j}) + (4 \cos 30°\mathbf{i}$$
$$- 4 \sin 30°\mathbf{j}) \times T_{DC}\mathbf{j} = (2 \cos 30°\mathbf{i} - 2 \sin 30°\mathbf{j})$$
$$\times \left[-\frac{50}{32.2}(a_G)_t\mathbf{i} + \frac{50}{32.2}(33.3)\mathbf{j} \right]$$

Expanding and simplifying gives a scalar equation in the $\mathbf{k}$ direction

$$-3.46T_{DC} = -189.6 + 1.55(a_G)_t \qquad\qquad (6)$$

Solving Eqs. (4), (5), and (6) simultaneously, we have

$$T_{BA} = T_{DC} = 47.5 \text{ lb} \qquad\qquad Ans.$$
$$(a_G)_t = 16.1 \text{ ft/sec}^2$$

so that again

$$\alpha = \frac{(a_G)_t}{r_{AB}} = \frac{16.1 \text{ ft/sec}^2}{3 \text{ ft}} = 5.37 \text{ rad/sec}^2 \qquad\qquad Ans.$$

Comparing the two solutions, we may conclude that it is generally easier to use a scalar approach rather than a vector approach to the solution of planar-kinetics problems. Furthermore, in this problem, it is more expedient to sum moments about the mass center of the body, thereby eliminating the inertia-force vector. In some problems, however, another point may be more suitable for moment summation, especially if many unknown forces are concurrent at the point.

Example 15-3

Because of the action of a 50-lb horizontal force, the 10-lb block moves along the horizontal support shown in Fig. 15–5a. The coefficient of friction between the block and the support is $\mu = 0.2$. If a 15-lb link is attached to the block at the pin connection A, determine the angle θ at which the link hangs from the vertical, the reaction components at the pin A, and the acceleration of the system.

Solution

Throughout the motion the link hangs at a constant angle θ from the vertical, so that the block and the link have the *same* acceleration **a.** The free-body and inertia-vector diagrams for each body are shown in Figs. 15–5b and 15–5c. The directions of the force components A_x and A_y have been assumed. In accordance with Newton's third law of motion, these components have equal magnitude and act in opposite directions on the free-body diagrams of the link and the block. Since the block is *moving* to the right, the frictional force is $F = \mu N = 0.2N$, and acts opposite to the motion of the block. Furthermore, assuming that the block does not tip, this force acts entirely at the base of the block, and the normal force **N** acts at an unknown distance x from the center of the block, such that $0 < x < \frac{1}{2}$ ft.

Applying the equations of motion to the block, we have

$$\stackrel{+}{\rightarrow}\Sigma F_x = m(a_A)_x; \qquad\qquad 50 - A_x - 0.2N = \frac{10}{32.2}a \qquad (1)$$

$$+\uparrow\Sigma F_y = m(a_A)_y; \qquad\qquad N - 10 - A_y = 0 \qquad (2)$$

$$\zeta + \Sigma M_A = 0; \qquad\qquad N(x) - \tfrac{1}{2}(0.2N) = 0 \qquad (3)$$

Applying the translational equations of motion to the link, we obtain

$$\stackrel{+}{\rightarrow}\Sigma F_x = m(a_G)_x; \qquad\qquad A_x = \frac{15}{32.2}a \qquad (4)$$

$$+\uparrow\Sigma F_y = m(a_G)_y; \qquad\qquad A_y - 15 = 0 \qquad (5)$$

Moments may be summed about any point on the link provided we also take into account the moment of the inertia vector $(15/32.2)$**a** about the

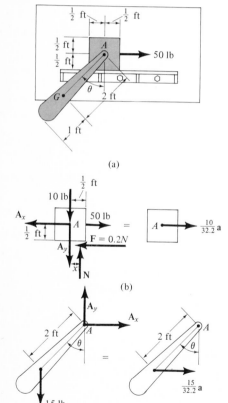

(a)

(b)

(c)

Fig. 15–5

point. It is convenient to sum moments about point A in order to eliminate the two unknown force components A_x and A_y at this point. In reference to the free-body and inertia-vector diagrams, Fig. 15-5c, we have

$$\zeta + \Sigma M_A = \Sigma(M_A)_{IV}; \quad 15(2 \sin\theta) = 2\cos\theta\left(\frac{15}{32.2}a\right) \tag{6}$$

Equations (1) through (6) contain six unknowns, N, x, A_x, A_y, θ, and a. Solving these equations, we have

$$N = 25.0 \text{ lb}$$
$$x = 0.1 \text{ ft}$$
$$A_x = 27.0 \text{ lb} \qquad \text{Ans.}$$
$$A_y = 15.0 \text{ lb} \qquad \text{Ans.}$$
$$\theta = 60.9° \qquad \text{Ans.}$$
$$a = 58.0 \text{ ft/sec}^2 \qquad \text{Ans.}$$

Since $x < \frac{1}{2}$ ft, the assumption that the block slides without tipping was correct.

Problems

Except when stated otherwise, assume that the coefficients of static and kinetic friction are equal, i.e., $\mu = \mu_s = \mu_k$.

15-1. The 3,500-lb car has a mass center located at G. Determine the normal reactions of the wheels on the road and the acceleration of the car if it is rolling freely down the incline. Neglect the weight of the wheels.

Prob. 15-1

15-2. The car has a center of gravity at G and rests on a horizontal surface for which the coefficient of friction is $\mu = 0.2$ between its back wheels and the surface. Determine the minimum distance at which it can reach a velocity of 60 mph without slipping on the surface. During the motion, the wheels in front are assumed to rotate freely. Neglect the weight of the wheels.

Prob. 15-2

15-3. A motorcyclist is rounding a horizontal curved road which has a radius of $\rho = 200$ ft. If the coefficient of friction between the tires and the road is $\mu = 0.65$, determine the maximum speed at which he may round the curve and the corresponding angle θ at which he must lean (measured from the vertical) so as not to tip over. Assume that ρ is measured to the center of gravity of both the rider and the motorcycle.

15-4. The motorcyclist is rounding a 60° banked curve which has a radius of $\rho = 50$ ft. Determine the speed at which he must travel and the angle θ at which he

must ride, so that he does not tip over. The coefficient of friction between the tires and the road is $\mu = 0.60$.

Prob. 15–4

Prob. 15–7

15-5. The 1-ft rectangular crate rests on the surface of the truck. The crate is uniform and weighs 20 lb. If the coefficient of friction between the crate and the truck is $\mu = 0.30$, determine the shortest distance s in which the truck can stop without causing the crate to tip or slide. The truck has an initial velocity of $v = 20$ ft/sec.

Prob. 15–5

15-6. If the string in Prob. 15–7 breaks and the 20-lb link is allowed to swing freely, determine the angle θ the link makes with the vertical if the frame is subjected to an acceleration of 15 ft/sec².

15-7. The 20-lb link AB is pinned to a moving frame at A and held in the vertical direction by means of a string BC which can support a maximum tension of 10 lb. Determine the maximum horizontal acceleration to which the link may be subjected just before the string breaks and the corresponding reaction at the pin, A.

15-8. The 300-lb refrigerator is being pulled up the 30° inclined plane by means of rope CD which is attached to a pulley at D. The center of gravity of the refrigerator is at G. If the angular acceleration of the pulley is 3 rad/sec², clockwise, determine the normal reactions of caster wheels at A and B on the incline. What is the maximum angular acceleration of the pulley at D without causing the refrigerator to tip over?

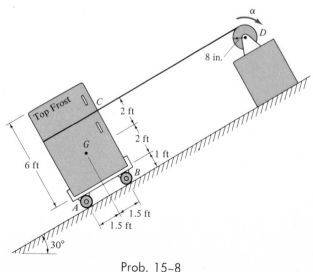

Prob. 15–8

15-9. Solve Prob. 15–8, assuming that the casters at *A* stick while the casters at *B* roll. The coefficient of kinetic friction between the stuck casters and the plane is $\mu_k = 0.40$.

15-10. The blocks *A* and *B* weigh 50 and 10 lb, respectively. If the magnitude of the force **P** is 100 lb, determine the normal force exerted by block *A* on block *B*. Neglect friction and the weights of the pulleys and bars of the triangular frame.

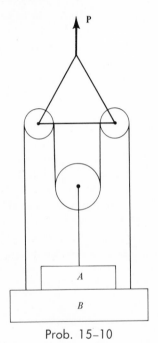

Prob. 15–10

15-11. If the 3,700-lb car in Prob. 15–12 is traveling around the banked curve at a velocity of 40 mph, determine the minimum coefficient of friction between the wheels and the pavement which will prevent the car from sliding up the curve. Assume that $\theta = 30°$.

15-12. The 3,700-lb car is rounding a banked curve having a radius of $\rho = 100$ ft and bank angle $\theta = 60°$. If the coefficient of friction between the wheels and the road is $\mu = 0.30$, determine the maximum speed at which the car may round the curve without sliding.

Prob. 15–12

15-13. If $\theta = 45°$, determine the minimum speed of the car in Prob. 15–12 so that it does not slip down the curve or overturn while traveling around the curve. Assume that the coefficient of friction between the pavement and the wheels is $\mu = 0.25$.

15-14. The jet aircraft shown weighs 154,000 lb and has a center of gravity at *G*. It is propelled forward by four engines, each engine providing the same constant thrust. At takeoff, starting from rest, it reaches a speed of 155 mph, traveling a distance of 3,000 ft. Determine (a) the reaction which the wheels exert on the runway during the takeoff, (b) the reactions if the plane is traveling at a constant speed of 15 ft/sec on the runway.

Prob. 15–14

15-15. The crank *AB* is turning with a constant angular velocity of 5 rad/sec. The uniform connecting rod *BC* weighs 20 lb and is pin connected at its end points. Determine the maximum and minimum vertical forces

641

which the pins exert on the ends of the rod during its motion.

Prob. 15-15

15-16. The rocket weighs 800 lb and has a center of gravity at G. It is supported on the monorail track at points A and B by rollers. If the engine provides a constant thrust T of 2,000 lb, determine the distance the rocket travels in 20 sec, starting from rest.

Prob. 15-16

15-17. Work Prob. 15-16, assuming that the rollers at A and B are replaced by runners for which the coefficient of friction between the runners and the rail is $\mu_k = 0.2$.

15-18. Compute the force developed in each of the supporting cables of the 10-lb triangular plate in Prob. 15-19 when $\theta = 0°$. Assume that the plate was released from rest at $\theta = 45°$.

15-19. The 10-lb triangular plate ABC is supported in the horizontal plane by means of three parallel cables and a rope at D. Each cable makes an angle of $\theta = 40°$ with the vertical. Determine the tension in each cable the instant after the rope at D is cut.

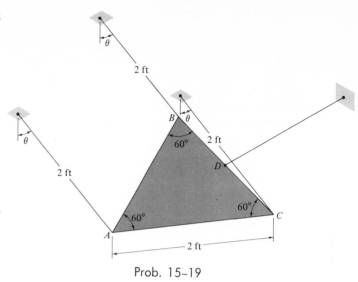

Prob. 15-19

15-20. The 80-lb T-frame is supported by a pin at D and a sliding roller at B. The tie bar EF constrains the links AB and CD to remain parallel during the motion. The center of gravity of the frame is at G. At the instant shown, the link AB has an angular velocity of 6 rad/sec and an angular acceleration of 3 rad/sec², both counterclockwise. Determine the horizontal and vertical components of reactions at B and D at this instant. Neglect the weight of the supporting rods.

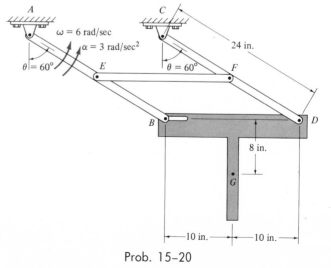

Prob. 15-20

642

15-21. Determine the maximum angular acceleration which may be given to the connecting links of the 20-lb box frame in Prob. 15-22 without causing the 10-lb crate to tip on the frame. Assume that the coefficient of friction between the frame and the crate is large enough to prevent slipping and that the crate is accelerated from rest in the position shown. Neglect the weights of the connecting links *AB* and *DC*.

15-22. The box frame *BEGD* weighs 20 lb and carries a uniform 10-lb crate. The coefficient of friction between the crate and the frame is sufficient to prevent slipping. What is the acceleration of the frame and the crate immediately after the rope *HI* is cut? Determine the normal force which the crate exerts on the frame at this

instant. Neglect the weight of the supporting rods *AB* and *CD*.

Prob. 15-22

15-4. Introduction to Mass Moment of Inertia

As shown in Sec. 15-2, the mass moment of inertia of a body represents an important property which is involved in the force analysis of any body subjected to rotational acceleration. We will briefly introduce this concept at this time in order to meet our needs for a further discussion of planar kinetics. A detailed treatment of the mass moment of inertia for a body is given in Chapter 20.

Consider the body shown in Fig. 15-6 which is rotating about the *AA* axis with an angular acceleration α. A small element of mass dm of the body, located at a perpendicular distance r from the axis, has a *tangential acceleration* of $a_t = r\alpha$. The magnitude of tangential force $d\mathbf{F}$ acting on this element and causing this acceleration is determined from Newton's second law of motion. Hence, $dF = dm\,a_t = r\alpha\,dm$. From Fig. 15-6 the moment of $d\mathbf{F}$ about the *AA* axis is $dM = r\,dF = r^2\alpha\,dm$. The moment of all these forces, acting on all the elements of the body, is determined by integration. Therefore $M = \int_m r^2\alpha\,dm$. Since α is the same for all radial lines in the body this term may be factored out of the integral sign, leaving an integral of the form

$$I = \int_m r^2\,dm$$

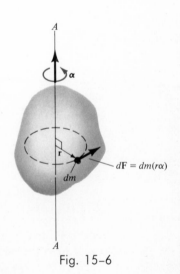

Fig. 15-6

643

This integral is termed the *mass moment of inertia* or simply the moment of inertia when there is no chance of confusing it with moment of inertia of an area.* Since the distance r is involved in the formulation, the moment of inertia is *unique* for each axis about which it is computed. From the form of the equation it is seen that the units of mass moment of inertia consist of mass times length squared. Since mass is measured in slugs, a common set of units for moment of inertia is slug-ft^2, ft-lb-sec^2, or m-kg$_f$-sec^2.

The moment of inertia represents an important property of the body and as shown here, it is involved whenever a force analysis is required of a body having an angular acceleration. Hence, the moment of inertia I_G is a measure of the resistance of a body to angular acceleration ($\Sigma M_G = I_G \alpha$) in the same way that mass m is a measure of the resistance to acceleration ($\Sigma F = ma$).

If the functional relation between the mass density ρ and the geometry of the body is known, we can write $dm = \rho \, dV$, and therefore, the moment of inertia of the body becomes

$$I = \int_V r^2 \rho \, dV$$

In the special case of ρ being a constant, this term may be factored out of the integral and the integration is then purely a function of geometry. In any case, the exact details of the integration are discussed fully in Sec. 20–3. For our present purposes, we will use *only the moments of inertia of the common shapes for which the integration has been computed.* The moments of inertia of these shapes are listed in Appendix D. Inspection of this appendix reveals that the moments of inertia are computed *only* about axes which pass through the mass center G of the body.

Provided, however, we know the moment of inertia I_G of a body about an axis *passing through the body's mass center,* the moment of inertia I of the body may be determined about an arbitrary *parallel axis* by using the *parallel-axis theorem.* This theorem may be stated mathematically as

$$I = I_G + md^2 \tag{15-9}$$

where m is the mass of the body and d is the perpendicular distance between the two *parallel axes.* The proof of this theorem is given in Chapter 20.

The parallel-axis theorem is often used for determining the moment of inertia of composite shapes when the moment of inertia I_G of each of the composite parts is either known (see Appendix D) or can be

---

*Area moments of inertia are of the form $I = \int_A r^2 dA$. See the discussion in Chapter 10, *Engineering Mechanics: Statics.*

computed. For example, if the body is constructed of a number of simple shapes such as disks, spheres, and rods, the moment of inertia of the body about any axis can be determined by adding together, algebraically, the moments of inertia of each of the composite shapes computed about this same axis. The parallel-axis theorem is needed for the calculations if the center of gravity of each composite part does not lie on the axis. Example 15–4 illustrates the procedure.

The moment of inertia of a body about a specified axis is occasionally reported using the *radius of gyration*. This length, k, which is usually measured in feet or meters, is defined from the relation

$$I = k^2 m \qquad \text{or} \qquad k = \sqrt{\frac{I}{m}} \qquad\qquad (15\text{–}10)$$

Note the similarity between the meaning of k in this formula and r in the equation $dI = r^2\, dm$, which defines the mass moment of inertia of an element about an axis. Consequently, if the mass of the body is imagined concentrated at a distance k from the axis, the moment of inertia of the body about the axis would be $I = k^2 m$. Occasionally in this book and elsewhere, the mass moment of inertia of a body about an axis is specified by reporting values of k and the weight W of a body. Since $m = W/g$, the moment of inertia for the body can be determined using Eq. 15–10.

Example 15–4

Compute the mass moment of inertia of the pendulum shown in Fig. 15–7a about an axis directed perpendicular to the page and passing through point O. The rod is slender and has a linear density of 16 lb/ft. The circular disk has an area density of 8 lb/ft².

Solution

The pendulum may be thought of as consisting of three composite parts, Fig. 15–7b; namely, the rod OA, *plus* a $\frac{1}{2}$-ft-radius disk, *minus* a $\frac{1}{4}$-ft-radius disk. The moment of inertia of the pendulum about point O can therefore be determined by computing the moment of inertia of each of these three composite parts about point O and then algebraically adding the results. The computations are performed by using the parallel-axis theorem in conjunction with the data listed in Appendix D and Fig. 15–7b.

The moment of inertia of the slender rod OA about an axis perpendicular to the page and passing through the center of gravity of the rod is $I_G = \frac{1}{12}ml^2$ (Appendix D). The perpendicular distance between this axis and the parallel axis passing through point O is 1 ft. (Both axes are perpendicular to the page.) Hence, for the rod,

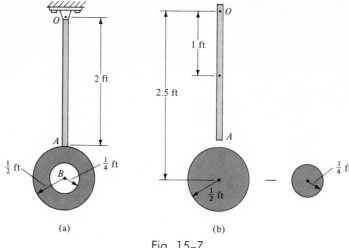

Fig. 15–7

$$(I_O)_r = \frac{1}{12}ml^2 + md^2$$

$$= \frac{1}{12}\left(\frac{16 \text{ lb/ft(2 ft)}}{32.2 \text{ ft/sec}^2}\right)(2 \text{ ft})^2 + \left(\frac{16 \text{ lb/ft(2 ft)}}{32.2 \text{ ft/sec}^2}\right)(1 \text{ ft})^2$$
$$= 1.33 \text{ slug-ft}^2$$

From Appendix D the moment of inertia of a thin disk about an axis perpendicular to the plane of the disk is $I_G = \frac{1}{2}mr^2$. The mass centers of *both* the $\frac{1}{2}$-ft-radius disk and the $\frac{1}{4}$-ft-radius disk are located at a distance of 2.5 ft from point O. Using the parallel-axis theorem, for the $\frac{1}{2}$-ft-radius disk, we have

$$(I_O)_d = \frac{1}{2}mr^2 + md^2$$

$$= \frac{1}{2}\left[\frac{\pi(\frac{1}{2}\text{ ft})^2(8 \text{ lb/ft}^2)}{32.2 \text{ ft/sec}^2}\right]\left(\frac{1}{2}\text{ ft}\right)^2 + \left[\frac{\pi(\frac{1}{2}\text{ ft})^2(8 \text{ lb/ft}^2)}{32.2 \text{ ft/sec}^2}\right](2.5 \text{ ft})^2$$
$$= 1.24 \text{ slug-ft}^2$$

For the $\frac{1}{4}$-ft-radius disk (hole),

$$-(I_O)_h = -\left\{\frac{1}{2}\left[\frac{\pi(\frac{1}{4}\text{ ft})^2(8 \text{ lb/ft}^2)}{32.2 \text{ ft/sec}^2}\right]\left(\frac{1}{4}\text{ ft}\right)^2 \right.$$

$$\left. + \left[\frac{\pi(\frac{1}{4}\text{ ft})^2(8 \text{ lb/ft}^2)}{32.2 \text{ ft/sec}^2}\right](2.5 \text{ ft})^2\right\} = -0.31 \text{ slug-ft}^2$$

The moment of inertia of the pendulum about the axis passing through point O is therefore the sum of the moments of inertia for the rod and the $\frac{1}{2}$ ft-radius disk less the $\frac{1}{4}$ ft-radius disk,

$$I_O = (I_O)_r + (I_O)_d - (I_O)_h$$
$$= 1.33 + 1.24 - 0.31$$
$$= 2.26 \text{ slug-ft}^2 \qquad \textit{Ans.}$$

15-5. Equations of Motion: Rotation of a Rigid Body About a Fixed Axis

Consider the rigid body (or slab) shown in Fig. 15–8a which is constrained to rotate about a fixed axis perpendicular to the page and passing through point O. Rotational motion is caused by the external force system acting on the body. At the *instant* shown, the angular acceleration of the body is α and the angular velocity is ω. Because of this rotation, the body's *center of mass G* is given an acceleration which can be represented by two components acting normal and tangent to the *circular path of motion*

(a)

(b)

Fig. 15–8

of this point, as shown in the figure. The tangential component of acceleration has a magnitude of $(a_G)_t = \alpha r_{OG}$ and acts in a tangential direction which is compatible with the angular acceleration $\boldsymbol{\alpha}$. The magnitude of the normal component is $(a_G)_n = \omega^2 r_{OG}$. This component is *always* directed from point G to point O.

The free-body and inertia-vector diagrams for the body are shown in Fig. 15–8b. The weight of the body, $W = mg$, and the pin reaction at O, $\mathbf{F}_O$, are included on the free-body diagram since they represent external forces acting on the body. The two inertia-force vectors shown on the inertia-vector diagram are associated with tangential and normal acceleration components of the body's mass center. These inertia vectors act in the same direction as the component accelerations and have magnitudes of $m(a_G)_t$ and $m(a_G)_n$. The inertia-couple vector acts in the same direction as $\boldsymbol{\alpha}$ and has a magnitude of $I_G\alpha$, where I_G is the moment of inertia of the body about an axis perpendicular to the page and passing through point G. From the derivation given in Sec. 15–2 (see Eqs. 15–6), the equations of motion which apply to this body may be written in the form

$$\begin{aligned} \Sigma F_n &= m(a_G)_n = m\omega^2 r_{OG} \\ \Sigma F_t &= m(a_G)_t = m\alpha r_{OG} \\ \Sigma M_G &= I_G\alpha \end{aligned} \qquad (15\text{--}11)$$

Although the moment equation here applies *only* at point G, other points may be chosen for summing moments, *provided* one accounts for the moments produced by the inertia-force components *and* the inertia couple about the point. In computing these moments, ΣM_{IV}, the inertia-force vector is treated in the same manner as a force; that is, it has the properties of a *sliding vector*. (Recall that $m\mathbf{a}_G$ is directly related to the resultant external force $\mathbf{F}_R$ by Eq. 15–1, i.e., $\mathbf{F}_R = \Sigma\mathbf{F} = m\mathbf{a}_G$.) In a similar manner, the inertia couple has the same properties as a couple. It is a *free vector* and may be applied to any point on the body. The inertia couple is a free vector since $\boldsymbol{\alpha}$ is a free vector. Thus, if moments of the external forces are summed about the arbitrary point A on the free-body diagram, Fig. 15–8b, they are equivalent to the moment summation about point A on the inertia-vector diagram. Hence, the required moment equation becomes

$$\zeta + \Sigma M_A = \Sigma(M_A)_{IV} = a(m(a_G)_n) - b(m(a_G)_t) + I_G\alpha$$

The dimensions a and b are defined in the figure.

If moments are summed about the *fixed point* O, the moment equation reduces to a simplified form. From Fig. 15–8b,

$$\zeta + \Sigma M_O = \Sigma(M_O)_{IV} = r_{OG}m(a_G)_t + I_G\alpha$$

Here r_{OG} is the perpendicular distance from O to the line of action of $m(\mathbf{a}_G)_t$. The inertia vector $m(\mathbf{a}_G)_n$ is *eliminated* in the moment summation since the line of action of this vector passes through O. Noting that $(a_G)_t = r_{OG}\alpha$, we may write the previous equation as

$$\zeta + \Sigma M_O = (I_G + mr_{OG}^2)\alpha \qquad (15\text{--}12)$$

From the parallel-axis theorem, Eq. 15–9, the term in parentheses represents the *moment of inertia of the body about the fixed axis of rotation* passing through point O. Denoting this term by I_O, the three equations of motion for the body become

$$\begin{aligned}
\Sigma F_n &= m(a_G)_n = m\omega^2 r_{OG} \\
\Sigma F_t &= m(a_G)_t = m\alpha r_{OG} \\
\Sigma M_O &= I_O\alpha
\end{aligned} \qquad (15\text{--}13)$$

The following examples illustrate the use of the equations of motion for solving problems involving the rotation of a rigid body about a fixed axis.

Example 15–5

The Charpy impact test is used in materials testing to determine the energy absorption characteristics of a material during impact. This test is performed by releasing the 60-lb pendulum shown in Fig. 15–9a, when $\theta = 45°$, and allowing it to fall freely to strike the specimen at S. The

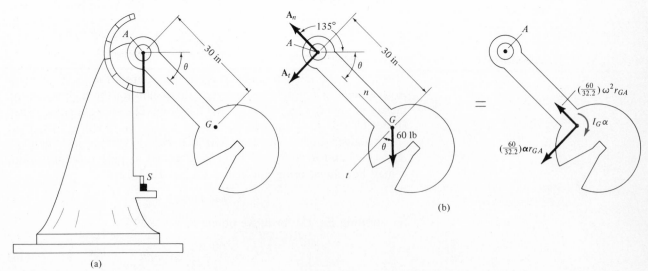

(a)

(b)

Fig. 15–9

pendulum has a mass center at G and a radius of gyration of $k_A = 27$ in. about the pin at A. Determine the reactive force at A (a) just after the pendulum is released, and (b) just before the pendulum strikes the specimen.

Solution

Part (a). The free-body and inertia-vector diagrams for the pendulum are shown in Fig. 15–9b, when the pendulum is in the general position θ. For convenience, the reaction components at A are shown acting in the n and t directions. The inertia-force vectors are applied at the center of mass G of the pendulum. Since the angular acceleration acts clockwise, $I_G\alpha$ acts clockwise and the tangential inertia force acts downward and to the left.

The moment of inertia of the pendulum about point A can be computed since the radius of gyration about A and the weight of the pendulum are known. Using Eq. 15–10,

$$I_A = k_A^2 m = \left(\frac{27 \text{ in.}}{12 \text{ in./ft}}\right)^2 \frac{60 \text{ lb}}{32.2 \text{ ft/sec}^2} = 9.43 \text{ slug-ft}^2$$

Since the moment of inertia about the *pivot point* (fixed point) is known, we will use Eqs. 15–13 for the solution. Summing moments about point A has an added advantage in that the reaction force at A is eliminated from the moment equation.

$$+\nwarrow \Sigma F_n = m\omega^2 r_{GA}; \qquad A_n - 60 \sin\theta = \frac{60}{32.2}\left(\omega^2 \frac{30}{12}\right) \qquad (1)$$

$$+\swarrow \Sigma F_t = m\alpha r_{GA}; \qquad A_t + 60 \cos\theta = \frac{60}{32.2}\left(\alpha \frac{30}{12}\right) \qquad (2)$$

$$\zeta+\Sigma M_A = I_A\alpha; \qquad 60 \cos\theta\left(\frac{30}{12}\right) = 9.43\alpha \qquad (3)$$

Note: Since the moment summation of the external forces about point A is equated to $I_A\alpha$, it is *not necessary* to account for the moment of the inertia-vectors $I_G\alpha$ and $m\alpha r_{GA}$ about this point. Do you understand why? (Refer to Eqs. 15–12 and 15–13.)

For a given angle θ, there are four unknowns in these three equations: A_n, A_t, ω, and α. Since the body is subjected to pure rotation, ω and α may be related using *kinematics* via Eq. 13–3, i.e.,

$$\omega \, d\omega = \alpha \, d\theta$$

Substituting Eq. (3) for α, we obtain

$$\omega \, d\omega = \frac{60(30)}{12(9.43)} \cos\theta \, d\theta \qquad (4)$$

Equations (1) through (4) may be used to obtain the complete solution to the problem for an arbitrary angle θ. In particular, when $\theta = 45°$, $\omega = 0$, since the pendulum is released from rest. In this case Eqs. (1) through (3) become

$$A_n - 60 \sin 45° = 0$$

$$A_t + 60 \cos 45° = \frac{60}{32.2}\left(\alpha\,\frac{30}{12}\right)$$

$$60 \cos 45°\left(\frac{30}{12}\right) = 9.43\alpha$$

Solving these equations, we have

$$\alpha = 11.25 \text{ rad/sec}^2$$
$$A_n = 42.4 \text{ lb}$$
$$A_t = 9.97 \text{ lb}$$

The reaction at A is therefore

$$F_A = \sqrt{(A_n)^2 + (A_t)^2}$$
$$= \sqrt{(42.4)^2 + (9.97)^2} = 43.5 \text{ lb} \qquad \textit{Ans.}$$

$$135° + \phi = 135° + \tan^{-1}\frac{9.97}{42.4} = 148.0° \qquad \textit{Ans.}$$

Part (b).   When $\theta = 90°$, the value of ω is determined using Eq. (4). Integrating this equation from $\theta_1 = 45°$, where $\omega_1 = 0$, to $\theta_2 = 90°$, where $\omega_2 = \omega$, yields

$$\int_{0°}^{\omega} \omega\,d\omega = \frac{60(30)}{12(9.43)}\int_{45°}^{90°} \cos\theta\,d\theta$$

$$\frac{\omega^2}{2}\bigg]_{0°}^{\omega} = 15.91[\sin\theta]_{45°}^{90°} = 15.91[1 - 0.707] = 4.66$$

Hence,

$$\omega^2 = 9.32 \text{ (rad/sec)}^2$$

Thus for $\theta = 90°$, Eqs. (1) through (3) become

$$A_n - 60 = \frac{60}{32.2}(9.32)\frac{30}{12}$$

$$A_t = \frac{60}{32.2}\left(\alpha\,\frac{30}{12}\right)$$

$$0 = 9.44\alpha$$

Solving, we obtain

$$\alpha = 0.0$$
$$A_n = 103.4 \text{ lb}$$
$$A_t = 0.0 \text{ lb}$$

Hence,

$$F_A = \sqrt{(A_n)^2 + (A_t)^2} = 103.4 \text{ lb} \qquad \qquad Ans.$$

at

$$\phi = 90° \qquad \qquad Ans.$$

Example 15–6

The rotor shown in Fig. 15–10a weighs 40 lb and has a radius of gyration of $k_G = 1.3$ ft about an axis passing through G. If a force of 10 lb is applied to the hand brake, determine the horizontal and vertical components of reaction at the pin B at the instant shown. The rotor has a counterclockwise angular velocity of 5 rad/sec at this instant. The coefficient of kinetic friction between the brake and the rotor is $\mu_k = 0.4$.

Solution

A free-body diagram of the brake is shown in Fig. 15–10b. Since the brake does not move, the laws of statics apply. Motion of the rotor causes the friction force $F = 0.4N$ to act to the *left* on the brake. The normal force may be obtained by summing moments about point A.

$$\zeta + \Sigma M_A = 0; \quad 0.4N(1) + N(1.5) - 10(4.75) = 0, \quad N = 25.0 \text{ lb}$$

Thus,

$$F = 0.4N = 0.4(25.0 \text{ lb}) = 10 \text{ lb}$$

Using these results the free-body and inertia-vector diagrams for the rotor are shown in Fig. 15–10c. Due to the frictional force, which acts to the *right* on the rotor, the rotor has a clockwise angular deceleration. Consequently, the inertia couple $I_G\alpha$ acts in a clockwise sense and the inertia force $m(a_G)_t = m\alpha r_{GB}$ acts to the left. (The motion $(a_G)_t$ must be compatible with α.) The inertia force $m(a_G)_n$ is always directed toward the center of rotation, point B. Applying the translational equations of motion yields

$$+\uparrow\Sigma F_n = m\omega^2 r_{GB}; \quad -25 + B_y - 40 = \frac{40}{32.2}(5)^2(1.5) \qquad (1)$$

$$\xrightarrow{+}\Sigma F_t = m\alpha r_{GB}; \quad -B_x + 10 = -\frac{40}{32.2}\alpha(1.5) \qquad (2)$$

The moment of inertia of the rotor about its mass center is determined from the radius of gyration and the weight of the rotor, i.e.,

$$I_G = mk_G^2 = \frac{40}{32.2}(1.3)^2 = 2.10 \text{ slug-ft}^2$$

Also, by the parallel-axis theorem, the moment of inertia of the rotor about point B is,

$$I_B = I_G + m(r_{GB})^2 = 2.10 + \frac{40}{32.2}(1.5)^2 = 4.89 \text{ slug-ft}^2$$

If moments are summed about the mass-center, Fig. 15–10c,

$$\zeta + \Sigma M_G = I_G\alpha; \qquad -B_x(1.5) + 10(2.5) = 2.10\alpha \qquad (3)$$

(a)

(b)

(c)

Fig. 15–10

If moments are summed about the fixed point B,

$$\zeta + \Sigma M_B = I_B \alpha; \qquad\qquad 10(1) = 4.89\alpha \qquad\qquad (4)$$

The term on the right side of Eq. (4) accounts for the moment of $I_G \alpha$ and $(40/32.2)\alpha r_{GB}$ about point B. Can you show this? (Refer to Eqs. 15–12 and 15–13.)

Solving Eqs. (1), (2), and (3) *or* Eqs. (1), (2), and (4), we have

$$\alpha = 2.04 \text{ rad/sec}^2$$

$$B_x = 13.81 \text{ lb} \qquad\qquad\qquad\qquad Ans.$$

$$B_y = 111.6 \text{ lb} \qquad\qquad\qquad\qquad Ans.$$

Example 15–7

The two gears shown in Fig. 15–11a are used to lift the 30-lb load. Gear B weighs 20 lb and has a radius of gyration of $k_B = 1.2$ ft. Gear A weighs 30 lb and has a radius of gyration of $k_A = 1.5$ ft. Determine the acceleration of the 30-lb load if a vertical force of 200 lb is applied to a cable wrapped around the drum on gear B, as shown in the figure.

Solution

The moments of inertia for each gear are

$$I_B = m_B k_B^2 = \frac{20}{32.2}(1.2)^2 = 0.894 \text{ slug-ft}^2$$

and

$$I_A = m_A k_A^2 = \frac{30}{32.2}(1.5)^2 = 2.10 \text{ slug-ft}^2$$

The free-body and inertia-vector diagrams of each gear and the block are shown in Fig. 15–11b. Why are these diagrams drawn? The mass centers of the gears are not subjected to a translational acceleration; hence, there is no inertia-force vector acting on the gears—only the inertia-couple vectors. Applying the rotational equation of motion about the mass center of each of the gears yields

$$\zeta + \Sigma M_A = I_A \alpha_A; \qquad\qquad T(1) - P(2) = -2.10\alpha_A \qquad\qquad (1)$$

$$\zeta + \Sigma M_B = I_B \alpha_B; \qquad\qquad 200(1) - P(1.5) = 0.894\alpha_B \qquad\qquad (2)$$

The block is subjected to translation only. Applying Newton's second law of motion to this body gives

$$+\uparrow \Sigma F_y = m(a_G)_y; \qquad\qquad T - 30 = \frac{30}{32.2} a \qquad\qquad (3)$$

Equations (1) through (3) contain five unknowns, T, P, a, α_A, and α_B.

(Applying the two translational equations of motion to each of the gears will not help in the solution, since these equations involve four other unknowns, A_x, A_y, B_x, and B_y.)

Using *kinematics,* we may obtain two more equations which relate α_B, α_A, and a. Since the points of contact between the gears at C, in Fig.

(a)

(b)

Fig. 15–11

15–11a, must have the same *tangential* components of acceleration, this requires that

$$(a_C)_t = r_B\alpha_B = r_A\alpha_A$$
$$= 1.5\alpha_B = 2\alpha_A$$

or

$$\alpha_B = 1.33\alpha_A \qquad (4)$$

Also, the acceleration of the cable supporting the 30-lb weight may be related to the angular acceleration of gear A by the equation

$$a = r\alpha_A$$
$$a = 1\alpha_A \qquad (5)$$

It is important that the *directions* of α_A, α_B, and $\mathbf{a}$, which were considered when deriving Eqs. (1)–(3), be the *same* as the directions considered when deriving Eqs. (4)–(5). This is necessary since we are seeking a simultaneous solution of equations. The results are:

$$P = 92.6 \text{ lb}$$
$$T = 77.7 \text{ lb}$$
$$\alpha_A = 51.2 \text{ rad/sec}^2$$
$$\alpha_B = 68.1 \text{ rad/sec}^2$$
$$a = 51.2 \text{ ft/sec}^2 \qquad \textit{Ans.}$$

Problems

15–23. The homogeneous disk weighs 50 lb. A 10-lb weight is attached to a cable wrapped around its rim, as shown. Determine the speed of the weight 2 sec after it is released from rest. Neglect the weight of the cable.

Prob. 15–23

15–24. The 50-lb uniform triangular plate is suspended by pins at points A and B. If the pin at A is suddenly removed, determine the immediate angular acceleration of the plate and the horizontal and vertical components of reaction at B. The moment of inertia of the plate about the pinned axis B is $I_B = 2.3$ slug-ft^2.

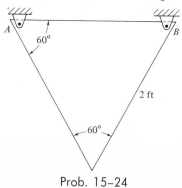

Prob. 15–24

15-25. A 1.5-ft-diameter flywheel weighs 30 lb and is pin-connected at its center. An inextensible cable is wrapped around the rim of the flywheel and when a 12-lb weight is attached to the cable, the weight attains a velocity of 6 ft/sec after moving downward a distance of 10 ft starting from rest. Determine the radius of gyration of the flywheel.

15-26. The 20-lb rod AB has its mass center at G. The radius of gyration of the rod about the mass center is $k_G = \frac{1}{2}$ ft. When the rod is in the position shown, it has an angular velocity of 6 rad/sec. Determine the angular acceleration of the rod at this instant and the magnitude of the reaction at pin A.

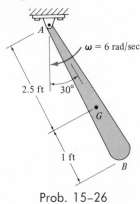

Prob. 15-26

15-27. The shaft consists of two cones and rotates about the horizontal axis which is supported by bearings at A and B. The shaft has a total weight of 80 lb and is at rest when it is suddenly acted upon by a moment of 20 lb-ft. Compute the angular acceleration of the shaft and the reactions at A and B.

Prob. 15-27

15-28. The pendulum consists of a 30-lb sphere and a 10-lb slender rod. Compute the reaction at the pin O just after the cord AB is cut.

Prob. 15-28

15-29. The 20-lb sphere is fixed to a shaft AB which is rotating with an angular velocity of $\omega = 400$ rad/sec. If a braking force is suddenly applied to the sphere by the link CD, as shown in the figure, determine the angular velocity of the sphere 1 sec later. The coefficient of friction between the brake and the sphere is $\mu_k = 0.4$. Neglect friction at the bearings A and B and the weight of shaft AB and the brake.

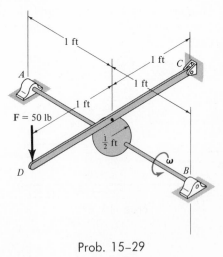

Prob. 15-29

15-30. The 60-lb disk has an initial angular velocity of 70 rad/sec. At a given instant, three braking pistons are forced up against the disk. Determine the angular veloc-

657

ity 3 sec later if $\mu_A = 0.2$, $\mu_B = 0.3$, and $\mu_C = 0.25$.

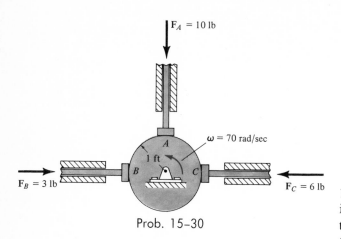

Prob. 15–30

15-31. The 60-lb sphere rests on the surface of a conveyor which is moving at a speed of 10 ft/sec. The sphere is prevented from translating by means of the smooth retaining walls at A and B. If the conveyor is suddenly stopped, determine the angular deceleration of the sphere. Assume that $\mu = 0.1$ between the conveyor and the sphere.

Prob. 15–31

15-32. The 30-lb cylinder is initially at rest when it is brought into contact with the wall and the rotor at A. The rotor has a constant angular velocity ω. If the coefficient of friction for the cylinder at its contacting surfaces is $\mu_k = 0.2$, determine the angular acceleration of the cylinder.

658

Prob. 15–32

15-33. The homogeneous slender rod weighs 80 lb and is pinned at O. It is supported by a cable at A. After the cable is cut, determine the reaction at O, (a) when the rod is still in the horizontal position, and (b) when the rod becomes vertical.

Prob. 15–33

15-34. Gears A and B weigh 20 lb each and are free to rotate about their centers. Determine a couple $\mathbf{M}$ which must be applied to gear C to give it an angular acceleration of 30 rad/sec². Gear C weighs 5 lb. All gears may be treated as circular disks.

Prob. 15–34

15-35. If a couple acting on gear C in Prob. 15–34 is 2 lb-ft, determine the angular acceleration of gears A and B.

15-36. At the instant shown, two forces act on the 30-lb slender rod which is pinned at O. Determine the magnitude of force F and the angular acceleration of the rod so that the horizontal reaction which the *pin exerts on the rod* is 5 lb directed to the right.

Prob. 15–36

15-37. The inertia-vector diagram representing the general rotational motion of a rigid body about a fixed axis is shown in the figure. Show that the inertia-couple $I_G\alpha$ may be eliminated by moving the inertia-force vectors $m(\mathbf{a}_G)_t$ and $m(\mathbf{a}_G)_n$ to point P, located a distance $r_{GP} = k_G^2/r_{OG}$ from the center of mass of the body. The point P is called the *center of percussion* of the body.

Prob. 15–37

15-38. Determine the distance h to the center of percussion P for the pendulum. The bob consists of a 20-lb plate, which is attached to a 10-lb slender rod. Compute

the reaction at O when the pendulum is struck by a 10-lb force F at the center of percussion, as shown in the figure.

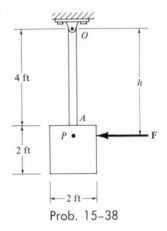

Prob. 15–38

15-39. The motor M supplies a couple of 200 lb-ft to its connecting hub. If the uniform beam OA weighs 20 lb and supports a force of 50 lb, determine the horizontal and vertical reactive components at O the instant the couple is applied. Neglect the weight of the pulley and cable.

Prob. 15–39

15-40. The 10-lb ring A is initially at rest when it is suspended from a rotor at B. The vertical wall at C is smooth. Determine the angular acceleration of the ring

if the rotor has a constant angular velocity of 20 rad/sec when the ring is placed on it. The coefficient of friction between the ring and the rotor is $\mu_k = 0.2$.

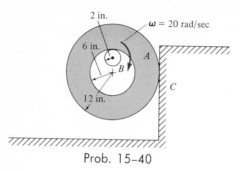

Prob. 15–40

15–41. The 10-lb disk D is subjected to a counterclockwise moment having a magnitude of $M = 10t$ lb-ft, where t is measured in seconds. Determine the angular velocity of the disk 2 sec after the moment is applied. The plate exerts a constant force of 100 lb on the disk. The coefficient of friction between the disk and the plate is $\mu = 0.2$.

Prob. 15–41

15–42. The 100-lb flywheel has a radius of gyration of $k_O = 6$ in. and is rotating clockwise with an angular velocity of 200 rpm. Determine the required force **P** which must be applied to the brake handle to stop the

660

flywheel in three revolutions. The coefficient of friction between the belt and the wheel is $\mu_k = 0.4$.

Prob. 15–42

15–43. The disk A turns with a constant clockwise angular velocity of 30 rad/sec. Disk B is initially at rest when it is brought into contact with A. Determine the time required for disk B to attain the same angular velocity as disk A. The coefficient of friction between the two disks is $\mu_k = 0.3$. Disk B weighs 60 lb. Neglect the weight of bar BC.

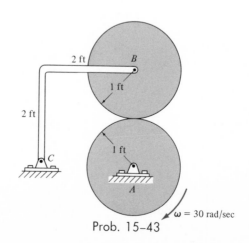

Prob. 15–43

15-44. Work Prob. 15–43 assuming that disk A rotates with a constant *counterclockwise* angular velocity of 30 rad/sec.

15-45. The 4-lb slender rod rotates about the z axis with an angular velocity of 3 rad/sec. Determine the axial force in the bar at a distance x from the axis of rotation.

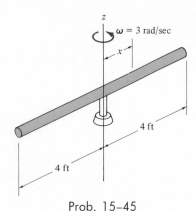

Prob. 15–45

15-6. Equations of Motion: Plane Motion of a Rigid Body

The equations of motion for a rigid body subjected to general plane motion were derived in Sec. 15–2. These equations apply only to rigid bodies which are symmetrical with respect to a fixed reference plane. With this assumption, the motion of the body is then represented by a slab having the same mass as the body. All the forces acting on the body are then projected into the plane of the slab.

The rigid body (or slab) shown in Fig. 15–12a is subjected to general plane motion caused by the externally applied force system (which includes the weight of the body). The free-body and inertia-vector diagrams for the body are shown in Fig. 15–12b. The inertia-force $m\mathbf{a}_G$ (shown dashed) has the same direction as the acceleration of the body's mass center. The inertia couple $I_G\alpha$ acts in the same direction as the angular acceleration of the body. If an x and y coordinate system is chosen as shown, the three equations of motion (Eqs. 15–6) which apply may be written as

$$\Sigma F_x = m(a_G)_x$$
$$\Sigma F_y = m(a_G)_y \qquad (15\text{-}14)$$
$$\Sigma M_G = I_G\alpha$$

(a)

(b)

Fig. 15-12

(a)

(b)

(c)

Fig. 15-13

The last of these equations may be replaced by a moment summation about *any point P* on or off the body provided the moments $\Sigma(M_P)_{IV}$ caused by the inertia vectors are taken into account. For example, the sum of the moments of the external forces about point D (free-body diagram), Fig. 15–12b, is equivalent to the moment summation of the inertia-vectors about point D (inertia-vector diagram). Hence,

$$\zeta + \Sigma M_D = \Sigma(M_D)_{IV} = I_G\alpha + c[m(a_G)_x] - b[m(a_G)_y]$$

The dimensions c and b are shown in the figure.

A special class of planar kinetic problems exists which deserves mentioning. These problems involve wheels, cylinders, or bodies of similar shape which roll along a plane surface. Due to the applied loadings, in some cases it may be difficult to determine whether or not the body slides as it rolls on the plane. Consider, for example, the *homogeneous* cylinder shown in Fig. 15–13a which has a weight $\mathbf{W}$ and is subjected to a known horizontal force $\mathbf{P}$. Because the cylinder is homogeneous, the *mass center G coincides with the centroid* of the cylinder. The free-body and inertia-vector diagrams for the cylinder are shown in Fig. 15–13b. In particular, note that the inertia-force vector $(W/g)\,\mathbf{a}_G$ is horizontal since the path of motion of point G is along a horizontal line. Applying the three equations of motion to the cylinder yields:

$$\xrightarrow{+}\Sigma F_x = m(a_G)_x; \qquad P - F = \frac{W}{g}a_G \qquad (15\text{-}15)$$

$$+\uparrow\Sigma F_y = m(a_G)_y; \qquad N - W = 0 \qquad (15\text{-}16)$$

$$\zeta + \Sigma M_G = I_G\,\alpha; \qquad Fr = I_G\,\alpha \qquad (15\text{-}17)$$

These *three equations* contain *four unknowns: F, N, α,* and a_G. If the

662

frictional force **F** is great enough to cause the wheel to roll along the surface *without slipping*, a relationship can be obtained between α and a_G using *kinematics*. To obtain this relationship, note that point G travels a distance $s = r\theta$ as the cylinder rotates through an angle θ, Fig. 15–13c. Differentiating this expression twice yields the kinematic equation

$$a_G = r\alpha \qquad (15\text{–}18)$$

The unknowns are obtained by solving *simultaneously* Eqs. 15–15 through 15–18.

Rather than considering moments about point G a more direct solution is obtained by summing moments about point IC (the instantaneous center of zero velocity). If this is done, we must account for the moments of both the force and inertia-vectors about this point, Fig. 15–13b, i.e.,

$$\zeta + \Sigma M_{IC} = \Sigma (M_{IC})_{IV}; \quad \Sigma M_{IC} = I_G \alpha + \left(\frac{W}{g} a_G\right)r$$

Using Eq. 15–18, and noting that $m = W/g$, we have

$$\Sigma M_{IC} = (I_G + mr^2)\alpha$$

From the parallel-axis theorem, the term in parentheses represents the moment of inertia of the cylinder about an axis perpendicular to the plane of motion and passing through the IC. Hence,

$$\Sigma M_{IC} = I_{IC}\alpha \qquad (15\text{–}19)$$

Applied to the cylinder, Fig. 15–13b, this equation becomes

$$Pr = I_{IC}\alpha \qquad (15\text{–}20)$$

If Eqs. 15–15, 15–16, 15–18, and 15–20 are used for the solution (instead of Eqs. 15–15 to 15–18), each equation may be solved *directly* for each unknown. In any case, when the solution is obtained, the assumption of no slipping must be *checked*.

Recall that no slipping occurs provided $F \leq \mu_s N$, where μ_s is the static coefficient of friction. If this inequality is satisfied, the problem is solved. However, if $F > \mu_s N$ the problem must be *reworked* since *both rolling and slipping occur*. In this case, α and a_G are *independent* so that Eq. 15–18 and Eq. 15–20 do not apply. Instead, the magnitude of frictional force **F** is *related* to the magnitude of the normal force **N**, using the coefficient of kinetic friction, μ_k, i.e.,

$$F = \mu_k N \qquad (15\text{–}21)$$

Equations 15–15, 15–16, 15–17 and 15–21 now form a complete solution to the problem.

The example just discussed has been constructed assuming the mass center of the cylinder coincides with its centroid. If this is not the case,

Fig. 15-14

it is necessary to use the kinematic equation for acceleration, Eq. 13–17, rather than Eq. 15–18.* Example 15–9 illustrates the procedure.

Example 15–8

A rectangular block is supported by a 1-ft slender rod as shown in Fig. 15–14a. The rod is welded centrally to the block and rests on a horizontal surface at A. The coefficients of static and kinetic friction between the rod and the surface are equal to $\mu = 0.25$. Determine the acceleration of point B on the block at the instant the 40-lb horizontal force is applied. The center of mass of the assembly is at G. Both the block and rod have a combined weight of 60 lb, and the radius of gyration is $k_G = \frac{2}{3}$ ft.

Solution

Since the radius of gyration and the weight are known, the moment of inertia of the assembly about its mass center is

$$I_G = mk_G^2 = \frac{60}{32.2}\left(\frac{2}{3}\right)^2 = 0.828 \text{ slug-ft}^2$$

The free-body and inertia-vector diagrams are shown in Fig. 15–14b. Why are these drawings needed? In particular, there is no component of inertia-force in the y direction since the acceleration of the assembly in this direction is zero. Specifically, the assembly cannot translate in this direction *and* there is no initial rotation ω of the assembly at the instant the 40-lb force is applied. (If $\omega \neq 0$, G would have a normal component

---

*A brief review of Secs. 13–8 and 13–9 may prove helpful in solving problems since, in general, computations for $\mathbf{a}_G$ and α require a kinematic analysis of acceleration.

of acceleration.) The inertia-force is assumed to be directed to the right and the inertia-couple acts clockwise.

Applying the equations of motion gives

$$\xrightarrow{+}\Sigma F_x = m(a_G)_x; \qquad -F + 40 = \frac{60}{32.2}a_G \qquad (1)$$

$$+\uparrow\Sigma F_y = m(a_G)_y; \qquad N - 60 = 0 \qquad (2)$$

$$\zeta+\Sigma M_G = I_G\alpha; \qquad F\left(\frac{5}{4}\right) - 40\left(\frac{3}{4}\right) = 0.828\alpha \qquad (3)$$

There are four unknowns in these three equations.

We will *first assume that no slipping occurs at A*. Hence, point A acts as a "pivot" and we may apply the kinematic equation

$$a_G = r_{AG}\alpha$$

or

$$a_G = 1.25\alpha \qquad (4)$$

Solving Eqs. (1) through (4) yields

$$N = 60.0 \text{ lb}$$
$$F = 27.5 \text{ lb}$$
$$a_G = 6.68 \text{ ft/sec}^2$$
$$\alpha = 5.35 \text{ rad/sec}^2$$

Testing the original assumption of no slipping, we require

$$F \leq \mu N$$

However,

$$27.5 \text{ lb} > 0.25(60 \text{ lb}) = 15 \text{ lb}$$

Since *slipping occurs* at the point of contact, we must rework the problem. For this case Eq. (4) does *not* apply. Instead, we must use the frictional equation $F = \mu N$, i.e.,

$$F = 0.25N \qquad (5)$$

Solving Eqs. (1) through (3) and (5) simultaneously yields

$$N = 60.0 \text{ lb}$$
$$F = 15.0 \text{ lb}$$
$$a_G = 13.4 \text{ ft/sec}^2$$
$$\alpha = -13.6 \text{ rad/sec}^2$$

Because of the negative sign, the angular acceleration of the assembly is counterclockwise, Fig. 15–14c. Knowing the acceleration of point G,

the acceleration of point B may be obtained by applying Eq. 13–17 to points B and G, i.e.,

$$\mathbf{a}_B = \mathbf{a}_G + (\boldsymbol{\alpha} \times \mathbf{r}_{B/G}) + \boldsymbol{\omega} \times (\boldsymbol{\omega} \times \mathbf{r}_{B/G})$$

$$a_B = 13.4 + 13.6(\tfrac{3}{4}) + 0$$

$$\angle^\theta \qquad \rightarrow \qquad \leftarrow$$

$$a_B = 3.20 \text{ ft/sec}^2 \rightarrow$$

Ans.

Example 15–9

The 50-lb wheel shown in Fig. 15–15a has a mass center at G and a radius of gyration of $k_G = \tfrac{3}{4}$ ft. If a 50-lb force is centrally applied to the wheel as shown, determine the acceleration of the center O of the wheel. The wheel is initially at rest and the coefficient of friction between the wheel and the horizontal plane is $\mu = 0.60$.

Solution

Using the radius of gyration and the weight, the moment of inertia of the wheel about the mass center is

$$I_G = mk_G^2 = \frac{50}{32.2}\left(\frac{3}{4}\right)^2 = 0.873 \text{ slug-ft}^2$$

The free-body and inertia-vector diagrams are shown in Fig. 15–15b. Note that since the mass center G of the wheel is *not* located at the geometric center O, there are two inertia-force components shown acting in the x and y directions on the inertia-vector diagram. The inertia-couple is assumed to act clockwise.

Applying the equations of motion gives

$$\xrightarrow{+} \Sigma F_x = m(a_G)_x; \qquad 50 \cos 30° - F = \frac{50}{32.2}(a_G)_x \tag{1}$$

$$+\uparrow \Sigma F_y = m(a_G)_y; \quad N + 50 \sin 30° - 50 = \frac{50}{32.2}(a_G)_y \tag{2}$$

$$\zeta + \Sigma M_G = I_G \alpha; \quad F(1) - N(\tfrac{1}{4}) - 50 \sin 30°(\tfrac{1}{4}) = 0.873\alpha \tag{3}$$

There are five unknowns in these three equations: F, N, $(a_G)_x$, $(a_G)_y$, and α.

If we *first assume no slipping occurs,* we can relate the acceleration components $(a_G)_x$ and $(a_G)_y$ and the angular acceleration α using *kinematics.* Since the wheel is assumed not to slip, the acceleration of the center of the wheel has a magnitude of $a_O = \alpha r = \alpha(1 \text{ ft}) = \alpha$. This vector, $\mathbf{a}_O = \{\alpha \mathbf{i}\}$ ft/sec^2 is shown on the kinematic diagram, Fig. 15–15c. Choosing point O as a base point and applying Eq. 13–17 between points G and O yields

$$\mathbf{a}_G = \mathbf{a}_O + \boldsymbol{\omega} \times (\boldsymbol{\omega} \times \mathbf{r}_{G/O}) + (\boldsymbol{\alpha} \times \mathbf{r}_{G/O})$$

or

$$(a_G)_x + (a_G)_y = \alpha + 0 + \alpha(\tfrac{1}{4})$$
$$\quad\;\rightarrow \qquad\quad \uparrow \qquad\quad \rightarrow \qquad\quad \uparrow$$

(a)

(b)

No slipping

(c)

Slipping

(d)

Fig. 15-15

Equating the respective horizontal and vertical components, we have

$$(a_G)_x = \alpha \tag{4}$$

$$(a_G)_y = \alpha/4 \tag{5}$$

Equations (1) through (5) may now be solved for the five unknowns, which yield

$$F = 24.3 \text{ lb}$$
$$N = 29.7 \text{ lb}$$
$$(a_G)_x = 12.2 \text{ ft/sec}^2$$
$$(a_G)_y = 3.05 \text{ ft/sec}^2$$
$$\alpha = 12.2 \text{ rad/sec}^2$$

Checking our original assumption that the wheel does not slip, it is necessary that

$$F \le \mu N$$

But

$$24.3 \text{ lb} > 0.60(29.7 \text{ lb}) = 17.82 \text{ lb}$$

Thus, the initial assumption is wrong, and *the wheel will start to roll and slide at the same time.* The problem must be reworked since Eqs. (4) and (5) are invalid. Instead,

$$F = 0.6N \tag{6}$$

The remaining equations are obtained using *kinematics*. Assuming that the center of the wheel has an acceleration of $\mathbf{a}_0$, Fig. 15–15d, Eq. 13–17 applied between points G and O becomes

$$\mathbf{a}_G = \mathbf{a}_0 + \boldsymbol{\omega} \times (\boldsymbol{\omega} \times \mathbf{r}_{G/O}) + (\boldsymbol{\alpha} \times \mathbf{r}_{G/O})$$
$$(a_G)_x + (a_G)_y = a_0 + 0 + \alpha(\tfrac{1}{4})$$
$$\quad\;\rightarrow \qquad \uparrow \qquad \rightarrow \qquad\quad \uparrow$$

Equating the respective horizontal and vertical components,

$$(a_G)_x = a_0 \tag{7}$$

$$(a_G)_y = \frac{\alpha}{4} \tag{8}$$

Solving Eqs. (1) through (3) and (6) through (8) gives

$$F = 15.8 \text{ lb}$$
$$N = 26.3 \text{ lb}$$
$$(a_G)_x = 17.7 \text{ ft/sec}^2$$
$$(a_G)_y = 0.848 \text{ ft/sec}^2$$
$$\alpha = 3.39 \text{ rad/sec}^2$$
$$a_0 = 17.7 \text{ ft/sec}^2 \qquad\qquad\qquad \textit{Ans.}$$

Example 15-10

The 30-lb slender rod AB shown in Fig. 15-16a has its mass center at G and its ends move on the smooth planes. End A is initially at rest when it is suddenly acted upon by a 2-lb horizontal force. Determine the normal reactions at A and B at this instant.

Solution

From Appendix D, the moment of inertia of the rod about its mass center is

$$I_G = \frac{1}{12} ml^2 = \frac{1}{12}\left(\frac{30}{32.2}\right)(3)^2 = 0.699 \text{ slug-ft}^2$$

(a)

(b)

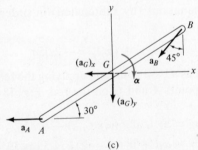

(c)

Fig. 15-16

The free-body and inertia-vector diagrams of the rod are shown in Fig. 15-16b. The direction of the inertia vectors have all been assumed. Applying the equations of motion yields

$$\xrightarrow{+} \Sigma F_x = m(a_G)_x; \qquad -2 - N_B \cos 45° = -\frac{30}{32.2}(a_G)_x \qquad (1)$$

$$+\uparrow \Sigma F_y = m(a_G)_y; \qquad N_A - 30 + N_B \sin 45° = -\frac{30}{32.2}(a_G)_y \qquad (2)$$

$$\zeta + \Sigma M_G = I_G \alpha; \qquad 2(1.5 \sin 30°) + N_A(1.5 \cos 30°)$$
$$- (N_B \cos 15°)1.5 = 0.699\alpha \qquad (3)$$

There are five unknowns in these three equations: N_A, N_B, $(a_G)_x$, $(a_G)_y$, and α. The remaining two equations will be obtained using *kinematics* to relate $(a_G)_x$ and $(a_G)_y$ to α.

Since the velocities at the ends A and B are zero at the instant considered, the angular velocity of the rod ω is also *zero*. Since points A and B have rectilinear motion, the *directions* of $\mathbf{a}_A$ and $\mathbf{a}_B$ are known. A kinematic diagram showing the accelerations of points A, B, and G is given in Fig. 15-16c. The directions of $(a_G)_x$, $(\mathbf{a}_G)_y$, and $\boldsymbol{\alpha}$ are *compatible* with the directions assumed for the inertia-force and inertia-couple vectors shown in Fig. 15-16b. (This is *necessary* since we will be seeking a simultaneous solution of equations.) Choosing point B as a base point and applying the relative acceleration equation between points B and G on the rod yields

$$\mathbf{a}_G = \mathbf{a}_B + \boldsymbol{\omega} \times (\boldsymbol{\omega} \times \mathbf{r}_{G/B}) + (\boldsymbol{\alpha} \times \mathbf{r}_{G/B})$$
$$-(a_G)_x\mathbf{i} - (a_G)_y\mathbf{j} = -(a_B \sin 45°)\mathbf{i} - (a_B \cos 45°)\mathbf{j} + 0$$
$$+ (-\alpha\mathbf{k}) \times (-1.5 \cos 30°\mathbf{i} - 1.5 \sin 30°\mathbf{j})$$

Expanding and equating the $\mathbf{i}$ and $\mathbf{j}$ components, we have

$$-(a_G)_x = -a_B \sin 45° - \alpha(1.5 \sin 30°)$$

and

$$-(a_G)_y = -a_B \cos 45° + \alpha(1.5 \cos 30°)$$

Combining these equations (by subtraction) in order to eliminate the unknown a_B yields

$$-(a_G)_x + (a_G)_y = -2.05\alpha \qquad (4)$$

Choosing point A as a base point and applying the relative acceleration equation between points A and G on the rod, Fig. 15-16c, we have

$$\mathbf{a}_G = \mathbf{a}_A + \boldsymbol{\omega} \times (\boldsymbol{\omega} \times \mathbf{r}_{G/A}) + \boldsymbol{\alpha} \times \mathbf{r}_{G/A}$$
$$-(a_G)_x\mathbf{i} - (a_G)_y\mathbf{j} = -a_A\mathbf{i} + 0 + (-\alpha\mathbf{k}) \times (1.5 \cos 30°\mathbf{i} + 1.5 \sin 30°\mathbf{j})$$

Expanding and equating the respective **i** and **j** components gives

$$-(a_G)_x = -a_A + 1.5 \sin 30° \alpha$$

and

$$-(a_G)_y = -1.5 \cos 30° \alpha = -1.30\alpha \qquad (5)$$

Solving Eqs. (1) through (5) simultaneously,

$$\alpha = 3.70 \text{ rad/sec}^2$$
$$(a_G)_x = 12.40 \text{ ft/sec}^2$$
$$(a_G)_y = 4.82 \text{ ft/sec}^2$$
$$N_A = 15.96 \text{ lb} \qquad \textit{Ans.}$$
$$N_B = 13.51 \text{ lb} \qquad \textit{Ans.}$$

Problems

15–46. The 30-lb wheel has a radius of gyration of $k_O = 8$ in. and rolls freely down the inclined plane. If the coefficient of friction between the wheel and the plane is $\mu = 0.2$, determine the acceleration of the wheel. Assume that $\theta = 20°$.

Prob. 15–46

15–47. Work Prob. 15–46 assuming that the wheel rolls freely down the plane when $\theta = 70°$.

15–48. Determine the maximum angle θ in Prob. 15–46 which will prevent slipping of the wheel as it rolls down the incline.

15–49. The 3-lb semicircular disk is released from rest at the instant shown. Determine the acceleration of point A the instant it is released. The coefficient of friction is large enough to prevent slipping. The moment of inertia of the disk about its mass center is $I_G = 0.0045$ slug-ft^2.

Prob. 15–49

15–50. The 20-lb gear rests on the surface of a gear rack which is suddenly given an acceleration of $a_R = 5$ ft/sec^2. If the centroidal radius of gyration of the gear is $k_O = 7$ in., determine the angular acceleration of the gear from rest at this instant.

Prob. 15–50

15-51. Determine the minimum coefficient of friction in Prob. 15-49 needed to prevent slipping of the disk on the contacting surface.

15-52. The triangular plate weighs 40 lb and is supported by a roller at A. If a force of $F = 70$ lb is suddenly applied to the roller, determine the acceleration of the center of the roller at the instant the force is applied. The plate has a moment of inertia about its center of gravity of $I_G = 0.85$ slug-ft^2. Neglect the weight and the size d of the roller.

Prob. 15-52

Prob. 15-54

15-55. The 12-lb cylindrical roller rests on a horizontal surface and is being pushed by the moving 4-lb piston with a force of $F = 2$ lb. If the cylinder rolls without slipping on the horizontal surface and the coefficient of friction between the piston and the cylinder is $\mu_k = 0.20$, compute the angular acceleration of the cylinder. The guides at A and B are smooth.

Prob. 15-55

15-53. Solve Prob. 15-52 assuming that the roller A is replaced by a slider block having a negligible mass. The coefficient of friction between the block and the horizontal plane is $\mu_k = 0.2$. Neglect the dimension d in the computation. The plate has a moment of inertia about its center of gravity of $I_G = 0.85$ slug-ft^2.

15-54. The 10-lb hoop has a radius of gyration of $k_O = 6$ in. It is given an initial angular velocity of 6 rad/sec when it is placed on a horizontal surface. If the coefficient of friction between the hoop and the surface is $\mu = 0.3$, determine the distance the hoop moves before slipping on the horizontal surface ceases.

15-56. Determine the smallest value of force F in Prob. 15-55 which can be applied to the piston without causing the cylinder to slip on the horizontal surface. Assume that $\mu = 0.3$ between the cylinder and the horizontal surface.

15-57. The 50-lb wheel has a rope wrapped around its inner hub and is released from rest on the inclined plane as shown. It is observed that the center of the wheel moves down the incline, and in 10 sec attains a velocity of 5 ft/sec. If the centroidal radius of gyration of the wheel is $k_O = 5$ in., determine the coefficient of friction of the inclined plane.

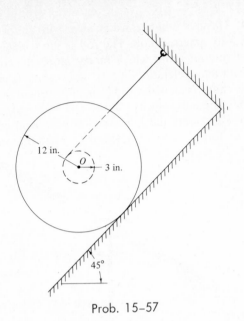

Prob. 15-57

15-58. The 16-lb bowling ball has a backspin as it moves along the horizontal surface such that its center is given a velocity of $v_0 = 10$ ft/sec. If the coefficient of friction between the floor and the ball is $\mu = 0.20$, determine the time at which the ball stops spinning. Assuming that after spinning has stopped the ball does not move, what was the initial angular velocity for the backspin?

$v_o = 10$ ft/sec

4.5 in.

Prob. 15-58

15-59. The freight elevator has a total weight of 200 lb with a center of gravity located at G. This weight includes the weight of the 50-lb disk pulley A. The vertical motion of the elevator is controlled by means of a motor at M. If the elevator has an upward acceleration of 3 ft/sec², determine the tension in all sections of the cable and the angular acceleration of the pulley at A. Assume that the cable does not slip on the pulley

surface. Neglect the mass of pulley B and the rotational effect of the motor.

Prob. 15-59

15-60. The two wheels C and D weigh 60 and 90 lb, respectively. The wheels are constructed so that the centroidal radius of gyration for each wheel is $k = 4$ in. Determine the angular acceleration of each wheel when the link is in the position shown. Neglect the weight of the link. Assume that the assembly does not slip on the plane.

Prob. 15-60

673

15-61. The uniform circular disk is 4 in. thick and is made of a material having a density of $\rho = 100$ lb/ft³. There is a 1.5-ft-radius semicircular hole in the disk, as shown in the figure. If the disk has an initial angular velocity of $\omega_1 = 1$ rad/sec, compute the angular acceleration of the disk at the instant shown and the normal force the disk exerts on the ground. Assume that slipping does not occur.

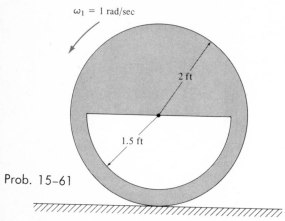

Prob. 15-61

15-62. What is the maximum angular velocity at which the disk in Prob. 15-61 can roll along the horizontal plane without leaving (hopping along) the surface?

15-63. The 20-lb block B is attached to the 10-lb disk pulley A. If a 20-lb force P is applied as shown, determine the acceleration of B and the angular acceleration of A. Neglect the mass of the pulley at C.

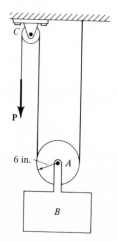

Prob. 15-63

674

15-64. The pendulum consists of a 5-lb uniform rod and a 3-lb spherical bob. The rod is pin-connected at B to a smooth collar which moves along the inclined pipe with an acceleration of 5 ft/sec² starting from rest. At this same instant the rod is in the vertical position and the angular velocity of the rod is 5 rad/sec, as shown. Determine the horizontal and vertical components of force acting on the rod at B at this instant.

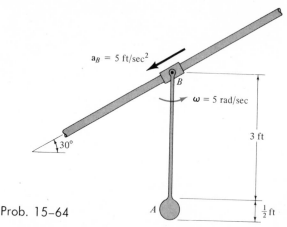

Prob. 15-64

15-65. The 20-lb uniform link AB is constrained so that its ends move along the grooved slots. Due to the force F the block at A has the velocity and acceleration at the instant shown. Determine the normal reactions which the blocks at A and B exert on the walls. Neglect the effects of friction and the mass of blocks A and B.

Prob. 15-65

15-66. The 30,000-lb rocket is fired such that each of its two engines provides a thrust of $T = 50,000$ lb. At a given instant, engine A suddenly fails to operate. Assuming that the rocket may be approximated by a slender rod 60 ft long from A to B, determine the angular acceleration of the rocket and the acceleration of its nose B.

Prob. 15-67

Prob. 15-66

15-68. The disk A weighs 3 lb and is supported along its periphery by a rope. Through the center of this disk there is a pin from which is suspended a 6-lb weight. If the pulley at B has a negligible mass, determine the acceleration of the 6-lb block and the angular acceleration of the disk.

15-67. The crankshaft has a constant angular velocity of 100 rad/sec. For all positions θ, the piston experiences a downward force of 2,000 lb because of a gas regulator at C. The connecting rod AB weighs 4 lb and has a centroidal radius of gyration of $k_G = 4$ in. Determine the force which the connecting rod exerts on the crankshaft at A when (a) $\theta = 0°$ and (b) $\theta = 180°$.

Prob. 15-68

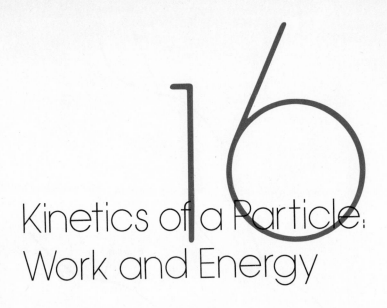

Kinetics of a Particle: Work and Energy

16–1. Work Done by a Force

In Chapter 14 Newton's second law of motion was used to solve problems in particle kinetics. This law relates the forces acting on a particle to the particle's mass and acceleration. For problems in which it becomes necessary to find the *velocity* or *displacement* of the particle, the kinematic equations given in Chapter 12 are required, in addition to Newton's second law. Problems such as these can be solved readily, however, by applying the principle of work and energy. Briefly this principle states that the work done by all the forces acting on the particle during a given displacement is equal to the change in kinetic energy of the particle.

In mechanics, a force does a differential amount of work when it moves through an infinitesimal displacement $d\mathbf{s}$ along a path. If β is the angle formed between the tails of the force $\mathbf{F}$ and displacement vector $d\mathbf{s}$, Fig. 16–1a, the *work dW* done by $\mathbf{F}$ is a *scalar quantity*, defined by the dot product

$$dW = \mathbf{F} \cdot d\mathbf{s} = F\,ds\cos\beta \qquad (16\text{–}1)$$

The work, as expressed by Eq. 16–1, may be interpreted in one of two ways: either as the product of the force magnitude F and the magnitude of the component of the displacement vector in the direction of $\mathbf{F}$, i.e. $ds\cos\beta$, Fig. 16–1b, or as the product of the displacement magnitude ds and the component of force magnitude in the direction of displacement, i.e., $F\cos\beta$, Fig. 16–1c. If $0° \leqslant \beta < 90°$, $\mathbf{F}\cos\beta$ and $d\mathbf{s}$ are in the *same direction* so that the work is *positive;* however if $90° < \beta \leqslant 180°$,

(a)

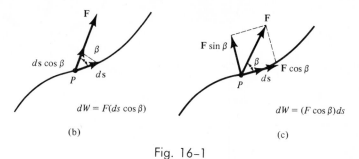

(b)

$$dW = F(ds \cos \beta)$$

(c)

$$dW = (F \cos \beta)ds$$

Fig. 16–1

(a)

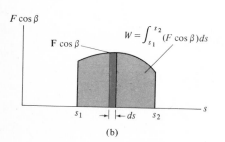

$$W = \int_{s_1}^{s_2} (F \cos \beta) ds$$

(b)

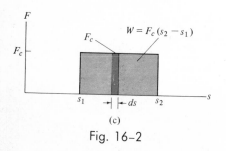

$$W = F_c (s_2 - s_1)$$

(c)

Fig. 16–2

the two vectors are in *opposite directions,* consequently the work is *negative.* There are two cases for which a force does zero work. As seen from Eq. 16–1, if **F** is *perpendicular* to the differential displacement $d\mathbf{s}$, $\beta = 90°$ so that the work done by **F** is zero. The second case occurs when the force is applied at a *fixed point* in which case $d\mathbf{s} = 0$.

The units of work are force times displacement (e.g., ft-lb or m-kg$_\text{f}$) the same units as those for moment. The two concepts are, however, in no way related. Moment is a vector quantity in which distance and force are measured perpendicular to one another, whereas work is a scalar such that force and distance are measured along the same line.

If the force undergoes a finite displacement along the path from s_1 to s_2, Fig. 16–2a, the work done by **F** can be determined by integrating Eq. 16–1.

$$W_{1-2} = \int_{s_1}^{s_2} \mathbf{F} \cdot d\mathbf{s} = \int_{s_1}^{s_2} (F \cos \beta)\, ds \qquad (16-2)$$

In general F is a function of displacement, i.e., $F = F(s)$. If the working component of the force ($F \cos \beta$) is plotted versus s, Fig. 16–2b, then the integral represented by Eq. 16–2 may be interpreted as the area under the curve between the points s_1 and s_2.

If the force **F** has a *constant magnitude* F_c and **F** always acts *tangent* to the path, then $\beta = 0°$ so that $F_c \cos \beta = F_c$. Equation 16-2 may be integrated directly, Fig. 16-2c, yielding

$$W = F_c(s_2 - s_1)$$

Two typical forces which do work and are frequently encountered in problems of particle kinetics are the work of a weight (or uniform gravitational force) and the work of a force exerted by an elastic spring. These cases will presently be discussed in detail. Consider a particle P having a weight w, which moves *down* along the path shown in Fig. 16-3. If

Fig. 16-3

its original elevation is h_2 and its final elevation is h_1, both measured with respect to the fixed horizontal reference plane or datum,* the work done by the *constant force*† w can be determined by using Eq. 16-2. As noted, the displacement component of ds in the direction of **w** is $-dh = ds \cos \beta$. Hence, applying Eq. 16-2, realizing that w is constant, we obtain

$$W = \int w \cos \beta \, ds = - \int_{h_2}^{h_1} w \, dh$$

or

$$W = w(h_2 - h_1) \qquad (16-3)$$

Thus, the work done by a weight force **w** is equal to the magnitude of the weight w times the difference in elevation of its vertical displacement. In this case the total work is *positive,* since the force w and the total

---

*The position of the datum plane is *arbitrary*. The results to be obtained indicate that the work done by the weight depends only upon the *difference in elevation*.

†The weight force is constant provided it is under the influence of a uniform gravitational field. This assumption is suitable for small differences in elevation. If the elevation change is significant, a variation of weight with elevation must be taken into account. See Prob. 16-53.

Fig. 16–4

displacement it undergoes $(h_2 - h_1)$ are in the same direction. If the weight is moved upwards from h_1 to h_2 the work is negative. Why?

The magnitude of force $\mathbf{F}$ required to stretch a linear elastic spring a distance x from its original unstretched position is

$$F = kx$$

where k is the spring stiffness. Thus, the work required to stretch a spring from an original (stretched) position x_1 to a farther position x_2, Fig. 16–4a, can be found by applying Eq. 16–2 and integrating along the path x.* Since the displacement $d\mathbf{x}$ is in the same direction as $\mathbf{F}$, we have

$$W = \int_{x_1}^{x_2} \mathbf{F} \cdot d\mathbf{x} = \int_{x_1}^{x_2} kx \, dx$$

or

$$W = \tfrac{1}{2}kx_2^2 - \tfrac{1}{2}kx_1^2 \tag{16–4}$$

This *same equation* is obtained when the spring is *compressed* from a distance x_1 to a farther distance x_2, both measured with respect to the unstretched position, Fig. 16–4b. Equation 16–4 represents the shaded

*We will neglect the mass of the spring, assuming it to be small compared to the body to which it is connected.

area under the curve $F = kx$ versus x, as shown in Fig. 16–4c. In both cases the work is positive since the force **F** moves in the same direction as displacement.

16–2. Principle of Work and Energy

Consider a particle moving along an arbitrary path and located at position P, a distance s from one end of the path, as shown in Fig. 16–5. The position P is located relative to the *inertial frame of reference xyz* by using the vector **r**. When the particle is at P, it has an instantaneous velocity **v** and it is subjected to a system of external forces, represented by the resultant $\mathbf{F}_R = \Sigma\mathbf{F}$. Since only the tangential components of force, i.e. $\Sigma F \cos \beta$ do work on the particle, let us consider writing Newton's second law of motion in the *tangential direction*. We have

$$\Sigma F_t = ma_t; \qquad\qquad \Sigma F \cos \beta = ma_t$$

During the displacement ds, $a_t = v\, dv/ds$ (Eq. 12–37). Hence,

$$\Sigma F \cos \beta\, ds = mv\, dv$$

Integrating both sides, assuming that at some initial time $t = t_1$ the particle has a position $s = s_1$ and a speed $v = v_1$, and at some later time $t = t_2$ that $s = s_2$ and $v = v_2$, Fig. 16–5, we have

$$\Sigma \int_{s_1}^{s_2} F \cos \beta\, ds = \tfrac{1}{2}mv_2^2 - \tfrac{1}{2}mv_1^2$$

Fig. 16–5

Using Eq. 16–2, we may write the final result as

$$\Sigma W_{1-2} = \tfrac{1}{2}mv_2^2 - \tfrac{1}{2}mv_1^2 \qquad (16-5)$$

This equation is known as the *principle of work and energy* for a particle. The scalar term on the left represents the work done by *all the forces acting on the particle* as the particle moves along the path. The two terms on the right side of Eq. 16–5, which are of the form $T = \tfrac{1}{2}mv^2$, represent the *kinetic energy* of the particle when the particle is located at the *final* and *initial points*, respectively, on the path. These terms are scalar quantities which represent the capacity of the particle to do work by virtue of its motion. So that Eq. 16–5 is dimensionally homogeneous, the kinetic energy has the same units as work, e.g., ft-lb or m-kg$_f$. Since the speed v is squared, *the kinetic energy will always be a positive quantity.*

Because Eq. 16–5 is determined from Newton's second law of motion, it is required that measurements of work and energy be made relative to an inertial frame of reference. We have defined such a frame of reference as having axes which are either fixed or translating with a constant velocity. Observers in *different* inertial frames of reference will therefore see a particle move with a different velocity and displacement; hence, these observers will not agree on the work done nor on the particle's change in kinetic energy. Even so, both observers will testify that Eq. 16–5 is valid; that is, total work done is equal to the change in kinetic energy.*

16–3. Method of Problem Solution

When applying the principle of work and energy to the solution of dynamics problems, it is convenient to rewrite Eq. 16–5 in the form

$$T_1 + \Sigma W_{1-2} = T_2$$

Here T_1 and T_2 represent the initial and final kinetic energies of the particle. In words, this equation states that

T_1	+	ΣW_{1-2}	=	T_2	
Initial kinetic energy of the particle		Total work done by all the forces acting on the particle		Final kinetic energy of the particle	(16-6)

---

*This assumes that both the forces acting on the particle and the particle have the *same displacement*. In general, this is not the case in collision or impact problems where permanent *deformations* are produced. See Chapter 18.

The kinetic energy terms are easily obtained when the speed of the particle is determined at the initial and final points of the path. To account for all the forces which do work on the particle, a free-body diagram of the particle should be drawn when the particle is located at an intermediate point along its path. For example, consider the particle of weight **w** which slides down along the rough path AB, Fig. 16–6a. From this diagram, it can be seen that the weight **w** does positive work and the frictional force **F** does negative work during the displacement. Why? There is no work done by the reactive force **N**, since this force always acts normal to the path of displacement. Knowing the speed of the particle at points A and B, Fig. 16–6b, and the geometry of the path, application of Eq. 16–6 is very direct.

There is an inherent advantage in using a work-energy approach to the solution of dynamics problems which involve *forces, velocities,* and *displacements.* To solve such problems using Newton's second law of motion necessitates first obtaining the particle acceleration by means of a *vector* force summation (sometimes involving the solution of simultaneous equations). Then, having found the acceleration, we must apply the kinematic equation $a_t\,ds = v\,dv$ to obtain a relation between acceleration and velocity. As noted from the derivation in Sec. 16–2, however, these two steps are already incorporated in the principle of work and energy. Application of this principle involves the velocities directly in the solution. Also, forces which do no work on the particle during the displacement are not considered when Eq. 16–6 is applied.

Since the principle of work and energy is a *scalar* equation, the terms involved are added algebraically rather than vectorially. It should be noted, however, that only *one equation* of work and energy may be written for each particle. This is in contrast to three scalar equations of motion which may be written for the same particle.

A decision as to the most direct procedure to be used for the solution of a particular type of problem should be made before attempting to solve the problem. The principle of work and energy cannot be used to determine accelerations, time intervals, or forces which do no work on a particle. There are cases, however, where the accelerations and forces are functions of the velocity. For such cases, it is usually easier to determine the velocity using the principle of work and energy, and then substitute this quantity into the equations of motion which relate force, mass, and acceleration.

(a)

(b)

Fig. 16–6

16–4. Principle of Work and Energy for a System of Particles

The principle of work and energy may be applied to a system of particles by considering each particle of the system separately. Consider, for

(a)

(b)

Fig. 16-7

example, the system of n particles shown in Fig. 16-7a. The arbitrary ith particle has a mass m_i and is subjected to internal forces $\mathbf{f}_{ij}$ and a *resultant* external force $\mathbf{F}_i$. The internal forces are reactive forces, developed between the ith particle and all other j particles of the system. The resultant external force $\mathbf{F}_i$ consists of gravitational, electrical, magnetic, or contact forces between adjacent bodies or particles *not included within the system* of n particles. At a given instant, the applied force system gives the ith particle a velocity of $\mathbf{v}_i$. The free-body diagram for the ith particle is shown in Fig. 16-7b. It is assumed that the particle travels through a displacement $d\mathbf{s}_i$ along the dotted path shown in the figure. Furthermore, the particle has an initial speed $(v_i)_1$ at $(s_i)_1$ and final speed $(v_i)_2$ at $(s_i)_2$. Applying the principle of work and energy to the ith particle, we obtain

$$\frac{1}{2}m_i(v_i)_1^2 + \int_{(s_i)_1}^{(s_i)_2} \mathbf{F}_i \cdot d\mathbf{s}_i + \int_{(s_i)_1}^{(s_i)_2} \mathbf{f}_{i1} \cdot d\mathbf{s}_i$$

$$+ \int_{(s_i)_1}^{(s_i)_2} \mathbf{f}_{i2} \cdot d\mathbf{s}_i + \cdots + \int_{(s_i)_1}^{(s_i)_2} \mathbf{f}_{ij} \cdot d\mathbf{s}_i + \cdots$$

$$+ \int_{(s_i)_1}^{(s_i)_2} \mathbf{f}_{in} \cdot d\mathbf{s}_i = \frac{1}{2}m_i(v_i)_2^2$$

Since work is a scalar quantity, the principle of work and energy for the entire system of n particles may be determined by writing similar equations for each particle and summing the results. This yields

$$\Sigma\frac{1}{2}m_i(v_i)_1^2 + \Sigma\int_{(s_i)_1}^{(s_i)_2} \mathbf{F}_i \cdot d\mathbf{s}_i + \Sigma\int_{(s_i)_1}^{(s_i)_2} \mathbf{f}_{ij} \cdot d\mathbf{s}_i = \Sigma\frac{1}{2}m_i(v_i)_2^2 \qquad (16\text{-}7)$$

The second and third terms on the left side of this equation represent, respectively, the work done by the external forces acting on, and internal forces acting within, the system. In cases where all the particles of the system are *connected by inextensible links or cables, or the particles are contained in a rigid body, the work created by the internal forces is zero.* This is because the internal forces occur in equal and opposite pairs, and all the particles are displaced by an equal amount. Under these conditions, Eq. 16-7 becomes

$$\Sigma\frac{1}{2}m_i(v_i)_1^2 + \Sigma\int_{(s_i)_1}^{(s_i)_2} \mathbf{F}_i \cdot d\mathbf{s}_i = \Sigma\frac{1}{2}m_i(v_i)_2^2$$

which may be represented in the form

$$\underset{\substack{\text{Total initial} \\ \text{kinetic energy} \\ \text{of the system}}}{\Sigma T_1} \quad + \quad \underset{\substack{\text{Total work} \\ \text{done by all} \\ \text{the external forces} \\ \text{acting on the system}}}{\Sigma W_{1-2}} \quad = \quad \underset{\substack{\text{Total final} \\ \text{kinetic energy} \\ \text{of the system}}}{\Sigma T_2} \qquad (16\text{-}8)$$

The following examples illustrate the methods for applying the principle of work and energy to the solution of problems in particle dynamics.

Example 16-1

The automobile shown in Fig. 16-8a is traveling up the 30° incline at a speed of 60 ft/sec. If the driver wishes to stop his car in a distance of 60 ft, determine the frictional force which must be supplied by the wheels at the pavement. The automobile weighs 4,000 lb.

(a)

(b)

Fig. 16-8

Solution

The free-body diagram of the auto is drawn first, Fig. 16–8b. Why? From this diagram it can be seen that the normal force **N** does no work since this force is never displaced in the normal direction. The friction force **F** is displaced 60 ft, and the 4,000 lb force is displaced 60 sin 30° ft. Both forces do *negative work* since they act in the opposite direction to their displacement. The principle of work and energy is therefore,

$$T_1 + \Sigma W_{1-2} = T_2$$

$$\frac{1}{2}\left(\frac{4{,}000}{32.2}\right)(60)^2 - 4{,}000(60 \sin 30°) - F(60) = 0$$

Solving this equation for F yields

$$F = 1{,}727 \text{ lb}$$

Ans.

Try working the problem using Newton's second law of motion in order to compare the solutions.

Example 16–2

A 20-lb block rests on a horizontal surface, as shown in Fig. 16–9a. The spring has a stiffness of $k = 50$ lb/ft and is initially compressed 1 ft from B to A. After the block is released from rest at A, determine (a) the velocity of the block when it passes point C and (b) the total distance d the block moves after coming to rest. The coefficient of kinetic friction between the block and the plane is $\mu_k = 0.2$. The block is not attached to the spring.

Solution

Part (a). Two free-body diagrams for the block are shown in Fig. 16–9b. The block moves under the influence of the spring force $F_s = kx$, along the 1-ft-long path AB, after which it continues to slide along the plane to point C. With reference to either free-body diagram, $\Sigma F_y = 0$; hence $N = 20$ lb. Only the spring and friction forces do work during the displacement. In particular, the spring force does positive work from A to B, whereas the frictional force does negative work from A to C. Why? Applying the principle of work and energy from A to C,

$$T_A + \Sigma W_{A-C} = T_C$$

$$\frac{1}{2}m_A(v_A)^2 + \int_0^1 kx\, dx + \int_0^2 (-0.2N)\, dx = \frac{1}{2}m(v_C)^2$$

$$0 + \frac{1}{2}(50)(x)^2 \Big|_0^1 - 0.2(20)x \Big|_0^2 = \frac{1}{2}\left(\frac{20}{32.2}\right)(v_C)^2$$

(a)

(b)

(c)

Fig. 16–9

Evaluating the limits of integration* we have

$$0 + 25 - 8 = 0.311(v_C)^2$$

Solving for v_C,

$$v_C = 7.40 \text{ ft/sec} \qquad \qquad Ans.$$

Part (b). When the block moves a distance d from its original position, the velocity $\mathbf{v}_D = 0$. The free-body diagrams are shown in Fig. 16–9c. Applying Eq. 16–6 between points A and D gives

---

*Integration is actually not needed to solve this problem if one realizes that the frictional force is *constant* during the motion, so that $W = F(s_2 - s_1)$, and the work done by the spring can be determined directly from Eq. 16–4, i.e., $W = \frac{1}{2}kx_2^2 - \frac{1}{2}kx_1^2$.

$$T_A + \Sigma W_{A-D} = T_D$$

$$\tfrac{1}{2}m(v_A)^2 + \int_0^1 kx\,dx + \int_0^d (-0.2N)\,dx = \tfrac{1}{2}m(v_D)^2$$

or

$$0 + \tfrac{1}{2}(50)(1)^2 - 0.2(20)d = 0$$

Thus,

$$d = 6.25 \text{ ft} \qquad\qquad \textit{Ans.}$$

Example 16-3

A block having a weight w is given an initial velocity of $v_1 = 1$ ft/sec when it is at the top of the smooth cylinder shown in Fig. 16-10a. If the cylinder has a radius of 10 ft, determine the angle θ_{max} at which the block leaves the surface of the cylinder.

Solution

The free-body diagram for the block is shown in Fig. 16-10b. *Only* the weight force **w** does work during the displacement. If the block is assumed to leave the surface when $\theta = \theta_{max}(N = 0)$ then the weight moves through a vertical displacement of $10(1 - \cos\theta_{max})$ ft as shown in the figure. Applying the principle of work and energy yields

$$T_1 + \Sigma W_{1-2} = T_2$$

$$\frac{1}{2}\frac{w}{g}(1)^2 + w(10)(1 - \cos\theta_{max}) = \frac{1}{2}\frac{w}{g}v_2^2$$

or

$$v_2^2 = 20g(1 - \cos\theta_{max}) + 1 \qquad\qquad (1)$$

There are two unknowns in this equation, θ_{max} and v_2. A second equation relating these two variables may be obtained by applying Newton's

(a)

(b)

(c)

Fig. 16-10

second law of motion in the *normal direction* to the forces acting on the free-body diagram.* The inertia-vector diagram is given in Fig. 16–10c. Thus,

$$+ \swarrow \Sigma F_n = ma_n; \qquad -N + w\cos\theta = \frac{w}{g}\frac{v^2}{10}$$

When the block leaves the surface of the cylinder at $\theta = \theta_{max}$, $\mathbf{N} = 0$ and $v = v_2$; hence,

$$\cos\theta_{max} = \frac{v_2^2}{10g} \tag{2}$$

Eliminating the unknown v_2^2 between Eqs. (1) and (2), using $g = 32.2$ ft/sec², gives

$$\cos\theta_{max} = 2(1 - \cos\theta_{max}) + 0.00311$$

Hence,

$$\cos\theta_{max} = 0.668$$

Solving for θ_{max}, we have

$$\theta_{max} = 48.1° \qquad\qquad\qquad Ans.$$

This problem has also been solved in Example 14–10. At this point, the reader should compare the two methods of solution. Since force, velocity, and displacement (θ_{max}) were involved in the solution, a work-energy approach is clearly advantageous.

Example 16–4

The blocks A and B shown in Fig. 16–11a weigh 10 and 100 lb, respectively. If the blocks are released from rest, determine the distance block A travels when the speed of B becomes 3 ft/sec. (Do you know why the principle of work and energy is suitable for solving this problem?)

Solution

This problem may be solved by considering blocks A and B separately and applying the principle of work and energy to each block. However, since the two blocks are connected together by an inextensible cable, we will consider the system of blocks A and B together.

If a free-body diagram of block B were drawn and the equations of vertical equilibrium applied, it would be found that a cable tension of 25 lb is required for vertical *equilibrium*. Since block A weighs 10 lb, which requires a cable tension of only 10 lb for vertical *equilibrium*, block B

---

*Recall that Newton's second law of motion applied in the *tangential direction* was used to derive the principle of work and energy.

(a)   (b)

(c)

Fig. 16–11

will begin to accelerate downward while block A accelerates *upward*. Using the methods of kinematics discussed in Sec. 12–10, it may be seen from Fig. 16–11b that at any given instant the total cable length l in the vertical direction may be expressed in terms of the displacement coordinates s_A and s_B as

$$s_A + 4s_B = l \qquad (1)$$

Taking the time derivative,

$$v_A = -4v_B \qquad (2)$$

The free-body diagram for the entire system is shown in Fig. 16–11c. Since the blocks start at rest, the initial velocity of the blocks is zero. From Eq. (2), their final velocities are $(v_B)_2 = 3$ ft/sec (downward) and $(v_A)_2 = 12$ ft/sec (upward). The cable force $\mathbf{T}$ and reactions $\mathbf{R}_1$ and $\mathbf{R}_2$, shown on the free-body diagram, do *no work*, since these forces represent the reactions at the supports and consequently do not move while the blocks are being displaced. From Eq. (1), a displacement s_A' (upward) of block A creates a *corresponding* displacement $s_B' = -\frac{1}{4}s_A'$ (downward) of block B. Hence, applying Eq. 16–8 gives

$$\Sigma T_1 + \Sigma W_{1-2} = \Sigma T_2$$

$$\frac{1}{2}m_A(v_A)_1^2 + \frac{1}{2}m_B(v_B)_1^2 - w_A s_A' + w_B s_B' = \frac{1}{2}m_A(v_A)_2^2 + \frac{1}{2}m_B(v_B)^2$$

$$0 + 0 - 10s_A' + 100\left(\frac{1}{4}s_A'\right) = \frac{1}{2}\left(\frac{10}{32.2}\right)(12)^2 + \frac{1}{2}\left(\frac{100}{32.2}\right)(3)^2$$

Solving this equation for s_A' gives

$$s_A' = 2.42 \text{ ft} \qquad\qquad Ans.$$

Problems

16–1. A 5,000-lb freight car is pulled along a horizontal track. If the car starts at rest and attains a velocity of 40 ft/sec after traveling a distance of 300 ft, determine the total work done on the car in this distance if the rolling frictional force between the car and track is 80 lb.

16–2. A 90-lb force is required to drag a 200-lb weight 60 ft along a rough incline at constant velocity. If the weight slides back down the incline, from a vertical height of 25 ft, determine the velocity it attains when it reaches the bottom.

16–3. A train weighing 200,000 lb is pulled up a 30° inclined plane by a steady drawbar pull of 1,000 lb. If the frictional resistance at the tracks is 10 lb per ton, and the train travels 1,000 ft up the incline before its speed is reduced to 10 ft/sec, determine the initial speed of the train.

16–4. The 30-lb block is subjected to the action of the two forces shown. If the block starts from rest, determine the distance it has moved when it attains a velocity of 30 ft/sec. The coefficient of friction between the block and the surface is $\mu = 0.2$.

Prob. 16-4

16-5. When a 15-lb cannon shell is fired from a 10-ft-long cannon, the resulting force acting on the shell, while in the cannon, varies in the manner shown. Determine the muzzle velocity of the shell as it leaves the barrel. Neglect the effects of friction inside the barrel.

Prob. 16-5

16-6. The 10-lb block slides along a smooth plane and strikes a *nonlinear spring* with a velocity of 12 ft/sec. If the spring has a resistance of $F = kx^2$, where $k = 10$ lb/ft^2, determine its velocity after it has compressed the spring 1 ft.

Prob. 16-6

16-7. The 20-lb block resting on the 30° inclined plane is acted upon by a 40-lb force. If its initial velocity is 5 ft/sec down the plane, determine its velocity after it has traveled a distance of 10 ft down the plane. How much time does it take for the block to travel this distance? The coefficient of friction between the block and the plane is $\mu = 0.2$.

Prob. 16-7

16-8. The small 2-lb ball starting from rest at A slides along the wire. If, during the motion, the ball is acted upon by a force $\mathbf{F} = 10\mathbf{i} + 6y\mathbf{j} + 2z\mathbf{k}$, where F is given in pounds; determine the velocity of the ball when it strikes the wall at B.

Prob. 16-8

16-9. The 10-lb block is dropped from rest 4 ft above the surface of a platform which is supported by a spring

with a stiffness of 10 lb/in. If the original length of the spring is 10 in. and if cables, attached to the platform, allow its extended length to be only 8 in., determine how high the block rises into the air after rebounding from the platform. Neglect the weight of the platform and assume that no energy is lost in the collision with the platform.

Prob. 16–9

16–10. The block assembly shown weighs 20 lb. The pulley is one unit which consists of an outer spool of 1-ft radius and an inner spool having a radius of $\frac{1}{2}$ ft. It is attached to the block by an axle through its center. If a 60-lb force **F** is applied to the outer cable, determine the speed that the block attains in moving a distance of 10 ft up the plane if it originally started from rest. The coefficient of friction between the block and plane is $\mu = 0.1$. Neglect the mass of the pulley.

692

Prob. 16–10

16–11. The 10-lb block B is given an initial velocity of 10 ft/sec downward. It moves 3 ft and stops. Determine the coefficient of friction between the block A and the inclined plane.

Prob. 16–11

16–12. The 30-lb block slides along the smooth inclined plane with an initial velocity of 10 ft/sec when it passes point A. Determine the kinetic energy of the block just before it strikes the spring. If the spring constant is $k = 5$ lb/in., determine how far the spring is compressed.

Prob. 16-12

16-15. The 25-lb block has an initial velocity of 10 ft/sec when it is midway between springs A and B. After striking spring B, it rebounds and slides across the horizontal plane and strikes spring A, etc. If the coefficient of friction between the plane and the block is $\mu = 0.4$, determine the total distance traveled by the block before the block comes to rest.

Prob. 16-15

16-13. Rework Prob. 16-12 assuming that the coefficient of friction between the block and plane is $\mu = 0.2$.

16-14. The 6-lb block is released from rest at A and moves down the smooth parabolic surface shown. Determine how far it compresses the spring.

Prob. 16-14

16-16. The 20-lb block slides in the smooth horizontal slot. When the block is at A, the springs are unstretched. If the block is given an initial velocity of 60 ft/sec to the right from this position, determine the maximum horizontal displacement s of the block.

Prob. 16-16

16-17. Determine the velocity of the 60-lb block A if the two blocks are released from rest and the 40-lb block B moves through a distance of 20 ft. The coefficient of friction between both blocks and the inclined plane is $\mu = 0.10$.

Prob. 16-17

16-18. The 2-lb box moves along the smooth curved ramp. If the velocity v_A of the box along the horizontal surface has a magnitude of 30 ft/sec, determine the velocity of the box and normal force acting on the ramp when the box is located at B and C. Assume the radius of curvature of the path at C is 5 ft.

Prob. 16-18

16-19. The 6-lb block moves within the frictionless slot. If the block starts at rest when the attached spring is in the unstretched position A, determine the constant horizontal force $\mathbf{F}$ which must be applied so that the block attains a velocity of 2 ft/sec when it reaches point B.

Prob. 16-19

16-20. The coefficient of friction between the 15-lb block and the surface is $\mu = 0.2$. The block is acted upon by a horizontal force of 20 lb and has a velocity of 20 ft/sec when it is at point A. Determine how much the block deforms the compound spring.

Prob. 16-20

694

16-21. The block shown weighs 2 lb and is given an initial velocity of 20 ft/sec when it is at A. If the spring has an unstretched length of 2 ft and a stiffness of $k = 100$ lb/ft, determine the velocity of the block when $s = 1$ ft.

Prob. 16–21

16-22. Determine the initial velocity of the block in Prob. 16–21 so that the maximum displacement of the block is $s = 4$ ft.

16-23. Determine the required horizontal velocity $\mathbf{v}_A$ of the box in Prob. 16–18 so that the box begins to leave the surface of the ramp when it reaches C. Assume the radius of curvature of the path at C is 5 ft.

16-24. If the 15-lb block in Prob. 16–20 is 1 ft from the wall and then released from rest, determine how far from the wall the block slides before coming to rest. The coefficient of friction is $\mu = 0.2$, and $P = 20$ lb, as shown in the figure.

Power is defined as the time rate of doing work. The power delivered to a body which is subjected to a force **F** while undergoing a displacement *d***s** during the time *dt* is

$$P = \frac{dW}{dt} = \frac{\mathbf{F} \cdot d\mathbf{s}}{dt}$$

or

$$P = \mathbf{F} \cdot \mathbf{v} \tag{16–9}$$

v represents the absolute velocity of the mass center of the body when the body is acted upon by the force **F**.

The term "power" provides a useful basis for determining the type of motor or machine which is required to do a certain amount of work in a given time. For example, two pumps may be able to empty the contents of a reservoir if given enough time; however, the pump having the larger power will complete the job first.

The units of mechanical and electrical power generally accepted in the engineering system of units are the *horsepower* (hp) and the *watt* (W) respectively. These units are defined as

$$1 \text{ hp} = 550 \text{ ft-lb/sec} = 76.04 \text{ m-kg}_f/\text{sec}$$
$$1 \text{ W} = 0.737 \text{ ft-lb/sec} = 1 \text{ joule/sec} = 0.102 \text{ m-kg}_f/\text{sec}$$

Thus,

$$1 \text{ hp} = 746 \text{ W}$$

The *mechanical efficiency* of a machine is defined as the ratio of the output of useful power created by the machine to the input of power supplied to the machine. Hence,

$$e = \frac{\text{power output}}{\text{power input}} \tag{16–10a}$$

If work is being done by the machine at a constant rate, then the efficiency may be expressed in terms of the ratio of output energy to input energy, i.e.,

$$e = \frac{\text{energy output}}{\text{energy input}} \tag{16–10b}$$

Since machines consist of a series of moving parts, frictional forces will always be developed within the machine. Because extra work or power

is needed to overcome these frictional forces, *the efficiency of a machine is always less than 1.* In all machines the transformation of mechanical energy into thermal energy owing to frictional forces is unavoidable. In some situations, however, this is desirable. For example, the kinetic energy of a moving vehicle is dissipated into thermal energy utilizing the frictional forces developed by the brakes.

Example 16–5

The jeep shown in Fig. 16–12*a* weighs 2,000 lb and has an engine running efficiency of $e = 0.80$. The coefficient of friction acting between the wheels and the pavement is $\mu = 0.25$. As the jeep moves forward, the wind creates a drag resistance on the jeep of $F_D = 0.6\, v^2$ lb, where v is the velocity in ft/sec. Assuming that the engine supplies power to *all the wheels,* determine the maximum horsepower that can be supplied by the engine to the jeep. Neglect the mass of the wheels.

Solution

The free-body and inertia-vector diagrams for the jeep are shown in Fig. 16–12*b*. The normal force **N** and frictional force **F,** shown on the free-body diagram, represent the *resultant forces* of all four wheels. In particular, the frictional force pushes the jeep forward. This force is created by the rotating motion of the wheels on the pavement.

(a)

(b)

Fig. 16–12

Applying Newton's second law of motion, we have

$$\stackrel{+}{\rightarrow}\Sigma F_x = ma_x; \qquad -F + 0.6v^2 = -\frac{2,000}{32.2}\frac{dv}{dt} \qquad (1)$$

$$+\uparrow\Sigma F_y = ma_y; \qquad N - 2,000 \text{ lb} = 0; \quad N = 2,000 \text{ lb}$$

At *maximum power* the drag resistance from the wind will *balance* the maximum frictional force which can be developed at the wheels, thereby preventing the jeep from accelerating further. When this happens the jeep is in *equilibrium* since it moves with *constant velocity*, i.e., $dv/dt = 0$. The frictional force reaches its maximum value when $F = 0.25N = 0.25$ (2,000 lb) = 500 lb. Hence, Eq. (1) becomes,

$$-500 + 0.6\ v^2 = 0$$

Solving for v, we get

$$v = \sqrt{\frac{500}{0.6}} = 28.9 \text{ ft/sec}$$

The power output of the jeep is created by the driving (frictional) force **F**. Thus, using Eq. 16–9,

$$P = Fv = (500)28.9 = 14,450 \text{ ft-lb/sec}$$

or

$$P = \frac{14,450}{550} = 26.3 \text{ hp}$$

Using Eq. 16–10a, the power supplied by the engine (power input) is therefore

$$\text{power input} = \frac{1}{e}\ (\text{power output}) = \frac{1}{0.8}(26.2)$$

$$\text{power input} = 32.8 \text{ hp} \qquad\qquad Ans.$$

Problems

16-25. A truck weighs 60,000 lb and has an engine which develops 360 hp. Determine the steepest grade the truck can climb with a constant speed of $v = 10$ mph.

16-26. The motor and driving device of a 1,300-lb elevator has an overall efficiency of 0.70. If the maximum velocity attained by the elevator is 10 ft/sec, determine the horsepower supplied by the motor.

16-27. An automobile weighing 4,000 lb travels up a 4° slope at a constant speed of 60 mph. Neglecting friction and wind resistance, determine the horsepower of the automobile.

16-28. The escalator moves upward with a speed of 2 ft/sec. If the steps are 8 in. high and 15 in. in length, determine the required horsepower of a motor needed to lift an average load of 180 lb per step.

Prob. 16-28

16-32. Determine the final velocity of the train in Prob. 16-31 if the wind resistance is $0.6v$ lb where v is the velocity in ft/sec.

16-33. The 100-lb crate is lifted up the 30° incline by using the pulley system and motor M. If the crate starts from rest and by constant acceleration attains a speed of 12 ft/sec in 30 ft, determine (a) the average horsepower developed by the motor and (b) the instantaneous power when the crate has moved 30 ft. Neglect friction along the plane.

16-29. A 150-lb man runs up the steps of the escalator in Prob. 16-28. When the escalator is not moving, the man can run up the steps in 15 sec. Determine the combined horsepower developed by the man and escalator if (a) the escalator is not operating and (b) the escalator is moving with a speed of 2 ft/sec.

16-30. A 2,000-lb jeep, starting from rest, develops a torque of 20 lb-ft for 30 sec at each of its 1.5-ft-diameter wheels. If rolling resistance of the jeep is 20 lb during its motion, determine the average power developed by the wheels.

16-31. An electrically powered train car draws 30,000 watts of power. If the car weighs 40,000 lb and starts from rest, determine the maximum speed it attains in 30 sec. The mechanical efficiency is $e = 0.8$.

Prob. 16-33

16-34. Compute the average power of the motor in Prob. 16-33 if the coefficient of friction between the plane and the crate is $\mu = 0.30$.

16-6. Conservative Forces

A particularly simple type of force is one which depends only on the position of the particle on which the force is acting and is independent of the velocity and acceleration of the particle. Furthermore, if the work done by this force in moving the particle from one point to another is independent of the path followed by the particle, this force is called a *conservative force.* For example, consider a body having a weight w and initially located at some fixed reference datum, Fig. 16-13. If the body is moved to some position P, a distance h above the datum along the

699

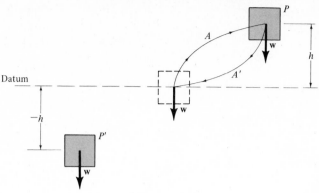

Fig. 16–13

arbitrary path *A*, the total work *W* done *on the body* by the force *w* is independent of the path and depends only on the vertical height *h*.* Thus,

$$W = -wh$$

The work is negative, since *w* acts in the direction opposite to displacement *h*. In a similar manner, the work done *on the body* by the weight in lowering the body a distance *h* back to the datum, along the arbitrary path *A'*, is

$$W = wh$$

In this case the work is positive. Why? Since the work done by the weight force is independent of the path and depends only upon the initial and final positions of the path, the *weight of a body is a conservative force.*

The *force* developed by an *elastic spring* is also a *conservative force.* The work *done on a body* by a spring force **F** which causes the spring to elongate or compress a distance *d* away from its equilibrium position, Fig. 16–14, is

$$W = \int \mathbf{F} \cdot d\mathbf{x} = -\int_0^{-d} kx\, dx = -\int_0^d kx\, dx = -\tfrac{1}{2}kd^2$$

The amount of work done *on the body* is thus dependent only upon the amount of extension *or* compression of the spring *d* and the stiffness *k*. In both cases the work is negative, since the spring exerts a force *on the body* which is opposite to its displacement.

In contrast to a conservative force, consider a frictional force exerted on a moving object by a fixed surface. The work done by the frictional

Fig. 16–14

*Due to the law of gravitation the weight of a body varies with elevation. Hence, if the elevation change is significant, a variation of weight with elevation must be considered. See Prob. 16–53.

force *depends upon the path*—the longer the path, the greater the work. When the block slides across a fixed surface to some point and then is returned to its original position, the frictional force reverses direction when the block is returned, so that instead of recovering the work done during the first displacement, it becomes necessary to do work on returning the block to its original position. Thus, a *frictional force is a non-conservative force.*

If work is done on a body by a conservative force, it gives the body *energy.* The energy stored in a body is known as *potential energy V.* For example, when the weight force *w*, Fig. 16–13, is *above* the datum at position *P*, it has a *positive* potential energy

$$V = wh$$

In this position, the force has the *capacity* for doing positive work *on the body* when the body is moved back to the datum. When the body is *below* the datum, at position *P'*, the potential energy is *negative,*

$$V = -wh$$

since the weight force does negative work *on the body* when the body is moved up to the datum plane.

The potential energy which an elastic spring stores in an attached body, when the spring is elongated or compressed a distance *d*, is

$$V = \tfrac{1}{2}kd^2$$

In the deformed position, the spring has the capacity for doing positive work *on the body* in returning the body back to the spring's original undeformed position, Fig. 16–14.

From this discussion, it may be concluded that the work done by a weight force is opposite to the potential energy gained or lost by the weight. For example, a weight force acting *on a body* does *negative work* when the body is moved above a datum; however, this causes a corresponding *increase* in the *potential energy* of the body. In a similar manner, a spring force does *negative work* when it acts *on the body* which compresses or elongates it; however, this causes an *increase* in the body's *potential energy.*

Potential energy may occur from types of forces other than the weight and the elastic spring restoring force. All that is required is that the work done by the force be independent of the displacement path and that the force itself be independent of velocity and acceleration. In general, when a particle is located at an arbitrary point (x, y, z) in space, its potential energy at this point can be measured using a potential energy function $V = V(x, y, z)$. The work done by a force in moving the particle from point (x_1, y_1, z_1) to point (x_2, y_2, z_2) is then measured by the *difference* in the potential energy function. Therefore,

$$W = V(x_1, y_1, z_1) - V(x_2, y_2, z_2) \qquad (16\text{--}11)$$

As an example, the potential energy function for a body acted upon by a spring has been determined as $V = \frac{1}{2}kx^2$, where x is the displaced position of the spring measured from the undeformed position. Using Eq. 16–11, the work done by the spring force acting on the block when the spring is displaced a distance x_1 to x_2 ($|x_2| > |x_1|$) is thus

$$W = \tfrac{1}{2}kx_1^2 - \tfrac{1}{2}kx_2^2$$

When the displacement path is infinitesimal, i.e., from point (x, y, z) to $(x + dx, y + dy, z + dz)$, Eq. 16–11 becomes

$$dW = V(x, y, z) - V(x + dx, y + dy, z + dz)$$

or

$$dW = -dV(x, y, z) \qquad (16\text{--}12)$$

If the force $\mathbf{F}$ and displacement $d\mathbf{s}$ are defined using rectangular coordinates, Eq. 16–1 becomes

$$dW = \mathbf{F} \cdot d\mathbf{s} = (F_x\mathbf{i} + F_y\mathbf{j} + F_z\mathbf{k}) \cdot (dx\mathbf{i} + dy\mathbf{j} + dz\mathbf{k})$$

or

$$dW = F_x\,dx + F_y\,dy + F_z\,dz \qquad (16\text{--}13)$$

Also, the total derivative of $V(x, y, z)$ may be written as

$$dV(x, y, z) = \frac{\partial V}{\partial x}\,dx + \frac{\partial V}{\partial y}\,dy + \frac{\partial V}{\partial z}\,dz \qquad (16\text{--}14)$$

Substituting Eqs. 16–13 and 16–14 into Eq. 16–12 yields

$$F_x\,dx + F_y\,dy + F_z\,dz = -\left(\frac{\partial V}{\partial x}\,dx + \frac{\partial V}{\partial y}\,dy + \frac{\partial V}{\partial z}\,dz\right)$$

Since changes in x, y, and z are all independent of one another, this equation is satisfied provided

$$F_x = -\frac{\partial V}{\partial x}, \qquad F_y = -\frac{\partial V}{\partial y}, \qquad F_z = -\frac{\partial V}{\partial z} \qquad (16\text{--}15)$$

or

$$\mathbf{F} = -\nabla V \qquad (16\text{--}16)$$

where ∇ (del) represents the vector operator $\nabla = (\partial/\partial x)\mathbf{i} + (\partial/\partial y)\mathbf{j} + (\partial/\partial z)\mathbf{k}$. Hence, if the force $\mathbf{F}$ satisfies Eqs. 16–15 or Eq. 16–16, $\mathbf{F}$ is a conservative force.

16–7. Conservation of Energy Theorem

Of all the laws in nature, the law of conservation of energy is probably the most important. In general, this law may be stated simply as follows: *Energy is always conserved when it is transformed from one form to another,* or, *energy can neither be created nor destroyed.*

There are many forms of energy in addition to the two forms of mechanical energy already discussed. These include thermal energy, electrical energy, chemical energy, and nuclear energy. In this book, however, we will be concerned only with the law of conservation of energy as applied to mechanical energy—which consists of potential and kinetic energy.

When a particle is acted upon by a conservative force, the principle of work and energy as expressed by Eq. 16–6 may be written in a different form. Equation 16–11 states that the work done on the particle by a conservative force may be written in terms of the difference in its potential energies. Substituting this equation into Eq. 16–6 yields

$$V_1 - V_2 = T_2 - T_1$$

or

$$T_1 + V_1 = T_2 + V_2 \qquad (16\text{–}17)$$

This equation is often referred to as the *conservation of energy theorem* for the mechanical energy of a particle. An equation similar to this may be written for a system of particles, which is based on Eq. 16–7 or Eq. 16–8. From Eq. 16–17 it may be seen that the *sum* of the kinetic and potential energy for the system remains *constant*. For example, consider the energy involved in dropping a ball of weight w from a height h above some ground datum, Fig. 16–15a. The potential energy of the ball is maximum before it is dropped, at which time its kinetic energy is zero. The total energy of the ball in its initial position is thus

$$E = T + V = 0 + wh = wh$$

When the ball has fallen a distance $h/2$, its speed can be determined using Eq. 12–5c. We obtain $v = \sqrt{2g(h/2)} = \sqrt{gh}$. The energy of the ball at the midheight position is therefore

$$E = V + T = w\frac{h}{2} + \frac{1}{2}\frac{w}{g}(\sqrt{gh})^2 = wh$$

Just before the ball strikes the ground, the potential energy of the ball is zero, and its speed is $v = \sqrt{2gh}$. Here again the total energy of the ball is

Potential Energy (max)
Kinetic Energy (zero)

Potential Energy and
Kinetic Energy

Potential Energy (zero)
Kinetic Energy (max)

(a)

height $= h$
$v = 0$

h

Datum

height $= 0$
$v = \sqrt{2gh}$

(b)

Fig. 16–15

$$E = V + T = 0 + \frac{1}{2}\frac{w}{g}(\sqrt{2gh})^2 = wh$$

When the ball comes in contact with the ground, the ball is deformed and, provided the ground is hard enough, the ball will rebound off the surface. It will reach a new height h' which will be *less* than the height h from which it was first released. The difference in height accounts for a potential energy loss, $E_l = w(h - h')$. Part of this energy loss has been converted into thermal energy which is used to raise the temperature of the ball and the surroundings at the moment the collision occurred. Other portions of the loss of energy have gone into deforming the ball and producing mechanical vibrations and noise. (Further details of the impact are discussed in Sec. 18–5.)

It is very important to remember that *only* problems involving *conservative force systems* may be solved using the conservation of energy theorem, Eq. 16–17. As stated previously, frictional forces are nonconservative. The work done by such forces is transformed into thermal energy used to heat up the surfaces of contact. This energy dissipates into the surroundings and may not be recovered. Since Eq. 16–17 applies only to a balance of mechanical energy, problems involving frictional forces should be solved using the principle of work and energy, Eq. 16–6.

To eliminate errors when using Eq. 16–17, two *energy diagrams* should accompany the problem solution. *After* a reference datum (used to measure the potential energy) has been established, these two diagrams should show the particle (or system of particles) in its initial and final position. The data necessary to calculate the mechanical energy of the particle should be labeled on these diagrams. For example, the necessary energy diagrams for the falling ball discussed earlier are shown in Fig. 16–15*b*. Further examples for applying the principle of conservation of energy to the solution of problems in particle dynamics follow.

Example 16–6

A small block having a weight w starts from rest at point A and slides down the smooth curved path shown in Fig. 16–16*a*. Determine the velocity of the block when it reaches points B and C.

Solution

The only force which does work on the body is the weight of the block. The normal reaction of the path on the block does no work since at any point the block does not displace normal to the path. The energy diagrams are shown in Fig. 16–16*b*. Why do we construct these diagrams? For convenience, the potential energy datum has been established through the center of mass of the block when the block is located at B. Applying Eq. 16–17 between points A and B yields

(a)

(b)

Fig. 16–16

$$V_A + T_A = V_B + T_B$$

$$wh_{BA} + 0 = 0 + \frac{1}{2}\frac{w}{g}(v_B)^2$$

$$v_B = \sqrt{2gh_{AB}} \qquad\qquad Ans.$$

Applying the energy equation between points A and C gives

$$V_A + T_A = V_C + T_C$$

$$wh_{BA} + 0 = wh_{BC} + \frac{1}{2}\frac{w}{g}(v_C)^2$$

Hence,

$$v_C = \sqrt{2g(h_{BA} - h_{BC})} \qquad\qquad Ans.$$

This same result may also be obtained by applying the conservation of
energy principle between points B and C, i.e.,

$$V_B + T_B = V_C + T_C$$

$$0 + \frac{1}{2}\frac{w}{g}(\sqrt{2gh_{BA}})^2 = wh_{BC} + \frac{1}{2}\frac{w}{g}(v_C)^2$$

Solving for v_C gives

$$v_C = \sqrt{2g(h_{BA} - h_{BC})} \qquad\qquad Ans.$$

The results of this problem indicate that the block attains a velocity which is *independent* of the path. This velocity is the *same* as that computed by Eq. 12–5c when the block *falls freely* from the vertical height of the path.

Example 16–7

The 10-lb rigid block shown in Fig. 16–17a has a downward velocity of 4 ft/sec when it is 5 ft from the ground. Determine the maximum deflection of the spring absorber after the block has fallen on it. Neglect the mass of the thin platforms A and B. Assume that no energy is lost in the collision.

Solution

The two energy diagrams are shown in Fig. 16–17b. For convenience the datum plane has been established at the ground. In the top position the block has both kinetic and potential energy. Why? In this position the center of gravity of the block is 5.5 ft above the datum. When the block strikes the platform, its kinetic energy is transferred into potential

Fig. 16–17

energy which is stored in the springs. In the bottom position the center of gravity of the block is $(y + 0.5)$ ft above the datum. At this elevation the springs undergo *maximum deflection*, so that the kinetic energy of the block is reduced to zero. Consequently the platforms are then y ft from the datum. Here it is *assumed* that the distance y is small enough to compress *all springs* of each of the two platforms, i.e., $|y|$ ft < 1 ft. With reference to Fig. 16–17b, applying Eq. 16–17, we have

$$V_1 + T_1 = V_2 + T_2$$

$$10 \text{ lb } (5.5 \text{ ft}) + \frac{1}{2}\left(\frac{10 \text{ lb}}{32.2 \text{ ft/sec}^2}\right)(4 \text{ ft/sec})^2 = 10 \text{ lb } (y + 0.5 \text{ ft})$$

$$+ \frac{1}{2}[4(75 \text{ lb/ft})](1.25 \text{ ft} - y)^2 + \frac{1}{2}[4(12.5 \text{ lb/ft})](1 \text{ ft} - y)^2 + 0$$

Simplifying gives

$$(175)y^2 - (415)y + 207 = 0$$

Solving,

$$y = 0.713 \text{ ft} \qquad \text{and} \qquad y = 1.658 \text{ ft}$$

The second root is not acceptable since platform A cannot deform for $|y| > 1.25$ ft. Therefore, since $y = 0.713$ ft < 1 ft, *all* eight springs are compressed as originally assumed. The springs supporting platform A are compressed a distance

$$\Delta_A = 1.25 - y = 1.25 - 0.713 = 0.537 \text{ ft} \qquad\qquad Ans.$$

And the springs supporting platform B compress

$$\Delta_B = 1 - y = 1 - 0.713 = 0.287 \text{ ft} \qquad\qquad Ans.$$

(a)

Example 16–8

A chain has a linear density of 2 lb/ft and hangs across the surface of a smooth drum, as shown in Fig. 16–18a. If it is released from rest in the position shown, determine the velocity of the chain when end A passes point B. Neglect the radius of the drum.

Solution

The two energy diagrams are shown in Fig. 16–18b. The datum has been located at the base of the drum. Since each link or particle of the chain moves with the same speed as its connecting links, the total energy for the *system of links* can be found by determining the mechanical energy for a given length of chain and adding (algebraically) the results. In particular, the potential energy of a chain length can be computed on the basis of knowing the position of the *mass center* of the chain length.

(b)

Fig. 16–18

(Recall that the mass center represents an average position for the mass of a body.) The mass center for the two lengths of chain on the right and left sides of the drum are located with respect to the datum as shown in the figure. In all cases the potential energy of a given length of chain is negative since the mass center of each length lies *below* the datum.
Applying the conservation of energy theorem gives

$$V_1 + T_1 = V_2 + T_2$$
$$-(2 \text{ lb/ft})(5 \text{ ft})(2.5 \text{ ft}) - (2 \text{ lb/ft})(10 \text{ ft})(5 \text{ ft}) + 0$$
$$= -(2 \text{ lb/ft})(15 \text{ ft})(7.5 \text{ ft}) + \frac{1}{2}\left(\frac{(2 \text{ lb/ft})(15 \text{ ft})}{32.2 \text{ ft/sec}^2}\right)v_2^2$$

So that

$$0.466v_2^2 = 100$$

Solving,

$$v_2 = 14.65 \text{ ft/sec} \qquad\qquad Ans.$$

Problems

16-35. The 6-lb ball is fired from a tube by a spring having a stiffness of $k = 20$ lb/in. Determine how far the spring must be compressed to fire the ball to a height of 8 ft at which point it has a velocity of 6 ft/sec.

Prob. 16-35

16-36. The 5-lb bob of the pendulum is released from rest when it is in the horizontal position shown. Determine the speed of the bob and the tension in the cord when the bob passes through its lowest position.

Prob. 16-36

16-37. Work Prob. 16-14 using the principle of conservation of mechanical energy.

16–38. Work Prob. 16–16 using the principle of conservation of mechanical energy.

16–39. The 5-lb block is released from rest at A and travels along the frictionless guide. Determine the speed of the block when it strikes the stop B. The spring has an unstretched length of 12 in.

Prob. 16–39

16–40. The toy car weighs 0.5 lb. Determine the minimum height h at which the car must be released to make the loop.

Prob. 16–40

16–41. Determine the stiffness k of the spring at the end of the track in Prob. 16–40 so that when the car is released from a height $h = 5$ ft, it rounds the loop, hits the spring, and compresses it 1 in., then rebounds and returns back to its original position. Assume that no energy is lost in the collision.

16–42. The 5-lb ball is shot from a spring device. The spring has a stiffness $k = 10$ lb/in. and the cables C and plate P keep the spring compressed 2 in. The plate is pushed back 3 in. from its initial position, so that $s = 30$ in. If it is then released from rest, determine the speed of the ball when it leaves the surface of the smooth inclined plane.

Prob. 16–42

16–43. Determine the stiffness k of the spring in Prob. 16–42 which is required to shoot the ball a distance $s = 30$ in. up the plane when the spring is pushed back 3 in. and released from rest. The cables C and plate P keep the spring compressed 2 in.

16–44. A 1-lb particle is projected from the earth's surface with an initial velocity of $\mathbf{v}$. Determine this "escape velocity" which will allow the particle to travel an infinite distance before its velocity is decreased to zero. Assume the radius of the earth to be 4,000 mi. Neglect the effect of air friction.

16-45. The 5-lb sphere is attached to a 2-ft cord and is fired from position A by a spring which is compressed 4 in. If the stiffness of the spring is $k = 100$ lb/in., determine the speed of the sphere and the tension in the cord when the sphere is at positions B and C. B is located at a point just *before* the entire cord becomes horizontal.

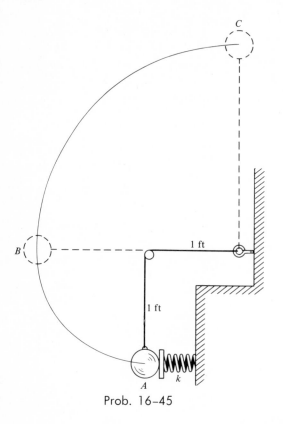

Prob. 16-45

16-46. If the spring in Prob. 16-45 is compressed 3 in. and released, determine the stiffness k so that (a) the speed of the sphere at B is zero and (b) the tension on the cord at C is zero.

16-47. A 2-lb ball is shot from a spring-loaded tube and follows the curved path shown. If the spring has a stiffness $k = 10$ lb/in. and it is compressed a distance of 3 in., determine the height h the ball is shot into the air after following the path. Neglect the effects of friction.

Prob. 16-47

16-48. Determine the speed and the force the ball exerts on the path in Prob. 16-47 when the ball passes points A and B. The spring is compressed 3 in. when the ball is released. Neglect the effects of friction.

16-49. How far should the spring in Prob. 16-47 be compressed so that when it is released the ball travels along the path and just reaches point C?

16-50. The chain has a total length of 5 ft and rests on a smooth surface. If the chain has a linear density of 1 lb/ft, determine the velocity of end B when it passes point A.

Prob. 16-50

16-51. The 1-lb homogeneous block is connected to the floor of a box having negligible weight. If the box is released from rest from the position shown, determine the maximum deflection of platform B. Neglect the weight of the platform and assume no energy loss in the collision.

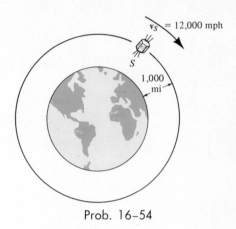

Prob. 16-51

16-52. Work Prob. 16-51 assuming that the box weighs 1 lb.

16-53. Show that if the gravitational force acting between two bodies is $F = G(Mm/r^2)$ (Eq. 14-3), the work done in moving the force from a point r_1 to a point r_2 is $W = GMm[(1/r_2) - (1/r_1)]$. Also prove that the gravitational force F is conservative.

16-54. The 300-lb satellite S is in a circular orbit around the earth. If the speed of the satellite is 12,000 mph when it is 1,000 mi above the surface of the earth, determine its speed when it is 1,500 mi from the earth's surface. Use the principle of conservation of energy. Assume the radius of the earth to be 4,000 mi. (*Hint:* See Prob. 16-53. Use $GM = 1.265 \times 10^{12}$ mi³/hr²).

Prob. 16-54

16-55. The blocks A and B weigh 10 and 30 lb, respectively. They are connected together by a light cord and ride in the frictionless grooves. Determine the speed of both blocks after they are released from rest and block A moves 6 ft along the plane.

Prob. 16-55

711

17

Planar Kinetics of Rigid Bodies: Work and Energy

17–1. Kinetic Energy of a Rigid Body

Rigid-body kinetics problems were discussed in Chapter 15, using Newton's second law of motion. In that chapter the discussion was limited to bodies which were subjected to motion in the plane. Such motions include translation, rotation about a fixed axis, and general plane motion. In an effort to keep the rotational equation of motion ($\Sigma M_G = I_G\alpha$) in a relatively simple form, it was necessary to assume that the body was *symmetrical* with respect to the fixed reference plane used to describe the motion. In this chapter we will discuss work and energy methods for rigid bodies subjected to motion in the plane. The results to be obtained will be rather general, however, since they are valid for rigid bodies of *any shape*. The principle of work and energy will be extended further in Chapter 21 to include the more complex rigid-body motion about a fixed point and general motion of a rigid body. As noted in Chapter 16, work and energy methods are particularly well suited for problems involving displacements and velocities.

Before discussing the principle of work and energy for a body we will consider a detailed analysis for obtaining the kinetic energy of a rigid body subjected to general plane motion. The special cases of translation and rotation about a fixed axis will then be discussed.

Consider the rigid body shown in Fig. 17–1 which is represented here

Fig. 17–1

by a *slab* moving in the reference plane of motion. The $x'y'$ coordinate system has its origin fixed in the body at the mass center G. The axes of this coordinate system translate or remain parallel to the x,y axes of the inertia frame of reference. An arbitrary ith particle of the body, having a mass Δm_i, is located at a distance $r_{i/G}$ from the center of mass. If, at the *instant* shown, this particle has a velocity $\mathbf{v}_i$,* the kinetic energy of the particle is

$$\Delta T_i = \tfrac{1}{2}\Delta m_i(v_i)^2 = \tfrac{1}{2}\Delta m_i(\mathbf{v}_i \cdot \mathbf{v}_i)$$

If the body is composed of n particles the kinetic energy of the body is determined by writing similar expressions for each particle and summing the results, i.e.,

$$T = \Sigma\tfrac{1}{2}\Delta m_i(\mathbf{v}_i \cdot \mathbf{v}_i)$$

The velocity of the ith particle may be measured relative to the mass center using the relative velocity equation, Eq. 13–12, i.e.,

$$\mathbf{v}_i = \mathbf{v}_G + \mathbf{v}_{i/G}$$

Here $\mathbf{v}_G$ is the velocity of the mass center and $\mathbf{v}_{i/G}$ is the relative velocity of particle i with respect to G. Using this equation the kinetic energy becomes

$$T = \Sigma\tfrac{1}{2}\Delta m_i(\mathbf{v}_G + \mathbf{v}_{i/G}) \cdot (\mathbf{v}_G + \mathbf{v}_{i/G})$$
$$= \tfrac{1}{2}(\Sigma \Delta m_i)v_G^2 + \mathbf{v}_G \cdot (\Sigma m_i\mathbf{v}_{i/G}) + \tfrac{1}{2}(\Sigma \Delta m_i v_{i/G}^2)$$

The first sum represents the total mass m of the body. The second sum is *zero* because $\Sigma m_i\mathbf{v}_{i/G} = d/dt(\Sigma m_i\mathbf{r}_{i/G}) = d/dt(\Sigma m_i\bar{\mathbf{r}}) = 0$, i.e., the position vector $\bar{\mathbf{r}}$ relative to G is equal to zero by definition of the mass center. Hence, we may write

$$T = \tfrac{1}{2}mv_G^2 + \tfrac{1}{2}\Sigma \Delta m_i v_{i/G}^2$$

By definition, the relative velocity $\mathbf{v}_{i/G} = \boldsymbol{\omega} \times \mathbf{r}_{i/G}$ (see Eq. 13–13), where $\boldsymbol{\omega}$ is the instantaneous angular velocity of the body. Thus, the magnitude of $v_{i/G} = \omega r_{i/G}$ so that

$$T = \tfrac{1}{2}mv_G^2 + \tfrac{1}{2}\omega^2\Sigma \Delta m_i r_{i/G}^2$$

Letting the particle mass $\Delta m_i \to dm$, the summation sign becomes an integral since the number of particles $n \to \infty$. Representing $r_{i/G}$ by a more generalized dimension r we have

$$T = \tfrac{1}{2}mv_G^2 + \tfrac{1}{2}\omega^2\int_m r^2\,dm$$

---

*The particle may also have accelerated motion, but this will not be considered in the following analysis.

The integral represents the mass moment of inertia I_G of the body about an axis (the z' axis) passing through G and directed perpendicular to the plane of motion. Our final result is therefore

$$T = \tfrac{1}{2}mv_G^2 + \tfrac{1}{2}I_G\omega^2 \qquad (17\text{--}1)$$

From this equation it is seen that the total kinetic energy of a body undergoing plane motion is the sum of the *translational* kinetic energy of the mass center of the body and the *rotational* kinetic energy of the body about its mass center. Both of these terms are always *positive* since the velocities are squared. Furthermore it may be verified that both terms have units of length times force, common units being ft-lb or m-kg$_f$.

In particular, when a rigid body is subjected to either rectilinear or curvilinear *translation,* the kinetic energy due to rotation is zero since $\omega = 0$. Each particle Δm_i of the body has translational kinetic energy $\Delta T_i = \tfrac{1}{2}\Delta m_i v^2$, Fig. 17–2. From Eq. 17–1, the total kinetic energy of the body is then

Fig. 17–2

$$T = \tfrac{1}{2}mv^2 \qquad (17\text{--}2)$$

where m is the mass of the body and v is the body's translational velocity at the instant considered.

When a body is *rotating about a fixed axis* with an angular velocity ω the magnitude of velocity of the mass center is $v_G = r_{G/O}\omega$, where $r_{G/O}$ is the distance from point O to the mass center G as shown in Fig. 17–3. Thus, computing the kinetic energy using Eq. 17–1, we have

$$T = \tfrac{1}{2}m(r_{G/O}\omega)^2 + \tfrac{1}{2}I_G\omega^2$$

or

$$T = \tfrac{1}{2}(I_G + mr_{G/O}^2)\omega^2$$

From the parallel-axis theorem (Eq. 15–9) the term in parentheses represents the moment of inertia I_O of the body about an axis (the z axis) passing through point O and directed perpendicular to the plane of motion. Hence, we can compute the kinetic energy of the body using the equation

Fig. 17–3

$$T = \tfrac{1}{2}I_O\omega^2 \qquad (17\text{--}3)$$

In a similar manner, the kinetic energy of a rigid body subjected to general plane motion may also be determined with respect to its instantaneous center of zero velocity, *IC.* At an *instant considered, the body appears to rotate about this point* and therefore the total kinetic energy of the body may be found using

$$T = \tfrac{1}{2}I_{IC}\omega^2 \qquad\qquad (17\text{–}4)$$

Here, I_{IC} represents the moment of inertia of the body about the instantaneous axis of zero velocity, that is, an axis passing through the instantaneous center of zero velocity and directed perpendicular to the plane of motion.

Since energy is a scalar quantity, the total kinetic energy for a system of *connected* rigid bodies is simply the sum of the kinetic energies of all the moving parts. Depending upon the type of plane motion, the kinetic energy of *each body* is found by applying Eq. 17–1 or the alternate forms Eqs. 17–2 through 17–4.* Once the kinetic energy of the body is computed, the principle of work and energy can then readily be applied.

Example 17–1

The 60-lb wheel shown in Fig. 17–4a has a radius of gyration of $k_G = \tfrac{1}{2}$ ft about its center of mass G. The connecting rod AB, having a mass center at D, weighs 10 lb. If the 1-lb slider block at B has a downward velocity of 10 ft/sec, determine the kinetic energy of the system at the instant shown. Assume that the wheel does not slip on the horizontal surface.

Solution

To compute the kinetic energy of the system, we must *first* determine both the angular velocity and the velocity of the mass center of the wheel and the connecting rod. This will be done by using the method of instantaneous centers (Sec. 13–7). Since the wheel does not slip, the instantaneous center IC for the *wheel* is at its point of contact with the ground. The velocity of point A is then directed horizontally, Fig. 17–4b. Knowing this, and the fact that $\mathbf{v}_B$ is directed downwards, we can determine the IC for the *rod*, Fig. 17–4c. From the figure the rod is 4 ft long, so that $r_{B/IC} = 4\cos 30°$ ft. Hence the angular velocity of the rod AB is

$$v_B = \omega_{AB}(r_{B/IC})$$
$$10 = \omega_{AB}(4\cos 30°)$$
$$\omega_{AB} = 2.89 \text{ rad/sec}$$

The magnitude of the velocity of the mass center D for the rod is therefore

$$v_D = \omega_{AB}(r_{D/IC}) \qquad\qquad (1)$$

Using the law of cosines for triangle $IC\text{-}B\text{-}D$ to determine the magnitude $r_{D/IC}$ (or by noting that triangle $A\text{-}IC\text{-}D$ is equilateral), we have

(a)

(b)

(c)

Fig. 17–4

---

*A brief review of Secs. 13–5 through 13–7 may prove helpful in solving problems since computations for kinetic energy require a kinematic analysis of velocity.

$$r_{D/IC} = \sqrt{(2)^2 + (4 \cos 30°)^2 - 2(2)(4 \cos 30°) \cos 30°} = 2 \text{ ft}$$

Then substituting into Eq. (1),

$$v_D = 2.89(2) = 5.78 \text{ ft/sec}$$

The magnitude of the velocity of point A on the rod is

$$v_A = \omega_{AB}(r_{A/IC})$$
$$= 2.89(4 \sin 30°) = 5.78 \text{ ft/sec}$$

Knowing this velocity, the angular velocity of the wheel ω_W can be determined with reference to Fig. 17-4b.

$$v_A = \omega_W(r_{A/IC})$$
$$5.78 = \omega_W(1.75)$$

Thus,

$$\omega_W = 3.30 \text{ rad/sec}$$

The magnitude of the velocity of the center of mass G of the wheel is thus

$$v_G = \omega_W(r_{G/IC})$$
$$= (3.30)\sqrt{(1)^2 + (1/4)^2}$$
$$= 3.40 \text{ ft/sec}$$

Using Appendix D, the moment of inertia of the rod about its center of mass D is

$$(I_{AB})_D = \frac{1}{12} m_{AB} l_{AB}^2 = \frac{1}{12}\left(\frac{10}{32.2}\right)(4)^2 = 0.414 \text{ slug-ft}^2$$

Knowing the radius of gyration and the weight, the wheel's moment of inertia is

$$(I_W)_G = m_W k_G^2 = \frac{60}{32.2}(1/2)^2 = 0.466 \text{ slug-ft}^2$$

The kinetic energies of the rod and wheel may now be computed using Eq. 17-1.

$$T_{AB} = \frac{1}{2} m_{AB} v_D^2 + \frac{1}{2}(I_{AB})_D \omega_{AB}^2 \qquad \text{(rod)}$$

$$= \frac{1}{2}\left(\frac{10}{32.2}\right)(5.78)^2 + \frac{1}{2}(0.414)(2.89)^2 = 6.92 \text{ ft-lb}$$

$$T_W = \frac{1}{2} m_W v_G^2 + \frac{1}{2}(I_W)_G \omega_W^2 \qquad \text{(wheel)}$$

$$= \frac{1}{2}\left(\frac{60}{32.2}\right)(3.40)^2 + \frac{1}{2}(0.466)(3.30)^2 = 13.31 \text{ ft-lb}$$

The kinetic energy of block B is determined by using Eq. 17-2, since the block is *translating*.

$$T_B = \frac{1}{2}m_B v_B^2 = \frac{1}{2}\left(\frac{1}{32.2}\right)(10)^2 = 1.55 \text{ ft-lb} \qquad \text{(block)}$$

The total kinetic energy of the system is therefore

$$T = T_{AB} + T_W + T_B$$
$$= 6.92 + 13.31 + 1.55$$
$$= 21.78 \text{ ft-lb} \qquad \qquad Ans.$$

Since the instantaneous centers for both the rod and wheel are known, Eq. 17-4 may have also been used for computing the kinetic energies T_{AB} and T_W. It is left as an exercise for the reader to show that one obtains the same results as those calculated here. You may also use the relative velocity equation, Eq. 13-13, in the solution.

Problems

17-1. The uniform slender rod has a mass m and is pinned at its end point. If the angular velocity of the rod is ω, show that the kinetic energy is the same when computed with respect to its mass center G or point A.

Prob. 17-1

17-2. The 20-lb disk D rolls down the inclined plane with an angular velocity of 12 rad/sec. The uniform connecting link AB weighs 5 lb. If the 40-lb wheel at C has a mass center at G and a radius of gyration $k_G = \frac{3}{4}$ ft, determine the kinetic energy of the system at the instant link AB becomes horizontal as shown. Both the wheel and the disk roll without slipping.

Prob. 17-2

17-3. The 30-lb cylinder has a counterclockwise angular velocity of 5 rad/sec when its center has a translational velocity of 20 ft/sec. Determine the kinetic energy of the cylinder at this instant.

Prob. 17-3

17–4. The uniform rectangular plate weighs 30 lb. If the plate is pinned at A and has an angular velocity of 3 rad/sec, determine the kinetic energy of the plate.

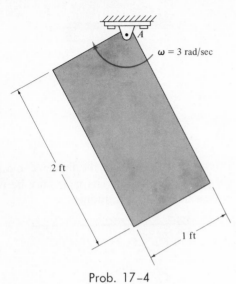

$\omega = 3$ rad/sec

2 ft

1 ft

Prob. 17–4

17–5. The mechanism consists of two rods AB and BC, which weigh 10 and 20 lb, respectively; and a 4-lb disk at C. The disk moves along the horizontal plane with a clockwise angular velocity of 4 rad/sec. If the disk does not slip on the plane, determine the kinetic energy of the system at the instant shown.

B

4 ft

2 ft

$r = 3$ in.
$\omega = 4$ rad/sec

A

C

Prob. 17–5

17–6. The 50-lb pulley has a centroidal radius of gyration of $k_O = \frac{2}{3}$ ft and is turning with a constant angular velocity of 20 rad/sec clockwise. Compute the total kinetic energy for the system. Assume that neither cable slips on the pulley.

$\omega = 20$ rad/sec

$\frac{1}{2}$ ft

1 ft

O

B | 30 lb

A | 20 lb

Prob. 17–6

17–7. Link AB of the mechanism has an angular velocity of 3 rad/sec. If links AB and CD each weigh 15 lb and link BC weighs 20 lb, determine the kinetic energy of the mechanism at the instant when $\theta = 30°$. Is the kinetic energy constant for a general angle θ?

A | $\omega_{AB} = 3$ rad/sec

D

θ

θ

2 ft

B

C

3 ft

Prob. 17–7

17–8. The 10-lb crank wheel has a radius of gyration of $k_O = \frac{3}{4}$ ft. The 3-lb connecting rod has a mass center at G. The piston weighs 2 lb. If the wheel has a clockwise angular velocity of 4 rad/sec, determine the kinetic energy of the system at the instant shown.

719

6.5 in.

6.5 in.

B

G

$\omega = 4$ rad/sec

A

O

5 in.

Prob. 17-8

17-9. Link AB has an angular velocity of 2 rad/sec. If each link is considered as a uniform rod with a linear density of 0.5 lb/in., determine the total kinetic energy of the links at the instant shown.

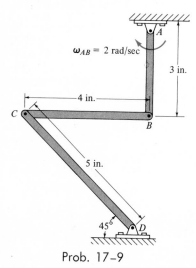

A

$\omega_{AB} = 2$ rad/sec

3 in.

4 in.

C

B

5 in.

45°

D

Prob. 17-9

17-10. Determine the kinetic energy for the system of three links. Links AB and CD each weigh 10 lb and link BC weighs 20 lb.

720

A $\omega_{AB} = 5$ rad/sec

1 ft

B

2 ft

C

1 ft

D

Prob. 17-10

17-11. From the data given in the figure, compute the kinetic energy of the system. The gear may be considered as a disk having a 6-in. radius.

4 lb

$v = 6$ ft/sec

6 in. 10 lb

$v = 3$ ft/sec

4 lb

Prob. 17-11

17-12. The 20-lb gear C is rotating counterclockwise with a constant angular velocity of 10 rad/sec. The link AB weighs 4 lb and is rotating clockwise at 2 rad/sec. If the 5-lb gear D revolves about C, determine the kinetic energy of the system. For the calculation assume that the gears are equivalent to disks having the radii shown.

$\frac{1}{4}$ ft

B

D

$\omega_C = 10$ rad/sec

A

$\omega_{AB} = 2$ rad/sec

C

$\frac{1}{2}$ ft

Prob. 17-12

17–13. Two 10-lb rectangular plates are welded to the shaft AB. If the shaft has an angular velocity of 6 rad/sec, determine the kinetic energy of both plates.

1 ft

2 ft

A

1 ft

ω = 6 rad/sec

B

2 ft

1 ft

Prob. 17–13

17–2. Work Done by a Force and a Couple

It was shown in Sec. 16–1 that if a particle is acted upon by a force **F**, the work done by this force in moving the particle along the path s can be expressed as

$$W = \int_s \mathbf{F} \cdot d\mathbf{s} = \int_s F \cos \beta \, ds$$

where β is the angle between the force vector and the differential path displacement $d\mathbf{s}$. Referring to Fig. 16–1, the work of a force may therefore be interpreted in one of the two following ways: Work is either equal to the magnitude of the component of the force in the direction of displacement ($F \cos \beta$) times the magnitude of displacement ds, or it is equal to the magnitude of the displacement component acting in the direction of the force ($ds \cos \beta$) times the force magnitude F.

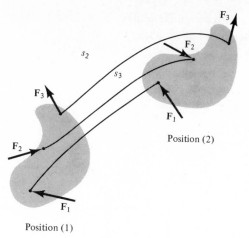

s_2

s_3

$\mathbf{F}_3$

$\mathbf{F}_2$

$\mathbf{F}_1$

Position (2)

$\mathbf{F}_2$

$\mathbf{F}_1$

Position (1)

Fig. 17–5

Consider now the work done by a system of three external forces*
acting on the rigid body shown in Fig. 17–5, as the body moves from
position (1) to position (2). Since a rigid body does not deform when subjected to the given loading, the total work done by the *internal forces*
which hold the particles of the body together is *zero*. When the body
moves along a path, the work done by the internal forces is again zero.
This is because all the internal forces occur in equal and opposite collinear
pairs so that the corresponding work done by each pair of internal forces
cancels. Therefore, the total work done is simply the algebraic addition
of the work done by each of the individual *external forces* acting on the
body. This work is

$$W = \int_{s_1} \mathbf{F}_1 \cdot d\mathbf{s}_1 + \int_{s_2} \mathbf{F}_2 \cdot d\mathbf{s}_2 + \int_{s_3} \mathbf{F}_3 \cdot d\mathbf{s}_3 \qquad (17\text{–}5)$$

Similar terms must be included in this equation if additional forces are
applied to the body. Each force is integrated over the path along which
the point of application of the force moves. In each application the
integration must take into account the variation of the force direction
and its magnitude as the force moves along its path.

There are some external forces which do no work when the body is
displaced. These forces are either applied at *fixed points* on the body
or they act in a direction which is *perpendicular* to the path of displacement of their point of application. Examples of forces which do no work
include the normal reaction acting on a block which moves along a
surface, the reactions at a pin support about which a body rotates, and

---

*External forces include contact forces between other bodies, and gravitational, electrical,
and magnetic forces.

the weight of a body when the center of gravity of the body moves in a *horizontal plane*.

Recall that a couple consists of a pair of equal and opposite noncollinear forces. *The work done by a couple (or moment) acting on a rigid body is computed only when the body is rotated.* Consider, for example, the couple moment $M = Fr$ acting on the rigid body shown in Fig. 17–6a. When the body undergoes a small displacement (to the dashed position) during an instant of time, the force $-\mathbf{F}$ acting at A moves through a displacement $d\mathbf{s}_A$ to point A', while the force $\mathbf{F}$ at B moves through a displacement $d\mathbf{s}_B$ to B'. To analyze the motion, the displacement of the body is separated into two independent motions consisting of a translation and a rotation. As shown in Fig. 17–6b, the entire body is first translated by an amount $d\mathbf{s}_A$ and then rotated about point A' through an angle $d\boldsymbol{\theta}$. To separate the motions in this manner, the displacement $d\mathbf{s}_B$ is divided into two components to represent each motion, i.e., $d\mathbf{s}_B = d\mathbf{s}_A + d\mathbf{s}_\theta$.

The total work done by the couple is the sum of the work done by the two forces during the translation plus the work done during the rotation. During the translation $d\mathbf{s}_A$ the work done by force $\mathbf{F}$ is canceled by the work of $-\mathbf{F}$ since the forces are equal in magnitude but opposite in direction. When the body is rotated about point A', $-\mathbf{F}$ does no work since A' does not move. Force $\mathbf{F}$ moves through a displacement $d\mathbf{s}_\theta$. Consequently, the total work is $dW = \mathbf{F} \cdot d\mathbf{s}_\theta = Fr\,d\theta$. Since the product Fr is equal to the magnitude M of the couple, then

$$dW = M\,d\theta$$

This result clearly shows that a couple does work *only* during a *rotation* and *not* during a *translation*. In the derivation the line of action of $d\boldsymbol{\theta}$

(a)   (b)

Fig. 17–6

is *parallel* to the line of action of **M**. This is *always the case* for *plane motion*, since **M** and $d\boldsymbol{\theta}$ are perpendicular to the plane of motion. For other types of motion, however, this is not the case. Therefore, in general, the component of $d\boldsymbol{\theta}$ in the direction of **M** is required; that is, the work done is defined by the dot product $dW = \mathbf{M} \cdot d\boldsymbol{\theta}$.

When the body is rotated through a finite angle θ, from θ_1 to θ_2, the work done by a couple is

$$W = \int_{\theta_1}^{\theta_2} \mathbf{M} \cdot d\boldsymbol{\theta} \qquad (17\text{-}6)$$

If **M** is *parallel* to $d\boldsymbol{\theta}$ (plane motion) and furthermore if **M** has a *constant magnitude* then

$$W = M(\theta_2 - \theta_1) \qquad (17\text{-}7)$$

where the angles θ_1 and θ_2 are measured in radians. If the body has an angular velocity $\omega = d\boldsymbol{\theta}/dt$, we may write Eq. 17-6 in the form

$$W = \int_{\theta} \mathbf{M} \cdot \boldsymbol{\omega}\, dt \qquad (17\text{-}8)$$

The dot product indicates that the work is done only by the component of **M** which acts in the direction of $\boldsymbol{\omega}$. For motion in the plane, **M** and $\boldsymbol{\omega}$ always have parallel lines of action so that

$$W = \int_{\theta} M\omega\, dt \qquad (17\text{-}9)$$

For applications, this equation may be integrated provided the moment **M** and angular velocity ω are expressed as functions of time. The limits on the integral are evaluated when θ is also a known function of time.

17-3. Principle of Work and Energy: Method of Problem Solution

With reference to Sec. 16-2, the principle of work and energy for a *particle* may be stated as follows:

$$\Sigma W_{1-2} = T_2 - T_1$$

The term on the left represents the work done by all the external forces acting on the particle while the particle moves from position s_1 to position s_2 along the path s. The terms on the right side of the equation represent the change in the kinetic energy of the particle.

Since a rigid body consists of a large number of particles, the principle of work and energy for the body is obtained by applying the principle of work and energy to each particle of the body and adding the results algebraically. Thus, the principle of work and energy for a rigid body is of the same form as for a particle, written as

$$\Sigma W_{1-2} = T_2 - T_1$$

When applying this equation, it is best to rewrite it in the form

$$
\begin{array}{ccc}
T_1 & + \quad \Sigma W_{1-2} \quad = & T_2 \\
\text{Initial} & \text{Total work done} & \text{Final} \\
\text{translational} & \text{by all the} & \text{translational} \\
\textit{and}\ \text{rotational} & \text{external forces} & \textit{and}\ \text{rotational} \\
\text{kinetic energy} & \text{and couples} & \text{kinetic energy} \\
\text{of the body} & \text{acting on the} & \text{of the body} \\
& \text{body} &
\end{array}
\qquad (17\text{--}10)
$$

The kinetic energy terms are obtained from the equations derived in Sec. 17–1 once the required moment of inertia, velocity, and angular velocity terms are computed. It is helpful to draw the *kinematic diagrams* of the body located at the initial and final positions of its path and to compute the required velocities from these diagrams. As stated with Eq. 17–10, the second term on the left represents the total work done by the externally applied forces and couples acting on the body when the body is moved from the initial position (1) to the final position (2). The work done by the forces may be computed from an equation similar to Eq. 17–5. The work done by the external couples is computed from Eq. 17–6 or Eq. 17–8. To account for all the forces and couples which do work on the body, a *free-body diagram* of the body should be drawn when the body is located at an intermediate point along its path.

When several rigid bodies are connected together, the principle of work and energy may be applied to each body separately or to the entire system of connected bodies, particularly when the bodies are pin-connected, connected by inextensible cables, or in contact with meshed gears. In all these cases the internal forces which hold the various members together do no work, and hence, are eliminated from the analysis.

It should be observed that Eq. 17–10 is a *scalar equation;* therefore, only one unknown may be obtained by using this equation when it is applied to a single rigid body. This is in contrast to three scalar equations of plane motion which may be written for the same body. (See Chapter 15.) The principle of work and energy cannot be used to determine accelerations, time intervals, or forces which do no work on the rigid body. There are cases, however, where accelerations and forces are functions of velocity. In such circumstances it is usually easier to determine the velocity using the principle of work and energy and then to substitute

this quantity into the equations of motion which relate force, mass, and acceleration.

Example 17-2

The 20-lb bar shown in Fig. 17-7a has a center of gravity at G. If the bar is given an initial clockwise angular velocity of $\omega_1 = 10$ rad/sec when $\theta = 90°$, compute the spring constant k so that the bar stops when $\theta = 0°$. What are the horizontal and vertical components of reaction at the pin A when $\theta = 0°$? The spring is required to deform 3 in. for $\theta = 0°$.

Solution

Two kinematic diagrams for the bar when $\theta = 90°$ (position 1) and $\theta = 0°$ (position 2) are shown in Fig. 17-7b. The kinetic energy of the bar may be computed with reference to either the point of rotation A or the center of mass G. If point A is used, we must apply Eq. 17-3. Using

Fig. 17-7

Appendix D and the parallel-axis theorem to compute the moment of inertia of the bar about point A we get

$$I_A = \frac{1}{12}ml^2 + md^2 = \frac{1}{12}\left(\frac{20}{32.2}\right)(2)^2 + \frac{20}{32.2}(1)^2 = 0.828 \text{ slug-ft}^2$$

Thus,

$$T_1 = \frac{1}{2}I_A\omega_1^2 = \frac{1}{2}(0.828)(10)^2 = 41.4 \text{ ft-lb}$$

and

$$T_2 = \frac{1}{2}I_A\omega_2^2 = \frac{1}{2}(0.828)(0)^2 = 0$$

These *same results* may be obtained using Eq. 17–1, which applies to point G. To show this note that

$$I_G = \frac{1}{12}ml^2 = \frac{1}{12}\left(\frac{20}{32.2}\right)(2)^2 = 0.207 \text{ slug-ft}^2$$

Then

$$T_1 = \frac{1}{2}m(v_G)_1^2 + \frac{1}{2}I_G\omega_1^2 = \frac{1}{2}\left(\frac{20}{32.2}\right)(10)^2 + \frac{1}{2}(0.207)(10)^2 = 41.4 \text{ ft-lb}$$

and

$$T_2 = \frac{1}{2}m(v_G)_2^2 + \frac{1}{2}I_G\omega_2^2 = \frac{1}{2}\left(\frac{20}{32.2}\right)(0)^2 + \frac{1}{2}(0.207)(0)^2 = 0$$

Having formulated the kinetic energy of the bar, we will now consider the work done by all the forces acting on the bar as the bar moves from position 1 to position 2. From the free-body diagram of the bar, Fig. 17–7c, the reactions A_x and A_y of the pin on the bar do no work since these forces do not displace. The 20-lb weight moves downward through a vertical height (displacement in the direction of the force) of 1 ft, hence the work is *positive*. The spring force F_s does *negative work on the bar*. Why? This force acts while the spring is being compressed 3 in. The work may be determined by either integrating the spring force $F_s = kx$ over the path of deformation, or by direct application of Eq. 16–4. Choosing the latter method,

$$W_s = -\frac{1}{2}kx^2 = -\frac{1}{2}k\left(\frac{3 \text{ in.}}{12 \text{ in./ft}}\right)^2 = -\frac{k}{32}$$

Applying the principle of work and energy, we write

$$T_1 + \Sigma W_{1-2} = T_2$$

$$41.4 + 20(1) - \frac{k}{32} = 0$$

Solving for k yields

$$k = 1,965 \text{ lb/ft} = 163.7 \text{ lb/in.} \qquad \textit{Ans.}$$

The pin reactions at $\theta = 0°$ must be determined by applying the equations of motion. (These forces cannot be obtained from the principle of work and energy. Why?) The free-body and inertia-vector diagrams are shown in Fig. 17–7d. When $\theta = 0°$, the magnitude of the spring force is $F_s = kx = (163.7 \text{ lb/in.})(3 \text{ in.}) = 491.1$ lb. The normal component of the inertia-force vector is zero, since the angular velocity of the bar is zero at this instant, that is, the bar is motionless. Applying the equations of motion, we have

$$\xrightarrow{+} \Sigma F_x = ma_n; \qquad\qquad -A_x = 0$$

$$+\uparrow \Sigma F_y = ma_t; \qquad A_y - 20 + 491.1 = -\frac{20}{32.2}(1)\alpha$$

$$\zeta + \Sigma M_A = I_A \alpha; \qquad 20(1) - 491.1(2) = 0.828\alpha$$

Solving these three equations gives

$$\alpha = -1,162 \text{ rad/sec}^2$$

$$A_x = 0.0 \text{ lb} \qquad\qquad \textit{Ans.}$$

$$A_y = 251 \text{ lb} \qquad\qquad \textit{Ans.}$$

The negative sign for α indicates that the bar will begin rotating counterclockwise just after coming to rest on the spring.

Example 17–3

The 30-lb rod shown in Fig. 17–8a is restrained so that its ends move within the grooved slots. The rod is initially at rest when $\theta = 0°$. If the slider block at B is suddenly acted upon by a horizontal force of $P = 5$ lb, determine the angular velocity of the rod at the instant $\theta = 45°$. Neglect the weight of blocks A and B. (Why is the principle of work and energy suitable for solving this problem?)

Solution

The kinetic energy of the rod may be determined using Eq. 17–1. To apply this equation it is first necessary to know the velocity v_G of the rod's mass center and the angular velocity ω. Two kinematic diagrams of the rod, when the rod is in the initial position 1 and final position 2, are shown in Fig. 17–8b. When the rod is in position 1, $T_1 = 0$. Why?

(a)

(b)

(c)

Fig. 17-8

In position 2 the instantaneous angular velocity is ω_2 and the velocity of the mass center is $(\mathbf{v}_G)_2$. Hence, the kinetic energy is

$$T_2 = \frac{1}{2}m(v_G)_2^2 + \frac{1}{2}I_G(\omega_2)^2$$

$$= \frac{1}{2}\left(\frac{30}{32.2}\right)(v_G)_2^2 + \frac{1}{2}\left[\frac{1}{12}\left(\frac{30}{32.2}\right)(4)^2\right](\omega_2)^2$$

$$= 0.466(v_G)_2^2 + 0.621(\omega_2)^2 \tag{1}$$

The two unknowns $(v_G)_2$ and ω_2 in this expression may be related via the instantaneous center of zero velocity of the rod, Fig. 17-8b. It is seen that as block A moves downward with a velocity of $(\mathbf{v}_A)_2$, block B moves horizontally to the left with a velocity of $(\mathbf{v}_B)_2$. Knowing these directions, the IC may then be determined as shown in the figure. Hence,

$$(v_G)_2 = r_{G/IC}\omega_2$$
$$= (2\tan 45°)\omega_2$$

$$(v_G)_2 = 2\omega_2$$

The kinetic energy of the rod when it is in position 2 may now be determined in terms of ω_2 using this relation. Hence, from Eq. (1),

$$T_2 = 0.466(2\omega_2)^2 + 0.621(\omega_2)^2$$
$$= 2.48(\omega_2)^2$$

Having formulated the kinetic energy, we may now consider the work done by all the forces acting on the rod as the rod moves from position 1 ($\theta = 0°$) to position 2 ($\theta = 45°$). The normal forces $\mathbf{N}_A$ and $\mathbf{N}_B$ shown on the free-body diagram, Fig. 17–8c, do no work as the rod is displaced. Why? The 30-lb weight is displaced a *vertical distance* of $(2 - 2\cos 45°)$ ft; whereas the 5-lb force moves a horizontal distance of $(4\sin 45°)$ ft. The work done by both of these forces is *positive*, since the forces act in the same direction as their corresponding displacement.

Applying the principle of work and energy gives

$$\{T_1\} + \{\Sigma W_{1-2}\} = \{T_2\}$$
$$\{0\} + \{30(2 - 2\cos 45°) + 5(4\sin 45°)\} = \{2.48(\omega_2)^2\}$$

Solving for ω_2 gives

$$\omega_2 = 3.58 \text{ rad/sec} \qquad\qquad Ans.$$

Example 17–4

The 50-lb wheel shown in Fig. 17–9a has a radius of gyration of $k_G = \frac{3}{4}$ ft about its center of gravity. If the wheel is subjected to a couple of 2 lb-ft and rolls on its inner hub without slipping, determine the angular speed of the wheel after the 20-lb block is released from rest and has fallen 1 ft. The spring has a stiffness of $k = 6$ lb/ft and is initially unstretched when the weight is released.

Solution

It is simpler to analyze the system of both the wheel and block together rather than treating each one separately. (A separate treatment involves two work equations and would introduce the work done by the unknown cable tension into the equations.) We will first obtain the kinetic energy when the system is in the initial position 1 and final position 2. The necessary kinematic diagrams are shown in Fig. 17–9b. Since the wheel and block are initially at rest, $T_1 = 0$. When the wheel is in position 2, the unknown velocities of the wheel and the block are ω_2, $(v_G)_2$, and $(v_B)_2$. The total kinetic energy of the system is therefore

$$T_2 = \frac{1}{2}m_{wh}(v_G)_2^2 + \frac{1}{2}I_G(\omega_2)^2 + \frac{1}{2}m_{bl}(v_B)_2^2$$

Fig. 17–9

Since $I_G = m_{wh}k_G^2$, we have

$$T_2 = \frac{1}{2}\frac{50}{32.2}(v_G)_2^2 + \frac{1}{2}\left[\frac{50}{32.2}\left(\frac{3}{4}\right)^2\right](\omega_2)^2 + \frac{1}{2}\frac{20}{32.2}(v_B)_2^2$$

$$= 0.776(v_G)_2^2 + 0.437(\omega_2)^2 + 0.311(v_B)_2^2 \qquad (1)$$

Using kinematics, the three unknowns ω_2, $(v_G)_2$, and $(v_B)_2$ may be related. Since the wheel does not slip as it rolls, the instantaneous center is located at the point of contact with the ground, Fig. 17–9b. The speed

of point B on the wheel moves with the same velocity as the block since the attached cable is inextensible. Therefore,

$$(v_B)_2 = r_{B/IC}\omega_2 = \tfrac{3}{2}\omega_2$$
$$(v_G)_2 = r_{G/IC}\omega_2 = \tfrac{1}{2}\omega_2$$

Using these relations, the total kinetic energy of both the wheel and block when they are in position 2 can now be expressed in terms of the single unknown ω_2. From Eq. (1), we get

$$T_2 = 0.776(\tfrac{1}{2}\omega_2)^2 + 0.437(\omega_2)^2 + 0.311(\tfrac{3}{2}\omega_2)^2 = 1.33(\omega_2)^2$$

Having formulated the kinetic energy, we will now consider the work done by all the forces acting on the system. Inspection of the free-body diagram, Fig. 17–9c, reveals that only the spring force $\mathbf{F}_s$, the 20-lb weight of the block, and the 2 lb-ft couple do work while the system moves from position 1 to position 2. The frictional force $\mathbf{F}$ does *no work*, since the wheel does not slip as it rolls. Note that during any instant of time the frictional force acts at a point having *zero velocity* (instantaneous center IC). Hence, for any small movement of the wheel, $dW = \mathbf{F} \cdot d\mathbf{s}_{IC} = F(v_{IC}\, dt) = 0$. Why is there no work done by forces $\mathbf{N}$ and $\mathbf{R}$? The work done by $\mathbf{F}_s$ is *negative*, since this force acts *opposite* to its displacement. This work may be computed by using Eq. 16–4, i.e., $-\tfrac{1}{2}kx^2$. Since the wheel does not slip, as the block moves downward 1 ft the wheel rotates $\theta = s/r_{B/IC} = 1\ \text{ft}/1.5\ \text{ft} = \tfrac{2}{3}$ rad, Fig. 17–9b. Hence the spring stretches $x = \theta(r_{D/IC}) = \tfrac{2}{3}$ rad (1 ft) $= \tfrac{2}{3}$ ft. The couple does work because of the rotation $\theta = \tfrac{2}{3}$ rad. Since this rotation is in the same direction as the couple, the work done is positive and can be computed using Eq. 17–7.

Using this data, applying the principle of work and energy to the system gives

$$\{T_1\} + \{\Sigma W_{1-2}\} = \{T_2\}$$
$$\{0\} + \{(2\ \text{lb-ft})(\tfrac{2}{3}\ \text{rad}) + (20\ \text{lb})(1\ \text{ft}) - \tfrac{1}{2}(6\ \text{lb/ft})(\tfrac{2}{3}\ \text{ft})^2\} = \{1.33(\omega_2)^2\}$$

or

$$20 = 1.33\omega_2^2$$

Thus,

$$\omega_2 = 3.88\ \text{rad/sec} \qquad\qquad Ans.$$

Problems

17-14. A 50-lb vertical force acts on the wheel. Determine the velocity of the 10-lb block B after the center of the wheel is raised 3 ft, starting from rest. Assume that the cable does not slip on the wheel. The radius of gyration of the wheel is $k_O = 8$ in.

P = 50 lb

20 lb

$r = 1$ ft

10 lb   B

Prob. 17-14

17-15. The wheel weighs 20 lb and has a radius of gyration of $k_C = 3$ in. If it is released from rest from the position shown, determine how far it must drop before the velocity of its center is increased to 20 ft/sec. The cable is wrapped around the inner hub of the wheel and unrolls as the wheel is dropped.

4 in.

C

2 in.

Prob. 17-15

17-16. The 100-lb disk and the surface of the conveyor belt have a coefficient of friction of $\mu_k = 0.2$. If the conveyor belt is moving with a velocity of 6 ft/sec when the disk is released and comes in contact with it, determine the number of revolutions the disk makes before it reaches a constant angular velocity. What is the force on link AB while the disk is slipping?

$\frac{1}{2}$ ft   A   B

$v_C = 6$ ft/sec

Prob. 17-16

17-17. Gears A and B each weigh 3 lb and have a radius of gyration of $k = 1.5$ in. about their centers. The outer gear ring C weighs 5 lb and has a radius of gyration of $k_O = 6.2$ in. If a torque of 20 lb-ft is applied to gear A, determine the angular velocity of gear B after the outer gear ring has rotated through one revolution starting from rest.

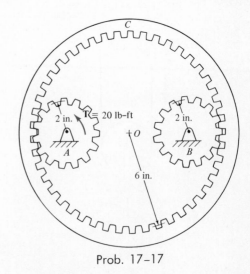

C

2 in.   T = 20 lb-ft   2 in.

O

A   B

6 in.

Prob. 17-17

733

17-18. The triangular plate weighs 50 lb. The angle $\theta = 30°$ when the rope AB is cut. Determine the velocity of point D on the plate when $\theta = 0°$. The moment of inertia of the plate about its mass center is $I_G = 0.3$ slug-ft².

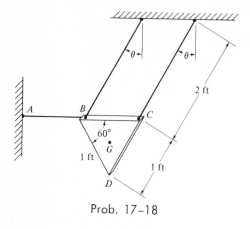

Prob. 17–18

17-19. The 20-lb block is raised by the 150-lb force applied to the cable. Assuming that the cable does not slip on the pulleys, determine the velocity of the block after it has been raised 5 ft starting from rest. The pulleys at A and B can be treated as 10- and 5-lb disks, respectively. The link CD weighs 2 lb.

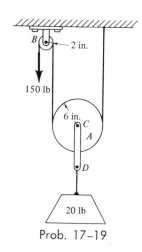

Prob. 17–19

17-20. Pulley A weighs 15 lb and has a radius of gyration of $k_O = 0.8$ ft. If the system is released from rest, determine the velocity of the center O of the pulley after the 10-lb block moves vertically 4 ft. Neglect the mass of the pulley at B.

Prob. 17–20

17-21. The 100-lb wheel has a radius of gyration of $k_O = \frac{2}{3}$ ft and is rotating at 20 rad/sec, when the force of $P = 60$ lb is applied to the brake handle. If the coefficient of friction between the brake and the wheel is $\mu_k = 0.35$, determine the number of rotations which the wheel makes before it stops. What are the horizontal and vertical components of reaction of A (a) while the wheel is stopping, and (b) when the wheel has stopped? Neglect the thickness of the brake pad B.

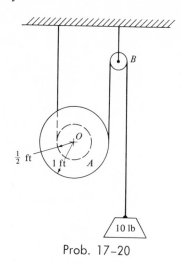

Prob. 17–21

734

17-22. The motor M supplies a torque $T = 60$ lb-ft which is used to rotate the pulley at A via the cable. The pulley weighs 25 lb and has a radius of gyration of $k_O = 0.8$ ft. Determine the velocity of the 10-lb weight after it reaches a height of 8 ft starting from rest. Neglect the weight of the pulley at B.

Prob. 17-22

17-23. The 10-lb cylinder is acted upon by a torque of $T = 6$ lb-ft. The 5-lb link AB connects the cylinder to a 10-lb block B. If the cylinder rolls without slipping, determine the angular velocity of the cylinder after it has moved 12 ft down the incline. Neglect the effect of friction on B in the analysis.

Prob. 17-23

17-24. A 20-lb cylindrical disk having a diameter of 3 ft is placed on a 30° inclined slope. Using the principle of work and energy, compute the unbalanced torque causing angular acceleration of the disk as the disk rolls down the incline.

17-25. Rod CB weighs 20 lb. A 4-lb semicircular disk is welded to the end of the rod at B. As the rod rotates, the spring always remains horizontal because of the roller support at A. If the rod is released when the spring is unstretched at $\theta = 30°$, determine the angular velocity of the rod when $\theta = 60°$. The rod and disk have a radius of gyration about point C of $k_C = 3.44$ ft.

Prob. 17-25

17-26. The 10-lb sphere starts from rest at $\theta = 0°$ and rolls without slipping down the cylindrical surface which has a radius of 10 ft. Determine the speed of the sphere when $\theta = 45°$.

Prob. 17-26

17-27. The motor supplies a constant torque of $T = 40$ lb-ft to the gear at C. If gear B weighs 10 lb and has a radius of gyration of $k_O = \frac{2}{3}$ ft, determine the distance s the 50-lb load must descend in order to increase its velocity from 1 to 10 ft/sec. The disk pulley at A weighs 5 lb. There is no slipping of the cord on the pulley.

Prob. 17-27

17-28. Rod ABC weighs 10 lb and the disk weighs 3 lb. The disk is rotating with a *constant* counterclockwise angular velocity of 4 rad/sec measured *with respect to the rod*. If point A has a downward speed of $v_A = 10$ ft/sec when the spring is unstretched at $\theta = 0°$, determine the angular velocity of the rod when $\theta = 30°$. The spring remains in the vertical position during the motion because of the smooth guide at D.

Prob. 17-28

17-29. The uniform rod in Prob. 17-1 is released from rest when $\theta = 90°$. Determine the reaction at A when $\theta = 0°$. The rod weighs w lb.

17-30. The disk at C weighs 12 lb and rolls on the horizontal surface without slipping. The links AB and BC each weigh 4 lb. When $\theta = 90°$, the elastic cable EC is unstretched. Determine the angular velocity of the disk when $\theta = 45°$ if the horizontal force $P = 50$ lb is suddenly applied to the pin at C when $\theta = 90°$.

Prob. 17-30

When the force system acting on the rigid body consists only of conservative forces, the conservation of energy theorem may be employed to solve a problem which would otherwise normally be solved using the principle of work and energy. The work done by a conservative force is *independent* of the path and depends only upon the initial and final points of the path. It was shown in Sec. 16-6 that this work may be expressed as the difference in potential energy measured relative to an arbitrarily selected horizontal reference or datum. A common example of a conservative force is the weight of a body. If a particle with mass Δm_i is located h_i ft *above* an arbitrary datum plane, the potential energy for the particle is

$$V = (\Delta m_i\, g)\, h_i$$

If we assume that the particle represents the *i*th particle of a rigid body, the potential energy for the rigid body composed of *n* particles is

$$V = \sum_{i=1}^{n} \Delta m_i\, g\, h_i$$

The summation becomes an integral as $\Delta m_i \to dm$ and consequently $n \to \infty$. Assuming that the acceleration of gravity *g* is constant, we have

$$V = g \int_m h\, dm = (mg)h_G = wh_G$$

Thus, the gravitational potential energy of a body may be determined by knowing the height h_G of its center of mass from the datum plane and the weight *w* of the body. In the above case, the potential energy is *positive* since the gravitational force has the ability to do work *on the body* when the body is moved *back* to the datum plane. If the body is located h_G *below* the datum, the potential energy is negative, $V = -wh_G$, since then the gravitational force does negative work *on the body* when the body is moved back to the datum.

The force developed in an elastic spring is another example of a conservative force. The potential energy which the spring stores in an attached body, when the spring is elongated or compressed a distance *d*, is

$$V = \tfrac{1}{2}kd^2$$

In the deformed position, the spring has the capacity for doing positive work *on the body* in returning the body back to the spring's original undeformed position.

In Sec. 16-7 the conservation of energy theorem was developed for

a particle. By successively applying this theorem to each of the particles of a rigid body and adding the results algebraically (since energy is a scalar quantity), we may write the conservation of energy theorem for a rigid body as

$$V_1 + T_1 = V_2 + T_2 \tag{17-11}$$

This theorem states that the *sum* of the potential and kinetic energy of the body is *constant* as the body moves from one position to another.* The kinetic energy terms are computed from the equations presented in Sec. 17–1 and the potential energy terms are computed from the gravitational and elastic forces acting on the body.

To eliminate errors in applying Eq. 17–11, two *energy diagrams* should accompany the problem solution. Once a reference datum (used to measure potential energy) is established, the two diagrams show the body in its initial and final position. The data necessary to calculate the potential and kinetic energy of the body should be labeled on these diagrams. The following examples illustrate the use of these diagrams for applying the principle of conservation of energy to solve planar-kinetics problems involving rigid bodies.

Example 17–5

The 10-lb rod AB, Fig. 17–10a, is confined so that its ends move in the horizontal and vertical slots. The spring has an unstretched length of $\frac{1}{4}$ ft, i.e., when $\theta = 0°$. Determine the velocity of the slider block B when $\theta = 0°$ if the block is released from rest when $\theta = 30°$. Neglect the mass of the slider blocks.

Solution

The two energy diagrams for the rod when the rod in its initial and final positions are shown in Fig. 17–10b. Why do we draw these diagrams? The datum plane used to measure the gravitational potential energy of the system is placed in line with the rod when $\theta = 0°$. When the rod is in position 1, the kinetic energy $T_1 = 0$. Why? Since the center of mass G is located *below the datum,* the gravitational potential energy is *negative.* The spring is stretched a distance of $\Delta_1 = (2 \sin 30°)$ ft when the rod is

---

*The theorem applies equally well to a system of smooth, pin-connected rigid bodies, bodies connected by inextensible cords, and bodies in mesh with other bodies. In these cases the forces acting at points of connection are *eliminated* from the analysis, since they occur in equal and opposite pairs and each pair of forces moves through equal distances when the system undergoes a small displacement.

(a)

(b)

(c)

Fig. 17–10

in this position. The potential energy stored in the spring is *positive,* since the spring has the capacity to do positive work *on the rod.* When the rod is in position 2, it has an angular velocity ω_2 and its mass center has a velocity $(v_G)_2$. In this position the potential energy of the rod is zero since the spring is unstretched, $\Delta_2 = 0$, and the center of mass G is located at the datum. Applying the conservation of energy theorem, we write

$$\{V_1\} + \{T_1\} = \{V_2\} + \{T_2\}$$

$$\left\{-wh + \frac{1}{2}k(\Delta_1)^2\right\} + \left\{\frac{1}{2}m(v_G)_1^2 + \frac{1}{2}I_G(\omega_1)^2\right\} = \left\{\frac{1}{2}k(\Delta_2)^2\right\}$$
$$+ \left\{\frac{1}{2}m(v_G)_2^2 + \frac{1}{2}I_G(\omega_2)^2\right\}$$

Substituting in the necessary data,

$$\left\{-10(1 \sin 30°) + \frac{1}{2}(50)(2 \sin 30°)^2\right\} + \{0 + 0\}$$

$$= \{0\} + \left\{\frac{1}{2}\left(\frac{10}{32.2}\right)(v_G)_2^2 + \frac{1}{2}\left[\frac{1}{12}\left(\frac{10}{32.2}\right)(2)^2\right](\omega_2)^2\right\}$$

or

$$0.155(v_G)_2^2 + 0.0518(\omega_2)^2 = 20 \qquad (1)$$

Using kinematics, we may relate $(v_G)_2$ to ω_2, as shown in Fig. 17–10c. At the instant considered, the instantaneous center of zero velocity for the rod is at point A; hence,

$$(v_G)_2 = (r_{G/IC})\omega_2$$
$$(v_G)_2 = (1)\omega_2 \qquad (2)$$

Substituting Eq. (2) into Eq. (1) and solving for ω_2, we have

$$\omega_2 = 9.83 \text{ rad/sec}$$

With reference to Fig. 17–10c, the velocity of the slider block at B is therefore

$$(v_B)_2 = (r_{B/IC})\omega_2 = 2(9.83)$$

or

$$(v_B)_2 = 19.66 \text{ ft/sec} \qquad \text{Ans.}$$

Example 17–6

The 20-lb homogeneous disk shown in Fig. 17–11a is attached to a uniform 15-lb rod AB. If the system is released from rest when $\theta = 60°$, that is, when the rod is vertical, determine the angular velocity of the rod when $\theta = 0°$. Assume that the disk rolls without slipping. Neglect friction along the guide and the weight of the slider block B.

Solution

The two energy diagrams for the rod and disk when they are located at their initial and final positions are shown in Fig. 17–11b. For convenience the datum plane (which is horizontally fixed) passes through point A_2, as shown in the figure. When the rod and disk are in the initial position 1, their weights have positive potential energy, since these forces are above the datum and therefore have the capacity to do work in moving the rod and disk. Furthermore, the entire system is at rest so that the kinetic energy is zero.

When the system is in the final position, the weight of the rod has positive potential energy, and the weight of the disk has zero potential

(a)

(b)

(c)

Fig. 17-11

energy. Why? The rod has an angular velocity of $(\boldsymbol{\omega}_R)_2$ and its mass center has a velocity of $(\mathbf{v}_G)_2$. Since the rod is *fully extended* in this position, the disk is momentarily at rest. Therefore $(\boldsymbol{\omega}_D)_2 = (\mathbf{v}_A)_2 = 0$.

Applying the conservation of energy theorem, we have

$$\{V_1\} + \{T_1\} = \{V_2\} + \{T_2\}$$

$$\{(2 + h)(w_R) + h(w_D)\} + \left\{\frac{1}{2}m_R(v_G)_1^2 + \frac{1}{2}I_G(\omega_R)_1^2 + \frac{1}{2}m_D(v_A)_1^2\right.$$

$$\left. + \frac{1}{2}I_A(\omega_D)_1^2\right\} = \{(h)w_R\} + \left\{\frac{1}{2}m_R(v_G)_2^2 + \frac{1}{2}I_G(\omega_R)_2^2\right.$$

$$\left. + \frac{1}{2}m_D(v_A)_2^2 + \frac{1}{2}I_A(\omega_D)_2^2\right\}$$

Substituting in the required data, yields

$$\{3(15) + 1(20)\} + \{0 + 0 + 0 + 0\} = \{1(15)\} + \left\{ \frac{1}{2}\left(\frac{15}{32.2}\right)(v_G)_2^2 \right.$$

$$+ \left. \frac{1}{2}\left[\frac{1}{12}\left(\frac{15}{32.2}\right)(4)^2\right](\omega_R)_2^2 + 0 + 0 \right\}$$

or

$$0.233(v_G)_2^2 + 0.311(\omega_R)_2^2 = 50 \qquad (1)$$

Using kinematics, a relation between $(v_G)_2$ and $(\omega_R)_2$ can be established. When the rod is in position 2, point A_2 represents the instantaneous center of zero velocity (IC) for the rod. Hence, from Fig. 17-11c,

$$(v_G)_2 = r_{G/IC}(\omega_R)_2$$
$$(v_G)_2 = 2(\omega_R)_2$$

Substituting this equation into Eq. (1) we get

$$1.243(\omega_R)_2^2 = 50$$

Thus

$$(\omega_R)_2 = 6.34 \text{ rad/sec} \qquad \qquad Ans.$$

Problems

17-31. The semicircular disk weighs 10 lb. If it is released from rest in the position shown, determine the velocity of point A when the disk has rotated counterclockwise 90°. Assume that the disk rolls without slipping on the surface. The moment of inertia of the disk about its mass center is $I_G = 0.0994$ slug-ft^2.

Prob. 17-32

Prob. 17-31

17-32. The elliptical cylinder weighs 50 lb and has a radius of gyration about its mass center of $k_G = 1.3$ ft. If it is released from rest in the position shown and rolls without slipping, determine the angular velocity of the cylinder when point A contacts the horizontal surface.

17-33. Solve Prob. 17-15 using the conservation of energy theorem.

17-34. Solve Prob. 17-20 using the conservation of energy theorem.

17-35. The 10-lb disk has a center of gravity at G and a radius of gyration of $k_O = 0.8$ ft. The cord is placed across the periphery of the disk and the weights are released when the disk is in the position shown. Determine the angular velocity of the disk when it has rotated 180°.

Prob. 17-35

17-36. The 20-lb disk rolls on the curved surface without slipping. The link AB weighs 5 lb. If the disk is released when $\theta = 0°$, determine the angular velocity of the disk when $\theta = 90°$.

Prob. 17-36

17-37. The 30-lb disk has a counterclockwise angular velocity of 3 rad/sec when it is in the position shown. If the rod AB weighs 10 lb and the slider block B weighs 2 lb, determine the angular velocity of the disk when the rod AB is in its lowest vertical position. The slider block slides on the smooth vertical pipe.

Prob. 17-37

17-38. Solve Prob. 17-25 using the conservation of energy theorem.

17-39. Solve Prob. 17-28 using the conservation of energy theorem.

17-40. Disks A and B weigh 15 and 25 lb, respectively. The connecting link CD weighs 5 lb. Determine the angular velocity of disk A just after the link falls through an angle of 180°. Assume that slipping does not occur between the two disks and that disk B is fixed from rotating.

Prob. 17-40

743

17-41. Pulley A weighs 100 lb and has a centroidal radius of gyration of $k_A = 7$ in. The disk pulley at C weighs 20 lb. Determine the velocity of the 50-lb weight after it has fallen 2 ft starting from rest. The rope does not slip on the pulley.

Prob. 17-41

17-42. The pendulum consists of a rod AB which weighs 10 lb and a disk which weighs 20 lb. The disk maintains a *constant* counterclockwise angular velocity of 2 rad/sec *relative to the rod*. Determine the angular velocity of the rod when $\theta = 90°$ if the rod is released from rest when $\theta = 0°$.

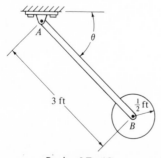

Prob. 17-42

17-43. The 20-lb disk at A rolls on the horizontal surface without slipping. Links AB and BC weigh 10 and 5 lb, respectively. If link BC has a counterclockwise angular velocity of 5 rad/sec in the position shown, determine the angular velocity of BC when it becomes horizontal.

Prob. 17-43

17-44. The 20-lb disk has a radius of gyration of $k_O = 4$ in. If the 50-lb block is suddenly applied to the cable which is attached to the disk as shown in the figure, determine the angular velocity of the disk after the block has fallen 1 ft. Assume that the disk rolls without slipping. Neglect the mass of the pulleys at A and B. The spring is initially unstretched.

Prob. 17-44

17-45. The trolley consists of two 5-lb cylinders A and B and *two* 2-lb links. If the trolley is released from rest in the position shown, determine its speed when the links become horizontal. Assume that the coefficient of friction at the surface is sufficient to prevent slipping.

Prob. 17–45

17–46. The rod weighs 3 lb/ft and is suspended from parallel cables at A and B. If the center of the rod has an angular velocity of 2 rad/sec about the zz axis at the instant shown, determine how high the center of mass rises at the instant when the rod stops swinging.

Prob. 17–46

17–47. The 10-lb plate shown is welded to a shaft which is supported horizontally by two smooth bearings at A and B. If the plate is released from rest from the position shown, determine the angular velocity of the shaft when the plate has rotated 180°. What are the vertical reactions at the bearings at this instant?

Prob. 17–47

17–48. The disk A is pinned at O and weighs 15 lb. A 1-ft rod weighing 2 lb and a 1-ft-diameter sphere weighing 10 lb are welded to the disk, as shown. The spring is originally stretched 1 ft. If the cord BC is cut, determine the velocity of point B on the sphere when the rod swings down to the vertical position.

Prob. 17–48

Kinetics of a Particle: Impulse and Momentum

18–1. Principle of Linear Impulse and Momentum of a Particle

Application of Newton's second law of motion proves to be the most direct method for obtaining either the acceleration of a particle or the resultant force acting on the particle during an instant of time. The principle of work and energy generally applies to problems in which the forces acting on the particle can be expressed in terms of displacements. For problems in which the *forces* can be expressed as functions of *time*, the principle of *impulse and momentum* often provides a direct solution. This method also provides the most suitable analysis for problems involving *impact*.

The principle of impulse and momentum is obtained from Newton's second law of motion by integrating this equation with respect to time. If a particle of mass m is subjected to several forces $\Sigma\mathbf{F}$, we can write Newton's second law of motion as

$$\Sigma\mathbf{F} = m\mathbf{a} = m\frac{d\mathbf{v}}{dt}$$

where $\mathbf{a}$ and $\mathbf{v}$ indicate the instantaneous acceleration and velocity of the particle. Since m is constant,

$$\Sigma \mathbf{F} = \frac{d(m\mathbf{v})}{dt} \qquad (18\text{--}1)$$

The vector $\mathbf{L} = m\mathbf{v}$ in this equation is defined as the *linear momentum* of the particle. Since m is a scalar, the linear momentum vector acts in the same direction as $\mathbf{v}$, and its magnitude mv has units of force-time, e.g., lb-sec or kg_f-sec. Equation 18–1 is an alternative formulation of Newton's second law of motion. This equation states that the resultant force acting on a particle equals the time rate of change of the linear momentum of the particle. Rearranging the terms in this equation, we have

$$\Sigma \mathbf{F}\, dt = d(m\mathbf{v})$$

Assuming that at $t = t_1$, $\mathbf{v} = \mathbf{v}_1$, and at $t = t_2$, $\mathbf{v} = \mathbf{v}_2$; then, integrating the left side of the equation between the limits t_1 and t_2 and integrating the right side from $\mathbf{v}_1$ to $\mathbf{v}_2$ yields

$$\Sigma \int_{t_1}^{t_2} \mathbf{F}\, dt = m\mathbf{v}_2 - m\mathbf{v}_1$$

The integral in this equation $\left(\int_{t_1}^{t_2} \mathbf{F}\, dt \right)$ is defined as the *linear impulse* of force $\mathbf{F}$. The *impulse* measures the effect of a force during the *time* the force acts. On the other hand, *work* measures the effect of a force by its relation to the *displacement* through which the force acts. Unlike work, which is a scalar, impulse is a vector quantity. The impulse vector acts in the same direction as the force, and its magnitude has units of force-time, e.g., lb-sec or kg_f-sec, the same units used to measure linear momentum. Provided all the forces acting on the particle can be expressed as functions of time, the impulse given to the particle may be determined by direct evaluation of the integral. If a force $\mathbf{F}$ acts in a *constant direction* during the time t_1 to t_2, the impulse acts in the direction of the force, and its magnitude can be represented by the area under the curve of force versus time, Fig. 18–1a. For example, when a force $\mathbf{F}_c$ acting on the particle is *constant* during the time interval $t = t_1$ to $t = t_2$, the resulting impulse becomes $\mathbf{I} = \int_{t_1}^{t_2} \mathbf{F}_c\, dt = \mathbf{F}_c(t_2 - t_1)$. The magnitude of this impulse is represented by the area under the curve shown in Fig. 18–1b.

Writing the previous equation in a different form we have

$$m\mathbf{v}_1 + \Sigma \int_{t_1}^{t_2} \mathbf{F}\, dt = m\mathbf{v}_2 \qquad (18\text{--}2)$$

This equation is referred to as the *principle of linear impulse and*

(a)

(b)

Fig. 18–1

Fig. 18-2

momentum for a particle. The principle states that the initial momentum of a particle plus the vector sum of all the impulses given to the particle during the time range t_1 to t_2 is equivalent to the final momentum of the particle. It should be noted that, in general, the particle's final momentum will be *different* from the particle's original momentum. The final momentum is determined by *vector addition* in accordance with Eq. 18-2. See Fig. 18-2. Since the principle of impulse and momentum is derived from Newton's second law of motion, both the initial and final momentum must be determined relative to an inertial frame of reference.

Provided the velocities v_1 and v_2 and the forces ΣF can be expressed by their Cartesian vector components, Eq. 18-2 may be written in the form

$$m(v_x)_1\mathbf{i} + m(v_y)_1\mathbf{j} + m(v_z)_1\mathbf{k} + \left(\Sigma \int_{t_1}^{t_2} F_x \, dt\right)\mathbf{i} + \left(\Sigma \int_{t_1}^{t_2} F_y \, dt\right)\mathbf{j}$$

$$+ \left(\Sigma \int_{t_1}^{t_2} F_z \, dt\right)\mathbf{k} = m(v_x)_2\mathbf{i} + m(v_y)_2\mathbf{j} + m(v_z)_2\mathbf{k}$$

Equating the respective **i, j,** and **k** components, we obtain the three scalar equations

$$m(v_x)_1 + \Sigma \int_{t_1}^{t_2} F_x \, dt = m(v_x)_2$$

$$m(v_y)_1 + \Sigma \int_{t_1}^{t_2} F_y \, dt = m(v_y)_2 \qquad (18\text{-}3)$$

$$m(v_z)_1 + \Sigma \int_{t_1}^{t_2} F_z \, dt = m(v_z)_2$$

These equations represent the principle of impulse and momentum for the particle in the x, y, and z directions, respectively.

In Sec. 14-4 it was shown that for a rigid body subjected to pure *translation,* the resultant force of all the forces acting on the body passes through the center of mass of the body. Thus, the translational motion

of the body may be analyzed as though it were a particle. By the same reasoning, the impulses acting on a translating rigid body also pass through the body's center of mass, and therefore the principle of impulse and momentum applies as well to a *translating rigid body*.

18-2. Method of Problem Solution

When applying the principle of impulse and momentum to the solution of dynamics problems, it is sometimes convenient to include a set of three diagrams along with the computations. Each of these diagrams corresponds to one of the three terms in Eq. 18-2. The *momentum-vector diagrams* are simply outlined shapes of the particle which indicate the correct direction and magnitude of the particle's momentum both before $(t = t_1)$ and after $(t = t_2)$ the impulse is applied. Like the velocity, the linear momentum always acts *tangent* to the path of motion. The magnitude and direction of each of the impulses given to the particle may easily be determined from the *impulse-vector diagram*. Similar to the free-body diagram, the impulse-vector diagram is an outlined shape of the particle showing *all* the *impulses* which act on the particle, when the particle is located at some intermediate point along its path. For the method of work and energy, it was necessary to consider *only* those force components which *move* in the direction of displacement. When applying the principle of impulse and momentum, however, *all forces* acting on the particle must be considered. This is necessary because there are cases in which an impulse may be exerted by a force which does no work on the particle. If the force $\mathbf{F}_c$ acting on the particle is *constant* for the time interval $(t_2 - t_1)$, the impulse applied to the particle is $\mathbf{F}_c(t_2 - t_1)$, acting in the same direction as $\mathbf{F}_c$. When the force direction or magnitude varies, the impulse must be determined by integration, in which case the impulse is indicated as $\int_{t_1}^{t_2} \mathbf{F} \, dt$.

As an example to illustrate the correct way of establishing the three diagrams, consider the particle shown in Fig. 18-3a, which has a weight W and moves along the smooth path. When the particle is at point A on the path, its velocity is $\mathbf{v}_A$, and it is acted upon by a constant horizontal force $\mathbf{F}$ for t seconds until it arrives at point B, where its velocity is $\mathbf{v}_B$. The correct momentum- and impulse-vector diagrams are shown in Fig. 18-3b. The momentum vectors act tangent to the path of motion at points A and B. The normal force acting on the particle does no work; however, it *does* create an impulse to the particle. The normal force $\mathbf{N}$ is a function of time, i.e., $\mathbf{N} = \mathbf{N}(t)$, since both the direction and magnitude of $\mathbf{N}$ vary during the time t the particle is in motion. Thus, the impulse given to

(a)

(b)

Fig. 18-3

the particle during this time is $\int_0^t \mathbf{N}(t)\,dt$. Forces $\mathbf{W}$ and $\mathbf{F}$ are constant during the motion, and therefore they create impulses of $\mathbf{F}t$ and $\mathbf{W}t$, as shown. Having established the three diagrams, we may apply Eq. 18–3 in a very direct manner to obtain the solution, since vector components can be determined from each of the diagrams. If motion is occurring in the xy plane, the third of Eqs. 18–3 is automatically satisfied. Thus, one may solve for at most two unknown scalar values on the diagrams shown in Fig. 18–3b. Of course if one of the unknown vectors shown on these diagrams is assumed acting with the wrong sense of direction, the resulting answer will be negative when Eqs. 18–3 are used for the solution.

Since there are many ways of applying the theory of dynamics to solve problems, a procedure should first be decided upon before attempting to solve a particular problem. For some problems, a combination of Newton's second law of motion and one of its two integrated forms, either the principle of work and energy or the principle of impulse and momentum, will yield the most direct solution. In particular, the following examples illustrate the method for applying the principle of impulse and momentum to the solution of problems.

Example 18–1

The block shown in Fig. 18–4a weighs 10 lb and is acted upon by a force having a variable magnitude $F = 0.2t$, where F is in pounds and

Fig. 18–4

t is in seconds. Compute the velocity of the block 4 sec after $\mathbf{F}$ has been applied. The initial velocity of the block is 1 ft/sec down the plane, and the coefficient of kinetic friction between the block and the plane is $\mu_k = 0.3$.

Solution

The momentum- and impulse-vector diagrams of the block are shown in Fig. 18–4b. Why are these diagrams drawn? Since the force $\mathbf{F}$ varies with time, the impulse created by this force is determined by integrating over the 4 sec time interval. The weight, normal force, and frictional force (which acts opposite to the direction of motion) are all *constant* so that the impulse created by each of these forces is simply the magnitude of the force times the 4 sec time interval.

Applying Eqs. 18–3 with respect to the xy axis shown gives

$$(+\nearrow) \qquad\qquad m(v_x)_1 + \Sigma \int_{t_1}^{t_2} F_x \, dt = m(v_x)_2$$

$$-\frac{10}{32.2}(1) - \int_0^4 0.2t \, dt + 0.3N(4) - 10(4) \sin 30° = -\frac{10}{32.2}(v_x)_2$$

$$-0.311 - 1.6 + 1.2N - 20 = -0.311(v_x)_2$$

or

$$(v_x)_2 = 70.6 - 3.86N \qquad\qquad (1)$$

Also,

$$(+\nwarrow) \qquad\qquad m(v_y)_1 + \Sigma \int_{t_1}^{t_2} F_y \, dt = m(v_y)_2$$

$$0 + N(4) - 10(4) \cos 30° = 0$$

Except for the constant multiple 4, this equation is representative of the summation of forces in the y direction. This is so since *no motion* occurs in this direction and all the force components acting are *constant*. Solving for N yields

$$N = 8.66 \text{ lb}$$

Substituting N into Eq. (1) and solving for $(v_x)_2$, we have

$$(v_x)_2 = 37.2 \text{ ft/sec} \qquad\qquad Ans.$$

Example 18–2

The 100-lb crate shown in Fig. 18–5a is moving to the right on a smooth

(a) (b)

Fig. 18–5

surface with a velocity of 3 ft/sec. Suddenly a force $F = 20$ lb, acting at an angle of $\theta = 45°$, is applied to the crate during a 10-sec time interval. Determine the velocity of the crate after the time period.

Solution

The momentum- and impulse-vector diagrams for the crate are shown in Fig. 18–5b. It has been assumed that during and after the sudden loading the crate remains on the path. Also, it is assumed that the crate continues to move to the right with a velocity $(v_x)_2$ after the time interval. The principle of impulse and momentum may be applied using the scalar Eqs. 18–3 with respect to the xy axes shown in the figure.

$(\xrightarrow{+})$
$$m(v_x)_1 + \Sigma \int_{t_1}^{t_2} F_x \, dt = m(v_x)_2$$

$$\frac{100}{32.2}(3) - 20(10) \cos 45° = \frac{100}{32.2}(v_x)_2 \tag{1}$$

$(+\uparrow)$
$$m(v_y)_1 + \Sigma \int_{t_1}^{t_2} F_y \, dt = m(v_y)_2$$

$$0 + N(10) - 100(10) + 20(10) \sin 45° = 0 \tag{2}$$

Solving Eq. (1) for $(v_x)_2$ yields

$$(v_x)_2 = -42.5 \text{ ft/sec} \qquad \textit{Ans.}$$

The negative sign indicates that the crate moves to the *left* after the 10-sec interval, rather than to the right as originally assumed. Equation (2) is representative of a force summation in the y direction. Why? Solving for N, we have

$$N = 85.9 \text{ lb}$$

Since N is positive, the crate *remains on the surface of the plane,* as it was originally assumed.

Problems

18-1. A 10-lb cannon ball is fired in the vertical direction with an initial velocity of 1,500 ft/sec. Determine how long it takes before its velocity is reduced to zero.

18-2. The particle P, weighing 3 lb, is acted upon by two forces. If the particle originally has a velocity of $v_1 = \{3i + j + 6k\}$ ft/sec, determine the speed after 10 sec.

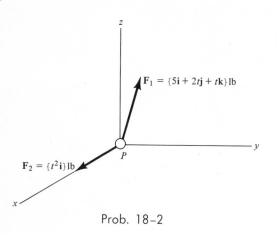

Prob. 18-2

18-3. The 2-lb block has an initial velocity of $v_1 = 10$ ft/sec acting in the direction shown. If a force of $F = \{5i + 2j\}$ lb acts on the block for $t = 5$ sec, determine the final velocity of the block. The xy plane is smooth.

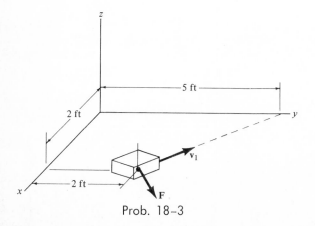

Prob. 18-3

18-4. A 5-lb coaster is given an initial velocity of 100 ft/sec up a 45° slope. Determine the time it will take to travel up the slope before it stops.

18-5. The 10-lb shell is fired from a cannon having an elevation of 100 ft. If an average force of 300 lb is exerted on the shell for 1 sec while it is being fired, determine the distance s measured from the point on the wall to where the shell strikes the ground.

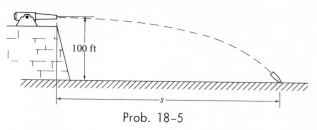

Prob. 18-5

18-6. The 60-lb hoisting block is operated by a motor M. Starting from rest, the supporting cable can be wound up by the motor so that it travels with a maximum velocity of 3 ft/sec in 5 sec. If the cable does not slip on the pulleys, determine the average force exerted on the cables during this time. Neglect the weights of each of the pulleys.

Prob. 18-6

18-7. Using the principle of impulse and momentum, prove that when the motor M in Prob. 18-6 is winding up the cable of the hoisting block at a constant velocity $v = 3$ ft/sec, the tension in the cable is 20 lb.

18-8. The 20-lb cabinet is subjected to the variable force $F = 3 + 2t$, where F is in pounds and t in seconds. If the cabinet is initially moving down the plane with a speed of 6 ft/sec, determine how long it will take the force to bring the cabinet to rest. **F** always acts parallel to the plane.

Prob. 18-8

18-9. A 20-lb weight slides down a 30° inclined plane with an initial velocity of 2 ft/sec. Determine the velocity of the weight in 10 sec if the coefficient of kinetic friction for the plane is $\mu_k = 0.25$.

18-10. A 50-lb block is moving up a 20° inclined plane with an initial velocity of 3 ft/sec. If the coefficient of kinetic friction between the block and the plane is $\mu_k = 0.1$, determine how long a 100-lb horizontal force must act on the block in order to increase the upward velocity of the block to 10 ft/sec. How far up the plane will the block travel while the force is acting?

18-11. A 10-lb block is originally at rest on a smooth horizontal surface when a horizontal force **F** is suddenly applied to the block. If the force varies with time as shown, determine the speed of the block after 10 sec.

Prob. 18-11

18-12. A 30-ton engine exerts a constant force of 20 tons on a train having three cars which have a total weight of 220 tons. If the rolling resistance is 10 lb per ton for both the engine and cars, determine how long it takes to increase the speed of the train from 20 to 30 mph. What is the driving force exerted by the engine wheels on the tracks?

18-13. The 20-lb block is moving to the right on a horizontal plane with a velocity of 4 ft/sec when it is acted upon by the horizontal force $F = 2e^{-t}$, where F is in pounds and t is in seconds. Determine the velocity of the block after the force acts for 2 sec. The force acts in the constant direction shown in the figure.

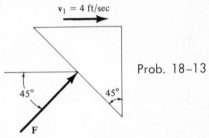

Prob. 18-13

18-14. A 20-lb block is moving with a velocity of $v = 10$ ft/sec, leftward, after it has been acted upon for 15 sec by a force **F** which varies in the manner shown. Determine the initial velocity of the block before the force was applied.

Prob. 18-14

755

18–15. The 10-lb slider block is moving to the right with a velocity of 10 ft/sec when it is acted upon by the forces F_1 and F_2. These loadings vary in the manner shown in the figure. Determine the velocity of the block after 6 sec. Neglect friction and the weight of the pulleys.

18–16. As a 5-lb sphere falls from rest through a liquid, the drag force exerted on the sphere is $F = 0.2v$, where v is the velocity in ft/sec and F is given in pounds. Using the differential form of the impulse and momentum principle ($F\,dt = m\,dv$), determine the time required for the sphere to attain one fourth its terminal velocity.

18–17. Work Prob. 18–16 assuming that the drag force is $F = 0.05v^2$, where v is the velocity in ft/sec and F is in pounds.

18–18. The 20-lb block is at rest on the inclined plane where the coefficient of friction between the block and the plane is $\mu = 0.3$. If the block starts from rest and is acted upon by a force $F = 3t$, where F is in pounds and t is in seconds, determine the time required for the block to attain a velocity of 5 ft/sec up the plane. The force always acts parallel to the plane. The stop block s prevents the block from initially moving down the plane.

Prob. 18–15

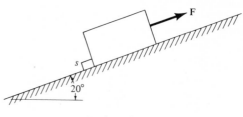

Prob. 18–18

18–3. Principle of Linear Impulse and Momentum for a System of Particles

Up to this point, the principle of impulse and momentum has been discussed only for a single particle. In the case of a system of n particles, the principle of impulse and momentum may be obtained by using Eq. 14–4. This equation may be rewritten as

$$\Sigma \mathbf{F}_i = \Sigma m_i \frac{d\mathbf{v}_i}{dt}$$

The terms are summed (vectorially) from $i = 1$ to $i = n$. In particular, the term on the left side represents the sum of external forces acting on

all the particles. On the right side $\mathbf{v}_i$ defines the instantaneous velocity of the arbitrary ith particle. Multiplying both sides of this equation by dt and integrating between the limits $t = t_1$, $\mathbf{v}_i = (\mathbf{v}_i)_1$, and $t = t_2$, $\mathbf{v}_i = (\mathbf{v}_i)_2$, yields

$$\Sigma \int_{t_1}^{t_2} \mathbf{F}_i \, dt = \Sigma m_i(\mathbf{v}_i)_2 - \Sigma m_i(\mathbf{v}_i)_1$$

or

$$\Sigma m_i(\mathbf{v}_i)_1 + \Sigma \int_{t_1}^{t_2} \mathbf{F}_i \, dt = \Sigma m_i(\mathbf{v}_i)_2 \qquad (18\text{--}4)$$

This equation represents the principle of linear impulse and momentum for a system of particles. It states that the initial linear momentum of the system plus (vectorially) the impulses of all the external forces acting on the system during the time period t_1 to t_2 is equal to the final linear momentum of the system.

The mass center of the system of n particles is defined by

$$m\mathbf{r}_G = \Sigma m_i \mathbf{r}_i$$

where $m = \Sigma m_i$ is the total mass of all the particles. Taking the time derivative of this equation, we have

$$m\mathbf{v}_G = \Sigma m_i \mathbf{v}_i$$

Thus, the total linear momentum of the system of particles is equivalent to the linear momentum of a particle of mass m and moving with a velocity of the mass center. Substituting this equation into Eq. 18–4, we have

$$m(\mathbf{v}_G)_1 + \Sigma \int_{t_1}^{t_2} \mathbf{F}_i \, dt = m(\mathbf{v}_G)_2$$

In words, the initial linear momentum of an aggregate particle of mass m, having an initial velocity of the center of mass of the system of particles, plus (vectorially) the external impulses acting on the system during the time interval t_1 to t_2 is equal to the final linear momentum of the aggregate particle.

18–4. Conservation of Linear Momentum

When the sum of the external impulses acting on a system of particles is zero, the principle of impulse and momentum reduces to a simplified form:

$$\Sigma m_i(\mathbf{v}_i)_1 = \Sigma m_i(\mathbf{v}_i)_2 \qquad (18\text{--}5)$$

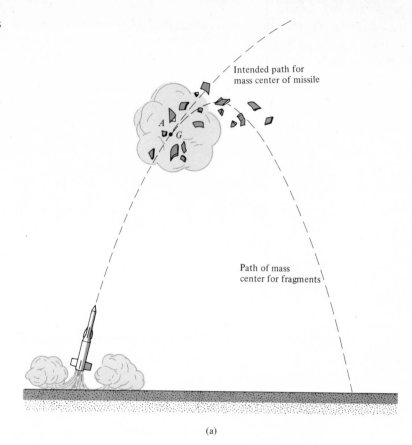

Intended path for
mass center of missile

Path of mass
center for fragments

(a)

$m\mathbf{v}_1$ $+$ $\mathbf{F}\Delta t$ $=$ $\Delta\ m\mathbf{v}_2$

$\mathbf{w}\ \Delta t \approx 0$

(b)

Fig. 18–6

Or, by using the definition of the center of mass,

$$(\mathbf{v}_G)_1 = (\mathbf{v}_G)_2$$

Equation 18–5 states that the total linear momentum for a system of particles remains *constant* throughout the time period t_1 to t_2 during which the external impulse acting on the system is zero. This principle is known as the *conservation of linear momentum*. It implies that the velocity $\mathbf{v}_G$ of the mass center for the system of particles remains constant.

There are essentially two cases for which the external impulse acting on a system of particles is zero, and yet the velocity (not the momentum) of all the particles may change over a period of time. Both cases may be explained by a simple example. Consider the missile shown in Fig. 18–6a which has an intended trajectory shown by the dotted path. When it reaches point A, it suddenly blows up. The fragments (or particles) of the missile travel out in all directions, as shown in the figure. Initially (before the explosion) the velocity of the system of fragments (or particles) was the same, since the missile was intact. Just after the explosion, however, the velocity of each fragment changed. Even so, the momentum of the system remains constant from *just before* to *just after* the explosion. There are two reasons for this. First, the fragments were blown apart *only* by impulses which occurred from equal but opposite *internal forces* acting on the system of particles. The total external impulse given to the system at the time of explosion is therefore zero. Second, the time for explosion was *very short*, so that any external force (such as gravity) which acts on the system and creates an impulse during the explosion time is very small and may be neglected in comparison to the internal explosive forces given to the system. For these reasons, the mass center G of the fragments will continue to follow the intended path of the missile for a short time, Fig. 18–6a. As the time after the explosion becomes *longer*, the impulse created by the gravity force (external force) becomes important and the path of the mass center of the fragments changes, as shown in the figure. This new path can be determined using Eq. 18–4.

If a study is made of the motion of just *one* of the fragments of the missile, we must use Eq. 18–2. For example, the fragment shown in Fig. 18–6b has an initial momentum $m\mathbf{v}_1$ just before the explosion. Here m is the mass of the fragment and $\mathbf{v}_1$ is the velocity of the missile. When the explosion occurs, the (external) explosive impulse $\mathbf{F} \, \Delta t$ acts on the fragment. During the same instant there is a much smaller (external) gravitational impulse $\mathbf{w} \, \Delta t$ also acting as shown in the figure. Since $F \gg w$ and the time considered, Δt, is short, the gravitational impulse may be neglected. The resulting momentum becomes $m\mathbf{v}_2$, which can be determined from Eq. 18–2. That is,

$$m\mathbf{v}_1 + \Sigma \int_{t_1}^{t_2} \mathbf{F} \, dt = m\mathbf{v}_2$$

becomes

$$m\mathbf{v}_1 + \mathbf{F} \, \Delta t = m\mathbf{v}_2$$

A *force* which is very large and acts for a short time such that it produces a significant change in momentum is called an *impulsive force*. In the previous example, the force of the explosion, $\mathbf{F}$, is an impulsive force. The impulse created by this force is $\mathbf{F} \, \Delta t$. By comparison, the weight

Recoil spring $k = 200$ lb/ft

(a)

$m_c(v_c)_1 = 0$ $m_s(v_s)_1 = 0$ $m_c(v_c)_2$ $m_s(v_s)_2$

$=$

(b)

Fig. 18-7

w of the fragment is a *nonimpulsive force*. This force creates no significant change in the momentum of the fragment during the short time Δt, so that it can be neglected in the analysis. In general, the weight of a body, the force of a spring, or any force which is *known to be small* compared to the impulsive force can be classified as a nonimpulsive force. *Reactive forces* which are *not known* cannot be classified as nonimpulsive, and must therefore be accounted for in the analysis.

Example 18-3

The 1,000-lb cannon, shown in Fig. 18-7a, fires a 10-lb shell with muzzle velocity of 1,500 ft/sec. If the cannon support is firmly fixed to the ground and the recoil of the cannon is absorbed by two springs, each having a stiffness of $k = 200$ lb/ft, determine the maximum deflection of the springs after the cannon is fired.

Solution

When the shell is fired, the cannon exerts an *impulsive force* on the shell which gives the shell a forward momentum. On the other hand, the shell exerts an equal but opposite *impulsive force* on the cannon. The recoil springs which are attached to both the cannon support and the cannon also exert an impulse to the cannon. However, for a small time Δt, just before to just after the firing, the recoil of the cannon is *very small*, so that the force exerted by the springs on the cannon is *nonimpulsive*.* Thus, if we consider the cannon and shell to be a single system,

*If the cannon was *firmly fixed* to its support (no springs) the reactive forces of the support on the cannon must be considered as impulsive forces in the analysis since the support is rigid.

momentum for the system is *conserved* since the impulsive forces are internal to the system. The impulse-vector diagram used to obtain the impulse term is therefore not needed. The momentum-vector diagrams of the system (cannon and shell) are shown in Fig. 18–7b. Applying Eq. 18–5 gives

$$(\xrightarrow{+}) \qquad m_c(v_c)_1 + m_s(v_s)_1 = m_c(v_c)_2 + m_s(v_s)_2$$

$$0 + 0 = -\frac{1,000}{32.2}(v_c)_2 + \frac{10}{32.2}(1,500)$$

The negative sign is included in the first term on the right-hand side, since the velocity of the cannon, $(v_c)_2$, is assumed to be opposite to the velocity of the shell after the cannon is fired, Fig. 18–7b. Solving for $(v_c)_2$, we have

$$(v_c)_2 = 15 \text{ ft/sec}$$

Since the cannon support is fixed to the ground, the kinetic energy of the cannon is transferred into stored potential energy in the two springs. Applying the conservation of energy theorem to the cannon, we write

$$(T_c)_2 + (V_c)_2 = (T_c)_3 + (V_c)_3$$

$$\frac{1}{2}m_c(v_c)_2^2 + 0 = 0 + 2\left[\frac{1}{2}kx^2\right]$$

$$\frac{1}{2}\left(\frac{1,000}{32.2}\right)(15)^2 + 0 = 0 + 2\left[\frac{1}{2}(200)x^2\right]$$

Solving for the spring deflection yields

$$x = 4.18 \text{ ft} \qquad\qquad Ans.$$

Example 18–4

The 160,000-lb tugboat T, shown in Fig. 18–8a, is used to pull the 20,000-lb barge B with a rope R. A 5,000-lb crate C rests on top of the barge. The coefficient of friction between the barge and the crate is $\mu = 0.3$. If the initial velocity of the tugboat is $(v_T)_1 = 10$ ft/sec while the rope is slack, determine the velocity of the tugboat *directly after* towing occurs if (a) the rope is inextensible, and (b) the rope is extensible so that towing takes place within a period of 1 sec. Initially the velocity of the barge is zero. Neglect the frictional effects of the water.

Solution

Part (a). When the rope is inextensible, the impulse created between the tugboat and the barge is *instantaneous*. As a result, the barge is pulled (or jerked) from under the crate so fast that the crate has effectively *zero velocity* and hence slips on the surface of the barge. The momentum-vector diagrams for the entire system (tugboat, crate, and barge) are

10 ft/sec

(a)

(b)

(c)

(d)

(e)

Fig. 18–8

shown in Fig. 18–8b. The impulses created by the rope R and the frictional force between the crate and the barge are internal to the system and therefore momentum of the system is conserved during the instant of towing. Applying Eq. 18–5 and noting that $(v_B)_2 = (v_T)_2$ we have

$$(\xleftrightarrow{\pm}) m_T(v_T)_1 + m_B(v_B)_1 + m_C(v_C)_1 = m_T(v_T)_2 + m_B(v_B)_2 + m_C(v_C)_2$$

$$\frac{160,000}{32.2}(10) + 0 + 0 = \frac{160,000}{32.2}(v_T)_2 + \frac{20,000}{32.2}(v_T)_2 + 0$$

Solving for the velocity of the tugboat yields

$$(v_T)_2 = 8.89 \text{ ft/sec} \qquad\qquad Ans.$$

This value represents the velocity of the tugboat *just after* towing. At this instant, the crate has zero velocity as explained previously. In time the

frictional force developed between the crate and the barge creates an impulse on the crate to give the crate the *same velocity* as the barge, i.e., when $v_C = v_B$ no sliding occurs.

Part (b). When the rope is allowed to stretch, the impulse given to the barge occurs within a finite time interval, in this case $t = 1$ sec. As a result, the crate is given an impulse via the frictional force *before* the tugboat attains its final velocity. We will *assume* that the frictional impulse (sliding) occurs for the *entire* 1 sec, during which time the tugboat gives the towing impulse to the barge. If this is the case, the momentum- and impulse-vector diagrams for the crate are shown in Fig. 18–8c. Applying Eq. 18–2 in the horizontal direction to the crate gives

$(\xrightarrow{+})$
$$m(v_{Cx})_1 + \Sigma \int_{t_1}^{t_2} F_x \, dt = m(v_{Cx})_2$$

$$0 + 0.3N(1) = \frac{5,000}{32.2}(v_C)_2 \qquad (1)$$

In the same manner, application of Eq. 18–2 in the vertical direction yields

$(+\uparrow)$
$$m(v_{Cy})_1 + \Sigma \int_{t_1}^{t_2} F_y \, dt = m(v_{Cy})_2$$

$$0 + N(1) - 5,000(1) = 0$$

or

$$N = 5,000 \text{ lb}$$

Substituting for N in Eq. (1) and solving for $(v_C)_2$ gives

$$(v_C)_2 = 9.66 \text{ ft/sec}$$

Knowing the final velocity of the crate, the *conservation of momentum principle* may now be applied to the *entire system* both before and after the *internal impulses* have been applied. (Note again that the frictional impulse given to the crate by the barge is equal and opposite to the friction impulse given to the barge by the crate.) The momentum-vector diagrams for the system are shown in Fig. 18–8d. Applying Eq. 18–5, noting that $(v_B)_2 = (v_T)_2$ after 1 sec, we have

$(\xrightarrow{+})$ $m_T(v_T)_1 + m_C(v_C)_1 + m_B(v_B)_1 = m_T(v_T)_2 + m_C(v_C)_2 + m_B(v_B)_2$

$$\frac{160,000}{32.2}(10) + 0 + 0 = \frac{160,000}{32.2}(v_T)_2 + \frac{5,000}{32.2}(9.66) + \frac{20,000}{32.2}(v_T)_2$$

Solving for $(v_T)_2$ gives

$$(v_T)_2 = 8.62 \text{ ft/sec}$$

Since $(v_T)_2 < (v_C)_2$, the assumption that the crate was sliding on the

surface of the barge for the entire 1 sec is *not* correct. (How could the crate possibly have a forward velocity of 9.66 ft/sec when the barge has a forward velocity of only 8.62 ft/sec!) We must therefore rework the problem, assuming that *before* the end of the 1-sec impulse the crate had *stopped sliding* on the surface of the barge, and therefore the barge, crate, and tugboat all move with the same final velocity, i.e., $(v_B)_2 = (v_C)_2 = (v_T)_2$. Once again all the impulses are internal, so that the conservation of momentum principle applies. The momentum-vector diagrams are shown in Fig. 18–8e. From the figure,

$$(\xleftrightarrow{}) \ m_T(v_T)_1 + m_C(v_C)_1 + m_B(v_B)_1 = m_T(v_T)_2 + m_C(v_C)_2 + m_B(v_B)_2$$

$$\frac{160{,}000}{32.2}(10) + 0 + 0 = \frac{160{,}000}{32.2}(v_T)_2 + \frac{5{,}000}{32.2}(v_T)_2 + \frac{20{,}000}{32.2}(v_T)_2$$

Solving for $(v_T)_2$ yields

$$(v_T)_2 = 8.65 \ \text{ft/sec} \qquad\qquad Ans.$$

Problems

18-19. In 5 sec, the 10-lb block A attains a velocity of 1 ft/sec after starting from rest. Determine the tension in the cord and the coefficient of friction between block A and the horizontal plane. Neglect the mass of the pulley.

18-20. The two blocks are released from rest. If the coefficient of friction between the plane and the blocks is $\mu = 0.15$, determine the velocity of the blocks after 4 sec. What is the tension developed in the cord? Neglect the mass of the pulley.

Prob. 18–19

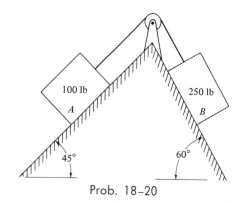

Prob. 18–20

18-21. Determine the velocity of each block 10 sec after the blocks are released from rest. Neglect the mass of the pulleys.

Prob. 18-21

18-22. Determine the tension in the cord of the blocks and pulley system in Prob. 18–21, 10 sec after the blocks are released from rest. Neglect the mass of the pulleys.

18-23. The two crates each weigh 20 lb. If the coefficient of friction between the plane and the crates is $\mu = 0.1$, determine the velocity of both crates after 5 sec. The crates are originally moving to the right with a speed of 2 ft/sec. The towing force acts parallel to the 20° plane and has a magnitude of $F = 20 + 2t + 5e^{-t}$, where t is in seconds and F is in pounds. Neglect the mass of the pulley.

Prob. 18-23

18-24. Each car of a three-car freight train weighs 30 tons. The train is pulled by a 50-ton engine. The rolling resistance of the engine and cars is 15 lb per ton. If it takes 3 min for the train to reach a speed of 40 mph, determine the force developed at the coupling between the engine and the train. The train travels in a horizontal direction with an initial speed of 5 mph.

18-25. A railroad car weighs 30,000 lb and is traveling horizontally at 30 ft/sec. At the same time another car weighing 10,000 lb is traveling 5 ft/sec in the opposite direction. If the cars meet and become coupled together, determine the speed of both cars after coupling. Compute the total kinetic energy both before and after coupling has occurred, and explain qualitatively what happened to the difference of these energies.

18-26. The 5-lb wooden block A is released from rest when $\theta = 45°$. When $\theta = 90°$, a 0.08-lb bullet B traveling at 2,000 ft/sec strikes the block. If the bullet becomes embedded in the block at this instant, determine the velocity of the bullet and block after impact.

Prob. 18-26

18-27. A rifle weighs 7 lb and has a barrel which is 2 ft long. If a bullet weighing 0.02 lb is fired from the gun with a muzzle velocity of 500 ft/sec, determine the recoil velocity of the gun.

18-28. A bullet having a weight of 0.025 lb and traveling with a horizontal velocity of 100 ft/sec is fired into a block weighing 2 lb. If the block is resting on a smooth plane, determine the speed of the block after impact if the bullet penetrates the block 3 in. What is the average resistance force of the bullet on the block?

18-29. The barge B weighs 30,000 lb and supports an automobile weighing 3,000 lb. If the barge is not tied to the pier P and someone moves the automobile to the other side of the barge for unloading, determine how

Prob. 18-29

far the barge moves away from the pier. Neglect the resistance of the water.

18-30. The barge weighs 45,000 lb and supports two automobiles A and B, which weigh 4,000 and 3,000 lb, respectively. If the automobiles are driven toward each other with a constant speed of 6 ft/sec relative to the barge, determine the speed of the barge just before they collide. Provided the two automobiles suddenly stop just before colliding, determine how far the barge moves.

Prob. 18-30

18-31. The 10-lb model train car has a speed of $v = 10$ ft/sec. If it strikes a 5-lb plate P, determine the impulse exerted on the plate if (a) the springs are perfectly rigid ($k \to \infty$), and (b) the springs each have a stiffness $k = 20$ lb/in. Neglect the weight of the wheels.

Prob. 18-31

18-32. Block A, weighing 100 lb, is released from rest at the position shown and slides freely down the smooth circular ramp onto a 20-lb cart. If the coefficient of friction between the cart and the block is $\mu_k = 0.2$ and $R = 4$ ft, determine the final velocity of the cart once the block comes to rest on the cart.

Prob. 18-32

18-33. Determine the smallest radius R of the smooth ramp in Prob. 18-32 so that the block slides completely across the surface of the cart and off the other side. What is the final velocity of the cart, just after the block has left the surface of the cart?

18-34. The unloading ramp weighs 100 lb. It is not tied down but is allowed to roll freely. A 200-lb crate is released from rest at A and allowed to slide to point B on the ramp. Determine the speed of the ramp when the crate reaches point B. Assume that the surface of contact is smooth.

Prob. 18-34

18-35. Work Prob. 18-34 assuming that the crate attains a maximum speed of 6 ft/sec measured relative to the ramp when the crate reaches point B on the ramp. For this case the ramp has a rough surface.

18-36. The 10-lb piece of putty P falls freely from a machine 3 ft into a 2-lb crate C which rides down the smooth 30° incline. If the crate is released from rest and the putty sticks in the crate upon impact, determine (a) the height h so that when both the crate and putty are released at the same instant, they meet at A, and (b) the velocity of the crate just after the putty falls into it.

18-37. The wooden pendulum bob A weighs 2 lb and is given an initial velocity v_1 when it is at $\theta = 90°$. A 0.1-lb bullet strikes the bob from below when $\theta = 0°$ with an initial velocity of 1,600 ft/sec and becomes embedded in the bob in 0.2 sec after the initial contact. Determine the maximum initial velocity v_1 of the bob so that when the impact occurs, the tension in the cord OC is zero. Assume that the bullet creates a constant force on the bob during the impact.

Prob. 18-37

Prob. 18-36

18-5. Impact

Impact occurs when two bodies collide with each other. It is necessary that the collision occurs during a very *short* interval of time, causing relatively large forces to be exerted between the bodies. The striking of a hammer upon a nail, or a golf club and ball are common examples of impact loadings.

To develop the necessary equations which describe the motion of particles subjected to impact loadings, we will consider two particles A

Central impact

(a)

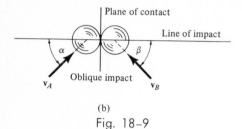

Oblique impact

(b)

Fig. 18-9

and B, Fig. 18–9, which have velocities $\mathbf{v}_A$ and $\mathbf{v}_B$, respectively. Collision occurs at the *plane of contact*. The line segment drawn perpendicular to this plane is called the *line of impact*. In general, there are two types of impact. *Central impact* occurs when the direction of motion of the mass centers of the two particles coincides with the line of impact, that is, there is a direct collision, Fig. 18–9a. When motion of one or both of the particles occurs at some angle with the line of impact, Fig. 18–9b, the impact is said to be *oblique impact*.

The principle of impulse and momentum provides the only practical means for analyzing the mechanics of impact. For simplicity in the analysis we will first consider the case involving the central impact of particles A and B, shown in Fig. 18–10a. These particles have mass m_A and m_B and initial velocities $(v_A)_1$ and $(v_B)_1$. Provided $(v_A)_1 > (v_B)_1$, collision will eventually occur. When this happens, due to the impact, both particles undergo a *period of deformation,* such that at maximum deformation they will have the same velocity $\mathbf{v}$, Fig. 18–10b. A *period of restitution* then occurs in which the particles will either return to their original shape or will remain somewhat permanently deformed. The "restitution" depends primarily upon the types of materials involved, and to a lesser extent, upon the size of the two particles as well as their relative velocities. After the end of the period of restitution, both particles will have final velocities $(\mathbf{v}_A)_2$ and $(\mathbf{v}_B)_2$, Fig. 18–10c.

The momentum- and impulse-vector diagrams for each phase of the collision are shown in Fig. 18–11. In particular, the deformation phase is shown in Figs. 18–11a, 18–11b, and 18–11c; and the restitution phase is shown in Figs. 18–11d, 18–11e, and 18–11f. It can be seen from the impulse-vector diagrams, Figs. 18–11b,e that the impulses created by forces for deformation ($\mathbf{P}$) and restitution ($\mathbf{R}$) are *equal* and *opposite* on each particle (a consequence of Newton's third law of motion). For this reason, the total linear momentum in the horizontal direction for the system of *both* particles is conserved; that is, the impulses are *internal* to the system. Hence, applying Eq. 18–5 and referring to the four momentum-vector diagrams* in Figs. 18–11a and 18–11f, we have

Fig. 18-10

*These diagrams represent the linear momentum conditions for the system shown in Figs. 18–10a and 18–10c.

768

Fig. 18-11

$(\xrightarrow{+})$ $\qquad$ $m_A(v_A)_1 + m_B(v_B)_1 = m_A(v_A)_2 + m_B(v_B)_2$ $\qquad$ (18-6)

There are two unknowns in this equation, $(v_A)_2$ and $(v_B)_2$.

A second equation relating $(v_A)_2$ and $(v_B)_2$ may be obtained by applying the principle of impulse and momentum to each particle *separately*. Referring to the momentum- and impulse-vector diagrams shown in Figs. 18-11a, 18-11b, and 18-11c, we obtain for the period of deformation

$(\xrightarrow{+})$ $\qquad$ $m_A(v_A)_1 - \int P \, dt = m_A(v)$ $\qquad$ (particle A) $\qquad$ (18-7)

and

$(\xrightarrow{+})$ $\qquad$ $m_B(v_B)_1 + \int P \, dt = m_B(v)$ $\qquad$ (particle B) $\qquad$ (18-8)

With reference to Figs. 18-11d, 18-11e, and 18-11f, for the period of restitution,

$(\xrightarrow{+})$ $\qquad$ $m_A(v) - \int R \, dt = m_A(v_A)_2$ $\qquad$ (particle A) $\qquad$ (18-9)

and

$(\xrightarrow{+})$ $\qquad$ $m_B(v) + \int R \, dt = m_B(v_B)_2$ $\qquad$ (particle B) $\qquad$ (18-10)

Combining Eqs. 18-7 and 18-9 yields

$$\frac{\int R \, dt}{\int P \, dt} = \frac{v - (v_A)_2}{(v_A)_1 - v} \qquad (18\text{-}11)$$

Likewise, from Eqs. 18–8 and 18–10, we have

$$\frac{\int R\,dt}{\int P\,dt} = \frac{(v_B)_2 - v}{v - (v_B)_1} \qquad (18\text{–}12)$$

We will now define the *coefficient of restitution, e,* as the ratio of the impulse of restitution to the impulse of deformation, i.e.,

$$e = \frac{\int R\,dt}{\int P\,dt} \qquad (18\text{–}13)$$

Since $\int P\,dt \geq \int R\,dt$, the value of e will vary from 0 to 1. Combining Eqs. 18–11 and 18–12 so as to eliminate the value of v and using the definition of e, gives

$$e = \frac{(v_B)_2 - (v_A)_2}{(v_A)_1 - (v_B)_1} = \frac{\text{Rel. vel. after impact}}{\text{Rel. vel. before impact}} \qquad (18\text{–}14)$$

In reference to Figs. 18–10c and 18–10a, the result indicates that the coefficient of restitution, e, is equal to the ratio of the *relative velocity* of separation for the particles *just after impact* to the *relative velocity* of separation of the particles *just before impact.*

Using Eq. 18–14, we therefore have a relatively simple means for experimental determination of the coefficient of restitution. In performing such experiments by measuring the relative velocities, it has been found that the factor e varies appreciably with impact velocity as well as with the size and shape of the colliding bodies. These differences occur because some of the initial kinetic energy of the bodies is transformed into heat energy as well as creating sound and elastic shock waves when the deformation occurs. For these reasons, the coefficient of restitution is reliable only when used under conditions which are close to those which were known to exist when measurements were made.

In particular, if the collision between the two particles is *perfectly elastic,* the impulse which deforms the particles $\left(\int P\,dt\right)$ is equal and opposite to the restitution impulse $\left(\int R\,dt\right)$. Hence, from Eq. 18–13, $e = 1$ for an elastic collision. Under these conditions it may be shown that the total kinetic energy of both particles before collision is equal to the total kinetic energy after collision. (See Prob.

18–40.) When $e = 0$, the impact is said to be *inelastic* or *plastic*. In this case there is no restitution impulse given to the particles and as a result, after collision occurs, both particles excessively deform and move together with a common velocity. In any case, provided the value of e is known, the final velocities $(v_A)_2$ and $(v_B)_2$ of the particles can be determined by solving Eqs. 18–6 and 18–14 simultaneously. In general therefore the solution of a problem involving the central impact of two particles requires writing the conservation of momentum of the system (Eq. 18–6) and using the coefficient of restitution to relate the relative velocities of the particles before and after the impact (Eq. 18–14). In applying these two equations here, it is recalled that particle B was located to the right of particle A and that both particles were initially moving to the right, Fig. 18–10. If, in application, this is not the case, a negative sign must account for particle motion to the left. (See Examples 18–5 and 18–6.)

When an *oblique impact* occurs, the particles move away from each other with velocities having *unknown directions* as well as *unknown magnitudes*. Therefore, four unknowns are present in the problem. As shown in Fig. 18–12a these unknowns may be represented as $(v_A)_2$, $(v_B)_2$, α_2, and β_2 [$(v_A)_1$, $(v_B)_1$, α_1, and β_1 are known]. If we establish the y axis along the plane of contact and the x axis along the line of central impact,

Fig. 18–12

the velocities may be resolved into components along the y axis and the x axis. If the particles are smooth (frictionless) then the impulsive forces occur *only in the x direction* since this direction is along the line of central impact, Fig. 18–12b. Hence, four independent equations can be written for the particles using the set of x and y velocity components. Specifically, for velocity components in the x direction, the relationship for the coefficient of restitution, Eq. 18–14, applies, and the x component of the total momentum for the system is conserved. Since *no impulse* is exerted on the particles in the y direction, that is, along the plane of contact, Fig. 18–12b, the y component of momentum for particle A is conserved and likewise, the y component of momentum for particle B is conserved. See Example 18–7.

Example 18–5

The bag A, having a weight of 5 lb, is released from rest at the position $\theta = 0°$, as shown in Fig. 18–13a. It strikes a 50-lb box B when $\theta = 90°$. If the coefficient of restitution between the bag and box is $e = 0.3$, determine the velocities of the bag and box just after the impact.

Solution

This problem involves central impact. Why? Before applying the mechanics of the impact however, we must first obtain the velocity of the bag just before it strikes the box. This can be done by applying the principle of work and energy to the bag, from $\theta = 0°$ to $\theta = 90°$. As shown in Fig. 18–13b, work is done only by the 5-lb weight, which moves through a vertical displacement of 3 ft. (The tension $\mathbf{T}$ does no work since this force is always perpendicular to the displacement.) Hence,

$$T_0 + \Sigma W_{0-1} = T_1$$

$$0 + 5(3) = \frac{1}{2}\left(\frac{5}{32.2}\right)(v_A)_1^2$$

Thus,

$$(v_A)_1 = 13.90 \text{ ft/sec}$$

The momentum diagrams of A and B, just before and just after the impact, are shown in Fig. 18–13c. (Why don't we need an impulse diagram?) Applying the conservation of momentum to the system, we have

$$m_B(v_B)_1 + m_A(v_A)_1 = m_B(v_B)_2 + m_A(v_A)_2$$

$$0 + \frac{5}{32.2}(13.90) = \frac{50}{32.2}(v_B)_2 + \frac{5}{32.2}(v_A)_2$$

or

(a)

(b)

(c)

Fig. 18–13

$$(v_A)_2 = 13.90 - 10(v_B)_2 \qquad (1)$$

Using the definition of the coefficient of restitution, in reference to Fig. 18–13c, gives

$$e = \frac{\text{Rel. vel. after impact}}{\text{Rel. vel. before impact}} = \frac{(v_B)_2 - (v_A)_2}{(v_A)_1 - (v_B)_1}$$

$$0.3 = \frac{(v_B)_2 - (v_A)_2}{13.90 - 0}$$

or

$$(v_A)_2 = (v_B)_2 - 4.17 \qquad (2)$$

Solving Eqs. (1) and (2) simultaneously, yields

$$(v_A)_2 = -2.53 \text{ ft/sec} \quad \text{and} \quad (v_B)_2 = 1.64 \text{ ft/sec} \qquad \textit{Ans.}$$

The negative sign for $(v_A)_2$ indicates that the bag moves to the *right* after impact instead of to the left, Fig. 18–13c.

Example 18–6

The 1-lb ball B shown in Fig. 18–14a is attached to a 1-ft-long elastic cord. If the cord is stretched $\frac{1}{2}$ ft and released on the smooth *horizontal* plane from the position shown, determine how far the cord is stretched after the ball rebounds from the wall. The stiffness of the cord is $k = 6$ lb/ft, and the coefficient of restitution for the ball is $e = 0.8$. The ball makes a central impact with the wall.

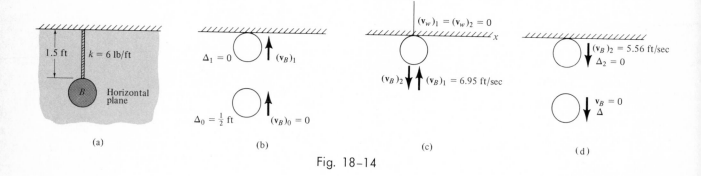

Fig. 18–14

Solution

In order to obtain the stretch in the cord after impact, it is necessary to study the mechanics of the impact. To do this we must first obtain the initial velocity of the ball *just before* it strikes the wall. This may be done by applying the conservation of energy principle to the ball. The energy diagrams for the ball just before it is released and just before it strikes the wall are shown in Fig. 18–14b. In the final position the cord is unstretched, so $\Delta_1 = 0$. Hence,

$$T_0 + V_0 = T_1 + V_1$$

$$\frac{1}{2}m(v_B)_0^2 + \frac{1}{2}k(\Delta_0)^2 = \frac{1}{2}m(v_B)_1^2 + \frac{1}{2}k(\Delta_1)^2$$

$$0 + \frac{1}{2}(6)\left(\frac{1}{2}\right)^2 = \frac{1}{2}\left(\frac{1}{32.2}\right)(v_B)_1^2 + 0$$

Solving for $(v_B)_1$ yields

$$(v_B)_1 = 6.95 \text{ ft/sec}$$

We shall now consider the interaction of the ball with the wall using the principle of impact. Note that the y axis represents the line of impact for the ball. Since the mass of the wall is essentially infinite, there is no need to write the conservation of momentum for the ball-wall system. Instead, the velocity of the wall is zero *both before* and *after impact,* Fig. 18–14c.

Using the definition of the coefficient of restitution, Eq. 18–14,

$$e = \frac{(v_B)_2 - (v_w)_2}{(v_w)_1 - (v_B)_1}$$

$$0.8 = \frac{(v_B)_2 - 0}{0 - 6.95}$$

Solving,

$$(v_B)_2 = -5.56 \text{ ft/sec}$$

The negative sign indicates that the ball rebounds *away* from the wall; that is, the ball moves in the negative y direction.

The maximum stretch Δ in the cord may be determined by applying the conservation of energy theorem to the ball. From the energy diagrams shown in Fig. 18–14d, we have

$$T_2 + V_2 = T + V$$

$$\frac{1}{2}m(v_B)_2^2 + \frac{1}{2}k(\Delta_2)^2 = \frac{1}{2}m(v_B)^2 + \frac{1}{2}k(\Delta)^2$$

$$\frac{1}{2}\left(\frac{1}{32.2}\right)(5.56)^2 + 0 = 0 + \frac{1}{2}(6)(\Delta)^2$$

Solving gives

$$\Delta = 0.400 \text{ ft} = 4.80 \text{ in.} \qquad \textit{Ans.}$$

Example 18–7

Two smooth wooden spheres A and B having a weight of 1 and 2 lb, respectively, collide with initial velocities shown in Fig. 18–15a. If the coefficient of restitution for the spheres is $e = 0.75$, determine the final velocity of each sphere after collision, and the loss in kinetic energy due to the collision.

Solution

The problem involves *oblique impact*. Why? In order to seek a solution, the x and y axes have been established along the line of impact and the

(a)

(b)

(c)

Fig. 18–15

plane of contact, respectively, Fig. 18–14a. By doing this the impulse which A and B exert on each other is *only* in the x direction. See Fig. 18–15b.

Resolving each of the initial velocities into x and y components, we have

$$(v_{Ax})_1 = 15 \cos 30° = 12.99 \text{ ft/sec}$$
$$(v_{Ay})_1 = 15 \sin 30° = 7.50 \text{ ft/sec}$$
$$(v_{Bx})_1 = -5 \cos 45° = -3.54 \text{ ft/sec}$$
$$(v_{By})_1 = -5 \sin 45° = -3.54 \text{ ft/sec}$$

Since the impulse (or impact) occurs only in the x direction (line of impact) we can apply the conservation of momentum for *both* spheres in this direction (Eq. 18–6) and Eq. 18–14, which defines the coefficient of restitution for the spheres. Thus, in reference to the momentum diagrams in Fig. 18–15b,

$(\xrightarrow{+})$ $\qquad m_A(v_{Ax})_1 + m_B(v_{Bx})_1 = m_A(v_{Ax})_2 + m_B(v_{Bx})_2$

$$\frac{1}{32.2}(12.99) + \frac{2}{32.2}(-3.54) = \frac{1}{32.2}(v_{Ax})_2 + \frac{2}{32.2}(v_{Bx})_2$$

$$(v_{Ax})_2 + 2(v_{Bx})_2 = 5.91 \qquad (1)$$

and,

$$e = \frac{(v_{Bx})_2 - (v_{Ax})_2}{(v_{Ax})_1 - (v_{Bx})_1} = \frac{\text{Rel. vel. after impact}}{\text{Rel. vel. before impact}}$$

$$0.75 = \frac{(v_{Bx})_2 - (v_{Ax})_2}{12.99 - (-3.54)}$$

$$(v_{Bx})_2 - (v_{Ax})_2 = 12.40 \qquad (2)$$

Solving Eqs. (1) and (2) for $(v_{Ax})_2$ and $(v_{Bx})_2$ yields

$$(v_{Ax})_2 = -6.29 \text{ ft/sec}$$

$$(v_{Bx})_2 = 6.10 \text{ ft/sec}$$

The momentum of *each sphere* is *conserved* in the y direction (plane of impact), since *no impulse* occurs in this direction. Hence, in reference to Fig. 18–15b,

$(+\uparrow)$ $\qquad m_A(v_{Ay})_1 = m_A(v_{Ay})_2$

Thus,

$$(v_{Ay})_2 = 7.50 \text{ ft/sec}$$

and

$(+\uparrow)$ $\qquad m_B(v_{By})_1 = m_B(v_{By})_2$

Thus,

$$(v_{By})_2 = -3.54 \text{ ft/sec}$$

Summing the vector components, we have

$$(v_A)_2 = \sqrt{(v_{Ax})_2^2 + (v_{Ay})_2^2} = \sqrt{(-6.29)^2 + (7.50)^2} = 9.79 \text{ ft/sec} \qquad \textit{Ans.}$$

$$\alpha_2 = \tan^{-1}\frac{(v_{Ay})_2}{(v_{Ax})_2} = \tan^{-1}\frac{7.50}{6.29} = 50.0° \qquad \alpha_2\diagdown^{(v_A)_2} \qquad \textit{Ans.}$$

$$(v_B)_2 = \sqrt{(v_{Bx})_2^2 + (v_{By})_2^2} = \sqrt{(6.10)^2 + (-3.54)^2} = 7.05 \text{ ft/sec} \qquad \textit{Ans.}$$

$$\beta_2 = \tan^{-1}\frac{(v_{By})_2}{(v_{Bx})_2} = \tan^{-1}\frac{3.54}{6.10} = 30.1° \qquad \diagdown_{\beta_2}\diagup_{(v_B)_2} \qquad \textit{Ans.}$$

These results are shown in Fig. 18–15c.

Knowing the velocities of each sphere both before and after the collision, the loss in kinetic energy becomes

$$\Delta T = T_1 - T_2$$

$$= \frac{1}{2}m_A(v_A)_1^2 + \frac{1}{2}m_B(v_B)_1^2 - \frac{1}{2}m_A(v_A)_2^2 - \frac{1}{2}m_B(v_B)_2^2$$

$$= \frac{1}{2}\left(\frac{1}{32.2}\right)(15)^2 + \frac{1}{2}\left(\frac{2}{32.2}\right)(5)^2 - \frac{1}{2}\left(\frac{1}{32.2}\right)(9.79)^2$$

$$- \frac{1}{2}\left(\frac{2}{32.2}\right)(7.05)^2$$

$$= 1.23 \text{ ft-lb} \qquad\qquad\qquad\qquad\qquad\qquad \textit{Ans.}$$

Problems

18–38. Blocks A and B weigh 5 and 10 lb, respectively. After striking block B, A slides 2 in. to the right, and B slides 3 in. to the right. If the coefficient of friction between the blocks and the surface is $\mu = 0.2$, determine the coefficient of restitution between the blocks. Block B is originally at rest.

Prob. 18–38

18–39. A 6-lb steel ball has an initial velocity of $v = 20$ ft/sec and strikes head on another steel ball that has a weight of 24 lb and is originally at rest. If the collision is perfectly elastic, determine the velocity of each ball after the collision and the impulse which the 6-lb ball imparts to the 24-lb ball.

18–40. If two spheres A and B are subjected to direct central impact and the collision is perfectly elastic ($e = 1$) prove that the kinetic energy before collision equals the kinetic energy after collision.

18–41. The three spheres each have an equal mass and are suspended from equal-length cables, as shown. Assuming that perfectly elastic impact occurs, describe the motion of each sphere when sphere A is released from rest at an angle θ.

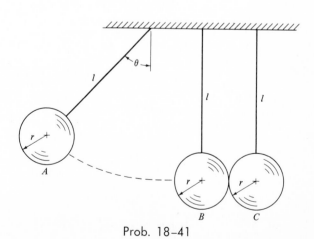

Prob. 18–41

18-42. The two blocks have a coefficient of restitution $e = 0.5$. Determine the velocities of each block after impact and the energy loss during impact.

Prob. 18-42

18-43. The ball A has a velocity of 5 ft/sec when it strikes ball B which is initially at rest. After the collision, ball B moves to the right with a velocity of 1 ft/sec and strikes ball C which is moving to the left with a velocity of 10 ft/sec. If the coefficient of restitution for all the balls is $e = 0.7$, determine the loss in kinetic energy which occurs after all the collisions. *Hint:* The weight of ball B is to be determined.

Prob. 18-43

18-44. The three balls each weigh 0.5 lb and have a coefficient of restitution of $e = 0.85$. If ball A is released from rest and strikes ball B and then ball B strikes ball C, determine the velocity of each ball after the second collision has occurred.

Prob. 18-44

18-45. An ivory ball is released from rest at a distance of 100 in. above a very large fixed metal surface. If the ball rebounds to a height of 60 in., determine the coefficient of restitution for the ball.

18-46. A basketball has a weight of 2 lb and is thrown with a speed of 40 ft/sec against a smooth rigid backstop at an angle of 60° with the backstop. If the collision is "perfectly elastic," what impulse is given to the backstop? What average resistance force does the ball exert on the backstop if the ball yields (dents) 2 in.?

18-47. The two spheres A and B weigh 3 and 5 lb, respectively. If they collide with the initial velocities shown, determine their velocities after impact. The coefficient of restitution is $e = 0.65$.

Prob. 18-47

18-48. The $\frac{1}{2}$-lb billiard ball is moving with a velocity of 5 ft/sec when it strikes the side of the pool table at A. If the coefficient of restitution between the ball and the side of the table is $e = 0.6$, determine the velocity

Prob. 18-48

of the ball after striking the table twice. Neglect the effects of friction while the ball is rolling and assume that the ball does not drop into one of the side or corner pockets.

18-49. The two hockey pucks A and B have equal masses. If they collide at point O and are deflected along the dotted paths, determine their speeds after impact. Assume that the ice surface over which they slide is smooth.

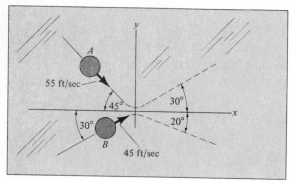

Prob. 18-49

18-50. Ball A weighs 2 lb and is traveling in the horizontal plane with a velocity of 3 ft/sec. Ball B weighs 11 lb and is initially at rest. If after the impact ball A has a velocity of 1 ft/sec directed along the positive x axis, determine the velocity of ball B after impact. How much kinetic energy is lost in the collision?

Prob. 18-50

18-51. The 8-lb sphere A is released from rest 10 ft from the surface of a flat plate P which weighs 6 lb. Determine the maximum compression in the spring if the impact is perfectly elastic.

Prob. 18-51

18-52. To test the manufactured properties of 2-lb steel balls, each ball is released from rest as shown and strikes a 45° inclined surface for which the coefficient of restitution is $e = 0.8$. Determine the distance s so that the ball strikes the horizontal plane at A. At what speed does each ball strike point A?

Prob. 18-52

780

18–53. Show that for an elastic collision (oblique impact) between two particles of equal mass, one of which is initially at rest, the recoiling angle between the particles is always a right angle.

18–54. Plates A and B each weigh 5 lb and slide along the frictionless guides. If the coefficient of restitution between the plates is $e = 0.7$, determine the maximum deflection of the spring when plate A moves with a velocity of 6 ft/sec before striking plate B. Plate B is originally at rest.

18–55. Work Prob. 18–54, assuming that the coefficient of restitution between the plates is (a) $e = 1$, and (b) $e = 0$.

Prob. 18–54

*18–6. Steady-Fluid Streams

Knowledge of the forces developed by steady moving fluid streams is of importance in the design and analysis of turbines, pumps, blades, and fans. The principle of impulse and momentum may be used to determine these forces. Consider, for example, the diversion of a steady stream of fluid (liquid or gas) by a fixed pipe, Fig. 18–16a. The fluid enters the

(a)

(b)

Fig. 18–16

pipe with a velocity $\mathbf{v}_A$ and exits with a velocity $\mathbf{v}_B$. The momentum- and impulse-vector diagrams for the fluid stream are shown in Fig. 18-16b. The force $\Sigma\mathbf{F}$, shown acting on the impulse-vector diagram, represents the resultant force of all the external forces acting on the fluid stream. It is this loading which gives the fluid stream an impulse whereby the original momentum of the fluid is changed in both its magnitude and direction. Since the flow is steady this force will be *constant* during the time interval dt. As shown in the figure, the fluid stream is in motion, and as a result a small amount of fluid, having a mass dm, enters the pipe with a velocity $\mathbf{v}_A$ at time t. Considering this element of mass and the mass of fluid in the pipe as a closed system, at time $t + dt$, a corresponding element of mass dm must leave the pipe with a velocity $\mathbf{v}_B$. The *average velocity* of the fluid stream, having mass m, *within* the pipe section is constant during the time interval dt. In Fig. 18-16b, its velocity is shown to be $\mathbf{v}$. Applying the principle of impulse and momentum to the fluid stream, we have

$$dm\,\mathbf{v}_A + m\mathbf{v} + \Sigma\mathbf{F}\,dt = dm\,\mathbf{v}_B + m\mathbf{v}$$

Solving for the resultant force yields

$$\Sigma\mathbf{F} = \frac{dm}{dt}(\mathbf{v}_B - \mathbf{v}_A) \qquad (18\text{--}15)$$

The term dm/dt is called the *mass flow* and indicates the constant amount of fluid which flows either into or out of the pipe per unit of time. $(\mathbf{v}_B - \mathbf{v}_A)$ represents the *vector difference* between the input and output velocity of the fluid stream. Provided motion of the fluid can be represented in the xy plane, it is usually convenient to express Eq. 18-15 in the form of two scalar equations:

$$\begin{aligned} \Sigma F_x &= \frac{dm}{dt}(v_{Bx} - v_{Ax}) \\[6pt] \Sigma F_y &= \frac{dm}{dt}(v_{By} - v_{Ay}) \end{aligned} \qquad (18\text{--}16)$$

The force summation in Eq. 18-15 or Eqs. 18-16 may easily be accounted for by accompanying the problem solution with a free-body diagram. Only the entrance and exit velocities of the fluid are required. These velocities represent the *relative velocities* of the fluid with respect to the system. For some problems, a kinematic diagram for the velocities will help in determining their values (refer to Example 18-9).

Example 18-8

Determine the reactive force which the pipeline contraction, shown in Fig. 18-17a, exerts on the fluid. The mass density of the liquid is $\rho_m = 2$

slug/ft³, and the static pressures at sections A and B are 50 and 75 lb/in.², respectively.

$v_A = 10$ ft/sec

$\frac{2}{3}$ ft $\frac{1}{2}$ ft $\frac{1}{3}$ ft

$v_B = 40$ ft/sec

(a)

Solution

The free-body diagram of the fluid passing through the contraction is shown in Fig. 18–17b. The forces acting on this free-body diagram consist of the static pressure forces $\mathbf{F}_A$ and $\mathbf{F}_B$, the weight of the fluid in the joint $\mathbf{W}$, and the resultant reactive force components $\mathbf{F}_x$ and $\mathbf{F}_y$ which the pipe exerts on the fluid.

The mass flow rate may be computed on the basis of knowing the velocities and density of the liquid, and the area of the cross section.

$F_A = p_A A_A$ $F_B = p_B A_B$

(b)

Fig. 18–17

$$\frac{dm}{dt} = \rho_m v_A A_A = \rho_m v_B A_B = (2 \text{ slug/ft}^3)(40 \text{ ft/sec})\pi \left(\frac{1}{6} \text{ ft}\right)^2$$

$$= 6.98 \text{ slug/sec}$$

Using Eqs. 18–16, we have

$(\xrightarrow{+})$
$$\Sigma F_x = \frac{dm}{dt}(v_{Bx} - v_{Ax})$$

$$p_A A_A - p_B A_B - F_x = \frac{dm}{dt}(v_B - v_A)$$

$$50(144)\pi \left(\frac{1}{3}\right)^2 - 75(144)\pi \left(\frac{1}{6}\right)^2 - F_x = 6.98(40 - 10)$$

$$F_x = 1,361 \text{ lb}$$

$(+\uparrow)$
$$\Sigma F_y = \frac{dm}{dt}(v_{By} - v_{Ay})$$

$$F_y - W = 0$$

$$F_y = W$$

The weight of the fluid W is equal to the mass density ρ_m times the product of gravitational acceleration g and the volume of liquid contained in the joint, i.e.,

$$F_y = \rho_m g V = (2 \text{ slug/ft}^3)\frac{32.2 \text{ lb}}{1 \text{ slug}}\left[\frac{1}{4}\pi \left(\frac{1}{3} \text{ ft}\right)^2 (1 \text{ ft})\right.$$

$$\left. -\frac{1}{4}\pi \left(\frac{1}{6} \text{ ft}\right)^2 \left(\frac{1}{2} \text{ ft}\right)\right] = 4.92 \text{ lb}$$

The magnitude of the resultant force which the contraction exerts on the fluid is thus

$$F = \sqrt{(F_x)^2 + (F_y)^2} = \sqrt{(1,361)^2 + (4.92)^2}$$

$$F \approx 1,361 \text{ lb} \qquad\qquad Ans.$$

A 1-in.-diameter water jet having a velocity of 80 ft/sec impinges upon a single moving blade, Fig. 18–18a. If the blade is moving at 20 ft/sec away from the jet, determine the magnitude of force which the blade is exerting on the water; $\rho_W = 62.4$ lb/ft³.

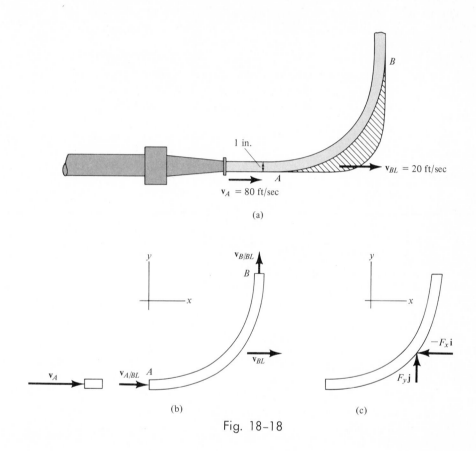

Fig. 18–18

Solution

The velocity terms to be used in Eq. 18–15 must be measured *with respect to the blade*, Fig. 18–18b. Hence, the relative velocity of the fluid at the blade entrance A is determined from the equation

$$\mathbf{v}_A = \mathbf{v}_{BL} + \mathbf{v}_{A/BL}$$

or

$$\mathbf{v}_{A/BL} = \mathbf{v}_A - \mathbf{v}_{BL} = 80\mathbf{i} - 20\mathbf{i} = \{60\mathbf{i}\} \text{ ft/sec}$$

This same relative flow velocity is directed vertically upward on the blade at B, as shown in the figure. Hence

$$\mathbf{v}_{B/BL} = (v_{A/BL})\mathbf{j} = \{60\mathbf{j}\} \ \text{ft/sec}$$

The free-body diagram of a section of fluid acting on the blade is shown in Fig. 18–18c. The weight of the fluid will be neglected in the calculations, since this force is small compared to the reactive components $\mathbf{F}_x$ and $\mathbf{F}_y$. Substituting the above results into Eq. 18–15 yields

$$\Sigma \mathbf{F} = \frac{dm}{dt}(\mathbf{v}_B - \mathbf{v}_A) = \rho_W(v_{A/BL})A(\mathbf{v}_{B/BL} - \mathbf{v}_{A/BL})$$

$$-F_x\mathbf{i} + F_y\mathbf{j} = \frac{62.4}{32.2}(60)\left(\pi\left(\frac{1}{2}\right)^2\right)(60\mathbf{j} - 60\mathbf{i})$$

Equating the respective $\mathbf{i}$ and $\mathbf{j}$ components gives

$$F_x = 91.3(60) = 5{,}480 \ \text{lb}$$
$$F_y = 91.3(60) = 5{,}480 \ \text{lb}$$

The magnitude of force of the blade on the water is therefore

$$F = \sqrt{(F_x)^2 + (F_x)^2}$$
$$= \sqrt{(5{,}480)^2 + (5{,}480)^2}$$
$$= 7{,}750 \ \text{lb} \qquad\qquad Ans.$$

Note that the water exerts an equal but opposite force on the blade.

*18–7. Propulsion with Variable Mass

In Section 18–6, we considered a case in which the amount of mass dm which entered and left a system was *constant*. There are, however, two other important cases involving mass flow. These are represented by a system which is either gaining or losing mass. The principle of impulse and momentum may be used to determine the forces acting on such a system. We will discuss each of these two cases separately.

Case I: A System Which Loses Mass. Consider a device which at an instant of time has a mass m and a forward velocity $\mathbf{v}$, Fig. 18–19a. At this same instant the device is expelling an amount of mass m_e. The flow velocity $\mathbf{v}_e$ of this lost mass is *constant* when the measurement is made slightly removed from the device. We will consider the *system* at an instant to include *both the mass m and the expelled mass m_e*, as shown by the dotted line in the figure. The momentum- and impulse-vector

(a)

(b)

Fig. 18–19

diagrams for the system are shown in Fig. 18–19*b*. During the time interval *dt*, the velocity of the system is increased from **v** to **v** + *d***v**. This increase in forward velocity of the system, however, does not change the velocity **v**_e of the expelled mass, since this mass moves at a constant speed once it has been ejected. To increase the velocity of the system during the time *dt*, an amount of mass dm_e has been ejected and thereby gained in the exhaust. The impulses are created by $\Sigma \mathbf{F}_s$, which represents the resultant of all the external forces which *act on the system* in the direction of motion. This force resultant *does not include* the force which causes the device to move forward, since this force (called a *thrust*) is *internal to the system;* that is, the thrust acts with equal magnitude but opposite direction on the mass *m* of the device and the expelled exhaust mass m_e*. Applying the principle of impulse and momentum to the system, in reference to Fig. 18–19*b*, we have

$$(\xrightarrow{+}) \qquad mv - m_e v_e + \Sigma F_s \, dt = (m - dm_e)(v + dv) - (m_e + dm_e)v_e$$

or

$$\Sigma F_s \, dt = -v \, dm_e + m \, dv - dm_e \, dv - v_e \, dm_e$$

The third term on the right side of this equation may be neglected since it is a "second-order" differential. Dividing by *dt* gives

$$\Sigma F_s = m\frac{dv}{dt} - (v + v_e)\frac{dm_e}{dt}$$

*$\Sigma \mathbf{F}_s$ represents the external resultant force *acting on the system,* which is different from $\Sigma \mathbf{F}$, the resultant force acting only on the device.

The relative velocity of the device as seen by an observer moving with the particles of the ejected mass is $v_{D/e} = (v + v_e)$. Thus, our final result is

$$\Sigma F_s = m\frac{dv}{dt} - v_{D/e}\frac{dm_e}{dt} \qquad (18\text{--}17)$$

where the term dm_e/dt represents the rate at which mass is being ejected.

A rocket is a typical example of a device which ejects its stored mass in order to increase its forward velocity. Consider, for example, the rocket shown in Fig. 18–20 which has a weight **W** and is moving upward against an atmospheric drag force $\mathbf{F}_D$. The system to be considered consists of the mass of the rocket and the mass of ejected gas m_e. Applying Eq. 18–17 to this system gives

$$(+\uparrow) \qquad -F_D - W = \frac{W}{g}\frac{dv}{dt} - v_{D/e}\frac{dm_e}{dt}$$

The last term of this equation represents the *thrusting force* **T** which the engine exhaust exerts on the rocket, Fig. 18–20. Recognizing that $dv/dt = a$, we may therefore write

$$(+\uparrow) \qquad T - F_D - W = \frac{W}{g}a$$

If a free-body diagram of the rocket is drawn, it becomes obvious that this latter equation is simply Newton's second law of motion for the rocket.

Case II: A System Which Gains Mass.

A device such as a scoop or a shovel may gain mass as it moves forward. Consider for example the device shown in Fig. 18–21a which at an instant of time has a mass m and is moving forward with a velocity **v.** At the same instant, the device is collecting a stream of mass m_i. The flow velocity $\mathbf{v}_i$ of this injected mass is constant and independent of the velocity **v.** It is required that $v > v_i$. The system to be considered at this instant includes both the mass of the device and the mass of the injected fluid, as shown by the dotted line in the figure. The momentum- and impulse-vector diagrams for the system are shown in Fig. 18–21b. With an increase in mass dm_i gained by the device, there is an assumed increase in velocity dv during the time interval dt. This increase is caused by the impulse created by the resultant of all the external forces *acting on the system* in the direction of motion, $\Sigma\mathbf{F}_s$. This force does not include the retarding force of the injected mass acting on the device. Why? Applying the principle of impulse and momentum to the system, we have

Fig. 18–20

(b)

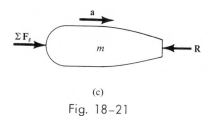

(c)

Fig. 18–21

$(\xrightarrow{+})\quad mv + m_i v_i + \Sigma F_s\, dt = (m + dm_i)(v + dv) + (m_i - dm_i)v_i$

Using the same procedure as in the previous case, we may write this equation as

$$\Sigma F_s = m\frac{dv}{dt} + (v - v_i)\frac{dm_i}{dt}$$

The relative velocity of the device as seen by an observer moving with the particles of the injected mass is $v_{D/i} = v - v_i$. Thus, our final result is

$$\Sigma F_s = m\frac{dv}{dt} + v_{D/i}\frac{dm_i}{dt} \qquad (18\text{–}18)$$

where dm_i/dt is the mass flow rate injected into the device. The last term in this equation therefore represents the magnitude of force **R** which the injected mass *exerts on the device*. Since $dv/dt = a$, Eq. 18–18 becomes

$$\Sigma F_s - R = ma$$

This equation represents Newton's second law of motion for the device (refer to Fig. 18–21c).

As in the case of steady flow, problems which are solved using Eqs. 18–17 and 18–18 should be accompanied by the necessary free-body diagram. With this diagram one can then determine ΣF_s *for the system* and isolate the force exerted on the device by the particle stream.

(a)

Example 18–10

The 35,000-lb jet airplane shown in Fig. 18–22a has a constant speed of 600 mph when it is flying along the straight-line path shown. Air enters the intake scoop S at the rate of 120 lb/sec. If the engine burns fuel at the rate of 1.2 lb/sec and the gas is exhausted relative to the plane with a velocity of 3,000 ft/sec, determine the drag force exerted on the plane by the air resistance.

(b)

Fig. 18–22

Solution

The jet plane represents a device which is propelled by *collecting* a fluid (liquid and gas) and *expelling* it by mechanical and thermal means with a velocity higher than the velocity at which it was collected. Such a device represents a *combination* of Cases I and II. When the principle of impulse and momentum is applied to the system consisting of the plane and portions of the intake and exhaust fluids, it may be shown that

$$\Sigma F_s = m\frac{dv}{dt} - v_{D/e}\frac{dm_e}{dt} + v_{D/i}\frac{dm_i}{dt} \tag{1}$$

Compare this equation with Eqs. 18–17 and 18–18. The terms are defined in the text.

The free-body diagram of the plane is shown in Fig. 18–22b. The forces which act *on the plane* consist of the engine thrust **T,** the force of the air stream at the intake **R,** the weight **W,** the drag force **D,** and the lift **L.** Recognizing that **T** and **R** are represented in Eq. (1) by the last two terms, respectively, the magnitude of the force resultant ΣF_s acting on the system in the *direction of flight* is then the sum of $-W\sin 10°$ and $-D$.

Since the plane is moving at constant velocity of $v = 600$ mph $= 880$ ft/sec, $dv/dt = 0$. The outside air is initially at rest, and therefore the relative velocity of the plane with respect to the injected air is $v_{D/i} = v$. From the problem data,

$$v_{D/e} = 3,000 \text{ ft/sec}$$

$$\frac{dm_i}{dt} = \frac{120}{32.2} = 3.73 \text{ slugs/sec}$$

$$\frac{dm_e}{dt} = \frac{120 + 1.2}{32.2} = 3.76 \text{ slugs/sec}$$

Substituting into Eq. (1) yields

(a)

(b)

(c)

Fig. 18–23

$$(+\nearrow) \qquad \Sigma F_s = m\frac{dv}{dt} - v_{D/e}\frac{dm_e}{dt} + v_{D/i}\frac{dm_i}{dt}$$

$$-35{,}000(\sin 10°) - D = 0 - 3{,}000(3.76) + 880(3.73)$$

or

$$D = 1{,}920 \text{ lb} \qquad\qquad Ans.$$

Example 18–11

A chain of length l, Fig. 18–23a, weighs γ lb/ft. Determine the magnitude of force $\mathbf{F}$ required to (a) raise the chain with a constant speed of v_c ft/sec, starting from rest when $y = 0$, and (b) to lower the chain with a constant speed of v_c ft/sec, starting from rest when $y = l$ ft.

Solution

Part (a). As the chain is raised, all the suspended links are given a sudden impulse downward by each added link which is lifted off the ground. Thus, we may consider the *suspended portion* of the chain as a device which is *gaining mass*. A free-body diagram of a portion of the chain which is located at an arbitrary height y above the ground is shown in Fig. 18–23b. The system to be considered is the length of the chain y which is suspended by $\mathbf{F}$ at any instant, including the next link which is about to be added but still at rest. The forces acting on this system exclude the force $\mathbf{P}$ which the added link exerts on the suspended portion of the chain. Hence, $\Sigma F_s = F - \gamma y$.

The velocity $\mathbf{v}_c$ of the chain at a given instant is equivalent to $\mathbf{v}_{D/i}$. Why? Since v_c is constant,

$$\frac{dy}{dt} = v_c \qquad \text{and} \qquad \frac{dv_c}{dt} = 0$$

Integrating, using the initial condition that $y = 0$ at $t = 0$, gives

$$y = v_c t$$

The mass of the system at any instant is thus

$$m = \frac{\gamma}{g}y = \frac{\gamma}{g}v_c t$$

The total mass to be injected is

$$m_i = \frac{\gamma}{g}(y - l) = \frac{\gamma}{g}(v_c t - l)$$

Thus, the *rate* at which mass is *added* to the suspended chain is

$$\frac{dm_i}{dt} = \frac{\gamma}{g}v_c$$

Applying Eq. 18–18 to the system, using this data we have

$(+\uparrow)$
$$\Sigma F_s = m\frac{dv_c}{dt} + v_{D/i}\frac{dm_i}{dt}$$

$$F - \gamma y = 0 + v_c\left(\frac{\gamma}{g}v_c\right)$$

Hence,

$$F = \gamma\left(y + \frac{v_c^2}{g}\right) \qquad Ans.$$

Part (b). When the chain is being lowered, the links which are expelled (given zero velocity) *do not* impart an impulse to the *remaining* suspended links. Why? Thus, if we consider the same system as in Part (a), we will not apply Eq. 18–17 to determine F. Newton's second law of motion may be applied to obtain the most direct solution. At time t the portion of chain still off the floor is y. The free-body and inertia-vector diagrams for a suspended portion of the chain are shown in Fig. 18–23c. Thus,

$+\uparrow\Sigma F = ma$
$$F - \gamma y = 0$$
$$F = \gamma y \qquad Ans.$$

(*Note:* In determining the *reaction on the ground* in Part (a), Newton's second law of motion should be applied for the solution. As the chain is raised, each link taken from the resting portion of the chain *does not* exert an impulse to the ground. However, as the chain is lowered, Part (b), the links added to the ground *do* cause an impulsive reaction. For this case the pile of links represents a device which is gaining mass and therefore Eq. 18–18 may be used.)

Problems

18–56. A plow located on the front of a locomotive scoops up snow at the rate of 10 ft³/sec and stores it in the train. If the locomotive is traveling at a constant speed of 10 mph, determine the resistance to motion caused by the shoveling. The density of snow is $\rho_s = 6$ lb/ft³.

18–57. The fireman weighs 150 lb and is holding a hose which has a nozzle diameter of 1 in. If the velocity of the water at discharge is 60 ft/sec, determine the total vertical force supported by the man's feet at the ground; $\rho_w = 62.4$ lb/ft³.

Prob. 18–57

18-58. The nozzle discharges water having a density of $\rho_w = 62.4$ lb/ft³ at a constant rate of 2 ft³/sec. The cross-sectional area of the nozzle at A is 4 in.² and at B the cross-sectional area is 12 in.². If the static pressure due to the water at B is 2 lb/in.², determine the magnitude of force which must be applied at section B to hold the nozzle in place.

Prob. 18-58

18-59. The nozzle discharges water at the rate of 60 ft/sec against a shield. If the cross-sectional area of the water stream is 1 in.², determine the force **F** required to hold the shield motionless. With what net velocity **v** does all the water leave the shield? $\rho_w = 62.4$ lb/ft³.

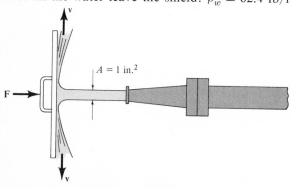

Prob. 18-59

18-60. What force is required to move the shield in Prob. 18-59 forward with a velocity of 10 ft/sec? Determine the net velocity at which the water leaves the shield.

18-61. A jet of water has a cross-sectional area of 5 in.². If it strikes the fixed blade with a speed of 60 ft/sec, determine the magnitude of force which the water exerts on the blade; $\rho_w = 62.4$ lb/ft³.

Prob. 18-61

18-62. The helicopter is hovering in midair. If the slipstream of the air passing through the rotor is outlined as shown, and the slipstream has a diameter of 18 ft where the velocity of the air is $v_a = 200$ ft/sec, determine the weight of the helicopter. The density of the air is $\rho_a = 0.076$ lb/ft³.

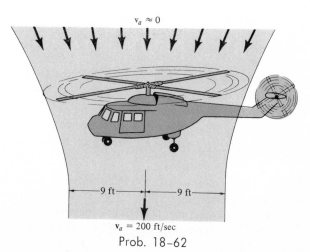

Prob. 18-62

18-63. The fan draws air through a vent with a speed of 12 ft/sec. If the cross-sectional area of the vent is 2 ft², determine the thrust on the blade. The density of the air is $\rho_a = 0.076$ lb/ft³.

Prob. 18-63

18-64. The small cart is being propelled up the incline with a constant velocity of 10 ft/sec by a water jet which has a velocity of 40 ft/sec. If the diameter of the water jet is 1 in., determine the weight of the cart. The density of the water is $\rho_w = 62.4$ lb/ft³.

Prob. 18-64

18-65. A rocket burns 2,000 lb of fuel in 2 min. If the velocity of the gas is 6,500 ft/sec with respect to the rocket, determine the thrust of the rocket engine.

18-66. The boat weighs 400 lb and is traveling forward on a river with a constant velocity of 45 ft/sec. The river

Prob. 18-66

is flowing in the opposite direction at 10 ft/sec. A tube is placed in the water, as shown. If the tube collects 100 lb of water in the boat in 2 min, determine the thrust **T** which is required to propel the boat forward. Neglect the drag effect of the water on the surface of the boat.

18-67. The nozzle has a diameter of 1.5 in. If it discharges water uniformly with a velocity of 120 ft/sec against the fixed blade, determine the vertical force exerted by the water on the blade. The density of water is $\rho_w = 62.4$ lb/ft³.

Prob. 18-67   Prob. 18-69

18-68. Work Prob. 18-67 assuming that the blade has a downward velocity of 40 ft/sec.

18-69. The second stage S_2 of the two-stage rocket weighs 2,000 lb (empty) and is launched from the first stage S_1 with a velocity of 3,000 mph. The fuel in the second stage weighs 1,000 lb. If it is consumed at the rate of 50 lb/sec and ejected at 8,000 ft/sec, determine the acceleration of the rocket at the instant the engine is fired. What is the rocket's acceleration after all the fuel is consumed? Neglect the effect of gravitation.

18-70. If the chain is lowered at a constant velocity of 2 ft/sec by the force **F**, determine the normal reaction on the floor as a function of time as the chain is lowered.

793

The chain has a linear density of 3 lb/ft and has a total length of 12 ft.

Prob. 18–70

18–71. The chain has a total length of 10 ft and a linear density of 2 lb/ft. Determine the magnitude of force **F** as a function of time which must be applied to the end of the chain to raise it with a constant velocity of 2 ft/sec. Initially the entire chain is at rest on the ground.

Prob. 18–71

18–72. The 10-ft chain rests within the fixed conical shell. As soon as the plate p is lowered, a small portion of the chain falls unobstructed through the opening at A. Determine the velocity **v** at which the plate must be lowered so that only the first link of the chain remains

touching the surface of the plate. The chain weighs 2 lb/ft.

Prob. 18–72

18–73. A rocket has an empty weight of 500 lb and carries 300 lb of fuel. If the fuel is burned at the rate of 15 lb/sec and ejected with a relative velocity of 4,400 ft/sec, determine the maximum speed attained by the rocket starting from rest. Neglect the effect of gravitation on the rocket.

Prob. 18–74

18–74. The rocket car has an empty weight of 400 lb and is fired from rest along the horizontal tracks. There is 300 lb of fuel stored in the car, which is consumed at a constant rate of 15 lb/sec and ejected with a relative velocity of 5,000 ft/sec. Determine the maximum speed attained by the car, starting from rest, if the frictional resistance due to the atmosphere is $F = 0.01v$, where v is the velocity given in ft/sec and F is measured in pounds.

18-75. The missile weighs 40,000 lb. The constant thrust provided by the turbojet engine is $T = 15,000$ lb. Additional thrust is provided by *two* rocket boosters B. The propellant in each booster is burned at a constant rate of 150 lb/sec, with a relative exhaust velocity of 3,000 ft/sec. If the mass of the propellant lost by the turbojet engine can be neglected, determine the velocity of the missile after the 4 sec burn time of the boosters. The initial velocity of the missile is 300 mph.

Prob. 18-75

18-76. Determine the velocity of the second stage of the rocket in Prob. 18-69 as a function of time t after the engine has fired. Neglect the effect of gravitation.

18-8. Angular Momentum of a Particle

The linear momentum of a particle has been defined as $\mathbf{L} = m\mathbf{v}$, where m is the mass of the particle and $\mathbf{v}$ represents the particle velocity. The *angular momentum* of the particle about any point is defined as the *moment of the linear momentum vector* about that point. For example, the angular momentum of the particle P shown in Fig. 18-24 about the origin O of the *inertial reference frame xyz* is defined by the vector cross-product

$$\mathbf{H}_O = \mathbf{r} \times m\mathbf{v} \qquad (18\text{-}19)$$

where $\mathbf{r}$ denotes the position vector drawn from point O to P. (This formulation is analogous to finding the moment of a force about a point.) The angular momentum vector acts in a *direction* which is perpendicular to the plane containing the position and the linear momentum vectors, in accordance with the right-hand rule. The *magnitude* of $\mathbf{H}_O$ is defined by the vector cross product, i.e., $H_O = rmv \sin \theta$, where θ is defined as the angle made between the tails of the two vectors, Fig. 18-24. Common units for this magnitude are lb-ft-sec or kg_f-m-sec.

Fig. 18-24

If $\mathbf{r}$ and $\mathbf{L} = m\mathbf{v}$ are expressed in terms of Cartesian components, then the angular momentum may be expressed in determinant form,

$$\mathbf{H}_O = \begin{vmatrix} \mathbf{i} & \mathbf{j} & \mathbf{k} \\ r_x & r_y & r_z \\ mv_x & mv_y & mv_z \end{vmatrix}$$

The moment about point O of the force system which acts on the particle may be related to the angular momentum of the particle using Newton's second law of motion. Since the mass of the particle is constant, this equation may be written in the form

$$\Sigma\mathbf{F} = m\frac{d^2\mathbf{r}}{dt^2}$$

Multiplying (vectorially) each side of this equation by the position vector $\mathbf{r}$ to obtain the moment $\Sigma\mathbf{M}_O$ of the external forces about point O, and noting that m is constant, we have

$$\Sigma\mathbf{M}_O = \mathbf{r} \times \Sigma\mathbf{F} = m\mathbf{r} \times \frac{d^2\mathbf{r}}{dt^2} \qquad (18\text{--}20)$$

By definition of the derivative of the cross product of two vectors, Eq. 12–14, we can write

$$\frac{d}{dt}\left(\mathbf{r} \times \frac{d\mathbf{r}}{dt}\right) = \frac{d\mathbf{r}}{dt} \times \frac{d\mathbf{r}}{dt} + \mathbf{r} \times \frac{d^2\mathbf{r}}{dt^2}$$

The first term on the right side of this equation is zero since the cross product of any two parallel vectors is zero. Hence, substituting into Eq. 18–20, noting that $\mathbf{v} = d\mathbf{r}/dt$, yields

$$\Sigma\mathbf{M}_O = m\frac{d}{dt}\left(\mathbf{r} \times \frac{d\mathbf{r}}{dt}\right) = \frac{d}{dt}(\mathbf{r} \times m\mathbf{v})$$

Using Eq. 18–19, the final result is

$$\Sigma\mathbf{M}_O = \frac{d\mathbf{H}_O}{dt} \qquad (18\text{--}21)$$

Therefore, *the moment about point O of all the forces acting on the particle is equal to the time rate of change of the angular momentum of the particle about point O.* This result is similar to Eq. 18–1, which can be written as

$$\Sigma\mathbf{F} = \frac{d\mathbf{L}}{dt} \qquad (18\text{--}22)$$

This equation states that *the sum of all the forces acting on the particle is equal to the time rate of change of the linear momentum of the particle.*

An equation having the same form as Eq. 18–21 may be derived for the system of n particles shown in Fig. 18–25. The forces acting on the arbitrary ith particle of the system consist of the resultant of the external forces $\mathbf{F}_i$, and the internal forces $\mathbf{f}_{ij}$ which act between the i and other j particles within the system. Expressing the moment of each of these forces about the origin O of the *inertial reference frame,* using the form of Eq. 18–21, we have

$$\mathbf{r}_i \times \mathbf{F}_i + \mathbf{r}_i \times \mathbf{f}_{i1} + \mathbf{r}_i \times \mathbf{f}_{i2} + \cdots + \mathbf{r}_i \times \mathbf{f}_{ij} + \cdots + \mathbf{r}_i \times \mathbf{f}_{in} = \frac{d(\mathbf{H}_i)_O}{dt}$$

Here $\mathbf{r}_i$ represents the position vector drawn from the origin O to the ith particle and $(\mathbf{H}_i)_O$ is the angular momentum of the ith particle about O. Similar equations can be written for each of the other particles of the system. When all results are summed (vectorially) the moment created by the internal force $\mathbf{f}_{ij}$ will cancel the moment of $\mathbf{f}_{ji}$—the force which the jth particle exerts on the ith particle. This is because of the principle of transmissibility, realizing that $\mathbf{f}_{ij}$ and $\mathbf{f}_{ji}$ are equal, opposite, and collinear. Hence, the moment created by each pair of internal forces will cancel, leaving the result

$$\Sigma\mathbf{M}_O = \frac{d\mathbf{H}_O}{dt} \qquad (18\text{–}23)$$

This equation states that *the sum of the moments about point O of the external forces acting on a system of n particles, is equal to the time rate*

Fig. 18–25

of change of the total angular momentum of the system of particles about point O. Recall that point O represents the origin of an inertial frame of reference. Since this frame is either *fixed* or allowed to *translate* with a *constant velocity,* the terms in Eq. 18–23 will be different for observers in different inertial frames of reference. Both observers will, however, conclude that Eq. 18–23 is valid for their own sets of measurements since this equation involves time rates of change of the motion.

When moments of the external forces acting on the system of particles are summed about the *mass center G* of the system, one again obtains the same simple form of Eq. 18–23 relating the moment summation $\Sigma \mathbf{M}_G$ to the angular momentum $\mathbf{H}_G$. To show this consider the system of n particles shown in Fig. 18–26, where xyz represents an inertial frame of reference and axes x', y', and z' *translate* with respect to this frame. The origin of the primed coordinates is located at the mass center G of the system of particles. Since this point is *accelerating,* the translating frame is *noninertial.* The angular momentum of the ith particle with respect to this frame is

$$(\mathbf{H}_i)_G = \mathbf{r}_{i/G} \times m_i \mathbf{v}_{i/G}$$

where $\mathbf{r}_{i/G}$ and $\mathbf{v}_{i/G}$ represent the instantaneous position and velocity of the ith particle with respect to the primed coordinates. Taking the time derivative gives

$$\frac{d(\mathbf{H}_i)_G}{dt} = \left(\frac{d\mathbf{r}_{i/G}}{dt}\right) \times m_i \mathbf{v}_{i/G} + \mathbf{r}_{i/G} \times m_i \left(\frac{d\mathbf{v}_{i/G}}{dt}\right)$$

By definition, $\mathbf{v}_{i/G} = d\mathbf{r}_{i/G}/dt$. Thus, the first term on the right-hand side of this equation is zero since the cross product of parallel vectors is zero. Also $\mathbf{a}_{i/G} = d(\mathbf{v}_{i/G})/dt$. Therefore we have

$$\frac{d(\mathbf{H}_i)_G}{dt} = \mathbf{r}_{i/G} \times m_i \mathbf{a}_{i/G}$$

Similar expressions can be written for the other particles of the system. When the results are summed for all n particles we get

$$\frac{d\mathbf{H}_G}{dt} = \Sigma(\mathbf{r}_{i/G} \times m_i \mathbf{a}_{i/G})$$

The relative acceleration for the ith particle is defined by the equation $\mathbf{a}_{i/G} = \mathbf{a}_i - \mathbf{a}_G$, where $\mathbf{a}_i$ and $\mathbf{a}_G$ represent the accelerations of the ith particle and point G measured with respect to the *inertial frame of reference.* Substituting, expanding, and using the distributive property of the vector cross-product, we have

$$\frac{d\mathbf{H}_G}{dt} = \Sigma(\mathbf{r}_{i/G} \times m_i \mathbf{a}_i) - (\Sigma m_i \mathbf{r}_{i/G}) \times \mathbf{a}_G$$

Fig. 18–26

By definition of the mass center, the sum $(\Sigma m_i \mathbf{r}_{i/G}) = (\Sigma m_i)\bar{\mathbf{r}}$ is equal to zero since the position vector $\bar{\mathbf{r}}$ relative to G is zero. Hence, the last term in the previous equation is zero. The product $m_i \mathbf{a}_i$ contained in the first term on the right may be replaced by the sum of the external forces $\mathbf{F}_i$ and internal forces $\mathbf{f}_{ij}$ acting on the particle. If this is done, the moment resultant of the internal forces for the entire system will be zero for the same reason as stated in the derivation of Eq. 18–23. Therefore, we are left only with the moment of the *external forces* acting on the system of particles. Denoting this value by $\Sigma \mathbf{M}_G = \Sigma \mathbf{r}_{i/G} \times \mathbf{F}_i$, the final result may be written as

$$\Sigma \mathbf{M}_G = \frac{d\mathbf{H}_G}{dt} \qquad (18\text{--}24)$$

This result represents one of the most important formulations in mechanics. As will be shown in Chapter 21, it forms the basis for obtaining the rotational equations of motion for a rigid body.

18–10. Principle of Angular Impulse and Momentum for a Particle, Conservation of Angular Momentum

Equation 18–21 may be rewritten in the form

$$\Sigma \mathbf{M}_O \, dt = d\mathbf{H}_O$$

Assuming that the angular momentum of the particle at time $t = t_1$ is $\mathbf{H}_O = (\mathbf{H}_O)_1$, and at time $t = t_2$, $\mathbf{H}_O = (\mathbf{H}_O)_2$, the above equation may be integrated between these limits to yield

$$\Sigma \int_{t_1}^{t_2} \mathbf{M}_O \, dt = (\mathbf{H}_O)_2 - (\mathbf{H}_O)_1$$

or

$$(\mathbf{H}_O)_1 + \Sigma \int_{t_1}^{t_2} \mathbf{M}_O \, dt = (\mathbf{H}_O)_2 \qquad (18\text{--}25)$$

The second term on the left side of this equation represents the *angular impulse* which is given to the particle. This term is computed on the basis of integrating, with respect to time, the moment of all the forces acting on the particle, over the time interval t_1 to t_2. The initial and final angular momenta $(\mathbf{H}_O)_1$ and $(\mathbf{H}_O)_2$ are defined as the moment of the linear momentum of the particle, Eq. 18–19, at times t_1 and t_2, respectively.

In a similar manner, using Eq. 18–23 for a system of n particles, we may write

$$\Sigma(\mathbf{H}_O)_{i_1} + \Sigma \int_{t_1}^{t_2} (\mathbf{M}_O)_i \, dt = \Sigma(\mathbf{H}_O)_{i_2} \qquad (18\text{--}26)$$

The angular impulse term in this expression may be written as

$$\text{ang imp} = \Sigma \int_{t_1}^{t_2} (\mathbf{M}_O)_i \, dt = \Sigma \int_{t_1}^{t_2} (\mathbf{r}_i \times \mathbf{F}_i) \, dt$$

where $\mathbf{r}_i$ is the position vector extending from the origin O to the ith particle, and $\mathbf{F}_i$ is the resultant external force acting on the ith particle.

Equations 18–25 and 18–26 represent the *principle of angular impulse and momentum* for a particle and a system of particles, respectively. These equations are analogous to Eqs. 18–2 and 18–4, which define the *principle of linear impulse and momentum*. Using impulse and momentum principles, we can therefore write *two vector equations* which define the motion of a particle; namely, Eqs. 18–2 and 18–25. These vector equations may be expressed as *six independent scalar equations* defining the linear and angular impulse and momentum of the particle along the *xyz* axes (inertial reference frame). If the particle is confined to move in the *xy* plane, *three independent scalar equations* may be written to express the motion. Two of these equations represent the principle of linear impulse and momentum in the *xy* plane, and the third equation represents the principle of angular impulse and momentum about the *z* axis. These three scalar equations are

$$m(v_x)_1 + \Sigma \int_{t_1}^{t_2} F_x \, dt = m(v_x)_2$$

$$m(v_y)_1 + \Sigma \int_{t_1}^{t_2} F_y \, dt = m(v_y)_2 \qquad (18\text{--}27)$$

$$(H_O)_1 + \Sigma \int_{t_1}^{t_2} M_O \, dt = (H_O)_2$$

It is recommended that a set of linear momentum- and impulse-vector diagrams be drawn before applying these equations to the solution of problems. A set of such diagrams for the particle P is shown in Fig. 18–27a. The linear-momentum diagrams indicate the momentum $m\mathbf{v}$ of the particle at times t_1 and t_2. Knowing the position vectors $\mathbf{r}_1$ and $\mathbf{r}_2$, the angular momentum can be computed on the basis of $\mathbf{H}_O = \mathbf{r} \times m\mathbf{v}$. Information given on the impulse-vector diagram is used to compute the x, y, and z components of each linear impulse given to the particle and the angular impulse about point O. Using vector notation, an angular impulse is computed on the basis of $\int_{t_1}^{t_2} \mathbf{r} \times \mathbf{F} \, dt$.

When the angular impulse acting on a particle is zero during the time

(a)

(b)

Fig. 18–27

period $t = t_1$ to $t = t_2$, Eq. 18–25 reduces to the following simplified form,

$$(\mathbf{H}_O)_1 = (\mathbf{H}_O)_2 \qquad (18\text{–}28)$$

This equation is known as the *conservation of angular momentum* for the particle. It states that when no angular impulse is applied to a particle during the time period t_1 to t_2, the angular momentum of the particle remains *constant* during the time period. In particular, if no external impulse is given to the particle, both the linear and angular momentum will be conserved. If, however, the particle is subjected only to a central force (Sec. 14–8) then angular momentum will be conserved but linear momentum will not be conserved. This special case is shown in Fig. 18–27*b*. The impulsive force $\mathbf{F}$ is always directed toward point O and hence the angular impulse created by this force about point O is always zero. (See Example 18–12.)

On the basis of Eq. 18–26, we can also write the conservation of angular momentum for a system of n particles, i.e.,

$$\Sigma(\mathbf{H}_O)_{i_1} = \Sigma(\mathbf{H}_O)_{i_2} \qquad (18\text{–}29)$$

where the summation includes all the particles.

The following Examples illustrate the use of these equations for solving problems.

The 10-lb block shown in Fig. 18-28a rests on a smooth horizontal surface, and is attached to an elastic cord having a stiffness of $k_c = 2$ lb/ft and unstretched length of 4 ft. The spring has a stiffness of $k_s = 15$ lb/ft and is initially compressed a distance of $\frac{1}{2}$ ft. When it is released, the block slides along the horizontal surface. Determine the component of velocity of the block acting perpendicular to the cord and the velocity of the block when the cord is stretched 1 ft.

(a)

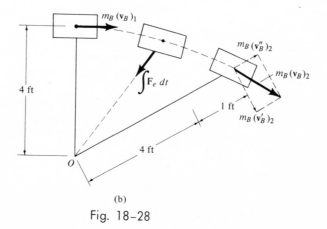

(b)

Fig. 18-28

Solution

The initial velocity of the block, just after the block is launched from the spring, can be determined by applying the conservation of energy

principle to the spring and block before, and just after the block has been launched. In this case, the potential energy stored in the spring, i.e., $\frac{1}{2}kx^2$, is converted into kinetic energy of the block, $\frac{1}{2}mv^2$. (During this time the small amount of potential energy which also goes into stretching the elastic cord when the block is moved $x = \frac{1}{2}$ ft will be neglected.) Thus,

$$T_0 + V_0 = T_1 + V_1$$

$$0 + \frac{1}{2}(15)\left(\frac{1}{2}\right)^2 = \frac{1}{2}\left(\frac{10}{32.2}\right)(v_B)_1^2 + 0$$

Solving for $(v_B)_1$ yields

$$(v_B)_1 = 3.47 \text{ ft/sec}$$

After the block has been launched, it follows the dashed path shown in Fig. 18–28b. The momentum- and impulse-vector diagrams for the block are also shown in this figure. Angular momentum about point O is *conserved,* since the moment of the linear impulse $\int F_e \, dt$ about point O is always zero. (F_e is a central force.) After the cable is stretched 1 ft, only the component of momentum $m_B(v_B')_2$ is effective in producing angular momentum of the block about point O. (Recall that angular momentum is the *moment* of linear momentum about a point, i.e., $H_O = r \times mv$.) Applying the conservation of angular momentum gives

$$(H_O)_1 = (H_O)_2$$

$$r_1 m_B(v_B)_1 = r_2 m_B(v_B')_2$$

$$4\left(\frac{10}{32.2}\right)(3.47) = 5\left(\frac{10}{32.2}\right)(v_B')_2$$

or

$$(v_B')_2 = 2.78 \text{ ft/sec} \qquad\qquad Ans.$$

The speed of the block, $(v_B)_2$, may be obtained by applying the conservation of energy principle before the block was launched and just after the cord has been stretched 1 ft.

$$T_0 + V_0 = T_2 + V_2$$

$$0 + \frac{1}{2}(15)\left(\frac{1}{2}\right)^2 = \frac{1}{2}\left(\frac{10}{32.2}\right)(v_B)_2^2 + \frac{1}{2}(2)(1)^2$$

Thus,

$$(v_B)_2 = 2.37 \text{ ft/sec} \qquad\qquad Ans.$$

(*Note:* It is possible to determine $(v_B'')_2 = \sqrt{(v_B)_2^2 - (v_B')_2^2}$, Fig. 18–28b. This component of velocity indicates the rate at which the cord is being stretched when the cord is 5 ft long.)

Example 18-13

The two identical blocks A and B, shown in Fig. 18–29a, each weigh 4 lb and are free to move along the frictionless horizontal rod DE. These two blocks are connected to the 2-lb block C which rides freely along rod GH. Holding block C so that it cannot travel up rod GH, a moment is applied to rod GH, which gives it an angular motion, until the velocities of blocks A and B reach 12 ft/sec as shown. When the moment is *removed* and block C is *released,* determine the tangential component of velocity of blocks A and B and the upward speed of C when A and B move 3 ft from the axis of rotation GH. Neglect the rotational effects caused by the rods and block C.

(a)

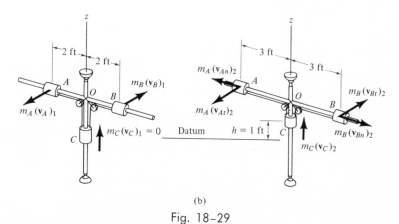

(b)

Fig. 18–29

Solution

Since the three blocks are connected by cords, all the impulses given to the system of the blocks and frame occur in equal but opposite pairs.

The angular momentum of the system is therefore *conserved* about point O. The momentum-vector diagrams are shown in Fig. 18–29b. The tangential velocities $(v_{At})_2$ and $(v_{Bt})_2$ are equal but opposite, since blocks A and B always travel along paths having the same radius of curvature. Applying the conservation of angular momentum to the system about point O yields

$$(H_O)_{A1} + (H_O)_{B1} = (H_O)_{A2} + (H_O)_{B2}$$

$$2\left(\frac{4}{32.2}\right)(12) + 2\left(\frac{4}{32.2}\right)(12) = 3\left(\frac{4}{32.2}\right)(v_{At})_2 + 3\left(\frac{4}{32.2}\right)(v_{At})_2$$

The terms represent the magnitude of the moments of the linear momentum about point O. In all cases this moment acts in the **k** direction, Fig. 18–29b. Solving for $(v_{At})_2$ gives

$$(v_{At})_2 = 8.0 \text{ ft/sec} \qquad \qquad \textit{Ans.}$$

$(v_{Bt})_2$ is the same; however, $(v_{Bt})_2$ acts in the opposite direction of $(v_{At})_2$, Fig. 18–29b.

The velocity of block C may be determined by applying the conservation of energy principle to the system. Locating the potential energy datum at the lowest position of block C, Fig. 18–29b, we have

$$\{T_1\} + \{V_1\} = \{T_2\} + \{V_2\}$$

$$\left\{ \frac{1}{2}m_A(v_A)_1^2 + \frac{1}{2}m_B(v_B)_1^2 + \frac{1}{2}m_C(v_C)_1^2 \right\} + \{0\}$$

$$= \left\{ \frac{1}{2}m_A(v_{At})_2^2 + \frac{1}{2}m_A(v_{An})_2^2 + \frac{1}{2}m_B(v_{Bt})_2^2 \right.$$

$$\left. + \frac{1}{2}m_B(v_{Bn})_2^2 + \frac{1}{2}m_C(v_C)_2^2 \right\} + \{W_C h\}$$

Substituting in the computed values of $(v_{At})_2$ and $(v_{Bt})_2$, realizing that $(v_{An})_2 = (v_{Bn})_2 = (v_C)_2$ because of the cords, we have,

$$\left\{ \frac{1}{2}\left(\frac{4}{32.2}\right)(12)^2 + \frac{1}{2}\left(\frac{4}{32.2}\right)(12)^2 + 0 \right\} + \{0\}$$

$$= \left\{ \frac{1}{2}\left(\frac{4}{32.2}\right)(8)^2 + \frac{1}{2}\left(\frac{4}{32.2}\right)(v_C)_2^2 + \frac{1}{2}\left(\frac{4}{32.2}\right)(8)^2 \right.$$

$$\left. + \frac{1}{2}\left(\frac{4}{32.2}\right)(v_C)_2^2 + \frac{1}{2}\left(\frac{2}{32.2}\right)(v_C)_2^2 \right\} + \{2(1)\}$$

Solving for $(v_C)_2$, we obtain

$$(v_C)_2 = 7.15 \text{ ft/sec} \qquad \qquad \textit{Ans.}$$

18-77. Determine the angular momentum $\mathbf{H}_O$ for the system of three particles about point O. All the particles are moving in the xy plane.

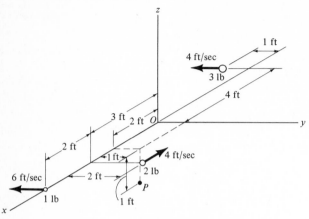

Prob. 18-77

18-78. Determine the angular momentum $\mathbf{H}_P$ of the system of three particles in Prob. 18-77 about the point P.

18-79. The 2-lb ball is attached to a cord as shown. When the ball is 3 ft from the hole, it is rotating around the hole such that its speed is 4 ft/sec. If the cord is suddenly pulled downward through the hole with a constant velocity of 1 ft/sec, determine the speed of the ball when the ball is 1 ft from the hole.

Prob. 18-79

18-80. The 1-lb slider block s moves outward from the center of a rod which has a negligible mass. The rod pivots about a pin and rotates in the *horizontal* plane. If the tangential component of velocity of the block at point A is 6 ft/sec and the block is forced to move outward with a constant velocity of 1 ft/sec relative to the rod, determine the tangential component of velocity after $t = 2$ sec from the position shown.

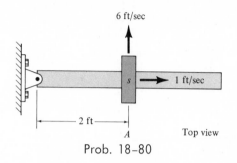

Prob. 18-80

18-81. The 10-lb shell is fired from a cannon with a muzzle velocity of 1,500 ft/sec. Determine the angular momentum of the shell about point O (a) when it is at the maximum height, point A, (b) at point B just before striking the ground, and (c) just after striking the ground.

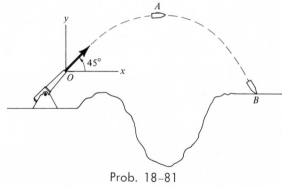

Prob. 18-81

18-82. The two slider blocks A and B weigh 3 lb each and slide freely outward along the horizontal rods. The radial component of velocity is always maintained at 2 ft/sec. When the blocks are located at the position

shown, their tangential velocity components are 16 ft/sec. The mass of the supporting frame is negligible and it is free to rotate about the vertical axis. Determine how long it takes for each block to attain a speed of 10 ft/sec.

Prob. 18–82

18-83. The eight 5-lb spheres are rigidly attached to the crossbar frame. The frame has a negligible weight. If a couple moment $M = (5t + 2)$ lb-ft, where t is in seconds, is applied as shown, determine the speed of each of the spheres in the inner and outer circles in 4 sec, starting from rest.

Prob. 18–83

18-84. The elastic cord, having an unstretched length of 2 ft and a stiffness of $k = 3$ lb/ft, is attached to a fixed point at A and a 6-lb block at B. If the block is given an initial velocity of $v_1 = 10$ ft/sec outward along the y axis, determine the velocity at which the block is moving toward point A the instant the cord becomes slack. How close does the block come within approaching point A after the cord has become slack? Assume that the xy plane is smooth.

Prob. 18–84

18-85. The 3-lb sphere located at A is released from rest and travels down the curved path. Knowing that the sphere exerts a normal force of 5 lb on the path when it reaches point B, determine the angular momentum of the sphere about the center of curvature, point O. The radius of curvature for the path at point B is indicated by ρ. Neglect any effects caused by rolling.

Prob. 18–85

18-86. The two spheres are rotating in a circular path with a constant speed. The cords are attached to a tubular shaft which causes the rotation. If the cord length

$l = 2$ ft is shortened to $l' = 1$ ft, by pulling the cords through the tube, determine the new diameter of the path d', such that the total angular momentum of the spheres remains constant. What is the tension in the cords in each case? Each sphere weighs 1 lb.

18-87. A small particle having a mass m is placed inside the vertical tube. The particle is displaced to the position shown and released. Using the principle of angular momentum, show that the motion of the particle is governed by the differential equation $d^2\theta/dt^2 + (g/R) \sin \theta = 0$.

Prob. 18-86

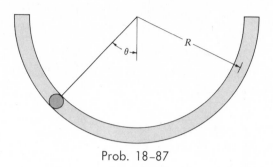

Prob. 18-87

19
Planar Kinetics of Rigid Bodies: Impulse and Momentum

19-1. Linear and Angular Momentum of a Rigid Body

The concepts of linear and angular momentum for a particle presented in Chapter 18 will be extended somewhat in order to determine momentum relationships which apply to a rigid body. Although the analysis will be quite general for translational motion, we will limit our discussion in this chapter to rotational motion of rigid bodies whose shape is symmetrical with respect to a fixed reference plane.* Because of this assumption, we may analyze the motion of the body by considering a slab having the same mass as the body, and projecting the forces acting on the body onto the plane of the slab. The three planar motions which we will consider are translation, rotation about a fixed axis, and general plane motion of a rigid body. The more general discussion of momentum principles applied to the spatial motion of rigid bodies is presented in Chapter 21.

The linear momentum $\mathbf{L}_i$ of a particle having a mass Δm_i and velocity $\mathbf{v}_i$ has been defined as $\mathbf{L}_i = \Delta m_i \mathbf{v}_i$. This vector has the same direction as the velocity of the particle. A common set of units used to measure the magnitude of linear momentum is lb-sec or kg_f-sec. Since a rigid body

---

*This assumption regarding the symmetry of the body was also necessary in deriving the rotational equations of motion in Chapter 15. However, recall that it was not required in deriving the kinetic energy expressions of Chapter 17.

consists of many particles, the linear momentum of the body may be obtained by summing (vectorially) the linear momenta of all n particles of the body. This sum is

$$\mathbf{L} = \Sigma\, \Delta m_i \mathbf{v}_i$$

The linear momentum of the body can be expressed in terms of the motion of the body's mass center G. By definition of the mass center,

$$(\Sigma\, \Delta m_i)\mathbf{r}_G = \Sigma\, \Delta m_i \mathbf{r}_i$$

where $\Sigma\, \Delta m_i = m$, the total mass of the body, and the position vector $\mathbf{r}_G$ locates the mass center with respect to the origin of an x, y, z inertia frame of reference. Taking the time derivative of this equation yields $(\Sigma\, \Delta m_i)\mathbf{v}_G = \Sigma\, \Delta m_i \mathbf{v}_i$. Hence, the linear momentum of the body can be written as

$$\boxed{\mathbf{L} = m\mathbf{v}_G} \qquad (19\text{--}1)$$

Hence, it can be seen that $\mathbf{L}$ has the same direction as $\mathbf{v}_G$, the velocity of the center of mass, and a magnitude of mv_G. In the derivation motion of the body has not been restricted in any way, so that Eq. 19–1 applies for *any motion* of the body.

The angular momentum of a particle has been defined as the moment of the linear momentum vector about a point. Using this concept, we may specify the line of action of the linear momentum vector $\mathbf{L}$ for each of the three types of plane motion of a rigid body. We will presently consider a detailed analysis for obtaining the angular momentum of a rigid body subjected to *general plane motion*. The special cases of translation and rotation about a fixed axis will then be discussed.

To simplify the analysis, the body will be assumed symmetric with respect to a fixed xy reference plane. Thus, the *general motion* of the body can be characterized by considering a slab of the body lying in the reference plane, as shown in Fig. 19–1a. The $x'y'$ coordinate system has its origin fixed in the body at the mass center G. The axes of this coordinate system translate (remain parallel) to the axes of the inertial xy frame of reference. At the instant shown the body has an instantaneous angular velocity $\boldsymbol{\omega}$ and its mass center has an instantaneous velocity of $\mathbf{v}_G$.* The velocity of the ith particle, located at $\mathbf{r}_{i/G}$ from the mass center, can be determined by using Eq. 13–13, i.e.,

$$\mathbf{v}_i = \mathbf{v}_G + \boldsymbol{\omega} \times \mathbf{r}_{i/G}$$

By the distributive property of the vector cross-product, the angular

(a)

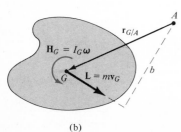

(b)

Fig. 19–1

*The body may also have accelerated motion but this is of no consequence in the following derivation.

momentum of the ith particle about the mass center G (z' axis) may therefore be written as

$$(\mathbf{H}_i)_G = \mathbf{r}_{i/G} \times \Delta m_i \mathbf{v}_i = \Delta m_i (\mathbf{r}_{i/G} \times \mathbf{v}_G) + \Delta m_i \mathbf{r}_{i/G} \times (\boldsymbol{\omega} \times \mathbf{r}_{i/G})$$

When similar expressions are written and summed for all n particles in the body, the total angular momentum becomes

$$\mathbf{H}_G = (\Sigma \, \Delta m_i \mathbf{r}_{i/G}) \times \mathbf{v}_G + \Sigma[\Delta m_i \mathbf{r}_{i/G} \times (\boldsymbol{\omega} \times \mathbf{r}_{i/G})]$$

By definition of the mass center, the sum $\Sigma \, \Delta m_i \mathbf{r}_{i/G} = (\Sigma \, \Delta m_i)\bar{\mathbf{r}}$ is equal to zero since the position vector $\bar{\mathbf{r}}$ relative to G is equal to zero. Hence, the first term on the right is equal to zero. The second term may be further simplified by noting that $\mathbf{r}_{i/G} \times (\boldsymbol{\omega} \times \mathbf{r}_{i/G})$ is a vector having a magnitude of $r_{i/G}^2 \omega$ and acting perpendicular to the plane of the slab. Realizing that $\mathbf{H}_G$ also acts in the same direction, we can write the previous equation in the scalar form

$$H_G = (\Sigma \, \Delta m_i r_{i/G}^2)\omega$$

Letting the particle size $\Delta m_i \rightarrow dm$, the summation sign becomes an integral since the number of particles $n \rightarrow \infty$. Representing $r_{i/G}$ by a more generalized dimension r, we have

$$H_G = \left(\int_m r^2 \, dm \right) \omega$$

The integral represents the moment of inertia I_G of the body about the z' axis which is perpendicular to the slab and passes through point G. Our final result, written in scalar form, is therefore

$$H_G = I_G \omega \tag{19-2}$$

Thus, in the case of general plane motion of a symmetrical rigid body (represented here as a slab) the *angular momentum* $\mathbf{H}_G$ of the body about the mass center is a *free vector*. The *direction* of this vector is the same as that of $\boldsymbol{\omega}$, that is, always perpendicular to the slab. The vector *magnitude*, $I_G \omega$, is commonly measured in units of ft-lb-sec or m-kg$_f$-sec.

To summarize our results, if a body is undergoing general plane motion, the resultant linear momentum of the body can be represented by the vector $\mathbf{L} = m\mathbf{v}_G$ and the resultant angular momentum of the body, *computed about the mass center*, is represented by $\mathbf{H}_G = I_G \boldsymbol{\omega}$. Note that $\mathbf{L}$ is created only by the translational motion of the body's mass center; whereas $\mathbf{H}_G$ is created only by the rotational motion $\boldsymbol{\omega}$. Since angular momentum is equal to the moment of linear momentum, the line of action of $\mathbf{L}$ must *pass through the mass center of the body* in order to preserve the correct magnitude of $\mathbf{H}_G$, Fig. 19-1b. Provided this is the case, $\mathbf{L}$ creates zero angular momentum about G so that $\mathbf{H}_G = I_G \boldsymbol{\omega}$ as defined

by Eq. 19–2. As a consequence, the angular momentum of the body can be computed about *any point* other than the mass center G provided one accounts for the angular momentum created by both $\mathbf{L}$ and $\mathbf{H}_G$ about the point. For example, if the angular momentum of the body is computed with respect to a fixed or moving axis passing through point A, Fig. 19–1b, one obtains

$$\mathbf{H}_A = (\mathbf{r}_{G/A} \times m\mathbf{v}_G) + I_G\boldsymbol{\omega}$$

which can be written in scalar form as

$$\zeta + H_A = b(mv_G) + I_G\omega$$

Here b is the perpendicular distance from point A to the line of action of the linear momentum vector $\mathbf{L} = m\mathbf{v}_G$ as shown in the figure. Hence, in the analysis it is seen that the linear momentum $\mathbf{L}$ and angular momentum $\mathbf{H}_G$ have the same vector properties as a force and couple.

In particular, when a rigid body is subjected to either rectilinear or curvilinear *translation,* $\mathbf{H}_G = 0$ since $\boldsymbol{\omega} = 0$. The angular momentum of the body about any point not lying on the line of action of $\mathbf{L}$ (which passes through the mass center) is computed simply by taking "moments" of $\mathbf{L}$ about the point. For example, consider the body shown in Fig. 19–2 which is translating with a velocity $\mathbf{v} = \mathbf{v}_G$. The body has a linear momentum $\mathbf{L} = m\mathbf{v}$. The angular momentum computed about point B is then $\mathbf{H}_B = \mathbf{r}_{G/B} \times m\mathbf{v}$. The magnitude of this vector may be expressed by $H_B = e(mv)$, where e is defined in the figure. The sense of direction is determined by the cross product, using the right-hand rule.

When the rigid body is subjected to *rotation about a fixed axis* passing through point O, Fig. 19–3, the angular momentum of the body computed with respect to point O reduces to a simplified form. From the general plane motion analysis, the rotational motion $\boldsymbol{\omega}$ about point O imparts an angular momentum $\mathbf{H}_G = I_G\boldsymbol{\omega}$ about point G and linear momentum $\mathbf{L} = m\mathbf{v}_G$ to the body. The line of action of the linear momentum passes through the mass center G, and, due to the rotational motion $\mathbf{v}_G = \mathbf{r}_{G/O} \times \boldsymbol{\omega}$, $\mathbf{L}$ is *perpendicular* to $\mathbf{r}_{G/O}$. Thus, as shown in the figure, we have

$$\zeta + H_O = I_G\omega + r_{G/O}(mv_G)$$

Since $v_G = r_{G/O}\omega$,

$$\zeta + H_O = [I_G + m(r_{G/O})^2]\omega$$

From the parallel-axis theorem, Eq. 15–9, the terms inside the brackets represent the moment of inertia I_O of the body with respect to point O. Thus, the *magnitude* of the angular momentum about point O becomes

$$H_O = I_O\omega \qquad (19\text{–}3)$$

$\mathbf{H}_O$ acts in the same direction as the angular velocity $\boldsymbol{\omega}$.

Fig. 19–2

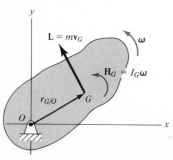

Fig. 19–3

In particular, if the body is rotating about a *fixed axis passing through its mass center G*, the linear momentum of the body is zero ($v_G = 0$), and the angular momentum of the body is simply $H_G = I_G\omega$.

19-2. Principle of Impulse and Momentum for a Rigid Body

The *principle of linear impulse and momentum* for a rigid body may be established directly from Eq. 15-1:

$$\Sigma F = ma_G = m\frac{dv_G}{dt}$$

or

$$\Sigma F = \frac{d}{dt}(mv_G)$$

since the mass m of the body is constant. Multiplying both sides of this equation by dt and integrating between the limits $t = t_1$, $v_G = (v_G)_1$, and $t = t_2$, $v_G = (v_G)_2$ yields

$$\Sigma \int_{t_1}^{t_2} F\, dt = m(v_G)_2 - m(v_G)_1 \qquad (19\text{-}4)$$

This result states that the impulses created by the *external force system* which acts on the body during the time interval t_1 to t_2 is equal to the change in the linear momentum of the body during the time interval.

The *principle of angular impulse and momentum* for the rigid body subjected to plane motion may be established directly from the rotational equation of motion. For example, if the body is subjected to *general plane motion*, the third of Eqs. 15-6 states that

$$\Sigma M_G = I_G\alpha = I_G\frac{d\omega}{dt}$$

or

$$\Sigma M_G = \frac{d}{dt}(I_G\omega)$$

since the moment of inertia of the body, I_G, is constant. Multiplying both sides of this equation by dt and integrating from $t = t_1$, $\omega = \omega_1$ to $t = t_2$, $\omega = \omega_2$, we obtain the result

$$\Sigma \int_{t_1}^{t_2} M_G\, dt = I_G\omega_2 - I_G\omega_1 \qquad (19\text{-}5)$$

In a similar manner, for *rotation about a fixed axis* passing through

point O, the third of Eqs. 15–13 ($\Sigma M_O = I_O \alpha$) when integrated becomes

$$\Sigma \int_{t_1}^{t_2} M_O \, dt = I_O \omega_2 - I_O \omega_1 \qquad (19\text{–}6)$$

Equations 19–5 and 19–6 state that the total angular impulse acting on the body during the time interval t_1 to t_2 is equal to the change in the angular momentum of the body during the time interval. In particular, the angular impulse considered is determined by integrating the moment about point G or O of all the external forces and/or couples applied to the body.

A special case of plane motion exists which deserves mentioning. This occurs for a body, such as a homogeneous sphere, cylinder, or wheel, which is *rolling on a surface without slipping and whose center of mass coincides with its centroid.* For this body, the moment equation $\Sigma M_{IC} = I_{IC} \alpha$ (Eq. 15–19) may be applied at the surface point of contact which is the instantaneous center of zero velocity, IC, for the body (refer to Sec. 15–6). This equation of motion may be integrated with respect to time as in the previous cases since the axis for computing the moment of the external forces *always* coincides with the instantaneous axis of zero velocity* at each instant during the motion; thus,

$$\Sigma \int_{t_1}^{t_2} M_{IC} \, dt = I_{IC} \omega_2 - I_{IC} \omega_1 \qquad (19\text{–}7)$$

where I_{IC} represents the moment of inertia of the body about the instantaneous axis of zero velocity. For some problems, application of this equation is particularly convenient, since the friction and normal contact forces at the IC are eliminated from the equation. See Example 19–2. If the rolling body does *not* have a coincident centroid and center of mass or slips at the contacting surface, Eq. 19–7 is *not* valid and one must apply Eq. 19–5 to express correctly the principle of angular impulse and momentum for the body.

Common units used to express the terms in Eqs. 19–5, 19–6, and 19–7 are ft-lb-sec or m-kg$_f$-sec. These units for angular momentum are *different* from those used to measure linear momentum, e.g., lb-sec or kg$_f$-sec. Thus, care should be taken *not* to mistakenly add these two quantities.

19–3. Method of Problem Solution

In Sec. 19–2, the principles of linear and angular impulse and momentum were established for a rigid body subjected to plane motion. When solving

---

* Recall that the instantaneous axis of zero velocity is perpendicular to the plane of motion and passes through the IC.

$$m(v_{Gx})_1 + \Sigma \int_{t_1}^{t_2} F_x \, dt = m(v_{Gx})_2$$

$$m(v_{Gy})_1 + \Sigma \int_{t_1}^{t_2} F_y \, dt = m(v_{Gy})_2 \qquad (19\text{–}8)$$

$$I_G \omega_1 + \Sigma \int_{t_1}^{t_2} M_G \, dt = I_G \omega_2$$

These three equations allow one to solve for, at most, three unknowns. If the body is rotating about a fixed axis, Eq. 19–6 may be substituted for Eq. 19–5 in the analysis. Or, for some problems involving plane motion it may be convenient to apply Eqs. 19–7. Equations 19–8 may also be applied to an entire system of connected bodies rather than to each body separately. Doing this eliminates the need to include reactive impulses which are *internal* to the system. If this is done, the resultant equations may then be written in symbolic form as

$$\left(\sum \substack{\text{System linear} \\ \text{momentum}} \right)_1 + \left(\sum \substack{\text{System linear} \\ \text{impulse}} \right)_{1-2} = \left(\sum \substack{\text{System linear} \\ \text{momentum}} \right)_2$$

$$\left(\sum \substack{\text{System angular} \\ \text{momentum}} \right)_1 + \left(\sum \substack{\text{System angular} \\ \text{impulse}} \right)_{1-2} = \left(\sum \substack{\text{System angular} \\ \text{momentum}} \right)_2$$

$$(19\text{–}9)$$

The system angular momentum and angular impulse must be computed with respect to the *same fixed reference point* for every element of the system when using these equations. See Example 19–1.

Before Eqs. 19–8 or 19–9 are applied to the solution of dynamics problems, it is suggested that you construct a set of momentum- and impulse-vector diagrams. As in the case for particles, each diagram corresponds to one of the three terms in Eqs. 19–8 or 19–9. An appropriate set of these three diagrams for general rigid body motion is shown in Fig. 19–4. The linear momentum vectors mv_G, Figs. 19–4a and 19–4c, are applied at the body's mass center. The angular momentum vectors $I_G \omega$ are free vectors, and therefore, like a couple, may be applied at any point on the body. When constructing the impulse-vector diagram, Fig. 19–4b, it is generally necessary to consider the impulses created by *all* forces and couples which act on the body. (This is *different* from the principle of work and energy where only forces which do work are considered.) If a force **F** or couple moment **M** acting on a rigid body is *constant* during the time interval t_1 to t_2, the integration of the impulse reduces to $\mathbf{F}(t_2 - t_1)$ and $\mathbf{M}(t_2 - t_1)$, respectively. This is the case for the weight force **W**, shown in Fig. 19–4b. When the force or couple moment

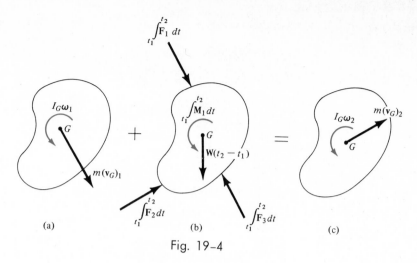

Fig. 19-4

varies with time, the impulse must be determined by integration, in which case the impulses are indicated as $\int_{t_1}^{t_2} \mathbf{F}(t)\,dt$ and $\int_{t_1}^{t_2} \mathbf{M}(t)\,dt$.

In general, a procedure to be used for the solution of a particular type of problem should be decided upon before attempting to solve the problem. The *principle of impulse and momentum* is generally most suitable for solving problems which involve *time* and *velocities*. As shown in Chapter 18, it also serves as the most practical method for solving problems involving impact. For some problems a combination of Newton's second law of motion and its two integrated forms, the principle of work and energy and the principle of impulse and momentum, will yield the most direct solution to the problem.

Example 19-1

The 20-lb pulley in Fig. 19-5a supports two weights A and B. The radius of gyration of the pulley is $k_O = 0.8$ ft. If weight B is initially moving downward with a velocity of 10 ft/sec, determine its speed in 6 sec. What are the horizontal and vertical components of reaction at the pin O in 6 sec?

Solution I

From *kinematics*, it is seen that the downward motion of block B causes the pulley to rotate clockwise and block A to move upwards. Since the radius of the pulley is $r = 1$ ft, at any instant of time the magnitude of the angular velocity of the pulley, ω, and the speed of blocks A and B may be related by the equations

$$\omega = v_A/1 \text{ ft} = v_B/1 \text{ ft}$$

Therefore at time t_1, when $(v_A)_1 = 10$ ft/sec,

$$\omega_1 = 10 \text{ rad/sec}, \quad (v_A)_1 = (v_B)_1 = 10 \text{ ft/sec}$$

And at time t_2,

$$(1 \text{ ft}) \omega_2 = (v_A)_2 = (v_B)_2$$

The momentum- and impulse-vector diagrams for each of the weights

Fig. 19–5

and the pulley are shown in Fig. 19–5b. Note that the momentum vectors act in a direction which is compatible with the motion just discussed. Because gravity is the cause of motion, all the impulsive forces acting on the impulse-vector diagram are constant throughout the 6-sec time interval. In particular, the tensions in the cable which supports weights A and B are $\mathbf{T}_A$ and $\mathbf{T}_B$. (The difference in cable tension is accounted for by the frictional force which is developed at the contacting surface between the cable and the pulley.)

Applying the *principle of linear momentum* to the weights A and B, noting that $(v_A)_1 = (v_B)_1 = 10$ ft/sec and $(v_A)_2 = (v_B)_2$, yields

$(+\uparrow)$ $m_A(v_A)_1 + \Sigma \int_{t_1}^{t_2} F_y \, dt = m_A(v_A)_2$ (weight A)

$$\frac{5}{32.2}(10) + T_A(6) - 5(6) = \frac{5}{32.2}(v_B)_2 \tag{1}$$

$(+\downarrow)$ $m_B(v_B)_1 + \Sigma \int_{t_1}^{t_2} F_y \, dt = m_B(v_B)_2$ (weight B)

$$\frac{25}{32.2}(10) + 25(6) - T_B(6) = \frac{25}{32.2}(v_B)_2 \tag{2}$$

The moment of inertia of the pulley about its fixed axis of rotation is determined using the radius of gyration and the weight of the pulley, i.e.,

$$I_O = mk_O^2 = \left(\frac{20}{32.2}\right)(0.8)^2 = 0.398 \text{ slug-ft}^2$$

Assuming that clockwise rotation is positive and applying the *principle of angular impulse and momentum* to the pulley about point O (Eq. 19–6) noting that $\omega_1 = 10$ rad/sec and $\omega_2 = (v_B)_2/1$ ft, we have

$(\curvearrowright +)$ $I_O\omega_1 + \Sigma \int_{t_1}^{t_2} M_O \, dt = I_O\omega_2$

$$0.398(10) - T_A(6)1 + T_B(6)1 = 0.398(v_B)_2 \tag{3}$$

Simplifying and solving Eqs. (1) through (3) simultaneously gives

$$(v_B)_2 = 100.3 \text{ ft/sec} \qquad Ans.$$
$$T_A = 7.33 \text{ lb}$$
$$T_B = 13.32 \text{ lb}$$

The reaction components at O may be obtained by applying the *principle of linear momentum* to the pulley, Fig. 19–5b,

$(\xrightarrow{+})$ $m_p(v_{Ox})_1 + \Sigma \int_{t_1}^{t_2} F_x \, dt = m_p(v_{Ox})_2$

$$0 + O_x(6) = 0$$

Hence,

$$O_x = 0 \qquad \textit{Ans.}$$

$(+\uparrow)$
$$m_p(v_{Oy})_1 + \Sigma \int_{t_1}^{t_2} F_y \, dt = m_p(v_{Oy})_2$$

$$0 + O_y(6) - 20(6) - T_A(6) - T_B(6) = 0$$

Substituting the numerical values of T_A and T_B into this equation yields

$$O_y(6) - 20(6) - 7.33(6) - 13.32(6) = 0$$

$$O_y = 40.7 \text{ lb} \qquad \textit{Ans.}$$

Solution II

A more direct solution to this problem may be obtained by applying the principle of impulse and momentum to the *entire system,* consisting of the two weights, the cable, and the pulley. The cable tensions are thereby eliminated from the analysis since they act as internal forces and hence cancel one another. The momentum-and impulse-vector diagrams for the system are shown in Fig. 19–5c. There are three unknowns, O_x, O_y, and $(v_B)_2$. (Note that $v_A = \omega(1 \text{ ft}) = v_B$.) Applying the principle of linear impulse and momentum to the system we have,

$$\left(\Sigma \begin{matrix} \text{System linear} \\ \text{momentum} \end{matrix}\right)_{x1} + \left(\Sigma \begin{matrix} \text{System linear} \\ \text{impulse} \end{matrix}\right)_{x(1-2)} = \left(\Sigma \begin{matrix} \text{System linear} \\ \text{momentum} \end{matrix}\right)_{x2}$$

$(+\uparrow)$
$$0 + \Sigma \int_{t_1}^{t_2} F_x \, dt = 0$$

$$0 + O_x(6) = 0 \qquad (4)$$

$$\left(\Sigma \begin{matrix} \text{System linear} \\ \text{momentum} \end{matrix}\right)_{y1} + \left(\Sigma \begin{matrix} \text{System linear} \\ \text{impulse} \end{matrix}\right)_{y(1-2)} = \left(\Sigma \begin{matrix} \text{System linear} \\ \text{momentum} \end{matrix}\right)_{y2}$$

$(+\uparrow)$
$$m_A(v_{Ay})_1 + m_B(v_{By})_1 + \Sigma \int_{t_1}^{t_2} F_y \, dt = m_A(v_{Ay})_2 + m_B(v_{By})_2$$

$$\frac{5}{32.2}(10) - \frac{25}{32.2}(10) + O_y(6) - 20(6) - 5(6) - 25(6) \qquad (5)$$

$$= \frac{5}{32.2}(v_B)_2 - \frac{25}{32.2}(v_B)_2$$

The principle of angular impulse and momentum will be applied with respect to point O. Accounting for the *moment* of both the momentum and impulses about this point, Fig. 19–5c, we have

$$\left(\sum \begin{array}{c}\text{System angular} \\ \text{momentum}\end{array}\right)_{O1} + \left(\sum \begin{array}{c}\text{System angular} \\ \text{impulse}\end{array}\right)_{O(1-2)} = \left(\sum \begin{array}{c}\text{System angular} \\ \text{momentum}\end{array}\right)_{O2}$$

$(\curvearrowleft +)$ $m_A(v_A)_1(r) + m_B(v_B)_1(r) + I_O\omega_1 - 5(6)r + 25(6)r$

$$= m_A(v_A)_2 r + m_B(v_B)_2 r + I_O\omega_2$$

Since $I_O = 0.398$ slug-ft^2, $(v_A)_1 = (v_B)_1 = 10$ ft/sec, $\omega_1 = 10$ rad/sec; and $(v_A)_2 = \omega_2$ (1 ft) $= (v_B)_2$, we obtain

$$\frac{5}{32.2}(10)(1) + \frac{25}{32.2}(10)(1) + 0.398(10) - 5(6)(1) + 25(6)(1) =$$

$$\frac{5}{32.2}(v_B)_2(1) + \frac{25}{32.2}(v_B)_2(1) + 0.398(v_B)_2 \quad (6)$$

After simplifying, solution of Eqs. (4) through (6) yields

$$O_x = 0 \qquad\qquad\qquad Ans.$$

$$O_y = 40.7 \text{ lb} \qquad\qquad Ans.$$

$$(v_B)_2 = 100.3 \text{ ft/sec} \qquad Ans.$$

Example 19–2

The 100-lb wheel shown in Fig. 19–6a has a radius of gyration of $k_G = 1.75$ ft. A cable is wrapped around the central hub of the wheel and a variable horizontal force having a magnitude of $P = (t + 3)$ lb is applied, where t is measured in seconds. If the wheel is initially at rest, determine its angular velocity in 10 sec. Assume that the wheel rolls without slipping.

Solution I

The momentum- and impulse-vector diagrams of the wheel are shown in Fig. 19–6b. Why are these diagrams needed? Since the wheel does not slip, the instantaneous center of zero velocity is at the point of contact with the ground. From the kinematics of rotation, the velocity of the mass center G can be expressed in terms of the wheel's angular velocity ω using the IC. This yields $(v_G)_2 = 2\omega_2$, shown from the momentum-vector diagram. The impulse created by the frictional force $\mathbf{F}$, shown on the impulse-vector diagram, acts to the left. This impulse must be determined by integration. (Note that F is *variable* since P is variable.)

The moment of inertia of the wheel about the mass center is

$$I_G = mk_G^2 = \left(\frac{100}{32.2}\right)(1.75)^2 = 9.51 \text{ slug-ft}^2$$

Using this data and applying Eqs. 19–8 to the wheel, in reference to Fig. 19–6b, we have

P = (t + 3) lb

(a)

(b)

Fig. 19-6

$(\xrightarrow{+})$ $m(v_{Gx})_1 + \Sigma \int_{t_1}^{t_2} F_x \, dt = m(v_{Gx})_2$

$0 + \int_0^{10} (t + 3) \, dt - \int_0^{10} F \, dt = \frac{100}{32.2}(2\omega_2)$

$(+\uparrow)$ $m(v_{Gy})_1 + \Sigma \int_{t_1}^{t_2} F_y \, dt = m(v_{Gy})_2$

$0 + N(10) - 100(10) = 0$

and

$(\curvearrowleft+)$ $I_G\omega_1 + \Sigma \int_{t_1}^{t_2} M_G \, dt = I_G\omega_2$

$0 + \int_0^{10} (t + 3)(1) \, dt + \int_0^{10} F(2) \, dt = 9.51\omega_2$

Simplifying these equations gives

$$80 - \int_0^{10} F \, dt = 6.21\omega_2 \qquad (1)$$

$$10N - 1,000 = 0 \qquad (2)$$

$$80 + 2 \int_0^{10} F \, dt = 9.51 \omega_2 \tag{3}$$

Eliminating the unknown $\int_0^{10} F \, dt$ between Eqs. (1) and (3) and solving for ω_2 yields

$$\omega_2 = 10.94 \text{ rad/sec} \qquad \textit{Ans.}$$

Although the normal force is not needed for the solution, it can be determined by solving Eq. (2), which gives

$$N = 100 \text{ lb}$$

Solution II

Since the wheel's centroid and center of mass *coincide* and the wheel *rolls without slipping*, Eq. 19–7 may be applied, whereby one obtains a direct solution for ω_2. Using the parallel-axis theorem, the moment of inertia of the wheel about the instantaneous center, *IC*, is

$$I_{IC} = I_G + mr_{G/IC}^2 \tag{4}$$

$$= 9.51 + \frac{100}{32.2}(2)^2 = 21.9 \text{ slug-ft}^2$$

Thus, in reference to the momentum- and impulse-vector diagrams, Fig. 19–6b, we have

$(\circlearrowleft +)$ $$I_{IC}\omega_1 + \Sigma \int_{t_1}^{t_2} M_{IC} \, dt = I_{IC}\omega_2$$

$$0 + \int_0^{10} (t + 3)3 \, dt = 21.9\omega_2 \tag{5}$$

Integrating and solving for ω_2 yields*

$$\omega_2 = 10.94 \text{ rad/sec} \qquad \textit{Ans.}$$

Example 19–3

The 30-lb disk A, shown in Fig. 19–7a, is initially at rest when it suddenly comes in contact with a 10-lb rotating disk B. The coefficient

---

*The term on the right side of Eq. (5) *includes* the moment of the angular momentum $I_G\omega_2$ and linear momentum $m(v_G)_2$ about IC. This is because I_{IC} was used for the computation, i.e., from Fig. 19–6b and Eq. (4),

$$H_{IC} = I_G\omega_2 + m(v_G)_2 r_{G/IC}$$
$$= I_G\omega_2 + m(r_{G/IC}\omega_2) r_{G/IC}$$
$$= (I_G + mr_{G/IC}^2)\omega_2 = I_{IC}\omega_2$$

(a)

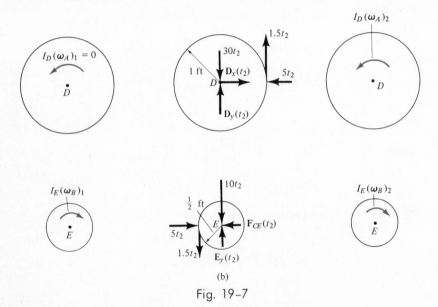

(b)

Fig. 19-7

of kinetic friction between the two disks is $\mu_k = 0.3$. The spring device at C creates a constant normal force between the two disks of 5 lb. If disk B is freely rotating with an initial angular velocity of $(\omega_B)_1 = 40$ rad/sec, determine how long it takes for slipping to stop between the two disks after they come in contact, and compute the angular velocities of both disks after slipping stops. (Why can this problem be solved using the principle of impulse and momentum?)

Solution

The momentum- and impulse-vector diagrams for each disk are shown in Fig. 19-7b. During the time interval $0 < t < t_2$, the disks are *slipping* and the constant frictional force acting between them is $F = \mu_k N = 0.3(5) = 1.5$ lb. This force and the 5-lb normal force create equal but opposite constant impulses between the disks as shown on the impulse-vector diagrams. When $t = t_2$, slipping between the disks has stopped,

and the angular velocity of both disks may be related. Since the circumference of disk B is one half that of disk A, $(\omega_A)_2 = (\omega_B)_2/2$.

The moment of inertia of each disk about its mass center is determined using Appendix D.

$$I_D = \frac{1}{2} m_A r_A^2 = \frac{1}{2}\left(\frac{30}{32.2}\right)(1)^2 = 0.466 \text{ slug-ft}^2$$

$$I_E = \frac{1}{2} m_B r_B^2 = \frac{1}{2}\left(\frac{10}{32.2}\right)\left(\frac{1}{2}\right)^2 = 0.0388 \text{ slug-ft}^2$$

Applying the equations of angular impulse and momentum to each disk gives

$(\zeta+)$
$$I_D(\omega_A)_1 + \Sigma \int_{t_1}^{t_2} M_D \, dt = I_D(\omega_A)_2$$

$$0 + 1.5t_2(1) = 0.466\left(\frac{(\omega_B)_2}{2}\right)$$

and

$(\uparrow+)$
$$I_E(\omega_B)_1 + \Sigma \int_{t_1}^{t_2} M_E \, dt = I_E(\omega_B)_2$$

$$0.0388(40) - 1.5t_2(\tfrac{1}{2}) = 0.0388(\omega_B)_2$$

Solving these two equations simultaneously, we have

$$t_2 = 1.55 \text{ sec} \qquad\qquad Ans.$$

$$(\omega_B)_2 = 10.0 \text{ rad/sec} \qquad\qquad Ans.$$

Hence,

$$(\omega_A)_2 = \frac{(\omega_B)_2}{2} = 5.0 \text{ rad/sec} \qquad\qquad Ans.$$

Problems

19-1. A 200-lb flywheel has a diameter of 3 ft and a radius of gyration about its center of $k_O = 1.2$ ft. A motor supplies a constant torque of 12 lb-in. to a 2-in.-diameter supporting shaft connected to the center of the flywheel. Bearing friction at the supporting shaft is 10 lb. Determine how long the torque must be applied to the shaft to increase the rotational speed of the flywheel from 4 to 15 rad/sec. If the motor is suddenly disengaged from the shaft once it is rotating at 15 rad/sec, determine how long it will take before the friction at the bearings stops the flywheel from rotating.

19-2. The shaft AB is supported at its ends by means of ball-and-socket joints. The slender supporting rods have a linear density of 1 lb/ft, and each of the four spheres has a diameter of $\frac{1}{4}$ ft and weighs 3 lb. Determine the angular momentum of the shaft if it is rotating with an angular velocity of 4 rad/sec.

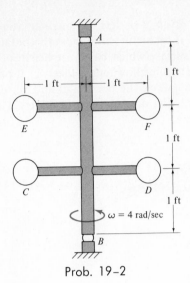

Prob. 19-2

19-3. Determine the angular momentum of the 30-lb plate about point A. The angular velocity of rod CD is 20 rad/sec.

Prob. 19-3

19-4. The pendulum consists of a uniform 6-lb rod AB and a 10-lb square plate. If the pendulum has a clockwise angular velocity of 10 rad/sec at the instant shown,

determine the linear and angular momentum of the pendulum.

Prob. 19-4

19-5. The 20-lb wheel has a radius of gyration of $k_O = 14$ in. If the wheel is released from rest and rolls down the plane without slipping, determine the velocity of point O 10 sec after the wheel is released.

Prob. 19-5

19-6. The block weighs 10 lb and has a radius of gyration about its mass center, G, of $k_G = 0.6$ ft. The kinetic energy of the block is 31 ft-lb when the block is in the position shown. If the block is rolling counterclockwise on the surface without slipping, determine the linear momentum of the block at this instant.

825

Prob. 19-6

19-9. The 25-lb disk has an angular velocity of 30 rad/sec. It is suddenly brought into contact with the horizontal surface which has a coefficient of friction of $\mu = 0.2$. Determine how long it takes for the disk to stop spinning. What are the horizontal and vertical components of reaction at A during this time? The link AB weighs 5 lb.

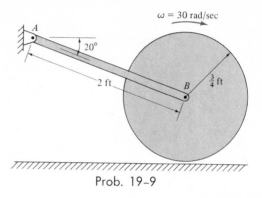

Prob. 19-9

19-7. The wheel weighs 100 lb and has a radius of gyration of $k_O = 1.5$ ft. A cord is wrapped around the inner hub of the wheel and is subjected to a force of $P = 2t + 3$, where P is in pounds and t is in seconds. If the wheel rolls without slipping, determine the speed of the center of the wheel 5 sec after the force is applied.

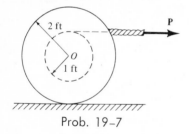

Prob. 19-7

19-8. The 100-lb wheel has a radius of gyration of $k_O = \frac{3}{4}$ ft. If the upper cable is subjected to a tension of $T = 50$ lb, determine the velocity of the center of the wheel in 3 sec, starting from rest. The coefficient of friction between the wheel and the surface is $\mu = 0.1$.

19-10. The 40-lb cylinder rests on an inclined surface having a coefficient of friction of $\mu = 0.1$. A cord is wrapped around the outer surface of the cylinder. If the cylinder is released from rest, determine the velocity of the center of the cylinder 1 sec after it is released.

Prob. 19-10

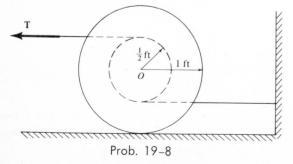

Prob. 19-8

19-11. The flywheel A weighs 30 lb and has a radius of gyration of $k_A = \frac{1}{2}$ ft. Disk B weighs 10 lb and is coupled to the flywheel by means of a V-belt which is

assumed not to slip at its contacting surface. A motor supplies a counterclockwise torque **T** to the flywheel. If $T = 50t$, where T is in lb-ft and t is in seconds, determine the angular velocity of the disk 5 sec after the motor is turned on. The spring support exerts a constant force of 80 lb acting to the right on B.

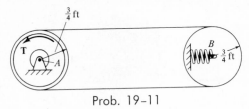

Prob. 19-11

19-12. Determine the amount of "backspin" ω which must be given to a 4-in.-diameter ball such that it stops spinning at the same instant that its forward velocity is zero. Initially, the ball is cast on the *rough surface* with a forward velocity of 2 ft/sec.

19-13. The bumper cushions on a pinball machine are 1 in. above their mounted surface, as shown. Determine the largest diameter ball which may be used so that, when a ball strikes the cushion, it rebounds without slipping. Assume that the cushion impact on the ball is horizontal.

Prob. 19-13

19-14. The 200-lb drum rolls along the incline, which has a coefficient of friction of $\mu = 0.2$. The radius of gyration of the drum is $k_O = 1.3$ ft. If the drum is released from rest, determine the maximum angle θ for the incline so that the drum rolls without slipping.

Prob. 19-14

19-15. Show that if a slab is rotating about a fixed axis perpendicular to the slab and passing through its mass center G, the angular momentum of the slab is the same about all points, fixed or moving.

19-16. The 30-lb flywheel A has a radius of gyration of $\frac{1}{3}$ ft. Disk B weighs 50 lb and is coupled to the flywheel by means of a belt which does not slip at its contacting surfaces. A motor supplies a clockwise torque **T** to the flywheel. If $T = 50t$ where T is in lb-ft and t is in seconds, determine the time required for the disk to attain an angular velocity of 60 rad/sec.

Prob. 19-16

19-17. The rigid body (slab) has a mass m and is rotating with an angular velocity ω about an axis passing through the fixed point O. Show that the momentum of all the particles composing the body can be represented by a single vector, having a magnitude of mv_G and acting through point P, which lies at a distance $r_{GP} = k_G^2/r_{OG}$ from the mass center G. The point P is called the *center of percussion*.

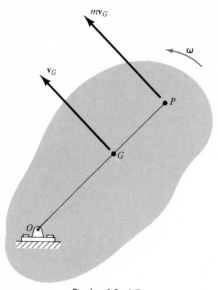

Prob. 19-17

of friction between the rack and the horizontal surface is $\mu = 0.2$. If the rack is initially moving to the left at a velocity of 2 ft/sec, determine the constant moment **M** which must be applied to the gear in order to increase the motion of the rack so that in 5 sec it will have a velocity of 6 ft/sec to the left.

Prob. 19-19

19-18. The 10-lb gear A rolls around the surface of the stationary outer gear rack. Motion occurs in the *horizontal plane*. If the 3-lb connecting link BC is subject to a torque of $T = 1$ lb-ft, determine the angular velocity of gear A in 30 sec starting from rest. The radius of gyration of gear A about point B is $k_B = 2$ in.

19-20. The 12-lb block is hoisted up the inclined plane using a motor M which supplies a clockwise torque **T** to the 5-lb disk. If $T = 3t$, where T is in lb-ft and t is in seconds, determine the speed of the block 4 sec after the motor is turned on. Neglect the mass of the pulley at A. The stop block s prevents the block from initially sliding down the plane. *Hint:* First determine the time it takes to overcome the static friction.

Top view

Prob. 19-18

Prob. 19-20

19-19. The 10-lb gear A has a radius of gyration of $k_O = 3$ in. The gear rack B weighs 5 lb. The coefficient

19-21. Work Prob. 19-20 assuming that $T = 0.9t^2$, where T is in lb-ft and t is in seconds.

19-22. The block and pulley system is released from rest. Using the data shown in the figure, determine the speed which block D attains 10 sec later. Assume that the cable does not slip on the pulleys.

$W_B = 5$ lb
$k_B = 2$ in.

3 in.

B

6 in.

A

$W_A = 10$ lb
$k_A = 4$ in.

10 lb

C

60 lb

D

Prob. 19-22

19-4. Conservation of Momentum

The principle of linear impulse and momentum for a rigid body or system of connected rigid bodies, discussed in Sec. 19-2, states that the linear impulse created by the external forces acting in a particular direction produces a time rate of *change* in the linear momentum of the system of bodies in the same direction. If the *linear impulse* acting in a given direction is *zero*, then the time rate of change of the linear momentum in that direction is zero, and therefore, the linear momentum in that direction is constant or conserved. The first of Eqs. 19-9 then reduces to the form

$$\left(\sum \frac{\text{System linear}}{\text{momentum}}\right)_1 = \left(\sum \frac{\text{System linear}}{\text{momentum}}\right)_2 \qquad (19\text{--}10)$$

This statement is called the *conservation of linear momentum*. When the system of rigid bodies is subjected to impulses which are small, Eq. 19-10 may be applied without inducing appreciable errors in the computations. Linear impulses are considered small when small forces act over very short periods of time. For example, the impulse created by the force of a tennis racket hitting a ball, during a short time interval Δt, is large;

however, the impulse of the weight of the ball is small and may therefore be neglected in the motion analysis during Δt.

In a similar manner, the angular momentum for a system of connected rigid bodies is conserved about the system's center of mass, G, or a point O lying on an axis of rotation of the system, when the angular impulses created by all the external forces acting on the system are zero or appreciably small when computed about these points. The second of Eqs. 19–9 then becomes

$$\left(\sum \frac{\text{System angular}}{\text{momentum}}\right)_1 = \left(\sum \frac{\text{System angular}}{\text{momentum}}\right)_2 \qquad (19\text{–}11)$$

This statement is called the *conservation of angular momentum*. It is often applied by a swimmer who executes a somersault after jumping off a diving board. By tucking his arms and legs in close to his chest, he decreases his body's moment of inertia and correspondingly increases his angular velocity ($I_G\omega$ must be constant). If he straightens out just before entering the water, his body's moment of inertia is increased and his angular velocity virtually stops. This example also illustrates that the angular momentum of a body may be conserved and yet the linear momentum is *not*. Such cases occur when the external forces creating the linear impulse pass through either the center of mass of the body or a fixed axis of rotation.

19–5. Impact

Impact occurs when two bodies collide with each other during a very short interval of time. During the collision very large equal but opposite impulsive forces are created between the two bodies at their point of contact. The concepts involving central and oblique impact of particles have been presented in Sec. 18–5.* We will presently discuss the eccentric impact of two bodies. *Eccentric impact* occurs when the line connecting the mass centers of the two bodies does not coincide with the line of impact. This type of impact most often occurs when one or both of the bodies is constrained to rotate about a fixed axis. Consider, for example, the collision between the two bodies A and B, shown in Fig. 19–8a, which collide at point C. Body B rotates about a fixed point O, whereas body A both rotates and translates. It is assumed that just before the collision occurs, body B is rotating counterclockwise at an angular speed of $(\omega_B)_1$, and the velocity of the contact point C, located on body A, is $(\mathbf{u}_A)_1$. Kinematic diagrams for both bodies just before the collision are shown

---

*It is suggested that this material be reviewed before working the problems.

Fig. 19–8

in Fig. 19–8*b*. Provided the bodies are smooth at point *C*, the *impulsive forces* they exert on each other *are directed along the line of impact*. Hence, the component of velocity of point *C* on body *B*, which is directed along the line of impact, is $(v_B)_1 = (\omega_B)_1 r$, Fig. 19–8*b*. Likewise, on body *A* the component of velocity $(\mathbf{u}_A)_1$ along the line of impact is $(\mathbf{v}_A)_1$. In order for a collision to occur, $(v_A)_1 > (v_B)_1$

During the impact an equal but opposite impulsive force **P** is exerted between the bodies which *deforms* their shape at the point of contact. The resulting impulse is shown on the impulse-vector diagrams for both bodies, Fig. 19–8c. Note that the impulsive force created at point C, on the rotating body B, creates impulsive pin reactions at the supporting pin O. On these diagrams it is assumed the impact creates forces which are much larger than the weights of the bodies. Hence, the impulses created by the weights are not shown since they are negligible compared to $\int P\, dt$ and the reactive impulses at O. When the deformation of point C is a maximum, point C on both the bodies moves with a common velocity **v** along the line of impact, Fig. 19–8d. A period of *restitution* then occurs in which the bodies tend to regain their original shapes. The restitution phase creates an equal but opposite impulsive force **R** acting between the bodies as shown on the impulse-vector diagram, Fig. 19–8e. After restitution the bodies move apart such that point C on body B has a velocity $(v_B)_2$ and point C on body A has a velocity $(u_A)_2$, Fig. 19–8f, where $(v_B)_2 > (v_A)_2$.

In general, a problem involving the impact of two bodies requires determining the *two unknowns* $(v_A)_2$ and $(v_B)_2$, assuming $(v_A)_1$ and $(v_B)_1$ are known (or can be determined using kinematics, energy methods, Newton's second law of motion, etc.) To solve this problem two equations must be written. The *first equation* generally involves application of *the conservation of angular momentum to the two bodies*. In the case of bodies A and B, we can state that angular momentum is conserved about point O since the impulses at C are internal to the system and the impulses at O create zero moment (or zero angular impulse) about point O. A *second equation* is obtained using the definition of the *coefficient of restitution*, Eq. 18–13. To establish a useful form of this equation we must first apply the principle of angular impulse and momentum about point O to bodies B and A separately. Combining the results, we then obtain the necessary equation. Proceeding in this manner, the principle of impulse and momentum applied to body B just before the collision to the time of maximum deformation (Figs. 19–8b, 19–8c, and 19–8d) becomes

$$(\zeta +) \qquad\qquad I_O(\omega_B)_1 + r\int P\, dt = I_O\omega \qquad\qquad (19\text{–}12)$$

Here I_O is the moment of inertia of body B about point O. Similarly, applying the principle of angular impulse and momentum from the instant of maximum deformation to the time just after the impact (Figs. 19–8d, 19–8e and 19–8f) yields

$$(\zeta +) \qquad\qquad I_O\omega + r\int R\, dt = I_O(\omega_B)_2 \qquad\qquad (19\text{–}13)$$

Solving Eqs. 19-12 and 19-13 for $\int P\,dt$ and $\int R\,dt$, respectively, and substituting into Eq. 18-13, we have

$$e = \frac{\int R\,dt}{\int P\,dt} = \frac{r(\omega_B)_2 - r\omega}{r\omega - r(\omega_B)_1} = \frac{(v_B)_2 - v}{v - (v_B)_1} \qquad (19\text{-}14)$$

In the same manner, we may write an equation which relates the magnitudes of velocities $(v_A)_1$ and $(v_A)_2$ of body A. The result is

$$e = \frac{(v_A)_2 - v}{v - (v_A)_1} \qquad (19\text{-}15)$$

Combining Eqs. 19-14 and 19-15, by eliminating the common velocity v, yields the desired result, i.e.,

$$e = \frac{(v_B)_2 - (v_A)_2}{(v_A)_1 - (v_B)_1} = \frac{\text{Rel. vel. after impact}}{\text{Rel. vel. before impact}} \qquad (19\text{-}16)$$

This equation is identical to Eq. 18-14, which was derived for the central impact occurring between two particles. It states that the coefficient of restitution, e (a measurable quantity), is equal to the ratio of the relative velocity of separation of the two bodies after impact (at the point of collision C) to the relative velocity of approach before impact. In deriving Eq. 19-16, we assumed that the point of contact C for both bodies moves to the right *both* before and after impact. If motion of the contacting point occurs to the left, the velocities of both bodies measured at point C are considered as negative quantities in Eq. 19-16.

As stated previously, when Eq. 19-16 is used in conjunction with the conservation of angular momentum, it provides a useful means of obtaining the absolute velocities of two colliding bodies both before and after collision.

Example 19-4

The 10 lb wheel shown in Fig. 19-9*a* has a centroidal radius of gyration of $k_G = \frac{1}{2}$ ft. Assuming the wheel does not slip, determine the minimum velocity $\mathbf{v}_G$ it must have to just roll over the obstruction at A.

Solution

Since no slipping occurs, the wheel essentially *pivots* about point A when contact occurs. The momentum-and impulse-vector diagrams for

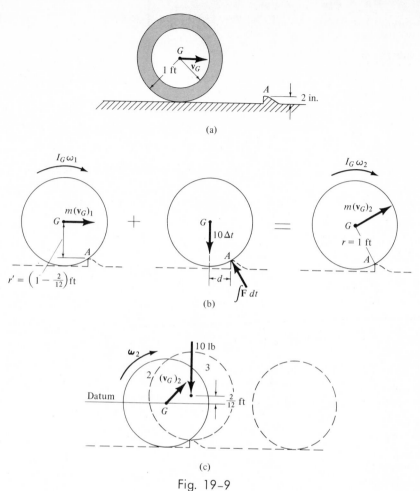

Fig. 19-9

the wheel are shown in Fig. 19-9b. These diagrams indicate, respectively, the momentum of the wheel *just before impact,* the impulses given to the wheel *during impact,* and the momentum of the wheel *just after impact.* In particular, two impulsive forces act on the wheel. (The normal force created by the horizontal plane is zero. Why?) During impact *only* the weight force creates an *angular impulse about point A.* However, since the time of impact is extremely small, this angular impulse can be neglected, i.e., $(10 \, \Delta t)d \approx 0$. Hence, angular momentum about point A is essentially *conserved.*

From *kinematics,* since no slipping occurs, the velocity of the mass center $\mathbf{v}_G$ is related to the angular velocity $\boldsymbol{\omega}$ by the equation $\omega = v_G/r$.

The moment of inertia of the wheel about G is determined from the radius of gyration and the weight of the wheel.

$$I_G = mk_G^2 = \left(\frac{10}{32.2}\right)\left(\frac{1}{2}\right)^2 = 0.0776 \text{ slug-ft}^2$$

The angular momentum when computed about point A must account for both the moment of the linear momentum, mv_G, about point A *and* the angular momentum $\mathbf{H}_G = I_G\omega$, which is due to the rotational motion of the wheel about G. Applying the conservation of angular momentum about A in reference to Fig. 19-9b, yields

$$(\mathbf{H}_A)_1 = (\mathbf{H}_A)_2$$

$$r'm(v_G)_1 + I_G\omega_1 = rm(v_G)_2 + I_G\omega_2$$

$$\left(1 - \frac{2}{12}\right)\left(\frac{10}{32.2}\right)(v_G)_1 + (0.0776)\left(\frac{(v_G)_1}{1}\right)$$

$$= (1)\left(\frac{10}{32.2}\right)(v_G)_2 + (0.0776)\left(\frac{(v_G)_2}{1}\right)$$

Thus,

$$(v_G)_2 = 0.867(v_G)_1 \tag{1}$$

The wheel will roll over the obstruction provided it passes the dotted position 3, shown on Fig. 19-9c. Hence, if $(v_G)_2$ (or $(v_G)_1$) is to be a minimum, it is necessary that the kinetic energy of the wheel at position 2 be equal to the potential energy at position 3. Constructing the datum through the initial position of the center of mass, as shown in the figure, and applying the conservation of energy theorem, we have

$$\{T_2\} + \{V_2\} = \{T_3\} + \{V_3\}$$

$$\left\{\frac{1}{2}\left(\frac{10}{32.2}\right)(v_G)_2^2 + \frac{1}{2}(0.0776)(\omega_2)^2\right\} + \{0\} = \{0\} + \left\{(10)\left(\frac{2}{12}\right)\right\}$$

Substituting $\omega_2 = (v_G)_2/1$ ft and Eq. (1) into this equation, and simplifying, yields

$$0.1167(v_G)_1^2 + 0.0292(v_G)_1^2 = 1.667$$

or

$$0.1459(v_G)_1^2 = 1.667$$

Thus,

$$(v_G)_1 = 3.38 \text{ ft/sec} \qquad \textit{Ans.}$$

Example 19-5

The 30-lb smooth slender rod shown in Fig. 19-10a is pinned at O and is initially at rest. If a 2-lb ball is thrown at the rod with a velocity

(a)

(b)

Fig. 19-10

of 50 ft/sec, as shown in the figure, determine the velocity of the ball just after impact, and compute the angle θ_{max} through which the rod swings. The coefficient of restitution between the ball and the rod is $e = 0.4$.

Solution:

The ball and rod will be considered as a complete system. The mo-

mentum-and impulse-vector diagrams for this system are shown in Fig. 19–10b. In particular, the momentum diagrams are drawn *just before and just after impact*. During impact, the ball and rod exchange equal but opposite impulses at A. As shown on the impulse-vector diagram, the impulses which are external to the system are due to the reactions at O and the weights of the ball and rod.

To obtain the velocities $(v_B)_2$, $(v_G)_2$, and ω_2, we will first apply the *principle of angular impulse and momentum with respect to point O*. By choosing this point, the angular impulses created by the reaction at O and the weight of the rod are *zero*. Furthermore, the radius of the ball, the contact time, and the weight of the ball are all small so that the angular impulse of the ball about point O is essentially *zero*. Hence, angular momentum is conserved about this point. The moment of inertia of the rod about its mass center is computed using Appendix D.

$$I_G = \frac{1}{12} ml^2 = \frac{1}{12}\left(\frac{30}{32.2}\right)(4)^2 = 1.24 \text{ slug-ft}^2$$

Hence, with reference to Fig. 19–10b, we have

$(\zeta+)$ $\qquad\qquad\qquad \Sigma(\mathbf{H}_O)_1 = \Sigma(\mathbf{H}_O)_2$

$m_B(v_B)_1(3 \text{ ft}) + m_R(v_G)_1(2 \text{ ft}) + I_G\omega_1$
$$= m_B(v_B)_2(3 \text{ ft}) + m_R(v_G)_2(2 \text{ ft}) + I_G\omega_2$$

$$\left(\frac{2}{32.2}\right)(50)(3) + 0 + 0 = \left(\frac{2}{32.2}\right)(v_B)_2(3) + \left(\frac{30}{32.2}\right)(v_G)_2(2) + 1.24\omega_2$$

or

$$9.31 = 0.1863(v_B)_2 + 1.863(v_G)_2 + 1.24\omega_2 \qquad (1)$$

The ball makes a direct collision with the rod at point A. Assuming that positive velocity components act to the right, we may write the coefficient of restitution equation (Eq. 19–16) in the form

$$e = \frac{(v_A)_2 - (v_B)_2}{(v_B)_1 - (v_A)_1}$$

where $(v_A)_1$ and $(v_A)_2$ represent the initial and final velocities of point A on the rod. Substituting into this equation, we have

$$0.4 = \frac{(v_A)_2 - (v_B)_2}{50 - 0}$$

or

$$(v_A)_2 - (v_B)_2 = 20 \qquad (2)$$

Since the rod pivots about point O, using kinematics, the velocities $(v_G)_2$, $(v_A)_2$, and ω_2 may be related to one another as shown in Fig. 19–10c. Hence,

$$(v_G)_2 = 2\omega_2 \tag{3}$$

$$(v_A)_2 = 3\omega_2 \tag{4}$$

Solving Eqs. (1) to (4) simultaneously,

$$\omega_2 = 2.36 \text{ rad/sec}$$
$$(v_G)_2 = 4.72 \text{ ft/sec}$$
$$(v_A)_2 = 7.08 \text{ ft/sec}$$
$$(v_B)_2 = -12.92 \text{ ft/sec}$$

The negative sign for $(v_B)_2$ indicates that the ball travels in a direction opposite to that shown in Fig. 19–10b.

The angle θ_{max} through which the rod swings may be computed by applying the conservation of energy principle to the rod. The datum is established through the initial position of the rod's center of mass, Fig. 19–10d. We require the initial kinetic energy of the rod to be converted entirely into potential energy. Thus,

$$\{T_1\} + \{V_1\} = \{T_2\} + \{V_2\}$$

$$\left\{\frac{1}{2}\left(\frac{30}{32.2}\right)(4.72)^2 + \frac{1}{2}(1.24)(2.36)^2\right\} + \{0\}$$

$$= \{0\} + \{30(2)(1 - \cos\theta_{max})\}$$

Solving for θ_{max},

$$\theta_{max} = 39.7° \qquad\qquad Ans.$$

Problems

19-23. A horizontal circular platform weighs 180 lb and is free to rotate about a vertical axis passing through its center. Initially the platform is rotating with an angular velocity of 2 rad/sec. A 60-lb dog suddenly jumps on the platform and runs in the opposite direction of rotation with a velocity of 30 ft/sec relative to the platform. Determine the angular velocity of the platform assuming that the dog runs along a circular path having a radius of $r = 10$ ft. The centroidal radius of gyration of the platform is $k = 4$ ft.

19-24. Disks A and B weigh 20 lb each. Two inextensible cables are attached to the 10-lb block and wrapped around the disks. The block is dropped $h = 2$ ft before the slack is taken up. If the impact is perfectly elastic, i.e., $e = 1$, determine the angular velocity of the disks just after impact.

Prob. 19–24

19-25. Solve Prob. 19–24 assuming that the impact is perfectly plastic, i.e., $e = 0$.

19-26. The 15-lb disk A is keyed to the vertical shaft and is rotating with it at 40 rpm. Disk B weighs 10 lb and is not rotating. Neglecting the mass of the axle, determine the angular velocity of both disks after disk B slides down the shaft, comes in contact with disk A, and slipping between the disks has stopped. What average frictional moment does disk B exert on disk A if slipping between the two disks occurs for 3 sec?

Prob. 19-26

19-27. The 15-lb square plate is rotating on the smooth surface with a constant velocity of 16 rad/sec. If a corner of the plate strikes the peg P, determine the new angular velocity of the plate if it starts to instantaneously rotate about the peg at the point of contact.

Prob. 19-27

19-28. The 10-lb slender rod AB is released from rest when it is in the horizontal position so that it begins to rotate clockwise. A 1-lb ball having a coefficient of restitution of $e = 1.0$ is thrown at the rod with a velocity of $v = 50$ ft/sec. The ball strikes the rod at the center of percussion P at the instant the rod is in the vertical position as shown. Determine the angular velocity of the rod just after the impact. See Prob. 19-17.

Prob. 19-28

19-29. The 15-lb rod AB is released from rest from the vertical position. If the coefficient of restitution between the floor and the cushion at B is $e = 0.7$, determine how high the end of the rod rebounds after impact with the floor.

Prob. 19-29

19-30. Two hampsters are placed on the surface of the 10-lb platform. The platform has a radius of gyration

839

of $k_o = 1.8$ ft and is free to rotate in the horizontal plane. The hampsters each weigh 1.4 lb and can run at a speed of 1 ft/sec relative to the platform. Determine the angular velocity of the platform if the hampsters run in circular paths in opposite directions on the platform.

Prob. 19–30

19-31. Work Prob. 19–30 assuming both hampsters run in the same direction along their paths.

19-32. The 4-lb sphere rolls without slipping along a horizontal plane with a velocity of 6 ft/sec. Neglecting any slipping and rebounding, determine the velocity of its center as it starts to roll up the inclined plane.

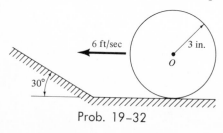

Prob. 19–32

19-33. The cylinder is at rest. If possible, determine the height h at which a bullet can strike the cylinder and cause it to roll without slipping.

Prob. 19–33

19-34. The pendulum consists of a rod AB which weighs 2 lb and a disk C which weighs 5 lb. If a 3-lb ball, having a speed of 60 ft/sec, strikes the pendulum at D, determine the angular velocity of the pendulum just after impact. The coefficient of restitution between the ball and the pendulum is $e = 0.6$.

Prob. 19–34

19-35. The 5-lb slender rod AB supports a 10-lb wooden block. A projectile weighing 0.2 lb is fired at

the block with a velocity of 1,000 ft/sec. If the block is initially at rest, and the projectile embeds itself into the block after impact, determine the angular velocity of the rod AB after the collision.

19-37. The rod AB weighs 20 lb and is hanging in the vertical position. A 5-lb block of ice, sliding on a smooth horizontal surface with a velocity of $v_0 = 12$ ft/sec, strikes the rod at B. Determine the velocity of the block immediately after the collision. The coefficient of restitution between the block and the rod is $e = 0.8$.

Prob. 19-35

Prob. 19-37

19-36. The 10-lb prismatic block is sliding on the smooth surface when the corner D hits a stop block S. Determine the minimum velocity v which the block can have which would allow it to tip over on its side so that it lands in the dotted position shown and has zero velocity.

19-38. The two disks each weigh 10 lb. If they are released from rest when $\theta = 30°$, determine θ after they collide and rebound from each other. The coefficient of restitution is $e = 0.75$. When $\theta = 0°$, the disks hang so that they just touch one another.

Prob. 19-36

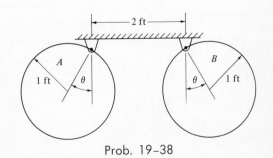

Prob. 19-38

20 Moments of Inertia for Mass

20-1. Definitions of Moments of Inertia for Mass and Radius of Gyration

The concept of the mass moment of inertia of a body was briefly introduced in Sec. 15–4 in order to develop the kinetic equations for plane motion of a rigid body. It will be shown in Chapter 21 that when a body is subjected to more complex motions, other mass moments of inertia are involved in the kinetic equations of motion. For this reason, a detailed study of the moments of inertia for mass is given in this chapter.

Consider the rigid body shown in Fig. 20–1. The *mass moment of inertia* for a differential element dm of the body about any one of the three coordinate axes is defined as the product of the mass of the element and the square of the shortest distance from the axis to the element. For example, as noted in the figure, $r_x = \sqrt{y^2 + z^2}$, hence the mass moment of inertia of the element about the x axis is

$$dI_{xx} = r_x^2 \, dm = (y^2 + z^2) \, dm$$

The moment of inertia for the entire body is determined by integrating this expression over the entire mass of the body. Thus,

$$I_{xx} = \int_m r_x^2 \, dm = \int_m (y^2 + z^2) \, dm \qquad (20\text{–}1a)$$

In a similar manner,

$$I_{yy} = \int_m r_y^2 \, dm = \int_m (z^2 + x^2) \, dm \qquad (20\text{–}1b)$$

$$I_{zz} = \int_m r_z^2 \, dm = \int_m (x^2 + y^2) \, dm \qquad (20\text{–}1c)$$

From these equations it is seen that the mass moment of inertia is *always a positive quantity,* since it is the sum of the product of the mass *dm*, which is always positive, and distances squared.

As defined in Sec. 15–4, the *radius of gyration* of the body is expressed in terms of the mass of the body and its mass moment of inertia. In this case,

$$k_x = \sqrt{\frac{I_{xx}}{m}}, \qquad k_y = \sqrt{\frac{I_{yy}}{m}}, \qquad k_z = \sqrt{\frac{I_{zz}}{m}}$$

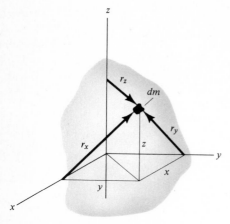

Fig. 20–1

The *mass product of inertia* for a differential element *dm* is defined with respect to a set of *two orthogonal planes,* as the product of the mass of the element and the perpendicular (or shortest) distances from the planes to the element. For example, with respect to the *xz* and *yz* planes, the mass product of inertia dI_{xy} for the element *dm*, shown in Fig. 20–1, is

$$dI_{xy} = xy \, dm$$

From this definition it is noted that $dI_{yx} = dI_{xy}$. Integrating over the entire mass, the product of inertia for the body may be expressed as

$$I_{xy} = I_{yx} = \int_m xy \, dm \qquad (20\text{–}2a)$$

In a similar manner,

$$I_{yz} = I_{zy} = \int_m yz \, dm \qquad (20\text{–}2b)$$

$$I_{xz} = I_{zx} = \int_m xz \, dm \qquad (20\text{–}2c)$$

Unlike the mass moment of inertia, which is always positive, the mass product of inertia may be positive, negative, or zero. The result depends upon the signs of the two defining coordinates, which vary independently from one another. In particular, if either one or both of the orthogonal planes are *planes of symmetry* for the mass, the *product of inertia* with respect to these planes will be *zero.* In such cases, elements of mass will occur in *pairs,* located on each side of the plane of symmetry. On one side of the plane the product of inertia for the element will be positive, while on the other side the product of inertia for the corresponding element will be negative, the sum therefore yielding zero. An example of this is shown in Fig. 20–2. In the first case, Fig. 20–2a, the *yz* plane is a plane of symmetry, and hence, $I_{xz} = I_{xy} = 0$. Computation for I_{yz} will yield a *positive* result, since all elements of mass are located using

only positive y and z coordinates. For the cylinder, with the coordinate axes located as shown in Fig. 20–2b, the xz and yz planes are both planes of symmetry. Thus, $I_{xz} = I_{yz} = I_{xy} = 0$.

The entire set of mass moments and products of inertia describes in a particular way the distribution of mass of a body relative to a given coordinate system having a specified orientation and point of origin. These nine quantities, six of which are independent of one another, can be grouped in an array which has the following form*:

$$\begin{pmatrix} I_{xx} & -I_{xy} & -I_{xz} \\ -I_{yx} & I_{yy} & -I_{yz} \\ -I_{zx} & -I_{zy} & I_{zz} \end{pmatrix}$$

This array is called an *inertia tensor*. In particular, the inertia tensor has a unique set of values when it is computed for *each location* of the origin and *inclination* of the coordinate axes. From the definition of mass moment and mass product of inertia, it is seen that each of the terms in the inertia tensor will have units of mass times length squared, a common set of units being slug-ft^2, ft-lb-sec^2 or m-kg$_f$-sec^2.

If we replace the subscripts x, y, and z in the inertia tensor by 1, 2, and 3, respectively, and let $I_{12} = -I_{xy}$, $I_{13} = -I_{xz}$, etc., the inertia tensor may be written in the "standard form":

$$I_{ij} = \begin{pmatrix} I_{11} & I_{12} & I_{13} \\ I_{21} & I_{22} & I_{23} \\ I_{31} & I_{32} & I_{33} \end{pmatrix}$$

When subscripts i and j are set equal to any combination of numbers 1, 2, and 3, all the terms of the inertia tensor are represented by I_{ij}. Since *two subscripts* are required to define all nine terms, the inertia tensor is referred to as a tensor of the *second order*. Though it has never been formally stated, a vector is actually a tensor of the *first order*. For example, the vector $\mathbf{A} = A_x\mathbf{i} + A_y\mathbf{j} + A_z\mathbf{k}$ may be represented by a column array, replacing x, y, and z by 1, 2, and 3, respectively. Thus,

$$A_i = \begin{pmatrix} A_1 \\ A_2 \\ A_3 \end{pmatrix}$$

A_i needs only the *single i subscript* to define the three vector components. In a similar manner, a scalar is defined completely using a single term; *no subscript* is needed, and hence, a scalar is a tensor of *zero order*. The inertia tensor is a second-order tensor not solely because it contains nine elements; rather it is defined as such because it transforms from one set

---

*The reason for using the negative signs for the products of inertia will become clear after studying Sec. 21–2.

(a)

(b)

Fig. 20–2

of coordinate axes to a second rotated set using a certain mathematical law for transformation. This method of transformation for I_{ij} is illustrated in Sec. 20–5. In the study of advanced strength of materials, one is generally introduced to the stress tensor and the strain tensor.

20–2. Transformation of Moments and Products of Inertia Between Parallel Coordinate Axes

Provided all the elements of the inertia tensor are known for a body relative to a given set of coordinate axes and point of origin, the elements of the inertia tensor for a different yet *parallel set of axes* may be computed using the parallel-axis and parallel-plane theorems. These two theorems are used to compute the mass moments and mass products of inertia for the body relative to two parallel axes and two parallel planes.

The *parallel-axis theorem* for mass moments of inertia was stated in Sec. 15–4. This theorem, which is similar to that for area moments of inertia,* will now be proven. Consider the body shown in Fig. 20–3, where the x', y', and z' axes have their origin located at the *mass center G* of the body, and the x, y, and z axes represent a corresponding set of parallel axes. The mass moment of inertia for the element of mass dm, with respect to the x axis, is

$$dI_{xx} = (y^2 + z^2)\, dm$$
$$= [(y' + y_G)^2 + (z' + z_G)^2]\, dm$$

Expanding this expression, and integrating each term to obtain the moment of inertia for the entire body, we have

$$I_{xx} = \int_m (y'^2 + z'^2)\, dm + (y_G^2 + z_G^2) \int_m dm$$

$$+ 2y_G \int_m y'\, dm + 2z_G \int_m z'\, dm$$

From Eq. 20–1a, the first integral represents the moment of inertia of the mass about the x' axis, i.e., $(I_{x'x'})_G$. The sum $(y_G^2 + z_G^2)$, in the second term, is equal to the square of the perpendicular distance $(r_G)_x$ from the x axis to the x' axis (see Fig. 20–3). The integral in this term is simply the mass m of the body. The last two terms in the above equation are equal to zero, since the origin of the primed coordinates passes through the mass center G. The parallel-axis theorem for the mass moment of inertia of the body about the x axis can therefore be written as

$$I_{xx} = (I_{x'x'})_G + m(y_G^2 + z_G^2) \tag{20–3a}$$

*See Chapter 10, *Engineering Mechanics: Statics.*

Fig. 20-3

This equation states that the mass moment of inertia of the body about the x axis, I_{xx}, is equal to the mass moment of inertia of the body about a *parallel x'* axis passing through the mass center of the body, $(I_{x'x'})_G$, plus the mass m of the body times the square of the distance between the x and x' axis. Similar expressions may be derived for finding I_{yy} and I_{zz}. These are

$$I_{yy} = (I_{y'y'})_G + m(x_G^2 + z_G^2) \tag{20-3b}$$

and

$$I_{zz} = (I_{z'z'})_G + m(x_G^2 + y_G^2) \tag{20-3c}$$

With reference to Fig. 20-3, the mass product of inertia dI_{xy} for the element of mass dm is

$$dI_{xy} = xy \, dm = (x' + x_G)(y' + y_G) \, dm$$

If we expand and integrate term by term, the product of inertia for the entire body becomes

$$I_{xy} = \int_m x'y' \, dm + y_G \int_m x' \, dm + x_G \int_m y' \, dm + x_G y_G \int_m dm$$

Since the origin of the primed axes is located at the mass center, $\int_m x' \, dm = \int_m y' \, dm = 0$. Hence the *parallel-plane theorem* for I_{xy} is

$$I_{xy} = (I_{x'y'})_G + m x_G y_G \tag{20-4a}$$

Here $(I_{x'y'})_G$ represents the mass product of inertia of the body relative to the $x'z'$ and $y'z'$ planes. Both planes contain the mass center, G. Also, x_G and y_G are the perpendicular distances from the $y'z'$ plane to the yz plane and from the $x'z'$ plane to the xz plane, respectively. Corresponding expressions for determining I_{xz} and I_{yz} are

$$I_{xz} = (I_{x'z'})_G + mx_G z_G \qquad (20\text{--}4b)$$

and

$$I_{yz} = (I_{y'z'})_G + my_G z_G \qquad (20\text{--}4c)$$

Because of the symmetry of the product of inertia, i.e., $I_{yx} = I_{xy}$, etc., all nine elements of the inertia tensor may be computed for a set of axes parallel to a given set using the transformation Eqs. 20–3 and 20–4.

20–3. Moments and Products of Inertia of Masses by Integration

Since mass extends in three dimensions, the mass moment and product of inertia can be found using either triple, double, or single integration. Usually single integration should be considered, because if this is possible, the limits on the integral are easy to establish. If the body consists of material having a variable mass density, $\rho_m = \rho_m(x, y, z)$, then the mass of the element may be expressed in terms of its density and volume as $dm = \rho_m \, dV$. Substituting into Eqs. 20–1 or 20–2, the moments of inertia are therefore computed using *volume elements* for the integration.

If differential-sized volume elements are chosen for integration, that is, elements having differential size in three directions, the *entire element*, because of its "smallness," lies at the *same distance* from the reference axis (or reference planes as in the case of product of inertia). In general, this is not true for elements having a differential size in one or two directions. These elements have a *finite* length or width and consequently portions of the entire element *do not* always lie at the same distance from the reference axis or planes. To perform the integration in such cases, it is first necessary to determine the moment or product of inertia of the *element* with respect to its own mass center, then to compute its moment or product of inertia about the reference axis or planes. This can be done using the parallel-axis or parallel-plane theorems. Examples 20–3 and 20–4 illustrate this procedure.

Example 20–1

Determine the inertia tensor for the homogeneous right circular cylinder shown in Fig. 20–4a with respect to the x, y, and z centroidal axes.

The mass density of the material is ρ_m.

Solution

The mass moment of inertia of the cylinder with respect to the z axis may be determined using a third-order differential-sized element, Fig. 20–4b. Using cylindrical coordinates, this element has a mass $dm = \rho_m\, dV = (\rho_m)r\, d\theta\, dr\, dz$, and is located at the arbitrary point $x = r\cos\theta$, $y = r\sin\theta$, and $z = z$. Hence, the moment of inertia of the element with respect to the z axis is

$$dI_{zz} = (x^2 + y^2)\, dm = (r^2\cos^2\theta + r^2\sin^2\theta)(\rho_m)r\, d\theta\, dr\, dz = \rho_m r^3\, d\theta\, dr\, dz$$

The simplification is possible since $\cos^2\theta + \sin^2\theta = 1$. Integrating over the entire volume, we have

$$I_{zz} = \rho_m \int_{-h/2}^{h/2}\left[\int_0^R\left(\int_0^{2\pi} d\theta\right)r^3\, dr\right]dz = \rho_m \int_{-h/2}^{h/2}\left(\int_0^R 2\pi\, r^3\, dr\right)dz$$

$$= \rho_m \int_{-h/2}^{h/2}\frac{\pi}{2}R^4\, dz = \frac{\rho_m\pi}{2}R^4 h$$

Since

$$m = \int_V \rho_m\, dV = \int_{-h/2}^{h/2}\int_0^R\int_0^{2\pi}(\rho_m)r\, d\theta\, dr\, dz = \rho_m\pi R^2 h \qquad (1)$$

then,

$$I_{zz} = \frac{1}{2}mR^2$$

The moment of inertia of the cylinder with respect to the x axis will be computed using the same third-order differential-sized element shown in Fig. 20–4b,

$$dI_{xx} = (y^2 + z^2)\, dm = (r^2\sin^2\theta + z^2)(\rho_m)r\, d\theta\, dr\, dz$$

Integrating over the entire volume, we obtain

$$I_{xx} = \rho_m \int_{-h/2}^{h/2}\left\{\int_0^R\left[\int_0^{2\pi} r(r^2\sin^2\theta + z^2)\, d\theta\right]dr\right\}dz$$

$$= \frac{\rho_m}{12}\pi R^2 h(3R^2 + h^2)$$

Using Eq. (1) gives

$$I_{xx} = \frac{m}{12}(3R^2 + h^2) \qquad (2)$$

In a similar manner,

$$I_{yy} = \frac{m}{12}(3R^2 + h^2)$$

(a)

(b)

Fig. 20–4

Since the xy, xz, and yz planes are all *planes of symmetry*,

$$I_{xy} = I_{yz} = I_{xz} = 0$$

The inertia tensor for the cylinder with respect to the centroidal axes shown in Fig. 20–4a is thus

$$\begin{pmatrix} \dfrac{m}{12}(3R^2 + h^2) & 0 & 0 \\[2mm] 0 & \dfrac{m}{12}(3R^2 + h^2) & 0 \\[2mm] 0 & 0 & \dfrac{1}{2}mR^2 \end{pmatrix} \qquad \textit{Ans.}$$

Fig. 20–5

Example 20-2

Compute the inertia tensor for the cylinder shown in Fig. 20–5 with respect to the x, y, and z axes.

Solution

In the previous example, the moments of inertia have been computed about axes passing through the mass center. Using this information along with the parallel-axis and parallel-plane theorems (Eqs. 20–3 and 20–4), we may compute the elements of the inertia tensor as follows:

$$I_{xx} = I_{x'x'} + m(y_G^2 + z_G^2)$$
$$= \frac{m}{12}(3R^2 + h^2) + m\left[(-R)^2 + \left(\frac{h}{2}\right)^2\right] = \frac{m}{12}(15R^2 + 4h^2)$$

$$I_{yy} = I_{y'y'} + m(x_G^2 + z_G^2)$$
$$= \frac{m}{12}(3R^2 + h^2) + m\left[0 + \left(\frac{h}{2}\right)^2\right] = \frac{m}{12}(3R^2 + 4h^2)$$

$$I_{zz} = I_{z'z'} + m(x_G^2 + y_G^2)$$
$$= \frac{1}{2}mR^2 + m[0 + (-R)^2] = \frac{3}{2}mR^2$$

$$I_{xy} = I_{x'y'} + mx_G y_G$$
$$= 0 + m(0)(-R) = 0$$

$$I_{xz} = I_{x'z'} + mx_G z_G$$
$$= 0 + m(0)\frac{h}{2} = 0$$

$$I_{yz} = I_{y'z'} + my_G z_G$$
$$= 0 + m(-R)\frac{h}{2} = -\frac{m}{2}Rh$$

Since the yz plane is a plane of *symmetry*, I_{xy} and I_{xz} could have been determined directly by inspection to be zero, without the need to apply the parallel-plane theorem.

The inertia tensor is therefore

$$
\begin{pmatrix}
\dfrac{m}{12}(15R^2 + 4h^2) & 0 & 0 \\[2mm]
0 & \dfrac{m}{12}(3R^2 + 4h^2) & \dfrac{m}{2}Rh \\[2mm]
0 & \dfrac{m}{2}Rh & \dfrac{3}{2}mR^2
\end{pmatrix}
\qquad Ans.
$$

Example 20-3

A solid is formed by revolving the shaded area shown in Fig. 20–6a about the y axis. Determine radius of gyration k_y of the solid. The mass density of the material is ρ_m slug/ft³.

(a)

Solution I

To obtain the radius of gyration, k_y, it is first necessary to determine the mass m and the mass moment of inertia I_{yy}. This will be done using a *thin cylindrical shell element* for integration as shown in Fig. 20–6b. The element intersects the curve at the arbitrary point (x, y). The mass of this element is

$$dm = \rho_m \, dV = \rho_m (2\pi x)(1 - y) \, dx$$

Hence, integrating with respect to x, from $x = 0$ to $x = 1$ ft, we obtain the entire mass.

$$m = 2\pi\rho_m \int_0^1 x(1 - y) \, dx = 2\pi\rho_m \int_0^1 x(1 - \sqrt{x}) \, dx = 0.628\rho_m \text{ slugs}$$

(b)

By inspection it is seen that *all parts of the entire element* lie at perpendicular distance x from the y axis. Hence, the *moment of inertia of the element* with respect to the y axis is:

$$dI_{yy} = x^2 \, dm = x^2(\rho_m)(2\pi x)(1 - y) \, dx$$

Hence, the moment of inertia of the solid becomes

$$I_{yy} = 2\pi\rho_m \int_0^1 x^3(1 - y) \, dx = 2\pi\rho_m \int_0^1 x^3(1 - \sqrt{x}) \, dx$$

$$= 0.1745\rho_m \text{ slug-ft}^3$$

The radius of gyration, k_y, is therefore:

$$k_y = \sqrt{\frac{I_{yy}}{m}} = \sqrt{\frac{0.1745\rho_m}{0.628\rho_m}} = 0.527 \text{ ft} \qquad Ans.$$

(c)

Fig. 20–6

Solution II

To illustrate a second method for integration, a *thin disk element* will be chosen as shown in Fig. 20–6c. This element intersects the curve at

the arbitrary point (x, y) and has a mass

$$dm = \rho_m \, dV = \rho_m(\pi x^2) \, dy$$

Hence, integrating with respect to y, from $y = 0$ to $y = 1$ ft, we obtain the entire mass.

$$m = \pi \rho_m \int_0^1 x^2 \, dy = \pi \rho_m \int_0^1 y^4 \, dy = 0.628 \rho_m \text{ slugs}$$

Although all portions of the element are *not* located at the same distance from the y axis, it is still possible to determine the moment of inertia dI_{yy} of the element about the y axis. In Example 20–1 it was shown that the moment of inertia of a cylinder about its longitudinal axis is $I = \frac{1}{2}mR^2$, where R and m are the radius and mass of the cylinder. Since the height of the cylinder is of no consequence in using this formula, the moment of inertia of the disk element, having a mass dm and radius x, Fig. 20–6c, is

$$dI_{yy} = \frac{1}{2}(dm)x^2 = \frac{1}{2}[\rho_m(\pi x^2) \, dy]x^2$$

Integrating, the moment of inertia of the solid becomes

$$I_{yy} = \frac{\pi \rho_m}{2} \int_0^1 x^4 \, dy = \frac{\pi \rho_m}{2} \int_0^1 y^8 \, dy = 0.1745 \rho_m \text{ slug-ft}^2$$

Therefore,

$$k_y = \sqrt{\frac{I_{yy}}{m}} = \sqrt{\frac{0.1745 \rho_m}{0.628 \rho_m}} = 0.527 \text{ ft} \qquad Ans.$$

Example 20–4

Determine the mass products of inertia I_{xy}, I_{xz}, and I_{yz} for the homogeneous prism shown in Fig. 20–7. The mass density of the material is ρ_m.

Solution

A rectangular element having a thickness dy is chosen for integration. The mass of this element is $dm = \rho_m \, dV = \rho_m hx \, dy$, having a center of gravity at point $G(x/2, y, h/2)$. The entire mass of the prism is

$$m = \int_m dm = \rho_m h \int_0^a (a - y) \, dy = \frac{\rho_m ha^2}{2}$$

Because of symmetry, the products of inertia of the *element* are zero for an $x'y'z'$ coordinate system with origin located at G and axes parallel to the x, y, and z axes. Hence, using the parallel-plane theorem to compute the product of inertia for the element with respect to the x, y, z axes, yields

$G(\frac{x}{2}, y, \frac{h}{2})$

$x + y = a$

Fig. 20–7

$$dI_{xy} = (dI_{x'y'})_G + dm\, x_G y_G$$

$$= 0 + (\rho_m h x\, dy)\frac{x}{2}y$$

For the entire prism, we have

$$I_{xy} = \frac{\rho_m h}{2}\int_0^a x^2 y\, dy = \frac{\rho_m h}{2}\int_0^a (a-y)^2\, y\, dy = \frac{\rho_m h a^4}{24} = \frac{m}{12}a^2 \quad Ans.$$

In a similar manner,

$$dI_{xz} = (dI_{x'z'})_G + dm\, x_G z_G$$

$$= 0 + (\rho_m h x\, dy)\frac{x}{2}\frac{h}{2}$$

so that

$$I_{xz} = \frac{\rho_m h^2}{4}\int_0^a x^2\, dy = \frac{\rho_m h^2}{4}\int_0^a (a-y)^2\, dy = \frac{\rho_m h^2 a^3}{12} = \frac{m}{6}ah \quad Ans.$$

And,

$$dI_{yz} = (dI_{y'z'})_G + dm\, y_G z_G$$

$$= 0 + (\rho_m h x\, dy)y\frac{h}{2}$$

$$I_{yz} = \frac{\rho_m h^2}{2}\int_0^a xy\, dy = \frac{\rho_m h^2}{2}\int_0^a (a-y)y\, dy = \frac{\rho_m h^2 a^3}{12} = \frac{m}{6}ah \quad Ans.$$

20–4. Moments and Products of Inertia of Composite Masses

When a body consists of a number of simple shapes such as spheres, rods, or cylinders, the mass moment or product of inertia for the body can be computed by adding together *algebraically* the mass moments or products of inertia for each shape, all computed with respect to the *same* axis or planes. The parallel-axis or the parallel-plane theorems are usually required for the computations. In this regard, the moments and products of inertia for a few common shapes are given in Appendix D.

The following example illustrates the procedure used for computing the mass moments and products of inertia for a composite body.

Example 20–5

Determine the mass moment of inertia I_{yy} and the product of inertia I_{xy} for the composite body shown in Fig. 20–8. The sphere weighs 30 lb, the slender rod weighs 10 lb, and the hollow cylinder has a density of 50 lb/ft³.

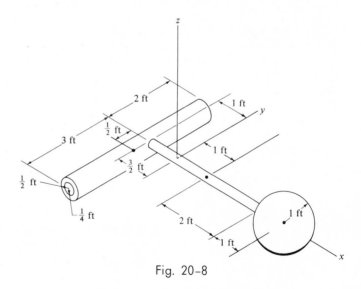

Fig. 20–8

Solution

Using the parallel-axis and parallel-plane theorems, and the data in Appendix D, we compute the required moments and products of inertia for the sphere, the slender rod, and the cylinder separately. The moment and product of inertia for the hollow cylinder is equivalent to that of

a solid cylinder of radius $\frac{1}{2}$ ft *less* that of a cylinder of radius $\frac{1}{4}$ ft. The computations are given in the following table:

	M *Mass* *(slug)*	I_{yy} *Moment of Inertia* *(slug-ft²)*	I_{xy} *Product of Inertia* *(slug-ft²)*
Sphere	$\dfrac{30}{32.2}$	$\dfrac{2}{5}\left(\dfrac{30}{32.2}\right)(1)^2 + \dfrac{30}{32.2}(4)^2 = 15.27$	$0 + \dfrac{30}{32.2}(4)0 = 0$
Slender Rod	$\dfrac{10}{32.2}$	$\dfrac{1}{12}\left(\dfrac{10}{32.2}\right)(4)^2 + \dfrac{10}{32.2}(1)^2 = 0.72$	$0 + \dfrac{10}{32.2}(0)0 = 0$
Cylinder (solid)	$\dfrac{50\pi(\frac{1}{2})^2(5)}{32.2} = 6.10$	$\dfrac{1}{2}(6.10)\left(\dfrac{1}{2}\right)^2 + 6.10\left(\dfrac{3}{2}\right)^2 = 14.48$	$0 + 6.10\left(-\dfrac{3}{2}\right)\left(-\dfrac{1}{2}\right) = 4.58$
Cylinder (hole)	$\dfrac{-50\pi(\frac{1}{4})^2(5)}{32.2} = -1.52$	$\dfrac{1}{2}(-1.52)\left(\dfrac{1}{4}\right)^2 + (-1.52)\left(\dfrac{3}{2}\right)^2 = -3.47$	$0 + (-1.52)\left(-\dfrac{3}{2}\right)\left(-\dfrac{1}{2}\right) = -1.14$

Thus,

$$I_{yy} = 15.27 + 0.72 + 14.48 - 3.47 = 27.0 \text{ slug-ft}^2 \qquad Ans.$$
$$I_{xy} = 0 + 0 + 4.58 - 1.14 = 3.44 \text{ slug-ft}^2 \qquad Ans.$$

Problems

20-1. Determine the mass moments of inertia I_x and I_y of the homogeneous paraboloid of revolution. The mass of the solid is 20 slugs.

20-2. Determine the mass moment of inertia of the homogeneous block with respect to its centroidal $\bar{x}$ axis. The total mass of the block is m.

Prob. 20-1

Prob. 20-2

20-3. Determine the mass moment of inertia of a homogeneous sphere with respect to a diametrical axis. The sphere has a radius a and a mass m.

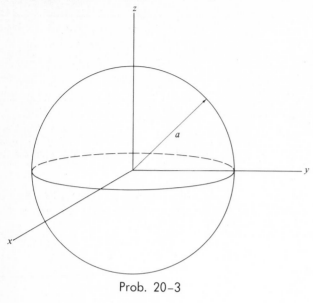

Prob. 20–3

20-5. Determine the mass moment of inertia I_y of the right circular cone, generated by revolving the line $z = 3y$ about the y axis. The total mass of the cone is m.

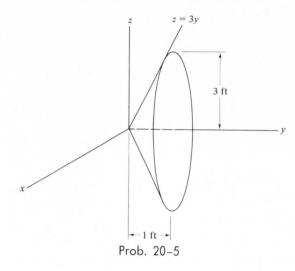

Prob. 20–5

20-4. An ellipsoid is formed by rotating the shaded area about the x axis. Determine the mass moment of inertia of this body with respect to the x axis. Express the result in terms of the mass m of the solid. Note that the boundary of the shaded area is defined by the equation $x^2/a^2 + y^2/b^2 = 1$.

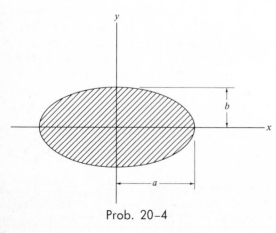

Prob. 20–4

20-6. Determine the mass moment of inertia of the hemispherical solid about the y axis. Express the result in terms of the mass m of the solid.

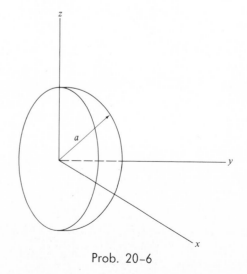

Prob. 20–6

20-7. Determine the radius of gyration k_y of the solid formed by revolving the shaded area around the x axis. The mass density of the material is ρ_m.

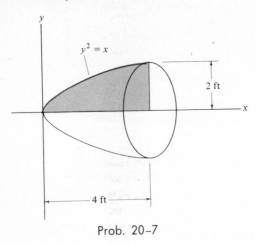

Prob. 20-7

20-8. Determine the mass moment of inertia I_y of the solid formed by rotating the shaded area about the y axis. The generated solid consists of material having a density of $\rho = 100$ lb/ft³.

Prob. 20-8

20-9. Determine the mass moment of inertia I_x for the solid formed by revolving the shaded area about the x axis. The density of the material is $\rho = 50$ lb/ft³.

Prob. 20-9

20-10. Determine the mass moment of inertia I_z and mass product of inertia I_{xz} for the slender rod. The mass of the rod is m.

Prob. 20-10

20-11. Determine the mass product of inertia I_{xy} for the homogeneous block in Prob. 20-2 with respect to the x and y axes passing through the base of the block. The total mass of the block is m.

20-12. Determine the radius of gyration k_{xy} for the solid in Prob. 20-7.

20-13. Determine the mass moment of inertia of the semicircular cylinder about (a) the x axis passing through the base and (b) the centroidal $\bar{x}$ axis. Express the results in terms of the mass m and radius a of the cylinder.

Prob. 20-13

20-14. Determine the radii of gyration k_x and k_y for the solid formed by revolving the shaded area about the x axis. The mass density of the material is ρ_m.

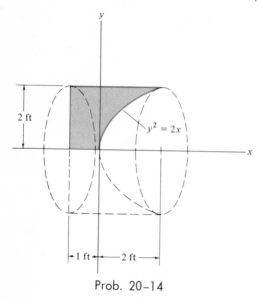

Prob. 20-14

20-15. Determine the mass moment of inertia of a thin circular ring about its central polar axis. The ring has a thickness of $\frac{1}{2}$ in. and an inner and outer radii of 3 ft and 5 ft, respectively. Express the result in terms of the mass m.

20-16. Determine the radii of gyration k_x and k_y for the solid formed by revolving the shaded area about the y axis. The mass density of the material is ρ_m.

858

Prob. 20-16

20-17. Determine the mass moment of inertia I_y and the mass product of inertia I_{xy} of the body formed by revolving the shaded region about the line $x = 5$ ft. The mass density of the material is ρ_m.

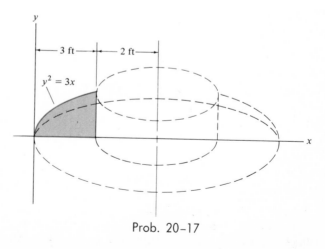

Prob. 20-17

20-18. Determine the mass moment of inertia of the torus about the z axis. The total mass of the torus is m.

Prob. 20-18

4 ft

1 ft

Prob. 20-20

20-19. Determine the mass moment of inertia of the composite solid with respect to the z and x axes. Each semicircular cylinder weighs 40 lb, and the rectangular block weighs 50 lb.

20-21. Determine the mass moment of inertia I_x for the composite assembly. The inner disk has a weight of 5 lb, the slender rods weigh 3 lb each, and the outer disk weighs 10 lb. The thin disks each have a thickness of $\frac{1}{2}$ in.

2 in.

12 in.

6 in.

Prob. 20-19

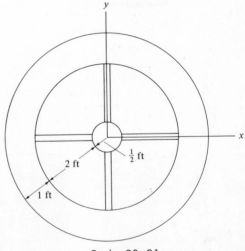

2 ft

$\frac{1}{2}$ ft

1 ft

Prob. 20-21

20-20. The pendulum consists of a sphere of weight 32.2 lb, securely attached to a slender rod of weight 16.1 lb. Determine the mass moment of inertia about an axis passing through the pin O and acting perpendicular to the page.

20-22. Determine the mass moment of inertia of the wheel with respect to the z axis. Each of the spokes weighs 3 lb, and the rim and hub weigh 10 and 2 lb, respectively.

Prob. 20-22

20-23. Determine the mass moment of inertia of the solid about the xx axis. The density of the material is $\rho = 100$ lb/ft³.

Prob. 20-23

20-24. Determine the radii of gyration k_x and k_{xy} for the bent rod. The rod has a linear density of 10 lb per foot of length.

Prob. 20-24

Prob. 20-25

20-26. The assembly is made of a material having a density of 400 lb/ft³. Determine the mass moment of inertia about the x axis and the mass product of inertia I_{xy}.

Prob. 20-26

*20–5. Transformation of Moments and Products of Inertia Between Inclined Coordinate Axes

We shall now develop the transformation equations required to obtain the elements of the inertia tensor for a set of axes which is *inclined* relative to a given set. In the discussion, it is assumed that both coordinate systems have the same origin. Consider the body shown in Fig. 20–9 for which the moments of inertia I_{xx}, I_{yy}, I_{zz}, I_{xy}, I_{xz}, and I_{yz} are known with respect to the *xyz* coordinate system. The mass moment of inertia $I_{x'x'}$ for the body is defined relative to the primed coordinate system as

$$I_{x'x'} = \int_m (y'^2 + z'^2)\, dm$$

Fig. 20–9

The vector **r** which locates the position of the mass element *dm* in the two coordinate systems may be written either as

$$\mathbf{r} = x\mathbf{i} + y\mathbf{j} + z\mathbf{k}$$

or

$$\mathbf{r} = x'\mathbf{i}' + y'\mathbf{j}' + z'\mathbf{k}'$$

Since the magnitude of this vector is preserved,

$$r^2 = x^2 + y^2 + z^2 = x'^2 + y'^2 + z'^2 \tag{20-5}$$

Thus,

$$y'^2 + z'^2 = x^2 + y^2 + z^2 - x'^2$$

so that

$$I_{x'x'} = \int_m (x^2 + y^2 + z^2 - x'^2)\, dm$$

The primed coordinate x' may be eliminated from this expression by realizing that

$$x' = \mathbf{i}' \cdot \mathbf{r} = \mathbf{i}' \cdot (x\mathbf{i} + y\mathbf{j} + z\mathbf{k}) = x(\mathbf{i}' \cdot \mathbf{i}) + y(\mathbf{i}' \cdot \mathbf{j}) + z(\mathbf{i}' \cdot \mathbf{k})$$
$$= xl_{x'x} + yl_{x'y} + zl_{x'z} \tag{20-6}$$

where $l_{x'x}$, $l_{x'y}$, and $l_{x'z}$ represent the *direction cosines* of the x' axis with the x, y, and z axes, respectively; e.g., $\mathbf{i}' \cdot \mathbf{i} = 1(1)\cos(x'x) = l_{x'x}$. Thus,

$$I_{x'x'} = \int_m [x^2 + y^2 + z^2 - (xl_{x'x} + yl_{x'y} + zl_{x'z})^2]\, dm$$

From the definition of a unit vector,

$$\mathbf{i}' = l_{x'x}\mathbf{i} + l_{x'y}\mathbf{j} + l_{x'z}\mathbf{k}$$

$$\mathbf{i'} \cdot \mathbf{i'} = l_{x'x}^2 + l_{x'y}^2 + l_{x'z}^2 = 1$$

Therefore,

$$I_{x'x'} = \int_m (x^2 + y^2 + z^2)(l_{x'x}^2 + l_{x'y}^2 + l_{x'z}^2)\,dm - (xl_{x'x} + yl_{x'y} + zl_{x'z})^2\,dm$$

$$= l_{x'x}^2 \int_m (y^2 + z^2)\,dm + l_{x'y}^2 \int_m (x^2 + z^2)\,dm$$

$$+ l_{x'z}^2 \int_m (x^2 + y^2)\,dm - 2l_{x'x}l_{x'y}\int_m xy\,dm$$

$$- 2l_{x'x}l_{x'z}\int_m xz\,dm - 2l_{x'y}l_{x'z}\int_m yz\,dm$$

Using the definitions of the mass moments and mass products of inertia, we have

$$I_{x'x'} = l_{x'x}^2 I_{xx} + l_{x'y}^2 I_{yy} + l_{x'z}^2 I_{zz} - 2l_{x'x}l_{x'y}I_{xy}$$
$$- 2l_{x'x}l_{x'z}I_{xz} - 2l_{x'y}l_{x'z}I_{yz} \qquad (20\text{--}7a)$$

The other moments of inertia of the body can be obtained in a similar manner, or by cyclic interchange of the x', y', and z' subscripts in Eq. 20–7a. The results are

$$I_{y'y'} = l_{y'x}^2 I_{xx} + l_{y'y}^2 I_{yy} + l_{y'z}^2 I_{zz} - 2l_{y'x}l_{y'y}I_{xy}$$
$$- 2l_{y'x}l_{y'z}I_{xz} - 2l_{y'y}l_{y'z}I_{yz} \qquad (20\text{--}7b)$$

and

$$I_{z'z'} = l_{z'x}^2 I_{xx} + l_{z'y}^2 I_{yy} + l_{z'z}^2 I_{zz} - 2l_{z'x}l_{z'y}I_{xy}$$
$$- 2l_{z'x}l_{z'z}I_{xz} - 2l_{z'y}l_{z'z}I_{yz} \qquad (20\text{--}7c)$$

Transformation equations for the products of inertia for the body can also be determined. For example, by definition, the product of inertia $I_{x'y'}$ for the body with respect to the $x'z'$ and $y'z'$ planes is

$$I_{x'y'} = \int_m x'y'\,dm$$

Using Eq. 20–6 and the fact that

$$y' = \mathbf{j'} \cdot \mathbf{r} = xl_{y'x} + yl_{y'y} + zl_{y'z}$$

where $l_{y'x}$, $l_{y'y}$, and $l_{y'z}$ represent the direction cosines of the y' axis with the x, y, and z axes, respectively, we have

$$I_{x'y'} = \int_m (x l_{x'x} + y l_{x'y} + z l_{x'z})(x l_{y'x} + y l_{y'y} + z l_{y'z})\, dm$$

$$= \int_m (x^2 l_{x'x} l_{y'x} + y^2 l_{x'y} l_{y'y} + z^2 l_{x'z} l_{y'z})\, dm$$

$$+ \int_m xy(l_{x'x} l_{y'y} + l_{x'y} l_{y'x})\, dm + \int_m yz(l_{x'y} l_{y'z} + l_{x'z} l_{y'y})\, dm$$

$$+ \int_m xz(l_{x'z} l_{y'x} + l_{x'x} l_{y'z})\, dm$$

Using Eq. 20–5, we can rearrange this equation in the form

$$I_{x'y'} = \int_m r^2(l_{x'x} l_{y'x} + l_{x'y} l_{y'y} + l_{x'z} l_{y'z}) - l_{x'x} l_{y'x} \int_m (y^2 + z^2)\, dm$$

$$- l_{x'y} l_{y'y} \int_m (x^2 + z^2)\, dm - l_{x'z} l_{y'z} \int_m (x^2 + y^2)\, dm$$

$$+ (l_{x'x} l_{y'y} + l_{x'y} l_{y'x}) \int_m xy\, dm + (l_{x'y} l_{y'z} + l_{x'z} l_{y'y}) \int_m yz\, dm$$

$$+ (l_{x'z} l_{y'x} + l_{x'x} l_{y'z}) \int_m xz\, dm$$

Since $\mathbf{i'}$ and $\mathbf{j'}$ are mutually perpendicular,

$$\mathbf{i'} \cdot \mathbf{j'} = (l_{x'x}\mathbf{i} + l_{x'y}\mathbf{j} + l_{x'z}\mathbf{k}) \cdot (l_{y'x}\mathbf{i} + l_{y'y}\mathbf{j} + l_{y'z}\mathbf{k}) = 0$$

or

$$\mathbf{i'} \cdot \mathbf{j'} = l_{x'x} l_{y'x} + l_{x'y} l_{y'y} + l_{x'z} l_{y'z} = 0$$

and therefore the first integral of the previous expression is zero. Using the definitions of the mass moments of inertia, we can write the final result as

$$\begin{aligned} I_{x'y'} = &-l_{x'x} l_{y'x} I_{xx} - l_{x'y} l_{y'y} I_{yy} - l_{x'z} l_{y'z} I_{zz} \\ &+ (l_{x'x} l_{y'y} + l_{x'y} l_{y'x}) I_{xy} + (l_{x'y} l_{y'z} + l_{x'z} l_{y'y}) I_{yz} \\ &+ (l_{x'z} l_{y'x} + l_{x'x} l_{y'z}) I_{xz} \end{aligned} \tag{20-8a}$$

By cyclic interchange of the primed subscripts in Eq. 20–8a, the products of inertia with respect to the other planes may be established. The results are

$$\begin{aligned} I_{x'z'} = &-l_{x'x} l_{z'x} I_{xx} - l_{x'y} l_{z'y} I_{yy} - l_{x'z} l_{z'z} I_{zz} \\ &+ (l_{x'x} l_{z'y} + l_{x'y} l_{z'x}) I_{xy} + (l_{x'y} l_{z'z} + l_{x'z} l_{z'y}) I_{yz} \\ &+ (l_{x'z} l_{z'x} + l_{x'x} l_{z'z}) I_{xz} \end{aligned} \tag{20-8b}$$

$$I_{z'y'} = -l_{z'x}l_{y'x}I_{xx} - l_{z'y}l_{y'y}I_{yy} - l_{z'z}l_{y'z}I_{zz}$$
$$+ (l_{z'x}l_{y'y} + l_{z'y}l_{y'x})I_{xy} + (l_{z'y}l_{y'z} + l_{z'z}l_{y'y})I_{yz} \qquad (20\text{-}8c)$$
$$+ (l_{z'z}l_{y'x} + l_{z'x}l_{y'z})I_{xz}$$

Thus, when the elements of the inertia tensor for the body are known at a point O with respect to the xyz coordinate axis, the elements of the inertia tensor for the body with respect to an inclined $x'y'z'$ coordinate axis, having the *same origin O,* can be determined using Eqs. 20-7 and 20-8.

*20-6. Ellipsoid of Inertia and Principal Axes of Inertia

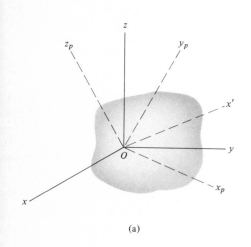

Using the transformation equations developed in Sec. 20-5, we can determine the moment of inertia of a body about *any* axis which is inclined with respect to a reference axis having the same origin. Let us now consider the nature of this variation for the mass moment of inertia $I_{x'x'}$ as the direction of the x' axis is changed, Fig. 20-10a. For each new inclination of the x' axis, the moment of inertia $I_{x'x'}$ can be computed using Eq. 20-7a. The results of this can be described graphically if we *plot* $O'P = 1/\sqrt{I_{x'x'}}$ in the direction of the x' axis. The locus of points P forms the surface shown in Fig. 20-10b. Confusion is avoided by plotting this surface using the $\xi\eta\zeta$ coordinate system which is *parallel* to the xyz axes shown in Fig. 20-10a. The equation of the surface can be determined using Eq. 20-7a. To show this, note that point P on the surface has coordinates (ξ, η, ζ), and the direction cosines of $O'P$ are the same as those for the axis x', since $O'P$ and Ox' are parallel. Thus,

$$l_{x'x} = \frac{\xi}{O'P} = \xi\sqrt{I_{x'x'}}$$

$$l_{x'y} = \frac{\eta}{O'P} = \eta\sqrt{I_{x'x'}}$$

$$l_{x'z} = \frac{\zeta}{O'P} = \zeta\sqrt{I_{x'x'}}$$

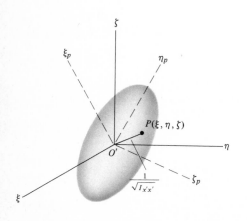

(b)

Fig. 20-10

Substituting these direction cosines into Eq. 20-7a and canceling the common factor $I_{x'x'}$ yields

$$I_{xx}\xi^2 + I_{yy}\eta^2 + I_{zz}\zeta^2 - 2I_{xy}\xi\eta - 2I_{xz}\xi\zeta - 2I_{yz}\eta\zeta = 1 \qquad (20\text{-}9)$$

This is the equation of an ellipsoid, and since the constants I_{xx}, I_{yy}, I_{zz}, I_{xy}, I_{xz}, and I_{yz} are included in the terms of this equation, the surface is called the *ellipsoid of inertia.* Representing the inertia tensor graphically at a point using the ellipsoid of inertia is similar to using an arrow to

represent a vector. Specifically, for each point in space, the size, shape, and orientation of the ellipsoid of inertia for the body will be different.

Equation 20–9, used to represent the surface of the ellipsoid, can be simplified if the ξ, η, and ζ coordinate axes are oriented along the three axes of symmetry of the ellipsoid. This special set of axes, called ξ_p, η_p, and ζ_p, is shown in Fig. 20–10b. The corresponding parallel set of x_p, y_p, and z_p axes for the body is shown in Fig. 20–10a. These axes are called the *principal axes of inertia* for the body. Using the ξ_p, η_p, and ζ_p axes, the ellipsoid of inertia may be represented as

$$I_x\xi_p^2 + I_y\eta_p^2 + I_z\zeta_p^2 = 1$$

where $I_x = I_{xx}$, $I_y = I_{yy}$, and $I_z = I_{zz}$ represent the *principal moments of inertia* for the body. (These values are computed with respect to x_p, y_p, and z_p axes, Fig. 20–10a.) By comparison with Eq. 20–9, it is seen that the *products of inertia for the body computed with respect to the principal axes of inertia are zero*.

From this analysis it may therefore be concluded that regardless of the shape of the body and the point considered, a set of principal axes may always be established at the point so that the products of inertia for the body are zero when computed with respect to these axes. When this is done, the inertia tensor for the body is said to be "diagonalized" and may be written in the simplified form

$$\begin{pmatrix} I_x & 0 & 0 \\ 0 & I_y & 0 \\ 0 & 0 & I_z \end{pmatrix}$$

Of these three principal moments of inertia, one will be a maximum and another a minimum of the moments of inertia for the body.

Mathematical determination of the direction of principal axes of inertia is somewhat complicated and is beyond the scope of this book. There are many cases, however, in which these axes may be determined by inspection. From the discussion in Sec. 20–1, if the coordinate axes, located at the point considered, are oriented such that *two* of the three orthogonal planes containing the axes are planes of *symmetry* for the body, then all the products of inertia for the body are zero with respect to the coordinate planes, and hence the coordinate axes are principal axes of inertia. For example, the x, y, and z axes shown in Figs. 20–2b or 20–4a represent the principal axes of inertia for the cylinder at the point O. This concept may be extended to include bodies which have partial rotational symmetry, such as the equilateral triangular plate shown in Fig. 20–11a. If the plate is rotated 120° about the x axis, it will coincide with both its original geometry and mass distribution. It follows that the ellipsoid of inertia for the plate remains the *same* for each 120° rotation. Hence, $O'P = O'Q = O'R$, Fig. 20–11b, so that the ellipsoid has an axis

(a)

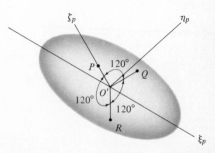

(b)

Fig. 20–11

of symmetry ξ_p which is parallel to the x axis (principal axis). The y and z axes, which are perpendicular to the x axis, will *also* represent principal axes, since these axes correspond to η_p and ζ_p for the ellipsoid. This same argument may be extended to include bodies having higher than three degrees of symmetry as illustrated here.

Example 20-6

Determine the inertia tensor for the homogeneous cylinder shown in Fig. 20–12 with respect to (a) the unprimed coordinate system and (b) the primed coordinate system. The mass of the cylinder is m.

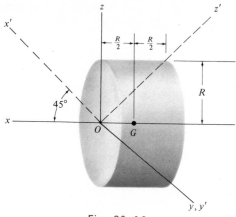

Fig. 20–12

Solution

Part (a). The inertia tensor can be computed with respect to the xyz axis using the data given in Appendix D together with the parallel-axis and parallel-plane theorems. Thus,

$$I_{xx} = \frac{1}{2}mR^2$$

$$I_{yy} = I_{zz} = \frac{1}{12}m(3R^2 + R^2) + m\left(\frac{R}{2}\right)^2 = \frac{7}{12}mR^2$$

$$I_{xy} = I_{xz} = I_{yz} = 0$$

Hence,

$$\begin{pmatrix} \frac{1}{2}mR^2 & 0 & 0 \\ 0 & \frac{7}{12}mR^2 & 0 \\ 0 & 0 & \frac{7}{12}mR^2 \end{pmatrix} \qquad Ans.$$

The x, y, and z axes are *principal axes* for the cylinder at point O since the products of inertia are all zero.

Part (b). The inertia tensor for the primed coordinates can be established by using the transformation equations for inclination with the moments of inertia previously calculated. From the geometry of the figure, the direction cosines for the axes are

$$l_{x'x} = \frac{1}{\sqrt{2}}, \qquad l_{x'y} = 0, \qquad l_{x'z} = \frac{1}{\sqrt{2}}$$

$$l_{y'x} = 0, \qquad l_{y'y} = 1, \qquad l_{y'z} = 0$$

$$l_{z'x} = \frac{-1}{\sqrt{2}}, \qquad l_{z'y} = 0, \qquad l_{z'z} = \frac{1}{\sqrt{2}}$$

From Part (a),

$$I_{xx} = \frac{1}{2}mR^2, \qquad I_{yy} = I_{zz} = \frac{7}{12}mR^2, \qquad I_{xy} = I_{xz} = I_{yz} = 0$$

Substituting these data into Eqs. 20–7 and 20–8 yields

$$I_{x'x'} = \left(\frac{1}{\sqrt{2}}\right)^2\left(\frac{1}{2}mR^2\right) + 0 + \left(\frac{1}{\sqrt{2}}\right)^2\left(\frac{7}{12}mR^2\right) - 0 - 0 - 0 = \frac{13}{24}mR^2$$

$$I_{y'y'} = 0 + (1)^2\left(\frac{7}{12}mR^2\right) + 0 - 0 - 0 - 0 = \frac{7}{12}mR^2$$

$$I_{z'z'} = \left(\frac{-1}{\sqrt{2}}\right)^2\left(\frac{1}{2}mR^2\right) + 0 + \left(\frac{1}{\sqrt{2}}\right)^2\left(\frac{7}{12}mR^2\right) - 0 - 0 - 0 = \frac{13}{24}mR^2$$

$$I_{x'y'} = 0$$

$$I_{x'z'} = -\left(\frac{1}{\sqrt{2}}\right)\left(\frac{-1}{\sqrt{2}}\right)\left(\frac{1}{2}mR^2\right) - 0 - \left(\frac{1}{\sqrt{2}}\right)\left(\frac{1}{\sqrt{2}}\right)\left(\frac{7}{12}mR^2\right)$$

$$+ 0 + 0 + 0 = -\frac{1}{24}mR^2$$

$$I_{z'y'} = 0$$

Changing the sign of $I_{x'z'}$, the inertia tensor for the primed coordinate system is therefore

$$\begin{pmatrix} \dfrac{13}{24}mR^2 & 0 & \dfrac{1}{24}mR^2 \\[2mm] 0 & \dfrac{7}{12}mR^2 & 0 \\[2mm] \dfrac{1}{24}mR^2 & 0 & \dfrac{13}{24}mR^2 \end{pmatrix} \qquad Ans.$$

Problems

20-27. Determine the mass moment of inertia of the cone assembly about the y' axis in terms of its mass moment of inertia about the y axis. The mass of each cone is m.

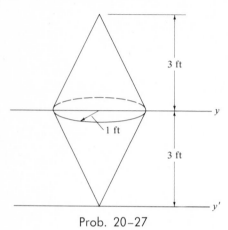

3 ft

1 ft

3 ft

Prob. 20-27

20-28. Determine the mass moment of inertia of the cube about the diagonal axis AB. The mass of the cube is m.

Prob. 20-28

20-29. Compute the mass moment of inertia of the disk about the axis of the shaft, AB. The disk has a total weight of 40 lb.

6 in.

30°

Prob. 20-29

20-30. Determine the inertia tensor for the cylinder with respect to the $x'y'z'$ coordinate system. The mass of the cylinder is m.

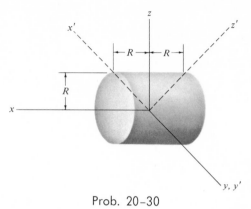

Prob. 20-30

20-31. Determine the inertia tensor for the cube with respect to (a) the xyz coordinate system and (b) the $x'y'z'$ coordinate system. The mass of the cube is m.

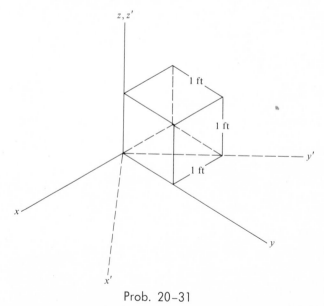

1 ft

1 ft

1 ft

Prob. 20-31

20-32. Determine the moment of inertia of both the 5-lb rod CD and the 10-lb disk about the axis of the shaft, AB.

Prob. 20-32

20-35. Determine the mass moment of inertia of the 5-lb circular plate about the axis of the rod OA.

Prob. 20-35

20-33. Determine the moment of inertia of the cone about the z' axis. The mass of the cone is m. The height of the cone is 2 ft.

Prob. 20-33

20-34. Compute the mass moment of inertia of the rod and plate assembly about the xx axis. The plates weigh 2 lb each and the two rods weigh 2 lb each.

Prob. 20-34

20-36. Compute the moment of inertia I_{xx} and product of inertia I_{yz} for the composite plate assembly. The plates have a density of 6 lb/ft².

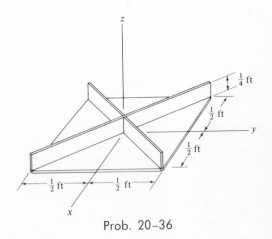

Prob. 20-36

20-37. Show that the sum of the mass moments of inertia for a set of orthogonal axes is independent of the orientation of the axes and thus depends only on the location of the origin.

20-38. Compute the mass moment of inertia of the composite body about the *AA* axis. The cylinder weighs 20 lb, and each hemisphere weighs 10 lb.

20-39. Compute the mass moment of inertia of the rod and thin ring assembly about the *OA* axis. The rods and ring have a linear density of 4 lb/ft.

Prob. 20-38

Prob. 20-39

21

Three-Dimensional Kinetics of a Rigid Body

*21–1. Introduction

The planar kinetics of a rigid body was discussed in Chapters 15, 17, and 19. In an effort to keep the rotational equation of motion and the equation of angular momentum in as simple a form as possible, the analysis was restricted to rigid bodies *symmetrical* with respect to a fixed reference plane. In this chapter we will present the more general kinetic analysis of a rigid body having an arbitrary shape and being subjected to any of the five types of motion discussed in Chapter 13. In general, as a body moves through space it has both translational and rotational motion. The kinetic aspects of the *translation* have been discussed in Chapter 15, where it was shown that the system of external forces acting on the body may be related to the acceleration of the body's mass center by the equation

$$\Sigma \mathbf{F} = m\mathbf{a}_G \qquad (21\text{–}1)$$

This equation is most general because it applies to a body of arbitrary shape undergoing any sort of motion. In this chapter we will emphasize primarily the rotational aspects of rigid body motion, since motion of the body's mass center, as exemplified by Eq. 21–1, is treated in the same manner as particle motion.

Six independent coordinates are needed to specify completely the location and orientation of a rigid body having general motion. For example, three Cartesian coordinates may be used to specify the location of a point in the body and, as will be shown in Sec. 21–6, three angles

(called *Euler angles*) may be used to specify the orientation of the body about this point. For each of these six coordinates, it is required that a set of six corresponding equations of motion be written in order to completely describe the motion of the body. The translational motion of the body is defined by the three scalar equations relating the external force system acting on the body to the acceleration of the body's mass center. These equations are defined by the three scalar components of Eq. 21-1. The three rotational equations relate the body's rotational motions to the moments created by the external forces about some point located either in or off the body. In the general sense, these "time differential" equations will be rather difficult to solve, since all the elements of the inertia tensor will be involved in the terms of these equations. Hence, we will have to exercise judgement in choosing the most suitable origin and orientation of the coordinate axes. Proper selection will allow us to solve many seemingly complicated problems in a rather direct fashion.

A derivation of the three rotational equations of motion requires a prior formulation of the angular momentum of the body. For this reason, the general angular momentum and kinetic energy expressions for the body will be presented first, Secs. 21-2 and 21-3. After gaining some familiarity in developing these expressions, the principles of impulse and momentum and work and energy will be at our disposal for solving rigid-body problems in three dimensions. The equations of rigid-body motion are developed in Sec. 21-4, followed by a general method for problem solution. Of special interest are problems involving the motion of an unsymmetrical body about a fixed axis, motion of a gyroscope (Sec. 21-7), and torque-free motion (Sec. 21-8).

*21-2. Angular Momentum of a Rigid Body

In this section we will develop the necessary equations used to determine the angular momentum of a rigid body about an arbitrary point located in the body. These formulations will provide the necessary means for developing both the principle of impulse and momentum and the equations of rotational motion for the rigid body.

Consider the rigid body in Fig. 21-1 which has a total mass m with center of mass located at G. The XYZ coordinate system represents an inertial frame of reference, and hence, its axes are fixed or translate with a constant velocity. The angular momentum as measured in this reference will be computed relative to the arbitrary point A located in the body. The position vectors $\mathbf{r}_A$ and $\boldsymbol{\rho}_A$ are drawn from the origin of coordinates to point A, and from point A to the ith particle P of the body. If the mass of particle P is Δm_i, the angular momentum is

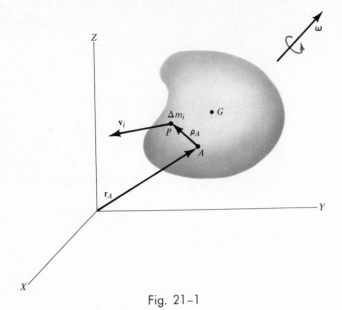

Fig. 21-1

$$\Delta \mathbf{H}_A = \boldsymbol{\rho}_A \times \Delta m_i \, \mathbf{v}_i$$

where $\mathbf{v}_i$ represents the velocity of the ith particle as measured from the X,Y,Z coordinate system. If the body also has an angular velocity $\boldsymbol{\omega}$ at the instant considered, $\mathbf{v}_i$ may be related to the velocity of A by using the kinematic equation

$$\mathbf{v}_i = \mathbf{v}_A + \boldsymbol{\omega} \times \boldsymbol{\rho}_A$$

Thus,

$$\Delta \mathbf{H}_A = \boldsymbol{\rho}_A \times (\mathbf{v}_A + \boldsymbol{\omega} \times \boldsymbol{\rho}_A) \, \Delta m_i$$
$$= (\boldsymbol{\rho}_A \, \Delta m_i) \times \mathbf{v}_A + \boldsymbol{\rho}_A \times (\boldsymbol{\omega} \times \boldsymbol{\rho}_A) \, \Delta m_i$$

For the entire body, summing all n particles,

$$\mathbf{H}_A = (\Sigma \boldsymbol{\rho}_A \, \Delta m_i) \times \mathbf{v}_A + \Sigma \boldsymbol{\rho}_A \times (\boldsymbol{\omega} \times \boldsymbol{\rho}_A) \, \Delta m_i$$

As $n \to \infty$, then $\Delta m_i \to dm$ and the summations become integrals:

$$\mathbf{H}_A = \left(\int_m \boldsymbol{\rho}_A \, dm \right) \times \mathbf{v}_A + \int_m \boldsymbol{\rho}_A \times (\boldsymbol{\omega} \times \boldsymbol{\rho}_A) \, dm \qquad (21\text{-}2)$$

There are two cases for which this equation reduces to a simplified form:

1. If A becomes a *fixed point* O on the body, Fig. 21-2a, then $\mathbf{v}_A = 0$ and Eq. 21-2 reduces to

(a)

(b)

(c)

Fig. 21-2

$$\mathbf{H}_O = \int_m \boldsymbol{\rho}_O \times (\boldsymbol{\omega} \times \boldsymbol{\rho}_O) \, dm \qquad (21\text{-}3)$$

2. If A is located at the *center of mass* G of the body, Fig. 21-2b, then $\int_m \boldsymbol{\rho}_A \, dm = 0$ and

$$\mathbf{H}_G = \int_m \boldsymbol{\rho}_G \times (\boldsymbol{\omega} \times \boldsymbol{\rho}_G) \, dm \qquad (21\text{-}4)$$

In general, A will be some point other than O or G, in which case Eq. 21-2 may nevertheless be simplified by noting that $\boldsymbol{\rho}_A = \boldsymbol{\rho}_G + \boldsymbol{\rho}_{G/A}$, Fig. 21-2c. Substituting into Eq. 21-2, we have

$$\mathbf{H}_A = \int_m (\boldsymbol{\rho}_G + \boldsymbol{\rho}_{G/A}) \times \mathbf{v}_A \, dm + \int_m (\boldsymbol{\rho}_G + \boldsymbol{\rho}_{G/A}) \times [\boldsymbol{\omega} \times (\boldsymbol{\rho}_G + \boldsymbol{\rho}_{G/A})] \, dm$$

or

$$\mathbf{H}_A = \left(\int_m \boldsymbol{\rho}_G \, dm \right) \times \mathbf{v}_A + (\boldsymbol{\rho}_{G/A} \times \mathbf{v}_A) \int_m dm + \int_m \boldsymbol{\rho}_G \times (\boldsymbol{\omega} \times \boldsymbol{\rho}_G) \, dm$$

$$+ \left(\int_m \boldsymbol{\rho}_G \, dm \right) \times (\boldsymbol{\omega} \times \boldsymbol{\rho}_{G/A}) + \boldsymbol{\rho}_{G/A} \times \left(\boldsymbol{\omega} \times \int_m \boldsymbol{\rho}_G \, dm \right) + \boldsymbol{\rho}_{G/A} \times (\boldsymbol{\omega} \times \boldsymbol{\rho}_{G/A}) \int_m dm$$

The first, fourth, and fifth terms on the right side are zero, since they contain the form $\int_m \boldsymbol{\rho}_G \, dm$, which is zero by definition of the mass center. Also, the third term on the right side is defined by Eq. 21-4. Thus,

$$\mathbf{H}_A = (\boldsymbol{\rho}_{G/A} \times \mathbf{v}_A)m + \mathbf{H}_G + \boldsymbol{\rho}_{G/A} \times (\boldsymbol{\omega} \times \boldsymbol{\rho}_{G/A})m$$

or

$$\mathbf{H}_A = \boldsymbol{\rho}_{G/A} \times [\mathbf{v}_A + \boldsymbol{\omega} \times \boldsymbol{\rho}_{G/A}]m + \mathbf{H}_G$$

Since

$$\mathbf{v}_G = \mathbf{v}_A + \boldsymbol{\omega} \times \boldsymbol{\rho}_{G/A}$$

then

$$\mathbf{H}_A = (\boldsymbol{\rho}_{G/A} \times m\mathbf{v}_G) + \mathbf{H}_G \qquad (21\text{-}5)$$

It is seen from this equation that the angular momentum of the body about the arbitrary point A consists of two parts—the moment of the

linear momentum $m\mathbf{v}_G$ of the body* about point A added (vectorially) to the angular momentum $\mathbf{H}_G$ of the body about the mass center. Equation 21-5 may also be used for computing the angular momentum of the body about a fixed point O in the body; the results, of course, will be the same as those computed using the more convenient Eq. 21-3.

To make practical use of Eqs. 21-3 through 21-5, the angular momentum must be expressed in terms of scalar components. For this purpose, it is convenient to choose a second set of axes x, y, and z having an arbitrary orientation relative to the XYZ axes, Fig. 21-2. Equations 21-3 and 21-4 are of the form

$$\mathbf{H} = \int_m \boldsymbol{\rho} \times (\boldsymbol{\omega} \times \boldsymbol{\rho}) \, dm$$

This form is also contained in Eq. 21-5. Expressing $\mathbf{H}$, $\boldsymbol{\rho}$, and $\boldsymbol{\omega}$ in terms of x, y, and z components, we have

$$H_x\mathbf{i} + H_y\mathbf{j} + H_z\mathbf{k} = \int_m (x\mathbf{i} + y\mathbf{j} + z\mathbf{k}) \times [(\omega_x\mathbf{i} + \omega_y\mathbf{j} + \omega_z\mathbf{k})$$
$$\times (x\mathbf{i} + y\mathbf{j} + z\mathbf{k}) \, dm$$

Expanding the cross products and combining terms yields

$$H_x\mathbf{i} + H_y\mathbf{j} + H_z\mathbf{k} = \left[\omega_x \int_m (y^2 + z^2) \, dm - \omega_y \int_m xy \, dm - \omega_z \int_m xz \, dm \right]\mathbf{i}$$
$$+ \left[-\omega_x \int_m xy \, dm + \omega_y \int_m (x^2 + z^2) \, dm - \omega_z \int_m yz \, dm \right]\mathbf{j}$$
$$+ \left[-\omega_x \int_m zx \, dm - \omega_y \int_m yz \, dm + \omega_z \int_m (x^2 + y^2) \, dm \right]\mathbf{k}$$

The integrals represent the moments and products of inertia of the body with respect to the x, y, and z coordinate axes. Equating the respective $\mathbf{i}$, $\mathbf{j}$, and $\mathbf{k}$ components, the three scalar components of the angular momentum may therefore be written as

$$\begin{aligned} H_x &= I_{xx}\omega_x - I_{xy}\omega_y - I_{xz}\omega_z \\ H_y &= -I_{xy}\omega_x + I_{yy}\omega_y - I_{yz}\omega_z \\ H_z &= -I_{zx}\omega_x - I_{yz}\omega_y + I_{zz}\omega_z \end{aligned} \qquad (21\text{-}6)$$

These equations represent the scalar form of the vector components of $\mathbf{H}_O$ or $\mathbf{H}_G$ (given in vector form by Eqs. 21-3 and 21-4). The angular momentum of the body about the arbitrary point A, other than the fixed point O or the center of mass G, may be expressed in scalar form by

---

*The linear momentum $\mathbf{L} = m\mathbf{v}_G$, where $\mathbf{v}_G$ is the velocity of the body's mass center, was derived in Sec. 19-1 for any general motion.

using Eq. 21–5, replacing $\rho_{G/A}$ and $\mathbf{v}_G$ in Cartesian component form, carrying out the cross-product operation, and substituting the components, Eqs. 21–6, for $\mathbf{H}_G$. In particular, when the rigid body is subjected to plane motion and the body is symmetrical with respect to the xy reference plane, then $\omega_x = \omega_y = 0$, and $I_{xz} = I_{yz} = 0$, so that Eqs. 21–6 reduce to the familiar form $H_z = I_{zz}\omega_z$, which represents Eq. 19–2 or Eq. 19–3.

The origin of the coordinate axes must be located at either point O or G when computing the moments and products of inertia of the body used in Eqs. 21–6. *The inclination of the axes about the origin is arbitrary.* Specifically, it has been pointed out in Chapter 20 that a different set of nine inertia terms exists for each orientation of the x, y, and z axes. These nine terms, when computed, represent the elements of the inertia tensor for the axes. Through Eqs. 21–6, the inertia tensor *transforms* the angular velocity $\boldsymbol{\omega}$ of the body into the angular momentum $\mathbf{H}$. To understand how this is done, replace the subscripts x, y, and z of the terms in Eqs. 21–6 by the numbers 1, 2, and 3, respectively, and write the inertia tensor in the standard form, as discussed in Sec. 20–1. The three Eqs. 21–6 may then be replaced by the single equation

$$H_i = \sum_{j=1}^{3} I_{ij}\omega_j \tag{21–7}$$

where for every $i = 1, 2,$ or 3, one obtains one of Eqs. 21–6. By inspection, the transformation is thus characterized by I_{ij}.

Although the inertia tensor *changes* with each orientation of the axes, the transformation of $\boldsymbol{\omega}$ into $\mathbf{H}$ as given by Eqs. 21–6 is the *same* regardless of the orientation. This is because the inertia ellipsoid for the body has a unique size, shape, and orientation at each point, and is *independent* of the orientation of coordinate axes used to describe it.

Equations 21–6 may be simplified further if the x, y, and z coordinate axes are oriented such that they become *principal axes of inertia* for the body at the point. In this case the products of inertia are zero, $I_{xx} = I_x$, etc., and the equations reduce to the simplified form

$$\begin{aligned} H_x &= I_x\omega_x \\ H_y &= I_y\omega_y \\ H_z &= I_z\omega_z \end{aligned} \tag{21–8}$$

Having presented the means for computing the angular momentum for the body, the principle of impulse and momentum, as discussed in Chapter 19, may be applied for solving kinetic problems when the body is subjected to any general motion. For this case six scalar equations are available. For example, three equations relate the linear impulse and

momentum of the body in the x, y, and z directions, and three equations relate the body's angular impulse and momentum about these axes. As was pointed out in Chapter 19, the principle of impulse and momentum is most suitable for solving problems involving time and velocities, and this method provides the only practical method for solving problems of impact.

*21–3. Kinetic Energy of a Rigid Body

The kinetic energy expressions of a rigid body subjected to planar motion have been formulated in Chapter 17. To apply the principle of work and energy to the solution of problems involving general rigid-body motion, it is first necessary to formulate an expression for the kinetic energy of the body in this more general sense.

Consider the rigid body in Fig. 21-3 which has a total mass m and center of gravity located at G. The kinetic energy of the ith particle P of the body having a mass Δm_i and velocity $\mathbf{v}_i$, measured relative to the inertial XYZ frame of reference, is

$$\Delta T_i = \tfrac{1}{2} \Delta m_i \, v_i^2 = \tfrac{1}{2} \Delta m_i \, (\mathbf{v}_i \cdot \mathbf{v}_i)$$

Provided the velocity $\mathbf{v}_A$ of an arbitrary point A on the body is known, the velocity $\mathbf{v}_i$ may be related to the velocity of A by the equation

$$\mathbf{v}_i = \mathbf{v}_A + \boldsymbol{\omega} \times \boldsymbol{\rho}_A$$

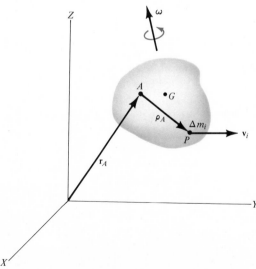

Fig. 21-3

where $\boldsymbol{\omega}$ is the instantaneous angular velocity of the body measured from the X, Y, and Z coordinate system, and $\boldsymbol{\rho}_A$ is a position vector drawn from point A to particle P. Substituting this expression for $\mathbf{v}_i$, we may write the kinetic energy for the particle as

$$\Delta T_i = \tfrac{1}{2} \Delta m_i (\mathbf{v}_A + \boldsymbol{\omega} \times \boldsymbol{\rho}_A) \cdot (\mathbf{v}_A + \boldsymbol{\omega} \times \boldsymbol{\rho}_A)$$

or

$$\Delta T_i = \tfrac{1}{2}(\mathbf{v}_A \cdot \mathbf{v}_A) \Delta m_i + \mathbf{v}_A \cdot (\boldsymbol{\omega} \times \boldsymbol{\rho}_A) \Delta m_i + \tfrac{1}{2}(\boldsymbol{\omega} \times \boldsymbol{\rho}_A) \cdot (\boldsymbol{\omega} \times \boldsymbol{\rho}_A) \Delta m_i$$

The kinetic energy for the entire body is obtained by summing the kinetic energies of all n particles of the body. Thus,

$$T = \tfrac{1}{2}\mathbf{v}_A \cdot \mathbf{v}_A \Sigma \Delta m_i + \mathbf{v}_A \cdot (\boldsymbol{\omega} \times \Sigma \boldsymbol{\rho}_A \Delta m_i) + \tfrac{1}{2}\Sigma(\boldsymbol{\omega} \times \boldsymbol{\rho}_A) \cdot (\boldsymbol{\omega} \times \boldsymbol{\rho}_A) \Delta m_i$$

Letting $n \to \infty$, then $\Delta m_i \to dm$, and the summations become integrals. Thus,

$$T = \tfrac{1}{2}m(\mathbf{v}_A \cdot \mathbf{v}_A) + \mathbf{v}_A \cdot \left(\boldsymbol{\omega} \times \int_m \boldsymbol{\rho}_A \, dm\right) + \tfrac{1}{2}\int_m (\boldsymbol{\omega} \times \boldsymbol{\rho}_A) \cdot (\boldsymbol{\omega} \times \boldsymbol{\rho}_A) \, dm$$

The last term on the right may be rewritten using the vector identity $\mathbf{a} \times \mathbf{b} \cdot \mathbf{c} = \mathbf{a} \cdot \mathbf{b} \times \mathbf{c}$, where $\mathbf{a} = \boldsymbol{\omega}$, $\mathbf{b} = \boldsymbol{\rho}_A$, and $\mathbf{c} = \boldsymbol{\omega} \times \boldsymbol{\rho}_A$. The final result is therefore

$$T = \tfrac{1}{2}m(\mathbf{v}_A \cdot \mathbf{v}_A) + \mathbf{v}_A \cdot \left(\boldsymbol{\omega} \times \int_m \boldsymbol{\rho}_A \, dm\right)$$

$$+ \tfrac{1}{2}\boldsymbol{\omega} \cdot \int_m \boldsymbol{\rho}_A \times (\boldsymbol{\omega} \times \boldsymbol{\rho}_A) \, dm \qquad (21\text{-}9)$$

This equation is rarely used for application because of computation involving the integrals. Simplification occurs, however, if the reference point A is either a fixed point O or the center of mass G.

1. If A is a *fixed point* O on the body, Fig. 21-2a, then $\mathbf{v}_A = 0$, and using Eq. 21-3, we can reduce Eq. 21-9 to

$$T = \tfrac{1}{2}\boldsymbol{\omega} \cdot \mathbf{H}_O \qquad (21\text{-}10a)$$

Given that $\boldsymbol{\omega} = \omega_x\mathbf{i} + \omega_y\mathbf{j} + \omega_z\mathbf{k}$ and $\mathbf{H}_O = (H_x)_O\mathbf{i} + (H_y)_O\mathbf{j} + (H_z)_O\mathbf{k}$, where the scalar components for $\mathbf{H}_O$ are defined by Eqs. 21-6, the kinetic energy may be expressed in an alternative form by substituting these expressions into Eq. 21-10a and carrying out the vector operations. The result is

$$T = \tfrac{1}{2}I_{xx}\omega_x^2 + \tfrac{1}{2}I_{yy}\omega_y^2 + \tfrac{1}{2}I_{zz}\omega_z^2 - I_{xy}\omega_x\omega_y - I_{yz}\omega_y\omega_z - I_{zx}\omega_z\omega_x \qquad (21\text{-}10b)$$

In particular, when the rigid body is rotating about a *fixed axis* which is perpendicular to the xy plane, then $\omega_x = \omega_y = 0$ and the above

equation reduces to the familiar form $T = \frac{1}{2}I_{zz}\omega_z^2$, which is essentially Eq. 17–3. Notice that this result is independent of the shape of the body.

If the x, y, and z coordinate axes, having origin at O, are oriented in the body such that these axes represent *principal axes of inertia* for the body, then

$$T = \frac{1}{2}I_x\omega_x^2 + \frac{1}{2}I_y\omega_y^2 + \frac{1}{2}I_z\omega_z^2 \qquad (21\text{--}10c)$$

2. If A is located at the *center of mass G* of the body, Fig. 21–2b, then $\int_m \boldsymbol{\rho}_A\, dm = 0$ and, using Eq. 21–4, we can reduce Eq. 21–9 to

$$T = \frac{1}{2}mv_G^2 + \frac{1}{2}\boldsymbol{\omega} \cdot \mathbf{H}_G \qquad (21\text{--}11a)$$

In a manner similar to that for a fixed point, the last term on the right side of this equation may be represented in scalar form. Requiring the origin of the x, y, and z coordinates to be located at the mass center, we obtain

$$T = \frac{1}{2}mv_G^2 + \frac{1}{2}I_{xx}\omega_x^2 + \frac{1}{2}I_{yy}\omega_y^2$$
$$+ \frac{1}{2}I_{zz}\omega_z^2 - I_{xy}\omega_x\omega_y - I_{yz}\omega_y\omega_z - I_{zx}\omega_z\omega_x \qquad (21\text{--}11b)$$

Note that when the body is subjected to *general plane motion* in the xy plane, $\omega_x = \omega_y = 0$ so that the above equation reduces to the form $T = \frac{1}{2}mv_G^2 + \frac{1}{2}I_{zz}\omega_z^2$, which is essentially Eq. 17–1.

If the x, y, and z axes are *principal axes of inertia* for the body at point G, then

$$T = \frac{1}{2}mv_G^2 + \frac{1}{2}I_x\omega_x^2 + \frac{1}{2}I_y\omega_y^2 + \frac{1}{2}I_z\omega_z^2 \qquad (21\text{--}11c)$$

The kinetic energy for the body may therefore be obtained using one of the preceding equations when the body is subjected to general motion. With these expressions the principle of work and energy, discussed in Chapter 17, may then be used as an aid to the solution of three-dimensional kinetic problems. Since both work and energy are scalar quantities only one equation may be written for each rigid body. Generally this equation is most suitable for the solution if the problem involves forces, velocities, and displacements.

Example 21–1

The bent rod in Fig. 21–4a has a linear density of 3 lb/ft. Initially it lies in the same horizontal plane and is released from rest when the end A is 4 ft above the hook at E. Assuming that the rod falls uniformly without rotation, determine the angular velocity of the rod and the

(a)

(b)

Fig. 21-4

(c)

velocity of end D just after A falls onto E. The hook at E provides a permanent connection for the rod because of the spring-lock mechanism S.

Solution

The velocity of the rod just *before* end A strikes the hook can be determined by applying the principle of work and energy (or the conservation of energy theorem). Since the rod does not rotate while it is falling, the kinetic energy Eq. 21-11b reduces to a simplified form involving only translation. Hence,

$$\{T_1\} + \{\Sigma W_{1-2}\} = \{T_2\}$$

$$\{0\} + \{3 \text{ lb/ft}(6 \text{ ft}) \, 4 \text{ ft}\} = \left\{\frac{1}{2}\left(\frac{3 \text{ lb/ft}(6 \text{ ft})}{32.2 \text{ ft/sec}^2}\right)(v_G)_1^2\right\}$$

Solving for $(v_G)_1$ gives

$$(v_G)_1 = 16.05 \text{ ft/sec}$$

The momentum and impulse vector diagrams for the rod are shown in Fig. 21-4b. During the short time Δt, the impulsive force **F** acting at A changes the momentum of the rod. (The impulse created by the rod's weight is small compared to **F** Δt, so that it is neglected.) Since the impulse is vertical, the angular momentum $(\mathbf{H}_G)_2$ imparted to the rod lies in the xy plane. Note that angular momentum of the rod is *conserved* about point A since the moment of the external impulse about point A is zero—**F** Δt passes through point A. Equations 21-5 must be used in computing the angular momentum of the rod since point A does *not* become a fixed point until *after* the impulsive interaction with the hook. Thus, applying the conservation of angular momentum,

$$\Sigma(\text{rod angular momentum})_1 = \Sigma(\text{rod angular momentum})_2$$

$$\mathbf{r}_{G/A} \times m(\mathbf{v}_G)_1 + 0 = \mathbf{r}_{G/A} \times m(\mathbf{v}_G)_2 + (\mathbf{H}_G)_2 \tag{1}$$

The scalar components of this equation may be determined knowing the position $(\bar{x}, \bar{y})$ of the mass center G and computing the moments of inertia of the rod relative to a set of coordinate axes x', y', and z' having an origin at the mass center. With reference to Fig. 21-4c, $\bar{y} = 1$ ft because of symmetry, and

$$\bar{x} = \frac{\Sigma \tilde{x} W}{\Sigma W} =$$

$$\frac{3 \text{ lb/ft(2 ft)}(-1 \text{ ft}) + 3 \text{ lb/ft(2 ft)}(-2 \text{ ft}) + 3 \text{ lb/ft(2 ft)}(-1 \text{ ft})}{(3 \text{ lb/ft}) 6 \text{ ft}} = -1.33 \text{ ft}$$

Thus,

$$\mathbf{r}_{G/A} = \{-1.33\mathbf{i} + 1\mathbf{j}\} \text{ ft}$$

The primed axes are *principal axes of inertia* for the rod because $I_{x'y'} = I_{x'z'} = I_{z'y'} = 0$. Since $(\mathbf{H}_G)_2$ has no component in the z' direction, only $I_{x'}$ and $I_{y'}$ must be computed. Using Appendix D and the parallel-axis theorem, we have

$$I_{x'} = 2\left[0 + \left(\frac{3 \text{ lb/ft(2 ft)}}{32.2 \text{ ft/sec}^2}\right)(1 \text{ ft})^2\right] + \frac{1}{12}\left(\frac{3 \text{ lb/ft(2 ft)}}{32.2 \text{ ft/sec}^2}\right)(2 \text{ ft})^2 = 0.435 \text{ slug-ft}^2$$

$$I_{y'} = 2\left[\frac{1}{12}\left(\frac{3 \text{ lb/ft(2 ft)}}{32.2 \text{ ft/sec}^2}\right)(2 \text{ ft})^2 + \left(\frac{3 \text{ lb/ft(2 ft)}}{32.2 \text{ ft/sec}^2}\right)(1.33 \text{ ft} - 1 \text{ ft})^2\right]$$

$$+ 0 + \left(\frac{3 \text{ lb/ft(2 ft)}}{32.2 \text{ ft/sec}^2}\right)(2 - 1.33)^2 = 0.248 \text{ slug-ft}^2$$

Substituting the computed data into Eq. (1), using Eqs. 21-8 to represent the scalar components of $(\mathbf{H}_G)_2 = H_x\mathbf{i} + H_y\mathbf{j} = I_{x'}\omega_x\mathbf{i} + I_{y'}\omega_y\mathbf{j}$, yields

$$(-1.33\mathbf{i} + \mathbf{j}) \times \left[\left(\frac{3(6)}{32.2}\right)(-16.05\mathbf{k})\right] = (-1.33\mathbf{i} + \mathbf{j}) \times \left[\left(\frac{3(6)}{32.2}\right)(-v_G)_2\mathbf{k}\right]$$
$$+ 0.435\omega_x\mathbf{i} + 0.248\omega_y\mathbf{j}$$

Expanding and equating the respective $\mathbf{i}$ and $\mathbf{j}$ components yields

$$0.435\omega_x - 0.559(v_G)_2 = -8.97 \tag{2}$$
$$0.248\omega_y - 0.745(v_G)_2 = -11.93 \tag{3}$$

We may obtain a third equation relating ω to $(v_G)_2$ by using *kinematics*. Realizing that after impact the rod rotates about the fixed point A, Eq. 13–22 may be applied, in which case

$$\mathbf{v}_G = \omega \times \mathbf{r}_{G/A}$$
$$-(v_G)_2\mathbf{k} = (\omega_x\mathbf{i} + \omega_y\mathbf{j}) \times (-1.33\mathbf{i} + \mathbf{j})$$

or

$$-(v_G)_2 = \omega_x + 1.33\omega_y \tag{4}$$

Solving Eqs. (2), (3), and (4) simultaneously gives

$$\omega_x = -3.31 \text{ rad/sec}$$
$$\omega_y = -7.63 \text{ rad/sec}$$
$$(v_G)_2 = 13.46 \text{ ft/sec}$$

Thus,

$$\omega = \{-3.31\mathbf{i} - 7.63\mathbf{j}\} \text{ rad/sec} \qquad \textit{Ans.}$$

and

$$(\mathbf{v}_G)_2 = \{-13.46\mathbf{k}\} \text{ ft/sec}$$

It should be noted that ω defines the direction of the instantaneous axis of rotation for the rod just after impact. (See Sec. 13–10.) The direction cosines for this axis are the components of the unit vector for ω, i.e., $\mathbf{u} = \omega/\omega$.

Applying Eq. 13–22, we find that the velocity of point D on the rod after impact is

$$\mathbf{v}_D = \omega \times \mathbf{r}_{D/A}$$
$$= (-3.31\mathbf{i} - 7.63\mathbf{j}) \times 2\mathbf{j}$$
$$= \{-6.62\mathbf{k}\} \text{ ft/sec} \qquad \textit{Ans.}$$

Example 21–2

The 10-lb gear A in Fig. 21–5a rolls on the fixed gear B. This is done by applying a 5 lb-ft torque to the vertical shaft CD, which allows A

Fig. 21–5

to turn freely about CE. Assuming that gear A starts from rest, determine the angular velocity of the vertical shaft after it has turned two revolutions. The axle CE weighs 2 lb. Neglect the weight of shaft CD, and assume that gear A is approximated by a thin disk.

Solution

The principle of work and energy may conveniently be applied to solve this problem. Why? Considering the vertical rod CD, the axle CE, and gear A as a system of connected bodies, we see that only the applied torque $\mathbf{T}$ does work. For two revolutions of shaft CD, this work is

$$\Sigma W_{1-2} = (5 \ \text{lb-ft})(4\pi \ \text{rad}) = 62.83 \ \text{ft-lb}$$

Since the gear is initially at rest, the initial kinetic energy of the system is zero. The final kinetic energy consists of both the kinetic energy of gear A and the axle CE. A kinematic diagram of the motion for both bodies is shown in Fig. 21–5b. The angular velocity of the vertical shaft CD is taken as $\boldsymbol{\omega}_{CD}$. Gear A has an angular velocity of $\boldsymbol{\omega}_{CE}$ about the axle CE. (Note that the axle itself is fixed from rotating with this angular velocity because of the constraint at C.) The angular velocity of the axle is therefore $\boldsymbol{\omega}_{CD}$. With some imagination, the gear may be thought of as a portion of a hypothetical massless extended body which is rotating about the *fixed point* C. The instantaneous axis of rotation for this body is along line CH because both points C and H on the body (gear) have zero velocity and must therefore lie on this axis. This requires that the components $\boldsymbol{\omega}_{CD}$ and $\boldsymbol{\omega}_{CE}$ be related by the equation

$$\frac{\omega_{CD}}{\frac{1}{4} \text{ ft}} = \frac{\omega_{CE}}{1 \text{ ft}} \qquad \text{or} \qquad \omega_{CE} = 4\omega_{CD}$$

The angular velocity of the gear at the instant considered is therefore

$$\boldsymbol{\omega}_A = -\omega_{CE}\mathbf{i} + \omega_{CD}\mathbf{k}$$

or

$$\boldsymbol{\omega}_A = -4\omega_{CD}\mathbf{i} + \omega_{CD}\mathbf{k} \qquad (1)$$

The x, y, and z axes in Fig. 21-5a represent *principal axes of inertia* for both the gear and the axle. Since point C is a fixed point of rotation for both bodies (gear A has been imagined to extend to this point), Eq. 21-10c may be applied to determine the kinetic energy. From Appendix D and the parallel-axis theorem, the mass moments of inertia of the gear and axle about point C are as follows:

for gear A,

$$I_x = \frac{1}{2}\left(\frac{10}{32.2}\right)\left(\frac{1}{4}\right)^2 = 0.00970 \text{ slug-ft}^2$$

$$I_y = I_z = \frac{1}{4}\left(\frac{10}{32.2}\right)\left(\frac{1}{4}\right)^2 + \frac{10}{32.2}(1)^2 = 0.315 \text{ slug-ft}^2$$

for axle CE,

$$I_x = 0$$

$$I_y = I_z = \frac{1}{12}\left(\frac{2}{32.2}\right)(1)^2 + \frac{2}{32.2}\left(\frac{1}{2}\right)^2 = 0.0207 \text{ slug-ft}^2$$

Thus, the kinetic energy is computed by using the equation

$$T = \tfrac{1}{2}I_x\omega_x^2 + \tfrac{1}{2}I_y\omega_y^2 + \tfrac{1}{2}I_z\omega_z^2$$

for gear A,

$$T_A = \tfrac{1}{2}(0.00970)(-4\omega_{CD})^2 + 0 + \tfrac{1}{2}(0.315)(\omega_{CD})^2 = 0.235\omega_{CD}^2$$

and for axle CE,

$$T_{CE} = 0 + 0 + \tfrac{1}{2}(0.0207)(\omega_{CD})^2 = 0.0104\omega_{CD}^2$$

Applying the principle of work and energy, we obtain

$$\{T_1\} + \{\Sigma W_{1-2}\} = \{T_2\}$$

$$\{0\} + \{62.83\} = \{0.235\omega_{CD}^2 + 0.0104\omega_{CD}^2\}$$

Therefore,

$$\omega_{CD} = 16.0 \text{ rad/sec} \qquad\qquad Ans.$$

Problems

21-1. Prove the triple scalar product identity $\mathbf{a} \cdot \mathbf{b} \times \mathbf{c} = \mathbf{a} \times \mathbf{b} \cdot \mathbf{c}$. *Hint:* Use Cartesian components.

21-2. The 20-lb circular disk is mounted on the shaft AB at an angle of $45°$ with the vertical. Determine the kinetic energy of the disk if the shaft is rotating at a speed of $\omega = 60$ rpm.

Prob. 21-2

21-3. Compute the magnitude of the angular momentum of the disk in Prob. 21-2 at the instant shown.

21-4. The uniform 5-lb rod AB is attached to two smooth collars at its end points by means of ball-and-socket joints. If collar A is moving downward at a speed of 2 ft/sec, determine the kinetic energy of the rod at the instant shown. Assume that the angular velocity of the rod is directed perpendicular to the axis of the rod.

Hint: See Prob. 13-72.

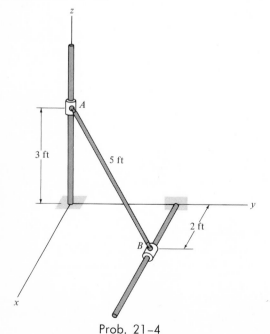

Prob. 21-4

21-5. Compute the angular momentum of rod AB in Prob. 21-4 at the instant shown.

21-6. Gear C has an angular velocity of $\boldsymbol{\omega}_C = \{18\mathbf{j}\}$ rad/sec. The radius of gyration of gear B and gear C is 2 in., and the radius of gyration of gear A is 6 in. Determine the magnitude of the total angular momentum of the system of three gears. Gear A weighs 10 lb, and gears B and C each weigh 2 lb.

Prob. 21-6

21-7. The 10-lb gear A is released from rest in the position shown and rolls along the fixed inclined gear C. Determine the maximum angular velocity of gear A after it is released. Neglect the weight of rod OB. The rod is connected to a ball-and-socket joint at O and fixed to the gear at B. Assume that gear A is a uniform disk having a radius of 1 ft.

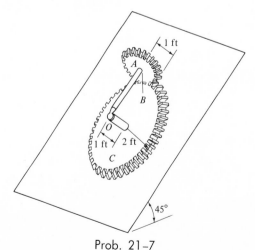

Prob. 21-7

21-8. Work Prob. 21-7 assuming that shaft OB weighs 2 lb.

21-9. The 25-lb square plate is suspended from a ball-and-socket joint at O. A 0.2-lb bullet is fired with a speed of 800 ft/sec into the plate at its corner and becomes imbedded in the plate. The bullet travels in a plane parallel to the xz plane at an angle of 45° with the horizontal as shown. Determine the angular momentum and the instantaneous axis of rotation of the plate just after impact.

Prob. 21-9

21-10. The linkage consists of the 5-lb uniform link OA, the 3-lb rod AB, and the 2-lb slider block B. If rod OA is released from rest when $\theta = 0°$, determine the angular velocity of link OA when $\theta = 90°$. The connections at A and B consist of ball-and-socket joints.

(b)

Prob. 21-10

21-11. The 10-lb circular plate is released from rest and falls horizontally onto the hook at S which provides a permanent connection. Determine the velocity of the mass center of the plate just after making the connection with the hook.

Prob. 21-11

21-12. The 4-lb uniform rod AB is attached at its ends by means of ball-and-socket joints. The joint at B is attached to a disk which is rotating at a constant angular speed of 2 rad/sec. Compute the kinetic energy of the rod at the position shown. Assume that the angular velocity of the rod is directed perpendicular to the axis of the rod. *Hint:* See Prob. 13-76.

(a)

Prob. 21-12

21-13. Compute the angular momentum of rod AB in Prob. 21-12 at the instant shown.

887

21-14. The 10-lb rectangular plate is free to rotate about the y axis because of the bearing supports at A and B. When the plate is balanced in the vertical plane, a 0.2-lb projectile is fired into it with a velocity of $v = \{-600\mathbf{i}\}$ ft/sec. Compute the angular velocity of the plate when the plate rotates 180°. If the projectile strikes corner D with this same velocity, instead of at C, does the angular velocity remain the same? Why?

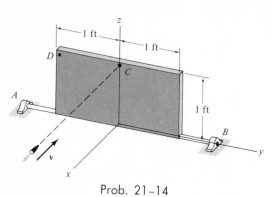

Prob. 21–14

21-15. The 6,000-lb spacecraft is initially moving with no angular velocity when it is struck by a 1-lb meteoroid. Compute the angular velocity of the spacecraft after impact. The meteoroid is traveling with a velocity of 6,000 ft/sec relative to the spacecraft before impact. The radii of gyration of the spacecraft are $k_x = k_z = 6$ ft and $k_y = 2$ ft.

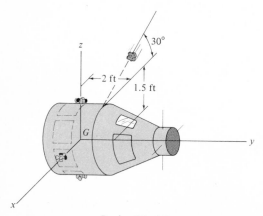

Prob. 21–15

21-16. The rod assembly is rotating with a constant angular velocity of $\omega = \{2\mathbf{k}\}$ rad/sec when the looped end at C encounters a hook at S, which provides a permanent connection. Determine the angular velocity of the assembly immediately after impact. The rod has a linear density of 5 lb/ft.

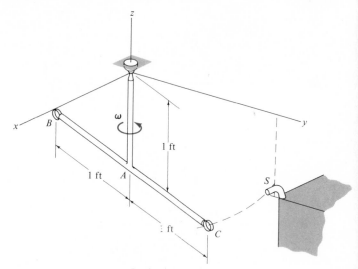

Prob. 21–16

888

*21–4. Equations of Motion of a Rigid Body

Having become familiar with the techniques used to describe both the inertia properties and the angular momentum of a body, we can now write the equations which describe the motion of the body in their most useful forms.

We have stated that we can best describe the translational motion of the body in terms of the absolute acceleration of its center of mass G, using the vector Eq. 21-1. The rotational motion of the body will be considered with respect to one of two points, either point O, which is fixed to an inertial reference and therefore does not accelerate, or the center of mass G of the body. As developed in Sec. 18-9, the equations

$$\Sigma \mathbf{M}_O = \frac{d\mathbf{H}_O}{dt} \qquad\qquad (21\text{-}12\text{a})$$

$$\Sigma \mathbf{M}_G = \frac{d\mathbf{H}_G}{dt} \qquad\qquad (21\text{-}12\text{b})$$

relate the moments of the external forces applied to a system of particles, to the angular momentum of the particles measured with respect to the inertial point O or the center of mass G for the system of particles. Since a rigid body is composed of many particles, these same equations may be used to relate the moments of external forces acting on the body to the angular momentum of the body about points O and G, respectively. The analysis given here is made with reference only to these two points, since, as was noted in Sec. 21-2, the angular momentum of the body may be expressed in a rather simple form. (Although it has never been formally shown here, the body's angular momentum, computed with reference to points which are accelerating toward the body's mass center at the instant considered, also reduces to this same simple form.)

The scalar components of the angular momentum $\mathbf{H}_O$ or $\mathbf{H}_G$ are defined by Eqs. 21-6 or, if principal axes of inertia are used either at point O or G, by the simpler Eqs. 21-8. It was pointed out in Sec. 21-2 that computation of these components is actually *independent* of the orientation of the xyz coordinate axes used to define these components. What is necessary is that the body's instantaneous angular velocity ω be measured with respect to an inertial frame of reference. In computing $\mathbf{H}_O$ or $\mathbf{H}_G$, the components of ω and the inertia tensor for the body are calculated with respect to the x, y, and z axes at the instant of time considered. Although the x, y, and z axes may have *any* angular velocity Ω, computations of the time derivatives, $d\mathbf{H}/dt$, as used in Eqs. 21-12, must account for the rotational change of the coordinate system. Specifically, the time derivative of vector $\mathbf{H}$ may be obtained by using Eq. 13-31. Hence Eqs. 21-12 become

$$\Sigma\mathbf{M}_O = \left(\frac{d\mathbf{H}_O}{dt}\right)_{x,y,z} + \boldsymbol{\Omega} \times \mathbf{H}_O$$

$$\Sigma\mathbf{M}_G = \left(\frac{d\mathbf{H}_G}{dt}\right)_{x,y,z} + \boldsymbol{\Omega} \times \mathbf{H}_G$$

(21–13)

where $\mathbf{H}$ is computed using Eqs. 21-6 or 21-8, and $(d\mathbf{H}/dt)_{x,y,z}$ is the rate of change of $\mathbf{H}$ with respect to the xyz reference.

Three cases exist for which it is desirable to define the orientation of the x, y, and z axes. With each orientation the rotational motion of the rigid body about point O or G is described by using Eq. 21–13. Obviously, the xyz reference should be chosen to yield the simplest set of moment equations for the solution of a particular problem. Let us now consider each of the three cases in detail.

Case I: ($\boldsymbol{\Omega} = 0$). If the body has general motion, Fig. 21–6a, the xyz frame may be chosen with origin at G, such that the axes only *translate* relative to the inertial XYZ frame of reference. Doing this would certainly simplify Eq. 21–13 since $\boldsymbol{\Omega} = 0$. However, the body would have a rotation $\boldsymbol{\omega}$ about these axes, and therefore, the moments and products of inertia of the body would have to be expressed as *functions of time*. In most cases this would be a difficult task, so that such a choice of axes has restricted value.

Case II: ($\boldsymbol{\Omega} = \boldsymbol{\omega}$). The x, y, and z axes may be chosen such that they are *fixed in and move with the body*. This case is shown in Fig. 21–6b for the body having a fixed point O. The moments and products of inertia for the body relative to these axes will be constant during the motion and may be computed using the methods of Chapter 20. Further simplification may result if the axes are oriented as *principal axes of inertia*. Since $\boldsymbol{\Omega} = \boldsymbol{\omega}$ in this case, Eqs. 21–13 become

$$\Sigma\mathbf{M}_O = \left(\frac{d\mathbf{H}_O}{dt}\right)_{x,y,z} + \boldsymbol{\omega} \times \mathbf{H}_O$$

$$\Sigma\mathbf{M}_G = \left(\frac{d\mathbf{H}_G}{dt}\right)_{x,y,z} + \boldsymbol{\omega} \times \mathbf{H}_G$$

(21–14)

We may express each of these vector equations as three scalar equations, utilizing Eqs. 21-6. Neglecting the subscripts O and G yields

$$\Sigma M_x \mathbf{i} + \Sigma M_y \mathbf{j} + \Sigma M_z \mathbf{k} = \left(I_{xx}\frac{d\omega_x}{dt} - I_{xy}\frac{d\omega_y}{dt} - I_{xz}\frac{d\omega_z}{dt}\right)\mathbf{i}$$

$$+ \left(-I_{yx}\frac{d\omega_x}{dt} + I_{yy}\frac{d\omega_y}{dt} - I_{yz}\frac{d\omega_z}{dt}\right)\mathbf{j}$$

$$+ \left(-I_{zx}\frac{d\omega_x}{dt} - I_{zy}\frac{d\omega_y}{dt} + I_{zz}\frac{d\omega_z}{dt} \right)\mathbf{k}$$
$$+ (\omega_x\mathbf{i} + \omega_y\mathbf{j} + \omega_z\mathbf{k}) \times (I_{xx}\omega_x - I_{xy}\omega_y - I_{xz}\omega_z)\mathbf{i}$$
$$+ (\omega_x\mathbf{i} + \omega_y\mathbf{j} + \omega_z\mathbf{k}) \times (-I_{xy}\omega_x + I_{yy}\omega_y - I_{yz}\omega_z)\mathbf{j}$$
$$+ (\omega_x\mathbf{i} + \omega_y\mathbf{j} + \omega_z\mathbf{k}) \times (-I_{xz}\omega_x - I_{yz}\omega_y + I_{zz}\omega_z)\mathbf{k}$$

(a)

(b)

(c)

Fig. 21–6

Computing the cross products, and expanding and equating the respective $\mathbf{i}$, $\mathbf{j}$, and $\mathbf{k}$ components yields

$$\Sigma M_x = I_{xx}\frac{d\omega_x}{dt} - (I_{yy} - I_{zz})\omega_y\omega_z - I_{xy}\left(\frac{d\omega_y}{dt} - \omega_z\omega_x\right) - I_{yz}(\omega_y^2 - \omega_z^2)$$

$$- I_{zx}\left(\frac{d\omega_z}{dt} + \omega_x\omega_y\right)$$

$$\Sigma M_y = I_{yy}\frac{d\omega_y}{dt} - (I_{zz} - I_{xx})\omega_z\omega_x - I_{yz}\left(\frac{d\omega_z}{dt} - \omega_x\omega_y\right) - I_{zx}(\omega_z^2 - \omega_x^2)$$

$$- I_{xy}\left(\frac{d\omega_x}{dt} + \omega_y\omega_z\right) \qquad (21\text{--}15)$$

$$\Sigma M_z = I_{zz}\frac{d\omega_z}{dt} - (I_{xx} - I_{yy})\omega_x\omega_y - I_{zx}\left(\frac{d\omega_x}{dt} - \omega_y\omega_z\right) - I_{xy}(\omega_x^2 - \omega_y^2)$$

$$- I_{yz}\left(\frac{d\omega_y}{dt} + \omega_z\omega_x\right)$$

Notice that for a rigid body symmetric with respect to the xy reference plane, and undergoing general plane motion in this plane, $I_{xz} = I_{yz} = 0$, and $\omega_x = \omega_y = d\omega_x/dt = d\omega_y/dt = 0$. Equations 21–15 reduce to the form $\Sigma M_x = \Sigma M_y = 0$, and $\Sigma M_z = I_{zz}\alpha_z$ (where $\alpha_z = d\omega_z/dt$), which is essentially the third of Eqs. 15–6 or 15–13 depending upon the choice of points G or O for summing moments.

If the x, y, and z axes are chosen as *principal axes of inertia,* the products of inertia are zero, $I_{xx} = I_x$, etc., and Eqs. 21–15 reduce to the form

$$\Sigma M_x = I_x\frac{d\omega_x}{dt} - (I_y - I_z)\omega_y\omega_z$$

$$\Sigma M_y = I_y\frac{d\omega_y}{dt} - (I_z - I_x)\omega_z\omega_x \qquad (21\text{--}16)$$

$$\Sigma M_z = I_z\frac{d\omega_z}{dt} - (I_x - I_y)\omega_x\omega_y$$

This set of equations is known historically as the *Euler equations of motion,* named after the Swiss mathematician Leonard Euler (1707–1783). Both sets of Eqs. 21–15 and 21–16 apply *only* for moments summed either about points O or G.

Case III ($\Omega \neq \boldsymbol{\omega}$). For some problems it is convenient to choose the x, y, and z axes having an angular velocity Ω, which is different from the angular velocity $\boldsymbol{\omega}$ of the body. This is particularly suitable for the analysis

of spinning tops and gyroscopes which are symmetrical about the spinning axis.* When this is the case, the moments and products of inertia remain constant during the motion. An example is given in Fig. 21–6c. Even though the disk is spinning with a constant rate ω, the moments of inertia of the disk about the (principal) x, y, and z axes are the same for *any* angular velocity Ω of the x, y, and z axes, provided Ω is collinear with ω.

Equations 21–13 are applicable for such a set of chosen axes. Each of these two vector equations may be reduced to a set of three scalar equations which are derived in a manner similar to Eqs. 21–15. Most often, the x, y, and z axes chosen are principal axes for the body. The *Euler equations* for the motion, analogous to Eqs. 21–16, then become

$$\Sigma M_x = I_x \frac{d\omega_x}{dt} - I_y \Omega_z \omega_y + I_z \Omega_y \omega_z$$

$$\Sigma M_y = I_y \frac{d\omega_y}{dt} - I_z \Omega_x \omega_z + I_x \Omega_z \omega_x \qquad (21\text{–}17)$$

$$\Sigma M_z = I_z \frac{d\omega_z}{dt} - I_x \Omega_y \omega_x + I_y \Omega_x \omega_y$$

where Ω_x, Ω_y, Ω_z represent the x, y, and z components of Ω, measured relative to an inertial frame of reference.

Any one of the sets of moment equations (Eqs. 21–15, 21–16, or 21–17) represent a series of three first-order nonlinear differential equations. These equations are coupled together since the angular velocity components are present in all the terms. Success in determining the solution of these equations for a particular problem, therefore, depends upon what represents the unknowns in these equations. Difficulty certainly arises when we attempt to solve for the unknown components of ω, given the external moments as functions of time. Further complications can arise if the moment equations are coupled to the three scalar equations of translation. This can happen because of the existence of kinematic constraints which relate the rotation of the body to its translation. For example, in the case of a hoop rolling without slipping, the angular velocity of the hoop is related to the acceleration of the hoop's mass center, thereby coupling the equations of motion. Problems necessitating the simultaneous solution of differential equations generally require using numerical methods with the aid of a computer. In many engineering problems, however, one is required to determine the applied moments acting on the body, given information about the motion of the body. In general, such problems may be solved directly, without the need to resort to computer techniques.

*A detailed discussion of such devices is given in Sec. 21–7.

Generally two methods are available for solving kinetic problems that involve the general motion of a rigid body. These methods consist either of the direct application of vector Eqs. 21–1 and 21–13 or of the application of the six scalar component equations, appropriate to the coordinate axes chosen for the problem. Usually, the first method has precedence over the second because the scalar equations are readily obtained from the compact form of the vector equations. When applicable, both methods of solution will be illustrated in the example problems.

A free-body diagram should accompany the solution of *all problems*. When correctly drawn, this diagram provides an important means of obtaining the external forces and moments applied to the body. For convenience, the *xyz* coordinate system should be drawn on the free-body diagram. The origin of this reference must be located either at the body's mass center *G*, or at a point *O*, considered fixed in an inertial reference frame and located in the body (or hypothetical massless extension of the body). The moment equations are simplified when the axes are inclined so that they represent principal axes of inertia of the body at point *O* or *G*.

Generally a kinematic diagram should also accompany the problem solution. This diagram provides a graphical aid for determining the acceleration of the body's mass center and for computing the components of the angular velocity. The angular velocity can usually be determined from the constraints of the body or from the given problem data.

If the equations of motion, along with the kinematic relationships, do not provide enough information for the complete solution of a particular problem, it may be possible to write frictional equations (if any) to relate some of the unknown quantities.

Example 21-3

The 30-lb flywheel shown in Fig. 21–7*a* forms an angle of 10° with a rotating shaft having negligible weight. The moment of inertia of the wheel about the *z* axis is $I_z = 1.5$ slug-ft^2, and the moments of inertia about the *x* and *y* axes are $I_x = I_y = 0.8$ slug-ft^2. If the shaft is rotating with a constant angular velocity of $\omega_{AB} = 30$ rad/sec, determine the reactions of the bearing supports *A* and *B* when the disk is in the position shown.

Solution I

The free-body diagram of the shaft is shown in Fig. 21–7*b*. The *xyz* coordinate system is *fixed in and rotates with the flywheel*. The origin of this coordinate system is located at the flywheel's center of mass *G*, which

(a)

(b)

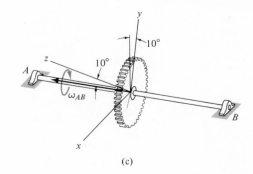

(c)

Fig. 21–7

is also a fixed point. By the choice of the axes, $\Omega = \omega$. As shown in Fig. 21–7c, the angular velocity ω of the flywheel is constant and is always directed along the axis of the shaft AB. This vector is measured in the XYZ inertial frame of reference and can be expressed in terms of the x, y, and z coordinates as

$$\omega = \{-30 \sin 10°\mathbf{j} + 30 \cos 10°\mathbf{k}\} \text{ rad/sec} \qquad (1)$$

Hence,

$$\omega_x = 0, \qquad \omega_y = -30 \sin 10°, \qquad \omega_z = 30 \cos 10°$$

and

$$\frac{d\omega_x}{dt} = \frac{d\omega_y}{dt} = \frac{d\omega_z}{dt} = 0$$

Also, since G is a fixed point,

$$(a_x)_G = (a_y)_G = (a_z)_G = 0$$

Applying the Euler equations of motion, Eqs. 21–16, yields

$$\Sigma M_x = I_x \frac{d\omega_x}{dt} - (I_y - I_z)\omega_y\omega_z$$

$$-(A_Y)(3) + (B_Y)(3) = 0 - (0.8 - 1.5)(-30 \sin 10°)(30 \cos 10°)$$

or

$$A_Y - B_Y = 35.91 \tag{2}$$

$$\Sigma M_y = I_y \frac{d\omega_y}{dt} - (I_z - I_x)\omega_z\omega_x$$

$$A_X(3) \cos 10° - B_X(3) \cos 10° = 0 + 0$$

or

$$A_X = B_X \tag{3}$$

$$\Sigma M_z = I_z \frac{d\omega_z}{dt} - (I_x - I_y)\omega_x\omega_y$$

$$A_X(3) \sin 10° - B_X(3) \sin 10° = 0$$

Again,

$$A_X = B_X$$

Applying Eq. 21–1 in scalar form, we have

$$\Sigma F_x = m(a_G)_x; \qquad A_X + B_X = 0 \tag{4}$$

$$\Sigma F_y = m(a_G)_y; \qquad A_Y + B_Y - 30 = 0 \tag{5}$$

$$\Sigma F_z = m(a_G)_z; \qquad 0 = 0$$

Solving Eqs. (2) through (5) simultaneously gives

$$A_X = B_X = 0 \qquad\qquad\qquad Ans.$$

$$A_Y = 32.95 \text{ lb} \qquad\qquad\qquad Ans.$$

$$B_Y = -2.95 \text{ lb} \qquad\qquad\qquad Ans.$$

Solution II:

Since the x, y, and z axes are chosen fixed to the flywheel, this problem may be solved by direct application of vector Eq. 21–14b. In using this equation, we must first know the components of the angular velocity ω and the angular momentum $\mathbf{H}_G$ of the flywheel, with respect to the x, y, and z axes. Using Eqs. 21–8 and Eq. (1), we write

$$H_x = I_x\omega_x = 0$$

$$H_y = I_y\omega_y = 0.8(-30 \sin 10°) = -4.18 \text{ lb-sec}$$

$$H_z = I_z\omega_z = 1.5(30 \cos 10°) = 44.3 \text{ lb-sec}$$

Thus,

$$\mathbf{H}_G = \{-4.18\mathbf{j} + 44.3\mathbf{k}\} \text{ lb-sec}$$

Since the x, y, and z frame of reference rotates with the same constant angular velocity as the flywheel, the angular momentum of the flywheel $\mathbf{H}_G$ remains *constant* with respect to an observer located in this rotating frame; hence, $(d\mathbf{H}_G/dt)_{x,y,z} = 0$, so that Eq. 21–14b becomes

$$\Sigma \mathbf{M}_G = \boldsymbol{\omega} \times \mathbf{H}_G$$

$$\mathbf{r}_{GA} \times \mathbf{F}_A + \mathbf{r}_{GB} \times \mathbf{F}_B = \boldsymbol{\omega} \times \mathbf{H}_G \qquad (6)$$

$$(-3 \sin 10°\mathbf{j} + 3 \cos 10°\mathbf{k}) \times (A_X\mathbf{i} + A_Y \cos 10°\mathbf{j} + A_Y \sin 10°\mathbf{k})$$
$$+ (3 \sin 10°\mathbf{j} - 3 \cos 10°\mathbf{k}) \times (B_X\mathbf{i} + B_Y \cos 10°\mathbf{j} + B_Y \sin 10°\mathbf{k}) \qquad (7)$$
$$= (-30 \sin 10°\mathbf{j} + 30 \cos 10°\mathbf{k}) \times (-4.18\mathbf{j} + 44.3\mathbf{k})$$

This equation is obviously rather cumbersome to expand. It appears more suitable to apply Eq. (6) in terms of components along the axes of the inertial XYZ frame of reference simply because all the vectors in Eq. (6), except $\mathbf{H}_G$, are easily expressed in these directions. Representing the direction of the X, Y, and Z axes by the unit vectors $\mathbf{I}$, $\mathbf{J}$, and $\mathbf{K}$, respectively, we may write

$$\boldsymbol{\omega} = \{30\mathbf{K}\} \text{ rad/sec}$$

$$\mathbf{H}_G = H_y(\cos 10°\mathbf{J} + \sin 10°\mathbf{K}) + H_z(\sin 10°\mathbf{J} + \cos 10°\mathbf{K})$$

Hence substituting into Eq. (6),

$$3\mathbf{K} \times (A_X\mathbf{I} + A_Y\mathbf{J}) + (-3\mathbf{K}) \times (B_X\mathbf{I} + B_Y\mathbf{J})$$
$$= (30\mathbf{K}) \times [-4.18(\cos 10°\mathbf{J} + \sin 10°\mathbf{K}) + 44.3(\sin 10°\mathbf{J} + \cos 10°\mathbf{K})] \qquad (8)$$

Expanding either Eq. (7) or (8) and equating the respective unit vector components yields

$$A_Y - B_Y = 35.91$$

and

$$A_X = B_X$$

which are the same as Eqs. (2) and (3) of the previous solution. Also, application of vector Eq. 21–1 yields the scalar Eqs. (4) and (5), previously obtained.

Example 21–4

The airplane shown in Fig. 21–8a is in the process of making a steady *horizontal* turn at the rate of $\omega_p = 0.5$ rad/sec. During this motion, the airplane's propeller is spinning at the rate of $\omega_s = 1,000$ rpm. For (a) a two-bladed propeller, Fig. 21–8b, and (b) a four-bladed propeller, Fig. 21–8d, determine the moments exerted on the propeller-shaft bearings

Fig. 21-8

when one of the blades is in the vertical position. For simplicity, assume that the weight of each blade of the propeller is 32.2 lb. The blade has a moment of inertia of 3 slug-ft² about an axis perpendicular to the blade and passing through its end point, and a negligible moment of inertia about an axis directed along the blade.

Solution

Part (a). The moments exerted on the engine bearings are obtained by applying the rotational equations of motion to the propeller. In applying these equations, we shall assume the *xyz* coordinate system to be *fixed* to the propeller blade in a manner shown in Fig. 21–8*b*. Hence $\Omega = \omega$. The axes chosen are thus principal axes of inertia for the two-bladed propeller, having an origin at the mass center G. The moments of inertia of the propeller with respect to these axes are

$$I_x = I_z = 2(3 \text{ slug-ft}^2) = 6 \text{ slug-ft}^2, \qquad I_y = 0$$

The angular acceleration components are computed from the time derivatives of the angular velocity components when the propeller is acting at the general angle θ from the vertical. As shown in Fig. 21–8*b*,

$$\omega_x = \omega_s; \qquad\qquad \frac{d\omega_x}{dt} = 0$$

$$\omega_y = -\omega_p \sin\theta; \qquad\qquad \frac{d\omega_y}{dt} = -\omega_p \cos\theta\frac{d\theta}{dt} = -\omega_p\omega_s \cos\theta$$

$$\omega_z = -\omega_p \cos\theta; \qquad\qquad \frac{d\omega_z}{dt} = \omega_p \sin\theta\frac{d\theta}{dt} = \omega_p\omega_s \sin\theta$$

$\omega_s = 1{,}000$ rev/min $(2\pi$ rad/rev$)(1$ min/60 sec$) = 104.7$ rad/sec, and $\omega_p = 0.5$ rad/sec; thus, for $\theta = 0°$,

$$\omega_x = 104.7 \text{ rad/sec}; \qquad\qquad \frac{d\omega_x}{dt} = 0$$

$$\omega_y = 0; \qquad\qquad \frac{d\omega_y}{dt} = -52.4 \text{ rad/sec}^2$$

$$\omega_z = -0.5 \text{ rad/sec}; \qquad\qquad \frac{d\omega_z}{dt} = 0$$

The free-body diagram of the propeller is shown in Fig. 21–8c. The effect of the connecting shaft on the propeller is indicated by the reactions $\mathbf{F}_R$ and $\mathbf{M}_R$. Substituting the computed quantities into Eqs. 21–16 gives

$$\Sigma M_x = I_x\frac{d\omega_x}{dt} - (I_y - I_z)\omega_y\omega_z$$

$$M_x = 0 \qquad\qquad\qquad\qquad Ans.$$

$$\Sigma M_y = I_y\frac{d\omega_y}{dt} - (I_z - I_x)\omega_z\omega_x$$

$$M_y = 0 - (6 - 6)(-0.5)104.7$$
$$= 0 \qquad\qquad\qquad\qquad Ans.$$

$$\Sigma M_z = I_z\frac{d\omega_z}{dt} - (I_x - I_y)\omega_x\omega_y$$

$$M_z = 0 \qquad\qquad\qquad\qquad Ans.$$

Part (b). The origin of the $x, y,$ and z coordinates is chosen at the mass center G of the four-bladed propeller shown in Fig. 21–8d. In this case, however, the axes *will not be fixed* with any one propeller blade, but rather will be fixed to the airplane, and thus turn at a constant rate $\Omega = \omega_p$. (The problem can, of course, be worked by fixing the axes to the propeller as in Part (a).) The choice of having $\Omega \neq \omega$ in the case of a four-bladed propeller is convenient since the $x, y,$ and z axes so chosen always represent *principal axes* of inertia for any angle θ of the propeller.*

---

*Refer to the discussion in Sec. 20–6 that pertains to Fig. 20–11.

Furthermore, the values of I_x, I_y, and I_z are *constant* with respect to time as the propeller rotates with respect to these axes. (This is *not* the case for a two-bladed propeller.) Thus, for any angle θ,

$$I_x = 4(3 \text{ slug-ft}^2) = 12 \text{ slug-ft}^2$$
$$I_y = I_z = 2(3 \text{ slug-ft}^2) = 6 \text{ slug-ft}^2$$

The angular velocity components for any angle θ of the propeller are

$$\Omega_x = 0, \qquad \Omega_y = 0, \qquad \Omega_z = \omega_p = 0.5 \text{ rad/sec}$$

and

$$\omega_x = -\omega_s = -1{,}000 \text{ rpm} = -104.7 \text{ rad/sec}; \qquad \frac{d\omega_x}{dt} = 0$$

$$\omega_y = 0; \qquad \frac{d\omega_y}{dt} = 0$$

$$\omega_z = \omega_p = 0.5 \text{ rad/sec}; \qquad \frac{d\omega_z}{dt} = 0$$

The free-body diagram is shown in Fig. 21–8e. Substituting into Eqs. 21–17, since $\Omega \neq \omega$, we have

$$\Sigma M_x = I_x \frac{d\omega_x}{dt} - I_y \Omega_z \omega_y + I_z \Omega_y \omega_z$$
$$M_x = 0 \qquad\qquad Ans.$$

$$\Sigma M_y = I_y \frac{d\omega_y}{dt} - I_z \Omega_x \omega_z + I_x \Omega_z \omega_x$$
$$M_y = 0 - 0 + 12(0.5)(-104.7)$$
$$M_y = -628 \text{ lb-ft} \qquad\qquad Ans.$$

$$\Sigma M_z = I_z \frac{d\omega_z}{dt} - I_x \Omega_y \omega_x + I_y \Omega_x \omega_y$$
$$M_z = 0 \qquad\qquad Ans.$$

Example 21–5

Two rods are welded to a shaft as shown in Fig. 21–9a. The rods and the shaft have a linear density of 2 lb/ft. A frictional moment **M**, developed in the bearings, causes the shaft to decelerate at 1 rad/sec² when it has an initial angular velocity of $\omega = 6$ rad/sec. Determine reactions at the bearings A and B, at the instant shown. What is the frictional moment **M** causing the deceleration?

Solution I

The free-body diagram of the shaft is shown in Fig. 21–9b. Since the

(a)

(b)

(c)

Fig. 21-9

assembly has a total rod length of 9 ft, the total weight is $W = (2 \text{ lb/ft})$
$(9 \text{ ft}) = 18 \text{ lb}$. This force is applied at the mass center G. The bearing
frictional moment is represented by $\mathbf{M}$, which is considered as a free-
vector in determining the *external reactions*, and therefore can be applied
at any point along the shaft. Because of the constraints, $\mathbf{a}_G = 0$. Applying
Eq. 21-1, we have

$$\Sigma \mathbf{F} = m\mathbf{a}_G; \qquad \mathbf{F}_A + \mathbf{F}_B - \mathbf{W} = 0$$
$$A_x\mathbf{i} + A_y\mathbf{j} + B_x\mathbf{i} + B_y\mathbf{j} - 18\mathbf{j} = 0$$

or

$$A_x + B_x = 0 \qquad\qquad (1)$$
$$A_y + B_y - 18 = 0 \qquad\qquad (2)$$

When applying the moment equations, the x, y, and z axes are chosen
such that they are *fixed in and rotate with the shaft*. The origin is located

at the center of mass G (also a fixed point), Fig. 21–9b. The moments and products of inertia for this set of axes are computed using Appendix D and the parallel-axis and parallel-plane theorems. The inertia terms which will be required for the solution are

$$I_{zz} = 0.166 \text{ slug-ft}^2$$
$$I_{xy} = I_{yx} = I_{yz} = I_{zy} = 0$$
$$I_{xz} = I_{zx} = 0.155 \text{ slug-ft}^2$$

Since points A and B are fixed at all times, the angular velocity $\boldsymbol{\omega}$ is always directed along the axis of the shaft. The angular acceleration $\boldsymbol{\alpha}$ therefore measures only a change in magnitude of $\boldsymbol{\omega}$. From the problem data,

$$\omega_x = 0, \qquad \frac{d\omega_x}{dt} = 0$$

$$\omega_y = 0, \qquad \frac{d\omega_y}{dt} = 0$$

$$\omega_z = 6 \text{ rad/sec}, \qquad \frac{d\omega_z}{dt} = -1 \text{ rad/sec}^2$$

Equations 21–15 must be used since $\boldsymbol{\Omega} = \boldsymbol{\omega}$ and the x, y, and z axes are *not* principal axes of inertia. Using the computed data, these equations reduce to the form

$$\Sigma M_x = -I_{zx}\frac{d\omega_z}{dt}$$
$$-A_y(2.5) + B_y(2.5) = -0.155(-1)$$
$$-A_y + B_y = 0.0620 \tag{3}$$

$$\Sigma M_y = -I_{zx}\omega_z^2$$
$$A_x(2.5) - B_x(2.5) = -0.155(6)^2$$
$$A_x - B_x = 2.23 \tag{4}$$

$$\Sigma M_z = I_{zz}\frac{d\omega_z}{dt}$$
$$-M = 0.166(-1)$$

$$M = 0.166 \text{ lb-ft} \qquad\qquad \textit{Ans.}$$

Solving Eqs. (1) through (4) simultaneously gives

$$A_x = 1.12 \text{ lb} \qquad\qquad Ans.$$

$$A_y = 8.97 \text{ lb} \qquad\qquad Ans.$$

$$B_x = -1.12 \text{ lb} \qquad\qquad Ans.$$

$$B_y = 9.03 \qquad\qquad Ans.$$

Since B_x is negative, this force acts in a direction opposite to that shown in Fig. 21–9b.

Solution II

The moment equations may *also* be established by direct application of vector Eq. 21–14. It is first necessary to compute the angular momentum $\mathbf{H}_G$ using Eqs. 21–6. Since

$$\boldsymbol{\omega} = \{6\mathbf{k}\} \text{ rad/sec}$$

then,

$$H_x = -I_{xz}\omega_z = -0.155\omega_z$$
$$H_y = 0$$
$$H_z = I_{zz}\omega_z = 0.166\omega_z$$

Therefore,

$$\mathbf{H}_G = H_x\mathbf{i} + H_y\mathbf{j} + H_z\mathbf{k}$$
$$= (-0.155\mathbf{i} + 0.166\mathbf{k})\omega_z$$

and

$$\left(\frac{d\mathbf{H}_G}{dt}\right)_{x,y,z} = (-0.155\mathbf{i} + 0.166\mathbf{k})\frac{d\omega_z}{dt}$$

Substituting $\omega_z = 6$ rad/sec and $d\omega_z/dt = -1$ rad/sec² into these equations and applying Eq. 21–14 yields

$$\Sigma\mathbf{M}_G = \left(\frac{d\mathbf{H}_G}{dt}\right)_{x,y,z} + \boldsymbol{\omega} \times \mathbf{H}_G$$
$$= 0.155\mathbf{i} - 0.166\mathbf{k} + (6\mathbf{k}) \times (-0.930\mathbf{i} + 0.996\mathbf{k})$$

or

$$\Sigma\mathbf{M}_G = 0.155\mathbf{i} - 5.58\mathbf{j} - 0.166\mathbf{k}$$
$$-M\mathbf{k} + (2.5\mathbf{k}) \times (A_x\mathbf{i} + A_y\mathbf{j}) + (-2.5\mathbf{k}) \times (B_x\mathbf{i} + B_y\mathbf{j})$$
$$= 0.155\mathbf{i} - 5.58\mathbf{j} - 0.166\mathbf{k}$$

Expanding and equating the respective **i**, **j**, and **k** components yields Eqs. (3) and (4) and the solution for M obtained previously.

From this analysis we see that the bearing reactions will rotate with the shaft and thus cause it to vibrate. Furthermore, these reactions are

proportional to the square of the angular speed (since $\Sigma M_y = -I_{zx}\omega_z^2$). In cases of high rotational velocities, such effects may cause severe damage to the supports and the shaft. As noted by the form of the moment equations obtained, such effects may be eliminated by "*balancing*" the shaft such that I_{zx} and all the other *products of inertia* for the shaft are *equal to zero*. This may be done either by removing the forked rods on the shaft or by adding two equivalent forked rods, symmetrically located about the mass center. Under these conditions, the shaft is said to be *dynamically balanced*.

Problems

21-17. The 20-lb disk is mounted eccentrically on the axis of shaft *AB*. If the shaft is rotating at a speed of 10 rad/sec, compute the reactions of the bearing supports when the disk is in the position shown.

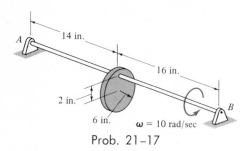

Prob. 21-17

21-18. The 20-lb disk is mounted on the horizontal shaft *AB* such that its plane forms an angle of 10° with the vertical. If the shaft *AB* rotates with an angular velocity of 3 rad/sec, determine the reactions developed at the bearings when the disk is in the position shown.

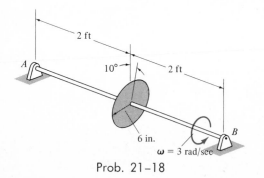

Prob. 21-18

21-19. The 4-lb bar rests along the smooth corners of a box. At the instant shown the box has an upward velocity of 5 ft/sec and an acceleration of 2 ft/sec². Determine the components of force which the walls exert on the bar.

Prob. 21-19

21-20. The 30-lb homogeneous rod *AB* is supported at *A* by a ball-and-socket joint and rests on the smooth surface of the flat cart at *B*. A smooth stop block is attached to the cart at *B*, which prevents the rod from

sliding off the cart. Determine the force components acting on the rod when the cart is accelerated at 10 ft/sec².

Prob. 21-20

21-21. The 40-lb block is mounted eccentrically on the rotating shaft *AB*. Determine the maximum reactions of the bearing supports as the block rotates.

Prob. 21-21

21-22. The 10-lb sphere is supported by a rod which is pin connected at *B* and rests against the vertical shaft at *C*. Determine the minimum angular speed ω of the shaft so that the reaction at *C* becomes zero. Neglect the weight of the connecting rod *BA*.

Prob. 21-22

21-23. The 20-lb square plate is mounted on the shaft *AB* so that the plane of the plate makes an angle of $\theta = 30°$ with the vertical. If the shaft is turning with an angular velocity of 25 rad/sec, determine the dynamic reactions, that is, the reactions due to the effects of the rotation, at the bearing supports *A* and *B* when the plate is in the position shown.

Prob. 21-23

21-24. The uniform hatch door weighs 20 lb and is supported in the horizontal plane by the bearings at *A* and *B*, and the stop block at *S*. If a force of $F = 60$

905

lb is suddenly applied to the door as shown, determine the vertical reactions at the bearings and the angular acceleration of the door.

Prob. 21-24

21-25. The 60-lb rectangular block rotates with a constant angular velocity of 6 rad/sec about the AB axis. The support at A is a smooth journal bearing which develops reactions normal to the shaft. The support at B is a smooth thrust bearing which develops reactions both normal and along the axis of the shaft. Determine the reaction components at A and B when the block is in the position shown.

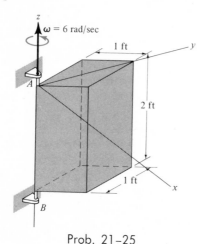

Prob. 21-25

21-26. Assume that the block in Prob. 21-25 has an angular acceleration of 2 rad/sec², acting in the same direction as ω. This acceleration is caused by a torque **T** applied to the block. Compute the magnitude and direction of **T**.

906

21-27. The 25-lb boom AB is pinned at A and held at B by means of a cable. The column CD is supported at the ends by means of ball-and-socket joints and is rotating with an angular velocity of 5 rad/sec. Determine the tension developed in the cable, and the magnitude of force developed at the pin A.

Prob. 21-27

21-28. Determine the tension in the cable in Prob. 21-27 if the supporting column has an angular acceleration of $\alpha = 3$ rad/sec² at the instant when $\omega = 2$ rad/sec. Assume that $\boldsymbol{\alpha}$ acts in the same direction as $\boldsymbol{\omega}$ shown in Prob. 21-27.

21-29. The 10-lb circular disk is mounted off-center on a shaft which is held by bearings at A and B. If the shaft is rotating at a speed of 7 rad/sec, determine the reactions at the bearing supports when the disk is in the position shown.

Prob. 21-29

21-30. The shaft AB rotates at a constant rate. The rod CD is attached to this fixed shaft, as shown. If two spheres weighing 6 lb each are attached to the ends of this rod, determine the reactions which occur at the bearing supports A and B at the instant shown. Neglect the weight of the shaft and rod. Due to the rotation the spheres have a speed of 8 in./sec.

Prob. 21-30

21-31. The 30-lb shaft AB is supported by a rotating arm. The support at A is a journal bearing, which develops reactions normal to the shaft. The support at B is a thrust bearing, which develops reactions both normal to the shaft and along the axis of the shaft. Neglecting friction, determine the reactions at these supports.

Prob. 21-31

21-32. The crankshaft is constructed from a rod which has a linear density of 12 lb/ft. Determine the compo-

nents of reaction at bearings A and B when the shaft is in the position shown.

Prob. 21-32

21-33. The sphere at B weighs 10 lb and the rod AB may be assumed to be weightless. If the rod is pinned at A to the vertical shaft which is rotating at $\omega = 7$ rad/sec, determine the angle θ for rotational equilibrium.

Prob. 21-33

21-34. If the vertical shaft in Prob. 21-33 is rotating with an angular acceleration of $\alpha = 6$ rad/sec^2, compute the angle θ when $\omega = 12$ rad/sec. α and ω act in the same direction.

907

21–35. The uniformly bent rod *ACD* has a linear density of 5 lb/ft and is supported at *A* by a pin and at *B* by a cable. If the vertical shaft rotates with a constant angular velocity of $\omega = 20$ rad/sec, determine the forces developed at these supports.

21–38. The rod *AB* has a linear density of $\frac{1}{2}$ lb/ft. It is fixed to the surface of a disk which rotates with a constant angular velocity of $\omega_D = 3$ rad/sec *relative* to rod *CD*. Rod *CD* has a constant angular velocity of $\omega_{CD} = 1$ rad/sec. Determine the reactions at the fixed support *A*, which the disk exerts on the rod.

Prob. 21–38

Prob. 21–35

21–36. The 25-lb disk *A* is fixed to the shaft *BCD*, which has negligible weight. Determine the torque **T** which must be applied to the vertical axis so that the shaft has an angular acceleration of $\alpha = 6$ rad/sec².

21–39. The shaft is rotating with a constant angular velocity of $\omega = 7$ rad/sec. There are four weights attached to the ends of the supporting rods. Assuming that the rods have negligible mass, determine the magnitude of W_C and W_D and the angles θ_C and θ_D which the supporting rods must make with the vertical, so that shaft *AB* is *dynamically balanced,* that is, so that *A* and *B* exert only vertical reactions on the shaft while the shaft is rotating.

Prob. 21–36

21–37. Work Prob. 21–36 assuming that rod *BCD* has a linear density of 2 lb/ft.

Prob. 21–39

21–40. Derive the scalar form of the rotational equation of motion for the x axis, when $\Omega \neq \omega$ and the moments and products of inertia of the body are not constant with respect to time.

21–41. Derive the scalar form of the rotational equation of motion for the x axis, when $\Omega \neq \omega$ and the moments and the products of inertia of the body are constant with respect to time.

21–42. Derive the Euler equations of motion for $\Omega \neq \omega$, i.e., Eqs. 21–17.

*21–6. Euler Angles

In defining the position of a rigid body at any given instant, it is necessary to specify the location of three noncollinear points in the body. Since the location of each point is defined in terms of three components of a position vector drawn to the point, a total of nine components (coordinates) is needed to specify the position of the body. These components are not entirely independent of one another; rather they are indirectly related to the geometry of the body. In total, six independent relations may be established. For dynamical purposes, however, it is more suitable to use another method for specifying these independent relations. This method consists of using three components of a position vector to specify the location of a point in the body, and three angular displacements to describe the orientation of the body about this point. Unfortunately, the three angular displacements cannot be represented as a vector, since we have shown in Sec. 13–10 that finite angular displacements do *not* obey the communitive law of vector addition. Provided, however, these displacements are made in a certain *orderly* fashion, the use of this method becomes practical for analytical work. Furthermore, when the angles are defined as the so-called *Euler angles,* this method is particularly well suited for defining the rotational motion of bodies having a fixed point. We will presently establish a method for defining the Euler angles; and in the remaining sections of this chapter we will use this method for the kinetic analysis of symmetrical tops and gyroscopes, and bodies subjected to torque-free motions.

To illustrate how the Euler angles are formed, reference is made to the conical top shown in Fig. 21–10. The top is fixed at point O and has an orientation relative to the inertial X, Y, and Z axes at some instant of time, as shown in Fig. 21–10d. To define this final position, a second set of x, y, and z coordinates will be needed. For purposes of discussion, assume that these coordinates are fixed in the top. Starting with the X,

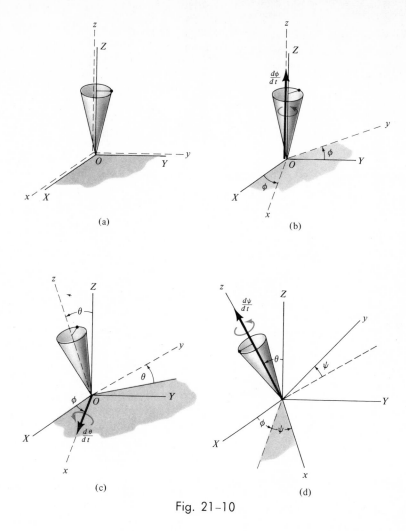

Fig. 21–10

Y, and Z and $x, y,$ and z axes in coincidence, Fig. 21–10a, the final position of the top is determined using the following three *orderly* steps:

1. Rotate the top about the Z (or z) axis through an angle ϕ ($0 \leq \phi < 2\pi$), Fig. 21–10b.
2. Rotate the top about the x axis through an angle θ ($0 \leq \theta \leq \pi$), Fig. 21–10c.
3. Rotate the top about the z axis through an angle ψ ($0 \leq \psi < 2\pi$) to obtain the final position, Fig. 21–10d.

The three angles, ϕ, θ, and ψ, so defined, are referred to as *Euler angles*.

When the top is in motion, in general, each of these three angles is changing with time. Although finite rotations are not vectors, infinitesimal

rotations are vectors (see Appendix E), and thus we may define the angular velocity $\boldsymbol{\omega}$ of the top in terms of the time derivatives of these angles. These angular velocity components, $d\phi/dt$, $d\theta/dt$, and $d\psi/dt$, are known as the *precession, nutation,* and *spin,* respectively. Their positive directions are shown in Fig. 21–10. Notice that they do *not* form an orthogonal set.

In solving problems in Sec. 21–5, you will recall that the x, y, and z axes may be chosen either to rotate with the body ($\boldsymbol{\Omega} = \boldsymbol{\omega}$) or to have an independent angular velocity ($\boldsymbol{\Omega} \neq \boldsymbol{\omega}$). Choosing $\boldsymbol{\Omega} \neq \boldsymbol{\omega}$ considerably simplifies the analysis of some problems. (For example, compare the methods of solution to Example 21–4, Parts (a) and (b).) If a body, rotating about a fixed point, is *symmetrical with respect to its axis of spin,* it is convenient to require the x, y, and z axes to follow the motion *only* in precession and nutation. Such is the case for the top shown in Fig. 21–11, which is symmetrical with respect to the z axis. The x, y, and z axes so chosen have only two components of angular velocity, $d\phi/dt$ and $d\theta/dt$, while the top has all *three* components of motion. Furthermore, the x, y, and z axes represent *principal axes of inertia,* for which the computed values of the moments of inertia are *constant.* Also, $I_x = I_y$. Because of these advantages, the analysis of the motion of *symmetrical* rotating bodies about a fixed point is considerably simplified. The differential equations defining the rotation can, in certain cases, be solved in closed form rather than by resorting to numerical methods and computer techniques for analyzing the motion. In Sec. 21–7 we will consider specifically the kinetic analysis of the gyroscope and symmetrical top. The Euler equations of motion will be used to do this. For these equations, it is first necessary to express the angular velocities $\boldsymbol{\omega}$ and $\boldsymbol{\Omega}$ in terms of Euler angle components along the x, y, and z axes. From the geometry of Fig. 21–11, realizing that the coordinate axes are subjected only to precession and nutation, while, in addition, the top (or symmetrical body) spins about the z axis, we obtain

$$\boldsymbol{\omega} = \omega_x\mathbf{i} + \omega_y\mathbf{j} + \omega_z\mathbf{k}$$

$$\boldsymbol{\omega} = \frac{d\theta}{dt}\mathbf{i} + \left(\frac{d\phi}{dt}\sin\theta\right)\mathbf{j} + \left(\frac{d\phi}{dt}\cos\theta + \frac{d\psi}{dt}\right)\mathbf{k} \qquad (21\text{–}18)$$

and

$$\boldsymbol{\Omega} = \Omega_x\mathbf{i} + \Omega_y\mathbf{j} + \Omega_z\mathbf{k}$$

$$\boldsymbol{\Omega} = \frac{d\theta}{dt}\mathbf{i} + \left(\frac{d\phi}{dt}\sin\theta\right)\mathbf{j} + \left(\frac{d\phi}{dt}\cos\theta\right)\mathbf{k} \qquad (21\text{–}19)$$

Fig. 21–11

*21-7. Gyroscopic Motion

We will presently develop the equations valid for analyzing the motion of a body which is symmetrical with respect to an axis and moving about a fixed point lying on the axis. These equations will then be used to study the motion of spinning tops and, a particularly interesting device, the gyroscope.

The motion will be analyzed using Euler angles such that the x, y, and z coordinate axes are subjected to precession ($d\phi/dt$) and nutation ($d\theta/dt$), having no spin. When the z axis represents the axis of symmetry for the body, as in the case of the top shown in Fig. 21–11, the axes represent *principal axes of inertia* of the body for any rotation (spin) of the body about these axes. Hence, the moments of inertia are constant and will be represented as $I_{xx} = I_{yy} = I$ and $I_{zz} = I_z$. The components of the angular velocity of the body and of the axes so chosen are defined by Eqs. 21–18 and 21–19, respectively. Since $\Omega \neq \omega$, the Euler Eqs. 21–17 must be used to establish the rotational equations of motion. Substituting into these equations the respective angular velocity, their corresponding time derivatives, and the moment of inertia components yields

$$\Sigma M_x = I\left[\frac{d^2\theta}{dt^2} - \left(\frac{d\phi}{dt}\right)^2 \sin\theta\cos\theta\right] + I_z\frac{d\phi}{dt}\sin\theta\left(\frac{d\phi}{dt}\cos\theta + \frac{d\psi}{dt}\right)$$

$$\Sigma M_y = I\left(\frac{d^2\phi}{dt^2}\sin\theta + 2\frac{d\phi}{dt}\frac{d\theta}{dt}\cos\theta\right) - I_z\frac{d\theta}{dt}\left(\frac{d\phi}{dt}\cos\theta + \frac{d\psi}{dt}\right)$$

$$\Sigma M_z = I_z\left(\frac{d^2\psi}{dt^2} + \frac{d^2\phi}{dt^2}\cos\theta - \frac{d\phi}{dt}\frac{d\theta}{dt}\sin\theta\right) \qquad (21\text{–}20)$$

The moment summation for these equations applies only at the fixed point O or the center of mass G of the body. Since the equations represent a coupled set of nonlinear second-order differential equations, in general a closed formed solution may not be obtained. It will therefore be necessary to plot the time functions of the Euler angles ϕ, θ, and ψ, and use numerical analysis and computer techniques for solving these equations.

A special case, however, does exist for which simplification of Eqs. 21–20 is possible. Commonly referred to as *steady precession,* it occurs when the nutation angle θ, precession $d\phi/dt$, and spin $d\psi/dt$, all remain *constant*. Equations 21–20 then reduce to the form

$$\Sigma M_x = -I\left(\frac{d\phi}{dt}\right)^2 \sin\theta\cos\theta + I_z\frac{d\phi}{dt}\sin\theta\left(\frac{d\phi}{dt}\cos\theta + \frac{d\psi}{dt}\right) \qquad (21\text{–}21)$$

$$\Sigma M_y = 0$$
$$\Sigma M_z = 0$$

Equation 21–21 may be simplified further by noting from Eq. 21–18 that $\omega_z = (d\phi/dt)\cos\theta + (d\psi/dt)$. Thus,

$$\Sigma M_x = -I\left(\frac{d\phi}{dt}\right)^2 \sin\theta\cos\theta + I_z\frac{d\phi}{dt}(\sin\theta)\omega_z$$

or

$$\Sigma M_x = \frac{d\phi}{dt}\sin\theta\left(I_z\omega_z - I\frac{d\phi}{dt}\cos\theta\right) \qquad (21\text{--}22)$$

It is interesting to note what effects the spin $d\psi/dt$ has on the moment about the x axis. Consider the spinning rotor shown in Fig. 21–12a.

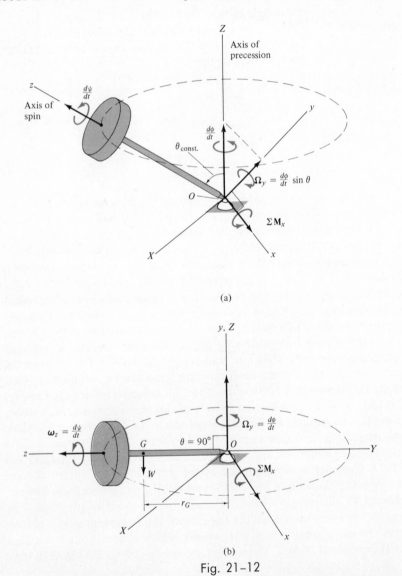

(a)

(b)

Fig. 21–12

Inspection of Eq. 21–21 reveals that the spin affects ΣM_x by an amount $I_z(d\phi/dt \sin\theta)\,d\psi/dt = (I_z\Omega_y)(d\psi/dt)$, while moments about the y and z axes are unaffected. This effect accounts for the characteristics exhibited by all spinning tops and gyroscopes. From the figure we note that as the rotor spins about the z axis and precesses about the Z axis, the angular velocity component Ω_y always remains *perpendicular* to the moment $\Sigma \mathbf{M}_x$. This moment, created by the spinning rotor, provides the necessary holding force which prevents the rotor from falling downward. As a consequence, the rotor precesses about the Z axis at a constant angular velocity $d\phi/dt$.

A more striking example of this effect is obtained if one considers the case where the spin, precession, and moment axes are mutually *perpendicular*. This situation requires $\theta = 90°$ and is shown in Fig. 21–12b. Equation 21–21 then reduces to the form

$$\Sigma M_x = I_z \frac{d\phi}{dt}\frac{d\psi}{dt}$$

or

$$\Sigma M_x = I_z\Omega_y\omega_z \qquad (21\text{–}23)$$

From the figure it is seen that the vectors $\Sigma \mathbf{M}_x$, Ω_y, and ω_z are mutually perpendicular to one another. Instinctively, one would expect the rotor to fall down under the influence of gravity! However, this is not the case at all provided the product $I_z\Omega_y\omega_z$ is correctly chosen to counterbalance the moment Wr_G of the rotor's weight about O. This unusual phenomenon of rigid body motion is often referred to as the *gyroscopic effect*.

Perhaps a more intriguing demonstration of the gyroscopic effect comes from studying the action of a *gyroscope* (frequently referred to as a *gyro*). A gyro is a rotor which spins at a very high rate about its axis of symmetry. This rate of spin is considerably greater than the precessional rate of rotation of the gyro about its transverse axis. Hence, for all practical purposes, the angular momentum of the gyro can be assumed directed along the axis of spin of the gyro. Thus, for the gyro rotor shown in Fig. 21–13a, $\omega_z \gg \Omega_y$, and the magnitude of the angular momentum about point O, as computed by Eqs. 21–8, reduces to the form $H_O = I_z\omega_z$. Since the magnitude and direction of $\mathbf{H}_O$ are always defined, the analysis is considerably simplified and can be used as a further aid to interpret Eq. 21–23. Consider now the motion of the gyro during an instant of time dt. Direct application of Eq. 21–12a requires the angular impulse $\Sigma \mathbf{M}_x\,dt$, created by the weight about the x axis, to be equal to the change in angular momentum $d\mathbf{H}_O$ of the rotor. Since $\mathbf{H}_O$ is constant in magnitude, only the directional change $d\mathbf{H}_O$ occurs, which is due to the precession of $\mathbf{H}_O$ about the Z axis. The vectors are shown in Fig. 21–13b. $d\mathbf{H}_O$ acts

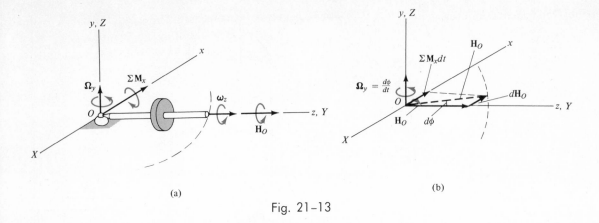

(a)   (b)

Fig. 21–13

in the direction of the positive x axis and has a magnitude of $dH_O = H_O\,d\phi = I_z\omega_z\,d\phi$. Thus,

$$\Sigma\mathbf{M}_O\,dt = d\mathbf{H}_O$$
$$\Sigma M_x\,dt = dH_O = I_z\omega_z\,d\phi$$

or

$$\Sigma M_x = I_z\omega_z\frac{d\phi}{dt}$$

which is equivalent to Eq. 21–23. Again it is repeated that the direction of the vectors $\Sigma\mathbf{M}_x$, $\mathbf{\Omega}_y$, and $\boldsymbol{\omega}_z$ (which acts in the same direction as $\mathbf{H}_O$) are all mutually perpendicular. This condition of orthogonality is in accordance with the cross product

$$\Sigma\mathbf{M}_x = \mathbf{\Omega}_y \times \mathbf{H}_O$$

which is obtained from Eq. 21–13 since $\mathbf{H}_O$ is constant. Notice that motion about a fixed point is different from the case of plane motion, in which the angular momentum and moment vectors are parallel.

When a gyro is mounted in gimbal rings, Fig. 21–14, it becomes free of external moments applied to its base. Thus, in theory, its angular momentum $\mathbf{H}$ will never precess but maintains its same fixed orientation when the base is rotated. This type of gyroscope is called a *free gyro* and is useful as a gyrocompass when the spin axis of the gyro is directed north. In reality, the gimbal mechanism is never completely free of friction, so that such a device is useful only for the local navigation of ships and aircraft. The gyroscopic effect is also useful as a means of stabilizing both the rolling motion of ships at sea and the trajectories of missiles and projectiles. As noted in Example 21–4, this effect is of significant importance in the design of shafts and bearings for rotors which are subjected to forced precessions.

Fig. 21–14

*21–8. Torque-free Motion

When the only external force acting on a body is its weight, the general motion of the body is referred to as *torque-free motion*. This type of motion is characteristic of planets, artificial satellites, and projectiles—provided the effects of air friction are neglected.

In order to describe the characteristics of this motion, the distribution of the body's mass will be assumed *axisymmetric*. The satellite shown in Fig. 21–15a is an example of such a body, where the z axis represents an axis of symmetry. The origin of the x, y, and z coordinates is located at the mass center G, such that $I_{zz} = I_z$ and $I_{xx} = I_{yy} = I$ for the body. Since the weight is the only external force present, summation of moments about the mass center is zero. From Eq. 21–12b, this requires the angular momentum of the body to be constant, i.e.,

$$\mathbf{H}_G = \text{const} \qquad (21\text{--}24)$$

At the instant considered, it will be assumed that the inertial frame of reference is oriented such that $\mathbf{H}_G$ acts along the positive Z axis and the y axis lies in the plane formed by the z and Z axes, Fig. 21–15a. The Euler angle formed between Z and z is θ, and therefore, with this choice of axes the angular momentum may be expressed as

$$\mathbf{H}_G = H_G \sin\theta\, \mathbf{j} + H_G \cos\theta\, \mathbf{k} \qquad (21\text{--}25)$$

Furthermore, using Eqs. 21–8, we have

$$\mathbf{H}_G = I\omega_x \mathbf{i} + I\omega_y \mathbf{j} + I_z \omega_z \mathbf{k} \qquad (21\text{--}26)$$

where ω_x, ω_y, and ω_z represent the x, y, and z components of the angular velocity of the body. Equating the respective $\mathbf{i}$, $\mathbf{j}$, and $\mathbf{k}$ components of Eqs. 21–25 and 21–26 yields

$$\omega_x = 0$$

$$\omega_y = \frac{H_G \sin \theta}{I}$$ (21-27a)

$$\omega_z = \frac{H_G \cos \theta}{I_z}$$

or

$$\boldsymbol{\omega} = \frac{H_G \sin \theta}{I}\mathbf{j} + \frac{H_G \cos \theta}{I_z}\mathbf{k}$$ (21-27b)

In a similar manner, equating the respective **i**, **j**, and **k** components in Eqs. 21-18 and 21-27b yields

$$\frac{d\theta}{dt} = 0$$

$$\frac{d\phi}{dt} \sin \theta = \frac{H_G \sin \theta}{I}$$

$$\frac{d\phi}{dt} \cos \theta + \frac{d\psi}{dt} = \frac{H_G \cos \theta}{I_z}$$

Solving we get

$$\theta = \text{const}$$

$$\frac{d\phi}{dt} = \frac{H_G}{I}$$ (21-28)

$$\frac{d\psi}{dt} = \frac{I - I_z}{II_z} H_G \cos \theta$$

(a)

(b)

Fig. 21-15

Thus, for torque-free motions of an axisymmetrical body, the angle θ formed between the angular momentum vector and the spin of the body remains constant. Furthermore, the angular momentum $\mathbf{H}_G$, precession $d\phi/dt$, and spin $d\psi/dt$ for the body remain *constant* at all times during the motion.

As shown in Fig. 21–15b, the body precesses about the Z axis, which is *fixed in direction,* while it spins about the z axis. These two components of angular velocity may be represented by a simple cone model, introduced in Sec. 13–10. The *space cone* defining the precession is *fixed* from rotating, while the *body cone rotates* around the space cone's outer surface without slipping. The radius of each cone is chosen such that the resultant angular velocity of the body is directed along the line of contact of the two cones. The line of contact represents the instantaneous axis of zero rotation for the body cone, and hence the angular velocity of both the body cone and the body must be directed along this line. From the construction, this angular velocity is $\boldsymbol{\omega}$. (Refer also to the discussion given in Example 13–15.) Since the spin is a function of the moments of inertia I and I_z of the body, Eq. 21–28, the cone model is satisfactory for describing the motion, provided $I > I_z$. Torque-free motion which meets these requirements is called *regular precession.* If $I < I_z$, the spin is negative and the precession is positive. This motion is represented by the satellite shown in Fig. 21–16 ($I < I_z$). The cone model may again be used to represent the motion. However, to preserve the correct vector addition in order to obtain the angular velocity $\boldsymbol{\omega}$, the space cone must be inside the body cone. This motion is referred to as *retrograde precession.*

Example 21–6

The 10-lb disk shown in Fig. 21–17a is spinning about its axis with a

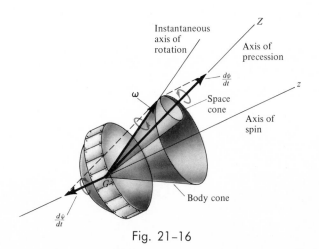

Fig. 21–16

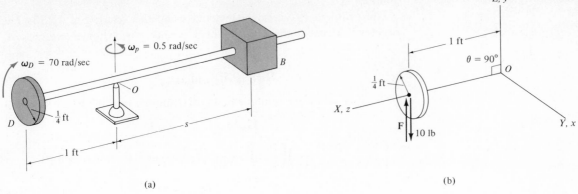

(a)   (b)

Fig. 21–17

constant angular velocity of $\omega_D = 70$ rad/sec. The counterweight at B weighs 5 lb, and by adjusting its position s one can change the angular velocity of precession of the disk about its vertical supporting shaft. Compute the distance s which will enable the disk to have a constant precessional velocity of $\omega_p = 0.5$ rad/sec about the pivot. Neglect the weight of the horizontal supporting shaft.

Solution I

The free-body diagram of the disk is shown in Fig. 21–17b. $\mathbf{F}$ represents the force reaction of the shaft on the disk. The origin for both the xyz and XYZ coordinate systems is located at point O, which represents a *fixed point* for the disk. (Although point O does not lie on the disk, imagine a massless extension of the disk to this point.) In the conventional sense, the Z axis is chosen along the axis of precession, and the z axis is along the axis of spin. Hence, $\theta = 90°$. Since the precession is *steady*, Eqs. 21–22 may be used for the solution. This equation reduces to

$$\Sigma M_x = \frac{d\phi}{dt} I_z \omega_z$$

which is essentially Eq. 21–23. Substituting in the required data gives

$$10\,\text{lb}(1\,\text{ft}) - F(1\,\text{ft}) = 0.5\,\text{rad/sec}\left[\frac{1}{2}\frac{10\,\text{lb}}{32.2\,\text{ft/sec}^2}\left(\frac{1}{4}\,\text{ft}\right)^2\right](-70\,\text{rad/sec})$$

$$F = 10.34\,\text{lb}$$

Using a free-body diagram of the shaft and block B, and summing moments about the pivot O, we require

$$(5\,\text{lb})\,s = 10.34\,\text{lb}(1\,\text{ft})$$

$$s = 2.07\,\text{ft} \qquad\qquad\qquad Ans.$$

Solution II

The problem may also be worked using the vector Eq. 21–13. The angular velocity of the disk for the x, y, and z coordinates are

$$\omega = \{0.5\mathbf{j} - 70\mathbf{k}\} \text{ rad/sec}$$
$$\boldsymbol{\Omega} = \{0.5\mathbf{j}\} \text{ rad/sec}$$

Using Eqs. 21–8, we obtain the angular momentum of the disk:

$$\mathbf{H}_O = I_y\omega_y\mathbf{j} + I_z\omega_z\mathbf{k}$$

$$= \left[\frac{1}{4} \left(\frac{10 \text{ lb}}{32.2 \text{ ft/sec}^2} \right) \left(\frac{1}{4} \text{ ft} \right)^2 + \frac{10 \text{ lb}}{32.2 \text{ ft/sec}^2} (1 \text{ ft})^2 \right] (0.5 \text{ rad/sec})\mathbf{j}$$

$$+ \frac{1}{2} \left(\frac{10 \text{ lb}}{32.2 \text{ ft/sec}^2} \right) \left(\frac{1}{4} \text{ ft} \right)^2 (-70 \text{ rad/sec})\mathbf{k}$$

or

$$\mathbf{H}_O = \{0.158\mathbf{j} - 0.679\mathbf{k}\} \text{ lb-sec}$$

Since $\mathbf{H}_O$ remains constant with respect to the xyz reference, Eq. 21–13 reduces to

$$\Sigma\mathbf{M}_O = \boldsymbol{\Omega} \times \mathbf{H}_O$$

Hence,

$$1 \mathbf{k} \times (-10\mathbf{j} + F\mathbf{j}) = 0.5\mathbf{j} \times (0.158\mathbf{j} - 0.679\mathbf{k})$$
$$F = 10.34 \text{ lb}$$

which is the same as that obtained in the previous solution.

Example 21–7

The top shown in Fig. 21–18a weighs 1 lb and is precessing about the vertical axis at a constant angle $\theta = 60°$ while it spins with an angular

(a)　　　　(b)

Fig. 21–18

velocity $\omega_s = 100$ rad/sec. Determine the precessional velocity ω_p. Assume that the axial and transverse moments of inertia of the top are 1.2×10^{-3} slug-ft^2 and 2.4×10^{-3} slug-ft^2, respectively, measured with respect to the fixed point O.

Solution I

The free-body diagram of the top is shown in Fig. 21–18b. Equation 21–21 may be used for the solution since the motion is a *steady precession*. As shown in Fig. 21–18b, the coordinate axes are established in the usual manner (the z axis along the axis of spin and the Z axis along the axis of precession), with the origin located at the fixed point O. Thus,

$$\Sigma M_x = -I\left(\frac{d\phi}{dt}\right)^2 \sin\theta \cos\theta + I_z\frac{d\phi}{dt}\sin\theta\left(\frac{d\phi}{dt}\cos\theta + \frac{d\psi}{dt}\right)$$

$$1\left(\frac{1.2}{12}\right)\sin 60° = -(2.4 \times 10^{-3})\left(\frac{d\phi}{dt}\right)^2 \sin 60° \cos 60°$$

$$+ (1.2 \times 10^{-3})\left(\frac{d\phi}{dt}\right)\sin 60°\left(\frac{d\phi}{dt}\cos 60° + 100\right)$$

or

$$\left(\frac{d\phi}{dt}\right)^2 - 200\left(\frac{d\phi}{dt}\right) + 166.7 = 0 \tag{1}$$

Solving this quadratic equation for the precession gives

$$\frac{d\phi}{dt} = 199.2 \text{ rad/sec} \quad \text{(high precession)} \qquad Ans.$$

and

$$\frac{d\phi}{dt} = 0.84 \text{ rad/sec} \quad \text{(low precession)} \qquad Ans.$$

Solution II

Using a vector approach, we can obtain the angular velocity of the top from Eqs. 21–18 and 21–19 or directly from Fig. 21–18a. In any case,

$$\omega = \frac{d\phi}{dt}\sin 60°\mathbf{j} + \left(\frac{d\phi}{dt}\cos 60° + 100\right)\mathbf{k}$$

$$\Omega = \frac{d\phi}{dt}\sin 60°\mathbf{j} + \frac{d\phi}{dt}\cos 60°\mathbf{k}$$

Using Eq. 21–8 yields

$$\mathbf{H}_O = (I_x\omega_x)\mathbf{i} + (I_y\omega_y)\mathbf{j} + (I_z\omega_z)\mathbf{k}$$

$$= (2.4 \times 10^{-3})\left(\frac{d\phi}{dt}\sin 60°\right)\mathbf{j} + (1.2 \times 10^{-3})\left(\frac{d\phi}{dt}\cos 60° + 100\right)\mathbf{k}$$

Since $\mathbf{H}_O$ is constant, relative to the xyz coordinate system, Eq. 21-13 reduces to the form

$$\Sigma \mathbf{M}_O = \mathbf{\Omega} \times \mathbf{H}_O$$

$$1\left(\frac{1.2}{12}\right)(\sin\theta)\mathbf{i} = \left(\frac{d\phi}{dt}\sin 60°\mathbf{j} + \frac{d\phi}{dt}\cos 60°\mathbf{k}\right)$$

$$\times\left[(2.4\times 10^{-3})\left(\frac{d\phi}{dt}\sin 60°\right)\mathbf{j} + (1.2\times 10^{-3})\left(\frac{d\phi}{dt}\cos 60° + 100\right)\mathbf{k}\right]$$

When expanded, each of the terms in this equation gives components only in the $\mathbf{i}$ (or x) direction. After simplification,

$$\left(\frac{d\phi}{dt}\right)^2 - 200\left(\frac{d\phi}{dt}\right) + 166.7 = 0$$

which is the same as Eq. (1) obtained previously.

Example 21-8

After passing a football, the motion is observed using a slow-motion projector. From the film, the spin of the football is seen to be directed 30° from the horizontal, as shown in Fig. 21-19a. Also, the football is seen to precess about the vertical axis at a rate $d\phi/dt = 3$ rad/sec. The ratio of the axial to transverse moments of inertia of the football is $\frac{1}{3}$, measured with respect to the center of gravity. Determine the magnitude of the initial spin and the angular velocity of the football. Neglect the effect of air resistance.

Solution

Since the weight of the football is the only force acting, the motion is torque-free. In the conventional sense, if we establish the z axis along the axis of spin and the Z axis along the precession axis, as shown in Fig. 21-19b, the angle $\theta = 60°$. Applying Eqs. 21-28 gives

$$\frac{d\phi}{dt} = \frac{H_G}{I}$$

or

$$H_G = 3I$$

Also,

$$\frac{d\psi}{dt} = \frac{I - I_z}{II_z}H_G\cos\theta$$

Since $I_z = I/3$,

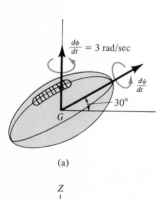

$\frac{d\phi}{dt} = 3$ rad/sec

$\frac{d\psi}{dt}$

30°

G

(a)

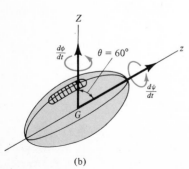

Z

$\frac{d\phi}{dt}$ $\theta = 60°$ z

$\frac{d\psi}{dt}$

G

(b)

Fig. 21-19

$$\frac{d\psi}{dt} = \frac{I - I/3}{I(I/3)}(3I)\cos 60°$$

Therefore, the spin becomes

$$\frac{d\psi}{dt} = 3 \text{ rad/sec} \qquad \qquad Ans.$$

The components of angular velocity are defined by Eqs. 21–27a:

$$\omega_x = 0$$

$$\omega_y = \frac{H_G \sin \theta}{I} = \frac{3I \sin 60°}{I} = 2.60 \text{ rad/sec}$$

$$\omega_z = \frac{H_G \cos \theta}{I_z} = \frac{3I \cos 60°}{I/3} = 4.50 \text{ rad/sec}$$

Thus,

$$\omega = \sqrt{(\omega_x)^2 + (\omega_y)^2 + (\omega_z)^2} = \sqrt{(0)^2 + (2.60)^2 + (4.50)^2}$$
$$= 5.20 \text{ rad/sec} \qquad \qquad Ans.$$

Problems

21–43. A thin rod is initially coincident with the Z axis when it is given three rotations defined by the Euler angles $\phi = 30°$, $\theta = 45°$, and $\psi = 60°$. If these rotations are given in the order stated, determine the final direction of the axis of the rod with respect to the X, Y, and Z coordinate axes. Is this direction also the same for any order of the rotations? Why?

21–44. Show that the angular velocity of a body, in terms of Euler angles ϕ, θ, and ψ, may be expressed as $\omega = (d\phi/dt \sin \theta \sin \psi + d\theta/dt \cos \psi)\mathbf{i} + (d\phi/dt \sin \theta \cos \psi - d\theta/dt \sin \psi)\mathbf{j} + (d\phi/dt \cos \theta + d\psi/dt)\mathbf{k}$, where $\mathbf{i}$, $\mathbf{j}$, and $\mathbf{k}$ are unit vectors directed along the x, y, and z axes shown in Fig. 21–10d.

21–45. The top consists of a disk which weighs 5 lb and a rod having a negligible weight. The top is spinning with an angular velocity of $\omega = 2,500$ rpm. Determine the steady-state precessional angular velocity of the rod when $\theta = 40°$.

Prob. 21–45

21–46. Work Prob. 21–45 when $\theta = 90°$.

21–47. The conical top weighs 4 lb and spins freely in the ball-and-socket joint at A with an angular velocity of $\omega = 250,000$ rpm. Compute the precession of the top about the axis of the shaft AB.

B

A

30° ω 9 in.

2 in.

Prob. 21–47

21–48. The driving armature of the engine of a ship may be approximated by the 150-lb cylinder having a radius of 8 in. If the armature is rotating with an angular velocity of $\omega_s = 3,000$ rpm when the ship turns at $\omega_T = 1$ rpm, determine the vertical reactions at the bearings A and B.

ω_T 2 ft

2 ft

2 ft

B

ω_s

8 in.

A

Prob. 21–48

21–49. The two disks A and C, each weighing 10 lb, are attached to the crossbar frame. The vertical axis

DE of the frame is subjected to an angular velocity of $\omega = 15$ rad/sec. Both disks roll on the horizontal surface without slipping. If the shaft *DE* is suddenly raised slightly so that the disks leave the surface, determine the gyroscopic bending effect exerted at B on the frame caused by the rotating disks. If the shaft *ABC* were elastic, would it bend upward or downward because of these moments?

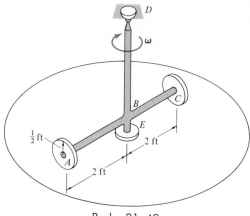

D

ω

B C

E

$\frac{1}{2}$ ft

2 ft

A

2 ft

Prob. 21–49

21–50. The 1-lb top is fixed to spin about point O. The top has a radius of gyration about its axis of symmetry of $k = 0.5$ in. About any transverse axis acting through point O, the radius of gyration is $k_t = 1.7$ in. If the precession is $\omega_p = 5$ rpm, determine the spin ω_s. The mass center is at G.

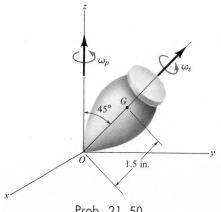

z

ω_p

ω_s

G

45°

O

1.5 in.

y

x

Prob. 21–50

924

21-51. The single-engine airplane is traveling at 150 mph when it enters a vertical curve. The propeller on the airplane weighs 20 lb and has a centroidal radius of gyration of 2.5 ft. When viewed from the front of the airplane, the propeller is turning clockwise at 2,000 rpm. Determine the gyroscopic bending moment which the propeller exerts on the bearings of the engine when the airplane is in the position shown.

$r = 300$ ft

Prob. 21–51

21-52. The projectile shown is subjected to torque-free motion. The transverse and axial moments of inertia for the projectile are I and I_z respectively. If θ represents the angle between the precessional axis Z and the axis of symmetry z, and β is the angle between the angular velocity ω and the z axis, show that β and θ are related by the equation $\tan \theta = (I/I_z) \tan \beta$.

Prob. 21–52

21-53. The gyroscope consists of a uniform 4-lb disk D which rotates in a spindle s having a negligible mass. The supporting frame weighs 1.2 lb and has a center of gravity at G. If the gyro is rotating at a speed of $\omega_D = 60$ rad/sec, determine the constant angular velocity ω_p at which the frame of the gyroscope precesses about the pivot point O. The frame moves in the horizontal plane.

Prob. 21–53

21-54. The 10-lb sphere rotates with an angular speed of $\omega_s = 1,500$ rpm about the axis of the horizontal rod. If the counterbalance block weighs 6 lb, determine the precession of rod DC. Neglect the weight of the shaft. The cross frame is supported by ball-and-socket joints at points C and D.

Prob. 21–54

21-55. The 3-lb disk is rotating about an axle at A with an angular velocity of $\omega_D = 6$ rad/sec. The axle is mounted on a platform which is rotating with an angular velocity of $\omega = 2$ rad/sec and an angular acceleration of $\alpha = 5$ rad/sec². Determine the moment which the reaction forces at the ends of the axle exert on the platform.

21-57. The 20-lb cylinder is rotating about shaft AB with a constant speed of $\omega = 7$ rad/sec. If the supporting shaft at C, initially at rest, is suddenly given an angular acceleration of $\alpha_C = 2$ rad/sec², determine the moment acting on the cylinder caused by the bearing forces at A and B.

Prob. 21-57

Prob. 21-55

21-58. Work Prob. 21-57 assuming that the cylinder is rotating about the z axis with an angular velocity of $\omega_C = \{3\mathbf{k}\}$ rad/sec when $\alpha_C = 2$ rad/sec².

21-56. The projectile weighs 4 lb and has axial and transverse radii of gyration of $k_z = 0.2$ in. and $k_t = 0.6$ in., respectively. If it is spinning at 5 rad/sec when it leaves the barrel of a gun, determine its angular momentum. Precession occurs about the Z axis.

Prob. 21-56

21-59. The motor weighs 50 lb and has a radius of gyration of 0.2 ft about the z axis. The shaft of the motor is supported by bearings at A and B. The shaft is turning at a constant rate of $\omega_s = 100$ rad/sec, while the frame has an angular velocity of $\omega = 2$ rad/sec. Determine the moment which the bearing forces at A and B exert on the motor because of this motion.

926

Prob. 21-59

Prob. 21-61

21-60. The space capsule has a known steady-state precession of two revolutions per hour about the z' axis. If the radius of gyration about an axis passing through the axis of symmetry is $k_z = 4$ ft, and about any transverse axis passing through the center of mass G is $k_t = 6$ ft, determine the rate of spin of the space capsule about its axis of symmetry.

Prob. 21-60

21-61. The rocket weighs 5,000 lb and has a radius of gyration of $k_z = 1.3$ ft and $k_y = 3.2$ ft. It is initially spinning about the z axis at $\omega = 2$ rad/hr. The meteoroid m strikes the rocket at A and creates an impulse of 100 lb-sec. Determine the axis of precession after the impact.

927

22
Vibrations

22-1. Simple Harmonic Motion

A *vibration* is the motion of a body or system of connected bodies which is randomly or uniformly repeated after a given interval of time. In engineering structures the occurrence of vibrations is widespread. Structures purposely designed to vibrate have been used for geological seismic investigations, to facilitate the packing of powdered materials such as sand or flour, and to determine the endurance or fatigue limits of machine members. Most often, however, the effects of vibrations are undesirable in engineering structures. Any vibration requires energy or power to produce it; therefore, the efficiency of machines is reduced. Vibrations also cause stress fatigue of materials, which hastens the time of their eventual failure.

In general, there are two types of vibration, free and forced. *Free vibration* occurs when the motion is maintained by restoring forces, such as the vibration of an elastic rod or the swinging motions of a pendulum. *Forced vibration* is caused by an external periodic or intermittent force applied to the system. Both these types of vibration may be either damped or undamped. *Undamped* vibrations can continue indefinitely because frictional effects are neglected in the analysis. Since in real life both internal and external frictional forces are present, the motion of all vibrating bodies is *damped*.

When the motion of a body is constrained so that it is allowed to vibrate in only one direction, it is said to have a *single degree of freedom*. A one-degree-of-freedom system requires only one coordinate to specify completely the position of the system at any time. In this book we will limit our discussion to one-degree-of-freedom systems, the simplest case of vibrating motion. The analysis of multidegree-of-freedom systems is

based on this simplified case and is thoroughly treated in textbooks devoted to vibrational theory.

The simplest type of vibrating motion with a single degree of freedom is undamped free vibration, represented by the model shown in Fig. 22–1a. The block has a mass m, rests on a frictionless surface, and is attached to a spring having a stiffness k. Vibrating motion is provided by displacing the block a distance x from its equilibrium position and allowing the spring to restore the block to its original position. As the spring draws the block back, the block will attain a velocity when it reaches its original position, and hence it will move out of equilibrium again. Since the surface on which the block moves is frictionless, oscillation of the block will continue indefinitely.

The equations which describe the motion of the block may be determined by using Newton's second law of motion. When the block is displaced a distance x to the right, the restoring force $\mathbf{F}$ is always *directed toward the equilibrium position*. (This holds true even when the block is displaced to the left of the equilibrium position.) The acceleration $\mathbf{a}$ is assumed to act in the direction of positive displacement. The free-body and inertia-vector diagrams for the block are shown in Fig. 22–1b. Applying Newton's second law of motion in the x direction, and noting that $a = d^2x/dt^2$, we have

$$\xrightarrow{+} \Sigma F_x = ma_x; \qquad\qquad -kx = m\frac{d^2x}{dt^2}$$

Rearranging the terms in this equation into a "standard form" gives

$$\frac{d^2x}{dt^2} + p^2x = 0 \qquad\qquad (22\text{–}1)$$

The constant p is called the *circular frequency*, and in this case,

$$p = \sqrt{\frac{k}{m}} \qquad\qquad (22\text{–}2)$$

(a)   (b)

Fig. 22–1

Fig. 22–2

Equation 22–1 may also be obtained by considering the block to be suspended, as shown in Fig. 22–2a. The displacement y is measured from the *equilibrium position*. The free-body and inertia-vector diagrams are shown in Fig. 22–2b. When the block is in the equilibrium position, the spring exerts a force of $W = mg$ upward on the block. When the block is displaced a distance y downward from this position, the magnitude of the spring force is $F = W + ky$. Applying Newton's second law of motion gives

$$+\uparrow \Sigma F_y = ma_y; \qquad W + ky - W = -m\frac{d^2y}{dt^2}$$

or

$$\frac{d^2y}{dt^2} + p^2y = 0$$

which is the same form as Eq. 22–1, where p is defined by Eq. 22–2.

When the motion of the system is defined such that the acceleration is proportional to the displacement, as shown by Eq. 22–1, the motion is called *simple harmonic motion*. Equation 22–1 is a second-order, linear, differential equation with constant coefficients. It can be shown, using the methods of differential equations, that the most general solution of this equation is

$$x = A \sin pt + B \cos pt \qquad (22\text{–}3)$$

where A and B represent constants of integration. The values of these constants are generally determined from the initial conditions of the problem. If we take the successive time derivatives of Eq. 22–3, it is possible to obtain the velocity and acceleration of the block:

$$v = \frac{dx}{dt} = Ap \cos pt - Bp \sin pt \qquad (22\text{–}4)$$

$$a = \frac{d^2x}{dt^2} = -Ap^2 \sin pt - Bp^2 \cos pt \qquad (22\text{–}5)$$

When Eqs. 22–3 and 22–5 are substituted into Eq. 22–1, indeed the differential equation is satisfied, and therefore Eq. 22–3 represents the true solution to Eq. 22–1.

As an example for determining the constants A and B, suppose in the general case, Fig. 22–1, that the block has been displaced a distance x_1 to the right from its equilibrium position, and given an initial (positive) velocity v_1—directed to the right. Substituting $x = x_1$ at $t = 0$ into Eq. 22–3 yields

$$x_1 = B$$

Since $v = v_1$ at $t = 0$, using Eq. 22–4, we have

$$\frac{v_1}{p} = A$$

If we substitute the values of A and B into Eq. 22–3, the equation describing the motion becomes

$$x = \frac{v_1}{p} \sin pt + x_1 \cos pt \tag{22–6}$$

It is seen for this general case that the solution consists of two parts. The first term is proportional to $\sin pt$ and depends upon the initial velocity of the block. The second term is proportional to $\cos pt$ and depends upon the initial displacement x_1.

Equation 22–3 may always be expressed in terms of simple sinusoidal motion. Let

$$A = C \cos \phi \tag{22–7}$$

and

$$B = C \sin \phi \tag{22–8}$$

where C and ϕ are new constants to be determined in place of A and B. Substituting into Eq. 22–3 yields

$$x = C \cos \phi \sin pt + C \sin \phi \cos pt$$

or

$$x = C \sin (pt + \phi) \tag{22–9}$$

If this equation is plotted on an x versus pt axis, the graph shown in Fig. 22–3 is obtained. The maximum displacement of the body from its equilibrium position is defined as the *amplitude* of vibration. From the figure (or Eq. 22–9) the amplitude is C. The *phase angle* is defined by the constant ϕ. This angle represents the amount by which the curve is displaced from the origin when $t = 0$. The constants C and ϕ are related

Fig. 22-3

to A and B by Eqs. 22-7 and 22-8. Squaring and adding these two equations, we obtain the amplitude

$$C = \sqrt{A^2 + B^2} \tag{22-10}$$

If we divide Eq. 22-8 by Eq. 22-7, the phase angle becomes

$$\phi = \tan^{-1}\frac{B}{A} \tag{22-11}$$

Since ϕ is a constant, the sine curve, Eq. 22-9, completes one *cycle*, and hence the cyclic motion of the block is repeated, in time $t = \tau$, so that

$$p\tau = 2\pi$$

or

$$\tau = \frac{2\pi}{p} \tag{22-12}$$

This length of time is called a *period*, Fig. 22-3. From Eq. 22-2, the period may be represented as

$$\tau = 2\pi\sqrt{\frac{m}{k}} \tag{22-13}$$

The *frequency* f is defined as the number of cycles completed per unit time. This is the reciprocal of the period:

$$f = \frac{1}{\tau} = \frac{p}{2\pi} \tag{22-14}$$

$$f = \frac{1}{2\pi} \sqrt{\frac{k}{m}} \tag{22-15}$$

The frequency is often expressed in terms of cycles/sec. This ratio of units is called a *hertz,* abbreviated as Hz.

22-2. Undamped Free Vibration

When a body or system of connected bodies is given an initial displacement from its equilibrium position and released, it will vibrate with a definite frequency known as the *natural frequency.* This type of vibration is called *free vibration,* since no external forces except gravitational or elastic forces act upon the body after the first displacement. Provided the *amplitude* of vibration remains *constant,* the motion is said to be *undamped.*

The undamped free vibration of a body has the *same* characteristics as simple harmonic motion, discussed in Sec. 22-1. The differential equation which describes the undamped free vibration of a body therefore has the *same form* as Eq. 22-1:

$$\frac{d^2x}{dt^2} + px = 0$$

Provided the value of the circular frequency p in this equation is known, it may be used to determine the period and natural frequency of the body as defined by Eqs. 22-12 and 22-14, respectively.

The theory for simple harmonic motion developed in Sec. 22-1 was based on the vibrating motion of the block and spring model shown in Fig. 22-1a or Fig. 22-2a. In the general case, the linear restoring force acting on the body will not be a single spring, rather it may consist of a series of composite springs or a gravitational force, as in the case of a simple pendulum. However, once the value of the circular frequency p has been determined for the body, the motion of the body may be characterized by using an *equivalent* block-and-spring model which has the same motion as the body; that is, we determine an *equivalent spring stiffness* k_{eq} for the model using Eq. 22-2—$k_{eq} = m_{eq}p^2$, where m_{eq} represents the *total mass* of the vibrating body or system of vibrating bodies. (See Example 22-4.)

The general procedure for determining p and the natural frequency of vibration of a rigid body or system of connected rigid bodies consists of displacing the body or system *slightly* from the equilibrium position and relating this displacement to the restoring forces and/or torques which act on the body. The forces and/or torques which tend to restore

(a)   (b)

Fig. 22–4

the body to the equilibrium position are generally due to the action of either a spring or gravity. The free-body and inertia-vector diagrams should be drawn for the body in the displaced position. We can then relate the restoring forces and torques acting on the body to the motions of the body by applying the equations of motion. Using the principles of kinematics, we obtain a differential equation having the same form as Eq. 22–1. The value of p is established after the resulting equation is rewritten in the "standard form," Eq. 22–1. The following examples illustrate this procedure.

Example 22–1

Determine the period of vibration for the simple pendulum shown in Fig. 22–4a. The bob weighs W lb and is attached to a cord having a length l.

Solution

Motion of the system will be related to the independent variable θ. The free-body and inertia-vector diagrams are shown in Fig. 22–4b. When the bob is displaced by an angle θ at time t, the restoring force acting on the bob is created by the weight component $W \sin \theta$. In particular, note that the tangential inertia-force vector is shown acting in the direction of *increasing* θ*. The bob undergoes a small displacement s, measured with respect to its equilibrium position, and therefore the tangential acceleration is $a_t = d^2s/dt^2$. Applying Newton's second law of motion to the bob†, we have

$$+ \nearrow \Sigma F_t = ma_t; \qquad\qquad - W \sin \theta = m \frac{d^2s}{dt^2}$$

---

*This was the format used in deriving the "standard form," Eq. 22–1. Refer to Figs. 22–1b and 22–2b.

†Application of Newton's second law of motion in the n direction is not needed since it involves the unknown tension T.

$$\frac{d^2s}{dt^2} + \frac{W}{m}\sin\theta = 0 \tag{1}$$

The arc length s may be related to θ by the equation $s = \theta l$. Given also that $W = mg$, Eq. (1) reduces to the form

$$\frac{d^2\theta}{dt^2} + \frac{g}{l}\sin\theta = 0 \tag{2}$$

The solution of this equation involves the use of an elliptic integral. For *small displacements*, however, $\sin\theta \approx \theta$, in which case Eq. (2) reduces to the much simpler form

$$\frac{d^2\theta}{dt^2} + \frac{g}{l}\theta = 0 \tag{3}$$

Comparing this equation with Eq. 22–1, which is the "standard form" for simple harmonic motion, we see that $p = \sqrt{g/l}$. Using Eq. 22–12, the period of time required for the bob to make one complete swing is therefore

$$\tau = \frac{2\pi}{p} = 2\pi\sqrt{\frac{l}{g}} \qquad\qquad Ans.$$

This interesting result indicates that the period depends only on the length of the cord and not on the mass of the pendulum bob.

The solution of Eq. (3) is given by Eq. 22–3, where p is defined as above and θ is substituted for x. The constants A and B may be determined, provided, for example, one knows the displacement and velocity of the bob at a given instant.

Example 22–2

The bent rod shown in Fig. 22–5a has a negligible mass and supports a 5-lb weight at its end. Determine the natural period of vibration for the weight.

Solution

The free-body and inertia-vector diagrams for the system are shown in Fig. 22–5b when the rod is displaced by a *small amount* θ from the equilibrium position. The spring exerts a force of $F_s = -k\delta + kx$ on the rod. The first term of this expression, $k\delta$, represents the static force needed to keep the rod BC in the horizontal position. The displacement δ represents the *static deflection* in the spring required for equilibrium. Since the mass of the bent rod is neglected and the rotation of the weight about its own mass center is small, only the translatory inertia of the 5-lb weight

(a)

(b)

(c)

Fig. 22–5

needs to be considered. To obtain the "standard form," Eq. 22–1, the inertia vector $(5/32.2)\mathbf{a}_y$ is assumed to act in the *upward direction* for $+\theta$ displacement, Fig. 22–5b. Moments will be summed about point B to *eliminate* the unknown reaction at this point. If we assume positive moments to be counterclockwise, then for small θ,

$$\zeta + \Sigma M_B = \Sigma(M_{IV})_B$$

$$-k\delta(2 \text{ ft}) + kx(2 \text{ ft}) + 5 \text{ lb}(4 \text{ ft}) = -\frac{5 \text{ lb}}{32.2 \text{ ft/sec}^2}a_y(4 \text{ ft})$$

The first term on the left-hand side of this equation, $k\delta(2 \text{ ft})$, represents the moment created by the spring force which is necessary to hold the weight in *equilibrium*, i.e., at $x = 0$. This moment is consequently equal and opposite to the moment 5 lb (4 ft) created by the 5-lb weight. These two terms, therefore, cancel in the previous equation, so that

$$kx(2 \text{ ft}) = -\frac{5 \text{ lb}}{32.2 \text{ ft/sec}^2} a_y(4 \text{ ft}) \tag{1}$$

From kinematics, the displacement of the spring and the 5-lb weight may be related to the angle θ, Fig. 22–5c. Since θ is small, $x = 2\theta$ ft and $y = 4\theta$ ft. Therefore, $a_y = d^2y/dt^2 = 4(d^2\theta/dt^2)$. Substituting into Eq. (1), we have

$$100 \text{ lb/ft}(2\theta \text{ ft})2 \text{ ft} = -\frac{5 \text{ lb}}{32.2 \text{ ft/sec}^2}\left(4\frac{d^2\theta}{dt^2} \text{ ft/sec}^2\right)(4 \text{ ft})$$

Rewriting this equation in "standard form" gives

$$\frac{d^2\theta}{dt^2} + (161.0)\theta = 0$$

Comparing with Eq. 22–1, we have

$$p^2 = 161.0$$
$$p = 12.7 \text{ Hz}$$

The natural period of vibration becomes

$$\tau = \frac{2\pi}{p} = \frac{2\pi}{12.7 \text{ Hz}} = 0.495 \text{ sec} \qquad Ans.$$

Example 22–3

The rectangular plate shown in Fig. 22–6a is suspended at its center from a rod having a rotational stiffness of $k = 1.5$ lb-ft/rad. Determine the natural period of vibration of the plate when it is given a small angular displacement θ in the plane of the plate. The density of the plate is $\rho = 120$ lb/ft³.

Solution

The free-body and the inertia-vector diagrams are shown in Fig. 22–6b. Since the plate is displaced in its own plane, the torsional *restoring* moment created by the rod is $M = k\theta$. This moment acts in the direction opposite to the displacement. The inertia-moment vector for the plate acts in the direction of *positive* θ. Why? Applying the equation of motion, we have

(a)

(b)

Fig. 22–6

$$\Sigma M_O = I_O \alpha; \qquad\qquad -k\theta = I_O \frac{d^2\theta}{dt^2}$$

or

$$\frac{d^2\theta}{dt^2} + \frac{k}{I_O}\theta = 0$$

Since this equation is in "standard form," the circular frequency is $p = \sqrt{k/I_O}$.

From Appendix D, the mass moment of inertia of the plate about an axis coincident with the rod is $I_O = \frac{1}{12}m(a^2 + b^2)$, where m is the mass of the plate. Thus,

$$I_O = \frac{1}{12}\left[\frac{120 \text{ lb/ft}^3(3 \text{ ft})(2 \text{ ft})(\frac{1}{2} \text{ ft})}{32.2 \text{ ft/sec}^2}\right][(2 \text{ ft})^2 + (3 \text{ ft})^2] = 12.1 \text{ slug-ft}^2$$

The natural period of vibration is therefore

$$\tau = \frac{2\pi}{p} = 2\pi\sqrt{\frac{I_O}{k}} = 2\pi\sqrt{\frac{12.1 \text{ slug-ft}^2}{1.5 \text{ lb-ft/rad}}} = 17.9 \text{ sec} \qquad Ans.$$

Example 22–4

A 10-lb weight is suspended from a cord which is wrapped around a 5-lb disk, as shown in Fig. 22–7a. The spring has a stiffness of $k = 20$ lb/ft. Determine the natural period of vibration for the system. If the vibrating motion of the system is represented by a block-and-spring

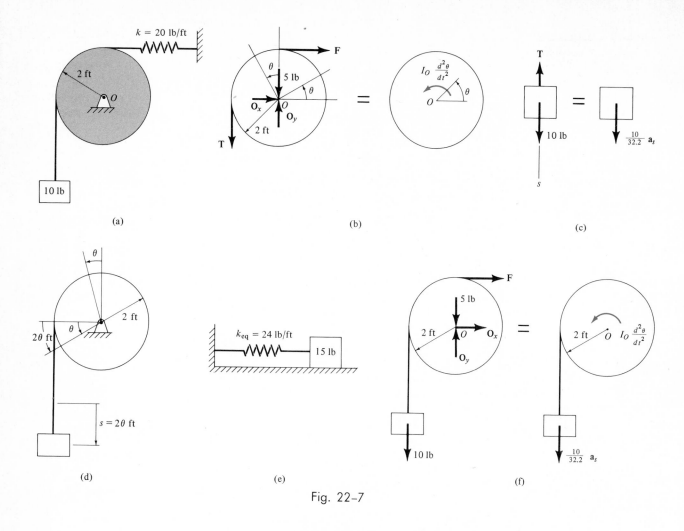

Fig. 22–7

model, determine the required stiffness of the spring, k_{eq}, and the mass of the block.

Solution I

   The system consists of the disk which undergoes a rotation defined by the angle θ, and the block which translates by an amount s. The free-body and inertia-vector diagrams for the disk and the block are shown in Figs. 22–7b and 22–7c. The inertia-moment vector $I_O d^2\theta/dt^2$ acts in the direction of increasing θ. Since we will be seeking a simultaneous solution of equations, it is necessary that the direction of θ be compatible with the direction of the inertia-force vector $(10/32.2)\mathbf{a}_s$, which acts in the direction of increasing s. From Appendix D, $I_O = \tfrac{1}{2}mr^2$, so that the equation of motion for the disk becomes

$$\zeta + \Sigma M_O = I_O \alpha;$$

$$-F(2 \text{ ft}) + T(2 \text{ ft}) = \left[\frac{1}{2}\left(\frac{5 \text{ lb}}{32.2 \text{ ft/sec}^2}\right)(2 \text{ ft})^2\right]\frac{d^2\theta}{dt^2} \qquad (1)$$

For the block,

$$+\downarrow\Sigma F_s = ma_s; \qquad -T + 10 \text{ lb} = \frac{10 \text{ lb}}{32.2 \text{ ft/sec}^2}a_s \qquad (2)$$

As shown on the kinematic diagram in Fig. 22–7d, a positive displacement θ of the disk causes the block to lower by an amount $s = 2\theta$ ft; hence, $a_s = d^2s/dt^2 = 2(d^2\theta/dt^2)$. When $\theta = 0°$, the spring force required for *equilibrium* of the disk is 10 lb, acting toward the right. For positive θ displacement, the spring force is therefore $F = (20 \text{ lb/ft})(2\theta \text{ ft}) + 10$ lb. Substituting these results into Eqs. (1) and (2), yields

$$-80\theta - 20 + 2T = \frac{10}{32.2}\frac{d^2\theta}{dt^2}$$

and

$$-T + 10 = \frac{20}{32.2}\frac{d^2\theta}{dt^2}$$

Combining these equations by eliminating the unknown cable tension T gives

$$\frac{50}{32.2}\frac{d^2\theta}{dt^2} + 80\theta = 0$$

or

$$\frac{d^2\theta}{dt^2} + 51.5\theta = 0 \qquad (3)$$

Hence,

$$p^2 = 51.5$$

or

$$p = 7.18 \text{ Hz}$$

Therefore,

$$\tau = \frac{2\pi}{p} = \frac{2\pi}{7.18} = 0.875 \text{ sec} \qquad \textit{Ans.}$$

The *total mass* of the *system* is equal to the mass of the block *and* the mass of the disk. Thus, $m_{eq} = 15 \text{ lb}/32.2 \text{ ft/sec}^2 = 0.466$ slug. The value of the spring stiffness is determined from the value of p as defined by Eq. 22–2. Therefore,

$$k_{eq} = p^2 m_{eq} = (7.18 \text{ Hz})^2(0.466 \text{ slug}) = 24 \text{ lb/ft}$$

The vibrating motion of the block-and-spring model, Fig. 22–7e, therefore, has the same motion characteristics as the original system.

Solution II

The resulting Eq. (3) derived in Solution I may be obtained directly by applying Newton's second law of motion to *both* the block and the disk. The free-body and inertia-vector diagrams for this two-body system are shown in Fig. 22–7f. Summing moments about point O yields

$$\zeta + \qquad 10 \text{ lb}(2 \text{ ft}) - F(2 \text{ ft}) = \frac{10 \text{ lb}}{32.2 \text{ ft/sec}^2} a_s(2 \text{ ft}) + I_O \frac{d^2\theta}{dt^2}$$

Using the kinematic relationship $a_s = 2(d^2\theta/dt^2)$ and noting that $I_O = \frac{1}{2}mr^2$, and $F = (20 \text{ lb/ft})(2\theta \text{ ft}) + 10 \text{ lb}$, we obtain

$$10(2) - 40\theta(2) - 10(2) = \frac{10}{32.2}\left(2\frac{d^2\theta}{dt^2}\right)2 + \left[\frac{1}{2}\left(\frac{5}{32.2}\right)(2)^2\right]\frac{d^2\theta}{dt^2}$$

or

$$\frac{d^2\theta}{dt^2} + 51.5 = 0$$

which is the same as Eq. (3).

Problems

22–1. When a 20-lb weight is suspended from a spring, the spring is stretched a distance of 4 in. Determine the natural frequency and the period of vibration for a 10-lb weight attached to the same spring.

22–2. A spring has a stiffness of 2 lb/in. If a 2-lb weight is attached to the spring and pushed 3 in. above its equilibrium position and released, determine the equation of motion.

22–3. A 6-lb weight is suspended from a spring having a stiffness of $k = 3$ lb/in. If the weight is given an upward velocity of 20 ft/sec when it is 2 in. above its equilibrium position, determine the equation which describes the motion and the maximum upward displacement of the weight measured from the equilibrium position.

22–4. A spring is stretched 3 in. by a 15-lb weight. If the weight is displaced 2 in. downward from its equilibrium position and given a downward velocity of 6 ft/sec, determine the equation which describes the motion. What is the phase angle?

22–5. If the weight in Prob. 22–4 is given an upward velocity of 6 ft/sec when it is displaced downward a distance of 2 in. from its equilibrium position, determine the equation which describes the motion. What is the amplitude of the motion?

22–6. A 2-lb weight is suspended from a spring having a stiffness of $k = 2$ lb/in. If the weight is pushed 1 in. upward from its equilibrium position and then released, determine the equation which describes the motion. What is the amplitude and the natural frequency of the vibration?

22–7. A pendulum has a 10-in.-long cord and is given a tangential velocity of 2 rad/sec downward, toward the vertical, from a position of $\frac{2}{10}$ rad from the vertical. Determine the equation which describes the angular motion.

22-8. Determine to the nearest degree the maximum angular displacement of the bob in Prob. 22–7 if the bob is initially displaced $\frac{1}{10}$ rad from the vertical and given a tangential velocity of 1 rad/sec away from the vertical.

22-9. Determine the frequency of vibration for the block-and-spring mechanisms.

(a)   (b)

Prob. 22–9

22-10. The 25-lb weight is fixed to the end of the rod assembly having negligible weight. If both springs are unstretched when the assembly is in the position shown, determine the natural period of vibration for the weight when the weight is displaced slightly and released.

Prob. 22–10

22-11. If the wire AB is subjected to a tension force of 20 lb, determine the equation which describes the motion when the 5-lb weight is displaced 2 in. horizontally and released.

Prob. 22–11

22-12. The thin hoop is supported by a knife-edge. Determine the period of oscillation for small amplitudes of swing.

Prob. 22–12

22-13. The semicircular disk weighs 20 lb. Determine the period of oscillation if it is displaced a small amount and released.

Prob. 22–13

943

22-14. The 6-lb weight is attached to the rods which have a negligible weight. Determine the frequency of vibration of the weight when it is displaced slightly and released.

$k = 5$ lb/ft

2 ft

6 lb

3 ft

Prob. 22–14

A

B

4 ft

2 ft

Prob. 22–16

22-15. If a 20-lb roller is displaced a small amount along the curved surface, determine the frequency at which it oscillates when it is released.

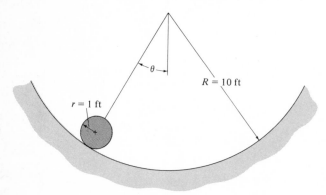

θ

$R = 10$ ft

$r = 1$ ft

Prob. 22–15

22-17. The U-tube has a cross-sectional area of 2 in.² and is partially filled with a liquid having a density of $\rho = 60$ lb/ft³. Determine the period of oscillation of the liquid when the valve V is opened quickly to allow the liquid to oscillate.

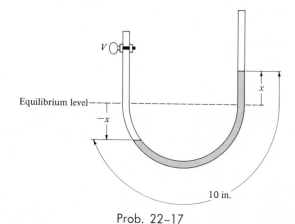

V

Equilibrium level

$-x$

x

10 in.

Prob. 22–17

22-16. The 20-lb rectangular plate has a natural period of vibration of $\tau = 0.3$ sec, as it oscillates around the axis of rod AB. Determine the torsional stiffness k, measured in lb-ft/rad, of the rod AB. Neglect the mass of the rod.

22-18. The pointer on the metronome weighs 0.5 lb. and supports a 1-lb weight W. The weight is positioned

944

a distance l from the pivot O of the pointer. A torsional spring exerts a restoring moment on the pointer having a magnitude of $M = 3\theta$, where θ represents the angle measured in radians and M is measured in lb-in. If the spring is untorqued when the pointer is in the vertical position, determine the natural period of vibration, τ, when $l = 3$ in.

Prob. 22-18

22-19. The 10-lb cylinder rolls without slipping on the horizontal surface as it oscillates about its equilibrium position. If the cylinder is displaced, such that it rolls 0.4 rad forward, determine the equation which describes the oscillatory motion of the cylinder when it is released.

Prob. 22-19

22-20. The 50-lb roller is attached to two springs. If the springs are always in tension when the roller is displaced a small amount and released, determine the period of vibration. The radius of gyration of the roller is $k_O = 1.5$ ft.

Prob. 22-20

22-21. The 20-lb disk is pinned at its center and supports the 50-lb weight. If the belt, which passes over the disk, is not allowed to slip on the surface of the disk, compute the period of oscillation of the system.

Prob. 22-21

22-22. The bent rod shown weighs 10 lb/ft and is constrained to rotate about the pin. If the rod is given a small displacement and released, determine the natural period of free vibration. The springs have zero tension force in them when the rod is in the position shown.

Prob. 22-22

22-24. The system shown consists of a spring with a stiffness of $k = 2$ lb/in. and an unstretched length of 10 in., a bar with a negligible mass, and a small sphere with a mass of 0.2 slug. Determine the frequency for small oscillations of the mass.

Prob. 22-24

22-23. The beam has a linear density of 50 lb/ft. If the end is displaced a small amount and released, determine the angular displacement of the beam as a function of time. Each spring has a stiffness of $k = 120$ lb/ft. The beam has a moment of inertia $I_0 = 266$ slug-ft^2.

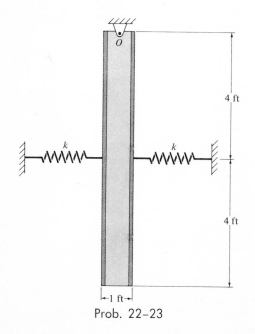

Prob. 22-23

In Sec. 22–2 we showed that the motion of a single-degree-of-freedom system subjected to free vibration may be represented by a differential equation having the standard form of Eq. 22–1. Furthermore, it was shown that the motion of the system is characterized by the block-and-spring model in Fig. 22–1a, or Fig. 22–2a, provided the mass m_{eq} of the block and the stiffness k_{eq} of the spring are correctly chosen. Thus, if we know the values of m_{eq} and k_{eq}, Eq. 22–1 may be written as

$$m_{eq}\frac{d^2x}{dt^2} + k_{eq}x = 0 \qquad (22\text{–}16)$$

Multiplying this equation by $v = dx/dt$ and *integrating* each term between the limits at $t = t_1$, $v = v_1$, $x = x_1$, and at $t = t_2$, $v = v_2$, $x = x_2$, we have

$$m_{eq}\int_{v_1}^{v_2} \frac{d^2x}{dt^2}\left(\frac{dx}{dt}\right) + k_{eq}\int_{x_1}^{x_2} x\left(\frac{dx}{dt}\right) = 0$$

or

$$m_{eq}\int_{v_1}^{v_2} \frac{dx}{dt}\left[\frac{d}{dt}\left(\frac{dx}{dt}\right)\right] + k_{eq}\int_{x_1}^{x_2} x\left[\frac{d}{dt}(x)\right] = 0$$

Hence,

$$\frac{m_{eq}}{2}\left[\left(\frac{dx}{dt}\right)^2\right]_{v_1}^{v_2} + \frac{k_{eq}}{2}[x^2]_{x_1}^{x_2} = 0$$

Substituting the limits and rearranging the terms yields

$$\tfrac{1}{2}m_{eq}(v_1)^2 + \tfrac{1}{2}k_{eq}(x_1)^2 = \tfrac{1}{2}m_{eq}(v_2)^2 + \tfrac{1}{2}k_{eq}(x_2)^2 = \text{const}$$

This equation represents the *conservation of energy* for the vibrating block-and-spring model. The terms define the kinetic and potential energies at the two points of time t_2 and t_1. For the arbitrary time t, the previous equation may be written in the more general form

$$\frac{1}{2}m_{eq}\left(\frac{dx}{dt}\right)^2 + \frac{1}{2}k_{eq}x^2 = \text{const} \qquad (22\text{–}17)$$

The differential equation of motion for the vibrating system may be obtained by using this form of the equation. For example, the total energy of the block-and-spring model shown in Fig. 22–1a, at *any instant*, consists of kinetic energy of the mass, $\tfrac{1}{2}m_{eq}(dx/dt)^2$, and potential energy of the spring, $\tfrac{1}{2}k_{eq}x^2$. The sum is therefore expressed by Eq. 22–17. *Differentiating* this equation with respect to time yields

$$m_{eq} \frac{dx}{dt} \frac{d^2x}{dt^2} + k_{eq}x\frac{dx}{dt} = 0$$

or

$$\frac{dx}{dt}\left(m_{eq}\frac{d^2x}{dt^2} + k_{eq}x\right) = 0$$

Since the system is vibrating, the velocity dx/dt cannot *always* be zero. Hence, the above equation is satisfied only if

$$m_{eq}\frac{d^2x}{dt^2} + k_{eq}x = 0$$

which is identical to Eq. 22–16.

For conservative force systems, which are subjected to reactive forces which do no work, determining the equation of motion by *time differentiating the energy equation* has an advantage. This is because the derivative of the energy equation yields a relation between the forces and the accelerations directly, without the necessity of having to dismember the system. For conservative force systems having a single degree of freedom, the time derivative of the energy equation will always be of the same form as Eq. 22–1.

The energy method also has an advantage over direct application of the equation of motion in that the circular frequency p may be computed *directly*. As an example, consider the total mechanical energy of the block-and-spring model when it is at *maximum displacement*. In this position the mass is temporarily at rest. The kinetic energy is zero, and the potential energy stored in the spring is a maximum. Therefore, Eq. 22–17 becomes $\frac{1}{2}k_{eq}x_{max}^2 = $ const. As the block passes through the equilibrium position, the kinetic energy of the block is a maximum and the potential energy of the spring is zero. Hence, Eq. 22–17 becomes $\frac{1}{2}m_{eq}(dx/dt)_{max}^2 = $ const. Since the vibrating motion of the block is *harmonic*, the *solution* for the displacement may be written in the form of Eq. 22–9:

$$x = C\sin(pt + \phi)$$

so that

$$x_{max} = C$$

Also,

$$\frac{dx}{dt} = Cp\cos(pt + \phi)$$

Hence,

$$\left(\frac{dx}{dt}\right)_{max} = Cp$$

From the conservation of energy principle ($T + V = $ const), the maximum stored energy in the spring is equal to the maximum kinetic energy of the block. Therefore,

$$\frac{1}{2}k_{eq}x^2_{max} = \frac{1}{2}m_{eq}\left(\frac{dx}{dt}\right)^2_{max} = \text{const}$$

or

$$k_{eq}C^2 = m_{eq}C^2p^2$$

Solving for p yields

$$p = \sqrt{\frac{k_{eq}}{m_{eq}}}$$

which is identical to Eq. 22–2. Knowing p we can compute the period τ and natural frequency f using Eqs. 22–12 and 22–14.

The following examples illustrate the method for obtaining both the period and natural frequency of vibration for a system employing the energy method.

Example 22-5

The small weight W, shown in Fig. 22–8a, is attached to the midpoint of a thin wire of length $2l$ and having a tension T. Determine the natural frequency of vibration for the system. Neglect the weight of the wire and assume that the tension T is much larger than the weight W.

Solution

A diagram of the weight, when the weight is located in the arbitrary position x, is shown in Fig. 22–8b. The kinetic energy is

$$T = \frac{1}{2}mv^2 = \frac{1}{2}\frac{W}{g}\left(\frac{dx}{dt}\right)^2$$

The component of tension force acting towards the equilibrium position is $(x/l)T$, assuming that the displacement x is small. The total restoring force is thus $F = 2(x/l)T$. The stiffness k of the system *in the direction of x* can be found using the equation

$$k = \frac{F}{x} = \frac{2(x/l)T}{x} = \frac{2T}{l}$$

Thus, the potential energy is

$$V = \frac{1}{2}kx^2 = \frac{1}{2}\left(\frac{2T}{l}\right)x^2$$

The total energy of the system is therefore

Fig. 22–8

$$T + V = \frac{1}{2}\frac{W}{g}\left(\frac{dx}{dt}\right)^2 + \frac{1}{2}\left(\frac{2T}{l}\right)x^2 = \text{const}$$

Taking the time derivative of this equation yields

$$\frac{1}{2}\frac{W}{g}2\left(\frac{dx}{dt}\right)\frac{d^2x}{dt^2} + \frac{1}{2}\left(\frac{2T}{l}\right)2x\frac{dx}{dt} = 0$$

which, after factoring out dx/dt and simplifying, reduces to the equation of motion

$$\frac{d^2x}{dt^2} + \frac{2Tg}{lW}x = 0$$

Comparing with the "standard form," Eq. 22–1,

$$p = \sqrt{\frac{2Tg}{lW}}$$

so that

$$f = \frac{1}{2\pi}\sqrt{\frac{2Tg}{lW}} \qquad\qquad Ans.$$

Example 22–6

A 10-lb weight is suspended from a cord wrapped around a 5-lb disk, as shown in Fig. 22–9a. If the spring has a stiffness of $k = 20$ lb/ft, determine the natural period of vibration for the system.

Solution I

The energy diagrams shown in Fig. 22–9b represent the position of the system for maximum kinetic energy and maximum potential energy, respectively. The datum is chosen at the elevation of the block when the system is in *equilibrium*. Hence, the maximum angular velocity of the disk and the maximum velocity of the weight occur when the block is located at the datum. Using kinematics, the velocity of the block is $v_{max} = 2$ ft(ω_{max}). Since the system undergoes simple harmonic motion, $v_{max} = Cp$, and therefore, $\omega_{max} = Cp/2$ ft.

The potential energy for the system is maximum when the system is temporarily at rest. At this instant the weight is displaced an amount $s_{max} = C$, while the spring is elongated by an amount $\delta_{st} + s_{max}$, where δ_{st} represents the *static deflection* of the spring. Since the static force in the spring is $F_{st} = 10$ lb $= (20$ lb/ft$)\delta_{st}$, the static deflection is $\delta_{st} = \frac{1}{2}$ ft. The principle of conservation of energy for the system may be written for the two positions shown in Fig. 22–9b:

① Maximum kinetic energy   ② Maximum potential energy

(a)   (b)

Fig. 22-9

$$\{T_1\} + \{V_1\} = \{T_2\} + \{V_2\}$$

$$\left\{\frac{1}{2}m_w(v_w)_1^2 + \frac{1}{2}I_D(\omega_D)_1^2\right\} + \left\{\frac{1}{2}k\delta_{st}^2\right\} = \left\{\frac{1}{2}m_w(v_w)_2^2 + \frac{1}{2}I_D(\omega_D)_2^2\right\}$$

$$+ \left\{\frac{1}{2}k(\delta_{st} + s_{max})^2 - ws_{max}\right\}$$

or

$$\left\{\frac{1}{2}\left(\frac{10 \text{ lb}}{32.2 \text{ ft/sec}^2}\right)(Cp)^2 + \frac{1}{2}\left[\frac{1}{2}\left(\frac{5 \text{ lb}}{32.2 \text{ ft/sec}^2}\right)(2 \text{ ft})^2\right]\left(\frac{Cp}{2}\right)^2\right\}$$

$$+ \left\{\frac{1}{2}(20 \text{ lb/ft})\left(\frac{1}{2} \text{ ft}\right)^2\right\} = \{0 + 0\}$$

$$+ \left\{\frac{1}{2}(20 \text{ lb/ft})\left(\frac{1}{2} \text{ ft} + C\right)^2 - 10 \text{ lb}(C)\right\}$$

Expanding and rearranging terms, this expression becomes

$$\frac{1}{8}\left(\frac{50}{32.2}\right)(Cp)^2 = 10(C)^2$$

Solving for p yields

$$p = \sqrt{\frac{80(32.2)}{50}} = 7.18 \text{ Hz}$$

Hence, the period of vibration is

$$\tau = \frac{2\pi}{p} = \frac{2\pi}{7.18} = 0.875 \text{ sec} \qquad Ans.$$

Solution II

When the weight and disk are displaced by an intermediate amount s and θ, respectively, where $s = 2\theta$, the kinetic energy is

$$
\begin{aligned}
T &= \frac{1}{2}m_w v_w^2 + \frac{1}{2}I_D\omega_D^2 \\
&= \frac{1}{2}\left(\frac{10 \text{ lb}}{32.2 \text{ ft/sec}^2}\right)\left(2\frac{d\theta}{dt}\right)^2 + \frac{1}{2}\left[\frac{1}{2}\left(\frac{5 \text{ lb}}{32.2 \text{ ft/sec}^2}\right)(2 \text{ ft})^2\right]\left(\frac{d\theta}{dt}\right)^2 \\
&= \frac{25}{32.2}\left(\frac{d\theta}{dt}\right)^2
\end{aligned}
$$

The potential energy is

$$
\begin{aligned}
V &= \frac{1}{2}k(\delta_{st} + s)^2 - ws \\
&= \frac{1}{2}20 \text{ lb/ft}(\delta_{st} + 2\theta \text{ ft})^2 - 10 \text{ lb}(2\theta \text{ ft})
\end{aligned}
$$

The total energy for the system is then

$$T + V = \frac{25}{32.2}\left(\frac{d\theta}{dt}\right)^2 + 10(\delta_{st} + 2\theta)^2 - 20\theta = \text{const}$$

Taking the time derivative of this equation yields

$$\frac{50}{32.2}\left(\frac{d\theta}{dt}\right)\frac{d^2\theta}{dt^2} + 40(\delta_{st} + 2\theta)\frac{d\theta}{dt} - 20\frac{d\theta}{dt} = 0$$

Recalling that $\delta_{st} = \frac{1}{2}$ ft, the previous equation reduces to the "standard form"

$$\frac{d^2\theta}{dt^2} + \frac{80(32.2)}{50}\theta = 0$$

So that

$$p = \sqrt{\frac{80(32.2)}{50}} = 7.18 \text{ Hz}$$

Thus,

$$\tau = \frac{2\pi}{p} = \frac{2\pi}{7.18} = 0.875 \text{ sec} \qquad Ans.$$

Problems

22-25. Using energy methods, determine the differential equation of motion of the 10-lb block. The horizontal surface is smooth. The springs are originally unstretched.

$k = 10$ lb/in.   10 lb $k = 5$ lb/in.

Prob. 22-25

22-26. The 20-lb disk is pin-connected at its midpoint. Determine the period of vibration of the disk if the springs have sufficient tension in them to prevent the belt from slipping on the surface of the disk.

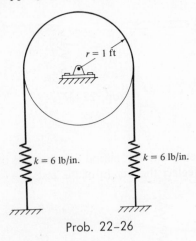

$r = 1$ ft

$k = 6$ lb/in. $k = 6$ lb/in.

Prob. 22-26

22-27. Determine the period of vibration of the 10-lb homogeneous semicircular disk.

$\frac{1}{2}$ ft

Prob. 22-27

22-28. Determine the natural frequency of vibration of the 20-lb disk. Assume that the friction force is great enough so that the disk does not slip on the surface of the plane while it is oscillating.

$k = 10$ lb/in.

1 ft

30°

Prob. 22-28

22-29. Solve Prob. 22–12 using energy methods.

22-30. Solve Prob. 22–15 using energy methods.

22-31. Solve Prob. 22–17 using energy methods.

22-32. Solve Prob. 22–19 using energy methods.

22-33. Determine the period of vibration of the pendulum. Consider the two bars to be slender, each having a linear density of 10 lb/ft.

O

2 ft

1 ft   1 ft

Prob. 22-33

22–34. If the disk weighs 20 lb, determine the natural frequency of vibration.

Prob. 22–34

Prob. 22–36

22–35. Determine the differential equation of motion of the 30-lb roller using the conservation of energy principle. Assume that the roller does not slip on the surface. The radius of gyration of the roller about its center of mass is $k_O = 1.5$ ft.

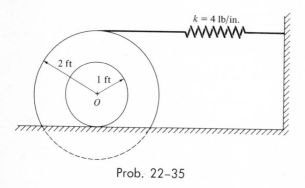

Prob. 22–35

22–37. The 5-lb weight is attached to a rod of negligible weight. Determine the natural frequency of vibration of the rod.

Prob. 22–37

22–38. Determine the period of vibration of the 10-lb weight. Neglect the weight of the rod.

Prob. 22–38

22–36. The slender rod has a linear density of 4 lb/ft and is supported in the horizontal plane by means of a ball-and-socket joint at A and a cable at B. Determine the natural frequency of vibration when the end B is given a small horizontal displacement and then released.

22–39. Solve Prob. 22–24 using energy methods.

22–4. Undamped Forced Vibrations

Undamped forced vibration is considered to be one of the most important types of vibrating motion in engineering work. The principles which describe the nature of this motion may, for example, be applied to the design of isolators. Isolators are necessary to prevent outside vibrations from reaching delicate instruments, as in the case of aircraft instruments. The principles of undamped forced vibrations may also be applied to the stress analysis of forces which cause vibration in machinery such as the forces caused by airplane engine vibrations acting within the fuselage of the airplane.

In general, *forced vibrations* arise from the application of an external periodic force or by a periodic excitation of the foundation of a system. Simplified models which represent these two cases are shown in Figs. 22–10a and 22–10c, respectively. (Frictional forces which cause *damping* of the system will be discussed in Sec. 22–5.)

In the first case, the periodic excitation force $F_O \sin \omega t$ is applied to the block, Fig. 22–10a. This force has a maximum magnitude of F_O and a *forcing frequency* ω. The free-body and inertia-vector diagrams for the block are shown in Fig. 22–10b. In particular, the distance x is measured

Fig. 22–10

from where the stretch in the spring is zero (the equilibrium position). Applying Newton's second law of motion, we have

$$\xrightarrow{+}\Sigma F_x = ma_x; \qquad F_0 \sin \omega t - kx = m\frac{d^2x}{dt^2}$$

or

$$\frac{d^2x}{dt^2} + \frac{k}{m}x = \frac{F_0}{m}\sin \omega t \qquad (22\text{--}18)$$

The model shown in Fig. 22–10c represents the case of periodic vibration of the block, which is caused by the harmonic movement $\delta_0 \sin \omega t$ at the support. The free-body and inertia-vector diagrams for the block in this case are shown in Fig. 22–10d. The coordinate x is measured from the point of zero displacement of the support, i.e., when the radius vector OA coincides with OB. Therefore, general displacement of the spring is $(x - \delta_0 \sin \omega t)$. Applying Newton's second law of motion, we have

$$\xrightarrow{+}\Sigma F_x = ma_x; \qquad -k(x - \delta_0 \sin \omega t) = m\frac{d^2x}{dt^2}$$

or

$$\frac{d^2x}{dt^2} + \frac{k}{m}x = \frac{k\delta_0}{m}\sin \omega t \qquad (22\text{--}19)$$

The forms of Eqs. 22–18 and 22–19 are identical. The solution of Eq. 22–18 will be discussed. The solution of Eq. 22–19 may be easily obtained from the solution of Eq. 22–18 by simply substituting the constant $k\delta_0$ for F_0.

Equation 22–18 is a second-order differential equation which is non-homogeneous. The total solution of this equation consists of a complementary solution x_c plus a particular solution x_p. The *complementary solution* is determined by setting the term on the right side of the equation equal to zero and solving the resulting homogeneous equation, which is equivalent to Eq. 22–1. The complementary solution is therefore the same as that for free vibration which was determined in Sec. 22–1. Therefore,

$$x_c = A \sin pt + B \cos pt$$

where p is the circular frequency defined by Eq. 22–2. The *particular solution* of Eq. 22–18 may be determined by assuming a solution of the form

$$x_p = C \sin \omega t \qquad (22\text{--}20)$$

where C is a constant. Taking the second derivative of Eq. 22–20 and substituting into Eq. 22–19, we have

$$-C\omega^2 \sin \omega t + \frac{k}{m}(C \sin \omega t) = \frac{F_0}{m} \sin \omega t$$

or

$$C\left(-\omega^2 + \frac{k}{m}\right) \sin \omega t = \frac{F_0}{m} \sin \omega t$$

This equation is satisfied for all time t provided

$$C = \frac{F_0/m}{\dfrac{k}{m} - \omega^2} = \frac{F_0/k}{1 - \left(\dfrac{\omega}{p}\right)^2}$$

Substituting this value for C into Eq. 22–20, we obtain the particular solution

$$x_p = \frac{F_0/k}{1 - \left(\dfrac{\omega}{p}\right)^2} \sin \omega t \qquad (22\text{–}21)$$

The total solution is therefore

$$x = x_c + x_p = A \sin pt + B \cos pt + \frac{F_0/k}{1 - \left(\dfrac{\omega}{p}\right)^2} \sin \omega t \quad (22\text{–}22)$$

This equation describes two types of vibrating motion of the block. The *complementary solution* defines the *free vibration*. This motion depends only upon the values of m and k and the constants A and B. These constants are determined from the initial conditions of the motion, as discussed in Sec. 22–1. The *particular solution* describes the motion of the block caused by the applied *forcing function* $F = F_0 \sin \omega t$. This periodic force F may also be expressed as $F_0 \cos \omega t$ or as a combination of sine and cosine functions (see Example 22–7). A graph of Eq. 22–22 is shown in Fig. 22–11a. We can see that the free vibration x_c is simply superimposed on the displacement caused by the forcing function. In Sec. 22–6 we shall show that when frictional effects are included in the solution the free vibration of the block will quickly damp out. When this occurs, the free vibration is then referred to as *transient*, and the forced vibration is called *steady state* since it is the only vibration which remains, Fig. 22–11b.

If the case of forced vibration induced by periodic excitation of the support had been considered, the particular solution of the differential equation, Eq. 22–19, would have been*

(a)

(b)

Fig. 22–11

*Substitute $k\delta_0$ for F_0 in Eq. 22–21.

$$x_p = \frac{\delta_0}{1 - \left(\dfrac{\omega}{p}\right)^2} \sin \omega t \qquad (22\text{-}23)$$

The *amplitude* of forced vibration, x_p, involving either a periodic force, Eq. 22–21, or support motion, Eq. 22–23, depends upon the *frequency ratio* ω/p. We will define the *magnification factor* MF as the ratio of the amplitude of steady-state vibration, $(x_p)_{\text{max}}$, to the static deflection F_0/k, which is caused by the amplitude of the periodic force F_0 or by the amplitude of the support movement $\delta_0 (F_0 = k\delta_0)$. From Eq. 22–21 or Eq. 22–23,

$$\text{MF} = \frac{(x_p)_{\text{max}}}{F_0/k} = \frac{(x_p)_{\text{max}}}{\delta_0} = \frac{1}{1 - \left(\dfrac{\omega}{p}\right)^2} \qquad (22\text{-}24)$$

This equation is graphed in Fig. 22–12. It is seen that when the frequency ratio is very small the MF is nearly unity, in which case the mass will be in phase with the force F_0 or displacement δ_0. If the force or displacement is applied with a frequency close to the natural frequency of the system, i.e., $\omega/p \approx 1$, the amplitude of vibration of the block becomes extremely large. This condition is called *resonance*, and in practice, resonating vibrations can cause tremendous stress and rapid failure of parts. When the force or displacement is applied at high frequencies $(\omega > p)$, the value of the MF becomes negative, indicating that the motion of the block is out-of-phase with the imposed force or displacement. Under these conditions, as the block is displaced to the right, the force or support displacement acts to the left, and vice versa. For extremely high frequencies $(\omega \gg p)$ the inertia of the mass prevents the block from following the force or displacement. As a result, the block remains almost stationary, and hence, the MF is approximately zero.

Example 22–7

The 20-lb block shown in Fig. 22–13a rests on a smooth surface and is acted upon by the forcing function $F = 3\cos 2t$, where F is given in pounds and t in seconds. At the same time, the support has a motion of $x_s = \frac{1}{12}\sin t$, where x_s is in feet and t in seconds. Determine the equation which describes the motion of the block if the block is initially at rest when $x = 0$.

Solution

The block is in the equilibrium position when the spring is unstretched at $t = 0$. When the block is located a distance x from this position, the spring is stretched $\Delta x = x - x_s$. The free-body and inertia-vector dia-

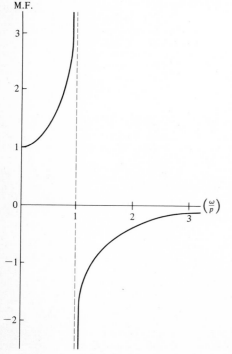

Fig. 22–12

(a)

(b)

Fig. 22–13

grams for the block in this position are shown in Fig. 22–13b. In particular, note from the equations of the forcing function and support displacement that $F_O = 3$ lb, $\omega = 2$ Hz and $\delta_0 = \frac{1}{12}$ ft. Applying Newton's second law of motion, we have

$$\overset{+}{\rightarrow} \Sigma F_x = ma_x; \qquad 3 \cos 2t - 100 \left(x - \frac{1}{12} \sin t \right) = \frac{20}{32.2} \frac{d^2x}{dt^2}$$

Rearranging terms and simplifying yields

$$\frac{d^2x}{dt^2} + 161.0x = 13.42 \sin t + 4.83 \cos 2t \qquad (1)$$

The solution of this equation consists of the complementary solution and the particular solution. The complementary solution is obtained by setting the terms on the right-hand side equal to zero. In Sec. 22–1, the solution of the resulting equation has been found (Eq. 22–3), and here it is of the form

$$x_c = A \sin 161.0t + B \cos 161.0t \qquad (2)$$

The particular solution is obtained by assuming a solution of the form

$$x_p = C \sin t + D \cos 2t \qquad (3)$$

where C and D are constants. Taking the second derivative of this equation and substituting into Eq. (1), we have

$$-C \sin t - 4D \cos 2t + (161.0)C \sin t + (161.0)D \cos 2t$$
$$= 13.42 \sin t + 4.83 \cos 2t$$

Equating the separate coefficients of the sine and cosine terms yields

$$-C + (161.0)C = 13.42$$
$$-4D + (161.0)D = 4.83$$

$$C = 0.0839 \text{ ft}, \qquad D = 0.0308 \text{ ft}$$

When we substitute these constants into Eq. (3), the particular solution is

$$x_p = 0.0839 \sin t + 0.0308 \cos 2t$$

This equation represents the displacement function for steady-state vibration. The total solution is therefore

$$x = x_c + x_p$$
$$= A \sin 161.0t + B \cos 161.0t + 0.0839 \sin t + 0.0308 \cos 2t \quad (4)$$

The constants A and B are determined from the initial conditions. Since $x = 0$ at $t = 0$,

$$0 = 0 + B + 0 + 0.0308; \qquad B = -0.0308$$

The velocity is

$$v = \frac{dx}{dt} = A(161.0) \cos 161.0t - B(161.0) \sin 161.0t$$
$$+ 0.0839 \cos t - 0.0616 \sin 2t$$

At $t = 0$, $v = 0$. Thus

$$0 = A(161.0) - 0 + 0.0839 - 0; \qquad A = -0.000521$$

Substituting the values of A and B into Eq. (4),

$$x = -0.000521 \sin 161.0t - 0.0308 \cos 161.0t$$
$$+ 0.0839 \sin t + 0.0308 \cos 2t \qquad Ans.$$

Problems

22–40. If the block-and-spring model is subjected to the impressed force $F = F_0 \cos t$, show that the differential equation of motion is $d^2x/dt^2 + (k/m)x = F_0/m \cos \omega t$, where x is measured from the equilibrium position of the block. What is the general solution of this equation?

22–41. The block-and-spring model shown is subjected to a periodic support displacement of $\delta = \delta_0 \sin \omega t$. Determine the equation of motion for the system and obtain its general solution. Define the displacement x measured from the static equilibrium position of the block when $t = 0$.

Prob. 22–40   Prob. 22–41

22-42. The spring shown stretches 6 in. when it is loaded with a 50-lb weight. Determine the equation which describes the motion of the weight when it is pulled 4 in. below its equilibrium position and released. The weight is subjected to the impressed force of $F = -7 \sin 2t$, where F is measured in pounds and t in seconds.

Prob. 22-45

Prob. 22-42

22-43. A 4-lb weight is attached to a vertical spring with a stiffness of $k = 10$ lb/ft. The weight is drawn downward a distance of 4 in. and released. If the support moves with an impressed displacement of $\delta = \sin 4t$, where δ is given in inches and t is measured in seconds, determine the equation which describes the motion of the system.

22-44. A 5-lb weight is suspended from a vertical spring having a stiffness of 50 lb/ft. An impressed force of $F = \frac{1}{4} \sin 8t$, where F is measured in pounds and t is given in seconds, is acting on the weight. Determine the equation of motion of the weight when it is pulled down 3 in. from the equilibrium position and released.

22-45. The 20-lb block is attached to a spring with a stiffness of 20 lb/ft. A force of $F = 6 \cos 2t$, where F is given in pounds and t is given in seconds, is applied to the block. Determine the maximum velocity of the block after frictional forces cause the free vibrations to dampen out.

22-46. The instrument box rests uniformly on a platform P, which in turn is supported by *four springs*, each spring having a stiffness of $k = 20$ lb/in. If the floor is subjected to a vibration of 200 cycles/min having a total vertical movement of 0.2 in., determine the amplitude of vertical motion of the platform and instrument box. The box and platform together weigh 60 lb.

Prob. 22-46

22-47. The light elastic rod supports a 10-lb sphere. When a 10-lb vertical force is applied to the sphere, the rod deflects $\frac{1}{2}$ in. If the wall oscillates with a harmonic frequency of 2 Hz and an amplitude of $\frac{1}{2}$ in., determine the amplitude of vibration of the sphere.

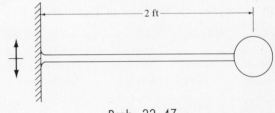

Prob. 22-47

22–48. The electric motor weighs 100 lb and is supported by *four springs,* each spring having a stiffness of 20 lb/ft. If the motor turns an eccentric disk D which is equivalent to a weight of 2 lb located $\frac{3}{4}$ ft from the axis of rotation, determine the angular rotation at which resonance occurs. Assume that the motor only vibrates in the vertical direction. (*Hint:* See the first part of Example 22–8.)

Prob. 22–50

Prob. 22–48

22–49. What is the amplitude of steady-state vibration of the motor in Prob. 22–48 when the angular speed of the disk is 150 rpm? (*Hint:* See the first part of Example 22–8.)

22–50. The electric motor is supported on a light horizontal beam. The motor turns an eccentric flywheel which is equivalent to an unbalanced 0.25-lb weight located 10 in. from the axis of rotation. If the static deflection of the beam is 1 in. because of the weight of the motor, determine the angular speed of the flywheel at which resonance will occur. The motor weighs 150 lb. (*Hint:* See the first part of Example 22–8.)

22–51. What will be the amplitude of steady-state vibration of the motor in Prob. 22–50 if the angular speed of the flywheel is 100 rpm? (*Hint:* See the first part of Example 22–8.)

22–52. Determine how much weight must be removed from the motor in Prob. 22–51 to produce an amplitude of vibration of 0.005 in.

22–53. The 500-lb trailer is pulled with a constant speed over the surface of a road which may be approximated by a cosine curve having an amplitude of $\frac{1}{4}$ ft and wave length of 10 ft. If the two springs s which support the trailer each have a stiffness of 50 lb/in., determine the speed v which will cause the greatest vibration of the trailer. Neglect the weight of the wheels.

Prob. 22–53

22–54. Determine the amplitude of vibration of the trailer in Prob. 22–53 if the speed $v = 5$ mph.

*22–5. Viscous Damped Free Vibrations

The vibration analysis considered thus far has not included the effects of friction or damping in the system. This has led to results which are therefore only in close agreement with the actual motion. Since all vibrations die out in time, the presence of damping forces should be included in the analysis.

Damping is generally attributed to the following three causes:

1. The internal frictional forces acting within the member which is being vibrated.
2. The frictional resistance which occurs between two solid contacting surfaces.
3. The resistance to the motion of a fluid in which the system vibrates, such as water, oil, or air.

Since the last case represents the most common occurrence of vibrational damping, its effects will be considered here.

Provided the body moves through a fluid at a low velocity, the resistance to motion is directly proportional to the speed. The type of force developed under these conditions is called a *viscous damping force*. The magnitude of this force may be expressed by an equation of the form

$$F = c\frac{dx}{dt} \tag{22–25}$$

The constant c is called the *coefficient of viscous damping* and has units of lb-sec/ft (kg_f-sec/m), provided the force F is measured in pounds (kilogram force) and the velocity dx/dt is measured in ft/sec (m/sec). Equation 22–25 pertains most often to the type of action which occurs in dashpots, lubricated bearings, and bodies moving with relatively low velocities through a fluid medium.

The appropriate block-and-spring model which includes the effects of viscous damping is shown in Fig. 22–14a. The *dashpot* which connects

(a)

(b)

Fig. 22–14

the block to the wall on the right provides the necessary viscous damping to the system. The effect of damping occurs when the piston P moves to the right or left within the enclosed cylinder. The cylinder contains a fluid, and the motion of the piston is retarded since the fluid must flow around or through a small hole in the piston. The dashpot is assumed to have a coefficient of viscous damping of c.

If the system is displaced a distance x from its equilibrium position, the resulting free-body and inertia-vector diagrams are shown in Fig. 22–14b. Both the spring force kx and the damping force $c(dx/dt)$ oppose the forward motion of the mass. Applying Newton's second law of motion, we have

$$\xrightarrow{+}\Sigma F_x = ma_x; \qquad -kx - c\frac{dx}{dt} = m\frac{d^2x}{dt^2}$$

or

$$m\frac{d^2x}{dt^2} + c\frac{dx}{dt} + kx = 0 \qquad (22\text{–}26)$$

This homogeneous differential equation has solutions of the form $x = e^{\lambda t}$, where e is the base of the natural logarithm and λ is a constant. The value of λ may be obtained by substituting the solution into Eq. 22–26, which yields

$$m\lambda^2 e^{\lambda t} + c\lambda e^{\lambda t} + ke^{\lambda t} = 0$$

or

$$e^{\lambda t}(m\lambda^2 + c\lambda + k) = 0$$

Since $e^{\lambda t}$ is always positive, solution to the above equation is possible provided

$$m\lambda^2 + c\lambda + k = 0$$

This quadratic equation may be solved for two values of λ.

$$\lambda_1 = -\frac{c}{2m} + \sqrt{\left(\frac{c}{2m}\right)^2 - \frac{k}{m}} \qquad (22\text{–}27)$$

and

$$\lambda_2 = -\frac{c}{2m} - \sqrt{\left(\frac{c}{2m}\right)^2 - \frac{k}{m}} \qquad (22\text{–}28)$$

The general solution of Eq. 22–26 is therefore a linear combination of exponentials which involves both of these roots.

Before discussing these cases, let us define the *critical damping coeffi-*

cient c_c as the value of c, which makes the radical in Eq. 22–27 and Eq. 22–28 equal to zero, i.e.,

$$\left(\frac{c_c}{2m}\right)^2 - \frac{k}{m} = 0$$

or

$$c_c = 2m\sqrt{\frac{k}{m}} = 2mp \qquad (22\text{–}29)$$

The value of p is the circular frequency defined by Eq. 22–2.

There are three cases of λ_1 and λ_2 which must be considered in the total solution. When $c > c_c$, the roots λ_1 and λ_2 are both real. The general solution of Eq. 22–26 may then be written as

$$x = Ae^{\lambda_1 t} + Be^{\lambda_2 t} \qquad (22\text{–}30)$$

Motion corresponding to this solution is nonvibrating. The effect of damping is so strong that when the block is displaced and released, it simply creeps back to its original position without oscillating. The system is said to be *overdamped*.

If $c = c_c$, then $\lambda_1 = \lambda_2 = -c_c/2m = -p$. This condition is known as *critical damping*. It represents the minimum condition of an overdamped system. Using the methods of differential equations, we may show that the solution to Eq. 22–26 for critical damping is

$$x = (A + Bt)e^{-pt} \qquad (22\text{–}31)$$

Finally, the third and most common case occurs when the system is *underdamped*. This occurs when $c < c_c$. In this case the roots λ_1 and λ_2 are complex, and it may be shown that the general solution of Eq. 22–26 can be written as

$$x = D[e^{-(c/2m)t} \sin{(p_d t + \phi)}] \qquad (22\text{–}32)$$

where D and ϕ are constants generally determined from the initial conditions of the problem. The constant p_d is called the *damped natural frequency* of the system. It has a value of

$$p_d = \sqrt{\frac{k}{m} - \left(\frac{c}{2m}\right)^2} = p\sqrt{1 - \left(\frac{c}{c_c}\right)^2} \qquad (22\text{–}33)$$

The ratio c/c_c is called the *damping factor*.

The graph of Eq. 22–32 is shown in Fig. 22–15. The initial limit of

$$De^{-(\frac{c}{2m})t}$$

$$x = D[e^{-(\frac{c}{2m})t} \sin (p_d t + \phi)]$$

Fig. 22–15

motion, D, diminishes with each cycle of vibration since it is confined within the bounds of the exponential curve. Using the damped natural frequency, we may write the period of damped vibration as

$$\tau = \frac{2\pi}{p_d} \tag{22–34}$$

*22–6. Viscous Damped Forced Vibrations

The most general case of single-degree-of-freedom vibrating motion occurs when the system includes both the effects of forced motion and induced damping. The analysis of this particular type of vibration is of practical value when applied to systems having significant damping characteristics.

If a dashpot is attached to the model shown in Fig. 22–10a, the differential equation which describes the motion becomes

$$m\frac{d^2x}{dt^2} + c\frac{dx}{dt} + kx = F_0 \sin \omega t \tag{22–35}$$

A similar equation may be written for a model having a forced displacement, Fig. 22–10c, which includes the effects of damping. In which

case, F_0 is replaced by $k\delta_0$. Since Eq. 22-35 is nonhomogeneous, the total solution is the sum of a complementary solution and a particular solution. The complementary solution is determined by setting the right side of Eq. 22-35 equal to zero and solving the homogeneous equation, which is equivalent to Eq. 22-26. The solution is therefore given by Eq. 22-30, Eq. 22-31, or Eq. 22-32, depending upon the values of λ_1 and λ_2. Because of friction, however, this solution will dampen out with time. Only the particular solution will remain. The *particular solution* describes the *steady-state vibration* of the system. Since the applied forcing function is harmonic, the steady-state motion will also be harmonic, and we will therefore assume a particular solution of the form

$$x_p = B \sin(\omega t - \phi) \tag{22-36}$$

Substituting this equation into Eq. 22-35, we may show that the two constants B and ϕ become

$$B = \frac{F_0/k}{\sqrt{\left[1 - \left(\frac{\omega}{p}\right)^2\right]^2 + \left(2\frac{c}{c_c}\frac{\omega}{p}\right)^2}} \tag{22-37}$$

and

$$\phi = \tan^{-1}\left(\frac{c\omega/k}{1 - \left(\frac{\omega}{p}\right)^2}\right) \tag{22-38}$$

The angle ϕ represents the phase difference between the applied force or support displacement and the resulting steady-state vibration of the system.

The *magnification factor* MF has been defined in Sec. 22-4 as the ratio of the amplitude of the forced vibration to the deflection caused by the force F_0 applied statically. From Eq. 22-36, the forced vibration has an amplitude of B; thus,

$$\text{MF} = \frac{B}{F_0/k} = \frac{B}{\delta_0} = \frac{1}{\sqrt{\left[1 - \left(\frac{\omega}{p}\right)^2\right]^2 + \left(2\frac{c}{c_c}\frac{\omega}{p}\right)^2}} \tag{22-39}$$

The MF is plotted in Fig. 22-16 versus the frequency ratio ω/p for various values of the damping factor c/c_c. It can be seen from this graph that the magnification of the amplitude increases as the damping factor decreases. Resonance obviously can occur only when the damping is zero, and the frequency ratio equals one.

M.F.

$\frac{c}{c_c} = 0$

$\frac{c}{c_c} = 0.10$

$\frac{c}{c_c} = 0.25$

$\frac{c}{c_c} = 0.50$

$\frac{c}{c_c} = 1.00$

$\left(\frac{\omega}{p}\right)$

Fig. 22–16

*22–7. Electrical Circuit Analogues

The characteristics of a vibrating system may be represented by an electric series circuit. Consider the circuit shown in Fig. 22–17a, which consists of an inductor L, a resistor R, and a capacitor C. When a voltage $E(t)$ is applied to the circuit, it causes a current of magnitude i to flow through the circuit. As the current flows past the inductor, the voltage drop is $L(di/dt)$, when it flows across the resistor the drop is Ri, and when it arrives at the capacitor the drop is $(1/C)\int i\,dt$. Since current cannot flow past a capacitor, it is only possible to measure the charge q acting on the

(a)　　　　　　　　　　　(b)

Fig. 22–17

Table 22–1 Electrical–Mechanical Analogues

Electrical		Mechanical	
Electric charge	q	Displacement	x
Electric current	i	Velocity	dx/dt
Voltage	$E(t)$	Applied force	$F(t)$
Inductance	L	Mass	m
Resistance	R	Viscous damping coefficient	c
Reciprocal of capacitance	$1/C$	Spring stiffness	k

capacitor. The charge may be related to the current by the equation $i = dq/dt$. Thus, the voltage drops which occur across the inductor, resistor, and capacitor may be written as $L\,d^2q/dt^2$, $R\,dq/dt$, and q/C, respectively. According to Kirchhoff's voltage law, the applied voltage balances the sum of the voltage drops around the circuit. Therefore,

$$L\frac{d^2q}{dt^2} + R\frac{dq}{dt} + \frac{1}{C}q = E(t) \qquad (22\text{--}40)$$

Consider now the model of a single-degree-of-freedom system, Fig. 22–17b, which is subjected to both a general forcing function $F(t)$ and damping. The equation of motion for this system was established in Sec. 22–6 and can be written as

$$m\frac{d^2x}{dt^2} + c\frac{dx}{dt} + kx = F(t) \qquad (22\text{--}41)$$

By comparison, we see that Eqs. 22–40 and 22–41 have the same form. Hence, mathematically, the problem of analyzing an electric circuit is the same as that of analyzing a vibrating mechanical system. The analogues between the two equations are given in Table 22–1.

This analogy has important application to experimental work, for it is much easier to simulate the vibration of a complex vibrating system using an electric circuit rather than to make an equivalent mechanical spring and dashpot model. Once we have determined the proper electrical circuits, we can study the motions of a vibrating system using an analogue computer.

Example 22–8

The 80-lb electric motor shown in Fig. 22–18 is supported by four springs, each spring having a stiffness of 5 lb/ft. If the rotor R is unbalanced such that its effect is equivalent to a 2-lb weight located 6 in. from the axis of rotation, determine the amplitude of vibration when the rotor is turning at 10 rad/sec. The damping factor is $c/c_c = 0.15$.

Fig. 22–18

Solution

The force which causes the motor to vibrate is the centrifugal force due to the unbalanced effect of the rotor. This force has a magnitude of

$$F_O = ma_n = mr\omega^2 = \frac{2 \text{ lb}}{32.2 \text{ ft/sec}^2}\left(\frac{6 \text{ in.}}{12 \text{ in./ft}}\right)(10 \text{ rad/sec})^2 = 3.11 \text{ lb}$$

The force which causes oscillation of the motor in the vertical direction may be expressed in the periodic form $F = F_O \sin \omega t$ where $\omega = 10$ rad/sec. Thus,

$$F = 3.11 \sin 10t$$

The stiffness of the entire system of four springs is $k = 4(5 \text{ lb/ft}) = 20$ lb/ft. Therefore, the natural frequency of vibration is

$$p = \sqrt{\frac{k}{m}} = \sqrt{\frac{20}{\dfrac{80}{32.2}}} = 2.84 \text{ Hz}$$

Since the damping factor is known, we may determine the steady-state amplitude by using Eq. 22–37:

$$B = \frac{F_O/k}{\sqrt{\left[1 - \left(\dfrac{\omega}{p}\right)^2\right]^2 + \left(2\dfrac{c}{c_c}\dfrac{\omega}{p}\right)^2}}$$

$$= \frac{3.11/20}{\sqrt{\left[1 - \left(\dfrac{10}{2.84}\right)^2\right]^2 + \left[2(0.15)\dfrac{10}{2.84}\right]^2}}$$

$$= 0.0136 \text{ ft} = 0.163 \text{ in.} \qquad\qquad Ans.$$

Problems

22–55. The 10-lb circular disk is attached to three springs and immersed in a fluid. Each spring has a stiffness of $k = 1$ lb/in. If the disk has a downward velocity of 10 ft/sec at the equilibrium position, determine the equation which describes the motion. Assume that fluid resistance acting on the disk furnishes a retarding force having a magnitude of $F = 0.8|v|$, where F is given in pounds and v is measured in ft/sec.

120°

Prob. 22–55

22-56. The 10-lb block is immersed in a thick liquid such that the damping force acting on the block has a magnitude of $F = |v|/2$, where F is given in pounds and v is measured in ft/sec. If the block is pulled down 3 in. and released, describe the motion of the block.

$k = 4$ lb/in.

10 lb

Prob. 22–56

22-57. A vertical spring has a stiffness of $k = 2$ lb/in. If a 5-lb weight is attached to the spring and is started from the equilibrium position with an upward velocity of 2 ft/sec, determine the position of the weight as a function of time. Assume that motion takes place in a medium which furnishes a retarding force F (pounds) having a magnitude numerically equal to four times the speed v of the weight, where v is measured in ft/sec.

22-58. The 20-lb block is subjected to the action of the harmonic force $F = 8 \cos 2t$, where F is in pounds and t is in seconds. Determine the steady-state motion of the block.

$k = 5$ lb/in.

$F = 8 \cos 2t$

$W = 20$ lb

$k = 5$ lb/in.

$c = 10$ lb–sec/ft

Prob. 22–58

22-59. A body having a mass of 0.5 slug is suspended from a spring having a stiffness of 0.6 lb/in. If a dashpot provides a damping force of 0.2 lb when the speed of the mass is 1 ft/sec, determine the period of free vibration.

22-60. The barrel of a cannon weighs 1,200 lb and returns to its initial position by means of a recuperator. The recoil of the cannon is 3 ft. Determine the required stiffness of each of the two springs mounted to the barrel and fixed to the base of the cannon so that the barrel recuperates without vibration. The damping coefficient is 2,000 lb-sec/ft.

22-61. Determine the differential equation of motion for the damped vibratory system shown. What type of motion occurs?

$k = 10$ lb/ft

100 lb

$c = 2$ lb–sec/ft

Prob. 22–61

22-62. A small 5-lb weight is suspended from a spring having a stiffness of 6 lb/in. The support at which the spring is attached is given a simple harmonic motion which may be expressed by $\delta = 1.5 \sin 2t$, where δ is in inches and t is in seconds. If the damping coefficient is 0.08 lb-sec/in., determine the phase angle ϕ of forced vibration.

22-63. Determine the magnification ratio of the spring and dashpot combination in Prob. 22–62.

22-64. The 200-lb electric motor is fastened to the midpoint of a simply supported light beam as shown. It is found that the beam deflects 2 in. when the motor is not running. The rotor turns an eccentric flywheel which is equivalent to an unbalanced weight of 1 lb located at 5 in. from the axis of rotation. If the rotor is turning at 100 rpm, determine the amplitude of steady-state vibration. The damping factor is $c/c_c = 0.20$.

Prob. 22–64

22-65. Determine the minimum amount of weight which can be fastened to the motor in Prob. 22-64 to produce a steady-state amplitude of 0.0362 in.

22-66. The bell-crank mechanism consists of a bent rod, having a negligible weight, to which is attached a 5-lb weight. Determine the critical damping coefficient c_c and the damping natural frequency for small vibrations about the equilibrium position.

Prob. 22–66

22-67. Draw the electric circuit which is analogous to the mechanical system shown. Determine the differential equation which describes the motion of the current in the loop.

Prob. 22–67

22-68. Determine the electrical circuit which is equivalent to the mechanical system shown.

Prob. 22–68

22-69. Determine the electrical circuit which is equivalent to the mechanical system shown. What is the differential equation which describes the motion of the current in the circuit?

$$F = F_O \cos \omega t$$

Prob. 22-69

22-70. Determine the mechanical analogue for the electrical circuit. What are the differential equations describing the mechanical and electrical systems?

$$E = E_O \sin \omega t$$

Prob. 22-70

Computer Program for the Simultaneous Solution of Linear Algebraic Equations

The application of the equations of static equilibrium or the equations of dynamic motion to the solution of problems involving three-dimensional force systems generally results in the necessity of solving a set of linear algebraic equations. The computer program listed in Fig. A–1 may be used to solve a series of n linear, nonhomogeneous, algebraic equations for n unknowns ($n \leq 21$). These equations are solved numerically using a Gauss elimination procedure. (Refer to a textbook on numerical analysis.)

The program is written in "Fortran IV" computer language, which is commonly used in most computer systems today. In using this program one must supply the *job control language cards (JCL)*, the *program statement cards,* and the *input data cards*. Personnel at the computer center should be able to lend assistance both in obtaining the proper *JCL* cards and in arranging these cards in the proper order in the program. *In general, the cards are arranged in the following manner: JCL cards, program statement cards, JCL cards, input data cards, JCL cards*. The *program statement cards* are composed of the program statements listed in Fig. A–1. Each of these statements must be punched on a separate computer card in the proper card columns, as indicated in the figure. The input data is recorded on cards which are contained at the end of the program.

1	2	3	4	5	6	7	Card column

```
C         PROGRAM FOR SOLVING SIMULTANEOUS EQUATIONS
          DIMENSION A(21,21),B(21),X(21),C(21,3)
          READ(5,1)RCH
        1 FORMAT(F10.4)
          N=RCH
          READ(5,2)((A(I,J),J=1,N),I=1,N)
        2 FORMAT(8F10.4)
          READ(5,2)(B(I),I=1,N)
          WRITE(6,3)
        3 FORMAT(35H1SOLUTION OF SIMULTANEOUS EQUATIONS//)
          WRITE(6,4)
        4 FORMAT(//13H THE A MATRIX/)
          DO 5 I=1,N
        5 WRITE(6,6)(A(I,J),J=1,N)
        6 FORMAT(1X,1PE16.7,6(2X,1PE16.7))
          WRITE(6,8)
        7 FORMAT(4X,2HX(,I2,2H)=,1PE16.7/)
        8 FORMAT(//13H THE B MATRIX/)
          DO 9 I=1,N
        9 WRITE(6,6)B(I)
          DO 10 I=1,N
       10 C(I,1)=0.0
          II=0
       11 AMAX=-1.0
          DO 16 I=1,N
          IF(C(I,1))16,12,16
       12 DO 15 J=1,N
          IF(C(J,1))15,13,15
       13 T=ABS(A(I,J))
          IF(T-AMAX)15,15,14
       14 IR=I
          IC=J
          AMAX=T
       15 CONTINUE
       16 CONTINUE
          IF(AMAX)27,32,17
       17 C(IC,1)=IR
          IF(IR-IC)18,20,18
       18 DO 19 J=1,N
          T=A(IR,J)
          A(IR,J)=A(IC,J)
       19 A(IC,J)=T
          II=II+1
          C(II,2)=IC
       20 P=A(IC,IC)
          A(IC,IC)=1.0
          P=1.0/P
          DO 21 J=1,N
       21 A(IC,J)=A(IC,J)*P
          DO 24 I=1,N
          IF(I-IC)22,24,22
       22 T=A(I,IC)
          A(I,IC)=0.0
          DO 23 J=1,N
       23 A(I,J)=A(I,J)-A(IC,J)*T
```

```
24 CONTINUE
   GO TO 11
25 IC=C(II,2)
   IR=C(IC,1)
   DO 26 I=1,N
   T=A(I,IR)
   A(I,IR)=A(I,IC)
26 A(I,IC)=T
   II=II-1
27 IF(II)25,28,25
28 DO 29 I=1,N
   X(I)=0.0
   DO 29 K=1,N
29 X(I)=X(I)+A(I,K)*B(K)
   WRITE(6,30)
30 FORMAT(///10H SOLUTIONS//)
   DO 31 I=1,N
31 WRITE(6,7)I,X(I)
   GO TO 34
32 WRITE(6,33)
33 FORMAT(///27H EQUATIONS CANNOT BE SOLVED)
34 STOP
   END
```

Fig. A-1

Each numerical element of data is punched anywhere within a field of 10 spaces (card columns) on the cards. Since a standard computer card contains 80 spaces, one can put at most *eight elements of input data on a card*. Each element *must* contain a decimal point, and these elements must be rounded off to, at most, four numeric characters after the decimal point.

Consider, for example, solving the following set of four linear algebraic equations for the unknowns x_1, x_2, x_3, and x_4.

$$A_{11}x_1 + A_{12}x_2 + A_{13}x_3 + A_{14}x_4 = B_1$$
$$A_{21}x_1 + A_{22}x_2 + A_{23}x_3 + A_{24}x_4 = B_2$$
$$A_{31}x_1 + A_{32}x_2 + A_{33}x_3 + A_{34}x_4 = B_3$$
$$A_{41}x_1 + A_{42}x_2 + A_{43}x_3 + A_{44}x_4 = B_4$$

The coefficients of these equations may be arranged in the following matrix form

$$[A][x] = [B]$$

or

$$\begin{bmatrix} A_{11} & A_{12} & A_{13} & A_{14} \\ A_{21} & A_{22} & A_{23} & A_{24} \\ A_{31} & A_{32} & A_{33} & A_{34} \\ A_{41} & A_{42} & A_{43} & A_{44} \end{bmatrix} \begin{bmatrix} x_1 \\ x_2 \\ x_3 \\ x_4 \end{bmatrix} = \begin{bmatrix} B_1 \\ B_2 \\ B_3 \\ B_4 \end{bmatrix}$$

To use the computer program for the solution of these equations, the input data (cards) must consist of the following:

1. Card 1 always contains the number of equations. In this case there are four equations, hence, the number 4.0 is punched anywhere within spaces 1 through 10 on the first card.
2. Cards 2 and 3 contain the A-matrix coefficients. (In all there are 16 coefficients.) Since there are 80 columns per card, and each coefficient is to be punched anywhere within a field of 10 spaces, the eight elements $A_{11}, A_{12}, A_{13}, A_{14}, A_{21}, A_{22}, A_{23}$, and A_{24} are punched *in this order* on the second card. The third card contains the remaining eight elements $A_{31}, A_{32}, A_{33}, A_{34}, A_{41}, A_{42}, A_{43}$, and A_{44}. (If there were more than 16 elements in the A-matrix, one would continue listing the data on cards 4, 5, etc.)
3. Card 4 contains the B-matrix coefficients. Each of these four elements is punched anywhere within a 10-space field on the card. They are entered in the following order: B_1, B_2, B_3, B_4.

When the program is received back from the computer center, the input data along with the answers (*output*) will be printed on paper. The answers are printed using "E-field" notation. This notation is similar to scientific notation. For example,

$$3.02 \text{ E } 02 = 3.02 \times 10^2 = 302.0$$
$$3.02 \text{ E } 00 = 3.02 \times 10^0 = 3.02$$
$$3.02 \text{ E} - 02 = 3.02 \times 10^{-2} = 0.0302$$

The following example illustrates the use of the program.

Example A–1

Solve the three equations

$$-x_1 - 2x_2 + 3x_3 = 2$$
$$2x_1 + x_2 + x_3 = -3$$
$$x_1 + 3x_2 + 4x_3 = 1$$

Solution:

Arranging the coefficients of three equations in the matrix form $[A][x] = [B]$ yields

$$\begin{bmatrix} -1 & -2 & 3 \\ 2 & 1 & 1 \\ 1 & 3 & 4 \end{bmatrix} \begin{bmatrix} x_1 \\ x_2 \\ x_3 \end{bmatrix} = \begin{bmatrix} 2 \\ -3 \\ 1 \end{bmatrix}$$

The input data is punched on four cards, as shown in Fig. A-2a. The first card contains the number of equations to be solved. The second and third cards contain the coefficients of the A-matrix, and the fourth card contains the B-matrix coefficients.

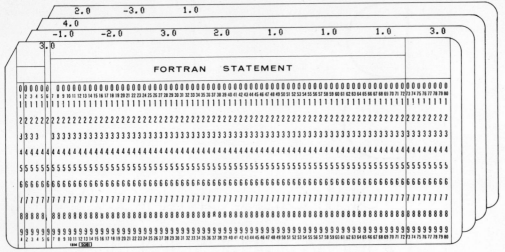

Fig. A-2a

SOLUTION OF SIMULTANEOUS EQUATIONS

THE A MATRIX

$$-1.0000000E\ 00 \qquad -2.0000000E\ 00 \qquad 3.0000000E\ 00$$
$$2.0000000E\ 00 \qquad 1.0000000E\ 00 \qquad 1.0000000E\ 00$$
$$1.0000000E\ 00 \qquad 3.0000000E\ 00 \qquad 4.0000000E\ 00$$

THE B MATRIX

$$2.0000000E\ 00$$
$$-3.0000000E\ 00$$
$$1.0000000E\ 00$$

SOLUTIONS

$$X(1)=\ -1.9285698E\ 00$$
$$X(2)=\ \ \ 5.0000006E-01$$
$$X(3)=\ \ \ 3.5714293E-01$$

Fig. A-2b

The printed output is shown in Fig. A-2b. Thus, the answers are

$$x_1 = -1.93 \qquad\qquad Ans.$$
$$x_2 = 0.50 \qquad\qquad Ans.$$
$$x_3 = 0.36 \qquad\qquad Ans.$$

B

Centroids of Line, Area, and Volume Elements

Circular arc segment $$\bar{x} = \frac{r \sin \alpha}{\alpha}$$	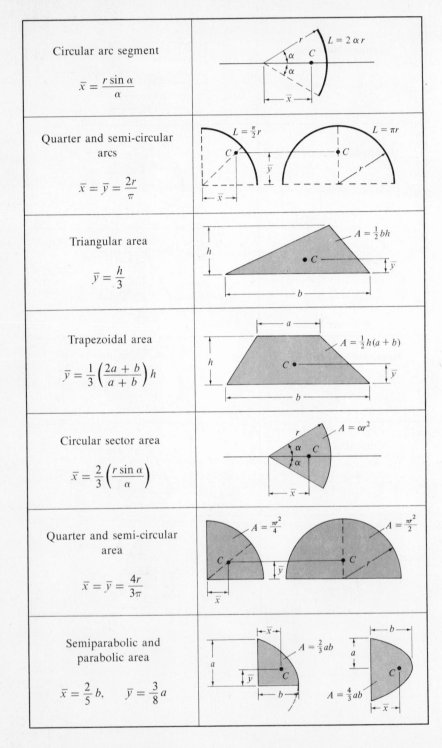
Quarter and semi-circular arcs $$\bar{x} = \bar{y} = \frac{2r}{\pi}$$	
Triangular area $$\bar{y} = \frac{h}{3}$$	
Trapezoidal area $$\bar{y} = \frac{1}{3}\left(\frac{2a+b}{a+b}\right)h$$	
Circular sector area $$\bar{x} = \frac{2}{3}\left(\frac{r \sin \alpha}{\alpha}\right)$$	
Quarter and semi-circular area $$\bar{x} = \bar{y} = \frac{4r}{3\pi}$$	
Semiparabolic and parabolic area $$\bar{x} = \frac{2}{5}b, \qquad \bar{y} = \frac{3}{8}a$$	

Semiparabolic area $\bar{x} = \dfrac{3}{4} b, \qquad \bar{y} = \dfrac{3}{10} a$	$A = \dfrac{ab}{3}$
Hemispherical surface $\bar{y} = \dfrac{r}{2}$	$A = 2\pi r^2$
Conical surface $\bar{y} = \dfrac{h}{3}$	$A = 2\pi r h$
Hemispherical volume $\bar{y} = \dfrac{3}{8} r$	$V = \dfrac{2}{3} \pi r^3$
Conical volume $\bar{y} = \dfrac{h}{4}$	$V = \dfrac{1}{3} \pi r^2 h$
Paraboloid of revolution volume $\bar{y} = \dfrac{h}{3}$	$V = \dfrac{1}{2} \pi r^2 h$

Table of Conversion Factors (FPS)→(MKGFS)

Fundamental Quantities:

Length 1 foot (ft) = 12 inches (in.) = 0.3048 meter (m) = 30.48 centi-meters (cm). The centimeter scale shown here approximates its actual size, and represents a comparison of length with the inch scale.

Force 1 pound (lb) = 0.4536 kilogram-force (kg_f). A weight equivalent to 1 kg_f is shown here in comparison to 1 lb of the same material.

Time The second (sec) is the standard unit of time in both systems of units.

Derived Quantity	Unit of Measurement (FPS)	To Convert (FPS) → (MKGFS) Multiply by	Unit of Measurement (MKGFS)
Moment of a Force Couple Torque	lb-ft	0.1383	kg_f-m
Linear load intensity Linear density	lb/ft	1.4882	kg_f/m
Surface load intensity Surface density Pressure (stress)	lb/ft^2	4.8825	kg_f/m^2
Volume load intensity Density	lb/ft^3	16.0187	kg_f/m^3
Area Moment of inertia	ft^4	0.008631	m^4
Mass Moment of inertia	ft-lb-sec^2	0.1383	m-kg_f-sec^2
Velocity	ft/sec	0.3048	m/sec
Acceleration	ft/sec^2	0.3048	m/sec^2
Work (energy)	ft-lb	0.1383	m-kg_f
Power	ft-lb/sec	0.1383	m-kg_f/sec
Impulse Momentum	lb-sec	0.4536	kg_f-sec
Frequency	Hz	1.0000	Hz
Angular velocity	rad/sec	1.0000	rad/sec
Angular acceleration	rad/sec^2	1.0000	rad/sec^2
Angular impulse Angular momentum	lb-ft-sec	0.1383	kg_f-m-sec

Mass Moments and Products of Inertia of Homogeneous Solids

Sphere

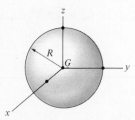

$$I_{xx} = I_{yy} = I_{zz} = \frac{2}{5}mR^2$$

$$I_{xy} = I_{xz} = I_{yz} = 0$$

$$V = \frac{4}{3}\pi R^3$$

Hemisphere

$$I_{xx} = I_{yy} = 0.259mR^2$$

$$I_{zz} = \frac{2}{5}mR^2$$

$$I_{xy} = I_{xz} = I_{yz} = 0$$

Cylinder

$$I_{xx} = I_{yy} = \frac{1}{12}m(3R^2 + h^2)$$

$$I_{zz} = \frac{1}{2}mR^2$$

$$I_{xy} = I_{xz} = I_{yz} = 0$$

$$V = \pi R^2 h$$

Semicylinder

$$I_{xx} = 0.0699mR^2 + \frac{m}{12}h^2$$

$$I_{yy} = 0.320mR^2$$

$$I_{zz} = \frac{1}{12}m(3R^2 + h^2)$$

$$I_{xy} = I_{xz} = I_{yz} = 0$$

Slender rod

$$I_{xx} = I_{yy} = \frac{1}{12}ml^2$$

$$I_{zz} = 0$$

$$I_{xy} = I_{xz} = I_{yz} = 0$$

Circular plate

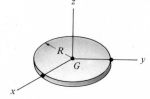

$$I_{xx} = I_{yy} = \frac{1}{4}mR^2$$

$$I_{zz} = \frac{1}{2}mR^2$$

$$I_{xy} = I_{xz} = I_{yz} = 0$$

Thin ring

$$I_{xx} = I_{yy} = \frac{1}{2}mR^2$$

$$I_{zz} = mR^2$$

$$I_{xy} = I_{yz} = I_{yz} = 0$$

Cone

$$I_{xx} = I_{yy} = \frac{3}{80}m(4R^2 + h^2)$$

$$I_{zz} = \frac{3}{10}mR^2$$

$$I_{xy} = I_{xz} = I_{yz} = 0$$

$$V = \frac{1}{3}\pi R^2 h$$

Rectangular block

$$I_{xx} = \frac{m}{12}(b^2 + h^2)$$

$$I_{yy} = \frac{m}{12}(a^2 + h^2)$$

$$I_{zz} = \frac{m}{12}(a^2 + b^2)$$

$$I_{xy} = I_{xz} = I_{yz} = 0$$
$$V = abh$$

Rectangular plate

$$I_{xx} = \frac{m}{12}b^2$$

$$I_{yy} = \frac{m}{12}a^2$$

$$I_{zz} = \frac{m}{12}(a^2 + b^2)$$

$$I_{xy} = I_{xz} = I_{yz} = 0$$

E

Infinitesimal Rotations

As stated in Chapter 13, finite rotations are not vectors, since they do not obey the communitive law of vector addition. When the rotations are infinitesimally small, however, they can be added in any manner, and therefore, infinitesimal rotations are vectors.

Fig. E-1

To prove this, consider point P, shown in Fig. E–1, located by position vector $\mathbf{r}$ from the *fixed point* O. We will consider vector $\mathbf{r}$ to be fixed in magnitude but not in direction; hence, physically, O and P may represent two points in a rigid body. A small change in $\mathbf{r}$ occurs because of a rotation $d\boldsymbol{\theta}$, therefore, we may write this change as

$$d\mathbf{r} = d\boldsymbol{\theta} \times \mathbf{r}$$

Using vector addition, we find that the final *position* of P is therefore

$$\mathbf{r}' = \mathbf{r} + d\mathbf{r} = \mathbf{r} + d\boldsymbol{\theta} \times \mathbf{r} \tag{E-1}$$

Two succesive infinitesimal rotations, $d\boldsymbol{\theta}_1$, then $d\boldsymbol{\theta}_2$, will be considered. These yield, respectively, the first position $\mathbf{r}_1$ and final position $\mathbf{r}_2$ of point P. Using Eq. D–1, we have

$$\mathbf{r}_1 = \mathbf{r} + d\mathbf{r}_1 = \mathbf{r} + d\boldsymbol{\theta}_1 \times \mathbf{r}$$

Then, since $d\mathbf{r}_2 = d\boldsymbol{\theta}_2 \times \mathbf{r}_1$, we obtain

$$\mathbf{r}_2 = \mathbf{r}_1 + d\mathbf{r}_2 = (\mathbf{r} + d\boldsymbol{\theta}_1 \times \mathbf{r}) + d\boldsymbol{\theta}_2 \times (\mathbf{r} + d\boldsymbol{\theta}_1 \times \mathbf{r})$$

or

$$\mathbf{r}_2 = \mathbf{r} + (d\boldsymbol{\theta}_1 + d\boldsymbol{\theta}_2) \times \mathbf{r} + d\boldsymbol{\theta}_2 \times (d\boldsymbol{\theta}_1 \times \mathbf{r})$$

The last term in this equation may be neglected since it represents a differential quantity of the second order (a *product* of two differentials). Therefore,

$$\mathbf{r}_2 = \mathbf{r} + (d\boldsymbol{\theta}_1 + d\boldsymbol{\theta}_2) \times \mathbf{r} \tag{E-2}$$

Comparing Eq. E–2 with Eq. E–1, we find that the total *displacement* of point P is

$$dr_1 + dr_2 = (d\theta_1 + d\theta_2) \times r \qquad (E-3)$$

Had the two successive rotations occurred in the order $d\theta_2$, then $d\theta_1$, the resulting *positions* would have been

$$r_1' = r + dr_1' = r + d\theta_2 \times r$$

Then,

$$r_2' = r_1' + dr_2' = (r + d\theta_2 \times r) + d\theta_1 \times (r + d\theta_2 \times r)$$

or

$$r_2' = r + (d\theta_2 + d\theta_1) \times r + d\theta_1 \times (d\theta_2 \times r)$$

Neglecting the last term in this equation, since it is a second-order differential quantity, we have

$$r_2' = r + (d\theta_2 + d\theta_1) \times r$$

Therefore, comparing with Eq. E–1, we see that the total displacement of point P is

$$dr_1' + dr_2' = (d\theta_2 + d\theta_1) \times r \qquad (E-4)$$

Since the vector cross product obeys the distributive law, comparison of Eqs. E–3 and E–4 proves that the resulting displacement $dr_1 + dr_2 = dr_1' + dr_2'$ is the *same* regardless of the order of the applied rotations $d\theta_1$ and $d\theta_2$. We may therefore conclude that infinitesimal rotations $d\theta$ are vectors since these quantities have both a magnitude and direction for which the order of (vector) addition is not important.

Answers to Even-Numbered Problems

12-2. 4.5 ft/sec².

12-4. $s = 110$ ft to the right of the origin, $v = 155$ ft/sec, $a = 148$ ft/sec².

12-6. 28.7 ft to the right of the origin.

12-8. 14.0 ft/sec.

12-10. 2.63 ft/sec.

12-12. 8.80 sec, 387 ft.

12-14. 32 ft/sec, 224 ft.

12-16. 34.1 sec.

12-18. Graph is the line $a = 2t$; $a = 14$ ft/sec².

12-20. $h = 218.4$ ft, $v = 5$ at $t = 0$, $v = -113.5$ at $t = 3.68$.

12-22. 3,333 ft, 500 ft/sec.

12-14. $|a_{max}| = 0.25$ ft/sec².

12-26. $t = 7.12$ sec
$v = 293.3$ at $t = 0$,
$v = 827.2$ at $t = 1.78$
$v = 293.3$ at $t = 7.12$.

12-28. $\{16t\mathbf{i} + 12\mathbf{j}\}$ ft/sec².

12-30. (a) $24t\mathbf{i} + 9t^2\mathbf{j} + 4t\mathbf{k}$.
(b) $32\mathbf{i} + 16\mathbf{j} + 5.33\mathbf{k}$.

12-32. (a) $16t^3\mathbf{i} + 18t^2\mathbf{j} + 10t^4\mathbf{k}$.
(b) $7.5\mathbf{i} + 7.0\mathbf{j} + 6.2\mathbf{k}$.

12-34. 101.1 ft/sec.

12-36. 31.4 ft, 90°

12-38. 497 ft.

12-40. 38.5 ft.

12-42. 18.3°, 66.1°.

12-44. 73.5 ft.

12-46. 0.62 sec $\leqslant t \leqslant 2.62$ sec.

12-48. 12.33 in./sec. 6 in./sec².

12-50. 6.20 ft/sec.

12-52. 33.5 ft/sec, 275 ft/sec².

12-54. 14.32 in./sec².

12-56. $\dfrac{d\mathbf{a}}{dt} = \left[\dfrac{d^3r}{dt^3} - 3\dfrac{dr}{dt}\left(\dfrac{d\theta}{dt}\right)^2\right.$
$\left. - 3r\dfrac{d^2\theta}{dt^2}\dfrac{d\theta}{dt}\right]\mathbf{u}_r + \left[r\dfrac{d^3\theta}{dt^3}\right.$
$+ 3\dfrac{d^2r}{dt^2}\dfrac{d\theta}{dt} + 3\dfrac{dr}{dt}\dfrac{d^2\theta}{dt^2}$
$\left. -r\left(\dfrac{d\theta}{dt}\right)^3\right]\mathbf{u}_\theta + \dfrac{d^3\theta}{dt^3}\mathbf{u}_z.$

12-58. 0.043 rad/sec.

12-60. 1.280 ft, 8.86 ft/sec².

12-62. 30,451 ft.

12-64. 70.7 ft/sec².

12-66. (0, −2 ft), 250 ft/sec².

12-68. 6.0 in./sec, 4.5 in./sec².

12-70. 6 ft/sec.

12-72. 97.2°.

12-74. 4.82 ft/sec, 1.80 ft.

12-76. $a_B = 9.11$ ft/sec²,
$v_{B/C} = 16.04$ ft/sec $\measuredangle^{51.3°}$
$a_{B/C} = 9.11$ ft/sec² $\measuredangle^{59.0°}$.

13-2. 1 rad/sec, 1 rad/sec².

13-4. 14.25 ft/sec, −0.25 ft/sec².

13-6. 1.8 in./sec ↑, 0.

13-8. 3 ft/sec
$\mathbf{a} = \{(-9 \sin 3t)\, \mathbf{u}_t + 4.5 \cos^2 3t)\, \mathbf{u}_n\}$ ft/sec².

13-10. $\omega_B = 2.4$ rad/sec ↰, $\omega_C = 1.6$ rad/sec. ↱

13-12. 10 rad/sec², 35.4 rad/sec, 35.3 rev.

13-14. 0.5 rad/sec ↰, 1.5 ft/sec ↓.

13-16. $v_B = 3.46$ ft/sec $\measuredangle^{30°}$, $v_C = 4$ ft/sec. →

13-18. 1.33 ft/sec. →

13-20. 15 in./sec $^{36.9°}\measuredangle$, $\omega_{CB} = 0$.

13-22. 4 ft/sec. ←

13-24. (a) $v = -e(\sin \phi)\omega - \dfrac{e^2(\sin 2\phi)\omega}{\sqrt{r^2 - e^2 \sin^2 \phi}}$.

  (b) $v = -e(\sin \phi)\omega - \dfrac{e^2(\sin 2\phi)\omega}{2\sqrt{(r_1 + r_2)^2 - e^2 \sin^2 \phi}}$.

  (c) $v = -e(\sin \phi)\omega$.

13-26. $\omega = 0$.

13-28. 368 ft/sec. $^{48.5°}\measuredangle$

13-30. 8 ft/sec. →

13-32. 6.71 ft/sec. $\measuredangle^{63.5°}$

13-34. 1.5 ft/sec ↓, 0.5 rad/sec. ↰

13-36. $\omega_{BC} = 1.50$ rad/sec ↰, $\omega_{CD} = 1.697$ rad/sec. ↰

13-38. 32.0 rad/sec. ↰

13-40. $v_A = 2.83$ ft/sec $\measuredangle^{45°}$, $v_B = 7.33$ ft/sec. →

13-42. 1 rad/sec ↰, 22 in./sec. $^{30°}\measuredangle$

13-44. $a_A = 480$ ft/sec² ↑, $a_B = 490.1$ ft/sec². $\measuredangle^{1.17°}$

13-46. $a_B = 960$ in./sec² ↑, $a_D = 5{,}120$ in./sec² ↑.

13-48. 7.07 rad/sec² ↴, 32.0 ft/sec². ↓

13-50. 0.231 rad/sec². ↱

13-52. 42.5 in./sec². ←

13-54. 113.8 in./sec². $\measuredangle^{71.6°}$

13-56. $\alpha_{AB} = 8.20$ rad/sec², ↴
$\alpha_{BC} = 4.78$ rad/sec². ↱

13-58. $a_A = 0.5$ ft/sec² ←,
$a_B = 63.6$ ft/sec². $\measuredangle^{2.25}$

13-60. $a_A = 108$ in./sec² ←, $a_B = 72$ in./sec². ←

13-62. 12 rad/sec, 12 ft/sec.

13-66. 36.4 rad/sec, 14.56 rad/sec².

13-68. (a) $\boldsymbol{\alpha} = \omega_s \omega_t \mathbf{j}$.
  (b) $\boldsymbol{\alpha} = -\omega_s \omega_t \mathbf{k}$.

13-70. $\mathbf{v}_A = \{20\mathbf{i} + 76.4\mathbf{j} - 43.6\mathbf{k}\}$ ft/sec,
$\mathbf{a}_A = \{70.9\mathbf{i} + 28\mathbf{j} - 8\mathbf{k}\}$ ft/sec².

13-72. $v_B = \{3\mathbf{i}\}$ ft/sec; no, the direction of ω must be specified.

13-74. $\omega_T = \sqrt{5}\omega_z$, $\alpha = 2\omega_z^2$.

13-76. $\mathbf{v}_A = \{0.667\mathbf{i}\}$ ft/sec,
$\mathbf{a}_A = \{-0.148\mathbf{i}\}$ ft/sec².

13-78. $\boldsymbol{\alpha} = \{0.375\mathbf{i} + 1.28\mathbf{j} + 2.64\mathbf{k}\}$ rad/sec²
$\mathbf{a}_C = \{-72.8\mathbf{i} + 38.3\mathbf{j}\}$ in./sec²
$\mathbf{a}_B = \{-3.25\mathbf{j}\}$ in./sec².

13-80. 12 ft/sec² at points B and C.

13-82. 17.78 ft/sec, 90.5 ft/sec².

13-84. 31 ft/sec $^{45°}\measuredangle$, 161.6 ft/sec². $^{40°}\measuredangle$

13-86. 1.5 rad/sec ↰, 9.5 rad/sec². ↱

13-88. 9.80 ft/sec $\measuredangle^{35.3°}$, 15.64 ft/sec². $^{14.1°}\measuredangle$

13-90. 25.3 ft/sec $\measuredangle^{63.4°}$, 73.8 ft/sec². $^{32.5°}\measuredangle$

13-92. $\mathbf{v}_P = \{-7.07\mathbf{i} + 7.07\mathbf{j} + 4.24\mathbf{k}\}$ ft/sec
$\mathbf{a}_P = \{-35.4\mathbf{i} - 48.1\mathbf{j} + 42.4\mathbf{k}\}$ ft/sec².

13-94. $\mathbf{v}_B = \{-\mathbf{i} + 18\mathbf{j} - \mathbf{k}\}$ ft/sec,
$\mathbf{a}_B = \{-8\mathbf{j} - 24\mathbf{k}\}$ ft/sec².

13-96. $\mathbf{v}_B = \{-425.8\mathbf{i} - 435.2\mathbf{j} - 30.9\mathbf{k}\}$ ft/sec
$\mathbf{a}_B = \{82.0\mathbf{i} - 141.5\mathbf{j} + 3{,}598\mathbf{k}\}$ ft/sec².

14-2. 76.4 lb, 12.3 ft/sec².

14-4. 8.05 ft/sec².

14-6. 3.58 ft/sec² ↑, 4.84 ft/sec². ↓

14-8. 344 lb.

14-10. 1.45 ft.

14-12. 15.36 ft/sec.

14-14. $N_A = 37,600$ lb, $T_{BA} = 32,600$ lb, $T_{CB} = 16,320$ lb.

14-16. 8.05 ft/sec², 12.5 lb.

14-18. (a) $a_A = a_B = 2.76$ ft/sec².
(b) $a_A = 67.6$ ft/sec², $a_B = 5.15$ ft/sec².

14-20. 23.1 ft/sec.

14-22. 15.0 ft/sec.

14-24. $s_{AB} = 396$ ft.

14-26. 1.00 sec.

14-28. 35.7 lb.

14-30. 12.04 ft/sec.

14-32. $32.2 \sin \theta = \left(\dfrac{d^2r}{dt^2} - 4r\right),$

$10 \cos \theta - N = \dfrac{40}{32.2}\dfrac{dr}{dt},$

$r = 4.03\,(\sinh \theta - \sin \theta).$

14-36. 557 ft.

14-38. 18,290 ft/sec.

14-40. $v_{\min} = 49.6$ ft/sec, $v_{\max} = 98.9$ ft/sec.

14-42. 5.32 ft/sec.

14-44. $\theta = 12.4°$, $\phi = 43.1°$.

14-46. $s = \dfrac{eVLl}{v_0^2\,wm}.$

14-48. $N_A = 11.8$ lb, $N_B = 29.5$ lb.

14-50. 2.52×10^4 ft/sec.

14-52. 1.214×10^5 ft/sec,
$1/r = 1.445 \times 10^{-13} \cos \theta + 1.874 \times 10^{-12}.$

14-54. 3.3×10^4 ft/sec.

14-56. 6.13×10^8 mi.

14-58. (a) 5.53×10^4 ft/sec.
(b) 4.84×10^8 mi.

14-60. 2.39×10^4 ft/sec.

15-2. 926 ft.

15-4. 29.9 ft/sec.

15-6. 25.0°.

15-8. $N_A = 17.5$ lb, $N_B = 242$ lb, $\alpha = 7.22$ rad/sec².

15-10. 33.3 lb.

15-12. Car slides $v = 116.7$ ft/sec.

15-14. (a) $F_A = F_B = 39,420$ lb, $F_C = 75,160$ lb.
(b) $F_A = F_B = 33,370$ lb, $F_C = 87,300$ lb.

15-16. 16,100 ft.

15-18. $T_A = T_B = T_C = 5.29$ lb.

15-20. $D_x = 147.5$ lb, $D_y = 32.2$ lb, $B_y = 150.2$ lb.

15-22. 22.8 ft/sec², 5.0 lb.

15-24. $\alpha = 21.7$ rad/sec², $B_x = 19.41$ lb, $B_y = 16.38$ lb.

15-26. 6.19 rad/sec² ↓, 73.2 lb. 59.1°↘

15-28. 6.09 lb. ↑

15-30. 55.8 rad/sec. ↱

15-32. 5.82 rad/sec². ↱

15-34. 1.31 lb-ft. ↺

15-36. 30.0 lb, 12.1 rad/sec. ↓

15-38. $O_x = 0$, $O_y = 30$ lb.

15-40. 5.15 rad/sec². ↓

15-42. 4.60 lb.

15-44. 2.02 sec.

15-46. 7.62 ft/sec². 20°↗

15-48. 33°.

15-50. 3.73 rad/sec². ↱

15-52. 166.2 ft/sec². →

15-54. 5.59 in.

15-56. 38.0 rad/sec². ↓

15-58. 66.6 rad/sec.

15-60. 22.3 rad/sec² ↙—both wheels.

15-62. 3.78 rad/sec.

15-64. $B_x = 0.219$ lb, $B_y = 20.8$ lb.

15–66. 23.0 ft/sec². $^{69.5°}$ ⬎

15–68. 11.17 ft/sec² ↑, 5.58 rad/sec². ⤵

16–2. 38.5 ft/sec.

16–4. 4.52 ft.

16–6. 11.07 ft/sec.

16–8. 89.9 ft/sec.

16–10. 20 ft, 59.9 ft/sec.

16–12. 347 ft-lb, 3.66 ft.

16–14. 7.6 in.

16–16. 11.97 in.

16–18. 24.0 lb.

16–20. 1.62 ft, $N_B = 7.18$ lb, $N_C = 1.18$ lb, $v_C = 16.0$ ft/sec.

16–22. 97.9 ft/sec.

16–24. 12.1 ft.

16–26. 33.8 hp.

16–28. 9.24 hp.

16–30. 4.06 hp.

16–32. 29.0 ft/sec.

16–34. 0.455 hp.

16–36. 19.66 ft/sec, 15 lb.

16–38. 12.0 ft.

16–40. 2.50 ft.

16–42. 5.67 ft/sec.

16–44. 3.69×10^4 ft/sec.

16–46. (*a*) 160 lb/ft, (*b*) 640 lb/ft.

16–48. $v_A = 19.45$ ft/sec, $N_A = 2.0$ lb
$v_B = 18.96$ ft/sec, $N_B = 23.7$ lb.

16–50. 14.58 ft/sec.

16–52. 9.54 in.

16–54. 9,900 mph.

17–2. 74.9 ft-lb.

17–4. 6.99 ft-lb.

17–6. 309 ft-lb.

17–8. 1.613 ft-lb.

17–10. 10.35 ft-lb.

17–12. 5.74 ft-lb.

17–14. 10.6 ft/sec.

17–16. 0.445 rev, 20 lb.

17–18. 4.15 ft/sec. →

17–20. 3.0 ft/sec. ↑

17–22. 67.5 ft/sec. ↑

17–24. 15.0 ft-lb.

17–26. 11.90 ft/sec.

17–28. 3.90 rad/sec. ⟲

17–30. 41.1 rad/sec. ⟳

17–32. 4.07 rad/sec. ⟳

17–34. 3.0 ft/sec. ↑

17–36. 19.13 rad/sec. ⟳

17–38. 2.01 rad/sec. ⟳

17–40. 23.7 rad/sec. ⟲

17–42. 4.79 rad/sec. ⟳

17–44. 9.48 rad/sec. ⟲

17–46. 2.24 in.

17–48. 6.12 ft/sec. ⬃$^{8.13}$

18–2. 4,260 ft/sec.

18–4. 4.39 sec.

18–6. 20.1 lb.

18–8. 4.64 sec.

18–10. 0.158 sec, 1.03 ft.

18–12. 6.07 sec, 4,800 lb.

18–14. 92.5 ft/sec.

18–16. 0.223 sec.

18–18. 9.88 sec.

18–20. 42.9 ft/sec, 114.6 lb.

18–22. 7.14 lb.

18–24. 2,940 lb.

18–26. 28.8 ft/sec. →

18–28. 1.23 ft/sec ←, 15.5 lb.

18-30. 0.115 ft/sec $\leftarrow$, 0.765 ft.

18-32. 13.4 ft/sec. $\leftarrow$

18-34. 25.4 ft/sec. $\leftarrow$

18-36. *a*) 0.188 ft. *b*) 6.96 ft/sec. $^{30°}\!\nearrow$

18-38. 0.0651.

18-42. $v_A = 1$ ft/sec $\leftarrow$, $v_B = 2.5$ ft/sec $\rightarrow$, $\Delta T = 0.815$ ft-lb.

18-44. $v_A = 1.043$ ft/sec $\rightarrow$, $v_B = 0.965$ ft/sec $\rightarrow$, $v_C = 11.90$ ft/sec. $\rightarrow$

18-46. 4.30 lb-sec, 446.8 lb.

18-48. 3.0 ft/sec. $^{45°}\!\nearrow$

18-50. 0.575 ft/sec $^{71.5°}\!\searrow$, 0.192 ft-lb.

18-52. 1.906 ft, 16.95 ft/sec.

18-54. 4.40 in.

18-56. 27.3 lb.

18-58. 303 lb.

18-60. 56.5 lb, $v_{\text{net}} = 0$.

18-62. 24,000 lb.

18-64. 35.4 lb.

18-66. 1.42 lb.

18-68. 260 lb.

18-70. $F = (6t + 0.373)$ lb.

18-72. $v = 4.63\sqrt{y}$.

18-74. 2,782 ft/sec.

18-76. $v = \left[8{,}000 \ln\left(\dfrac{93.2}{93.2 - 1.55t} \right) + 4{,}400 \right]$ ft/sec.

18-78. $\mathbf{H} = \{0.559\mathbf{i} - 0.248\mathbf{j} + 1.925\mathbf{k}\}$ ft-lb-sec.

18-80. 3.0 ft/sec.

18-82. 3.27 ft, 0.633 sec.

18-84. 6.44 ft/sec, 1.681 ft.

18-86. $d = 2$ ft, $T = 1.15$ lb; $d = 1.55$ ft, $T = 1.59$ lb.

19-2. 1.66 ft-lb-sec.

19-4. 8.85 lb-sec, 18.33 ft-lb-sec.

19-6. 4.63 lb-sec. $\leftarrow$

19-8. 26.8 ft/sec.

19-10. 20.4 ft/sec.

19-12. 30 rad/sec. $\curvearrowright$

19-14. 25°

19-16. 1.035 sec.

19-18. 250 rad/sec. $\curvearrowright$

19-20. 31.2 ft/sec.

19-22. 33.3 ft/sec.

19-24. 15.13 rad/sec.

19-26. 3.23 rad/sec, 0.1673 ft-lb.

19-28. 3.49 rad/sec. $\curvearrowleft$

19-30. 0.0355 rad/sec.

19-32. 5.43 ft/sec.

19-34. 18.13 rad/sec.

19-36. 5.96 ft/sec. $\rightarrow$

19-38. 22.4°.

20-2. $\frac{1}{12}m\,(a^2 + h^2)$.

20-4. $\frac{2}{5}mb^2$.

20-6. $\frac{2}{5}ma^2$.

20-8. 790 slug-ft^2.

20-10. $I_z = \frac{1}{3}mL^2$, $I_{xz} = 0$.

20-12. $k_{xy} = 0$.

20-14. $k_x = 1.528$ ft, $k_y = 1.415$ ft.

20-16. $k_x = 1.80$ ft, $k_y = 2.35$ ft.

20-18. $m(R^2 + \frac{3}{4}a^2)$

20-20. 28.1 slug-ft^2.

20-22. 0.335 slug-ft^2.

20-24. $k_x = 3.49$ ft, $k_{xy} = 1.82$ ft.

20-26. $I_x = 324$ slug-ft^2, $I_{xy} = 0$.

20-28. $\frac{1}{6}m$.

20-30. $\begin{pmatrix} 0.542mR^2 & 0 & 0.0416mR^2 \\ 0 & 0.583mR^2 & 0 \\ 0.0416mR^2 & 0 & 0.542mR^2 \end{pmatrix}$

20-32. 0.429 slug-ft^2.

20-34. 0.1035 slug-ft^2.

20-36. $I_{xx} = 0.0292$ slug-ft^2, $I_{yz} = 0$.

20-38. 1.128 slug-ft^2.

21-2. 1.148 ft-lb.

21-4. 0.336 ft-lb.

21-6. 0.468 ft-lb.-sec.

21-8. 12.4 rad/sec.

21-10. 9.83 rad/sec.

21-12. 0.589 ft-lb.

21-14. 38.6 rad/sec.

21-16. $\omega = \{-0.363\mathbf{j} + 0.363\mathbf{k}\}$ rad/sec.

21-18. $F_A = 5.02$ lb, $F_B = 4.92$ lb.

21-20. $A_x = 0$, $A_y = -4.66$ lb, $A_z = B_z = 15.0$ lb, $B_y = 4.66$ lb.

21-22. 3.97 rad/sec.

21-24. $A_y = 66.3$ lb, $B_y = -38.7$ lb, $\alpha_x = -72.4$ rad/sec^2.

21-26. $\mathbf{T} = \{0.621\mathbf{k}\}$ lb-ft.

21-28. 14.70 lb.

21-30. $F_A = 5.90$ lb, $F_B = 6.10$ lb.

21-32. $A_x = B_x = -37.3$ lb, $A_z = 23.1$ lb, $B_z = 62.1$ lb.

21-34. 83.6°.

21-36. 21.0 lb-ft.

21-38. $\mathbf{F}_A = \{-0.0447\mathbf{j} + 1.0\mathbf{k}\}$ lb, $\mathbf{M}_A = \{0.0413\mathbf{i}\}$ lb-ft.

21-40. $\Sigma M_x = \dfrac{d}{dt}(I_x\omega_x - I_{xy}\omega_y - I_{xz}\omega_z)$

$\qquad - \Omega_z(I_y\omega_y - I_{yz}\omega_z - I_{yx}\omega_x)$

$\qquad + \Omega_y(I_z\omega_z - I_{zx}\omega_x - I_{zy}\omega_y).$

21-42. $\Sigma M_x = I_x\dfrac{d\omega_x}{dt} - I_y\,\Omega_z\omega_y + I_z\Omega_y\omega_z.$

21-46. 2.21 rad/sec.

21-48. $A_y = 69.3$ lb, $B_y = 80.7$ lb.

21-50. 4,430 rad/sec.

21-52. $\tan\theta = (I/I_z)\tan\beta$

21-54. 0.820 rad/sec.

21-56. 1.97×10^{-4} ft-lb-sec.

21-58. $M_x = 0$, $M_y = -0.408$ lb-ft
$M_z = 0.433$ lb-ft.

22-2. $x = -\frac{1}{4}\cos 19.7t.$

22-4. $x = 0.529\sin 11.34t + 0.167\cos 11.34t$
$\phi = 72.5°.$

22-6. 0.083 ft, 3.13 sec.

22-8. 10.8°.

22-12. $2\pi\sqrt{\dfrac{2r}{g}}.$

22-10. 0.910 sec.

22-14. 0.549 Hz.

22-16. 90.8 lb/rad.

22-18. 1.04 sec.

22-20. 4.98 sec.

22-22. 0.463 sec.

22-24. 0.331 Hz.

22-26. 0.292 sec.

22-28. 1.81 Hz.

22-30. 0.246 Hz.

22-32. $\theta = 0.4\cos 20.7t.$

22-34. 4.85 Hz.

22-36. 1.105 Hz.

22-38. 1.11 sec.

22-40. $x = A\sin pt + B\cos pt + \dfrac{F_0/m}{(k/m - \omega^2)}\cos\omega t.$

22-42. $x = 0.0186\sin 8.02t + 0.333\cos 8.02t - 0.0746\sin 2t.$

22-44. $x = 0.00278\sin 17.9t + 0.250\cos 17.9t$
$\qquad + 0.00624\sin 8t.$

22-46. 0.20 in.

22-48. 17.6 rad/sec.

22-50. 19.7 rad/sec.

22-52. 121.4 lb.

22-54. 4.13 in.

22-56. $x = 0.251e^{-0.805t} \sin(12.40t + 86.3°)$.

22-58. $x = 0.0671 \cos(2t - 9.66°)$.

22-60. 13,420 lb/ft.

22-62. 1.54°.

22-64. 0.027 in.

22-66. $c_c = 5.46$ lb-sec/ft, $p_d = 4.40$ Hz.

22-68. 2 resistors, R; 1 capacitor, C; 1 inductor, L.

22-70. $m\dfrac{d^2x}{dt^2} + kx = F_0 \sin \omega t$

$$L\dfrac{d^2q}{dt^2} + 1/Cq = E_0 \sin \omega t.$$

Index